BIBK c.1

LIBRARY
44376
TA 705
A8

ATOMIC ENERGY
OF
CANADA LIMITED
WHITESHELL

PRINCIPLES OF ENGINEERING GEOLOGY

PRINCIPLES OF ENGINEERING GEOLOGY

P. B. ATTEWELL and I. W. FARMER
University of Durham

LONDON
CHAPMAN AND HALL

A Halsted Press Book
JOHN WILEY & SONS, INC., NEW YORK

First published 1976
by Chapman and Hall Ltd
11 New Fetter Lane, London EC4P 4EE

Reprinted 1979

© *1976 J. E. Attewell and L. C. Attewell*

Typeset by Preface Ltd, Salisbury, Wilts
Printed in Great Britain by
Fletcher & Son Ltd, Norwich

ISBN 0 412 11400 3

All rights reserved. No part of this book
may be reprinted, or reproduced or utilized in
any form or by any electronic, mechanical or
other means, now known or hereafter invented,
including photocopying and recording, or in
any information storage and retrieval system,
without permission in writing from the Publisher.

Distributed in the U.S.A.
by Halsted Press, a Division
of John Wiley & Sons, Inc., New York

Library of Congress Cataloging in Publication Data

Attewell, P   B
   Principles of engineering geology.

  1. Engineering geology. I. Farmer, Ian
William, joint author. II. Title.
TA705.A87   1975   624'151   75-20012
ISBN 0-470-03641-9

Contents

	Preface	xi
	Symbols	xvii
1	*Composition of Rocks*	1
1.1	Origin and geological classification of rocks	1
1.2	Rock forming minerals	7
1.3	Clay minerals	16
1.4	Base exchange and water adsorption in clay minerals	20
1.5	Mineralogical identification	25
2	*Rock Particles and Particle Systems*	30
2.1	Rock particle classification	30
2.2	Typical rock particle systems	33
2.3	Physical properties of particulate systems	42
2.4	Permeability of particulate systems	45
2.5	Representation of stress in a soil mass	48
2.6	Effective stress	56
2.7	Frictional properties of rock particles	60
2.8	Soil deformation – drained granular media	66
2.9	Soil strength – drained granular media	75
2.10	Soil strength and deformation – clay soils	81
2.11	Pore pressure parameters	88
2.12	Rate of porewater pressure dissipation	92
2.13	The critical state concept	97
2.14	Limiting states of equilibrium	100
3	*Clays and Clay Shales*	104
3.1	Interparticle attraction and repulsion	105
3.2	Sediment formation and clay fabrics	109
3.3	Unstable clay fabrics	117
3.4	Glacial and periglacial clays	122

3.5	Depth − strength profiles	125
3.6	Macrostructure of overconsolidated clays and clay shales	130
3.7	Engineering influence of discontinuities in clay shales	143
3.8	Classification of clay shales	146
3.9	Consolidation and diagenetic considerations	150
3.10	Physical breakdown of shales	153
3.11	Suction pressure	157
3.12	Swelling pressure	162
3.13	Chemical and mineralogical analyses of clays	167
3.14	Relationship between mineralogy, geochemistry and geotechnical properties of clays and clay shales	175
4	*Rock as a Material*	182
4.1	Uniaxial strength	184
4.2	Uniaxial short-term deformation	194
4.3	Deformation mechanisms in rock	199
4.4	Complete stress − strain characteristics of rock in uniaxial compression	206
4.5	Effect of rate and duration of loading	210
4.6	Deformation and failure of rocks in triaxial compression	218
4.7	Failure criteria for rocks	224
4.8	Yield criteria	229
4.9	Rock dynamics	232
4.10	Wave transmission through rocks	234
4.11	Wave attenuation	239
4.12	Rock as a construction material	244
5	*Preferred Orientation, Symmetry Concepts and Strength Anisotropy of some Rocks and Clays*	250
5.1	Studies of the orientation density distribution of clay minerals and other associated minerals	251
5.2	X-ray texture goniometry	252
5.3	Symmetry concepts	260
5.4	Deformation paths	263
5.5	Deformation ellipsoid	263
5.6	Randomization	266
5.7	Symmetry elements and sub-fabrics	271
5.8	Crystallographic plane multiplicities and symmetry	272

5.9	Engineering influence of intrinsic anisotropy	285
5.10	Comparative degree of intrinsic anisotropy — mechanical evidence from rock experimentation	288
5.11	Intrinsic strength anisotropy of brittle and semi-brittle rocks comprising a dominant clay mineral control	289
5.12	Intrinsic anisotropy and sedimentation	300
5.13	Anisotropy of clay shales	302
5.14	Clay strength anisotropy	302
6	*Rock Discontinuity Analysis*	315
6.1	The engineering interest in discontinuities	315
6.2	Genesis and modification of fissures and slickensides	317
6.3	Controls on fissuring and fissure patterns	319
6.4	Classification of discontinuities	320
6.5	Character of discontinuities	326
6.6	Test specimen size—strength relationships	326
6.7	Stereographic representation of discontinuity data	328
6.8	Direct and inverse transformations from polar to equatorial angles	329
6.9	Linear orthogonal transformations	333
6.10	Eulerian angles	336
6.11	Discontinuity survey techniques	336
6.12	Analysis of discontinuity data	344
6.13	Influence of gouge material and surface roughness characteristics of discontinuities	352
6.14	Distributions	355
6.15	Orientation density distribution of discontinuities	364
6.16	Discontinuity shear stability in a polyaxial stress field	369
6.17	Shear strain energy concepts	376
6.18	Preliminary consideration of certain types of discontinuity structure in two dimensions	385
6.19	Statistics of scanlines through discontinuity distributions	388
6.20	Continuity	394
6.21	Preliminary shear stability analysis of discontinuities at the foundation interface of an earth or rock-fill dam	398
6.22	Stability of jointed rock in the foundation of an arch dam	409
6.23	Stability of a discontinuous clay surrounding an unlined tunnel	426

7	*Site Investigation*	427
7.1	Preliminary investigation	429
7.2	Aerial photographs	437
7.3	Terrain evaluation for highway projects	442
7.4	Geophysical exploration techniques	448
7.5	Seismic refraction surveying	453
7.6	Site exploration	457
7.7	Borehole logging	465
7.8	Sampling and testing	475
7.9	Site investigation reports	483
7.10	Mechanical tests *in situ*	484
7.11	Field monitoring techniques	503
7.12	Use of field seismic techniques in engineering geology	512
7.13	Analysis of ground vibrations	514
7.14	Marine geotechnical exploration	529
7.15	Mining subsidence	534
7.16	Probability theory in site investigation	547
7.17	What is 'safety' in soil and rock mechanics?	557
8	*Groundwater*	560
8.1	Types of subsurface water	560
8.2	Groundwater flow	565
8.3	Seepage forces	577
8.4	Drainage and drain wells	580
8.5	Permeability tests – rock	585
8.6	Permeability tests – soils	591
8.7	Economic exploitation of groundwater	598
8.8	Ownership of groundwater and permitted abstractions	601
8.9	Groundwater exploration	601
8.10	Regional investigations	614
8.11	Simulation of groundwater regimes	618
8.12	Well losses	626
8.13	Improving aquifer yield	627
8.14	Groundwater quality	627
9	*Stability of Soil Slopes*	632
9.1	Planar slides	633
9.2	Circular failure surfaces	635
9.3	Slope stability case histories	645

9.4	Simple wedge method of analysis	661
9.5	Use of design curves	672
9.6	Pore pressure ratio	674
9.7	Clay slopes and shear strength parameters	675
9.8	Slope angle measurements in clays and clay shales	683
9.9	Classification of gravitational mass movements in clay	688
9.10	Rock breakdown and landform development	697
9.11	Geomorphological classification of slope profile development	704
9.12	General methods of preventing slope failure	705
9.13	Highway slopes	708
9.14	Protection against coastal erosion	714
10	*Rock Slope Stability*	720
10.1	Geomorphological classification of rock slope instabilities	720
10.2	Classification of rock masses	730
10.3	Character of joints in rock masses	738
10.4	Engineering recognition of rock failure modes	743
10.5	Surface roughness of joints	749
10.6	Discontinuity roughness classification	753
10.7	Planar sliding and the friction cone concept	758
10.8	Instability on intersecting joint planes	765
10.9	Influence of discontinuity orientation distributions	787
10.10	Seismic influences on stability with respect to sliding	792
10.11	Instability caused by block overturning	797
10.12	General rock slope design curves	803
10.13	Slopes in highway cuttings and embankments	809
11	*Ground Improvement*	814
11.1	Shallow compaction	817
11.2	Deep compaction	821
11.3	Pre-loading and consolidation	826
11.4	Sand drains	830
11.5	Grout treatment	836
11.6	Fissure grouting	851
11.7	Hydrofracture	855
11.8	Cavity grouting	863
11.9	Electro-chemical stabilisation	866
11.10	Groundwater freezing	871

11.11	Bentonite suspension	874
11.12	Ground anchors	879
12	*Water Resources, Reservoirs and Dams*	887
12.1	Water requirements in England and Wales	888
12.2	Planning of water resources	889
12.3	Conjunctive use schemes	895
12.4	Flood and dam design parameters	897
12.5	Channel protection	900
12.6	Design capacity of a storage reservoir	904
12.7	Air-photo interpretation for catchment development	907
12.8	Geological influences upon the selection of reservoir sites	908
12.9	Foundation investigations	911
12.10	Water movement into and out of a reservoir	914
12.11	Synthetic flow generation techniques	917
12.12	Dam foundations	918
12.13	Classification of dam types according to their purpose, construction and foundation geology	922
12.14	Long term stability of earth dams	944
12.15	Dam seismicity	945
	References	969
	Supplementary References	1022
	Author Index	1025
	Subject Index	1035

Preface

'Engineering geology' is one of those terms that invite definition. The American Geological Institute, for example, has expanded the term to mean 'the application of the geological sciences to engineering practice for the purpose of assuring that the geological factors affecting the location, design, construction, operation and maintenance of engineering works are recognized and adequately provided for'. It has also been defined by W. R. Judd in the McGraw-Hill Encyclopaedia of Science and Technology as 'the application of education and experience in geology and other geosciences to solve geological problems posed by civil engineering structures'. Judd goes on to specify those branches of the geological or geo-sciences as surface (or surficial) geology, structural/fabric geology, geohydrology, geophysics, soil and rock mechanics. Soil mechanics is firmly included as a geological science in spite of the perhaps rather unfortunate trends over the years (now happily being reversed) towards purely mechanistic analyses which may well provide acceptable solutions for only the simplest geology.

Many subjects evolve through their subject areas from an interdisciplinary background and it is just such instances that pose the greatest difficulties of definition. Since the form of educational development experienced by the practitioners of the subject ultimately bears quite strongly upon the corporate concept of the term 'engineering geology', it is useful briefly to consider that educational background.

Engineering geologists have usually received a basic training in either a geological or engineering discipline and there seems to be a popular acceptance of the potential advantages and disadvantages of both forms of training. Klaus John (1974) has summarized quite admirably the general feeling: 'They (geologists) prefer to approach a problem intuitively, indirectly, and in general qualitative terms, often preferring the problem to the results. Complexities are emphasized, simplifications are only hesitatingly accepted' and, in the case of engineers, '(they) are trained to be analytical, to depend on theory, and rely on numerical data, on abstractions of natural conditions ... often carried to excess with a tendency to unduly simplify in order to be able to numerically analyse a problem because, due to training and environment, (they) are dominated by their orientation toward results.' There is also a widespread feeling that a university training often tends to be given and received in a form and in an environment that is rather remote from the professional world where major design decisions have to be taken under pressures that are of both an economic and time-dependent nature.

In Great Britain, it would appear that most of the entrants to the profession of engineering geology arrive with a background in geological science. With that in mind, it could be argued that the fundamental objectives of an engineering geology education should be:

(a) to provide an adequate analytical and numerical framework within which the observational skills of the professional geologist can be harnessed to produce information that will be immediately useful in engineering design procedures;
(b) to emphasize the geological, physical and structural variants inherent in both undisturbed and excavated earth materials and to indicate the manner in which the inevitable departures from an ideal material behaviour may constrain and cause to be modified any formalized process of design and construction of earth and rock structures; and
(c) to underline the importance of water and particularly groundwater as an economic asset to be conserved and as a problem in ground engineering design.

To these fundamental objectives should perhaps be added the express wish that the engineering geologist at all times maintain a socio-environmental awareness of the possible ramifications of his professional work and advice.

It is with the three primary objectives in mind that the authors have attempted to structure the book by isolating a series of main headings which, it is felt, cover in a reasonable manner the very wide field of interest that is vested in the term 'engineering geology'.

It should be stated that during the early planning of this book it soon became apparent that there were at least two possible ways of tackling the work. One option, which had a certain attraction, would have been to present what might well have developed into a simple text on Quarternary and structural geology combined with a compendium of engineering geology case histories and supported by some very basic and undemanding analysis. Readers would have been expected to absorb the broad experience of others and, by interpolation and extrapolation, to use that experience in order to solve their own particular problems. Necessarily, such an approach to the subject would be qualitative and broadly descriptive, it would probably encompass numerous examples of rather simple geological structures, and it would no doubt be very acceptable to many geologists. On the other hand, one may question the requirement for a formalized collation of case histories when the proceedings of most of the engineering geology and rock mechanics conferences provide such case history expositions in easily-accessible and easily-readable form.

An alternative option, which is in line with the arguments expressed earlier and which is the one adopted for this book, attempts to treat many of the more basic engineering geology concepts at a rather greater analytical depth and where possible uses case history evidence to back-up those concepts. It attempts to indicate to the reader how he can solve his own problems more immediately by a knowledge of some of the underlying principles moderated by an appreciation of the physical nature of the materials with which he is dealing. The present method of presentation therefore integrates many of the contributory disciplines in both the geological and engineering sciences upon which engineering geology must draw, but it retains those disciplines for a service function rather than allowing them to dictate the form of that presentation.

Much of the text is written at university undergraduate level and, it is hoped, will be particularly appropriate reading for students in the final year of a first degree course in geological science. Some of the material should be of interest and use to students engaged in higher degree taught courses and to first degree civil engineering students. But since the text is geologically orientated throughout, students in these latter categories would need to supplement their reading with much more extensive study in the subjects of soil mechanics, rock mechanics, rock engineering and hydrology.

In writing and organizing the text, the authors have drawn to some extent on their experience of lecturing to higher degree students in engineering geology, to geology undergraduates of all levels, and to civil engineering science students, mainly, but not exclusively, in the University of Durham. However, certain lecture material that would normally be presented to M.Sc. students in Engineering Geology and to the B.Sc. students in Engineering Science has been omitted from the text on the grounds that it has rather less direct geological relevance. Typical examples of such omissions would be greater in-depth studies of stress analysis and dynamics, excavation techniques and support systems. Temptations to engage in discussions on aspects of foundation engineering have usually been firmly resisted, and there has been a conscious decision not to include either formalized descriptions of geological structures or general concepts in structural geology on the grounds that these subjects and their implied relationship to engineering geology are adequately covered in other books. Similarly omitted are the simple trigonometrical constructions used to resolve information on geological maps since, as before, these procedures are dealt with elsewhere. Mathematical demands on the reader have been set generally at a low level and rarely in the book do they rise above pre-university standard. On a very few occasions, the going does become a little heavier and probably of more specialist research interest at the present time, but the treatment of the subject is such that those particular sections may be ignored if necessary by the average reader without incurring any penalty throughout the subsequent text. Quantities are usually expressed in S.I. units but in most instances an equivalent alternative is given at an appropriate point in the text. Where imperial units have been used the authors justify their decision on the grounds both that S.I. units have not yet been adopted universally and also that it is really no major task to convert one to the other.

Symbols are defined in the text when used and are also separately listed in the book. Because of the extent of the subject coverage many of the symbols are re-used one or more times, occasionally in the same chapter. Care has been taken, however, to ensure that the reader is not faced with ambiguities that might be attributed to the allocation of the same symbols to different parameters.

There are twelve chapters in the book and they have been arranged in such a way as to minimize the degree of forward reference as the subject is developed. Chapter 1 examines the composition of rocks and earth materials generally and highlights the importance of silicates and clay minerals in engineering geology. Chapter 2 considers the physical properties of rock particles and covers the basic mechanics of particle systems or soils. This development is extended in Chapter 3 to a rather more detailed study of clays and clay shales, the latter with particular reference to mineralogy and chemistry and to the physical processes influencing breakdown. Chapter 4 looks at rock as a material in a materials science context and Chapter 5 examines some of the controls on and implications of intrinsic anisotropy in rock and clay. Chapters 6 and 7 form a particularly important part of the book. They cover respectively the subjects of rock discontinuities and site investigation techniques – the major interface between classical geology and civil engineering. These chapters include practical examples of discontinuity analysis and site appraisal. Chapter 8 is devoted to groundwater in both its engineering and economic roles. It includes a general theoretical appraisal of flow regimes and permeability measurement, aquifer abstraction and a consideration of water requirements. The remaining chapters are generally illustrative. Chapters 9 and 10 cover respectively soil and rock slope engineering problems with an emphasis on case history description and subordinate analytical backing. Chapter 11 describes in some detail various ground improvement techniques such as consolidation, grouting and freezing, the successful adoption of which is quite intimately dependent upon the geological character of the ground to be treated. The final Chapter 12 is concerned with dams, reservoirs and water resources.

  The authors wish to record help and encouragement from several sources.

  They extend very warm thanks to Dr. R. K. Taylor, a colleague of several years' standing, for many enlightening discussions on varied topics in engineering geology. The work on clay shales in Chapter 3 in particular has benefited from his experience in that field. Special thanks must also be accorded to Mr. J. P. Woodman for his work with P.B.A. in the formulation and solution of rational discontinuity models – a subject which is introduced in Chapter 6 – and to both he and Dr. J. C. Cripps for discussions on the use of decision theory in site investigation.

  The work of Engineering Geology M.Sc. Advanced Course students at Durham is acknowledged at appropriate points in the text and the authors express their appreciation of this contribution. They are also grateful to Dr. D. M. Hirst for reading through and commenting on Chapters 1 and 5 and to Dr. J. A. Hudson for performing the same operation on Chapter 4. The authors also feel that the final product has benefited from a first draft reading by Professor T. H. Hanna to whom they also express their thanks. Needless to say, they accept full responsibility for any inevitable errors that remain.

The authors wish jointly to acknowledge their debt to Dr. Albert Roberts, formerly of the University of Sheffield, England and the University of Nevada, U.S.A. He encouraged and supported their early work at the University of Sheffield and they will always be grateful for the research opportunities and the research environment that he provided.

In acknowledging the facilities at Durham University the authors have also appreciated the encouragement and support given by Professors G. M. Brown and M. H. P. Bott of the Department of Geological Sciences and by Professor G. R. Higginson of the Department of Engineering Science in the University. P.B.A. has also appreciated the tangible support provided during the development of Engineering Geology at Durham University by Professor Sir Kingsley Dunham, formerly Head of the Department of Geology in the University and Director of the Institute of Geological Sciences.

Much of the research described in the text was possible only with outside support. Singly and jointly the authors acknowledge the following bodies for grants and contract support: Natural Environment Research Council (U.K.), Transport and Road Research Laboratory, Department of the Environment (U.K.), European Research Office, U.S. Army.

Technical assistance with some of the previously unpublished work outlined in the text was provided by Messrs. A. Swann, C. B. McEleavey and P. A. Kay. Messrs. G. Dresser and J. Clayton were responsible for some of the photographic printing.

Finally, the onerous task of typing — and sometimes interpreting — the draft manuscript was performed most patiently and efficiently by Mrs. P. Farmer and Mrs. A. Taylor. Very special thanks are extended to them.

August 1974
P. B. Attewell
I. W. Farmer

Symbols

Symbol	Meaning	Chapter Reference
A	Porewater pressure parameter	2
A	Total capillary discharge area	2
A_i	Asperity contact area	2
$\bar{A}$	Porewater pressure parameter	2
A	Cross-sectional area of cylindrical rock specimen	4
A	Creep constant	4
A	Wavefront area	4
A	Area factor	6
$A*$	Predicted value in probability analysis	7
A_i	Mutually exclusive events in probability analysis	7
A	Shape factor in flow equation	8
$A_{1...n}$	Strip areas in soil slopes for r_u evaluation	9
A	Total area of shear plane	10
A_j	Total joint area within shear plane	10
A	Cross-sectional area of river channel	12
B	Porewater pressure parameter	2,9
$\bar{B}$	Porewater pressure parameter	2,9
B	Constant in strength anisotropy criterion	5
B	Width of strip footing or foundation	7
B	An event in probability analysis	7
B	Maximum width of a slide	9
C	Areal coefficient	2
C	Porewater pressure parameter	2
C	Torsion strain in vane tests	7
C_R	Cone resistance in standard penetration tests	7
C	Capillary constant	8
C	Drawdown constant	8
C	Electrical capacitance in groundwater analogue models	8
C	Explosive charge factor	11
C	Constant in blasting compaction law	11
C	Constant in pressure grouting uplift analysis	11
C_h	Coefficient of consolidation due to radial flow	11
C	Coefficient in Creager flood discharge formula	12

xviii *Symbols*

Symbol	Meaning	Chapter Reference
D_r	Relative density	2
D	Diameter of rock cylinder	4
D	Maximum thickness of a slide	9
D	Depth parameter in Taylor's slope design curves	9
D'_e	Underground excavation span parameter	10
D	Dielectric constant	11
D	Depth of water in river channel	12
E	Young's modulus of elasticity	2,4,7
E_R	Attraction and repulsion energy between plates	3
E_m	Deformation modulus from Boussinesq rigid punch test	7
E_d	Deformation modulus from plate (central hole) jacking test	7
E_s	Secant modulus	7
E_r	Elastic modulus from pressure chamber test	7
E_r	Energy ratio in vibration	7
E	Potential difference	11
E_i	Potential gradient	11
δE	Grout input energy	11
δE_s	Stored strain energy in rock and grout fluid	11
δE_r	Irrecoverable energy during hydrofracture	11
E_n	Evaporation parameter for month n in synthetic flow generation	12
F	Dimensionless strength ratio	6
F	Force dimension in dimensional analysis	7
F	Factor of safety	9,10
F_R	Pull-out resistance of a grouted anchor	11
G_w	Specific gravity of water	2
G_s	Specific gravity of solids	2
G	Rigidity modulus	4,7
G	Dimensionless parameter in limit equilibrium analysis on stereographic projection	6
H	Half thickness of a consolidating layer	2
H_c	Height of collapse rock	7
H	Vertical distance between standing piezometric level and base of abstraction or test borehole	8
H	Slope height in Taylor's slope design curves	9
H	Trough-to-crest wave height	9
H	Slope height in rock slope stability analysis	10
H_w	Height of water table in slope	10
H	Depth of wedge in diaphragm wall analysis	11
H_1	Height of dam from crest to top of foundation	12

Symbol	Meaning	Chapter Reference
H_2	Impounded head of water from base of dam	12
I_B	Brittleness index	4
I	Normalized diffracted X-ray intensity	5
I	Electrical current in resistivity investigation	7
I_f	Influence factor	7
$I_{\alpha,\beta}$	Edge formed by planes p_α, p_β on stereographic projection	10
J_n	Joint structure number in Tunnelling Quality Index equation	10
J_r	Joint roughness number in Tunnelling Quality Index equation	10
J_a	Joint alteration number in Tunnelling Quality Index equation	10
J_w	Joint water reduction factor in Tunnelling Quality Index equation	10
K_o	Coefficient of earth pressure at rest	2,3,7
K_a	Coefficient of active earth pressure	2, 11
K_P	Coefficient of passive earth pressure	2,7
K	Bulk modulus	4,7
K_s	Stiffness of rock specimen on unloading	4
K_m	Unloading stiffness of rock testing machine	4
K	Constant in Griffith crack stability criterion	5
K	σ_3/σ_1 ratio	5,11
K, K_1, K_2	Dimensionless parameters in limit equilibrium analysis on stereographic projection	6
K	Multiplying factor in rock slope stability analysis	10
L	Length of cylindrical rock test specimen	4
L	Length standardization multiplying factor for discontinuity scanlines	6
L	Length dimension in dimensional analysis	7
L	Beam span	7
L	Length of flow path in seepage analysis	8
L	Length along a borehole	8
L	Leakage parameter in non-steady state drawdown under leaky aquifer conditions	8
L	Maximum length of a slide up-slope	9
L	Length of cylindrical injection source	11
L	Path length in stabilisation	11
L	Effective length of grouted anchor	11
L	Width of wedge	11

xx *Symbols*

Symbol	Meaning	Chapter Reference
M	Critical state friction constant	2
M	A strength ratio	6
M	A fraction between 0 and 1	6
M	Total moment (Σm) of rotationally-shearing soil in slope stability analysis	9
M	Catchment area	12
N	Normal force	2,9
N	Number of capillary tubes	2
N	Number of fatigue cycles	4
N	Accumulated microseismic activity	4
N	Number of poles to discontinuities	6
N	A fraction between 0 and 1	6
N	A constant	6
N_γ	Bearing capacity factor	7
N_q	Bearing capacity factor	7
N	Number of blows in penetration testing	7
N	Number of prior distributions	7
N_f	Number of flow lines	8
N_D	Number of equipotential lines	8
N	Taylor stability number in soil slope stability analysis	9
N	Number of fissures per metre	11
P	Uniaxial force on a rock specimen	4
P	Probability function	6,7
P	Projection point of pole on a sphere	6,10
P	Total force applied in plate bearing test	7
P	A resultant vector in friction circle method of analysis	9
P	Designated as the applied normal force on a plane in stereographic analysis of rock slope stability	10
P	Myers percentage rating for floods	12
$P_{n,n-1}$	Precipitation indices	12
Q	Total volume flow rate	2,8,11,12
Q	Activation energy in Arrhenius equation	4
Q_{ult}	Ultimate bearing capacity	7
Q^{-1}	Material friction parameter (seismic wave transmission)	7
Q	Tunnelling Quality Index	10
R	Gas constant in Arrhenius equation	4
R	Angle of refraction of a wave at a boundary	4
R	Radius of projection sphere	6

Symbol	Meaning	Chapter Reference
R_e	Reliability of an estimate on an exponential utility function	7
R_q	Reliability of an estimate on a quadratic utility function	7
R	Drawdown radius (cone of depression) in pumping tests	8
R	Reciprocal of transmissivity $(1/T)$	8
R	Resultant vector of weight and porewater force in slope stability analysis	9
R	Skempton's 'residual factor'	9
R_α, R_β	Normal reactions to forces on planes p_α, p_β in rock slope stability analysis	10
R	Radius of grout penetration	11
R	Electrical resistance	11
R_s	Skin friction resistance of a grouted anchor	11
R_p	End-bearing resistance of a grouted anchor	11
R_n	Rainfall in month n (synthetic flow generation)	12
s	Asperity shear resistance	2
S	Saturation	2
S_o	Specific surface area per unit volume of particles	2, 11
S_c	Particle crushing strength and unconfined compressive strength of rock	4, 7, 10, 11
S_t	Tensile strength	4, 7
S_{AR}	Compressive strength of rock specimen having aspect ratio AR	4
S	Fatigue strength	4
S_{RT}	Ultimate compressive strength at room temperature	4
S_u	Ultimate compressive strength at any temperature	4
S_s	Strength of rock in direct (double) shear	4
S	Storage coefficient of an aquifer	8
S_a, S_y	Boulton's aquifer storage coefficients	8
S	Shear force along a plane	9
S_c	Cohesive element of shear force	9
S_s	Shear strength of soil during rotational failure of embankment caused by seismicity	12
S	Slope of stream or river	12
T_v	Time factor in one-dimensional consolidation	2
T	Surface tension	3
T	Absolute temperature in Arrhenius equation	4
T	Time dimension in dimensional analysis	7
T	Time lag in permeability	8
T	Transmissivity of an aquifer	8

xxii Symbols

Symbol	Meaning	Chapter Reference
T	Weight resolved into a plane of shear	9
T_h	Time factor in radial (sand drain) consolidation	11
V_w	Volume of water in pores of a soil	2,8
V_v	Volume of voids in a soil	2
V_s	Volume of solids in a soil	2
V	Specimen bulk volume	2,4
V	Electrical voltage in resistivity investigation	7
V	Cost parameter in probability analysis	7
δV	Volume of grout fluid injected per unit time during hydrofracture	11
W_w	Weight of water	2
W_s	Weight of solids	2
W	Energy in compression testing	4
W_i	Input energy to a wave	4
W_s	Stored energy in a wave	4
W_k	Kinetic energy in a wave	4
W_o	Source energy of a diverging wave disturbance	4
W_r	Energy per unit area of wavefront at distance r from source	4
W	Weathering standardizing multiplying factor for discontinuity scanlines	6
W	Total shear strain energy due to shear movements along discontinuities	6
W	Energy released by explosive (charge weight)	7,11
W	Weight of soil or rock in slope stability analysis	9,10
W	Weight of wedge in diaphragm wall analysis	11
W	Width of river channel at water surface	12
X	Slope angle function in rock slope design curves	10
Y	Slope height function in rock slope design curves	10
Z	Zeta potential	11
a	Graphical ordinate intercept in p-q space	2
a_w	Volume ratio of water phase	2,11
a	Area of new cracks per unit volume of rock	4
a	Area on plane of projection	6
a	Inter-electrode spacing for resistivity investigation	7
a_1	Radius of central hole in plate (bearing test)	7
a_2	Radius of plate	7
a	Internal radius of pressure chamber	7

Symbols xxiii

Symbol	Meaning	Chapter Reference
a	Ground acceleration	7,12
a_p	Radius of piezometer	8
a	Borehole radius	8,11
a_o	Radius of grout injection source	11
a	Capillary radius	8,11
a_w	Volume ratio of water phase	11
a	Catchment area	12
a_n	Constant of proportionality in synthetic flow generation	12
a	Lever arm parameter in Newmark's seismic stability analysis	12
b	Effective principal stress difference ratio	4
b'	Saturated thickness of an aquitard	8
b	Slice width in 2-dimensional soil slope stability analysis	9
b	Constant in synthetic flow generation equation	12
b	Distance parameter in Newmark's seismic stability analysis	12
c	Cohesion	2,4,11
c_i	Cohesion at asperity contact point	2
c_u	Undrained shear strength	2,9,11
c_v	Coefficient of consolidation	2,11
c	Crack half-length	4
c_{p1} (or c_1)	P-wave (or wave) velocity in medium 1	4,7,10
c_{p2} (or c_2)	P-wave (or wave) velocity in medium 2	4,7
c_c	Crack length in anisotropic failure criterion	5
c_e	Effective cohesion along a partially discontinuous surface	10
d	Clay mineral particle/plate diameter	1
d	Capillary diameter	2
d_o	Particle diameter parameter	2
d	Interparticle distance	3
d	Vane diameter	7
d	Jacking plate diameter	7
d	Thickness of block in rotational stability analysis	10
d_e	Drain spacing	11
d_w	Drain diameter	11
d_p	Maximum particle diameter	11
d	Anchor diameter	11
d	Induced dynamic distortion in a structure	12

xxiv Symbols

Symbol	Meaning	Chapter Reference
e	Void ratio	2
f	Frequency	7, 12
f_o	Janbu correction factor in soil slope stability analysis	9
f	Wave fetch (coastal protection)	9
h	Water head	2,8
h_c	Capillary water height above water table	2,8
h_w	Head of water above dam foundations or block	6,10
h	Height of vane	7
h_t	Seam thickness	7
h_a	Artesian head	8
h	Height of block of rock in rotational stability analysis	10
h_g	Grout hydraulic injection head	11
h_s	Head losses due to shear resistance	11
i	Hydraulic gradient	2,8,11
i	Angle of incidence of a wave at a boundary	4
i_c	Angle of critical incidence in seismic refraction	7
i	Standard deviation	7
i	Current in analogue simulation of groundwater	8
i	Average angle of incidence of surface undulations in direction of shear displacement	10
i	Slope angle in rock slope design curves	10
k	Coefficient of permeability	2,8
k	Equal area projection parameter	6
k	Constant	7
k_s	Constant expressing sampling and testing cost	7
k_t	Constant expressing the degree of loss due to error	7
k_e	Effective permeability of anisotropic strata	8
$k_{\alpha(\beta)}$, $k_{\beta(\alpha)}$	Constants in rock slope stability analysis	10
k_e	Equivalent permeability of fissured strata	8,11
k_h	Coefficient of horizontal permeability	8,11
k_v	Coefficient of vertical permeability	8,11
k_e	Electro-osmotic coefficient of permeability	11
l	Length of principal axes of deformation ellipsoid	5
l	Quantum number of Legendre polynomial equation	5
l	Length of a discontinuity intersection on a plane	6
l	Slice length parameter along a shear plane	9

Symbol	Meaning	Chapter Reference
l	Length of chord of a failure arc in slope stability analysis	9
l	Base width of dam	12
m	Moisture content	2
m	Poisson's number	2
m_v	Volume compressibility of a soil	2,11
m	Constant	4,10
m	Quantum number in Legendre polynomial equation	5
m	Fractional height of a water table above a reference surface	9
m	Moment of a force along a slope failure arc	9
m_α	Parameter in soil slope stability analysis	9
n	Porosity	2,4,11
n	Exponent in creep equation	4
n	Number of variables in probability analysis	7
n	Fraction of vertical force (W) exerted horizontally (specifically by seismic waves)	7,9,10,12
n_α	Parameter in soil slope stability analysis	9
n	Constant in rock slope stability analysis	10
n^*	Ratio between sand-drain spacing and diameter	11
p	Half sum of principal stresses	2
p	Suction pressure	3
p_x	Event probability	6
p	Scanline angle parameter	6
$p_{\alpha,\beta}$	Designation of a plane	10
p_a	Average injection pressure acting in a fissure	11
p_o	Internal grout pressure in a borehole	11
p_r	Fluid pressure at radius r	11
p	Basic forcing frequency	12
q	Half principal stress difference	2
q_u	Yield strength or unconfined compressive strength	2,6,9
q	Scanline direction constant	6
q	Bearing pressure	7
q	Flood discharge per unit area	12
r	Radius or radial distance	1,3,4,6,7,8,9,11
r_u	Pore pressure ratio	9

xxvi *Symbols*

Symbol	Meaning	Chapter Reference
s	Average specific surface area of clay minerals	1
s	Material parameter	6
s	Settlement	7
s	Drawdown as a function of radius from pumped well	8
s_t	Total drawdown due to pumping	8
s_d	Dynamic shear strength	12
t	Clay mineral thickness	1
t	Consolidation time	2
t	Temperature	3
t	Time	4,7,8,11
t	Beam thickness	7
t	Torque at failure in vane test	7
$t_{n,n+1}$	Interslice shear forces in slope stability analysis	9
u	Porewater pressure	2,8,11
u_a	Pore pressure change under hydrostatic consolidation	2
u_b	Pore pressure change due to deviator stress	2
u_g	Pressure of gas phase in partially-saturated soil	2
u_w	Pressure of water phase in partially-saturated soil	2
u_e	Excess porewater pressure in consolidation	2
u	Surface or particle displacement	7,12
u	Parameter in Theis equation	8
u_a, u_y	Parameters in Boulton's equation	8
v	Particle, ground surface, or water discharge velocity	2,4,7,8
v	Voltage in analogue simulation of groundwater	8
w	Weighting constant used in computer fabric analysis	6
w	Weight of slice of soil in slope stability analysis	9
x	Specimen displacement axis in graph of uniaxial compression test	4
x_1, x_2	Coordinate axes	6,7
x_i	Projection of measured discontinuity length l_i on to horizontal axis	6
x	Value of a variable in probability analysis	7
$\bar{x}$	Mean value of a variable	7
y	Geophone distance from source of seismic pulse	7
y_c	Critical distance in seismic refraction	7

Symbol	Meaning	Chapter Reference
z	Interatomic charge	1
z	Depth	2,4,7,8,11
z_i	Projection of measured discontinuity length l_i on to vertical axis	6
z_t	Depth of tension crack	10
z_w	Depth to water table and of water in tension crack	10
Γ	Ordinate of the critical state line	2
Δ	Volumetric strain	2,4
Δ	Vibration intensity	7
Λ	Defines projection domain of discontinuity length onto a reference axis (p)	6
Ξ	Defines projection domain of discontinuity length onto a reference axis (θ)	6
Ξ	Dimensionless area factor in seepage flow	11
Σ	Summation symbol	2,6,9
Φ	Function in dimensional analysis	7
Ψ	Electrical potential	3
Ψ	Thermal diffusivity	11
α	Effective hydraulic radius	2,11
α	Slope of p-q envelope	2
α	Soil-water contact angle	3,8
α	Variable parameter in extended Tresca law	4
α	Wave attenuation coefficient	4
α	Direction cosine angle	6
α	Bedding or shear plane inclination to horizontal (passive wedge in soil slope stability analysis)	7,9
α	Reciprocal of delay index (leaky water-table aquifer condition)	8
α^*	Analogue model mesh length	8
β	A reflection angle in wave transmission	4
β	Angle formed by an anisotropic plane and a major principal stress direction	5
β	Dip angle of pole to a discontinuity from the vertical and also direction cosine angle	6
β_i	Measured dip to horizontal of discontinuity intersection with scanline measurement plane	6

xxviii *Symbols*

Symbol	Meaning	Chapter Reference
β	Embankment slope angle	8
β	Shear plane inclination to horizontal (neutral wedge in soil slope stability analysis)	9
β	Angle in Newmark's dynamic stability analysis	12
γ	Bulk unit weight	2,4,7,8,9,10
γ_d	Dry unit weight	2,4
γ_f	Fluid unit weight	2
γ_b	Submerged unit weight	2,8,11
γ_w	Unit weight of water	2,3,6,8,9,11
γ_s	Unit weight of solids	2
γ_c	Unit weight of concrete	6
γ_r	Unit weight of rock	6
γ	Direction cosine and equatorial angle on projection	6
γ_k	Density of collapse rock	7
γ	Shear plane inclination to horizontal (active wedge in soil slope stability analysis)	9
γ_g	Density of grout suspension	11
δ	Equatorial angle (small circle) on projection	6
$\boldsymbol{\delta}$	Mathematical function	7
δ_z	Displacement of rock at depth z beneath plate	7
δ	Fissure thickness	8,11
δ	Displacement along undulating surface	10
δ_o	Fissure thickness before injection extension	11
δ_r	Fissure thickness at edge of extension front	11
δ_a	Average fissure width extension	11
$\epsilon_{x,y,z}$	Strain	2,4,5,10,11
ϵ_o	Elastic strain	4
ϵ_c	Long-term creep strain	4
ϵ_L	Lateral strain	4
$\epsilon_{\alpha,\beta}$	Azimuth angle between two planar full-dip directions	10
ζ	Angle defining surface of no finite longitudinal strain	5
η	Fluid viscosity	2,8,11
η	Latitude angle of poles to crystallographic planes in texture-goniometry	5
η	Eulerian angle in transformation of axes	6
η_e	Fraction of coal extraction from a seam	7
η	Storativity ratio parameter in Boulton's non-steady state water table type curves	8
η_p	Bingham viscosity of grout	11

Symbols xxix

Symbol	Meaning	Chapter Reference
η_g	Grout viscosity	11
η_w	Water viscosity	11
η_a	Apparent viscosity	11
θ	Angle defining plane in principal stress space	2
θ	Azimuth angle of poles to crystallographic planes in texture-goniometry	5
θ	Azimuth angle of a discontinuity pole (direction of planar full dip)	6
θ	Scanline direction angle referenced to an arbitrary axis	6
θ	Angle in construction of parabola	8
θ	Wedge angle	11
λ	Gradient of the critical state compression line	2
λ	Wavelength of transmitted waves	4
λ	Angle formed by a shear failure plane and a major principal stress direction	5
λ	Eulerian angle in transformation of axes	6
λ	Wavelength of undulations on discontinuous surfaces	10
μ	Coefficient of friction (tangent of friction angle)	2,4
μ_c	Friction coefficient along a crack	5
ν	Poisson's ratio	2,4,7,10
ν_o	'Elastic' Poisson's ratio	4
ν_c	Crack dilation portion of ν	4
ξ	Designation of a crystal face	5
ξ	Eulerian angle in transformation of axes	6
ξ	General shear plane angle to horizontal	6
ξ_α, ξ_β	Inclination angles of joint shear planes to horizontal in rock slope stability analysis	10
ξ_I	Dip angle from vertical of p_α, p_β planar intersection line $I_{\alpha,\beta}$ in rock slope stability analysis	10
ρ	Specific surface energy per unit length of crack	4
ρ	Density	2,4,7,10,12
ρ	Electrical resistivity	7,11
σ	Stress (also σ_x, σ_y, σ_z along coordinate axes x, y, z)	2,4
σ	Hydrostatic (all-round) stress	2
$\sigma_{1,2,3}$	Principal stress	2,4
σ_n	Normal stress on a plane	2,4
σ'_{vm}	Maximum historical overconsolidation pressure	2
σ'_{vo}	Initial vertical geostatic pressure on a soil	2

Symbol	Meaning	Chapter Reference
σ_t	Tensile stress concentration	4
σ_{3L}	Limiting confining pressure	4
σ_o	Critical stress value in distortional strain energy law	4
σ	Pressure in a wave	4
σ_i	Stress in incident wave	4
σ_r	Stress in wave reflected from a boundary	4
σ_{tr}	Stress in wave transmitted through a boundary	4
σ_{nc}	Normal stress on a crack	5
σ_t	Tangential stress around a borehole	11
σ_r	Radial stress around a borehole	11
σ_{to}	Tangential stress around a hole with zero internal pressure	11
σ_R	Radial pressure distribution around a borehole with seepage flow of injectant	11
σ_T	Tangential pressure distribution around a borehole with seepage flow of injectant	11
τ	Shear stress	2,4,7,9,11,12
τ_A	Shear resistance of asperities	2
τ_f	Shear stress at failure (function of normal pressure)	4,12
τ_r	Residual shear strength	4,9
τ_c	Shear stress along a crack	5
τ_p	Peak shear strength	9
$\bar{\tau}$	Mean shear strength	9
τ_s	Shear strength of grout	11
ϕ	Friction angle	2,4,5,6,7,9,10,11,12
ϕ_i	Inter-particle friction angle	2
ϕ_p	Peak friction angle	2,9
ϕ_r	Residual friction angle	2,9
ϕ	Particle diameter	9
ϕ_s	Skin or wall friction in anchor stability	11
χ	An event in fracture analysis	6
χ	Angular measurement on stereographic projection for rock slope stability analysis	10
$\psi_{\alpha,\beta}$	Azimuthal angle for planar full dip direction	10
ψ_I	Azimuthal angle for planar intersection line direction	10
ω	Circular frequency	12

1 ~ Composition of Rocks

Rock composition is a subject that is adequately covered in most texts on petrology and mineralogy and its study forms a basic part of degree courses in the geological sciences. A detailed treatment of the subject area is difficult to justify in a book devoted specifically to engineering geology, but there are some aspects of rock composition which do have an important bearing on the engineering properties of rocks and of rock particle systems. These affect both microstructures and massive forms. The most important effects derive from the specific physical properties of some minerals, and particularly some clay minerals which are able to absorb large quantities of water and have structures which tend to render the material anisotropic. These properties and the way in which they influence the engineering behaviour of rocks and clays are considered in greater detail in Chapters 3 and 5. In the present chapter, the composition of rocks together with the basic physical properties and structures of the more important minerals are considered and defined.

1.1 Origin and geological classification of rocks

A model earth comprises a series of concentric, near-spherical layers (Figure 1.1), the main divisions being the crust, the mantle and the inner and outer cores. These are divided and subdivided by discontinuities representing changes in material properties which may be detected through changes in their sonic ($P-$) wave velocity (see Section 4.10). The solid parts of the earth's surface are the *crust* (up to 60 km thick in mountain ranges) which comprises layers of sedimentary and granitic rocks on a basalt base, and the *lithosphere* part of the mantle. This may be up to 50 km thick and comprises ultrabasic peridotite. Below the lithosphere is the *rheosphere* comprising partly fused peridotite. Most earth movements are believed to originate in the lithosphere.

2   Principles of Engineering Geology

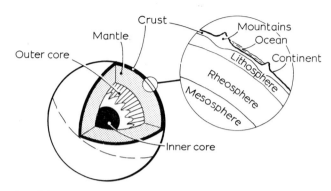

Properties of earth layers

Zone	Discontinuity	Material	Thickness km	Density Mg m^{-3}	P-wave velocity km s^{-1}
Crust (1.5% by volume)	Sial	Sedimentary rocks and granite	Oceans 10 Continents 30	2.7	6.5
	Sima	Basalt	Mountains 60	2.9	6.9
Mantle (82.3% by volume)	Lithosphere	Peridotite	50	3.3	8.1
	Rheosphere	Fused peridotite	200	3.3	7.8
	Mesosphere	Dense peridotite	750	4.3↓	10.7↓
	Lower	Dense peridotite	2000	5.7	13.6
Core (16.2% by volume)	Outer	Liquid Ni–Fe	2000	9.7–11.8	8.1–10.3
	Inner	Solid Ni–Fe	1400	14–16	11.2
			~6400		

Figure 1.1   Structure and composition of the earth.

The main interest of the geologist, and more particularly of the engineer, is centred on the rocks in the earth's crust. These are normally classified under three main headings, depending upon their method of formation. Thus, *igneous* rocks (comprising 98 per cent of the earth's crust) are formed by cooling of molten rock magma, *sedimentary* rocks by breakdown through weathering of an existing rock mass, and *metamorphic* rocks by alteration of existing rocks under conditions of high temperature and/or pressure. These classifications represent to a certain extent the basic *cyclic* processes to which rocks are subjected, and which have an important bearing on their mechanical properties. A typical history of cyclic change might follow several paths starting, since the original rocks in the earth's crust were almost certainly molten, as a primary magma, and possibly progressing through local fusion to an alternative magmatic state:

Primary magma
   ↓   cools
igneous rock
   ↓   weathering
sediment
   ↓   diagenesis
sedimentary rock
   ↓   heat and pressure
metamorphic rock
   ↓   heat and pressure
secondary magma
   ↓   cools
igneous rock

Each step will be accompanied by significant physical and chemical change, although the actual elements in the rock will remain essentially the same. This can be illustrated by examining the relative composition of igneous and typical sedimentary rocks in Table 1.1. Apart from the observation that relatively few elements are present in quantity, it is evident that shales, sandstones and igneous rocks have quite similar average elemental compositions, although often widely differing mechanical properties depending principally on their mode of formation, degree of weathering and state of diagenesis. Thus, although most rocks are elementally similar, they are mineralogically and texturally diverse.

Table 1.1 Average proportions of elements in igneous and typical sedimentary rocks by weight (*after* Hurlbut, 1971)

Igneous		Shales		Sandstones		Limestones	
O	46.60	O	49.00	O	51.00	O	49.00
Si	27.72	Si	27.28	Si	36.75	C	11.35
Al	8.13	Al	8.19	Ca	3.59	Mg	4.77
Fe	5.00	Fe	4.73	Al	2.53	Ca	3.45
Ca	3.63	K	2.70	C	1.38	Si	2.42
Na	2.83	Ca	2.23	K	1.10	Al	0.43
K	2.59	C	1.53	Fe	0.99	Fe	0.40
Mg	2.09	Mg	1.48	Mg	0.71		
Ti	0.44	Na	0.97	Na	0.33		

Table 1.2 Classification of Igneous Rocks

(a) *Mineralogical*

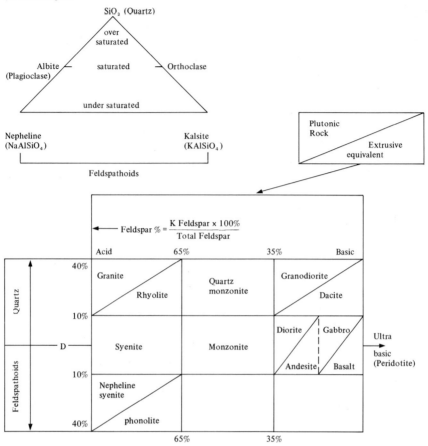

(b) *Mode of occurrence*

	Acid > 66% SiO_2	*Intermediate* 52–66% SiO_2	*Basic* < 52% SiO_2	*Ultrabasic*
Extrusive (Volcanic)	Rhyolite	Andesite	Basalt	
Minor Intrusive (Hypabyssal) Dykes, Sills, etc.	Quartz Porphyry	Porphyrite	Dolerite	
Major Intrusive (Plutonic)	Granite	Diorite	Gabbro	Picrite Peridotite

← Silica and Alkalis increase (Na_2O, K_2O)

Increase in Lime, Magnesia, Iron Oxide →

Most igneous rock classifications are based on mineralogy and mode of occurrence. Some of these are suggested in Table 1.2.

Both the mineral content and the grain size tend to be an indication of the rate of cooling. The classification has, however, limited application in engineering except as a means of identification, since by their very method of formation, igneous rocks comprise an intimate agglomeration of closely locked particles. These particles have a high surface contact area, minimum voids and strong mechanical bonds, and form a competent, largely homogeneous and isotropic rock material. The result is that, despite interesting evidence that grain size and quartz content can affect strength (see Merriam *et al*, 1970), igneous rocks in an unweathered state tend to be (apart from widely separated shrinkage joints) basically sound engineering materials.

Sedimentary rocks also tend to be classified in terms of texture and mineral content, as well as origin, which differentiates mainly between mechanical sedimentation of particles and chemical precipitation from solution.

A simplified classification in Table 1.3 suggested by Krynine (1948) differentiates mechanical sediments essentially between *sandstones* of varying grain size comprising mainly quartz, and *shales* comprising mainly feldspars and clay minerals.

The only common precipitates are carbonates. As in the case of igneous rocks, this classification has little direct geotechnical application or interest. It does, however, isolate one important factor which affects the behaviour of sediments (or soils) in particular and sedimentary rocks in general; that is the presence of clay minerals caused by the alteration of feldspars and other silicates

Table 1.3 Classification of sedimentary rocks

		← Predominantly Quartz		Predominantly Feldspar + clay mineral →
MECHANICAL SEDIMENTS	COARSE	Quartz conglomerate		
	MEDIUM	Sandstone and	Greywacke	Arkose
	FINE	Quartzite	Micaceous and Chloritic shales	Shales/mudstones
CHEMICAL PRECIPITATES		Limestone, dolomite, evaporites, chert		

Table 1.4 Textural classification of important sediments and sedimentary rocks

Texture	Rock	Identifying component	Characteristic features
CLASTIC	Till, Tillite Boulder Clay	Rock fragments and clay	Large size range, rocks in clay matrix
	Conglomerate, Gravel Boulders	Rounded rock and mineral fragments	More than 50% grains > 2 mm; < 25% clay
	Breccia	Angular rock and mineral fragments	More than 50% grains > 2 mm; < 25% clay
	Sand, Sandstone Gritstone Arkose Greywacke	Sand size rock particles	More than 50% grains < 2 mm; > 0.06 mm; < 25% clay
	Silt Siltstone	Silt size rock particles	More than 50% grains < 0.06 mm; < 25% clay
	Clay, Shale Mudstone Claystone	Clay minerals	More than 25% clay
	Marl Marlstone	Clay minerals, calcite	Fine-grained 25% to 75% calcite in clay matrix
	Limestone	Calcite grains	More than 50% calcite grains; < 25% clay
	Tuff	Volcanic material	Fine-grained (< 2 mm)
CRYSTALLINE	Limestone Chalk	Calcite	> 50% carbonate > 50% calcite
	Dolomite	Dolomite	> 50% carbonate > 50% dolomite
BIOFRAGMENTAL	Limestone	Fossil structures	> 50% coral, algal, foraminifera remains
	Peat, Coal Lignite	Partly or fully carbonised plant remains	Fibrous to compact carbonaceous material
	Diatomite	Fossil remains	> 50% diatom remains

during the weathering process. These can have a significant effect on texture and mechanical properties and may be of fundamental importance in engineering geology.

A rather more detailed sedimentary rock classification, of greater geotechnical interest and based principally on texture, is illustrated in Table 1.4. It is essentially Krynine and Judd's (1957) adaptation of a classification by Mielenz (1948) and forms a useful basis for the

Table 1.5 Textural classification of metamorphic rocks

Texture	Rock	Identifying features
GRANULAR	Hornfels Quartzite	Fine-grained, predominantly quartz particles
	Marble	Fine to coarse grained – particles of limestone or dolomite
BANDED	Gneiss	Elongated to platy mineral grains – with compositional banding
FOLIATED	Schist Serpentinite Slate Phyllite	Finely-foliated rocks with high proportion of phyllosilicates

detailed mechanical property description of sediments and sedimentary rocks in later Chapters.

Classification of metamorphic rocks is of only minor significance in engineering, but for completeness is included in Table 1.5. The simplest basis for classification is in terms of structure or texture.

1.2 Rock-forming minerals

Most rocks (and rock particles) comprise a mechanically-bonded aggregate of mineral particles. *Minerals* are usually defined as naturally occurring, inorganic crystalline compounds or elements. The total number of minerals in existence is limited to about 2000 by the stability of chemicals in the earth's environment. Of this number, only a few are present in large proportions.

According to Palache *et al* (1944) minerals may be classified on the basis of chemical composition into 12 main sub-divisions;

Native elements	Halides	Phosphates
Sulphides	Carbonates	Sulphates
Sulpho-salts	Nitrates	Tungstates
Oxides (ex silica)	Borates	Silicates

Of these, the most common are *silicates* which make up about 99 per cent of the earth's crust. The remaining minerals, although numerically and economically of considerable importance, are of lesser significance in engineering geology. This is mainly because, with the exception of carbonates, they do not usually appear in sufficient quantities to affect the physical or mechanical properties of commonly occurring rocks and soils.

The few essential rock-forming primary minerals can be grouped into feldspars, feldspathoids, silica minerals, olivine, pyroxenes, amphiboles and micas (Pirsson and Knopf, 1948). Relatively small amounts of chlorite, epidote, magnetite, ilmenite, and apatite are commonly included. With the exception of quartz, the essential rock-forming mineral groups are represented by a series of solid solutions the compositions of which are determined by the physicochemical environment under which they are formed (Pirsson and Knopf, 1948; Deer *et al*, 1966). The relative distribution of the essential minerals along with their characteristic texture forms the basic criterion in rock identification.

Weathering, subsequent sedimentation and in part the physical and chemical nature of the parent rocks determine the resultant mineral composition of the inorganic sedimentary rocks. Weathering phenomena are considered at appropriate points in subsequent chapters.

The physical properties of individual mineral crystals depend principally on the magnitude of the forces holding the atomic structure together. In terms of mechanical properties, the stronger these forces, the greater is the strength and rigidity of the mineral crystal. An indication of relative element-oxygen bond strengths in some common rock-forming minerals is given in Table 1.6.

The concept of strength and rigidity is, of course, vital to any development of engineering geology. One simple way of describing strength is in terms of surface hardness, the most common index being the Moh scale of hardness:

Softest	1.	Talc
	2.	Gypsum
	3.	Calcite
	4.	Fluorspar
	5.	Apatite
	6.	Orthoclase feldspar
	7.	Quartz
	8.	Topaz
	9.	Corundum
Hardest	10.	Diamond

Table 1.6 Relative bond strengths (*after* Nicholls, 1963)

Bond	Relative bond strength*
Si–O	2.4
Ti–O	1.8
Al–O	1.65
Fe^{3+}–O	1.4
Mg–O	0.9
Fe^{2+}–O	0.85
Mn–O	0.8
CaO–O	0.7
Na–O	0.35
K–O	0.25

*Relative bond strength = $\dfrac{2z}{(r - 1.40)^2}$
where z = charge on the metallic ion
r = radius of the metallic ion, Å

This scale is based on the relative abrasiveness of minerals, increasing orders in the scale being determined in terms of the ability of one mineral to scratch another. It is a useful scale for determining the relative toughness of minerals but might seem to have minor use in determining the mechanical properties of rocks since strength does not always correlate directly with hardness. On the other hand, the Schmidt hammer hardness test is being increasingly used for rapidly indexing intact rock via a correlation with its compressive or tensile strength.

The fundamental principles of interatomic bonding to the limit of present-day knowledge are explained in any modern textbook on physical chemistry. Most simply, atoms comprise a nucleus surrounded by layers of electrons, and the chemical properties of an element are determined by the number of electrons in the outer shell. These are known as valence electrons. In the case of common rock-forming elements, the electron number is:

    Oxygen   6
    Silicon   4
    Aluminium   3
    Iron   3
    Calcium   2
    Potassium   2
    Magnesium   2
    Sodium   1

However, all atoms tend towards a stable outer shell comprising eight electrons, as is the case with inert gas atoms. Other atoms can achieve this state either by losing electrons in the outer shell and leaving the atom with a nett positive charge (for example, Si^{4+}, Al^{3+}, Na^+) or by gaining electrons and leaving the atom with a nett negative charge (O^{2-}). Charged atoms are known as ions, positively charged atoms as *cations* and negatively charged atoms as *anions*. Alternatively, stability can be obtained between adjacent atoms by the phenomenon of electron sharing.

Interatomic bonding occurs either as a result of electrostatic attraction between oppositely charged ions (known as *ionic* bonding) or of electron sharing between atoms (known as *covalent* bonding). As would be expected, covalent bonds are stronger and more stable than ionic bonds.

The relative proportions of ionic and covalent bonding in a mineral crystal, and the magnitude of the attractive forces in ionic bonding, depend on the size and charge of the ions involved. In general, large, low valency ions which can co-ordinate large numbers of compatible ions on to their surface area tend to ionic bonding, whilst small, high valency ions tend to covalent bonding. Sizes and charges of the *common* mineral-forming ions in silicates are included in Table 1.7. It should be noted that in aqueous solution the ionic radii will change due to hydration.

Table 1.7 Charge and size of silicate-forming elements (*after* Hurlbut, 1971)

Ion	Ionic radius Å	Ratio: Ionic radius / Oxygen radius	Co-ordination number with respect to oxygen	Position in Silicate Formula
Si^{4+}	0.42	0.30	4	Si
Al^{3+}	0.51	0.36	4–6	Y (or Si substitute)
Fe^{3+}	0.64	0.46	6	Y
Mg^{2+}	0.66	0.47	6	Y
Ti^{4+}	0.68	0.49	6	Y
Fe^{2+}	0.74	0.53	6	Y
Na^+	0.97	0.69	8	X
Ca^{2+}	0.99	0.71	8	X
K^+	1.33	0.95	8–12	X
O^{2-}	1.40	1.00	–	O

Table 1.1 gives the proportion by weight of elements in igneous and sedimentary rocks. In *volumetric terms*, approximately 60 per cent of the atoms in the earth's crust are oxygen, 20 per cent silicon, 7 per cent aluminium and 2 per cent each of iron, calcium, sodium, potassium and magnesium. Exact proportions can be obtained by dividing the percentage by weight of the elements forming igneous rock by their atomic weights. Virtually all other atoms exist in insignificant quantities. It is possible, therefore, to build up a picture of most of the earth's crust as a network of oxygen ions, strongly bonded together (covalent bonds) by small highly-charged silicon (or possibly aluminium) ions. Into this structure, larger cations and associated anions (such as hydroxyls) will be weakly attached by ionic bonds.

The general chemical formula for this network may be written for most common elements in the form:

$$X_n \ Y_m \ (Si_p \ O_q) \ Z_r$$

where X represents large cations (Na, Ca, K),
Y represents medium sized cations (Al, Fe, Mg, Ti, Mn),
and Z represents anions (OH^-, Cl^-, F^-).

n, m, p, q, r, are integers, n, m, r being determined by the conditions for electrical neutrality and p and q defining the basic *silicate structure*. In this, an aluminium ion may be substituted for a silicon ion, as in the case of feldspars. The contents of both the X and Y groups, which both contain ions of a similar size, are largely interchangeable through ionic substitution, the main limitation, apart from size, being the need for electrical neutrality in the crystal structure.

The most important part of the network is the *silicate structure* which determines to a large extent the physical properties of the silicate crystals, and thereby to an indirect extent the mechanical properties of the rock or soil made up from the crystals. The silicate structure forms the basis of most silicate mineral classifications (Table 1.8).

Silicate structures comprise a series of linked oxygen and silicon atoms. The fundamental silicon-oxygen unit, determined by maximum coordination of silicon and oxygen atoms, comprises a tetrahedron (Figure 1.2a) formed by four oxygen ions (radius 1.4Å) grouped around a single silicon ion (radius 0.42Å). Since conditions for full covalent bonding are not satisfied by the

Table 1.8 Classification of silicates

Arrangement of tetrahedra	Oxygen : Silicon ratio and unit	Shared oxygen atoms per silicon atom	Important Sub-groups
Single (Nesosilicates)	4 : 1 $(SiO_4)^{4-}$	0	Olivines Garnets Aluminium Silicates
Double (Sorosilicates)	3½ : 1 $(Si_2O_7)^{3-}$	1	Epidotes
Ring (Cyclosilicates)	3 : 1 $(SiO_3)^{2-}$	2	Beryls
Chain – single (Inosilicates)	3 : 1 $(SiO_3)^{2-}$	2	Pyroxenes
Chain – double	2¾ : 1 $(Si_4O_{11})^{1½-}$	2½	Amphiboles
Sheet (Phyllosilicates)	2½ : 1 $(Si_2O_5)^{1-}$	3	Clay minerals Micas Chlorites Septechlorite
Framework (Tectosilicates)	2 : 1 $(SiO_2)^0$	4	Quartz Feldspars

outer electrons in this grouping, the bond which holds the tetrahedron together will be partly covalent and partly ionic (in fact approximately 50:50). The covalent bond will, of course, be much stronger and more stable than the ionic bond and will be evenly distributed between oxygen atoms in the tetrahedron. Thus, if another silicon ion becomes available for covalent bonding, the interatomic forces will be sufficient to break the ionic bond and effectively create a different silicate structure. As originally proposed by Bragg and summarized by Berman (1937), a series of structural units can be constructed by assuming varying degrees of covalent bonding symbolized by a sharing of oxygen atoms between adjacent tetrahedra (Figure 1.2b–e). In this way, the magnitude of bonding energy between atoms will increase, being a minimum in the case of a single tetrahedron and a maximum in the case of sheet or framework structures.

A detailed treatment of silicate structures is beyond the scope of

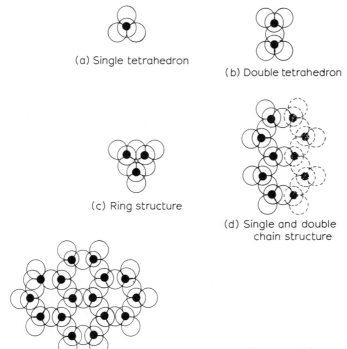

Figure 1.2 Representation of silicate structures.

the present book and can be obtained from any mineralogy text. For completeness, however, the main structural features of the more important silicates are summarized below.

Olivines consist of separate groups of SiO_4 tetrahedra held together by the attraction of positively charged ions (cations) to give overall electrical neutrality. Typical forms are *olivine* $(MgFe)_2 SiO_4$, *forsterite* $Mg_2 SiO_4$, and *fayalite* $Fe_2 SiO_4$.

Pyroxenes have a structure containing long chains of SiO_4 tetrahedra, each tetrahedron sharing two of its oxygens with neighbouring tetrahedra. The residual negative charge is balanced by cations between the chains which also serve to hold the chains together. In all pyroxenes, chains run parallel to the vertical crystallographic axis. Pyroxenes are classified according to crystal structure as orthorhombic (e.g. *hypersthene* $(Mg,Fe)_2 Si_2 O_6$) and monoclinic (e.g. *augite* $(Ca,Mg,Fe,Al)_2 (Si,Al)_2 O_6$).

Amphiboles are related to the pyroxenes, forming double chains, again being held together by cations, but in addition containing OH groups. In other words, they are hydrous silicates. The structure contains 'bands' of linked SiO_4 tetrahedra, each band comprising in effect two pyroxene chains united by shared oxygens. The unit of pattern contains Si_4O_{11}, but to avoid the necessity of splitting an atom, this is doubled to Si_8O_{22}, each unit being associated with two hydroxyls. This amounts to a total negative charge of 14, which is neutralized by the same cations as occur in pyroxenes. As in the case of pyroxenes there are orthorhombic (e.g. *anthophyllite* $(Mg,Fe)_7(Si_8O_{22})(OH)_2$) and monoclinic (e.g. *hornblende* $(Ca,Mg,Fe, Na, Al)_{3-4}(Al,Si)_4O_{11}(OH)$ and *tremolite* $Ca_2Mg_5(Si_4O_{11})_2(OH)_2$) types. Asbestos is a fibrous form of amphibole.

Micas or phyllosilicates are chemically distinct from pyroxenes and amphiboles insofar as the alkali elements (Na,K,Li) are important and calcium is absent. The atoms are arranged in extended sheets with the SiO_4 tetrahedra linked continuously at three corners, three out of four oxygens always being shared. The atoms in the sheets are linked in a hexagonal plan, the unit pattern containing Si_4O_{10}, but as in the amphiboles, in order to avoid half atoms it is convenient to double this to Si_8O_{20}. The essential hydroxyls are contained in the plane containing the unshared oxygens, four being associated with each 'doubled' unit. Al^{3+} may substitute for silicon and each unit contains two trivalent aluminiums, the standard formula being $(OH)_4(Al_2Si_6)O_{20}$ with cations to a total charge of 14. The micas are actually double sheets, with the unshared oxygen atoms facing each other. They are monoclinic and the main sub-groups are:

Muscovite	$K_2Al_4(Al_2Si_6)O_{20}(OH)_4$
Biotite	$K_2(Mg,Fe)_6(Al_2Si_6)O_{20}(OH,F)_4$
Phlogopite	$K_2(Mg_6)(Al_2Si_6)O_{20}(OH,F)_4$
Lepidolite	$K_2(Li,Al)_6(Al_2Si_6)O_{20}(OH,F)_4$

Chlorites, like micas, are built up from sheet structures and are monoclinic. The sheets in chlorite, however, are polar and all the free oxygens point the same way. Chemically, they contain more water than micas and do not contain any alkali metals. A typical formula is $(Mg,Fe)_{10}Al_2(Al_2Si_6)O_{20}(OH)_{16}$. Serpentine minerals are closely related to chlorite.

Clay minerals are considered in more detail in the following Section. They also have a sheet lattice similar to mica, although the sheet stacking is very variable. They can normally only be identified by X-ray methods (see Section 1.5). All are basically hydrous alumino-silicates.

Feldspars and *quartz* are the most stable forms of silicate structure. It is possible to have each SiO_4 tetrahedron sharing *all* its oxygens with adjacent tetrahedra giving a three-dimensional framework. This produces a structure which is electrically neutral and no cations are needed in the lattice. The formula is SiO_2 and the mineral is *quartz*. If, however, some of the Si^{4+} is replaced by Al^{3+}, the structure will no longer be electrically neutral and cations will be required to stabilize it. This is basically the structure of feldspars. Thus, replacing every fourth Si by Al in Si_4O_8 gives $(Al\ Si_3O_8)^-$; neutralizing with K^+ gives *orthoclase* $(K\ Al\ Si_3O_8)$ and with Na^+ gives *albite* $(Na\ Al\ Si_3O_8)$. Similarly, replacing every other Si by Al gives $(Al_2Si_2O_8)^-$ and neutralizing with Ca^{2+} gives *anorthite* $(Ca\ Al_2Si_2O_8)$. Ca^{2+} and Na^+ have similar radii, with a consequent continuous variation in composition between albite and anorthite. These soda-lime feldspars are known as the *plagioclase* feldspar series.

There is sometimes a relationship between the conditions of formation of igneous rocks and the type of silicate structure formed. As a general point, the simpler structural forms tend to crystallize out from the molten magma at *higher* temperatures than the more complex and stable structures. If the rate of cooling is rapid, basic igneous rocks with a preponderance of olivines and pyroxenes tend to be formed. On the other hand, if the rate of cooling is delayed, reaction between these simpler forms and the still-molten magma fraction may lead to the formation of igneous rocks with a preponderance of the more stable silicates such as amphiboles, micas, feldspars and quartz. But this is a rather simplistic approach.

A more confidently predictable process of chemical change would be expected during weathering of igneous rocks, and this is confirmed by the absence of most single, double, ring and chain structure silicates in sediments. These comprise almost exclusively sheet and framework minerals, most typically quartz, which is largely resistant to chemical change, and clay minerals, primarily hydrous aluminium silicates resulting from the further alteration through weathering of feldspars and micas. These are the constituents of most non-calcareous rocks and sediments or soils. Mechanically speaking, the *quartz* is relatively stable; the *clay mineral*, depending on its ability to absorb water, is potentially unstable, and its presence can have a substantial effect on the engineering properties of soils and some types of rock.

1.3 Clay minerals

Clay minerals are essentially micro-crystalline, hydrous aluminium, (occasionally magnesium or iron) sheet silicates, having a layered flaky structure. *Clay soils*, which will be considered later in Chapter 3, are an aggregate of clay mineral and non-clay mineral particles, the properties of which are determined largely but not exclusively by the properties of the clay minerals.

The atomic structure of clay minerals (see Grim, 1962) is usually formalized in terms of the Pauling model for *sheet silicates.* This model incorporates two basic structural units, the first comprising a two-layer sheet made up from units of six hydroxyl (or oxygen) ions in octahedral coordination with aluminium, iron or magnesium ions (Figure 1.3), and the second comprising a sheet silicate structure. When magnesium or ferrous ions are present in the former structure, all the positions in the basic octahedral unit are satisfied and the structural unit is known as *brucite*. This unit forms the basic structure of the sheet minerals, serpentine, talc, biotite, phlogopite and vermiculite. When aluminium or ferric ions are present, in order to obtain electrical neutrality, every third anion space is empty. This unit is known as *gibbsite*, and forms the basic structure of muscovite and some clay minerals.

In sheet silicate structural layers, the silica-hydroxyl tetrahedra are arranged in expanding hexagonal sheets so that the base of each tetrahedron is in the same plane and the tip points in the same direction. The basic sheet silicate mineral crystal is formed by the attachment of these oxygen ions to the brucite or gibbsite sheet structure, either through covalent bonding, hydrogen bonding or

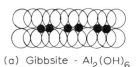

(a) Gibbsite - $Al_2(OH)_6$

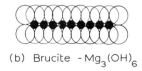

(b) Brucite - $Mg_3(OH)_6$

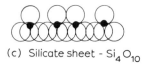

(c) Silicate sheet - Si_4O_{10}

Figure 1.3 Representation of sheet silicate structural units.

exchangeable cation linkage. The type and continuity of bonding determines the stability of the crystal.

A typical example of a clay mineral group formed from alternating sheets of gibbsite and silica tetrahedra are the *kandites* where the basic crystal results from the combination of the tips of the tetrahedra and one of the layers of the gibbsite sheet. The reason for this is readily apparent, since the longitudinal dimensions of the individual structural units and the thicknesses of the sheets are similar. In the common layer, the shared hydroxyls become oxygen atoms and the bonding is effectively covalent. The mineral *kaolinite* having the structural formula $(OH)_8 Al_4 O_{10}$ is an example.

Actual crystals are formed from a series of basic crystal double layers held together by hydrogen bonding between the hydroxyls and oxygens in adjacent units. This bonding is strong, and although the surfaces may be considered as cleavage planes, the tendency to cleavage is small, and well-formed kaolinite crystals tend to be relatively large, stable, and difficult to disperse. Typically, kaolinite crystals are pseudo-hexagonal, $0.3-4\mu m$ in diameter and $0.5-2\mu m$ thick (Figure 1.4).

In nature, many kaolinite particles are less well-formed and the hydrogen bonding between crystal units is weaker. This permits adsorption of water and consequent finer dispersion along the crystal unit cleavage. There is also a possibility in these poorly-formed crystals of a small amount of isomorphous substitution of the aluminium by iron or titanium, so resulting in a unit charge deficiency and reduction of crystal size.

A similar structural composition to kaolinite is found in the *halloysites*, the main difference being that halloysites normally occur in a partly or fully hydrated form having a structural formula $(OH)_8 Al_4 Si_4 O_{10} \cdot nH_2 O$, where n may be equal to zero, two or four. $nH_2 O$ represents a single molecular layer of water between the kaolinite crystal unit layers. Halloysite crystals, however, usually occur as hollow rods (Figure 1.5) probably due to the weakness of the bonding between layers which accentuates the slight difference in curvature of the gibbsite and silicate sheets. Their stability depends on the degree of hydration, with partly hydrated forms ($n = 2$) being the least stable.

Probably the most important clay minerals in an engineering context are the *montmorillonites* (Figure 1.6) and *illites*. Both are formed from structural units comprising a central gibbsite octahedral

Figure 1.4 Scanning electron micrograph of small kaolinite crystals forming a larger crystallite unit (*after* Bohor and Hughes, 1971; reproduced by permission of the Clay Minerals Society).

sheet sandwiched between two silicate sheets so that the tips of the silica tetrahedra penetrate both the hydroxyl layers of the gibbsite. The montmorillonite crystals are formed by successive layers of these units, held together by extremely weak bonding between oxygen atoms in the adjacent units. This structural configuration makes the cleavage a potentially dominant feature with the space between the structural units being prone to penetration by water and other polar molecules. Montmorillonite crystals can thereby accommodate thick interlayer sheets of adsorbed water in the structure. The thickness of these sheets depends to a certain extent on the type of exchangeable cation present, particularly its size and the strength of the ionic bonding exerted between the cations and the oxygen atoms. This bonding is lowest (and hence dispersion is greatest) when sodium is the exchangeable linking cation.

Although the basic structural formula for montmorillonite is $(OH)_4 Al_4 Si_8 O_{20} n H_2 O$, this will invariably be altered by substitution

Figure 1.5 Scanning electron micrograph (x10 000) of halloysite (*after* Eswaran, 1972; reproduced by permission of the author).

of various elements such as magnesium for aluminium in the gibbsite structure, aluminium for silicon in the tetrahedra and by the presence of soluble, linking cations. *Illite* is in many ways a special case of montmorillonite except that some of the silicons are replaced by aluminium, and potassium ions are present between crystal unit layers. The structure, which takes the formula $(OH)_4 Al_4 K_2 (Si_6 Al_2) O_{20}$, is stable in well-formed varieties of the mineral but is unstable in poorly-crystallized varieties.

Other important clay minerals are: *allophane*, an irregular and variable structural agglomeration of octahedra and tetrahedra; *chlorite*, a series of brucite and sheet silicate layers similar to serpentine; *vermiculite* and *attapulgite*, based on double-chain structures. Some of the main features of the important clay minerals are summarized in Figure 1.7 after Lambe and Whitman (1969). A tabular classification of clay minerals and some notes on that classification may be found in Fayed and Attewell (1965).

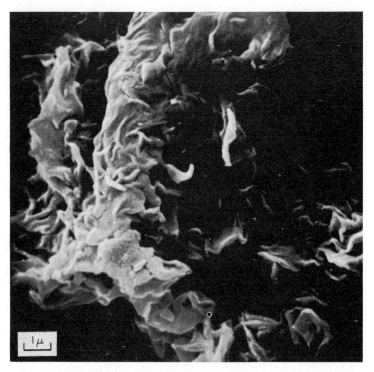

Figure 1.6 Scanning electron micrograph of rock sample of Wyoming bentonite (*after* Bohor and Hughes, 1971; reproduced by permission of the Clay Minerals Society).

1.4 Base exchange and water adsorption in clay minerals

Compositional variation through ionic or isomorphous substitution within the clay mineral crystal lattice (particularly prevalent in montmorillonites and vermiculites) of, say, trivalent aluminium for quadrivalent silicon, can leave the structural unit with a nett negative charge. Substitution also reduces the crystal size and alters its shape. Exposed hydroxyl groups and broken surface bonds can also lead to a nett negative charge on the structural units.

The presence of this nett negative charge means that soluble (also possibly insoluble) cations can be attracted or adsorbed on to the surface of clay mineral structural units without altering the basic structure of the clay mineral. These cations can be exchanged for other soluble cations if the ionic environment changes. The most common soluble cations are Na^+, K^+, Ca^{2+}, Mg^{2+}, H^+ and NH_4^+. There may also be some cases where positive nett charges caused by broken

Composition of Rocks 21

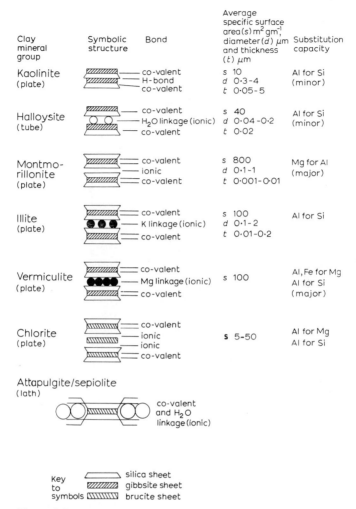

Figure 1.7 Major clay mineral features (*after* Lambe and Whitman, 1969).

bonds at particle surfaces can attract exchangeable anions, but these have minor engineering significance. Cation exchange capacity (CEC) does, however, have major significance in determining clay mineral properties, particularly the facility with which they adsorb water.

CEC, measured in terms of milliequivalents of the atomic weight of solvent per gram, varies widely for various types of clay mineral (Table 1.9), depending mainly on whether attractive forces result from broken bonds (kaolinite, halloysite) or from lattice substitutions (montmorillonite, vermiculite). Thus, kaolinites have a low

Table 1.9 Cation Exchange Capacity (Milliequivalents per 100 gm) *after* Grim (1953)

Kaolinite	3–15
Halloysite 2H$_2$O	5–10
Halloysite 4H$_2$O	10–40
Montmorillonite	80–150
Illite	10–40
Vermiculite	100–150
Chlorite	10–40
Sepiolite/Attapulgite	20–30

cation exchange capacity, adsorbing small amounts of cations together with associated water on to the surfaces of mineral particles. Montmorillonites, on the other hand, have a high cation exchange capacity, adsorbing soluble cations (and associated water) on to the structural unit interlayer surfaces. Potentially, therefore, the montmorillonite and vermiculite structures can adsorb much larger amounts of water, with obvious implications regarding stability and plasticity of a rock or soil containing these minerals.

The water adsorbed on to the structural unit interlayer *surfaces* is physically in a different state from ordinary liquid water or porewater and water at particle surfaces. Grim (1953) quotes evidence to the effect that from three to ten (or more in some circumstances) molecular layers of water may form a hexagonal net tied to the interlayer surface of the structural units by the bond between free hydrogen atoms in the water layer and oxygen atoms in the silica tetrahedra. The thickness of this layer (8 to 28 Å) may be two to three times the thickness of the structural layer of the clay mineral crystal and can only be removed readily and completely by application of energy to the clay mineral, usually by heating at 100°C.

The soluble ions adsorbed with the water on to the interlayer surface can affect the adsorbed water arrangement in several ways (see Grim, 1953). Principally, they act as a bond of varying strength holding the structural layers together and controlling the thickness of adsorbed water. Their effectiveness will depend on size and charge. Thus Na$^+$, K$^+$ will tend to be weak and clay-water systems containing these ions, such as sodium montmorillonites or *bentonites*, will be capable of adsorbing large amounts of water. Ca^{2+}, Mg^{2+}, on the other hand, will have stronger links and clay-water systems containing them will possess substantially lower water contents. Inclusion of Fe^{3+} or Al^{3+} would, of course, reduce the water content and plasticity and this is in fact the basis of the electro-osmotic method of clay stabilisation (see Section 11.9) where dispersion of iron or aluminium anodes into the soil provides a source of Fe^{3+} and Al^{3+} ions in the soil to replace more weakly bonded ions. In general, the replacing power of exchangeable cations increases with increasing valency and bond strength. A frequently quoted order (Grim, 1953) of replacing power is sodium, potassium, calcium, magnesium, ammonia, hydrogen, iron, aluminium. Although this order may vary with different clay mineral combinations and experimental conditions, it is generally true that sodium can be replaced by virtually anything except lithium, calcium and magnesium are replaced only with difficulty, and aluminium and iron are virtually irreplaceable. Grim quotes *pH value* as a useful parameter for assessing *in-situ* the nature of the exchangeable cation, being 9 for sodium, 7.5 for calcium and less than 7 for hydrogen, when aluminium or iron are invariably present in the exchange position.

The effect of exchangeable ion composition and water content on clay mineral properties is less easily identified. Most simply, this may be considered in terms of plasticity, defined for the purposes of engineering in terms of the *Atterberg* plastic state limits. The *plastic limit* (BS 1377 of the British Standards Institution) is the minimum moisture content at which the clay exhibits obvious plasticity, by being capable of forming threads 1/8 in (3 mm) in diameter without breaking when rolled by the palm of the hand on a glass plate. The *liquid limit* is the minimum moisture content at which the soil will begin to *flow* when subjected to small impacts in a liquid limit device designed by Casagrande (1932a). In each case, samples are prepared from dried, powdered clay mixed with water.

Experimental results in Table 1.10 suggest that the plastic limit reduces in the commoner clay minerals in the order attapulgite, montmorillonite, halloysite ($4H_2O$), illite, halloysite ($2H_2O$), and kaolinite, with slightly higher plastic limits occurring when the higher valency exchangeable ions are present. The exception is the case of montmorillonite where the reverse is the case.

Bearing in mind the inexact nature of the tests and the probable presence of variable impurities and variable crystal structure in many clays, the data are indicative of the relative amounts of water adsorbed during the short period of the test on to the particle surfaces of each mineral type. In view of this, the plastic limit can be related quite closely to specific surface area (directly) and to particle size (indirectly). Since particle size is dependent on the perfection of the mineral crystals, a high degree of scatter in the Atterberg limits for minerals with similar exchangeable ions might be expected.

Table 1.10 Plasticity of clay minerals (*after* White, 1955 quoted by Grim, 1962)

Mineral	Exchangeable ion	Plastic Limit percentage	Liquid Limit percentage
Kaolinite	Li	28–33	37–67
	Na	26–32	29–52
	K	28–38	35–69
	Ca	26–36	34–73
	Mg	28–31	39–60
	Fe	35–37	56–59
Halloysite ($2H_2O$)	Li	37	49
	Na	29	56
	K	35	57
	Ca	38	65
	Mg	47	65
Halloysite ($4H_2O$)	Li	47	49
	Na	54	56
	K	55	57
	Ca	58	65
	Mg	60	65
Montmorillonite	Li	59–80	292–638
	Na	54–93	280–710
	K	57–98	108–660
	Ca	63–81	123–510
	Mg	51–73	128–410
	Fe	73–75	140–290
Illite	Li	38–41	63–89
	Na	34–53	59–120
	K	40–60	72–120
	Ca	36–45	69–100
	Mg	35–46	71–95
	Fe	46–49	79–110
Attapulgite	Li	103	226
	Na	100	212
	K	104	161
	Ca	124	232
	Mg	109	179
	H	150	270

Attapulgite, in particular, tends to small unit dispersion when subjected to disturbance. Plastic limits for montmorillonites, particularly sodium, tend to be distorted by the thixotropic properties of the clay-water system.

The liquid limit is also strongly related to the specific surface area and, in addition, is affected by water adsorbed on to *interlayer* surfaces of structural units of more easily dispersed clay minerals. The liquid limit is therefore more strongly affected by the basic structure and the exchangeable ion present, being particularly high in the case of Li^+-, Na^+-, K^+- montmorillonite. In general, liquid limits decrease in the order montmorillonites, attapulgite, illite, halloysite, kaolinite, and in these groups with increasing exchangeable ion valency in the case of montmorillonites and possibly attapulgite, but with decreasing exchangeable ion capacity in other minerals.

As in the case of the plastic limits, the liquid limits for some montmorillonites are almost meaningless because of the thixotropic properties of the clay. In particular, increases in shear strength with time (the standard liquid limit has a shear strength equivalent of approximately 0.7 kN m^{-2}) in highly thixotropic, bentonitic clays can make the liquid limit almost totally time-dependent. They do, however, indicate the extremely high capacity for water adsorption of montmorillonite clays and by implication show how a relatively small amount of montmorillonite can rapidly alter the water adsorption capacity of a clay soil particularly where a low valency exchangeable ion is present. As will be shown later, this can have a critical effect on soil or rock properties. Accurate identification of small quantities of clay mineral is therefore an important factor in determining the engineering properties of rocks and soils.

1.5 Mineralogical identification

The minerals in a fine-grained polymineralic aggregate are most readily identified by X-ray diffraction methods. Geology students will be familiar with the basic theory of X-ray diffraction, of single crystal goniometry, of powder photography and diffractometry, of the use of A.S.T.M. index cards, and of some of the less routine deductive approaches to identification. It is also accepted that they will be familiar with the law of rational indices for the specification of crystal form. Without this quite basic knowledge, it will be difficult for the reader to understand the arguments set out in the remainder of this section and some of the explanations in Sections

3.13 and 5.2. For any reader unfamiliar with X-ray diffraction, and for the easy reference of others, the standard text by Klug and Alexander (1954) is recommended.

Mineralogical composition in most rocks may be determined by direct optical observation under the petrological microscope. The mineralogy of argillaceous rocks and clays must be determined by X-ray methods because the individual crystal sizes are very often below the range of the resolving power of the optical microscope. At the simplest level, we might only require to obtain just a very general indication of the presence of quartz in a clay together with an identification of the type of clay mineral from observation of diffraction peaks at the low 2θ (high dÅ) end of the Bragg angle spectrum. At a slightly higher level, we might require to evaluate quantitatively the free quartz to total clay mineral content, since this ratio can be correlated directly with the shear strength of the clay (see for instance Figure 3.17). Finally, at an even higher level, we might require to provide a complete quantitative identification of all resolvable minerals in a clay in cases where some indication of the genetic history of the clay would be useful or where, for example, the pre-disposition of the clay (or clay shale) to weathering and breakdown is questioned.

Prepared surfaces of stiffer *intact* clay or shale may be X-rayed, but more usually the sample takes the form of a powder in a cavity or on a smear mount. In the latter cases, resulting directly from the inequant platy form of the clay mineral crystal, it is virtually impossible to present a randomly-orientated powder to an X-ray beam, and there will therefore be clay mineral peak enhancement with respect to the peaks of other more equant minerals against which the clay mineral content might be balanced. Similarly, the preparation of a smooth diffracting surface on an intact specimen leads to preferred orientation effects unless special techniques are adopted at the preparation stage.

There are several chemical treatments available for assisting the identification process in polymineralic materials. They may be summarized as: 'acidising' (acid-soluble weight loss), heating (increased chlorite 14 Å basal reflection with loss of other orders), glycolation (expansion Na–montmorillonite basal spacing from about 12.5 Å, and Ca–, Mn–montmorillonite from about 15.5 Å, to a series at 17 Å, 8.5 Å, 5.7 Å, 4.2 Å, 3.4 Å, 2.8 Å . . .). Reference should be made to Schultz (1964) for full details of these procedures.

In some instances where illite is the dominant clay mineral present in a sample, there is sometimes a marked 'tail' present on the low-angle side of the 10 Å peak. This is taken to indicate the presence of mixed-layer clay minerals, and the attenuated tail to the peak tends to create a problem during the quantitative assessment of illite percentage. In some instances, following sample treatment with ethylene glycol, a 10 Å illite peak broadens towards the high angle side. This may be interpreted as an indication of disordered mixed-layer minerals. It may be noted that disordering is often a function of weakening; Jørgensen (1965), for example, has found that dioctahedral illite weathers to a mixed-layer vermiculite.

Klug and Alexander (1954) have detailed the procedure to be adopted for the quantitative assessment of the minerals present in a diffraction sample. Ideally, the sample should be 'doped' or 'spiked-up' with a known amount of pure mineral standard of suitable crystallite size to project sharp diffraction lines close to the unknown mineral lines to be measured but not too close to become superimposed on other lines. Internal standards for quartz could be calcium fluoride (CaF_2) or nickel oxide (NiO). A good internal standard for clay minerals is boehmite. Two diffractograms for the laminated clay considered in Chapter 5 are shown in Figures 1.8 and 1.9, the latter trace showing two boehmite calibrating lines.

It is found that the weight fraction x_1 of a mineral of unknown weight in an original sample is proportional to the ratio of the diffractogram intensities I_1/I_s where subscript l refers to the unknown mineral and subscript s refers to the spiking mineral. The constant of proportionality would be found by mixing together known proportions of the spiking mineral and a pure sample of the 'unknown' mineral and so generating a calibration line. Having determined the constant from the slope of the line, it is then a matter of measuring off intensities from spiked-up diffractograms of the type in Figure 1.9.

Differential thermal analysis (DTA) is a further aid to clay mineral identification. A sample of the material under study is heated at a controlled rate (10°C per minute) alongside an inert substance (alumina), both exothermic and endothermic peaks being noted on a chart recorder. Gillott (1968) has described the major thermal effects

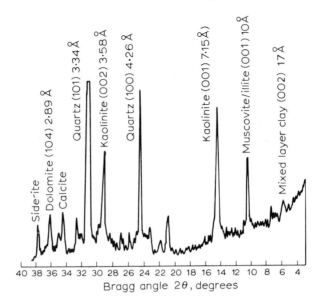

Figure 1.8 X-ray diffractogram of a sandy parting in laminated clay (*see* Figure 5.35). CoK$_\alpha$ radiation.

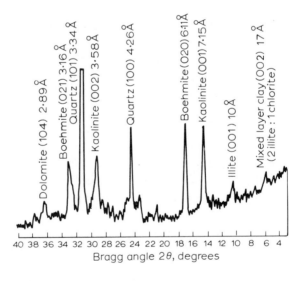

Figure 1.9 X-ray diffractogram of laminated clay 'spiked-up' with 10% by weight of boehmite (*see* Figure 5.35). CoK$_\alpha$ radiation.

on the clay minerals as: (a) at 50° to 200°C endothermic peaks appear due to the loss of surface water; (b) at 450° to 700°C endothermic peaks also develop due to the loss of hydroxyl ions which, as has been shown, form an integral part of the alumino-silicates; and (c) at temperatures greater than 800°C endothermic peaks appear due to final breakdown of the clay mineral structure and exothermic peaks develop due to the crystallization of new phases. In practice, there are often problems which result from the carbon burn-off peak blanketing the recorder traces.

2 ~ Rock Particles and Particle Systems

A fundamental basis of engineering geology is the difference in the response to stress that is exhibited, on the one hand, by massive rock (comprising virtually all the earth's crust) and, on the other, by the rock particles which form its upper surface, for it is upon and in this latter material that most engineering structures are built. In practice, the amplitudes of many of the natural stress fields and imposed stress situations encountered in near-surface rock are small and the possibility of failure of *competent* rocks is only really a major consideration in deep mining excavations (see Woodruff, 1966; Obert and Duvall, 1967). The emphasis in engineering geology must therefore be directed principally towards the weaker, well-fissured and jointed rocks and to deposits of rock particles, or *soils* to use the engineering term, resulting from the action of geological processes on the original intact rock formations.

In the present chapter, some of the fundamental properties of rock particle systems are considered.

2.1 Rock particle classification

The resultant effect of erosion and weathering is to reduce an original massive rock to a series of particles of various sizes, shapes and mechanical properties. These range from large, rounded boulders or angular blocks which may be up to several metres wide, down through gravels, sands and grits to the finest clay mineral fragments. The size and shape of particles depend (with the exception of clay minerals) principally on the weathering environment (climate, temperature), the extent and duration of energy input during transportation and destruction, and on the type of transport mechanism. Thus, a volcanic sand subjected to high energy input over a short period may comprise small and angular grains, whereas a river sand, subjected to varying degrees of abrasion, principally by rolling, over a prolonged time period may take the form of small and rounded grains.

Table 2.1 Rock particle classification

Size fraction	Sub-division	Rock particle diameter (mm)			
		M.I.T.	U.C.S.	Wentworth-Udden	U.S. Dept. Agric.
Boulder (block)		> 200	> 300	> 256	
Cobble		200–60	300–150	256–64	
Gravel	Coarse (pebble)	60–20	150–18	64–4	
	Medium	20–6			
	Fine (granule)	6–2	18–4.76	4–2	2–1
Sand	Coarse	2–0.6	4.76–2		1–0.5
	Medium	0.6–0.2	2–0.42	2–0.06	0.5–0.25
	Fine	0.2–0.06	0.42–0.074		0.25–0.10
	Very fine				0.10–0.05
Silt	Coarse	0.06–0.02	< 0.074 (non plastic)	0.06–0.004	0.05–0.002
	Medium	0.02–0.006			
	Fine	0.006–0.002			
Clay		< 0.002	< 0.074 (plastic)	< 0.004	< 0.002

Several classifications of rock particles based on their *size* have been proposed. Ideally, these should be logarithmic, cyclic and decimal with divisions and subdivisions occurring at geometric midpoints. In practice, few satisfy this criterion (Table 2.1) except the widely accepted Massachusetts Institute of Technology system adopted by the British Standards Institution (1957). Some rather noticeable differences between this and the other classifications emphasize the need for standardization. The U.C.S. classification (Wagner, 1957) is sometimes used in soil mechanics, the Wentworth–Udden classification (Pettijohn, 1957) is used in geology and the U.S. Department of Agriculture classification is used in agriculture.

Rock particle size (or texture) is an important factor in determining the properties of particulate systems, each major fraction having a specific effect on the properties of the (soil) system. For instance, boulders, blocks and cobbles have a stabilizing effect because of their size and weight when included in natural deposits and man-made fills such as rockfill.

Similarly, coarser sands and gravels are relatively stable if well-drained, and they are not affected significantly by moisture. Fine sands, silts and clays, on the other hand, have a tendency to instability when wet. In addition, clays have properties, particularly

Table 2.2 Particle shape classification (*after* Pettijohn, 1957)

Shape	Class	Roundness	Description
	Angular	0–0.15	Little or no evidence of wear
	Sub-angular	0.15–0.25	Maintains original form but shows definite signs of wear
	Sub-rounded	0.25–0.40	Considerable wear; tips of corners worn off and surface area reduced
	Rounded	0.40–0.60	Original faces removed, but some flat surfaces remaining
	Well-rounded	0.60–1	No faces, edges, corners; entire surface comprising well-rounded curves

expansion, associated with their clay mineral composition which may cause considerable instability. The presence of organic matter, such as peat, will also reduce stability by decreasing the density and increasing the compressibility of the material.

Thus, particle *size* is an important indicator of the potential stability of a soil; high in coarse-grained, well-drained aggregates, low in fine-grained, low-permeability layers. Particle *shape* is rather less important, although in sands, gravels and boulders, stability is related to a certain extent to irregularity, and shapes can often affect particle size measurement which is invariably based on square mesh sieve analysis.

In practice, most *large* (silt size and above) rock particles tend to be equidimensional to a greater or lesser extent, and subsequent analyses based on equivalent particle sphericity can be relatively easily justified. These more equant particles mainly derive from *quartz* or unaltered *feldspars*, and depending on the amount of wear will vary between an angular blocky particle and a perfect sphere. Pettijohn (1957) analyses varying degrees of roundness of particles, defining roundness in an approximate manner as the ratio between the average radius of curvature of projections on a particle to its equivalent sphere radius. He suggests five classifications which are listed in Table 2.2.

2.2 Typical rock particle systems

Rock particle systems rarely exist as single-size assemblages of particles. Usually they form a well- or poorly-graded collection which crosses several formal size, mineral or shape classifications. Whereas, therefore, a general size classification can be used to describe a soil quite approximately, an accurate description requires a quantitative statement of each range of particle size present. Such a classification is usually presented as a *particle size distribution* curve in which the cumulative weight percentage passing a particular mesh size is plotted in the manner of Figure 2.1.

Grading curves are obtained from dry and wet sieve analyses down to 0.06 mm (silt size fractions) and hydrometer or sedimentation analyses for finer fractions. The unavoidable loss of accuracy arising from the different definitions of particle size imposed by a square mesh sieve, and by Stokes' law applied to a falling particle, is tolerable within the limits of accuracy usually required for particle size distributions. Several typical particle size distributions are illustrated in Figure 2.1.

Two useful quantities can be obtained from such curves. The first of these is the D_{10} particle size, which is the particle diameter below which 10 per cent of the particles are finer. This is equated by Terzaghi and Peck (1967) among others to the *effective (or equivalent) spherical diameter* of a bed of uniform particles having the same permeability. As such it can be used to estimate flow characteristics through the Kozeny equation quoted in Section 2.4. A second quantity is the *uniformity coefficient* (C.U.), which is the ratio between the D_{60} particle size, below which 60 per cent of particles are finer, and the D_{10} particle size. A low uniformity coefficient indicates a single size soil; a high coefficient indicates a well-graded soil. For instance, in Figure 2.1 the approximate values are:

	D_{10} size	D_{60} size	C.U.
	mm	mm	
Uniform gravel	0.80	1.3	1.6
Medium sand	0.15	0.31	2.1
Sandy gravel	0.20	1.5	7.5
Silty sand	0.06	0.56	9.3
Sandy clay	0.001	0.028	28

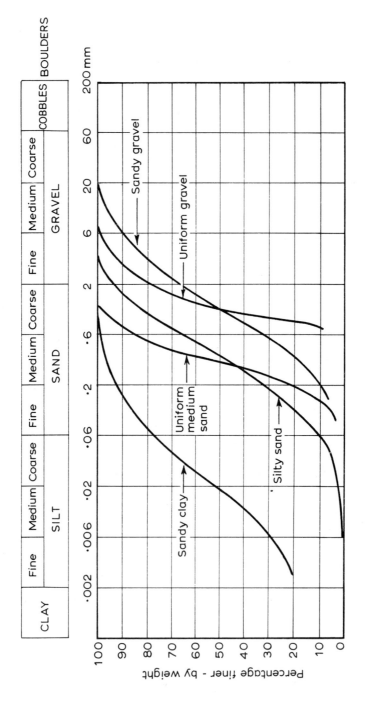

Figure 2.1 Particle size distribution chart.

It can be seen, therefore, that a D_{10} (or D_{60}) size *and* a uniformity coefficient can usefully classify a granular soil. In the case of clay soils, it is generally agreed that plasticity is a better index property. The Atterberg limits for clay minerals have been discussed in Section 1.4. These limits define the limiting moisture contents at which the soil exists in a plastic, as distinct from a solid, semi-solid or liquid state, and are dependent on the clay mineral content and related particle size of a clay soil. They have been used by Skempton (1953c) to define a quantity known as the *activity* of a soil, where

$$\text{soil activity} = \frac{\text{plasticity index \%}}{\text{\% clay size fraction (<0.002 mm)}}$$

The *plasticity index* P.I. is the difference between the liquid and plastic limits and indicates the moisture content range over which the soil remains plastic. Skempton (1953c) showed that activity was a valid concept and demonstrated a relationship between P.I. and the clay fraction for several clays and clay minerals. Typical activities quoted by Skempton range from 0.38 for kaolinite to 7.2 for Na-montmorillonite. The *liquidity index* (L.I.) is a particularly important parameter defining the deformation and flow properties of a clay at its natural moisture content:

$$\text{L.I.} = \frac{m\% - \text{P.L.\%}}{\text{P.I.\%}}$$

Various methods of identifying typical soil systems have been suggested. The difficulty of identification can be appreciated if some of the more commonly quoted materials are described as in Table 2.3.

Some of these terms, through common usage, can be quite specific and together with previously defined gravels, sands, silts, and clays may be used to give a reasonably sound description of a soil. Most, however, are adopted to describe too wide a range of materials to be of much use as the basis of a formal identification method. For instance, the terms *boulder clay* and *till* are virtually synonymous.

Possibly the best formal identification method suggested so far is the U.C.S. system (Casagrande, 1948; Wagner, 1957) upon which most soil identification systems are based or to which they are related. This method, listed in Table 2.4, classifies a series of typical

Table 2.3 Descriptions of some common soil materials

Boulder clay	an unstratified glacial clay containing various unsorted rock particles of all sizes.
Diatomaceous earth	a fine, porous, siliceous fossil soil, rather like flyash, and having the same use in concrete technology as a cement replacing pozzolan.
Gumbo	a fine plastic clay.
Loam	a mixture of sand, silt, clay and humus.
Laterite	a weathered red soil, largely derived from hydrated ferric oxide.
Loess	a wind-blown unstratified deposit of fine clay and silt.
Marl	usually describes a soft calcareous clay.
Peat	an organic, highly compressible vegetable deposit formed in swamp conditions.
Till	an unstratified glacial clay.
Volcanic ash	an unstratified fine-grained volcanic soil.

soils on the basis of *grading* and *plasticity*. The main requirement for such a classification is that each group of soils should respond in a similar mechanical manner. Wagner compares the important properties (permeability, strength, compressibility and workability) and the utility (dams, canals, foundations, highways) of each group in Table 2.5 graded in order of desirability from 1 (high) to 14 (low).

The findings confirm some earlier suggestions to the effect that the coarser the material, the greater generally is its strength and the finer the material, the worse are its engineering properties. The presence of organic materials in fine clay soils exacerbates the situation.

The use of Casagrande's (1948) A-line or plasticity chart classification of clays is worthy of special mention. The A-line (Table 2.4), having an equation P.I. = 0.73 (L.L. − 20) on a plasticity index (P.I.) versus liquid limit (L.L.) plot, is sometimes used to differentiate inorganic clays, which usually lie above the line, from organic clays which usually lie below the line. It is based, according to Terzaghi and Peck (1967), on traditional boring log classifications, and can be improved by further divisions representing degrees of plasticity (inorganic clays) or compressibility (organic clays) at liquid limits of

NOTES TO TABLE 2.4 (see next page)

[1] *Boundary classifications:* — Soils possessing characteristics of two groups are designated by combinations of group symbols. For example $GW–GC$, well graded gravel-sand mixture with clay binder.
[2] All sieve sizes on this chart are U.S. standard.

Field identification procedure for fine grained soils or fractions
These procedures are to be performed on the minus No. 40 sieve size particles, approximately 0.4 mm. For field classification purposes, screening is not intended; simply remove by hand the coarse particles that interfere with the tests.

Dilatancy (Reaction to shaking):
After removing particles larger than 0.4 mm size, prepare a pat of moist soil with a volume of about 2000 mm^3. Add enough water if necessary to make the soil soft but not sticky.

Place the pat in the open palm of one hand and shake horizontally, striking vigorously against the other hand several times. A positive reaction consists of the appearance of water on the surface of the pat which changes to a livery consistency and becomes glossy. When the sample is squeezed between the fingers, the water and gloss disappear from the surface, the pat stiffens and finally it cracks or crumbles. The rapidity of appearance of water during shaking and of its disappearance during squeezing assist in identifying the character of the fines in a soil.

Very fine clean sands give the quickest and most distinct reaction whereas a plastic clay has no reaction. Inorganic silts, such as a typical rock flour, show a moderately quick reaction.

Dry Strength (Crushing characteristics):
After removing particles larger than 0.4 mm size, mould a pat of soil to the consistency of putty, adding water if necessary. Allow the pat to dry completely by oven, sun or air drying, and then test its strength by breaking and crumbling between the fingers. This strength is a measure of the character and quantity of the colloidal fraction contained in the soil. The dry strength increases with increasing plasticity.

High dry strength is characteristic for clays of the CH group. A typical inorganic silt possesses only very slight dry strength. Silty fine sands and silts have about the same slight dry strength, but can be distinguished by the feel when powdering the dried specimen. Fine sand feels gritty whereas a typical silt has the smooth feel of flour.

Toughness (Consistency near plastic limit)
After removing particles larger than 0.4 mm size, a specimen of soil about 2000 mm^3 in size, is moulded to the consistency of putty. If too dry, water must be added and if sticky, the specimen should be spread out in a thin layer and allowed to lose some moisture by evaporation. Then the specimen is rolled out by hand on a smooth surface or between the palms into a thread about one-eighth inch in diameter. The thread is then folded and re-rolled repeatedly. During this manipulation the moisture content is gradually reduced and the specimen stiffens, finally loses its plasticity, and crumbles when the plastic limit is reached.

After the thread crumbles, the pieces should be lumped together and a slight kneading action continued until the lump crumbles.

The tougher the thread near the plastic limit and the stiffer the lump when it finally crumbles, the more potent is the colloidal clay fraction in the soil. Weakness of the thread at the plastic limit and quick loss of coherence of the lump below the plastic limit indicate either inorganic clay or low plasticity, or materials such as kaolin-type clays and organic clays which occur below the A-line.

Highly organic clays have a very weak and spongy feel at the plastic limit.

Table 2.4 Unified Classification System for soils (*after* Wagner, 1957)

Field identification procedures (Excluding particles larger than 76 mm and basing fractions on estimated weights)							Group symbols[1]	Typical names
Coarse grained soils (More than half of material is larger than No. 200 sieve size[2] (0.08 mm)) (The No. 200 sieve size is about the smallest particle visible to naked eye)	Gravels (More than half of coarse fraction is larger than No. 4 sieve size) (For visual classification, 8 mm size may be used as equivalent to the No. 4 sieve size)	Clean gravels (little or no fines)	Wide range in grain size and substantial amounts of all intermediate particle sizes				GW	Well graded gravels, gravel-sand mixtures, little or no fines
			Predominantly one size or a range of sizes with some intermediate sizes missing				GP	Poorly graded gravels, gravel-sand mixtures, little or no fines
		Gravels with fines (appreciable amount of fines)	Non-plastic fines (for identification procedures see *ML* below)				GM	Silty gravels, poorly graded gravel-sand-silt mixtures
			Plastic fines (for identification procedures, see *CL* below)				GC	Clayey gravels, poorly graded gravel-sand-clay mixtures
	Sands (More than half of coarse fraction is smaller than No. 4 sieve size)	Clean sands (little or no fines)	Wide range in grain sizes and substantial amounts of all intermediate particle sizes				SW	Well graded sands, gravelly sands, little or no fines
			Predominantly one size or a range of sizes with some intermediate sizes missing				SP	Poorly graded sands, gravel sands, little or no fines
		Sands with fines (appreciable amount of fines)	Non-plastic fines (for identification procedures, see *ML* below)				SM	Silty sands, poorly graded sand-silt mixtures
			Plastic fines (for identification procedures, see *CL* below)				SC	Clayey sands, poorly graded sand-clay mixtures
Fine grained soils (More than half of material is smaller than No. 200 sieve size (0.08 mm))	*Identification procedures on fraction smaller than No. 40 sieve size (0.4 mm)*			Dry strength (crushing characteristics)	Dilatancy (reaction to shaking)	Toughness (consistency near plastic limit)		
	Silts and clays liquid limit less than 50			None to slight	Quick to slow	None	ML	Inorganic silts and very fine sands, rock flour, silty or clayey fine sands with slight plasticity
				Medium to high	None to very slow	Medium	CL	Inorganic clays of low to medium plasticity, gravelly clays, sandy clays, silty clays, lean clays
				Slight to medium	Slow	Slight	OL	Organic silts and organic silt-clays of low plasticity
	Silts and clays liquid limit greater than 50			Slight to medium	Slow to none	Slight to medium	MH	Inorganic silts, micaceous or diatomaceous fine sandy or silty soils, elastic silts
				High to very high	None	High	CH	Inorganic clays of high plasticity, fat clays
				Medium to high	None to very slow	Slight to medium	OH	Organic clays of medium to high plasticity
Highly organic soils			Readily identified by colour, odour, spongy feel and frequently by fibrous texture				Pt	Peat and other highly organic soils

ormation required for cribing soils	Laboratory classification criteria			
ve typical name; indicate proximate percentages of sand and vel; maximum size; angularity, face condition, and hardness of e coarse grains; local or geological me and other pertinent descriptive ormation; and symbol in rentheses r undisturbed soils add information stratification, degree of mpactness, cementation, moisture nditions and drainage aracteristics ample: Silty sand, gravelly; about 20% hard, angular gravel particles 12 mm maximum size; rounded and subangular sand grains coarse to fine, about 15% non-plastic fines with low dry strength; well-compacted and moist in place; alluvial sand; (SM)	Determine percentages of gravel and sand from grain size curve Depending on percentage of fines (fraction smaller than No. 200 sieve size), coarse grained soils are classified as follows: Less than 5% — GW, GP, SW, SP More than 12% — GM, GC, SM, SC 5% to 12% — Borderline cases requiring use of dual symbols	$CU = \dfrac{D_{60}}{D_{10}}$ Greater than 4 $CC = \dfrac{(D_{30})^2}{D_{10} \times D_{60}}$ Between 1 and 3 Not meeting all gradation requirements for *GW*		
		Atterberg limits below 'A' line, or PI less than 4	Above 'A' line with P.I. between 4 and 7 are *borderline* cases requiring use of dual symbols	
		Atterberg limits above 'A' line, with P.I. greater than		
		$CU = \dfrac{D_{60}}{D_{10}}$ Greater than 6 $CC = \dfrac{(D_{30})^2}{D_{10} \times D_{60}}$ Between 1 and 3 Not meeting all gradation requirements for *SW*		
		Atterberg limits below 'A' line, or P.I. less than 4	Above 'A' line with PI between 4 and 7 are *borderline* cases requiring use of dual symbols	
		Atterberg limits show 'A' line with PI greater than 7		
ive typical name; indicate degree nd character of plasticity, amount nd maximum size of coarse grains; olour in wet condition, odour if any, ocal or geological name, and other ertinent descriptive information, nd symbol in parentheses or undisturbed soils add information n structure, stratification, onsistency in undisturbed and emoulded states, moisture and rainage conditions xample: Clayey silt, brown; slightly plastic; small percentage of fine sand; numerous vertical root holes; firm and dry in place; loess; (ML)	Use grain size curve in identifying the fractions as given under field identification	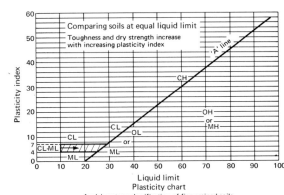 Plasticity chart for laboratory classification of fine grained soils		

Table 2.5 Engineering use chart (*after* Wagner, 1957)

Typical names of soil groups	Group symbols	Important properties			
		Permeability when compacted	*Shearing strength when compacted and saturated*	*Compressibility when compacted and saturated*	*Workability as a construction material*
Well-graded gravels, gravel-sand mixtures, little or no fines	GW	pervious	excellent	negligible	excellent
Poorly graded gravels, gravel-sand mixtures, little or no fines	GP	very pervious	good	negligible	good
Silty gravels, poorly graded gravel-sand-silt mixtures	GM	semi-pervious to impervious	good	negligible	good
Clayey gravels, poorly graded gravel-sand-clay mixtures	GC	impervious	good to fair	very low	good
Well-graded sands, gravelly sands, little or no fines	SW	pervious	excellent	negligible	excellent
Poorly graded sands, gravelly sands, little or no fines	SP	pervious	good	very low	fair
Silty sands, poorly graded sand-silt mixtures	SM	semi-pervious to impervious	good	low	fair
Clayey sands, poorly graded sand-clay mixtures	SC	impervious	good to fair	low	good
Inorganic silts and very fine sands, rock flour, silty or clayey fine sands with slight plasticity	ML	semi-pervious to impervious	fair	medium	fair
Inorganic clays of low to medium plasticity, gravelly clays, sandy clays, silty clays, lean clays	CL	impervious	fair	medium	good to fair
Organic silts and organic silt-clays of low plasticity	OL	semi-pervious to impervious	poor	medium	fair
Inorganic silts, micaceous or diatomaceous fine sandy or silty soils, elastic silts	MH	semi-pervious to impervious	fair to poor	high	poor
Inorganic clays of high plasticity, fat clays	CH	impervious	poor	high	poor
Organic clays of medium to high plasticity	OH	impervious	poor	high	poor
Peat and other highly organic soils	Pt	–	–	–	–

Relative desirability for various uses

	Rolled earth dams			Canal sections		Foundations		Roadways			
								Fills			
...mo- ...eous ...bank- ...nt	Core	Shell	Erosion resist-ance	Com-pacted earth lining	Seepage im-portant	Seepage not im-portant	Frost heave not possible	Frost heave possible	Sur-facing
	—	1	1	—	—	1	1	1	3
	—	2	2	—	—	3	3	3	—
	4	—	4	4	1	4	4	9	5
	1	—	3	1	2	6	5	5	1
	—	3 if gravelly	6	—	—	2	2	2	4
	—	4 if gravelly	7 if gravelly	—	—	5	6	4	—
4	5	—	8 if gravelly	5 erosion critical	3	7	8	10	6
3	2	—	5	2	4	8	7	6	2
6	6	—	—	6 erosion critical	6	9	10	11	—
5	3	—	9	3	5	10	9	7	7
8	8	—	—	7 erosion critical	7	11	11	12	—
9	9	—	—	—	8	12	12	13	—
7	7	—	10	8 volume change critical	9	13	13	8	—
10	10	—	—	—	10	14	14	14	—
—	—	—	—	—	—	—	—	—	—

20, 30, and 50, thus:

Liquid limit	Soil type
<20	cohesionless
20–30	low plasticity and compressibility
30–50	medium plasticity and compressibility
>50	high plasticity and compressibility

An example of the A-line applied to Tyne clays is shown in Figure 3.17b.

2.3 Physical properties of particulate systems

In nature, rock particle systems (or soils) are complex assemblages of rock particles of various shapes, sizes and mineralogical compositions, compacted to various densities. They are also, by definition, two, three or four *phase* mixes having intergranular spaces filled with air, water or possibly ice. Normally, soils may be regarded as three-phase systems (Figure 2.2), with the soil as the solid phase. The voids between the soil skeleton contain water (liquid phase) and air (gaseous phase). Various relationships between the phases can be derived to define the main physical properties of the soil. These would also apply to any rock comprising a three-phase system, or for that matter a two-phase system since the air volume is usually assumed weightless and is included in the voids (V_v).

If W and V are the total weight and volume respectively of a sample, and the subscripts w, s and v refer respectively to the water, solid and void phases, the main relationships are:

(a) *Volume*

$$\text{Porosity } n = \frac{V_v}{V},$$

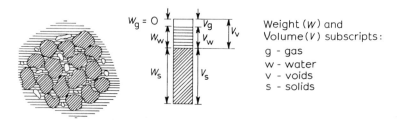

Figure 2.2   Three phase soil system.

Void ratio $e = \dfrac{V_v}{V_s}$,

Hence: $n = \dfrac{e}{1+e}$, $e = \dfrac{n}{1-n}$

Saturation $S = \dfrac{V_w}{V_v} = \dfrac{W_w}{\gamma_w V_v}$

(b) *Weight*

Moisture content $m = \dfrac{W_w}{W_s}$

Bulk unit weight $\gamma = \dfrac{W}{V}$; Bulk density $\rho = \dfrac{W}{V_g}$

Dry bulk unit weight $\gamma_d = \dfrac{W_s}{V} = \dfrac{\gamma}{1+m}$

Submerged (buoyant) bulk unit weight $\gamma_b = \gamma - \gamma_w$

where γ_w is the unit weight of the porewater

(c) *Specific gravity*

Specific gravity (water) $G_w = 1$ at $4°C$

Specific gravity (solids) $G_s = \dfrac{\gamma_s}{\gamma_w}$

Hence,

$$G_s m = Se = \dfrac{V_w}{V_s}$$

These relationships are fundamental index properties used by engineers and geologists as a simple means of describing soils and porous rocks. Obviously, the magnitude of porosity, unit weight or void ratio will depend upon the grading and the degree of compaction of the soil. This can be illustrated most simply by considering the case of uniform spherical particles, an unreal example of rock fragments of equal radius and absolute roundness. In this case, the void volume will depend solely on the packing of the particles (Figure 2.3). Thus, cubic-packed particles will form a less dense assemblage than will

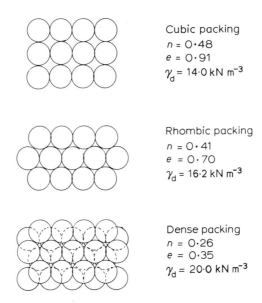

Figure 2.3 Arrangements of uniform spheres. Unit weights are based on a specific gravity of 2.70 for glass.

rhombic or hexagonally-packed particles which will in turn be less dense than the densest packing state in which spheres are always resting against three or four other spheres. Even in the densest configuration, however, there is still 26 per cent pore space in the particulate system. In order to reduce and ultimately to eliminate this porespace, smaller particles must be introduced into it; in other words a *graded* material is needed. In the case of a well-graded assemblage, the unit weight depends largely on the degree of compaction of the soil. Compaction (which is discussed later in its engineering sense) can usefully be characterized in terms of the *relative density* (D_r)* of dry soil:

$$D_r = \frac{e_{max} - e}{e_{max} - e_{min}} = \frac{\gamma_{d\,max}}{\gamma_d}\left(\frac{\gamma_d - \gamma_{d\,min}}{\gamma_{d\,max} - \gamma_{d\,min}}\right)$$

$$= \frac{(n_{max} - n)(1 - n_{min})}{(n_{max} - n_{min})(1 - n)}$$

(2.1)

*This is a term widely used in soil mechanics (see Terzaghi and Peck, 1967) to describe the looseness of a soil, and should not be confused with specific gravity.

where *max.* and *min.* for e and n refer respectively to the loosest and densest states of the soil. Typical D_r values are suggested by Lambe and Whitman (1969):

0–15 per cent	very loose
13–35 per cent	loose
35–65 per cent	medium
65–85 per cent	dense
85–100 per cent	very dense

These values depend to a certain extent on the methods used to achieve loose and dense states, usually some form of dry pouring and wet rodding or vibrating. Under such conditions, typical void ratios range between 0.2 (minimum) to 1.2 (maximum), porosities vary from 20 to 50 per cent, and dry unit weights vary from 12 to 22 kN m^{-3} for granular soils. This extreme dependence on relative density means that it is difficult to quote accurate data for hypothetical soils. Ideally, such data should be accompanied by information on the relative density of the soil or on the specific gravity of the soil phase. The situation is even more varied in clay soils. Saturated bentonitic clays containing a substantial sodium montmorillonite content can have void ratios varying between 25 at low pressures and low values at high pressures or when artifically dried.

2.4 Permeability of particulate systems

The rate at which a fluid, usually water, is able to flow through a particulate system can have an important effect on the properties of that system. For instance, in the specific case of flow through soils, permeability can determine the rate of leakage beneath or through a dam and the rate at which a saturated soil will be allowed to *compress* under load. This latter effect, introducing the concepts of consolidation, will be considered in detail later. The emphasis in the present section is on factors affecting rate of flow, expressed in terms of permeability.

In general, the voids in a particulate system may be considered interconnected and continuous, and this holds for most soils however small their particle size. The implication is that a soil can be treated, to a first approximation, as a bundle of capillaries each of *uniform* diameter (d). Poiseuille's law for laminar fluid flow through a single tube states (after Taylor, 1948) that:

$$Q = \frac{\pi d^4 \gamma_f}{128\eta} \frac{\delta P}{\delta L} \quad \text{or} \quad v = \frac{d^2 \gamma_f}{32\eta} \frac{\delta P}{\delta L} \tag{2.2}$$

where Q is the steady state volume flow rate,

v is the discharge velocity,

$\frac{\delta P}{\delta L}$ is the pressure gradient along the tube,

γ_f is the fluid unit weight,

and η is the fluid viscosity

Thus, for the analogous situation of a bundle of N tubes covering a discharge area A where $A = \frac{N\pi d^2}{4n}$, and n is the soil porosity, equation 2.2 becomes:

$$Q = \frac{N\pi d^4}{128} \frac{\gamma_f}{\eta} \frac{\delta P}{\delta L} \tag{2.3}$$

and

$$v = \frac{Q}{A} = \frac{nd^2}{32} \frac{\gamma_f}{\eta} \frac{\delta P}{\delta L} \tag{2.4}$$

Then, substituting $k = \frac{nd^2 \gamma_f}{32\eta}$

gives

$$v = k \frac{\delta P}{\delta L} = ki \tag{2.5}$$

which is the governing equation known as Darcy's Law for flow through a porous medium where i is the *pressure or hydraulic gradient* and k is the *coefficient of permeability*. Coefficient k has the same dimensions as velocity flow and is constant for a particular soil (represented by n,d) and fluid (represented by γ_f, η). For the usual case of water flow through soils at temperatures in the region 5° to 10°C (η is dependent on temperature), typical k values fall within the ranges:

Clay	$k < 10^{-9}$ m s^{-1}.
Silts	$10^{-9} < k < 10^{-7}$ m s^{-1}.
Fine sands	$10^{-7} < k < 10^{-5}$ m s^{-1}.
Coarse sands	$10^{-5} < k < 10^{-2}$ m s^{-1}.
Gravels	$10^{-2} < k$ m s^{-1}.

Clays and some silts are therefore virtually impervious, whilst very rapid flow rates can occur through coarse sands and gravels subjected to relatively low hydraulic gradients. Nevertheless, for all practical purposes, the assumption of laminar rather than turbulent flow is valid.

Examination of relative permeabilities indicates a relationship between the coefficient of permeability and *particle size*. This relationship can be formalized (maintaining the capillary analogy) by considering in its turn the relationship between particle size and the dimension of the proposed capillary openings in a particulate system, which itself can be generalized in terms of effective hydraulic radius α where:

$$\alpha = \frac{2 \times \text{Volume of unit capillary}}{\text{Surface area of unit capillary}} \qquad (2.6)$$

Since, by definition, the porosity n is the pore volume per unit volume of material, and the *specific surface area* S_0 per unit volume of a particle system is the surface area of particles per unit volume of particles, then for a bunch of capillaries:

$$\alpha = \frac{2n}{S_0(1-n)} = \frac{2e}{S_0} \qquad (2.7)$$

and substituting $\alpha = \dfrac{d}{2}$ in equation 2.4 gives

$$v = \frac{n^3}{(1-n)^2} \frac{\gamma_f}{2\eta S_0^2} \frac{\delta P}{\delta L} \qquad (2.8)$$

Since S_0 is approximately equal to $6/d_0$, where d_0 is the diameter of equivalent spherical particles with the same surface area, the above

Table 2.6 Permeability and soil size relationship

k m s^{-1}	d_0 mm	Size classification
1.10 × 10^{-1}	6	Medium-fine gravel
1.22 × 10^{-2}	2	Coarse sand–fine gravel
1.10 × 10^{-3}	0.6	Coarse–medium sand
1.22 × 10^{-4}	0.2	Medium–fine sand
1.10 × 10^{-5}	0.06	Medium silt-fine sand
1.22 × 10^{-6}	0.02	Coarse–medium silt

equation may be written:

$$v = \frac{1}{72} \frac{n^3}{(1-n)^2} \frac{\gamma_f d_0^2}{\eta} \frac{\delta P}{\delta L} \qquad (2.9)$$

This equation is sometimes known as the Kozeny–Carman equation for flow through a porous medium. In fact, in order to accommodate the uneven flowpath, an empirical modification to allow for increased friction losses is needed. The particular form (Scott, 1963a; Raffle quoted by Farmer, 1969) is usually:

$$v = \frac{1}{180} \frac{n^3}{(1-n)^2} \frac{\gamma_f d_0^2}{\eta} \frac{\delta P}{\delta L} \qquad (2.10)$$

This is similar to the Darcy equation (equation 2.5) and relates k to d_0 in the form:

$$k = \frac{1}{180} \frac{n^3}{(1-n)^2} \frac{\gamma_f d_0^2}{\eta} \qquad (2.11)$$

Putting $n = 0.3$, permeability and particle size can be related in Table 2.6 for water flow. Although the assumption of uniform particle size is implied in the reasoning, Terzaghi and Peck (1967) suggest a similarity between d_0 and D_{10} particle sizes which, particularly for loose, reasonably uniform, cohesionless sands, provides an adequate basis for the calculation of permeability.

2.5 Representation of stress in a soil mass

Soil mechanics may be defined as the study of the reaction of naturally-occurring masses of fine rock particles and clay mineral assemblages to various naturally and artificially created stress fields. The inexact nature of soils virtually ensures that soil mechanics remains an uneasy mixture of theoretical analysis and empirical data.

Rock Particles and Particle Systems

The treatment in the present book is necessarily limited and aims only to present some of the fundamental concepts essential to the study of engineering geology. Engineering design, which is the ultimate *raison d'être* of soil mechanics, is covered in the standard works, particularly Taylor (1948), Terzaghi (1948), Terzaghi and Peck (1967), Scott (1963b) and Lambe and Whitman (1969) which are recommended for further reading.

Consider first of all that the soil comprises an assemblage of particles and voids, that the voids are dry and that the pore pressure in them is atmospheric. Under load, the forces in the particulate system are transmitted through discrete contact points. Figure 2.4 shows that because of the limited size of the particles in relation to the area on which the external and/or body force acts, the resultant stress will be properly represented as force per total area rather than as force per contact area. This distinction is important because the contact stresses will be several orders of magnitude greater than the average stress transmitted over an area of the whole assemblage. Thus, for a given overall loading, the density and severity of the individual pressure contacts will control the degree of soil compression. This is considered in Section 2.8. Compression as a function of the average stress may be interpreted in terms of various deformation and yield criteria, some of which are considered briefly in Chapter 4 in the context of a stiffer rock framework.

In a real soil, interconnections between voids may be tortuous and they may be partly or fully saturated. Drainage and pressure equalization with the atmosphere may therefore be inhibited and compression under load may be restricted by the support offered by the pore phase.

It follows that the interparticle stresses will be correspondingly reduced. On the other hand, if the soil is in a state of desiccation, then the void pressure will be below atmospheric and the particle framework will be required to support rather higher local stresses with a corresponding increase in compression. These important concepts are developed further in the present chapter and in Chapter 3.

If, temporarily, we consider the particle assemblage as an equivalent continuum, accepting the possible presence and implications of a void infilling phase but ignoring local interparticle behaviour, the stress in the body may be described most simply by nine components (Figure 2.5):

50 Principles of Engineering Geology

$$\begin{matrix} \sigma_x & \tau_{xy} & \tau_{xz} \\ \tau_{yx} & \sigma_y & \tau_{yz} \\ \tau_{zx} & \tau_{zy} & \sigma_z \end{matrix}$$

which represent the resolved stress vectors normal and tangential to the yz, xz and zy planes as represented by the surfaces of an elemental cube. If the material is in static, internally irrotational equilibrium then $\tau_{xy} = \tau_{yx}$, $\tau_{zy} = \tau_{yz}$ and $\tau_{zx} = \tau_{xz}$.

This stress tensor can further be resolved (see Jaeger, 1969 or Jaeger and Cook, 1969 for a full theoretical treatment) into three orthogonal planes on each of which the stress is uniquely normal and the shear stresses are zero. These planes are called the *principal planes*, and the stresses acting on them – designated $\sigma_1, \sigma_2, \sigma_3$ – are called the *principal stresses*. Of these, the greatest (σ_1) is called the *major principal stress*, the smallest (σ_3) is called the *minor principal stress* and the intermediate (σ_2), the *intermediate principal stress*.

When the magnitudes and directions of the principal stresses are known, it is convenient in any stress analysis to use these as reference parameters. In the case of ground stresses in the undisturbed remote body of a soil or rock mass it is conventional, and usually correct, to assume that the three principal stresses act in a vertical and two horizontal orthogonal directions and that the horizontal geostatic stresses are equal to one another. The stress situation is therefore axi-symmetric about the vertical, an assumption that is quite sound where the ground surface and bedding are horizontal and the ground properties are uniform in a horizontal direction. Soils generally satisfy these conditions. There would then be no shear stresses acting

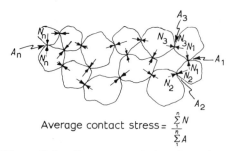

Average contact stress = $\dfrac{\sum\limits_1^n N}{\sum\limits_1^n A}$

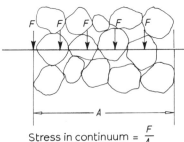

Stress in continuum = $\dfrac{F}{A}$

Figure 2.4 Stress transmission through a granular soil (porespace at atmospheric pressure).

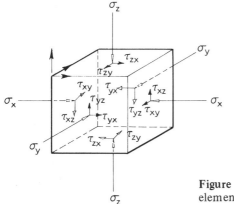

Figure 2.5 Stress components on an elemental cube.

on horizontal or vertical planes and the vertical geostatic stress σ_z at any depth z will be equal to the pressure of the soil column γz above that depth. The horizontal geostatic stress σ_x under these same conditions will be uniform in all directions and depending on the loading history of the soil will usually be less, but may be greater, than the vertical stress. Any pressurized porewater which may be present in a saturated soil will be supporting a proportion of the total stress, this support (u) being the same independent of direction. We may therefore denote the stresses actually supported by the solid skeleton as $\sigma'_z (= \sigma_z - u)$ and $\sigma'_x (= \sigma_x - u)$. Parameters σ'_z and σ'_x are designated *effective stresses* and they will be considered in greater detail in Section 2.6.

The ratio σ'_x/σ'_z for the ground in which lateral strain has been inhibited is known as the *coefficient* of *earth pressure at rest* and is designated K_0. During the accumulation of a detritus overload in a deep sedimentary basin, there is a limited facility for lateral adjustment, with a consequential enhancement of lateral pressure. When overburden is denuded, lateral pressure relaxation occurs only very slowly. K_0 may therefore range both below and above unity.

Visualizing vertical pressure at a point in an undisturbed soil mass as a major principal stress, we may write (provided that $K_0 < 1$):

$$\sigma_1 = \sigma_z = \sigma_v = \sigma'_v + u = \gamma z \qquad (2.12)$$

in order to cover the range of symbols commonly used. The two

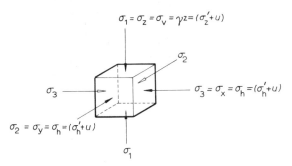

Figure 2.6 Representation of principal ground stresses under simple gravitational loading.

equal horizontal minor principal stresses (see Figure 2.6) may be written

$$\sigma_3 = \sigma_2 = \sigma_x = \sigma_y = \sigma_h = \sigma'_h + u = K_0 \sigma'_v + u \qquad (2.13)$$

Since the stress fields in soil mechanics are usually compressive, a positive sign is used to denote compression.

For most purposes, therefore, stress at a point in a horizontally-layered soil (or rock) may be modelled 2-dimensionally, so conveniently eliminating much of the complicated geometry involved in solving 3-dimensional problems. There are obviously ground stress situations, particularly in rock engineering and structural geology, where the presence of an intermediate principal stress cannot be ignored and also where a major principal stress acts horizontally. There is also a growing tendency to acknowledge in analytical soil mechanics the 3-dimensional nature of soil slope failures. Nevertheless, a 2-dimensional approach is satisfactory for most soil mechanics problems. In some rock engineering problems, however, and particularly those associated with slope stability, it is necessary to analyse 3-dimensionally. The necessary operations are outlined in Chapters 6 and 10.

Specifically, therefore, we are interested in a plane containing the major and minor principal stresses and in the stress situation of that plane. As shown in Figure 2.7, these principal stresses can be resolved as normal (σ_n) and shear (τ) stresses acting on any plane normal to the plane containing the major and minor principal stresses. The plane is inclined at an angle θ to the σ_x or σ_3 direction and the

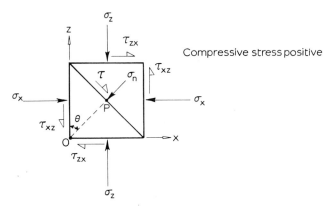

Figure 2.7 Two-dimensional stress field. Consider stresses in the xz plane representing part of the stress situation at O. Then the force across any plane through O, of area a, and having a normal OP inclined at an angle θ to OZ will be:

$$\sigma_n a = (\sigma_z \cos\theta - \tau_{zx} \sin\theta)a \cos\theta + (\sigma_x \sin\theta - \tau_{xz} \cos\theta)a \sin\theta$$

and the force along the plane by:

$$\tau a = (\sigma_z \sin\theta + \tau_{zx} \cos\theta)a \cos\theta - (\sigma_x \cos\theta + \tau_{xz} \sin\theta)a \sin\theta$$

Since $\tau_{zx} = \tau_{xz}$ these become:

$$\sigma_n = \sigma_z \cos^2\theta + \sigma_x \sin^2\theta - 2\tau_{xz} \sin\theta \cos\theta$$
$$\tau = (\sigma_z - \sigma_x)\sin\theta \cos\theta - \tau_{xz}(\sin^2\theta - \cos^2\theta)$$

Substituting the principal stress axes for the x,z axes and putting $\sigma_z = \sigma_1$, $\sigma_x = \sigma_3$, and $\tau_{xz} = \tau_{zx} = 0$ gives:

$$\sigma_n = \sigma_1 \cos^2\theta + \sigma_3 \sin^2\theta = \frac{\sigma_1 + \sigma_3}{2} + \frac{\sigma_1 - \sigma_3}{2} \cos 2\theta$$

$$\tau = (\sigma_1 - \sigma_3)\sin\theta \cos\theta = \frac{\sigma_1 - \sigma_3}{2} \sin 2\theta$$

normal to the plane is inclined at angle θ to the σ_z or σ_1 direction, whence:

$$\sigma_n = \frac{\sigma_1 + \sigma_3}{2} + \frac{\sigma_1 - \sigma_3}{2} \cos 2\theta \qquad (2.14)$$

$$\tau = \frac{\sigma_1 - \sigma_3}{2} \sin 2\theta \qquad (2.15)$$

This stress situation can be represented graphically in two simple ways: as a *Mohr circle* diagram or as a *p–q diagram*.

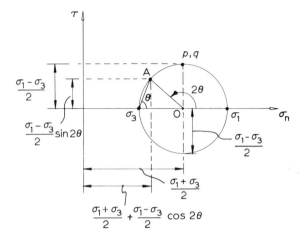

Figure 2.8 Mohr circle representation.

In the former case, equations 2.14 and 2.15 describe a circle of radius $(\sigma_1 - \sigma_3)/2$ and centre $(\sigma_1 + \sigma_3)/2$ on axes representing τ and σ_n (Figure 2.8). The periphery of the circle describes the stress situations in all the planes normal to the plane containing the major and minor principal stresses and having *normals* inclined at an angle θ to the direction of *major* principal stress. In Figure 2.8, OA represents a *plane* at an angle θ to the *minor* principal stress direction, having a normal at an angle θ to the major principal stress direction and being subjected to a normal stress given by equation 2.14 and a shear stress given by equation 2.15.

Thus, given the principal stresses and their directions it is possible to determine the stresses in any other direction. By similar construction, given the normal stress and shear stress on any two planes, the magnitudes and directions of the principal stresses can be found.

On a p–q diagram, the state of stress is represented by a single 'top' point having co-ordinates $p = (\sigma_1 + \sigma_3)/2$ and $q = (\sigma_1 - \sigma_3)/2$, and representing the plane of maximum shear stress on the Mohr circle construction. As such, it is no substitute for the Mohr circle representation if stress situations in various directions are to be determined. It is, however, extremely useful for illustrating the loading history of a soil or for representing the successive states of stress at various depths in the ground. It also lends itself more readily to least squares linear regression analysis for subsequent back substitution of the c, ϕ shear strength parameters into the τ, σ_n (Mohr) plane (see Section 2.9).

Use of the $p-q$ representation, as popularized by Lambe (1967) — see also Lambe and Whitman (1969) — is most simply illustrated by considering successive states of stress below (say) ground surface, where the σ'_x/σ'_z ratio remains constant. In this case, the *stress path* with increasing depth may be simply represented as in Figure 2.9 by a straight line rather than as a series of Mohr circle constructions. If as, for example, in the London Clay, the earth pressure at rest coefficient varies as a function of depth (Skempton, 1961a), then the $p'-q$ relationship will be non-linear as in Figure 2.10.

Similarly, the stress path during various stages of a triaxial test when the imposed vertical effective stress σ'_v is increased after initial application of an effective radial stress σ'_r (which subsequently remains constant) can be simply represented as a $p'-q$ diagram, but only with difficulty as a series of Mohr circles (see Figure 2.9). The triaxial test is considered in Sections 2.8, 2.9 and 2.10.

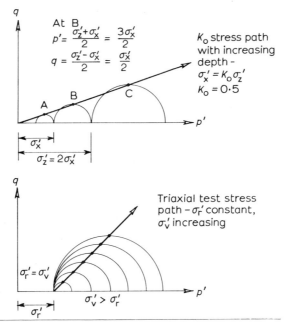

Figure 2.9   Representation of stress paths. A, B, C are 'top points'.

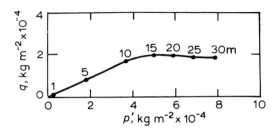

Figure 2.10 Undisturbed stresses at different depths in the London Clay. Calculations are based on Skempton's (1961a) variation of K_0 with depth.

2.6 Effective stress

The general subject area of continuum mechanics is beyond the scope of the present book. It is worth mentioning, however, that boundary value problems can be solved in soils as in any other engineering material by substituting, in the controlling equations, experimentally determined values of material constants relating stress and strain. Where soils do differ from most other engineering materials is that they are dominantly two phase systems, comprising a relatively flexible particle matrix, the pores of which are filled with a relatively incompressible liquid, namely water. When a soil is saturated with free rather than fully adsorbed water, the porewater has a positive pressure, and when it is stationary exerts a pressure equal to the hydraulic head of free water.

The porewater pressure, u, at a point O (Figure 2.11) in a soil mass at a depth z below the ground surface and a depth h below the water table will then be given by:

$$u = \gamma_w h \qquad (2.16)$$

where γ_w is the unit weight of water.

Any porewater pressure will act in all directions in the porespace and will exert an uplift or buoyancy effect on the soil skeleton surrounding the porespace, so that the *effective stress* acting through the soil skeleton at O will be equivalent to the difference between the total ground stress and the pressure of the porewater:

$$\sigma' = \sigma - u \qquad (2.17)$$

where σ' denotes effective stress.

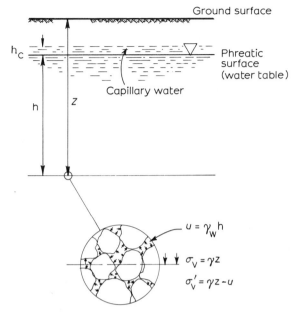

Figure 2.11 Representation of porewater pressure at a point.

The concept of effective stress was developed by Terzaghi and is fundamental to soil mechanics. The stresses in all directions will be affected, thus:

$$\sigma'_v = \sigma_v - u = \gamma z - \gamma_w h \qquad (2.18)$$

$$\sigma'_h = \sigma_h - u = K_0 \sigma'_v \qquad (2.19)$$

The coefficient of earth pressure at rest K_0 (see Section 2.5) in an undisturbed saturated soil represents actual interparticle contact stress conditions and will therefore always be expressed in terms of effective stress.

In the case of a *partially saturated* soil, the effective stress equation must be modified to allow for pore pressure exerted by the air phase. Bishop et al (1960) suggest an expression:

$$\sigma' = \sigma - [u_g - a_w(u_g - u_w)] \qquad (2.20)$$

where u_g is the pore pressure exerted by the air, and a_w is related to the degree of saturation.

Equations 2.17 and 2.20 are valid for most problems in soil mechanics. They are, however, experimentally derived and it is interesting to examine briefly the fundamental principles of effective stress. The simplest approach is to assume that effective stress is equal to the intergranular stress in a saturated soil:

$$\sigma' = \sigma - (1 - a)u_w \qquad (2.21)$$

where a is the intergranular contact area per gross area of material.

Equation 2.21 suggests that if a were high in a particulate material, then the concept of effective stress would be invalid. Skempton (1961b) showed that this was not necessarily true and developed and demonstrated relationships between effective stress, shear resistance and compressibility:

$$\sigma' = \sigma - (1 - a \tan \phi_p / \tan \phi_m) u_w \qquad (2.22)$$

$$\sigma' = \sigma - (1 - m_p / m_m) u_w \qquad (2.23)$$

where ϕ_p, ϕ_m are respectively the particle and mass friction angles, and m_p, m_m are the respective compressibilities.

In a particle system where a is small or where $\phi_m \gg \phi_p$ and $m_m \gg m_p$, these equations approximate to equation 2.17. In a denser, stiffer-frame material such as rock, the importance of the effective stress concept is reduced.

Above the water table (Figure 2.11), porewater is held in the soil structure by capillary forces. The surface tension of the water exerts an attractive force between capillary sidewalls, and the porewater pressure in this case will be *negative*. Negative or suction porewater pressures exist in all unsaturated soils (see Section 3.11) but are most usually associated with the *capillary zone*.

Terzaghi and Peck (1967) quote an empirical equation for the height of this zone above the water table:

$$h_c = \frac{C}{e D_{10}} \qquad (2.24)$$

where D_{10} is the D_{10} particle size (see Section 2.2),
      e is the void ratio,
and C is an areal coefficient depending on size and impurities.

In Table 2.7, Lane and Washburn (1946) quote experimental values of h_c (which includes a partly-saturated upper part of the capillary zone) for various soils.

Table 2.7 Capillary parameters for some particulate materials having different particle sizes (*after* Lane and Washburn, 1946)

Soil	D_{10} mm	e	h_c m
Silt	0.006	0.95	3.6
Fine Sand	0.03	0.36	1.6
Medium Sand	0.02	0.57	2.4
Coarse Sand	0.11	0.27	0.8
Silty Gravel	0.06	0.45	1.1
Fine Gravel	0.30	0.29	0.2
Sandy Gravel	0.20	0.45	0.3
Coarse Gravel	0.82	0.27	0.05

The magnitude of the suction porewater pressure in the capillary zone will be equal to:

$$u = -\gamma_w h_c \qquad (2.25)$$

The *physical* effect of saturation on soil strength can be illustrated most simply as a Mohr circle (Figure 2.12) or p–q graph where p, q, τ and σ_n are expressed in *total* rather than effective stress terms. The relationship between the two states may be expressed:

$$p = \frac{\sigma_1 + \sigma_3}{2} = \frac{(\sigma_1' + u) + (\sigma_3' + u)}{2} = \frac{\sigma_1' + \sigma_3'}{2} + u = p' + u \qquad (2.26)$$

$$q = \frac{\sigma_1 - \sigma_3}{2} = (\sigma_1' + u) - (\sigma_3' + u) = \frac{\sigma_1' - \sigma_3'}{2} = q' \qquad (2.27)$$

In Figure 2.12, circle A is the *total* stress circle for a triaxial test on a dry cohesionless soil or the *effective* stress circle for a test on the same soil saturated with water at a pore pressure u. Circle B is the total stress condition for the saturated soil.

The importance of effective stress in soil mechanics is readily apparent. If the soil is dry or is allowed to drain then $u = 0$ and the total and effective stresses are equal. If the soil is saturated or partly saturated and drainage is inhibited, the effective stress will be less than the total stress and the deformation and strength characteristics of the soil will be altered.

A major physical factor affecting drainage is the permeability of the soil (Section 2.4). If this is high, as in rockfills, gravels, sands and

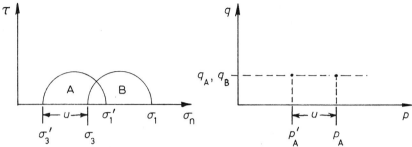
Figure 2.12 Graphical representation of effective stress.

most granular soils, then drainage will be rapid. If permeability is low, as in most clay soils, then drainage will be slow. This is particularly important in soil testing where the existence and magnitude of porewater pressure will depend upon the rate at which the soil is loaded. If this rate is higher than the rate at which the soil can drain, then the effective stresses in and the shear strength of the soil will be reduced. This phenomenon is reflected in the two major types of test – *undrained* (or quick) tests where drainage is prevented, and *drained* (or slow) tests where porewater pressure is allowed to dissipate. More will be said on this subject in later sections.

2.7 Frictional properties of rock particles

In considering the fundamental aspects of frictional behaviour of particulate systems, assume first that pore pressures are atmospheric and that the particle contacts alone control deformation under stress. It is now intended to extend the earlier comments concerning interparticle compressions to the idea of interparticle shear.

In a well-cemented rock continuum, resistance to deformation may involve fundamental processes such as distortion and dislocation of the bonds holding together the constituent crystal structure. Within a discrete particulate system, although individual particles may deform in this way, the deformation and strength of the system as a whole will be controlled by the interaction between individual particles, with the principal deformation mechanism being sliding between particles. The mechanical properties of a particulate system will therefore depend largely on the *frictional* properties of the inter-particle contact points. This is of fundamental importance in soil mechanics and is considered in detail by Lambe and Whitman (1969).

The laws of friction for unlubricated or dry surfaces are summarized by Bowden and Tabor (1967). There are three basic statements:

(a) The frictional resistance (S) between two bodies is directly related to the normal force (N) acting over the contact area, the relating constant μ being known as the *coefficient of friction:*

$$S = \mu N$$

or

$$S = N \tan \phi_i \tag{2.28}$$

where ϕ_i is the angle of inter-particle friction;
(b) The frictional resistance and coefficient of friction are independent of the surface area of contact for constant normal forces;
(c) Frictional resistance is independent of rubbing velocity.

These laws hold at low normal pressures and velocities. In the case of lubricated surfaces they must be modified, the most important change being that when a layer of lubricant is present between two surfaces, then the frictional resistance is independent of the normal pressure. If *shear* resistance is substituted for frictional resistance in this statement, it is possible immediately to grasp the effect of inserting a plastic clay layer at a rock joint contact surface, or between rock particles (as, for example, in a boulder clay).

The magnitude of the coefficient of friction is determined partly by the mechanical and partly by the surface properties of the contacting materials. Most material surfaces and all rock and mineral surfaces (Figure 2.13) are uneven, however smooth they may appear to be, so contact is invariably limited to *asperities* on the material surface (Figure 2.14). Since the contact area ΣA_i represented by these asperities is rather limited, they will be subjected to relatively high normal stresses which will cause deformation and ultimately brittle failure until the normal stress at an asperity N_i/A_i equals the yield strength q_u of the material. The normal force, N, will therefore be related through q_u to the contact area:

$$N = \sum_{i=1}^{i=n} N_i = q_u \sum_{i=1}^{i=n} A_i \tag{2.29}$$

In order for sliding to take place at the surface, the shear resistance τ_A of the asperities (a combination of interlocking and chemical bonding under high stress) must be overcome. The force (S)

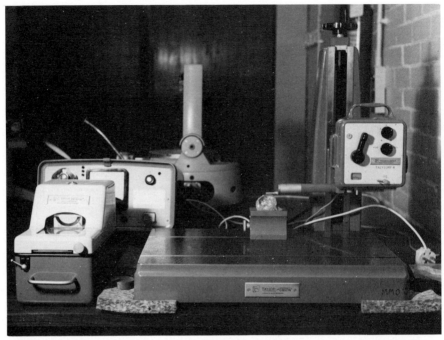

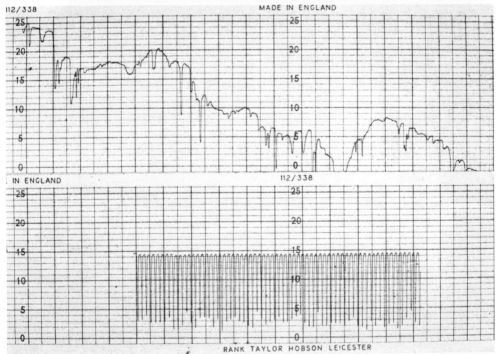

required to overcome friction may therefore be represented by:

$$S = \sum_{i=1}^{i=n} S_i = \tau_A \sum_{i=1}^{i=n} A_i \qquad (2.30)$$

and combining equations 2.29 and 2.30:

$$\frac{S}{N} = \frac{\tau_A}{q_u} = \mu \qquad (2.31)$$

where τ_A is the shear strength of the bonded asperities.

The shear strength of many rocks is generally quoted as half the compressive yield strength, suggesting a value for μ in the region of 0.5. Some experimental values quoted by Horn and Deere (1962) for saturated minerals are given in Table 2.8.

Because the particle surfaces are uneven, the presence or absence of water at atmospheric pressure has little effect on the coefficient of friction except in so far as it may weaken the asperities.

Although these and other quoted results must be treated with caution, sufficient additional information exists to suggest that, in general, sheet silicates have interparticle angles of friction in the region of 10° and framework silicates and calcite have interparticle angles of friction in the range 25° to 35°. These angles will, of course, depend on the cleanliness of the mineral surface, since a relatively small amount of contaminant can act as a highly effective lubricant.

Figure 2.13
(a) Talysurf instrument for measuring the roughness of a quartz crystal face that has been smoothed-down using alumina powder.

(b) Roughness records from 'Talysurf'. The *upper record* is of the smoothed quartz crystal face and there is a 100 000-times vertical scale magnification with respect to the actual record. Each vertical scale small division is equivalent to an asperity height of 0.2×10^{-6} mm. The record span of twenty-five divisions therefore covers a topographic range of 5×10^{-6} mm. The centre line average (CLA) for this record was 0.024 μm, with the quartz being below the line and air being above. There is a twenty-times magnification on the horizontal scan axis of the original record: 1 small division is equivalent to 0.25 mm, meaning that an approximate crystal face length of 8 mm has been scanned.

The *lower record* is that of a roughness standard gauge for instrument check and calibration. Each vertical small division represents 2×10^{-4} mm (10 000-times magnification on the original record). The peak–peak displacement is 97 μ inches or 24×10^{-4} mm. Horizontal scale calibration is as for upper record.

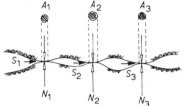

Figure 2.14 Frictional contact at asperities.

When a mass of rock particles is loaded, shear failure between particles will occur when the ratio between the nett tangential forces and the nett normal forces resulting from the load application at particle contacts exceeds at any point the interparticle coefficient of friction. Thus, in a mass of particles where sliding is the only failure mechanism, the shear resistance of the mass could be equated directly to the coefficient of friction of the rock particles. This is not in fact found to be true, for shear failure in a particulate material must also involve particle motion in the mass. If the particle mass has sufficient interlocking, as represented by a relative density greater than about 50 per cent, then the rolling motion of particles up and over each other results in *dilation* (see Rowe, 1962) or expansion in volume of the particle mass. This in turn means that considerable extra shear resistance is generated and the nett tangential force required to cause failure will be much higher. Thus, depending on the *density* of packing, the coefficient of friction of a mass of particles will be higher than the interparticle coefficient of friction. The question of dilation in the context of shear motion along asperities on discontinuous rock surfaces is considered in Section 10.5.

Table 2.8 Frictional properties of some minerals (*after* Horn and Deere, 1962)

	μ	$\phi_i = \tan^{-1} \mu$
Quartz	0.42–0.51	$23°-27°$
Calcite	0.60–0.68	$30°-34°$
Feldspar	0.76–0.77	$37°$
Muscovite	0.22–0.26	$13°-15°$
Biotite	0.13	$8°$
Chlorite	0.22	$13°$
Serpentine	0.29–0.48	$16°-26°$

The concept of interparticle friction as a major mechanical property determinant in particle systems can be applied to all soils but must be modified for soils containing clay minerals and water. Adhesion due to molecular, or more probably ionic, bonding develops at all interparticle contact points, and this is aided by any negative void pressures. But, in the case of blocky or angular particle systems, limited contact areas ensure that this overall effect is minor compared with the influence of frictional resistance. However, in the case of sheet minerals, the platy texture and smallness of the particles, allied to the enhanced facility for the development of negative pore pressures especially in overconsolidated soils, means that attraction between clay particles may contribute additional shear resistance to the mass. This resistance in termed *cohesion* and is independent of applied normal stresses.

Thus, the force required to overcome friction in equation 2.30 must be modified for a soil with cohesion:

$$S = \sum_{i=1}^{i=n} S_i + \sum_{i=1}^{i=n} c_i A_i \qquad (2.32)$$

where c_i is the cohesion at each contact.

Equation 2.31 then becomes:

$$\frac{S - cA}{N} = \mu \quad \text{or} \quad S = cA + \mu N \qquad (2.33)$$

where c represents the cohesive stress holding the particle system together.

It is possible, therefore, to distinguish *mechanically* between two types of rock particle systems, or soils. One of these comprises granular materials unbonded or *loosely* bonded, the reactions of which to load are determined principally by interparticle friction and interlocking. These are usually termed *cohesionless* soils. The other type containing clay minerals usually have poor frictional resistance but *may* have an initial cohesive resistance between particles which must be overcome before frictional resistance is mobilized. These are termed *cohesive* soils.

The influence of negative pore pressures has been mentioned with respect to cohesion, but of course the whole question of the effect of pore pressures, both positive and negative, is quite fundamental to the interpretation of the behaviour of soil in compression and shear.

2.8 Soil deformation – drained granular media

Soil deformation is a complex process, controlled by mechanisms ranging from interparticle sliding and crushing in drained granular media to porewater pressure dissipation and flow in poorly draining clay soils. The simple approach, based on laboratory testing and used to outline fundamental principles in the present chapter, should not divert the reader from acknowledging the difficulties inherent in estimating deformations in materials of widely varying geological character.

Two types of laboratory test are commonly used to study the compression of soils, an *oedometer* and a laboratory *triaxial test*. In an oedometer test, a vertical stress is applied to a soil sample laterally confined in a rigid container. The sample size can range from 75 mm. to 250 mm. and the equipment (Figures 2.15 and 2.16) can vary from a simple pressure pot to a drained cell equipped to measure lateral pressures. Because the sample is constrained in a horizontal direction, the measured axial strain is equal to the volumetric strain, and the ratio between horizontal and vertical effective pressures is equivalent to a coefficient of earth pressure at rest, K_0. The test does tend, therefore, to simulate a state of stress at depth in a soil subjected only to self-weight loading.

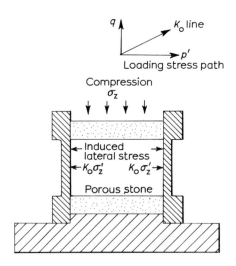

Figure 2.15 Oedometer or consolidation cell.

Figure 2.16 Rowe consolidation cell.

Stress-strain features of *granular* soils and aggregates (including rockfill) during confined compression have been examined by various workers. Typical compression curves such as in Figure 2.17 follow three stages:

(a) a near linear stage A in which deformation is a result of interparticle sliding and locking mechanisms and elastic deformation of particles,
(b) a strongly curvilinear stage B in which deformation results from fracturing and breakdown, initially of particle contact points, later of particles, accompanied by sliding and locking of the

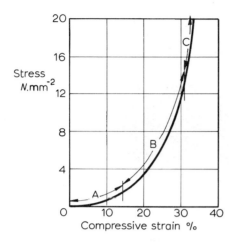

Figure 2.17 Stages in the uniaxially confined compression of 12.7 to 19 mm sandstone aggregate (*after* Farmer and Attewell, 1973).

degraded particles into a dense array,

(c) a near-linear but much steeper curve C where the now much denser aggregate acts as a nearly-elastic material with high internal contact area and reduced interparticle contact stresses.

Plotted in the more usual form as void ratio e against effective vertical stress σ_v' (see Figure 2.29 for a clay soil) the general equation of the graph in Figure 2.17 may be written (Terzaghi and Peck, 1967):

$$e = e_0 - C_c \log_{10}[(\sigma_{vo}' - \delta\sigma_v')/\sigma_{vo}'] \qquad (2.34)$$

where C_c is known as the compression index, and the subscript 'o' specifies initial conditions.

Experiments by the authors (Figure 2.18) suggest that the effect of particle crushing strength (S_c) and average particle diameter (d_m) can be quantified for materials of constant e_0 in the form:

$$\epsilon = \Delta = K_1 + K_2 \ln\left(\frac{\sigma}{S_c} d_m^{\frac{1}{2}}\right) \qquad (2.35)$$

where K_1, K_2 are constants.

The results of confined uniaxial compression studies are useful in determining compression and also compaction characteristics of drained soils and aggregates used for highway subgrades and other loadbearing foundations. They can also provide the drainage-controlled consolidation characteristics of clay soils (Section 2.12).

The *laboratory triaxial test* (Figures 2.19, 2.20) extends the scope of the uniaxially confined compression test by allowing different

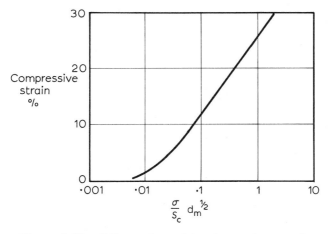

Figure 2.18 Effect of particle size and strength on stress-strain characteristics of dolerite, limestone and sandstone aggregates in sizes ranging from 1.6 to 19 mm (*after* Farmer and Attewell, 1973).

vertical and lateral pressures to be applied to the test specimen which is located between two vertical loading platens. A standard specimen size is 1½ in (38 mm) or 4 in (100 mm) diameter with a 2 or 3:1 aspect ratio. It is surrounded by a water-filled pressure vessel, and is protected by a flexible membrane which separates the confining water from the specimen.

In a drained test, the specimen is first subjected to an all-round confining pressure ($\sigma_1 = \sigma_2 = \sigma_3 = \sigma_c$) through the water medium and it undergoes volumetric strain. The vertical pressure σ_v on the end of the cylindrical specimen is then increased in order to introduce a shear stress to the specimen. End pressure is usually allowed to increase until the specimen fails in shear. The principal parameters operative during the triaxial test may be defined, after Wroth (1972) as:

Mean effective principal stress = $\frac{1}{3}(\sigma'_1 + 2\sigma'_3)$ (2.36)

Deviator stress = $(\sigma'_1 - \sigma'_3)$ (2.37)

Volumetric strain = $\Delta = (\epsilon_1 + 2\epsilon_3)$ (2.38)

Shear strain = $\frac{2}{3}(\epsilon_1 - \epsilon_3)$ (2.39)

A granular specimen will almost invariably be in a 'disturbed' drained state. A specimen with some clay mineral content and having some degree of saturation may require a drainage facility.

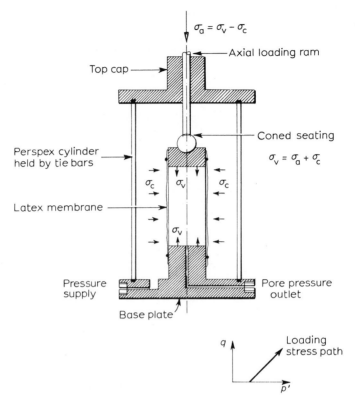

Figure 2.19 Triaxial test.

Whereas in the uniaxially confined compression test the overall lateral pressure varies thoughout the test and is dependent upon the applied end pressure, a technical feature of the triaxial test is that the confinement is both constant and independently applied. Dilation and a discrete macroscopic shear facility are a feature of the latter test but not of the former.

Figure 2.21 illustrates the main characteristics of a typical deviator stress-axial strain curve and of a resultant volumetric strain-axial strain curve for a dense and for a loose medium sand. The stress-strain curves have a non-linear decreasing slope leading to a peak stress level in the dense sand and approaching an eventual ultimate or residual stress level at a constant volume in both cases.

The volumetric-axial strain curves indicate that at low confining pressures and at some particular density an increasing deviator stress will result in increasing specimen volume as vertical strains are accommodated by lateral expansion of the granular material.

Figure 2.20 Laboratory triaxial cell for 1.5 in. (38 mm) diameter cylindrical specimen.

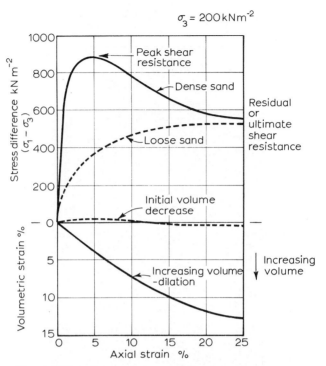

Figure 2.21 Stress-strain curves for dense (100% *relative density* D_r) and loose (20% D_r) Fort Peck sand (*after* Taylor, 1948 – extended by Lambe and Whitman, 1969).

Depending particularly on density, this phenomenon is known as dilation and has been discussed in Section 2.7.

A further interesting feature of triaxial compression is the peak level of stress that can be accepted by dense soils. This limit is manifest as a maximum shear stress along one of the possible shear planes in the specimen and the 'peak' shear resistance is a result of interlocking of particles on this failure plane. Thus the *peak* shear resistance of the soil is determined partly by the interparticle friction and partly by interlocking. In fact tests in the authors' laboratory (Papageorgiou, 1974) in which attempts were made to reduce friction by lubrication indicated that interlocking may be the dominant factor.

Once the interlocking has been broken down, however, shear resistance falls rapidly, reverting to the *ultimate* shear resistance provided mainly by interparticle friction. In loose soils where interlocking is less of a factor affecting shear resistance, there is a gradual build up to the ultimate strength.

A general conclusion is that the higher the initial density, or the lower the void ratio, of the granular material, the greater will be the peak shear resistance, but that further straining of most granular materials will eventually result in an ultimate shear resistance related to the interparticle friction, irrespective of initial density. Other effects of density, stress difference and particle characteristics on stress-strain behaviour of much coarser rockfill material are shown for a comprehensive series of tests described by Marachi *et al.* (1972). These tests were performed on three materials. The characteristics of two of these tests are summarized in Figures 2.22, 2.23, the materials (one loose, one dense) being:

(a) a weak, poorly-graded angular crushed shale (C.U. = 10; e_o = 0.45)
(b) a well rounded and graded beach gravel (C.U. = 30; e_o = 0.22).

The tests were performed on a 36 in (0.9 m) diameter specimen comprising <6 in (150 mm) material, and they illustrate several important aspects of stress-strain behaviour:

(i) at higher effective confining pressures, dilatancy is suppressed and a dense particulate material (Figure 2.23) tends to act in the manner of a loose particulate material (Figure 2.22), exhibiting compression towards a constant volume (see Section 2.13);
(ii) the slope of the stress-strain curve reduces and ultimately changes sign with increasing strain;
(iii) the strain at failure, the shear strength, and the slope of the rising curve all increase with increasing confining pressure;
(iv) a well-graded and rounded particle system has superior mechanical properties to one that is poorly-graded and angular.

The graphs also illustrate the extreme difficulties inherent in quoting deformation moduli for particle systems, depending as they do on the loading conditions. It is possible, using plate bearing tests or laboratory tests, to estimate a modulus of deformation for a particular stress condition – usually the secant modulus (Figure 2.24) between two deviator stress extremes or to 50% failure stress. Typical initial tangent moduli from seismic tests vary within an order of magnitude of 10^5 kN m^{-2} for most soils.

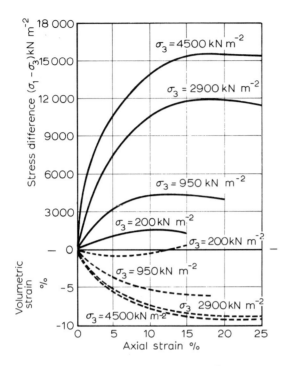

Figure 2.22 Triaxial stress-strain curves for −6 in weak angular crushed shale in 36 in diameter specimen (*after* Marachi *et al*, 1972). Negative volumetric strain denotes compression.

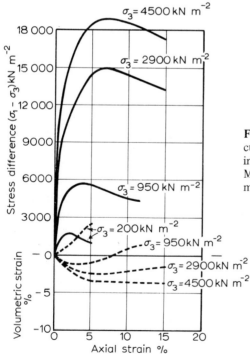

Figure 2.23 Triaxial stress-strain curves for −6 in well rounded gravel in 36 in diameter specimen (*after* Marachi *et al*, 1972). Negative volumetric strain denotes compression.

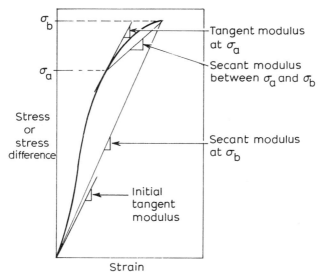

Figure 2.24 Representation of deformation modulus for a typical stress-strain curve.

2.9 Soil strength — drained granular media

A viable criterion of failure is more easily based upon soil strength than it is upon direct unacceptable deformation.

In most engineering materials, strength may be quite easily defined in terms of a limiting stress (usually uniaxial) at which the material yields (yield strength) or fails (ultimate strength). Rocks are mainly brittle materials in which strength is usually defined as the peak stress to be tolerated in uniaxial compression (compressive strength). Peak stress is associated with a *rapid* increase in strain rate to fracture and a consequential fall in stress. Because of their finely particulate nature, soils, as we have seen previously, can normally be evaluated only in a triaxially-confined state, their properties being determined primarily by intergranular movement. In such a case, strength is implicitly synonymous with *shear resistance*. As such, the strength of soils can be investigated either through a series of triaxial tests to failure or by direct shear tests in a *shear-box* (Figure 2.25). The former type of test creates a triaxial stress situation in which any shear failure develops along a failure plane determined by the relative amplitudes of the three principal stresses and the soil properties. The latter allows a constant normal stress, variable shear stress and large strain to be imposed *directly* on an intact, fissured or pre-cut plane.

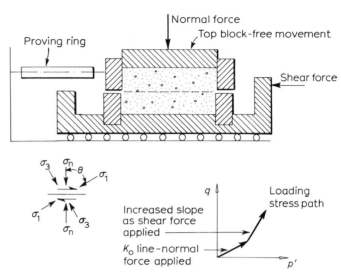

Figure 2.25 Shear box test, the outer box being filled with water to a level above the shear plane to allow equalization of porewater pressures during shear.

In addition to a lack of close control on moisture movement, the shear box has one major drawback. This is due to the kinematic constraints imposed by the system which, together with a certain mechanical flexibility, makes estimation of principal stress and strain rate directions difficult (see Figure 2.26).

Figure 2.22 shows that the peak stress difference level of a cohesionless material in a triaxial test is related to the confining pressure. Although in this case the σ_1/σ_3 ratio at high stress levels decreases with increasing confining pressure (due to decreased interlocking caused by interparticle crushing) the results imply that for the lower confining pressures ($<$950 kNm^{-2}) at shear *failure*

$$\frac{\sigma_1 - \sigma_3}{\sigma_1 + \sigma_3} = \frac{q_f}{p_f} = \tan \alpha = \text{constant} \qquad (2.40)$$

Since the material is drained, total stresses are equivalent to effective stresses.

This type of p_f, q_f relationship at failure holds for most drained granular soils confined at moderate pressures. The presence of a small amount of moisture in a granular soil may introduce an apparent cohesion between particles by capillarity, and a similar effect may be related to dilatancy as such a soil shears towards a constant volume

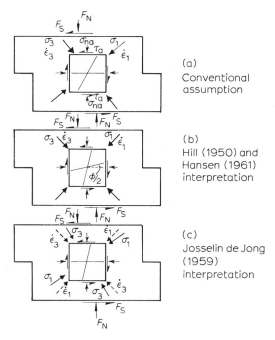

Figure 2.26 Interpretations of shear box test stress (σ) and strain rate ($\dot{\epsilon}$) directions (*after* Morgenstern and Tchalenko, 1967c). F_N and F_S are the applied normal force and shear force respectively.

state from a condition of high initial packing density. This form of cohesion is in physical contrast to the intrinsic cohesion characteristic of dense clay soils and which is considered further in Section 2.10. But with cohesion, the relationship takes the form:

$$q_f = a + p_f \tan \alpha \qquad (2.41)$$

This is the general equation of the K_f line or *failure envelope* of a soil constrained by low confining pressures. It is strain rate dependent and it represents the strength or shear resistance of the soil when subjected to triaxial stress. The linear relationship is, however, only an approximation, and at higher confining pressures and in coarser and more dense materials a curved envelope (Figure 2.27a) is more representative of the true failure criterion.

The failure condition is more commonly (Figure 2.27b) represented as a limiting τ–σ_n plot in which the failure envelope is a line tangential to the Mohr circles representing σ_1 and σ_3 at failure. In

78 *Principles of Engineering Geology*

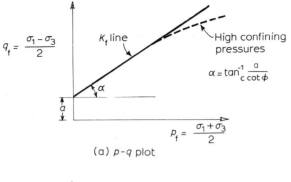

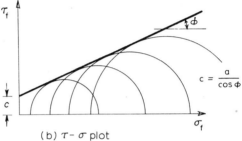

Figure 2.27 Representation of failure envelope.

For a cohesionless soil

$$\sin\phi = \frac{\sigma_1 - \sigma_3}{\sigma_1 + \sigma_3} = \frac{q_f}{p_f} = \tan\alpha = \frac{\sigma_1/\sigma_3 - 1}{\sigma_1/\sigma_3 + 1}$$

$$\sigma_1/\sigma_3 = \frac{1 + \sin\phi}{1 - \sin\phi} = \frac{1 + \tan\alpha}{1 - \tan\alpha} = \tan^2\left(45° + \frac{\phi}{2}\right) = \tan^2\theta$$

For a cohesive soil

$$c\cos\phi = q - p\sin\phi = q - p\tan\alpha = a$$

this case, the general equation of the failure envelope may be written:

$$\tau_f = c + \sigma_f \tan\phi \tag{2.42}$$

where c is the *cohesion*,
and ϕ is the coefficient of *internal friction* of the soil, a parameter that is related to the previously-mentioned coefficient of sliding friction.

This relationship is known as the Mohr–Coulomb criterion of failure for soils and is generally assumed to be a straight line at lower confining pressures. But it may become curvilinear at higher confining pressures as yield dominates over discrete shearing. c and ϕ are related quite simply to a and α, indicated in Figure 2.27.

The failure envelope can be built up from a series of three or preferably more triaxial tests conducted on soil specimens confined at different pressures, or by direct shear tests. The σ_1, σ_3 or p, q values at peak stress or some other specified failure level are then plotted as a p–q graph or as a series of Mohr circles, and a best fit line drawn either by eye or by a least squares regression analysis. A further most important level of shear stress that occupies a place in design analysis is the *residual shear stress* that is still tolerated by the failure surface at *large strains*. In this residual condition, the soil strains at constant volume and at constant shear stress. But such a state can only be investigated at large shear displacements using the direct shear method (Figure 2.25). Since shear stresses and normal stresses are applied directly in shear box tests, peak and residual failure envelopes may be plotted quite conventionally as graphs on the τ, σ_n axes. It should be noted that a small shear strain is a necessary condition for the mobilization of peak shear strength.

Taken as a reasonable approximation of the failure characteristics of soils, Mohr peak and residual failure envelopes define several aspects of soil behaviour at limiting equilibrium. These include:

(a) the range of stress states to which the soil may be subjected without failure, as represented by p, q co-ordinates below the K_f line or Mohr circles not intersecting the failure envelopes;
(b) the stress states where the soil is most likely to fail, as represented by p, q co-ordinates above the K_f line or Mohr circles intersecting the failure line.
(c) the inclination of the plane on which failure occurs, as represented by the position of the point of tangency between the failure envelope and the Mohr circle denoting the stress state at failure. The angle $\theta = 45° + \phi/2$ (Figure 2.8) represents the angle between the *failure plane* and the direction of minor principal stress. In nature, the shear failure plane on a large scale may be curved. Limit equilibrium analyses of rotational failure in soil slopes are considered briefly in Chapter 9.

The actual magnitude of the ϕ and c parameters depends on the physical and mineralogical properties of the soil, but principally on the void ratio, particle friction characteristics, particle size, shape and grading, overconsolidation ratio and the soil-water content. The effect of soil water is considered more specifically in the following sections together with the effect of *loading rates*. In addition, the c value, which is affected particularly by test conditions, is considered later in the rather more important context of overconsolidated clays.

In dry, cohesionless soils, ϕ will depend partly on the amount of interlocking, as represented by the initial void ratio and particle size, shape and grading, and on the *coefficient of sliding friction* (ϕ_i) between particles. The first point to note from Table 2.9 is that ϕ, both for peak stress (ϕ_p) in loose and dense soils and for residual stress (ϕ_r) is greater than the previously quoted surface friction values of ϕ_i (Table 2.6) for minerals, so indicating the effect of interlocking in all but the loosest soils. It is also interesting to note that whereas ϕ_p increases with decreasing initial void ratio (Lambe and Whitman, 1969, suggest a near-linear inverse relationship in Figure 2.28), ϕ_r is reasonably constant for a variety of materials and, with some reservations, may be considered a material property under the constant volume conditions that exist rather than as a specific bulk material parameter that is density-dependent. In that case ϕ_r should be close to the Horn and Deere (1962) fundamental ϕ_i values for particle surfaces (see Table 2.8), and may be as low as 8–9° for some clay soils.

The effect of *particle size* on ϕ is less certain. In general, ϕ appears to increase with decreasing particle size (see Rowe, 1962), but since the initial void ratio also tends, in granular soils, to decrease with decreasing particle size, this may not be very significant. Similarly, a well-graded sand, apart from having the probability of a lower void ratio, will almost certainly have better interlocking properties than

Table 2.9 Peak and residual friction angles for some cohesionless materials

	$\phi_p°$		$\phi_r°$
	Loose	Dense	
Inorganic silt	27–31	31–34	26–30
Fine sand	27–32	31–35	27–30
Uniform fine round grain sand	26–29	29–34	30–35
Sand and gravel	35–40	40–50	30–35

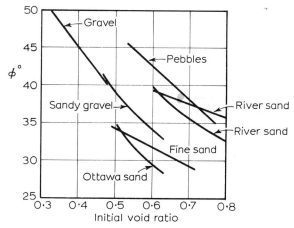

Figure 2.28 Effect of void ratio on ϕ (*after* Lambe and Whitman, 1969).

will a single size sand and will consequently have a greater ϕ value. Interlocking may also be a feature of angular rather than rounded soils.

The actual soil content may also affect the magnitude of ϕ. The presence of clay in the soil will lower the tendency to interlock, and the reduced coefficient of surface friction may substantially reduce ϕ and impart some cohesion.

The effect of moisture in small quantities, causing weakening of angularities – the Rehbinder (1948) effect of surface softening – and possibly lubricating some contacts, may also be a factor militating against high friction. The real significance of water in a soil, however, lies in its effect on the overall state of stress. This is a more demanding problem in the context of slow-draining clay soils.

2.10 Soil strength and deformation – clay soils

Because they are easily-drained, and excess porewater pressures are rapidly dissipated, the effective stress-strain behaviour of dry and of saturated *granular* soils is virtually identical. Deformation mechanisms, considered in a previous section, involve (in hydrostatic or uniaxial confinement) densification through intergranular slip at low pressures and through fracture at higher pressures. In triaxial compression, initial densification is followed by further compression in loose soils and by dilation at low normal compression in dense soils.

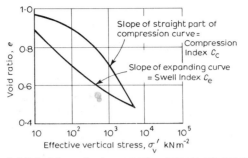

(a) Uniaxially confined compression-drained laminated clay

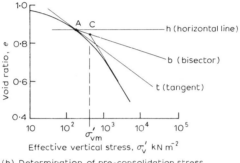

(b) Determination of pre-consolidation stress

Figure 2.29 Stress-strain characteristics of clay soils.

The stress-strain behaviour of *clay* soils, tested under *drained* conditions ($u = 0$ or $u = \sigma_b$, where σ_b is a constant back pressure which is applied to prevent dispersion of dissolved air from the pore water) in uniaxial confinement in an oedometer (Figure 2.29a), is similar to that of granular soils. The results are usually plotted as e versus σ_v' on linear-log axes, the slopes of the loading and unloading curves being used to obtain a compression index (C_c) and a swell index (C_e). C_c is defined in equation 2.34 for granular soils and can be related to the liquid limit (Terzaghi and Peck, 1967) to give a quick estimate of settlement in clay soils. C_e provides an indication of the volume increase due to removal of pressure. It should not be confused with swelling in clay soils due to wetting, a phenomenon considered later in Chapter 3. Other important compression indices which can be obtained from uniaxially-confined tests are the

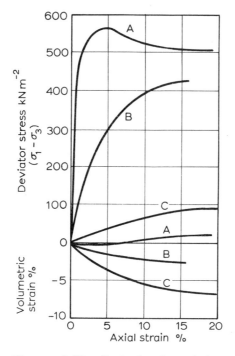

Figure 2.30 Drained stress-strain curves for: (A) Overconsolidated clay (London Clay) $\sigma_3 = 420\,\text{kNm}^{-2}$, $c_u = 200\,\text{kNm}^{-2}$, (B) Normally consolidated clay (Tyneside lake clay) $c_u = 70\,\text{kNm}^{-2}$, $\sigma_3 = 350\,\text{kNm}^{-2}$, (C) Soft clay $\sigma_3 = 50\,\text{kNm}^{-2}$, $c_u = 20\,\text{kNm}^{-2}$. Negative volumetric strain denotes compression.

coefficient of compressibility a_v and the *coefficient of volume compressibility* m_v, where

$$m_v = \frac{a_v}{1+e_0} = \frac{e_0 - e}{\delta\sigma_v'(1+e_0)} \tag{2.43}$$

The dependence of natural clay soil properties on fabric and stress history means that if the clay soil is disturbed during sampling or testing, its properties can be altered. Some alteration as a result of normal handling and change in scale is inevitable, but tolerable, except in clays with variable layered structures (see Section 3.3). If, however, the clay is remoulded, then its compressibility and to some extent its shear strength can be radically changed.

One method of determining the stress history of a soil is from the uniaxially confined compression test plotted as compressive strain or void ratio against the logarithm of effective stress. Known as Casagrande's method (Casagrande, 1936), it involves (Figure 2.29b) locating the point of maximum curvature (A) and drawing lines tangential (At) to and horizontally (Ah) through this point. The line bisecting the angle between these lines (Ab) will intersect the backward extension of the linear part of the curve at C, which is the pre-consolidation stress σ'_{vm}. Then, the *overconsolidation ratio* (OCR) $\sigma'_{vm}/\sigma'_{vo}$ gives an indication of the stress history of the sample, with due allowance made for fabric changes due to scale, sampling, handling and testing. σ'_{vo} is the estimated present vertical effective stress on the sample.

The stress-strain behaviour of clay soils tested in *drained* triaxial compression is also related to the overconsolidation ratio (Figure 2.30). *Overconsolidated* clays, which have at some time in their geological history been subjected to high overburden stress later removed by erosion, are usually stronger, stiffer and denser than *normally-consolidated* or *sensitive* clays, and in drained triaxial compression possess several stress-strain characteristics associated with dense, granular soils. These include a post-peak residual stress and a dilatant property following initial compression. Normally consolidated and soft clays usually possess similar characteristics to loose sand (see Figure 2.21).

The *strength* properties of clay soils defined at either *peak* or *residual* stress level under *drained* test conditions bear some similarity to those of cohesionless granular soils. Expressed in effective stress terms, the residual failure envelope takes the form:

$$\tau_f = \sigma'_f \tan \phi'_r \qquad (2.44)$$

where ϕ'_r is the effective residual or ultimate friction angle.

This characteristic relationship has been studied in detail by Skempton (1964) who has shown that it holds for all types of clays from soft-sensitive to heavily overconsolidated, and that the magnitude of ϕ'_r is related to the clay fraction (Figure 2.31a). There are also indications of a relationship between ϕ'_r and the plasticity index (see Voight, 1973 and Figure 2.31b) but since it could be argued that this index is an indicator of clay content in clay soils (see Section 2.2), such a relationship would be expected. In some soils

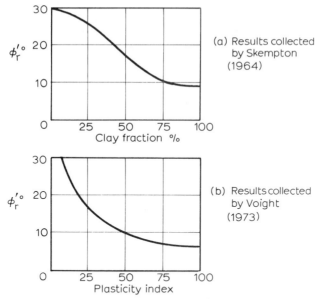

Figure 2.31 Relationship between ϕ'_r and clay content of clay soils.

considerable shear strain at low strain rates (using a ring or reversing shear box) in order completely to dissipate any excess pore pressures is needed before a residual state is acquired. A very stiff clay may be ultrasonically disaggregated and placed in the shear box at its liquid limit for a more rapid estimate of residual strength.

In view of the large strains that are necessary, it would seem that a true residual state of shear might have limited design validity. However, there is evidence that for soft, *normally* consolidated clays, a drained failure envelope based on *peak stress* at very low shear rates also follows the pattern of a frictional material, having negligible cohesion:

$$\tau_f = \sigma'_f \tan \phi'_p \qquad (2.45)$$

The effect of overconsolidation and pre-consolidation on a clay soil is to densify the soil by pushing the particles closer together. Strength will be increased as a result of an increasing inter-particle contact area. Such an operation only rarely imparts to the clay any significant unconfined compressive strength, but it invariably improves its shear resistance at lower confining pressures.

The resulting curved failure envelope can, for most practical design

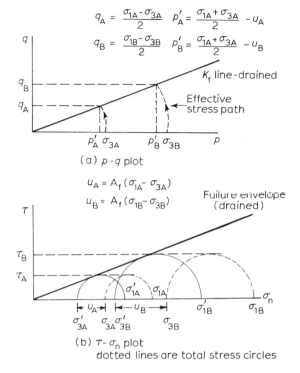

Figure 2.32 Consolidated undrained loading to failure.

purposes, be represented as a straight line:

$$q = a' + p' \tan \alpha' \quad \text{or} \quad \tau = c' + \sigma_n' \tan \phi' \quad (2.46)$$

where ϕ' is usually similar to the friction angle for a normally consolidated clay, particularly at higher effective stresses, and c' has values usually below about 25 kN m^{-2} (Lambe and Whitman, 1969) depending on the overconsolidation ratio.

If a consolidation pressure greater than the pre-consolidation stress is applied to an overconsolidated clay, then the clay should generate a drained failure envelope similar to that of a normally consolidated clay when sheared, with c' equal to zero.

Alternatively, if the effective normal pressure on a potential shear failure plane can be estimated, then the c', ϕ' shear strength parameters can be read off a curved envelope at that normal pressure from the slope and axis intercept of the tangent to the envelope at that point.

Whilst the results of *drained* tests are technically satisfying because they suggest that a similar (frictional) mechanism determines the properties in compression of *all* soils, they sometimes have minor practical significance, involving prolonged load application times and strain capacities which may be inapplicable on a field scale. In practice, in many foundation problems and particularly those involving imposed consolidation through pre-loading, it is more useful to have an indication of the *undrained soil strength* c_u obtained from rapid triaxial tests on undrained samples.

In simple (or unconsolidated) *undrained* tests on *saturated* clays and also loose sands, the resulting failure envelope should appear as a $c(c_u)$ value, with $\phi = \phi_u = 0$. In other words, once an *initial* effective stress condition for failure in an undrained state has been satisfied, further total stress increments will be transmitted as porewater pressure. This implies that for a soil with a particular stress history, the undrained strength will be related to the K_f line or failure envelope based on the drained condition and on the degree of *initial* consolidation applied to the soil under test. Thus in Figure 2.32, q_A, q_B (or τ_A, τ_B) represent the undrained shear strength of a soil subjected to initial consolidation pressures σ_{3A} and σ_{3B} respectively.

The *consolidated undrained* test can therefore be used to give a rapid estimation of strength under *in situ* loading conditions. In dense sands and possibly in heavily overconsolidated clays, dilation of the specimen leading to a negative porewater pressure along a shear plane can lead to cavitation of porewater and simulation of drained conditions in the rest of the material unless a high initial total stress is applied. In partly-saturated soils, an initial drained condition will gradually revert to an undrained condition, so forming a curved envelope in the τ, σ_n plane and ϕ will be greater than zero. In such a case, careful simulation of expected field conditions is required.

It cannot be emphasized too strongly that unless the rate of loading and straining in a *drained* triaxial or direct shear test is less than or equal to the rate of porewater dissipation (see Section 2.12), then the induced porewater pressures will change the shear resistance of the soil, so giving a τ, σ_n relationship somewhere between the drained and undrained states. There will be an apparent cohesion and reduced friction angle which may prove to be misleading. For a fuller treatment of the many variables affecting the mechanical response of clay soils to drained and particularly undrained loading, the reader is referred to Lambe and Whitman (1969, pp. 423–463).

2.11 Pore pressure parameters

The existence of a pressure in the voids of a soil obviously inhibits the mobilization of maximum shear strength because it restricts frictional contact between the structural elements of the material. If the pore pressures can be dissipated by drainage, then, for example, an embankment, be it a natural slope or a constructed earth dam, will gain in strength.

There are three general cases. First, with good drainage or no water present originally, the application of additional all-round external pressure will tend to decrease the volume of the soil and strengthen it. Second, if the material is incompletely saturated and drainage is restricted either partially or totally, then under high external all-round pressure the air will go into solution and there will be some volume decrease. Third, if the soil is fully saturated and drainage is inhibited, there will be minimal volume change and the principal effective stresses $\sigma_1'\ (=\sigma_1 - u)$, $\sigma_2'\ (=\sigma_2 - u)$ and $\sigma_3'\ (=\sigma_3 - u)$ will be independent of the external principal stresses.

In the worst case of no drainage, it is useful to relate the changes in pore pressure to changes in the principal stresses to which any element of soil is subjected. This is normally done through pore pressure parameters A and B (Skempton, 1954). It should be remembered that a pore pressure will act hydrostatically even when the external pressure on the clay element comprises both a hydrostatic, or non-deviatoric, component equal to $\frac{1}{3}(\sigma_1 + \sigma_2 + \sigma_3)$ and a deviatoric component equal to $\sigma_1 - \frac{1}{3}(\sigma_1 + \sigma_2 + \sigma_3)$.

First, let the all-round, isotropic confinement on the soil be σ and assume that σ increases to $\sigma + \delta\sigma$ by preventing drainage. Then,

$$\text{the } B \text{ porewater pressure parameter} = \frac{\delta u_a}{\delta\sigma} \qquad (2.47)$$

where δu_a is the pore pressure change, and $\delta\sigma \equiv \delta\sigma_3 \equiv \delta\sigma_c$.

As an example, suppose that the all-round confining compression on an element of soil were to be raised from zero to 138 kN m^{-2} and that the measured pore pressure changed as a consequence from -90 kN m^{-2} to 14 kN m^{-2}. The B parameter would then be 0.75. If the soil had been saturated, the low compressibility of the porewater would have produced a B parameter of 1. On the other hand, if dry air had filled the voids, the B parameter would have been zero due to the high compressibility of the air. It follows that in *soils, the magnitude of B indicates the degree of saturation*.

Suppose now that an element of soil *in situ* is loaded triaxially but axi-symmetrically, that is $\delta\sigma_2 = \delta\sigma_3$. This three-dimensional loading can be considered to be made up from an isotropic stress of $\delta\sigma_3$ plus a deviator stress of $\delta\sigma_1 - \delta\sigma_3$. Thus, in this case, the change in pore pressure due to the increment of deviator stress is

$$\delta u_b = \tfrac{1}{3} B(\delta\sigma_1 - \delta\sigma_3) \qquad (2.48)$$

if the material deforms elastically. However, since the stress-strain curve for a soil skeleton is non-linear, it is necessary to replace the fraction with another *pore pressure parameter A* such that

$$\delta u_b = AB(\delta\sigma_1 - \delta\sigma_3) \qquad (2.49)$$

Thus, the total pore pressure increment

$$\delta u = \delta u_a + \delta u_b = B[\delta\sigma_3 + A(\delta\sigma_1 - \delta\sigma_3)] \qquad (2.50)$$

and in a saturated compressible soil where $\delta u_a = \delta\sigma_3$ and $B = 1$:

$$A = \frac{\delta u - \delta\sigma_3}{\delta\sigma_1 - \delta\sigma_3} \qquad (2.51)$$

Some B and A pore pressure parameter values are given respectively in Tables 2.10 and 2.11. A further *pore pressure parameter C*, which is applied to uniaxially confined compression in an oedometer and which is therefore directly related to consolidation, is defined as:

$$C = \frac{\delta u}{\delta\sigma_v} \qquad (2.52)$$

In a laboratory triaxial test simulation of shear stress development under undrained conditions in the field, pore pressure parameter B would be measured before any drained consolidation and AB (written as $\bar{A}$) would be measured during the application of deviator stress. The appropriate equation would then be

$$\delta u = B\delta\sigma_3 + \bar{A}(\delta\sigma_1 - \delta\sigma_3) \qquad (2.53)$$

and the parameter at failure would be written as $\bar{A}_f$.

Equation 2.53 can be divided throughout by $\delta\sigma_1$ and a new parameter $\bar{B}$ introduced:

$$\bar{B} = \frac{\delta u}{\delta\sigma_1} = B\left[\frac{\delta\sigma_3}{\delta\sigma_1} + A\left(1 - \frac{\delta\sigma_3}{\delta\sigma_1}\right)\right] \qquad (2.54)$$

The *parameter A* can be determined from a consolidated un-

drained triaxial test on a specimen subjected initially to a hydrostatic stress σ_3 where $u = 0$ (Figure 2.33a). σ_3 remains constant during the test, so that $\delta\sigma_3 = 0$. In that case, at any stage during the test

$$A = \frac{\delta u}{\delta \sigma_1} \tag{2.55}$$

Table 2.10 *B* pore pressure parameter values

Material	B-value	% Saturation	Reference
Sandstone	0.286	100	
Granite	0.342	100	
Marble	0.550	100	From values of
Concrete	0.582	100	compressibility given by
Dense sand	0.9921	100	Skempton (1961b) and
Loose sand	0.9984	100	computed by Lambe and
London Clay	0.9981	100	Whitman (1969)
Gosport clay (normally consolidated)	0.9998	100	
Boulder clay	0.69	93	Measured by Skempton
Boulder clay	0.33	87	(1954); reference Lambe
Boulder clay	0.10	76	and Whitman (1969)

Table 2.11 *A* pore pressure parameter values (100% saturation)

Material	A-value	Reference
Very loose fine sand	2 to 3 (at failure)	
Sensitive clay	1.5 to 2.5 (at failure)	
Normally consolidated clay	0.7 to 1.3 (at failure)	Lambe and Whitman (1969) from values given by Bjerrum
Lightly overconsolidated clay	0.3 to 0.7 (at failure)	
Heavily overconsolidated clay	−0.5 to 0 (at failure)	
Very sensitive soft clays	1 (for foundation settlement)	
Normally consolidated clays	0.5 to 1 (for foundation settlement)	From Skempton and Bjerrum (1957) quoted by Lambe and Whitman (1969)
Overconsolidated clays	0.25 to 0.5 (for foundation settlement)	
Heavily overconsolidated sandy clays	0 to 0.25 (for foundation settlement)	

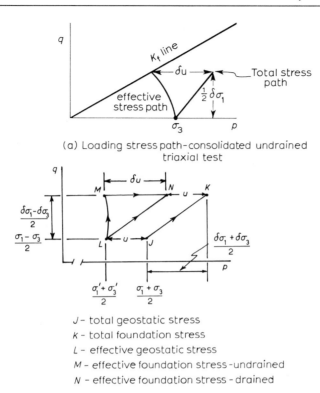

Figure 2.33 Determination and use of the A parameter.

Values of A at failure are usually quoted. The significance of high and low or negative A values lies in the reaction of the soil skeleton to shear. Thus, a loose sand or soft clay skeleton will tend to collapse on loading, and some of the effective stress acting on the specimen will be transferred to porewater pressure. In the case of an overconsolidated clay or dense sand, there will be a tendency to expand when subjected to shear, possibly resulting in a negative porewater pressure.

This latter phenomenon is important, for example, in the analysis of the stability of slopes in overconsolidated clays. Initially, under the influence of negative porewater pressures in shear zones, there will be a tendency for water movement towards a shear plane, so reducing stability. But drainage will improve with time. It follows that if the slope can be retained in the immediate, post-excavation period, there will be a good chance of its remaining stable thereafter unless conditions change. Slopes are discussed in more detail in Chapter 9.

Although A is related to soil properties, it is not necessarily constant but rather varies with changes in time, stress and strain. This is readily apparent, since δu and the effective stresses in the soil will be related to the same variables. This means that if A is required for a particular application, say primary consolidation under an embankment, the test on which A is based should be subjected to the same basic stress-strain-time considerations.

The main practical application of A is to estimate initial excess porewater pressures induced in a soil by a change in total stress conditions. It is possible to estimate with reasonable accuracy by Boussinesq theory (using the elastic analogy of soil in compression) the sub-surface stress increments imposed on the existing geostatic stress situation by a foundation (Figure 7.10). If A is known, then the porewater pressure increment can be calculated and the relative stability of the foundation estimated by relating the loading stress path to the K_f line. In Figure 2.33b, the effective undrained stress path LM will be followed if the rate of loading is rapid compared with the consolidation rate (Section 2.12). Slower rates of loading allowing partial drainage will cause the path to approach the drained effective stress path LN more closely.

The use and significance of the pore pressure parameters will be considered further with respect to soil slope stability in Chapter 9.

2.12 Rate of porewater pressure dissipation

Consolidation (or primary consolidation) describes the compression of a saturated soil where the rate and amount of compression is controlled principally by the rate at which water can flow from the soil and the rate at which porewater pressure is allowed to dissipate. There is also further *secondary compression* associated with consolidation which, according to Terzaghi and Peck (1967), is probably due to the presence of high viscosity adsorbed water between clay mineral grains which delays normal compression processes. The result is to delay consolidation and introduce unquantifiable long-term compression which can be serious particularly if a soil comprises a

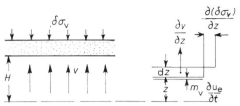

Figure 2.34 Representation of one-dimensional consolidation.

series of thin compressible layers and/or *primary* consolidation is completed rapidly (see Bjerrum, 1967).

In practice, the term 'consolidation' is invariably used to describe the compression of low permeability *clays* or *silty soils*. The practical effects of consolidation will be an increased density and ultimately increased shear strength as water is removed from the soil. The theoretical and empirical treatments of consolidation based on Terzaghi's classical *theory of one-dimensional consolidation* enable prediction of the rates at which porewater pressures induced in a saturated soil by an applied load are dissipated. Since consolidation is important in both a geological and geotechnical sense it merits rather more than a cursory mention.

Terzaghi's (1943) theory proposes quite simply that the time (t) for porewater pressure dissipation in a compressible layer bounded by two rigid permeable layers (Figure 2.34) is *directly proportional* to the volume (V_w) of water expelled from the soil by an applied vertical stress increment and is *inversely proportional* to the flow velocity (v) of the expelled water, that is:

$$t \propto \frac{V_w}{v} = \frac{\delta\sigma_v m_v H}{k_v (\delta\sigma_v / \gamma_w H)} = \frac{m_v \gamma_w H^2}{k_v} \qquad (2.56)$$

where $\delta\sigma_v$ is the vertical stress increment,

$m_v = \dfrac{\Delta}{\delta\sigma_v}$ is the volume compressibility of the soil,

Δ is the volumetric strain $= \dfrac{\delta H}{H} = \dfrac{\delta V}{V}$ for one-dimensional consolidation,

H is the half-thickness of the compressible layer,

V is the sample volume,

$\dfrac{\delta\sigma_v}{\gamma_w H}$ is the hydraulic gradient,

and k_v is the Darcy coefficient of vertical permeability.

The assumptions inherent in such an approach are: (i) all pore voids are saturated and all additional vertical loads are initially carried by the porewater as excess porewater pressure, (ii) compressive strains are small and occur in a *vertical* direction only, (iii) porewater and soil particles are incompressible, (iv) k_v, v and m_v remain constant throughout the consolidating layer, and (v) secondary compression is ignored.

Equation 2.56 summarizes some of the factors affecting consolidation. The actual rate of consolidation, subject to the assumptions summarized above, can be computed by considering the flow of water through an element of compressible soil of thickness dz and subjected to an excess hydraulic gradient

$$\frac{\partial(\delta\sigma_v)}{\partial z} = -\frac{\partial u_e}{\partial z} \qquad (2.57)$$

where u_e is the excess porewater pressure at z, the vertical position in the compressed layer.

If the flow obeys Darcy's law, then:

$$v = k_v i = -\frac{k_v}{\gamma_w}\frac{\partial u_e}{\partial z} \qquad (2.58)$$

and if the rate at which water leaves the layer is the same as the rate at which the volume of the layer decreases through compression, then:

$$\frac{\partial v}{\partial z} = m_v \frac{\partial(\delta\sigma_v)}{\partial t} = -m_v \frac{\partial u_e}{\partial t} \qquad (2.59)$$

Combining these two equations gives

$$m_v \frac{\partial u_e}{\partial t} = \frac{-\partial v}{\partial z} = \frac{k_v}{\gamma_w}\frac{\partial^2 u_e}{\partial z^2} \qquad (2.60)$$

or

$$\frac{\partial u_e}{\partial t} = c_v \frac{\partial^2 u_e}{\partial z^2} \qquad (2.61)$$

where $c_v = \dfrac{k_v}{m_v \gamma_w}$ is the *coefficient of consolidation* of the soil.

If c_v is assumed constant, this differential equation then expresses porewater pressure as a function of vertical position and time, and it can be solved for a given set of boundary conditions. For the one-dimensional case, these are: (a) at $t = 0$, $u_e = u_0 = \delta\sigma_v$ at any distance z; (b) at any time t, at $z = H$, $u_e = 0$; (c) at any time t, at $z = 0$, $du_e/dz = 0$; (d) at $t = \infty$, $u_e = 0$ at any distance z.

The solution, using Fourier series and which is given in full in Taylor (1948), takes the form:

$$u_e = u_0 \sum_{m=0}^{m=\infty} \frac{4}{\pi(2m+1)} \sin\left[\frac{\pi}{2}(2m+1)\frac{z}{H}\right]$$

$$\exp - \left[T_v \frac{\pi^2}{4}(2m+1)^2\right] \quad (2.62)$$

where m is a dummy variable,
and $T_v = c_v t/H^2$ is an independent dimensionless variable known as the *time factor*.

This relationship can conveniently be portrayed graphically in Figure 2.35 which plots z/H against $1 - u_e/u_0$ for different values of T_v. In this case, $1 - u_e/u_0$ represents the degree of pore pressure dissipation: zero when $u_e = u_0 = \delta\sigma_v$ immediately after application of pressure; 100 per cent when $u_e = 0$. From this graph, it can be seen that flow (dv/dz) is zero at $z = 0$, rising to a maximum at $z = H$ at the top and bottom of the layer. No flow occurs, therefore, across the layer centre line. Significant changes in volume only occur at mid-depth ($z = 0$) when $T_v > 0.05$. When $T_v > 0.15$, the form of the relationship is very nearly a sine curve, indicating that the first term of the series in Equation 2.62 ($m = 1$) is a reasonable approximation of the whole expression.

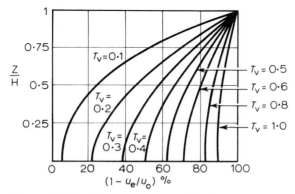

Figure 2.35 Percentage consolidation at various depths and time factors.

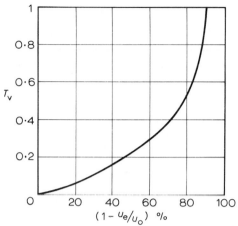

Figure 2.36 Relationship between average consolidation for $z = 0$ to $z = H$ and T_v.

If the average pore pressure dissipation $1 - (u_e/u_0)$ from Figure 2.35 is plotted for each value of T_v, it can be seen from Figure 2.36 that approximately 90 per cent of the porewater pressure is dissipated at $T_v = 1$ and, by extrapolation of the steeply-rising curve, that 99 per cent is dissipated at $T_v = 3$. In practice (and also for convenience), it is assumed that 90 per cent consolidation is close enough to complete consolidation for engineering purposes, whence:

$$T_v = 1 = \frac{c_v t}{H^2} \quad \text{and} \quad t_{90} = \frac{H^2}{c_v} = \frac{m_v \gamma_w H^2}{k_v} \qquad (2.63)$$

This time for complete consolidation is an exact reproduction of equation 2.56, having a constant of proportionality of unity. It is an interesting illustration of the fact, noted by Lambe and Whitman (1969), that some intuitively derived equations may ultimately be correct.

A value for c_v may be obtained by observing the rate of deformation of a laboratory sample in an oedometer or consolidation cell (see Table 11.4) designed to simulate the one-dimensional consolidation conditions assumed in the theory. Results from these tests are usually plotted as compression against $t^{\frac{1}{2}}$, or log t. c_v can be computed from equation 2.63 and Figure 2.36 if the degree of consolidation at any time t can be estimated, and provided that the sample is representative of the soil mass. Various methods available for this estimation are outlined in Lambe (1951) and Ackroyd (1953). Alternatively, c_v may be determined from a measurement of permeability and an estimate of compressibility.

The process of consolidation is synonymous with increasing the *strength* of the clay soil as the particles are pushed closer together and the water content is reduced. In other words, the actual effect of consolidation on a soil is equivalent to increasing the consolidation pressure in a *consolidated undrained* test (Figure 2.32).

In practice, a knowledge of consolidation theory can be used to optimize consolidation rates and consequent increases in strength. From Figure 2.36 it is evident that about 30 per cent of consolidation occurs in 10 per cent of the time necessary for 90 per cent consolidation. Obviously, therefore, complete consolidation can be obtained much faster if a consolidating layer *in situ* (or a laboratory specimen) is loaded in increments. A reasonable practical procedure is outlined in Section 11.3.

2.13 The critical state concept

The conventional treatment of soil mechanics followed in the previous sections has shown that the drained stress-strain behaviour of all rock particle systems ranging from the coarsest rockfill to the finest clay tends to follow a unified pattern. There may be exceptions to this pattern — particularly overconsolidated clays and some partly saturated soils where interparticle attractive forces may induce real or apparent cohesion — but most soils tend to behave in compression as granular, frictional materials.

Two basic types of deformation in laboratory tests have been considered. The first, uniaxially confined compression, better described as consolidation in clay soils, in which the specimen is rigidly confined in a lateral direction, shows that the void ratio of a specimen reduces in proportion to the logarithm of the effective vertical stress σ'_v (see equation 2.34):

$$e = e_0 - C_c \log_{10}[(\sigma'_{vo} - \delta\sigma'_v)/\sigma'_{vo}]$$

This type of deformation is illustrated for coarse aggregate in Figure 2.17 and for clays in Figure 2.29.

The second type of deformation, in a shear box or triaxial cell, involves the shearing of a specimen either directly or indirectly through differential stress application. From these tests, the relationship between the differential stress ($2q$) and effective normal stress ($2p'$) at which the specimen reaches a plastic state can be determined in the form (equation 2.40):

$$q = p' \tan \alpha'$$

This type of deformation is illustrated in Figure 2.21 for sand, Figures 2.22, 2.23 for rockfill and Figure 2.30 for clays.

Examination of these Figures indicates that the actual controlling constants in equations 2.34, 2.40 depend upon both the nature of the material particles and the character of their assembled state. In an attempt to unify the behavioural description of soil deformation, Schofield and Wroth (1968) developed the concept of the *critical state*.

Two basic statements of 'frictional flow' define the critical state. The first is compatible with the linear relationship in p', q space, as considered earlier:

$$q = Mp' \qquad (2.64)$$

where M is the critical state friction constant and is an expression of the power input required to drive the material in shear. If, from a series of triaxial tests, it is found that the particular friction constant for the material is less than M at yield, then the specimen was relatively loose initially, less power is needed to shear it, and during shear it has become more dense. Conversely, if the friction constant is greater than M, then more power is needed and the specimen has dilated. Indirectly, therefore, volume is shown to be critically important and that at some critical or ultimate stage in the shear process, 'flow' will continue without further change in q, p', or volume.

It has been shown that the volume property of a soil can be investigated more directly through the oedometer (Figure 2.15). The second critical state equation may be written:

$$V = \Gamma - \lambda \ln p' \qquad (2.65)$$

where

$$V = 1 + e \qquad (2.66)$$

and is the *specific volume* of unit volume of solids.

Γ is the ordinate of the critical state line in V, p' space and λ is the gradient of the compression line.

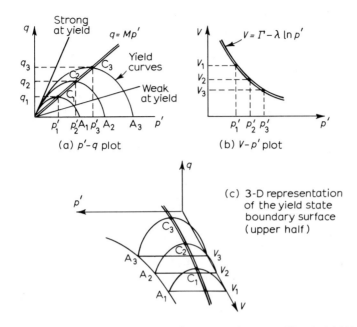

Figure 2.37 Critical state lines (*after* Schofield and Wroth, 1968).

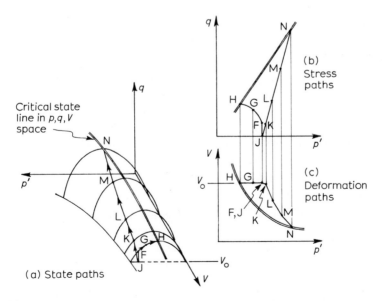

Figure 2.38 Projections of state paths for drained and undrained triaxial tests (*after* Schofield and Wroth, 1968).

Then since the V values associated with critical state p', q values must lie on a unique uniaxially-confined compression curve (Figure 2.37) there must be a unique curve of critical states, as represented by equations 2.65, 2.66 and an envelope of yield states in p', q, V space.

Schofield and Wroth (1968) show how the yield surface in Figure 2.37c can be used to construct state paths for drained and undrained triaxial tests on a hypothetical clay soil (Figure 2.38). The projection of these state paths as stress and deformation paths in p', q and p', V space can be used to predict deformation. In both state paths, J represents the original state of consolidation, JK and JF the reversible deformation to first yield and KLMN and FGH the test paths to the critical state.

Although to a certain extent the critical state approach is a re-statement of the classical approach to soil mechanics with emphasis on deformation rather than 'strength' or 'failure', it does also extend the scope of classical soil mechanics. For instance, the p', q, V envelope can be used to estimate the likely increase in shear resistance associated with a given degree of compaction or consolidation. More generally, the effective stress and specific volume of a soil in any state can be represented by a single point in p', q, V space and its likely reactions under stress referred to a closely-defined critical state reference. In soil testing it can be shown that the plasticity index is a critical state property, and a knowledge of the concept can be used to refine and develop testing techniques. For further information, the reader is referred to Schofield and Wroth (1968).

2.14 Limiting states of equilibrium

Unless the geostatic state of stress is hydrostatic, the stresses occurring naturally in the ground can be represented as a normal stress and shear stress on a plane inclined to the principal stress directions. These latter are assumed in turn to lie vertically and (uniformly) horizontally and are related through the coefficient of earth pressure at rest K_0 (see Section 2.5).

In normally-consolidated granular soils, K_0 is usually less than 1 and is represented most simply by Jaky's equation (see Lambe and Whitman, (1969)

$$K_0 = 1 - \sin \phi' \qquad (2.67)$$

For normally-consolidated clays, Lambe and Whitman (1969): quote Hendron and Alpan:

$$K_0 = 0.95 - \sin \phi' \qquad (2.68)$$
$$= 0.19 + 0.233 \log \text{P.I.} \qquad (2.69)$$

where P.I. is the plasticity index (Section 2.2).

The effect of *overconsolidation* in both sands and clays is to increase K_0 to values which may be as high as 2.5 to 3 at overconsolidation ratios in the region of 30 to 35 (see Skempton 1961a and Figure 3.33).

The K_0 line, representing the initial state of stress at various depths, will lie between the two K_f lines which define the *limiting state of equilibrium* of a soil on a p–q diagram. (Figure 2.39.) These may be defined as the range of σ'_1 and σ'_3 values (σ'_v and σ'_h values in practice) within which the shear strength of the soil exceeds the maximum shear stress on the potential failure plane. If the original at-rest stress state is point A, then failure will occur: (a) if σ'_v remains constant and σ'_h is reduced, in other words if the soil expands laterally (point B); (b) if σ'_v remains constant and σ'_h is increased, in other words if the soil is compressed laterally (point C).

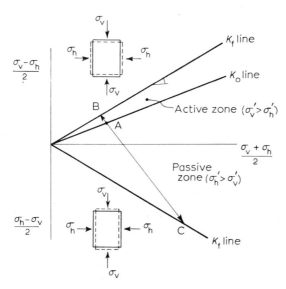

Figure 2.39 Representation of loading stress paths to active and passive states.

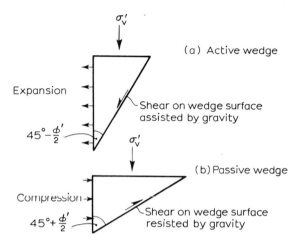

Figure 2.40 Active and passive wedges.

These situations usually occur in natural soils subjected to *constant effective overburden pressure* and varying degrees of confinement due to excavation or containment. The concept of limiting states of equilibrium has wide application in foundation engineering and excavation, and was first propounded by Rankine (1857). The extreme states of stress are known as the *active state* σ'_{ha} in the expanding case and the *passive state* σ'_{hp} in the compressing case (Figure 2.40).

The ratio of horizontal to vertical pressure at the point of failure is known respectively as the *coefficient of active earth pressure*, K_a, where

$$K_a = \frac{\sigma'_{ha}}{\sigma'_v} = \frac{\sigma'_3}{\sigma'_1} = \frac{1 - \sin \phi'}{1 + \sin \phi'} = \tan^2\left(45 - \frac{\phi'}{2}\right) = \frac{1 - \tan \alpha'}{1 + \tan \alpha'} \quad (2.70)$$

and the *coefficient of passive earth pressure*, K_p, where

$$K_p = \frac{\sigma'_{hp}}{\sigma'_v} = \frac{\sigma'_1}{\sigma'_3} = \frac{1 + \sin \phi'}{1 - \sin \phi'} = \tan^2\left(45 + \frac{\phi'}{2}\right) = \frac{1 + \tan \alpha'}{1 - \tan \alpha'} \quad (2.71)$$

Thus, for various ϕ' values, K_p and K_a together with K_0 (the latter for normally-consolidated soils) may be specified in Table 2.12. This concept must be based on effective stress.

Once the active pressure (generally developed at a strain of <0.5

Table 2.12 Some earth pressure coefficients for different friction angles.

ϕ'	K_o	K_a	K_p
10°	0.83	0.70	1.42
20°	0.66	0.49	2.04
30°	0.50	0.33	3.00
40°	0.36	0.22	4.60

per cent in normally-consolidated soils) is mobilized, this then is the *minimum* pressure which will act in a horizontal direction in a laterally *expanding* soil. Similary, the passive pressure, developed at a much greater strain (2 to 50 per cent mobilized at 0.5 per cent strain), represents the maximum pressure which can act in a horizontal direction in a soil subjected to lateral *compression*. Outside these limits, the lateral pressure will remain constant irrespective of strain; in other words, the soil will have assumed a plastic state.

An important application of this earth pressure concept lies in retaining wall design. The earth pressure tending to move the wall into the excavation can readily be seen to be an active pressure. Movement is resisted by any tie stresses and by a passive pressure in the soil restraining the buried toe. Design of a wall requires a knowledge of the operative pressures, a point which is illustrated by Golder *et al.* (1970) in an interesting case history. Four eminent civil engineers estimated the performance of a braced sheet pile wall which was retaining 18 m of gravel fill (8 m), silt (6 m) and till (4 m). The phreatic surface was 2 m from the top of the wall and a significant feature of the calculations was that the estimated pore pressures acting on the wall were between 3 and 4 times the estimated active effective stresses.

3 ~ Clays and Clay Shales

At the root of theoretical soil mechanics, outlined in the previous chapter, is the assumption that soils act as a mechanical continuum the reaction of which to stress is modified by an idealized particle structure and by the presence of pore water. This may be an acceptable approximation with respect to granular and cohesionless soils. However, in the case of soils with a significant clay mineral fraction, mechanical properties may be significantly modified by the soil structure, deriving partly from its compaction and partly from its depositional and geological stress history. The interpretation of site investigation and laboratory test data to determine the effect of soil structure and fabric on *in situ* soil properties is one of the principal functions of the engineering geologist.

The engineering terms *clay* and *clay shale* used to define a soil/rock with a high clay mineral content are, to a certain extent, synonymous. Essentially, the term 'clay' would be used to describe an 'unconsolidated' sediment whilst the term 'clay shale' would refer to stiff overconsolidated material with some tendency towards fissility. The 'spectrum' of clays is broadly defined by the British Standards Institution Code of Practice CP 2001:1957 in Table 3.1.

Clay shales, on the other hand, are usually regarded as geological materials that occupy an intermediate geotechnical position between soils and rocks. They possess a degree of intactness, continuity and strength not enjoyed by soils or clays. However, they comprise a dominant clay mineral content which introduces a much lower overall strength and a degree of strength anisotropy not possessed by rocks in general. Furthermore, although clay shales are usually replete with macroscopic discontinuities (fissures, partings, slickensides) which serve to lower the mass strength of the deposit, a preponderance of such discontinuities is of limited length and cannot be confused with systematic jointing associated with sedimentary rocks.

Although there is a temptation to treat shales, mudstones and

Table 3.1 Undrained shear strengths of clay soils

Consistency	Field indication	Strength kN m^{-2}
Very stiff	Brittle or very tough	>150
Stiff	Cannot be moulded in the fingers	75–150
Firm	Can be moulded in the fingers by firm pressure	40–75
Soft	Easily moulded in the fingers	20–40
Very soft	Extrudes between fingers when squeezed in the fist	<20

other argillaceous deposits as 'soft rocks', and there is logic in such an approach, a rather different treatment will be adopted here. A more detailed analysis of discontinuity aspects of the problem is treated in Chapter 6 but it is felt desirable to introduce the subject in single case history form in the present chapter. However, since the intact strength and material properties are evaluated through the use of soil mechanics equipment and techniques, these matters have been included in Chapter 2.

3.1 Interparticle attraction and repulsion

The attractive and repulsive forces which are responsible for much of the real and apparent *cohesion* in mineral particle systems have been described by various authors (see for instance, Lambe, 1960; Moore, 1968; Morgenstern, 1969; Mitchell, 1969). The magnitude of these forces, which occur in all sediments, is related to the characteristic electrostatic properties of the particle surfaces, and the associated electrolyte environment (Rosenqvist, 1955, 1958, 1959, 1960). They are only really significant however in fine-grained particulate systems, and specifically in systems which are fine enough to simulate the behaviour of colloids.* Even here it has been shown (see Mitchell, in Morgenstern 1969) that interparticle forces have very little effect except in the most lightly consolidated systems. The basic conclusion in practice must be, therefore, that whilst interparticle forces may affect the microstructure of a sediment during sedimentation, they have a less significant effect on its subsequent engineering behaviour.

*Colloids are usually defined as crystalline particles having a diameter 0.001 mm (1 μm) or by Lambe (1953) as a particle whose specific surface area is so great that particle behaviour is controlled by *surface* rather than mass energy. The classical text, which is essential reading for the student of colloidal physics, is the book by Verwey and Overbeek (1948).

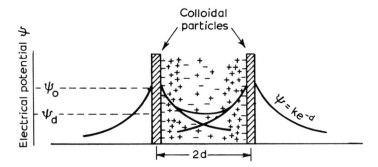

Figure 3.1 Forces between two particles in close proximity.

Attractive forces in colloidal suspensions generally result from:
(a) ionic bonding between negatively charged surfaces (generally resulting from isomorphous substitution of a high valency atom by a low valency atom and interparticle electrolyte cations (see Section 1.2);
(b) hydrogen bonding between oxygen atoms and hydroxyl groups formed by the partial dissociation of water molecules; and
(c) non-polar Van der Waals – London attractive forces, dependent on distance but generally independent of particle and electrolyte properties and therefore invariant in strength.

Repulsive forces may be set up as a result of ion or mineral hydration (long term) or as a result of osmotic pressures caused by the overlapping of surface double (Stern) layers of adsorbed water during compression (Figure 3.1). The magnitude of these forces is a function of the chemistry of the soil water system, the electrical potential Ψ at unit distance from the particle surface decreasing with increasing ion concentrations in the surrounding electrolyte.

Lambe (1953) was one of the first to study the detailed interaction of repulsive and attractive forces in the formation of sediments. Two hypothetical inter-particle total potential curves (Figure 3.2) are based on the summation of an exponential decrease in repulsion ($+E_R$) with distance (d) and a decrease in attraction ($-E_R$) with the square of distance. The first assumes a critical distance apart at which double layer repulsive forces are greater than the attractive forces. The second assumes a situation in which attractive forces are always greater than repulsive forces. The former describes a *weak* electrolyte concentration where the double layer is

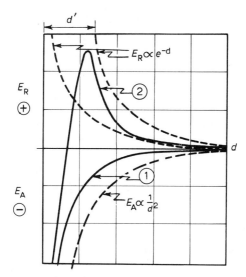

Figure 3.2 Total potential curves
(1) E_A always greater than E_R
(2) $E_R > E_A$ at critical distance d'.

more effective at greater distances than the Van der Waals – London attractive forces, so implying sedimentation of individual particles – a *dispersed* system. The second may describe a stronger electrolyte concentration where *flocculated* particles may accumulate, so leading to more rapid formation of a more open sediment. Ions have a flocculation capacity in a decreasing order of power: trivalent, divalent, monovalent, K^+ and NH_4^+ being exceptions since they are preferentially adsorbed.

The three basic sediment structures shown in Figure 3.3 are usually quoted as typifying most clay systems. These comprise

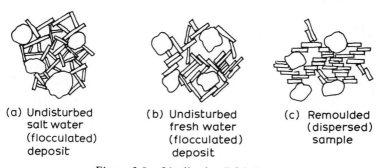

(a) Undisturbed salt water (flocculated) deposit

(b) Undisturbed fresh water (flocculated) deposit

(c) Remoulded (dispersed) sample

Figure 3.3 Idealized soil fabrics.

108 *Principles of Engineering Geology*

dispersed structures in which the clay particles (or plates) have *face-to-face* contact and flocculated structures having *edge-to-edge* or *edge-to-face* contact. These structures are not easily explained by colloid theory, but it would appear that when interparticle distances are small under flocculation controls, other electrical forces act on the clay particles. In particular, the positively charged edge surfaces (the positive charge resulting from broken sheets) are attracted towards the negatively charged plate surfaces. This can result in edge-to-face and edge-to-edge structures of greater domain complexity (see Figure 3.4 and also refer to Collins and McGown, 1974), so

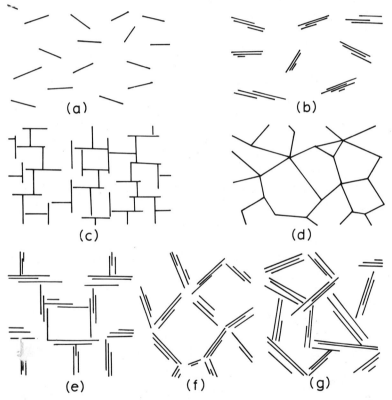

Figure 3.4 Modes of particle associations in clay suspensions (*after* Van Olphen, 1963).
(a) dispersed and deflocculated
(b) aggregated but deflocculated (face-to-face associations)
(c) edge-to-face flocculated but dispersed
(d) edge-to-edge flocculated but dispersed
(e) edge-to-face flocculated and aggregated
(f) edge-to-edge flocculated and aggregated
(g) edge-to-face and edge-to-edge flocculated and aggregated

creating a soil microstructure the form of which can have considerable bearing on its permeability anisotropy, its shear resistance, and its consolidation behaviour. These three factors, however, are also a function of any more-equant habit silty particles present in the sediment for they will exert an independent control on both the structural openness and the frictional properties of the material.

3.2 Sediment formation and clay fabrics

Most soils are formed by transportation and sedimentation of weathering products. Some of the mineralogical aspects of weathering are considered later in this chapter, but for a useful engineering geology paper on the subject, reference may be made to Fookes *et al* (1971). Other soils are formed *in situ* by progressive weathering. These are known as residual soils and they are formed in areas that have experienced a tropical climate in the past or at the present time (see, for example, Lumb, 1965; Ruddock, 1967; Wesley, 1973; Wallace, 1973).

Under tropical conditions of extreme wetting and drying with high desiccation rates being promoted through evapo-transpiration associated with luxuriant vegetation, the weak groundwater solutions resulting from leaching of the rocks during the rainy season are increasingly concentrated until the dissolved minerals are deposited out of solution. Of these ex-solution products, aluminium and iron hydroxide are not re-dissolved at the onset of the next rainy season and remain *in situ* as lateritic deposits (see, for example, Grant (1974) for an appraisal of Australian laterites and 'lateritic' gravels). *Bauxite* (high-grade aluminium hydroxide) used as a commercial source of aluminium, is a residual soil resulting from weathering of rock products containing aluminium and is found in Guyana and the Southern States of the U.S.A. Laterites, comprising a high proportion of hydrated ferric oxide, are commonly found in tropical countries, such as Western and Northern Australia, Africa, India and South America, and form an important source of road stone.

The most common forms of transportation, water, ice, air and gravity, will each have a unique effect on the structure and particularly on the texture of the resulting sediment. In general, the transportation mechanism predetermines the *texture* (that is, the fineness and uniformity) of a sediment in two ways: by further weathering associated with transportation and by the sorting action of transportation.

Thus, typical sediments transported by *gravity* will be subject to some reduction through impact, forming angular particles which will tend to be poorly-sorted and generally large. Sediments which have been transported by *ice* will be unsorted and may show signs of abrasion through grinding. On the other hand, *wind-blown* deposits will be subject to considerable abrasion and reduction and will contain characteristically rounded grains which will be well-sorted through variable velocity gradients into uniform layers and zones.

The majority of sediments are *water*-transported, and here the primary bed-load of degraded and progressively degrading larger particles is accompanied by a suspended load of finer clay minerals, silt and organic particles resulting partly from the chemical action of the water. Deposition is again controlled by the flow velocity of the water. Thus, reductions in flow velocity at places where, for instance, a river widens will lead to settling out from suspension of finer and finer suspended fractions. In addition, any temperature increase will lead to reductions in the viscosity and solubility of the suspension/solvent water. These two conditions are conducive to the sedimentation out of suspension of *clay mineral* and organic fractions.

The arrangement, size and orientation (or *fabric*) of particles in the deposited sediment will depend on, and in turn indicate, the conditions existing at the time of deposition. In the case of larger particles above silt size, which have been transported on or near to the river bed, this will be wholly dependent on velocity gradients and mainly a question of particle size. In the case of clay mineral particles, size is less important, and fabric orientation will depend principally on the depositional environment.

When sedimentation takes place in a fresh water, low electrolyte environment (typical of an inland lake — lacustrine deposit) sedimentation rates will be low, and a high degree of particle alignment will occur. Silt and larger clay particles sediment out of suspension individually, while smaller particles settle out in flocs which are large enough to overcome Brownian motion. With increasing salinity as the river mouth meets the sea, particles remaining in suspension will tend to flocculate, forming a sediment with a high void ratio and water content. In both fresh and salt water sediments, included silt fractions will tend to provide the soil with a loose, more open honeycomb texture (Casagrande, 1932b) and, in the case of fresh water deposits, to reduce the degree of preferred orientation of the platy clay minerals. Concentrations of NaCl greater than 0.1 to 0.2 molar

Clays and Clay Shales 111

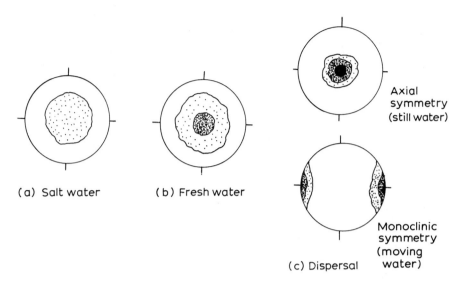

Figure 3.5 Equal area projections of idealized soil fabrics; horizontal plane of projection.

are generally sufficient to flocculate a clay. Typical, but idealized soil fabrics are shown in Figure 3.3 drawn generally after Lambe and Whitman (1968). A fully dispersed fabric can only be obtained after *remoulding* or consolidation and may not even be achieved then.

The dispersed soil micro-structure is strongly axi-symmetric. This is illustrated in equal area projections (poles to clay mineral basal planes) corresponding to the three cases in Figure 3.3 which are sketched in Figure 3.5. In the dispersed state the axis of fabric symmetry is parallel to the axis of sedimentation. Gillott (1968) has pointed out that sediments deposited by a flowing medium are likely to show monoclinic symmetry of fabric rather than axial symmetry. In this case, orientation transitions of fabric concentrations (perhaps via an intermediate fabric girdle) should indicate something of the dynamic controls on sediment formation.

A genetic classification of clays suggested by Skempton and Hutchinson (1969) is included in Table 3.2. This classifies clays according to the processes which have produced them.

Some recent work on the fabric/microstructure of clay soils has been carried out by Barden (1972a,b) and Barden and Sides (1971) using a scanning electron microscope to prepare a series of micrographs of clay soils of engineering interest from different parts of the world.

112 *Principles of Engineering Geology*

The major conclusion of the analyses, which confirms the work of soil scientists such as Brewer (1964) and Olsen (1962), is that whilst the idealized picture proposed by Lambe of flocculated and dispersed structures is basically correct, a closer approximation would be to regard all clay soils (both natural and artificial) as comprising an aggregate of soil particles (either peds, crumbs or

Table 3.2 Genetic classification of clays

Process	Comments
(a) Weathering *in situ*	Residual clays form as a result of deep, weathering of rocks, particularly igneous rocks, in *tropical climates*. Clays are also formed in *temperate climates* from the weathering of argillaceous rocks. Clay soils result from the more shallow weathering of competent rocks on slopes in relatively young valleys. The former type of clay is a stiff clay but after a slope failure in the material under heavy rainfall it would probably be classified as a soft intact clay. The second type would be grouped with the stiff clays and the third would probably be classified as a soft fissured clay.
(b) Sedimentary	These clays deposited under fresh or saline water would be classified as soft intact clays. The 'quick' clays, which are extremely sensitive to mechanical disturbance, fall under this heading. Sedimentary clays can be both normally-consolidated and over-consolidated. Of the latter type, the greater the clay mineral content, the more likely is the clay to be structurally discontinuous. The stiff, intact over-consolidated clays will usually have a higher silt content.
(c) Glacial	*Boulder clays* or *clay tills* are typical glacial clays and would usually be placed in the stiff, intact category. The softer intact material would have formed under a relatively small thickness of ice or under semi-buoyant ice. *Glacial lake clays* formed in temporary lakes during the ice age, although similar in engineering properties to fluvial lacustrine deposits, are usually more coarsely stratified (varved) than are recent lake deposits, and they may contain fine sand.
(d) Periglacial	Soliflucted clays are often stiff and contain slickensides but some of these clays are devoid of significant discontinuities. They will have been remoulded during their progress down-slope.
(e) Transported through landsliding processes	With little internal disturbance and discrete shearing of clay units, the picture would be one of slipped masses or '*slump blocks*'. *Mudflows*, on the other hand, would have produced intense remoulding and softening and an obliteration of any original texture. *Colluvial clays* comprise the products of weathering, earthflows and multiple sliding and have a moderate to stiff consistency.

clusters), each made up from clay plates rather than as a colloidal suspension made up from single plate units. This picture fits in far better with the generally accepted view of soils as *particulate* media, controlled uniformly by interparticle slip when subjected to deformation in a *drained* condition.

Barden's approach does not deny the existence of strongly orientated clay soil fabrics, but accepts them as stratified granular structures rather than as single plate structures. Similarly, flocculated structures are seen as irregular aggregates of soil peds with a tendency to side-face configuration between platy clay crystals inside the ped. Typical micrographs from Barden (1972) are shown in Figures 3.6 to 3.8 inclusive. These include *lightly overconsolidated post-glacial marine clays* (Figure 3.6), *lightly overconsolidated fresh or brackish water clays* (Figure 3.7), and *heavily overconsolidated stiff fissured clays* (Figure 3.8). The *recent marine clays* (Figure 3.6), in general, have weak, flocculated, open structures formed from flaky angular particles with silt inclusions which appear to determine the openness of the structure. Consolidation and remoulding is minimal and the clays will have high compressibility accompanied in most cases by low permeability because of the limited particle size. In addition, they will probably be fairly *sensitive*, depending on the degree to which the original salt content has been leached by ground water.

The flocculating action of a salt has been well established by leaching experiments (Rosenqvist, 1960; Kenney *et al*, 1968). After leaching, the clay tends to maintain its structure until it is remoulded, after which the particles remain dispersed. Suppose that a flocculated clay is sedimented from a salt water suspension, and that having developed only rather weak 'diagenetic bonds' it is subsequently flushed through by river water. Foundation engineering works in the clay might then be sufficiently disturbing to cause some collapse of the loose 'card house' structure, so creating foundation and slope instability. This is the situation that exists with respect to 'quick' clays in the Drammen area of Norway (Bjerrum, 1967; Torrance, 1974). In a similar manner, Søderblom (1966) quotes the case of phosphate effluents seeping from a waste dump in northern Sweden down into a clay of normal sensitivity. Some three years later, the clay was in a quick condition as a result of cation exchange processes. A phenomenon such as this lends practical evidence for the possibility of sensitive clay stabilisation through the injection of a dispersant (see Section 11.9).

It should be noted that remoulding in the case of a non-leached

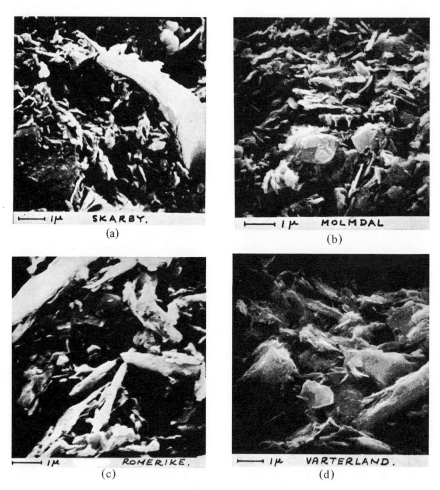

Figure 3.6 Electron micrographs of normally or lightly overconsolidated recent marine clays (*after* Barden, 1972; reproduced by permission of the Geological Society of London).
(a) Swedish marine clay from Skarby – very plastic leached illitic clay of high sensitivity.
(b) Swedish marine clay from Molmdal – similar to (a) but very high sensitivity.
(c) Norwegian marine clay from Romerike – leached illitic clay with zones of quick clay surrounded by clay of low sensitivity.
(d) Norwegian marine clay from Vaterland – unleached illitic clay.

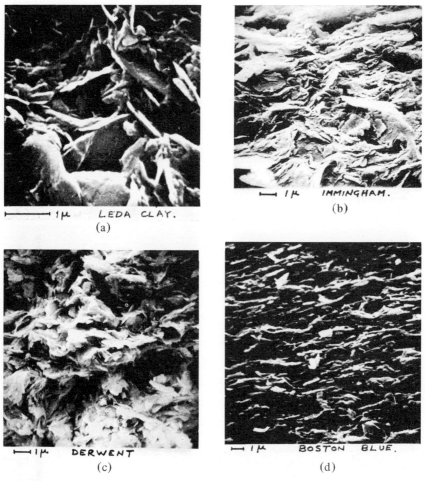

Figure 3.7 Electron micrographs of normally or lightly overconsolidated recent fresh or brackish water clays (*after* Barden, 1972; reproduced by permission of the Geological Society of London).
(a) Leda Clay, Canada – medium plasticity illitic clay of high sensitivity, deposited in brackish water and leached by fresh water.
(b) Immingham Clay, U.K. – unleached brackish water esturine clay of low sensitivity.
(c) Laminated clay, Derwent, U.K. – plastic freshwater clay.
(d) Boston Blue Clay – soft brackish water glacial clay.

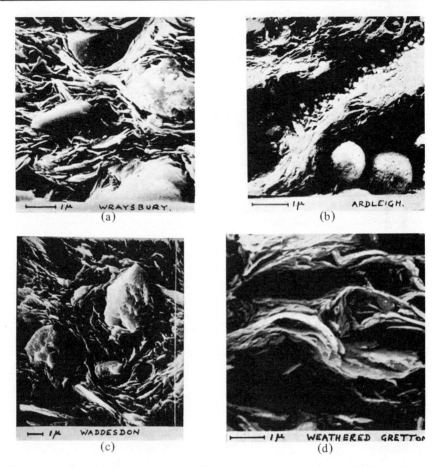

Figure 3.8 Electron micrographs of heavily overconsolidated, stiff fissured clays (*after* Barden, 1972; reproduced by permission of the Geological Society of London).
(a) London Clay at Wraysbury
(b) London Clay at Ardleigh
(c) Kimmeridge Clay at Waddesdon
(d) Upper Lias Clay at Gretton

clay is difficult, since much of the original structure is likely to be regained after the remoulding process is completed.

The *recent freshwater clays* (Figure 3.7) vary in structure from the strongly orientated Boston Blue Clay to the flocculated Leda Clay. With these materials, the degree of flocculation is dependent upon the salt content of the depositional environment and the sensitivity of the flocculated soils to the electrokinetic potential of the pore fluid. It is also dependent on the cementing media in the case of

Leda Clay. Dispersed soils usually have low permeability, low sensitivity and low compressibility.

The *earlier, heavily overconsolidated* clays (Figure 3.8) may or may not have collapsed from an originally unflocculated state, but all demonstrate a high degree of near-horizontal orientation which may be slightly modified by weathering. The basic properties are low compressibility and sensitivity and low material permeability. The overall properties may, however, be controlled more by fissures resulting from later erosion of overburden and subsequent weathering, and by the pronounced anisotropy of the clay, than from the actual microstructure. The effect of fissures and anisotropy on clay properties are considered later, but one important aspect of soil fabric should be immediately apparent. This is that in recent, lightly consolidated soils, the microstructure, as determined by the electrochemical environment during deposition, can have an important effect on soil properties. Thus in flocculated soils, compressibility will be high and sensitivity probably high. Both are a property of the microstructure, whereas in dispersed and overconsolidated soils, compressibility and sensitivity will be low and soil properties will be related to the soil *macrostructure.*

These remarks refer of course to conventionally deposited and consolidated soils. Some soils, specifically Keuper marls, some silt and loess deposits, and some expanding clays, have quite unusual fabrics and often spectacular reactions when wetted. The relatively unstable fabric/microstructure of these soils deserves special study.

3.3 Unstable clay fabrics

The *sensitivity* of a clay may be defined as the ratio between natural and remoulded strength. Although most clay soils are sensitive to a greater or lesser degree, their reactions to load are generally consistent and predictable on the basis of accepted soil mechanics testing procedures. Where the reactions of soils differ radically, this can usually be explained in terms of the fabric or microstructure (including clay mineral content), and particularly in the response of this microstructure to changes in water content.

An obvious example is a clay of high sensitivity (say between 8 and 16) resulting from a honeycomb structure, or from leaching of saltwater in a well-flocculated recent clay. Typical of these are the Scandinavian clays (Figure 3.6). In a partly saturated state the clay will compress under load, but in a saturated state, load will be transferred to the water in the porous structure, so creating a 'quick' condition (Höeg et al, 1969) in which the clay will behave as a viscous fluid rather than as a deformable solid.

A similar but not directly analogous condition occurs in some aeolian (wind-blown) soils of which the most readily identifiable are dune sands and loess. The former, of limited grain size (fine or medium sand), has no cohesive strength, moderately high permeability, and moderate compressibility. The latter has received quite detailed study and reference may be made to Turnbull (1948), Holtz and Gibbs (1951), Denisov (1953), Pitcher et al (1954), Clevenger (1956), Gibbs and Holland (1960), Krinitzsky and Turnbull (1967), Lutton (1969), Fookes and Best (1969) and Muchowski and Sztyk (1974).

Loess has a metastable structure generally in equilibrium with its environment. Denisov (1953) suggested that the loesses in Europe and Russia were formed under conditions of cold and dry climates of arid steppes or semi-deserts during glacial periods of the Pleistocene. Loess-like clayey soils with only a low degree of potential settlement, or which do not settle at all, he considered to have formed during interglacial periods when humidity was higher. If the internal moisture conditions change and/or there is an increasing overburden load on the loess, then if overall stability is maintained the loess will settle in a controlled manner and assume a terminal state of stability. Loess which has so consolidated has the properties of any other ordinary water-lain silt (Gibbs and Holland, 1960).

Structurally, loess consists mainly of angular particles of silt or fine sand with a small amount of clay binder. True loess in a natural state has a characteristic structure formed by the remnants of small vertical root holes that endow the material with a permeability anisotropy. Although low density, dry, natural loessial soils have a moderately high strength and can stand vertically in cuttings, they readily undergo structural collapse upon excessive wetting and disturbance and, in fact, collapse may occur by quite dramatic undercutting at the base of a wall. Such undercutting, often promoted by ponded water at the base and leading to more general collapse, is accommodated adjacent to highways in North America by the provision of adequate 'shoulder'. Loessial deposits are a feature of the land bordering the Mississippi River (Figure 9.53) and for an excellent engineering geological appraisal of the material, the reader is referred to the report by Lutton (1969).

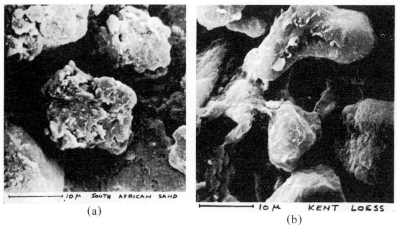

Figure 3.9 Electron micrographs of collapsing soils (*after* Barden, 1972; reproduced by permission of the Geological Society of London).
(a) Aeolian red sand, South Africa.
(b) Kent Loess.

South African red sand and Kent loess (Figure 3.9) are both original Pleistocene sands in which silt-size quartz and, in the South African sand, feldspars were deposited in a loose, open structure. Subsequent changes in climatic conditions have led to weathering of the feldspars to kaolinite in the South African sand and this has tended to cement the particles together. In the Kent loess, the cement is mainly calcareous. The presence of the cementing medium has prevented complete collapse of the soil during subsequent natural consolidation, but if the soil is saturated when loaded, the subsequent weakening of the intergranular bond can allow slip and ultimately *collapse*. The phenomenon of *collapsing soils* (Dudley, 1970, reviews this in detail) is mainly associated with aeolian deposits cemented by weathering or percolation, but it can occur in intensely desiccated sedimentary deposits containing swelling clays and in clay soils compacted artificially in a dry state. Barden (1972) quotes the case of Tucson silty clay from Arizona (Figure 3.10) which swells under moderate pressures and collapses under high pressures. The inference is that any soil, cemented or flocculated but having a high porosity, can collapse if conditions are created which substantially weaken the existing soil skeleton. This process can be assisted by various natural or artificial weathering mechanisms which can change the properties of a soil radically over a short or long period of time. For instance, as

120 *Principles of Engineering Geology*

Figure 3.10 Electron micrograph of silty clay, from Tucson, Arizona (*after* Barden, 1972; reproduced by permission of the Geological Society of London).

with loess, collapse can be induced by mechanical action combined with flooding.

A common example of natural collapse in Britain is the *Keuper Marl*, a red Upper Triassic fine-grained mudstone described by Davis (1968) and Barden and Sides (1970). The Keuper Marl probably originated through deposition of water-borne and wind-blown particles in highly saline inland seas. The resulting unweathered fabric (Figure 3.11) is of variable composition, but on average comprises 20 per cent silt-size quartz particles and 80 per cent clay mineral which is flocculated into silt-size peds or particles. Depending on the degree of weathering, the clay mineral peds can break

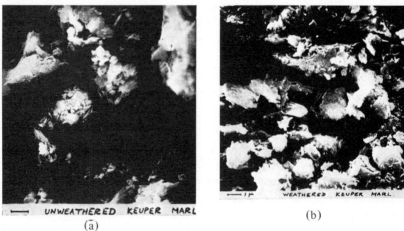

Figure 3.11 Electron micrographs of (a) unweathered and (b) weathered Keuper Marl (*after* Barden, 1972; reproduced by permission of the Geological Society of London).

down to give up to 50 per cent of clay size (<0.002 mm) particles, thus creating a variable structural pattern of primary and secondary clay peds and quartz particles surrounded by clay plates orientated parallel to the quartz surfaces.

Despite high overconsolidation stresses, the original flocculated structure is maintained in an unweathered or only partly-weathered Keuper Marl, in which case the strength and permeability will be high. Advanced weathering can, however, lead to virtual collapse of the structure with a reduction of strength to that of a soft clay and also a reduction in permeability by several orders of magnitude. Keuper Marl can therefore occur as a strong mudstone or as a soft clay, the difference depending solely on the fabric. Chandler (1969) has summarized the various weathering zones in the deposit.

Another type of soil having an important fabric is *peat* or *Muskeg* shown in Figure 3.12. Peat has a highly compressible amorphous or fibrous organic structure. Its interest as a material arises because its secondary consolidation due to long-term pore drainage or plastic deformation of the skeleton is of greater significance than primary consolidation resulting from drainage of the main skeleton. Since both types of consolidation are a direct reflection of the structure, the fabric assumes an unusual importance in any design problem. Reference may be made to Hanrahan (1954, 1964), Barden (1968), Skempton and Petley (1970), and to Berry and Poskitt (1972).

We may regard instability as an issue not only in collapsing soils but also in soils which swell and contract significantly in response to pore moisture changes. Swelling and suction pressure phenomena in montmorillonitic and interlayered illitic soils are considered later in

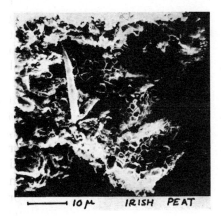

Figure 3.12 Electron micrograph of Irish peat (*after* Barden, 1972; reproduced by permission of the Geological Society of London).

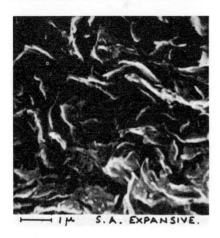

Figure 3.13 Electron micrograph of an expansive clay from South Africa (*after* Barden, 1972; reproduced by permission of the Geological Society of London).

the present chapter. Black Cotton soils present this type of engineering problem (Palit, 1953; Ranganatham, 1961).

An example of the swelling clay structure quoted by Barden (1972) is illustrated in Figure 3.13. This is a South African expansive clay described by Jennings (1953), and Barden describes the 'crinkled lettuce leaf effect' as a characteristic of montmorillonite clays when viewed under the scanning electron microscope.

3.4 Glacial and periglacial clays

Glacial soils (or drifts) comprise mainly *boulder clays* (or *tills*). These are formed by direct ice action and are deposited as moraines at the glacier end, or as more widespread layers beneath ice sheets. *Periglacial* soils are not formed directly by ice action, but are formed under associated climatic conditions. They usually occur as loess or windblown soils (see Section 3.3) or as residual soils resulting primarily from freeze-thaw weathering. In Britain where the furthest extent of Pleistocene ice followed a line roughly from the Severn to the Humber, glacial soils occur in Wales and the north and periglacial soils in the south.

Boulder clays usually comprise a stiff clay matrix formed from rock flour created by the grinding action of the moving ice. This matrix contains boulders of varying size, angularity and quantity, carried and subsequently deposited by the main ice mass. Boulder clays may also include sand and gravel layers and pockets deposited from sub-glacial streams. Typical size distributions, illustrated in Figure 3.14a, are variable and generally form a poor basis for engineering classification. McGown (1971) has shown that the slope of the size distribution curves usually exhibits a split or gap grading which occurs around the sand size fraction. This is particularly apparent in a log-probability plot. The type of distribution is quite typical of comminution products, and McGown suggests that in most cases the fractions on either side of the split can be separated for identification and testing purposes. He also shows that when till particle size distributions are split, a high degree of conformity is found between tills of the same parent rock type and glacial history. This point is illustrated in Figure 3.14b.

Since the fine soil fraction may range down to silt or clay size, its engineering properties will show a much greater variation than those

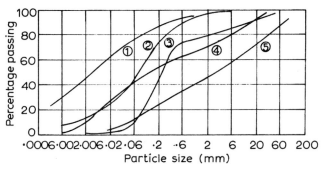

(a) Typical gradings of morainic soils
(1) Valparaiso, USA (2) Berlin
(3) Glen Kingie, Scotland (4) Ontario, Canada
(5) Ville la Salle, Canada

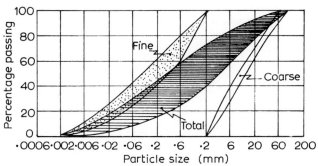

(b) Range of coarse and fine fractions from morainic soils - Corrour Forest, Scotland

Figure 3.14 Gradings of moraine deposits (*after* McGown, 1971).

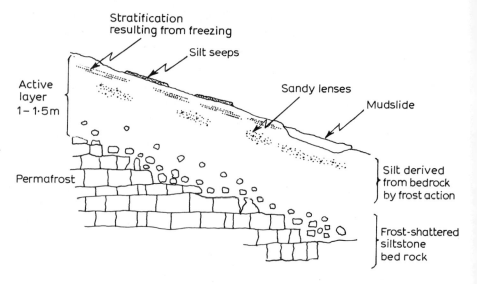

Figure 3.15 Slope solifluction deposit in permafrost zone – diagrammatic section from trial pit records (*after* Chandler, 1972b).

of the coarse fraction. McGown suggests therefore that the properties of the fine soil fraction should form the basis for any engineering classification of tills.

Residual *periglacial* soils usually occur as solifluction deposits and are frequently characterized by low density, variable permeability and porosity, and variable grading. An example investigated by Chandler (1972b) in Spitzbergen is illustrated in Figure 3.15. Sheet movements of solifluction deposits on low angled slopes, resulting in mudslide types of slope failure (see Section 9.9), are an important engineering problem. Chandler suggests several reasons for observed high artesian pressures causing mudflows in this case. These include (a) surface freezing of partially thawed upper layers in the spring; (b) silt accumulations (silt seeps) at the surface, effectively creating an impermeable blanket above coarser layers; (c) creation of ice lenses and segregation of coarse and fine layers, which will be aligned parallel to the ground surface, by freezing and thawing; (d) breakdown and fissuring of near-surface bedrock by frost action to create zones of high permeability.

Of these, the one having most universal relevance may be (c), and it appears probable that many mudflows in periglacial soils, often occurring some time after periglacial conditions have ameliorated, may result from variable permeability in the near-surface active layers.

3.5 Depth/strength profiles

A freshly-deposited 'unconsolidated' marine sediment will contain between 40 and 100 per cent water depending on its particle size and mineral content (see Keller, 1969). More specifically, a freshly-deposited medium sand will contain approximately 45 per cent water (equivalent to the porosity in a saturated soil), a silt 50 per cent to 70 per cent, a clay mud 80 per cent to 90 per cent (depending on dispersion), and a colloid mud up to 100 per cent water. The organic content will depend on the particle size, ranging from less than 1 per cent in a medium sand to more than 10 per cent in a clay mud.

Obviously, in this state both sands and clay will have a near-zero relative density, and high porosity and permeability. In mechanical terms they will also possess minimum strength and maximum compressibility, a result of the low inter-particle contact area. Fortunately, structures are rarely required in practice to rest on recent sediments but, where they are so required, a long process of drainage and consolidation is needed before sufficient bearing capacity to support the structure is obtained. Soils are usually consolidated to a certain degree by the natural process of increasing sedimentation which gradually builds up an increasing thickness of cover. The forces associated with this process compact the lower sands and consolidate the lower clays, increasing the density through slip and interlocking in the former case, and by re-orientation of clay particles in the latter case.

The result of this natural consolidation is a decrease in porosity and permeability, and an increase in shear strength and stiffness. This is illustrated by results obtained by Krinitzsky (1970) from tests on undisturbed cores obtained from drillings in homogeneous soft clay deposits of the Lower Mississippi deltaic plain. He shows in Figure 3.16 that gradual decreases in water content with depth are accompanied by almost linear increases in shear strength.

These results refer specifically to samples of Balize-Plaquemines interdistributary sediments and prodelta sediments from a single borehole. Individual samples varied in the former case from massive fat clays with high organic contents to thinly bedded clay silts, some with cross bedding, and in the latter case from moderately-to-highly

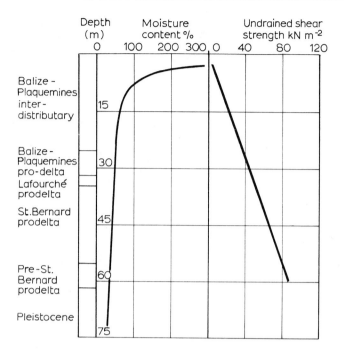

Figure 3.16 Depth-moisture content and depth-strength profiles, Mississippi Valley sediments (*after* Krinitzsky, 1970).

fractured calcareous fat clays, the lower of which exhibited brittle failure during testing.

The full test series described by Krinitzsky included most of the fine-grained sedimentary soils found in alluvial and deltaic deposits of the Lower Mississippi Valley, and a summary of undrained shear strengths indicates significant variations in the shear-strength/depth profile, particularly at shallow depths. This feature was investigated by Krinitzsky who used X-rays to examine the microstructure of the soils. He concluded that, apart from depth, the greatest influence on strength variability was the presence of fissures or other structural discontinuities in the *test specimen* unless these occurred at orientations nearly normal to the axis of major compression. Other features which affected strength were irregular masses of *organic material* and the presence of roots which tended to improve strength through desiccation-assisted consolidation. It must also be emphasized that the shear strength of a clay, in addition to its Atterberg limits, is quite intimately related to the percentage quartz in the

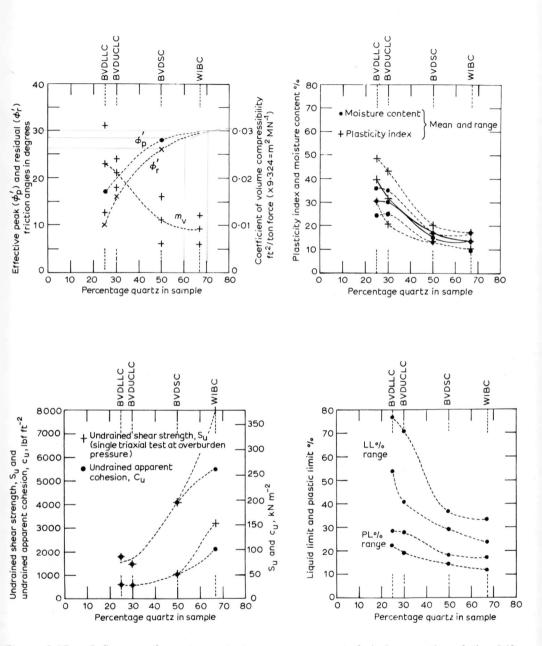

Figure 3.17a Influence of quartz content upon some geotechnical properties of the drift sequence on Tyneside, England. Mean and range of m_V shown. (*After* Boden, 1969).
Buried Valley Deposits Lower Laminated Clay (BVDLLC)
Buried Valley Deposits Upper Coarsely Laminated Clay (BVDUCLC)
Buried Valley Deposits Stony Clay (BVDSC)
Western Ice Boulder Clay (WIBC)

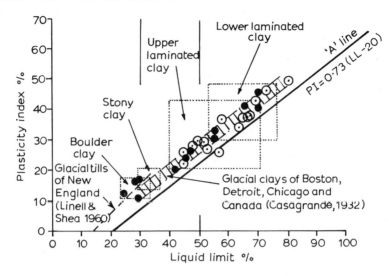

Figure 3.17b Casagrande plasticity chart classification of the four Tyneside (England) clays plotted in Figure 3.17a. In addition to the ranges of these laminated, stony and boulder clays are plotted other glacial clays and tills. The solid circles are representative of glacial lake clays of the Great Lakes (Wu, 1958). The open circles (Marshall, 1969) refer to a range of laminated clays sampled at different depths in the Team Valley (pre-glacial valley of the River Wear in N.E. England). See also Figure 9.16b.

material (see Figure 3.17a). The influence of clay composition on Atterberg limits has been discussed in a general way in Section 2.2 and Table 2.4, and for the clays specified in Figure 3.17a, the Casagrande plasticity chart is shown in Figure 3.17b.

The basic conclusion is, however, that soil *strength* is a property of the *soil constituents*, that is, of relative clay or quartz content at the simplest level, and of *depositional history, post-depositional history* and *structural geology* of the soil. This is illustrated in the geotechnical profile case-histories 1 to 7 drawn in Figures 3.18–24 and summarized in Table 3.3. Some of these are utilized by Lambe and Whitman (1969) in a similar soil-profile case-history summary.

The main points illustrated by these profiles are:

(a) In weakly consolidated (that is, recent) soils, strength increases, and water content and permeability decrease with increasing depth in the sediment (see cases 4 to 7);

Clays and Clay Shales 129

(b) In overconsolidated soils, water content and permeability are reasonably constant with depth, and strength — although it may increase with depth — is more a property of *weathering* with consequent increase in fissure density and effectiveness than of consolidation (see cases 1 to 3).

In the case of granular soils, degree of compaction and hence relative density and strength increase with depth in loosely compacted soils and depend on post-depositional history in the case of well-compacted soils. Additional factors such as cementation and clay content may, however, affect the soil response to stress and deformation.

Table 3.3 Depth/strength profile case histories

Case no.	Description	Depositional history	Post-Depositional history	Figures and reference
1	Upper Lias Clay (fissured heavily overconsolidated Jurassic clay-shale)	Marine deposits up to 300 m thick overlain by 1000 m overburden	Uplift and erosion of overburden	Figure 3.18 Chandler (1972a)
2	London Clay (fissured over-consolidated Eocene Clay)	Marine deposits up to 300 m thick with 300 m overburden	Uplift and erosion of up to 400 m overburden	Figure 3.19 Marsland (1972a)
3	London Clay	,,	,,	Figure 3.20 Skempton and Henkel (1957)
4	Boston Blue Clay (over–to–normally-consolidated Pleistocene Clay)	Glacial clay sedimented in marine water. 25 m thick	Uplift and re-submersions means that upper clay is stronger than lower clays	Figure 3.21 in Lambe and Whitman (1969)
5	Thames Estuary Clay (normally-consolidated recent clay)	Estuarine deposits	Normal consolidation	Figure 3.22 Skempton and Henkel (1953)
6	Norwegian Marine Clay (normally-consolidated recent clay)	Glacial clay deposited under marine conditions	Uplift and fresh-water leaching	Figure 3.23 Bjerrum (1954)
7	Norwegian Marine Clay	,,	,,	Figure 3.24 Bjerrum (1954)

3.6 Macrostructure of overconsolidated clays and clay shales

The most dominant structural features of stiff, overconsolidated clays and clay shales, and those features which can exert the most influence on their geotechnical behaviour, are their fissility and the presence of discontinuities. A strong fissility characterizes a true shale; for example, the Kimmeridge Clay in Kimmeridge Bay east of Weymouth on the south coast of England comprises horizons of highly bituminous 'paper shale' interbedded with more massive cementstone bands, the latter exhibiting large scale fissuring patterns but lacking the very fine fissility of the paper shale bed. Discontinuities, their genesis, measurement and geotechnical analysis, are considered in Chapter 6. Geological dictionaries define fissility as being a special case of jointing by fracture or by flowage, with the 'rock' capable of being split into thin sheets. Thus, not only may an anisotropic fissility plane actually parallel the original bedding and be a function only of quite moderate sedimentary overload pressures but it may also represent a fabric reaction to subsequent tectonic pressures on a regional or local basis. In the latter case, a pronounced fissility may be regarded as a very low-grade metamorphic feature. Discontinuities tend also to reflect regional structural trends particularly if their fabric trends are compatible with the joint directions in adjacent rocks. It is possible, therefore, to draw the overall conclusion that on a macroscopically discernible scale overconsolidated clays and clay shales have their geotechnical responses regionally conditioned. For this reason, it is useful to look at one such example.

A section of coast line in the south of England and the Isle of Wight was surveyed by Reeves (1973) for discontinuity trends in the stiff clay and clay shale beds. Stratigraphically, these beds are of Eocene, Cretaceous, and Jurassic age, their formational position in the stratigraphical column for the Hampshire Basin being shown in Figure 3.25. Structural and sample location maps together with extracts from processed stereonets of discontinuity orientations are given in Figures 3.26 to 3.29.

In a freshly excavated state, the Lias would be considered a true clay shale. Where it outcrops between Bridport and Lyme Regis towards the western end of the area in Figure 3.26 it tends to degrade very readily into a clay/mud flow and it is therefore not possible to measure discrete structural elements in it. The Lias, in fact, tends to be of much more variable lithology than the other

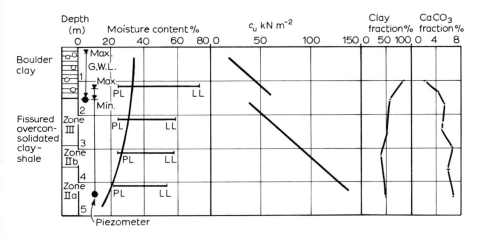

Figure 3.18 Geotechnical profile – Upper Lias Clay at Gretton (*after* Chandler, 1972a).

materials to be described subsequently and some cliff sections are formed almost entirely of slumped material. Since this section of the coast has a long history of erosion and landslipping, the Lias there is not really amenable to discontinuity measurements.

The Oxford Clay at Furzy Cliffs near Osmington, east of Weymouth, is the next oldest deposit of clay shale in the column of Figure 3.25. This 150 m thick deposit lies above the Cornbrash, the measuring and sampling areas A and B in Figure 3.26

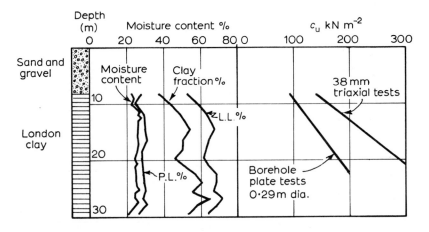

Figure 3.19 Geotechnical profile – London Clay at Wraysbury (*after* Marsland, 1972a).

132 *Principles of Engineering Geology*

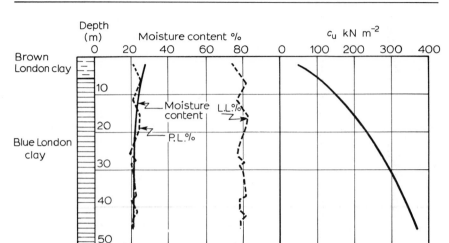

Figure 3.20 Geotechnical profile – London Clay at Paddington (*after* Skempton and Henkel, 1957).

being located towards the top of the sequence where the clays are bluish-grey and the fossils (typically *Gryphaea bilobata*) are pyritised. Sites A and B are located along the same cliff section some 100 metres apart.

These Oxford Clay cliffs are degrading much more rapidly than are the younger Kimmeridge Clay cliffs. Examination of the disintegration mode suggests that the clay fails in a more 'massive' manner

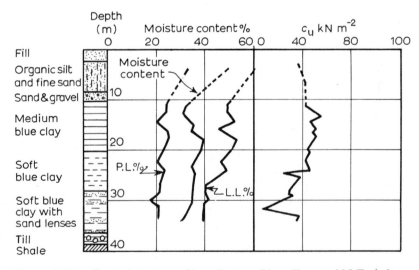

Figure 3.21 Geotechnical profile – Boston Blue Clay at M.I.T. (*after* Lambe and Whitman, 1969).

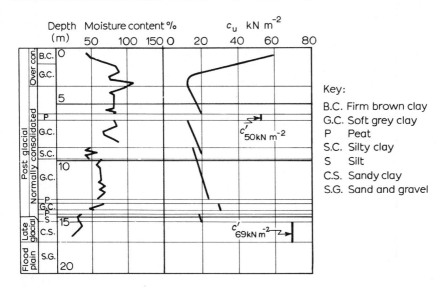

Figure 3.22 Geotechnical profile – Thames Estuary clay at Shellhaven (*after* Skempton and Henkel, 1953).

with block failures rather than the separation of individual fissure fragments. Many small fissures are slightly curved but on a scale of 4 m or more, a linear continuity can usually be discerned. Reference to the stereonets for sites A and B suggests a minor concentration of discontinuities, sub-horizontal, against the trend of the anticline, and tending somewhat to condition the material towards a planar shear

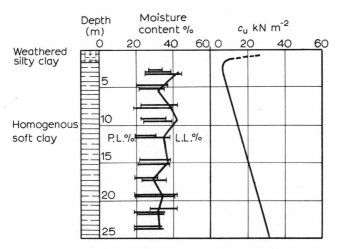

Figure 3.23 Geotechnical profile – Norwegian marine clay at Drammen – high sensitivity (*after* Bjerrum, 1954).

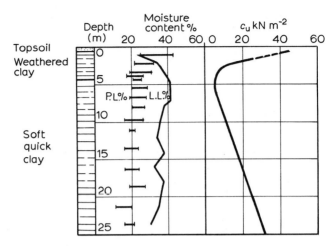

Figure 3.24 Geotechnical profile – Norwegian marine clay at Manglerud – low sensitivity (*after* Bjerrum, 1954).

type of failure. There is also a more extensive N/S and E/W conjugate set of discontinuities which is probably a result of the deformation phases that the deposit has been through but which could also be due to the lateral extension of discontinuities caused by the continuous extension of the cliff by sea erosion. The curvilinear nature of some of the discontinuity surfaces would suggest that they have been caused by drying out of the clay rather than by the removal of pre-consolidation loads.

Between the Oxford Clay and the Kimmeridge Clay outcropping further to the east lie the Corallian Beds. The Corallian represents shallower, warm-water marine deposits in contrast to the Oxford Clay below and the Kimmeridge Clay above which are both deeper-water marine deposits. The whole of the Jurassic in Southern England was one huge marine cycle but with many different marine lithologies represented.

In contrast to the grey, blocky clay shale of the Oxford Clay, the Kimmeridge Clay can be more accurately described as a true shale. Horizons of highly bituminous paper shale are interbedded at 10 to 20 metre intervals with more massive cementstone bands, the latter exhibiting large scale fissure patterns but, as noted earlier, lacking the very fine fissility of the former. Sites C, D, E in Figures 3.26, 3.29 occur in and around Kimmeridge Bay which is the type locality for this deposit. These sites are respectively on the eastern limb of an

PERIOD		UNIT DEPTH	FORMATION	TYPE OF DEPOSIT	ASSOCIATED WITH THE ALPINE OROGENY AND TETHIAN GEO-SYNCLINE
QUATERNARY	RECENT		Blown sand and shingle beaches; alluvium; tufa; peat	Superficial deposits	
	PLEISTOCENE		Raised beach		
			River terraces (gravels)		
			Coombe deposits		
			Angular flint gravel		
			Plateau gravel and brickearth		
Uncomformity F3			Clay with flints	Unconformity	Main Alpine orogeny F3
TERTIARY	OLIGOCENE	83m	Hamstead Beds		
		33m	Bembridge Beds		
		33m	Osborne Beds		
		50m	Headon Beds		
	40 myrs				
	EOCENE	66m	Barton Beds		
		200m	Bracklesham Beds		
		267m	Bagshot Beds	Thick, completely marine clay formation	Folding, uplift and erosion of Chalk before further deposition F2
		117m	London Clay	Non-marine deltaic sands and clays	
		33m	Reading Beds		
		850m+drift			
Unconformity F2	60 myrs				
	CRETACEOUS	500m	Chalk		
			Upper Greensand and Gault	Fresh water deltaic sands and lacustrine clays-prolonged phase of lake expansion (top ⅓ Weald Clay)	Folding along E/W axes, faulting and erosion F1
F1	Unconformity		Lower Greensand		
		500m	Wealden Beds		
	135 myrs				
MESOZOIC	JURASSIC		Purbeck Beds		
			Portland Beds		
		550m	Kimmeridge Clay	Deep water marine	
			Corallian Beds		
		170m	Oxford Clay and Kellaways Beds	Deep water marine	
			Cornbrash		
			Great Oolite series		
			Inferior Oolite		
	180 myrs		Lias		

Figure 3.25 Stratigraphy of the beds on the south coast of England (*after* Reeves, 1973).

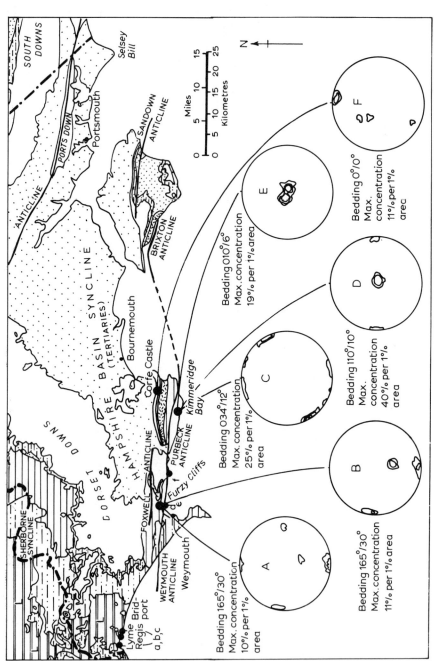

Figure 3.26 Structure and stratigraphy of the south coast of England in the vicinity of the Isle of Wight. Discontinuity measurement points and extracts from processed equal area fabrics are shown. Discontinuity orientations are contoured at 5% intervals with the exception of fabric D. Bedding orientations are given as dip direction (clockwise) and dip amplitude. Projections are upper hemisphere. (*After* Chatwin, 1960, *British Regional Geology – The Hampshire Basin and adjoining areas*, 3rd Edition, I.G.S., 1960).

anticlinal structure exposed in the cliffs at the eastern end and at the western end of the Bay. Site E measurements are referred to a strata section consisting entirely of paper shales. The projection indicates a strong planar fissility virtually parallel to the bedding. Sites C and D were located in a wave-cut platform below each cliff section and formed on a cementstone horizon. Fabric C is controlled by the orientation of the sub-vertical fissures in the cementstone but fabric D combines the near-vertical fissures in the cementstone with the planar fissures paralleling the bedding in the fissile shale. In this location, therefore, fissure fabric is controlled by the variable lithology. It has been estimated that a possible 1400 m of overlying deposits were present for some time during the post-depositional history of the Kimmeridge. Vertical off-loading by erosional processes following the three post-Jurassic periods of tectonic activity has been responsible for the fissility parallel to the bedding in the shale horizons. Fissures in the well-cemented, harder horizons may have resulted from folding during the tectonic events rather than from vertical stress release.

Other clay shale horizons in the south of England are best studied on the Isle of Wight (Figure 3.27). It is also possible in this way to assess the influence on the Isle of Wight of the anticlinal structures which formed the Isle of Purbeck on the mainland (see Figure 3.26).

The next clay shale in the Mesozoic succession is the Weald Clay of the south-east coast of the Isle of Wight. There the Wealden Beds comprise fresh-water deltaic sands and lacustrine clays formed during a prolonged phase of lake expansion at the beginning of the Cretaceous period. The Weald clays forming the top one-third of the Wealden Beds are very variable in character, often outcropping as marls or clays rather than as true clay shales. Lignite occurring as a pine raft and found near site H (Figure 3.29) suggests that the deposits are deltaic.

The beginning of the Cretaceous period was marked by an uplift of the surrounding higher land and probably also by a climatic change since it was accompanied by a revival of active rivers. These rivers, draining a land area to the west, spread their detritus over southern England and northern France in the form of extensive delta-flats. When sedimentation decreased through the lowering of the surrounding land masses, large fresh-water lagoons were formed and the deposits of these waters gave rise to the finely laminated shales.

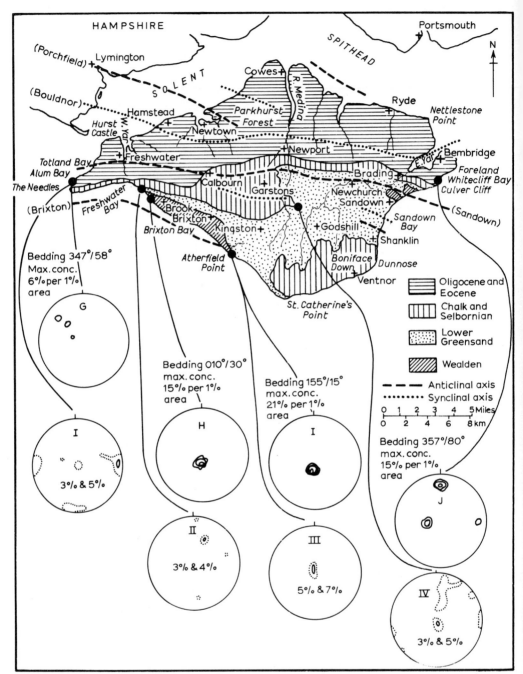

Figure 3.27 Structure and stratigraphy of the Isle of Wight. Discontinuity measurement points and extracts from processed equal area fabrics are shown. Lower four fabrics are lower hemisphere projections *after* Fookes and Denness (1969). The upper four are upper hemisphere projections *after* Reeves (1973), the contours being at 5% intervals. (*After* Chatwin, 1960, *British Regional Geology – The Hampshire Basin and adjoining areas*, 3rd Edition, I.G.S., 1960).

Clays and Clay Shales

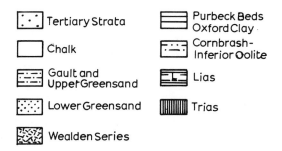

Figure 3.28 Data for Figures 3.26, 3.27.

Reference to Figures 3.27 and 3.29 will show that the upper section of the Wealden series is brought to the surface by two anticlines. Two distinct divisions may be recognized: a lower division comprising the Weald Marls or variegated clays and an upper division of Wealden shales. The Weald Marls are of freshwater origin but in the highest beds there is evidence of marine incursion. At the end of Wealden times, earth movements became more appreciable and eventually the sea, advancing from the south, covered most of the area and deposited the Lower Greensand. The Cretaceous period of marine deposition ended with the laying down of the extensive chalk formations.

Although sites H and I (Figures 3.27 and 3.29) are described as being in Weald Clay, the material can more accurately be described as a clay-marl even though it possesses an extensive fissure system typical of overconsolidated clay shales. The discontinuity fabrics for both sites show sub-horizontal concentrations with little obvious bedding influence. Site H is on the steeper northern limb of the Brixton Anticline and site I is just to the south of the suggested crestline of this structure. It will be noted that the Weald Clay fabric for the Fookes and Denness site III closely matches in orientation if not in intensity the site I fabric (the former is lower hemisphere and must therefore be rotated through an angle of 180° in the plane of the projection for comparison purposes with the latter fabric). The Fookes and Denness site II fabric is in the Gault Clay and cannot therefore be compared with fabric site H. However, the relationships between the discontinuity fabrics and the regional structure touched upon thus far in the present section are expanded in Fookes and Parrish (1969).

The Oxford, Kimmeridge, and Weald clays, all conforming in variable degrees to the definition of a clay shale, possess extensive,

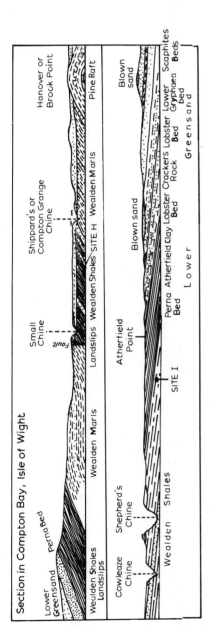
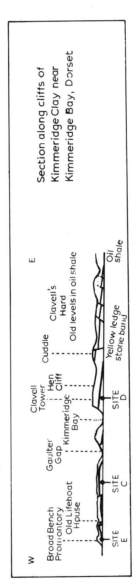

Figure 3.29 Coastal cross-sections which include some of the discontinuity measurement sites. (*After Chatwin, 1960, British Regional Geology – The Hampshire Basin and adjoining areas*, 3rd Edition, I.G.S., 1960).

complex fissure systems. They were all deposited conformably with the intermediate strata and were not affected by any large-scale tectonism until mid-Cretaceous times. This period of earth movement (F1 — Figure 3.25) was one of two main precursors to the main Alpine movements of the Miocene period.

After the folding, uplift and erosion of the Chalk land surface at the end of the Cretaceous period, the 30 m or so of the fluviatile Reading Beds were deposited before the onset of the marine incursion which laid down the London Clay. The London Clay of the Hampshire Basin was deposited in shallow seas and contrasts with the more uniform and deeper-water clays of the London Basin. In the Isle of Wight, the clay departs rather more severely from the concept of a clay-shale, its sand content increasing from east to west across the section.

On either side of the Isle of Wight, the Eocene beds lying above the profound unconformity with the Chalk are nearly vertical and dip towards the north. Of some 120 m of London Clay exposed at Alum Bay (compared with some 20 to 24 m in the Purbeck district) the upper 50 m are sandy and unfossiliferous and may be grouped with the overlying Bagshot Beds. Site G is near the base of the London Clay at the southern end of Alum Bay (Figure 3.29) and site J is in Whitecliff Bay on the east coast of the Island.

The discontinuity patterns shown for sites G and J are not amenable to obvious interpretation in terms of structural trends. Because the Eocene beds have only been subjected to a single major phase of folding (the main Alpine/late Tertiary episode) the fabric pattern might be expected to be less complicated than the pattern in, for example, the Kimmeridge Clay (sites C, D, E). Since this is not the case it might appear that the older, more compacted, better cemented and lithified the deposit is, the more basic is the discontinuity symmetry pattern. Minor readjustments and structure 'accommodations' would have a correspondingly greater effect on a weaker sediment.

Fissure concentrations in the London Clay of Alum Bay (site G) are weak and generally inclined to both the bedding plane and exposure surfaces. They do not seem to relate to the Chalk discontinuities recorded at the south of Alum Bay (Site I) by Fookes and Denness (1969). Site J in Whitecliff Bay seems to present a rather more logical discontinuity pattern in that the 15 per cent per 1 per cent area main concentration tends to parallel the bedding plane. The other two >5 per cent concentrations are very approximately orthogonal to each other and to the former major concentration and are of such an attitude as to possibly predispose the clay to a wedge-type of shear failure (see Chapter 10). Fortunately, the optimum wedge-shear facility seems generally to be into the slope (NW).

There is the possibility that the rather more dispersed nature of the discontinuity concentrations at site G and the lower degree of fabric symmetry at site J could arise from the fact that both sites are located in steeply dipping strata near to the axes of a number of anticlines. In the case of site G, possible deep-seated slope failures of a large section of the Alum Bay cliffs could be contributing factors. Slope failures are visible in the Reading Beds at Alum Bay but not in the London Clay.

Discontinuity information with respect to an outcrop of fissured Upper Chalk is also shown in Figure 3.26 (site F). At this location the deposit is horizontally-bedded near to the crest of the Foxwell Anticline (bearing $082°$ at this point) outcropping to the east of Corfe Castle, the structure being an extension of the exhumed Chalk erosion surface adjacent to site G.

On-site observation indicates an extensive near-vertical set of fissures together with further low-angle sets, a fabric pattern which is reproduced on the stereonet. The slight sub-vertical dip to the north arises because the site is on the north side of the crest of the anticline nearer to the steeper northern limb. But taking the fissure fabric as a whole, there is no immediately obvious correlation with the fabrics of sites G (London Clay) and I (Gault Clay), most probably because of the earth movements that took place at the end of the Cretaceous and pre-dating the London Clay. It will be noted, however, that the site F major concentration (upper hemisphere projection) is compatible with the Fookes and Denness site I concentration at the south pole of the lower hemisphere projection.

Overlying the London Clay, the rest of the Eocene and Oligocene succession is approximately 640 m thick in the Isle of Wight. All the Tertiary deposits dip steeply on the northern limb of the main east-west anticline. This feature is a constitutive element in a series of folds which are the result of earth movements that reached their climax in Miocene times. During the same period, crustal movements were taking place on the continent of Europe, resulting in uplift of the Alps. Events in southern England are only peripheral to the main

Alpine orogeny. The style of the folds in southern England is that of a swarm of short, parallel, en echelon anticlines, mostly asymmetrical with the steeper northern limbs and gentler southern limbs trending approximately E–W. A figure of about 1220 m has been calculated for the amplitude of the main asymmetric anticline of the Isle of Wight.

To the west of the coastal area in Figure 3.26 the older series of anticlines and synclines formed during the Cretaceous period are conspicuous in the area near Weymouth. The earth movements that gave rise to these structures occurred before Upper Cretaceous times, as proved by the fact that in places the Jurassic and Wealden strata were tilted and eroded before the Upper Greensand and the Chalk were laid down. On the Dorset coast at Ringstead Bay, highly-inclined Kimmeridge, Portland, and Purbeck Beds were overlain by Upper Greensand with a very slight dip. These intra-Cretaceous earth movements (post-Wealden and pre-Gault in age) affected the Oxford Clay at sites A and B and the Kimmeridge Clay at sites C, D. E. Subsequently, the main Alpine folding has modified the early minor fold structures in these deposits.

3.7 Engineering influence of discontinuities in clay shales

This subject is considered in much greater detail and also in a more general manner in Chapter 6 but it is useful to introduce it at this stage and in the light of what has gone earlier in this present chapter.

We may expect that the orientation of fissures in stiff clays and clay shales will be related to the stress field that created the regional and local structure and to the joint orientations in adjacent brittle rock. But if there is later tectonism, the weaker material may respond by further fissuring while more brittle but strong contiguous rock may not react at all. It may not always be easy to resolve the two discrete fabrics. Fissure orientations may also be related both to the direction of stress relief due to erosion of overlying strata and to the bedding planes and other 'accommodation' features of the original deposition process. If fissure orientation in stiff clays and clay shales can be interpreted with respect to geological structure, then it may be possible in some instances to forecast the manner in which the ground will react to changes in external or self-loading conditions without necessarily having to resort to *in situ* testing.

It may be argued that the presence of a high fissure density in a stiff clay or clay shale weakens the material and predisposes it to shear along planes that are generally compatible with directions of maximum resolved shear stress in the ground. There is, therefore, the concept of shear sliding exploiting the lower shear strengths of fissure faces and linking up individual fissures when they are sufficiently close to one another and suitably orientated. This type of problem and the role of the stereographic projection in its resolution are factors that are discussed in Chapter 6.

There are two major problems associated with any attempt to compute on a theoretical basis any quantitative modifications to the mass strengths of stiff fissured clays using techniques developed in rock mechanics. In the first place, the fissures are usually insufficiently extensive or identifiable to permit actual testing in order to determine fissure surface friction properties as in the case of rocks. In the second place, the presence of fissures on a small scale in laboratory test specimens may lead to scatter on a scale which may under certain circumstances invalidate the concept of a *soil material strength* as a useful index property.

The obvious inference is that for accurate and meaningful descriptions of stiff fissured clays in terms of strength and deformation, *in situ* large scale test techniques as described in Chapter 7 are required to replace conventional laboratory testing. In the case of London Clay, this proposition has been examined in detail by the British Building Research Establishment (see Marsland, 1971a,b; 1972a) which has carried out a series of large-scale plate tests and *in situ* shear tests.

The underlying philosophy of a plate-bearing test, which is usually carried out by incrementally loading a rigid circular plate in a test pit or unlined borehole, is that the plate simulates a simple foundation loading situation which can easily be analysed in terms of the stress distribution or assumed failure mechanism beneath the plate in order to compute the modulus of deformation or the shear strength of the rock or soil.

In the work described by Marsland, results of *in situ* tests on 0.87 m and 0.29 m diameter plates were compared with the results of laboratory triaxial test and laboratory penetration (small plate, 5.5 mm diameter) tests. The penetration tests to failure probably provide a measure of the *material strength* of the clay, and these were generally two to three times higher than the laboratory triaxial test results and up to six times higher than the 0.87 m plate results depending on depth (Figure 3.30). Somewhat surprisingly, the

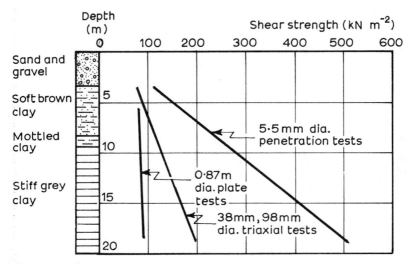

Figure 3.30 Effect of test method on shear strength – London Clay at Chelsea (*after* Marsland, 1971a).

E-values tend to follow a reverse trend (Marsland, 1971b) having a higher magnitude in the larger scale plate tests (Figure 3.31). This apparent anomaly is attributed partly to sample disturbance and partly to different stress levels applied in the laboratory tests. Comparable laboratory and *in situ* test results are achieved only if laboratory consolidated triaxial tests, in which the specimens are

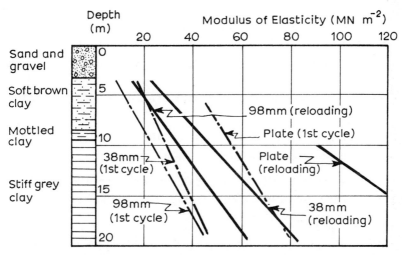

Figure 3.31 Effect of test method on modulus of elasticity – London Clay at Chelsea (*after* Marsland, 1971b).

initially subjected to their original *in situ* stress situation, are performed. In the case of direct shear tests on the clay, less difference is found between the shear strength measured *in situ* and in the laboratory. This result probably represents a confirmation of the general complexity of fissure fields and the limited spatial extent of many individual fissures.

A basic conclusion that may be drawn from Marsland's work on the typical stiff fissured clay is that its shear strength and deformational properties are a manifestation of complex inter-relationships between fissure structure and *depth*, and that ideally some form of *in situ* testing is a prerequisite of accurate design.

3.8 Classification of clay shales

At the weaker end of the shale spectrum, Terzaghi (1936) was the first to attempt a classification of clays via a 3-fold division:

(a) soft intact clays free from joints and fissures,
(b) stiff, intact clays also free from joints and fissures,
(c) stiff, fissured clays.

Consideration of genetic (Table 3.2) and compositional characteristics of clays leads to Underwood's (1967) classification in Table 3.4 which assists in the development of his evaluation of shales on an engineering basis (Table 3.5). Bjerrum (1967) based his 3-fold classification on bond strength:

(a) overconsolidated clays (i.e. overconsolidated plastic clays with weak, or no bonds),
(b) clay shales (i.e. overconsolidated plastic clays with well-developed diagenetic bonds),
(c) shales (i.e. overconsolidated plastic clays with strongly-developed diagenetic bonds).

Other terms often encountered in the literature which mean virtually the same as 'weakly-bonded shale' are:

(a) overconsolidated clay shale (Scott and Brooker, 1968),
(b) clay shale (Johnson, 1969; Fleming *et al*, 1970), and
(c) stiff, fissured clay (Chandler, 1970).

Scott and Brooker (1968) define an overconsolidated clay shale as: 'A sedimentary deposit composed primarily of silt- and clay-sized particles dominated by members of the montmorillonite group of

Table 3.4 A geological classification of shales (*after* Underwood, 1967; Fleming *et al*, 1970).

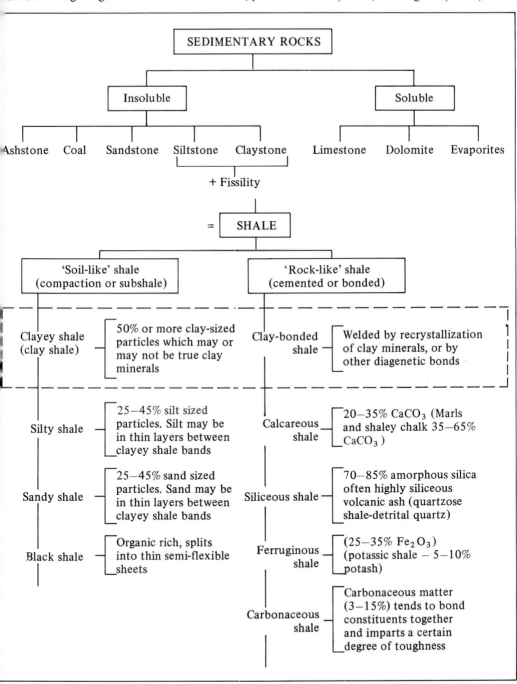

Table 3.5 Underwood's (1967) engineering evaluation of shales

Physical properties			Probable in situ behaviour						
Laboratory tests and in situ observations	Average range of values		High pore pressure	Low bearing capacity	Tendency to rebound	Slope stability problems	Rapid slaking	Rapid erosion	Tunnel support problems
	Unfavourable	Favourable							
Compressive strength, lbf in^{-2} (× 6.895 kN m^{-2})	50–300	300–5000	yes	yes					
Modulus of elasticity, lbf in^{-2} (× 6.895 kN m^{-2})	20,000–200,000	2×10^5 – 2×10^6		yes					yes
Cohesive strength, lbf in^{-2} (× 6.895 kN m^{-2})	5–100	100–>1500			yes	yes			yes
Angle of internal friction, degrees	10–20	20–65			yes	yes			yes
Dry density, lb ft^{-3} (× 0.016 Mg m^{-3})	70–110	110–160	yes					yes?	
Potential swell, percentage	3–15	1–3			yes	yes			yes
Natural moisture content, percentage	20–35	5–15	yes			yes		yes	
Coefficient of permeability m s^{-1}	10^{-7} – 10^{-12}	$>10^{-7}$	yes			yes	yes		

Predominant clay minerals	Montmorillonite or illite	yes			yes			
	Kaolinite or chlorite							
Activity ratio = $\frac{\text{plasticity index}}{\text{clay content}}$	0.75–>2.0				yes			
	0.35–0.75							
Wetting and drying cycles	Reduces to grain sizes					yes		
	Reduces to flakes						yes	
Spacing of rock defects	Closely spaced		yes		yes		yes?	yes
	Widely spaced							
Orientation of rock defects	Adversely orientated		yes		yes			yes
	Favourably orientated							
State of stress	Greater than existing overburden			yes	yes			yes
	About equal to existing overburden							

149

clay minerals, and these deposits have been subjected to consolidation loads in excess of those provided by present overburden'. A similar definition was provided by Fleming *et al* (1970) who described clay shales as: 'materials of sedimentary origin composed largely of silt- and clay-sized particles, which may or may not be slightly cemented by foreign agents, such as iron oxide, calcite, silica, and which have been subjected to consolidation loads greatly in excess of their present overburden loads'. They go further, however, in explaining the effect of slaking and grain size on these materials such that: 'The material is composed principally of clay minerals and pieces of intact material tending to slake when exposed to cyclic wetting and drying'. This behaviour contrasts with materials cemented to the extent that they do not slake when exposed to cyclic wetting and drying; these are termed 'siltstone or claystone', depending on particle gradation.

In the absence of a truly definitive materials classification, the term 'weakly-bonded shale' may be taken to mean an overconsolidated, stiff fissured clay with a high proportion of constituent clay minerals, although it could be argued that the London Clay, while satisfying the latter description would not be considered a shale, albeit weakly-bonded. It would probably be wise to reserve the term 'weakly-bonded shale' for such materials as the Japanese Tertiary shales (Nakano, 1967) and many of the classical North American and Canadian shales of the Missouri River and its tributaries.

In addition to this material classification, Underwood (1967) also listed a range of engineering properties characteristic of shales. These properties are reproduced in Table 3.5.

3.9 Consolidation and diagenetic considerations

One can conclude, with Johnson (1969), that the only feature common to all the definitions of 'clay shale' is the quite high degree of overconsolidation coupled with the other chemical and physical alterations which a sediment undergoes during and after burial. On these points, Bjerrum (1967) has considered the importance of diagenesis in the development of interparticle bonds in sediments.

Reference to Figure 3.32 from Fleming *et al* (1970) provides a useful graphical representation of consolidation. During the loading of a sediment as it becomes progressively more deeply buried in a sedimentary basin, it undergoes a reduction in pore volume and a consequential decrease in water content (syneresis). Platy clay

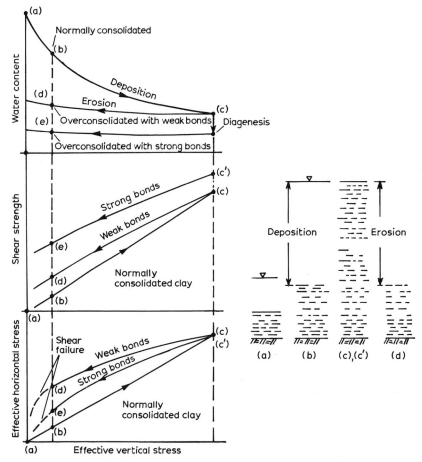

Figure 3.32 Graphical representation of the geological history of an overconsolidated clay (*from* Fleming *et al*, 1970, *after* Skempton, 1964 and Bjerrum, 1967).

minerals are re-orientated more intensely at right angles to the vertical compression axis. Bjerrum's (1967) concept includes the formation of 'diagenetic bonds' during the period from completion of consolidation up to the time of erosion (see Table 3.6 with respect to some N. American clay shales). This period may be sufficiently long for the minerals to re-crystallize under suitable pressure and temperature gradients, for adhesion to be induced, or for some cementing medium to be precipitated in the intercrystalline interstices. The more strongly developed these bonds, the smaller is the increase in the water content on unloading through erosion because

Table 3.6 Estimated thickness of sediment removed from different groups of clay shale in N. America

Clay shale group	Thickness removed	
Colorado	1900 ft	579 m
Claggett	1800 ft	549 m
Pierre	600 ft	183 m
Bearpaw	2200 ft*	671 m
Fort Union	950 ft	290 m

*Bearpaw took 25 million years to accumulate. Uplift and erosion lasted for some 49 million years. Assuming that the Ice Age began 1 million years ago, then these deposits were re-loaded for approximately 900 000 years and unloading has occurred within the last 100 000 years. (Note that all five formations are sequences of marine shales, although the Colorado and Fort Union series contain non-marine members with thin bentonite seams or lenses developed at various horizons. All the shales incorporate facies variations; for example, thin sandstones, siltstones and limestones occur in the Colorado group, lignites are developed in the Fort Union, and calcareous shales are not uncommon in other formations. All the host sediments were deposited in, or near to, a shallow sea which covered the Great Plains area in Upper Cretaceous and Lower Tertiary times. Western seaboard variations, marine transgressions and regressions, gave rise to non-marine deposits. The bentonites seem to correlate with local intermittent volcanic activity, most of the source sediment originating from a landmass which today form the Rocky Mountains.) (*After* Fleming *et al*, 1970.)

the expansion facility of the clay has been inhibited. From his one-dimensional consolidation experiments on remoulded samples of clay, Brooker (1967) suggested that the energy accepted by a sediment during consolidation could be partitioned into three components: work expended in consolidation (partially recoverable); elastic deformation (recoverable on release of constraint); work expended in the formation of diagenetic bonds (partially recoverable, depending upon bond strength).

Loss of overconsolidation may well be complicated by later geological events. In Britain, for example, Jurassic clay shales are commonly affected by glacial phenomena. In America, near Bismarck,

N. Dakota, Lane (1960) suggests that downcutting of the Missouri River may have relaxed horizontal pressures in the Fort Union clay shale series. Attewell and Taylor (1973), as a result of some studies of the Ampthill Clay (Upper Jurassic clay shale, Corallian age, Melton, North Yorkshire, England), have suggested that loss of overconsolidation in the clay could be due to three interacting factors: rebound or volume expansion on removal of load (recoverable strain energy), which will also give rise to jointing and fissuring; valley bulging, which has certain features in common with Peterson's (1958) 'time-rebound'; the affinity of certain expandable mixed layer clay minerals for water.

Researchers such as Bjerrum (1967) and Brooker (1967) have argued that in order for the clays to expand in a vertical direction, the changes in vertical effective stress are larger than those for the effective horizontal stress. Since clays with strong diagenetic bonds cannot expand so readily, the horizontal effective stresses to which they are subjected are less than those in weakly-bonded shales. Thus, it follows that the destruction of bonds through weathering processes leads to the development of high horizontal stresses in the weathered zone at ground surface.

In the case of overconsolidated clays, Peterson (1954) has shown that *in situ* horizontal stresses may be one and a half times the vertical stress and in London Clay Skempton (1961a) and Bishop *et al* (1965) found that the earth pressure at rest coefficient K_0 varied with depth from about 1.65 at 30 metres to greater than about 2.5 at 7 metres (see Figure 3.33). Brooker concluded that lower plasticity clays were more prone to accommodate higher horizontal stresses. It would appear that the horizontal-to-vertical effective stress ratio at rest, as determined by the bond strength, is dependent not only on the amount of strain energy absorbed but also on the duration of load application, the constituent mineralogy and the ambient temperature.

3.10 Physical breakdown of shales

The larger scale breakdown of rock and soil in the context of slope morphology is considered elsewhere (see Section 9.11) but it is useful to mention some of the structural and mineralogical features of the weaker argillaceous rocks which facilitate physical breakdown, noting at the same time that it is not always easy to differentiate strictly between physical and chemical changes.

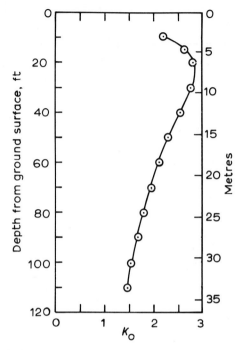

Figure 3.33 Variation of the earth pressure at rest coefficient, K_0 with depth below the ground in London Clay (*after* Skempton, 1961a).

(a) *Influence of mineralogy*

Minerals can conveniently be divided into two groups:

(i) *Detrital:* those which have been transported into the basin of deposition. Typical minerals are quartz, micas and derivatives, kaolinite, chlorite, feldspars and calcite.
(ii) *Non-detrital:* those which have grown within the sediment after deposition, for example, siderite ($FeCO_3$) and pyrite (FeS_2).

Of the former group, quartz is the most resistant to breakdown and any particle may well have experienced numerous cycles of erosion, transportation and deposition. Most clay minerals will have achieved some stability in the face of weathering and kaolinite is the most stable of the three clay minerals, being free (in British Coal Measures shales at least) of both ferrous and ferric iron, CaO, MgO and MnO. A chlorite may be iron-rich and although often a minor component it can tend to be a sensitive mineral in a weathering scheme (Loughman, 1969). Although Attewell and Taylor (1971) reported an increase in height of the chlorite basal plane peak with weathering in a British Coal Measures shale, this was interpreted by Taylor as a possible orientation effect of small, poorly crystalline platelets collapsing on glycolation. Other iron present in a shale may combine with sulphur to form pyrite (FeS_2) and part may be attributed to siderite ($FeCO_3$) or secondary iron oxides such as limonite or lepidocrocite which may be the end products formed by the weathering of Fe-rich minerals. The mineral pyrrhotite ($FeS_{(0.99-1.14)}$) weathers easily and can cause expansion and disintegration of mechanically weak rocks, for example, metamorphic anthracite or graphite-bearing slates (Martna, 1970). Pyrite weathering is accelerated through contact with pyrrhotite due to electro-chemical interaction between minerals, but pyrite on its own, Martna claims, is rather a stable mineral. Ankerite, a complex carbonate embodying, for example, FeO, MnO, MgO and CaO may be present as may TiO_2 as rutile and P_2O_5 as calcium phosphate. Inevitable organic matter may, however, also be an additional source of P_2O_5 and sulphur.

There is little direct evidence from the work of Attewell and Taylor (1971) on Coal Measures rocks that the non-detrital minerals play a dominant role in the breakdown process. Certainly on a relatively short-term basis (10,000 years or so) physical disintegration is a much more important issue.

(b) *Influence of sedimentary structure*

Bedding or stratification gives rise to a plane of weakness along which water may more easily pass and so create zones of preferred softening. In some mudstones, the stratification takes the form of varve-like units each of which represents a change of sedimentological regime. The Mansfield Marine shale, for example, studied by Spears (1969) comprises alternating lighter and darker laminae with an illitic clay mineral fraction and organic content only marginally higher in the latter laminae compared with the former. But under the petrological microscope, the clay minerals appear to be isotropic in the darker laminae, suggesting a floc-type of clay mineral lath structure, and a discernible propensity for swelling was attributed to interparticle water uptake into this open structure. In a similar manner, the non-marine Lumley mudstone (High Main coal seam roof rock, County Durham, England and examined further in Section 5.11) exhibits a 4 per cent swelling within the darker laminae (Taylor and Spears, 1969). Compositional variations between laminae were much greater in this latter rock (dark — clay and organic-rich; light — quartz and feldspar-rich) due to more rapid non-marine sedimentation of the component minerals.

Slickensides are other structural features in, for example, seat earths of coal seams, and they precondition the parent material for breakdown when subjected to cycles of wetting and drying. Joints in rocks perform a similar function, the weaker the rocks the higher generally is the joint frequency. Shales and mudstones also fracture in a polygonal manner during desiccation due probably to negative pore pressures and pore pressure gradients emanating from discrete centres in the material (Christiansen, 1970). These desiccation polygons are very similar in appearance to the ice polygons seen from the air in tundra regions (for example, Brooks Range, Alaska). Such desiccation fractures, together with any jointing and laminations and aided by any stress relief effects created by erosional denudation, can rapidly reduce most clay shales and mudstones to an aggregate in a few months, the actual length of time being a function of wet-dry (slaking) cyclic frequency together with the compositional mineralogy and fabric of the rock.

(c) *Intra- and inter-particle effects*

In a low to moderate temperature environment, most of the minerals in clays and shales are chemically stable. Non-detrital pyrite oxidizes irreversibly and clay minerals such as montmorillonite expand when water is accepted into the layered structure. Such *intraparticle* swelling is a function of the amount and properties of the water and the nature of the interlayer cation (Na^+ — montmorillonite is particularly prone to dissociation in the presence of water). Montmorillonite may occur as a separate phase or as a component in a mixed layer illite — montmorillonite mineral. The presence of the expandable mineral is registered on an X-ray diffraction trace as an attenuated tail on the high dÅ side of the 10 Å mica peak or as a discrete reflection. It can be proved quite readily by slaking tests that the total clay mineral content of a rock is not the only parameter conditioning breakdown; monomineralic kaolinite-rich rocks show negligible disposition to breakdown.

(d) *Moisture movement into shales*

Natural slopes in a temperate environment tend to be quite highly weathered within a superficial zone, the weathered material undergoing significant variations in seasonal strength. Field experiments described by Attewell and Taylor (1971) showed that the ultimate bearing capacity of a British laminated mudstone in the summer was almost double that in the winter, the large variations in strength being related to small increases in porosity (29.5 per cent to 34 per cent) and an associated moisture intake of less than 10 per cent. Moisture movement into the interstices of a rock, creating a potential for swelling, weakening, and ultimate disintegration, occurs under the influence of suction pressure or, more specifically, the surface tension of the water at the air-water interface.

3.11 Suction pressure

Soil suction is a physical expression of the negative pressure (below atmospheric) which causes water to be retained within the pores of a soil or rock when the sample is free from external mechanical pressure. Certain external loadings may also create an internal pressure deficiency but this is termed, by convention, a *negative pore pressure* to distinguish it from a *suction pressure*.

Soil suction is measured on a pF scale (Schofield, 1935) and chemistry students will note from Figure 3.34 its conceptual similarity with the pH scale. The potentially very high values of suction pressure will also be noted from Figure 3.34.* It will also be appreciated that in an undisturbed, saturated soil at equilibrium, the initial porewater pressure just balances the soil suction. Raising the external compressive pressure raises the internal pore pressure to a somewhat lower level which recognizes the existence of the suction pressure and in the simple saturated case the suction pressure is numerically equal to the effective stress applied to the soil skeleton.

More generally, with unsaturated, incompressible materials, the soil suction is equal to some fraction of the applied stress less the porewater pressure. That fraction may be 0.5 for a silty clay or about 0.15 for a sandy clay (Capper and Cassie, 1969). It is also most important to appreciate that as the moisture content of a soil decreases, its pF and hence its shear strength increases significantly

---

*Soil in equilibrium with free water has a pF of about zero. Oven—dried soil has a pF of almost 7.

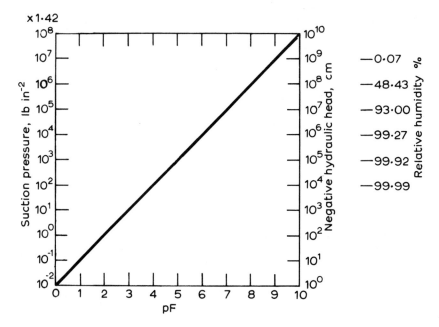

Figure 3.34 Soil suction.

(Croney and Coleman, 1961). Shear strength of a clay can often be measured by vane test and related to the suction (see equations of Dumbleton and West, 1970).

In very wet climates, the water table is near to the ground surface and this determines the equilibrium distribution of suction pressure and moisture content. In very dry climates the moisture content of the upper metre or so of soil is sometimes determined by the atmospheric humidity and under these circumstances the suction can be resolved from the average humidity (Croney and Coleman, 1961; see also Russam, 1959). During a summer drought there will be a nett soil moisture deficit created only marginally through the extraction of water from the ground by the root action of trees. The roots will tend to grow towards covered areas of ground, so creating some settlement through the moisture extraction process. Such settlement can be estimated from the shrinkage curves (soil suction/moisture content relationship) for that particular soil (Croney and Lewis, 1949).

A somewhat inverse problem to the latter, and one which also carries engineering overtones, arises in very cold climates when soil freezing can take place. With ice and water present together in the interstices of a soil, Schofield (1935) has shown that a suction pressure of 98.7 lbf in^{-2} (0.68 MN m^{-2}) is created for every degree Farenheit below freezing point (32°F). There will therefore be a quite considerable migration of soil moisture into the freezing zone under the action of this negative pressure but if the soil is quite desiccated (high suction pressure) no ice will form until the temperature has fallen in accordance with the relationship:

$$pF = 4.1 + \log_{10} t \qquad (3.1)$$

where t is the depression of the freezing point.

After ice has formed, segregation of additional ice can occur if water is available in an adjacent unfrozen zone at a lower suction pressure. All types of soil can be potentially liable to frost heave by segregation and an obvious remedial measure involves reducing the available water in a zone that is prone to freezing.

Soil suction is commonly measured in the laboratory through the use of two standard pieces of apparatus which are available commercially and which are described in the soil mechanics literature (see also Dumbleton and West, 1968). The *suction plate* measures suctions in the range of about 5 mm of mercury to 1 atmosphere while the *pressure membrane* measures suction in the range 1 to 100 atmospheres.

Suction curves are shown in Figure 3.35 for three different Coal Measures rocks (from Philpott, 1970) and for hard chalk (from Lewis and Croney, 1965). These curves are characteristic of incompressible materials. The initial vertical portion of the drying curves for the argillaceous materials indicates that the specimen is able to accept considerable suction before pore water is removed. A change in slope of the curve indicates that drainage begins when the necessary air-entry suction is achieved (about pF 2, 2 and 2.3 in Figures 3.35a, b, c respectively). Between pF 3 and pF 6 all three materials show a steady decrease in moisture content with suction but between a pF of 6 and 7 (the latter, the oven-dry condition) the steepening of the curves shows that the pore spaces have nearly emptied of water. The wetting curves can be similarly interpreted, and the hysteresis loops between the curves simply indicate that the pore spaces empty at different suction pressures from those at which they fill. Smaller pore spaces in the mudstone (Figure 3.35c) are responsible for the more extended vertical limbs of the curves at low pF. On the other hand, the suction pressure curves for the chalk (Figure 3.35d) are more sigmoidal and encompass a lower overall pF range. This means that once drainage has begun in the chalk, it continues rapidly.

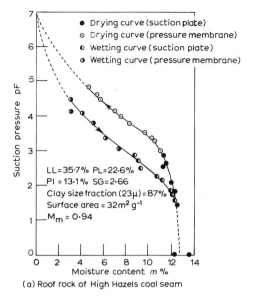
(a) Roof rock of High Hazels coal seam

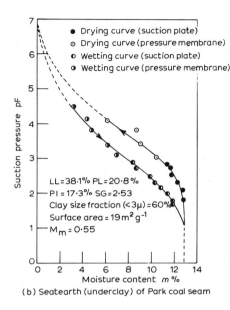
(b) Seatearth (underclay) of Park coal seam

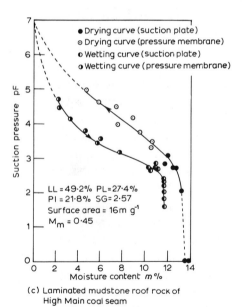
(c) Laminated mudstone roof rock of High Main coal seam

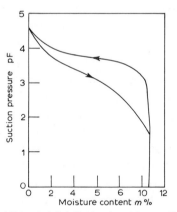
(d) Hard chalk (after Lewis and Croney, 1965)

Figure 3.35 Some suction pressure curves for different materials (generally *after* Philpott, 1970). M_m is the moisture content of a soil when the surface of the constituent particles is completely covered by a uni-molecular layer of water. (a) Roof rock of High Hazels coal seam. (b) Seatearth (underclay) of Park coal seam. (c) Laminated mudstone roof rock of High Main coal seam. (d) Hard Chalk (*after* Lewis and Croney, 1965).

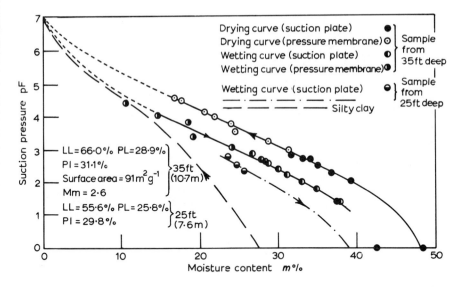

Figure 3.36 Suction pressure curves for laminated clay from Tyneside, Northern England (generally *after* Philpott, 1970).

Under increasing pF, as water is driven out of a clay by the incoming air, there will be some shrinkage of the material skeleton before the air can offer support. This decrease in the void ratio will contribute to the hysteresis between the drying and wetting curves. With a more compressible soil (Figure 3.36), air entry is rather more inhibited as shrinkage develops over the entire suction range.

Figure 3.37 (after Philpott, 1970) shows that there is a generally decreasing steepness of the pF curves for several undisturbed clays and clay shales as the natural moisture content and the liquid limit increase. But there is insufficient firmness in the data to conform with the contention of Dumbleton and West (1970) that there is a linear relationship between moisture content and liquid limit at constant suction pressure.

Surface area of mineral constituents

Two major theories have been proposed to account for the sorption of liquids and vapours by porous solids. One theory is based on the thermodynamics of the multi-phase system, the adsorbed material being regarded as a continuum. The other theory is based on statistical mechanics and takes into account the molecular structure of the adsorbed liquid or vapour. From this molecular theory, it is

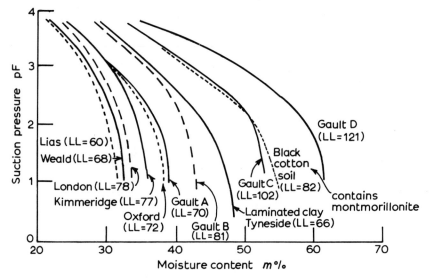

Figure 3.37 Suction pressure curves for several clays and clay shales.

possible to determine the surface area of the constituent mineral crystals which constitute the clays considered earlier and it will be noted from Figure 3.38 that the more active expanding clay mineral has the greatest surface area.

3.12 Swelling pressure

Swelling of saturated and partly-saturated expansive clays creates a problem not only in terms of terminal breakdown, but also at an intermediate level in the form of heave on structural foundations. Under laboratory test conditions, the specimen of clay is usually permitted to reach full saturation at the end of the soaking period, so exhibiting a full range of swelling behaviour under a given load or swelling pressure under conditions of no permitted volume change. In nature, however, a clay may be well protected by pavements, buildings or embankments and in such cases the moisture movement accompanied by heave takes place as a result of potential gradients which themselves arise from differences in soil water suction with depth. The relationship between swell pressure and suction for a Ca-montmorillonite expansive clay from Israel has been studied experimentally by Kassiff and Shalom (1971).

The actual magnitude of the swelling of a clay in the presence of water (and hence, indirectly, the potential size of the engineering

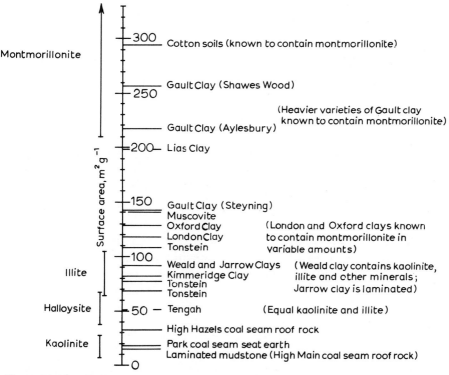

Figure 3.38 Surface areas of several materials tested in Engineering Geology Laboratories, Durham University (*after* Philpott, 1970; *see also* Road Research Note RN27/63).

problem in the material) depends upon a number of factors among which are: (i) nature and quantity of clay minerals present; (ii) presence and type, or absence, of matrix cement; (iii) exchangeable ions of the clay minerals; (iv) electrolyte content of the water phase; (v) particle size and void size distribution and the extent of the adsorbing surface; (vi) internal structural arrangement of the soil skeleton; (vii) actual moisture content; (viii) applied pressure.

According to Barshad (1955) there are two fundamental types of swelling: intramicellar or interlayer swelling which is the swelling of the crystal lattice, and intermicellar swelling which is a reflection of the adsorption of water molecules between individual clay minerals. Intramicellar swelling occurs in montmorillonite, vermiculite and certain hydrous micaceous minerals which can only be identified by using X-ray diffraction methods. Intermicellar swelling must be physically measured.

Barshad (1955) has discussed in detail the historical development of interlayer expansion and it becomes evident that, given a suitable amount of the right types of cations, the major controlling factor on swelling magnitude is the total surface area of clay mineral particles which is available for adsorbing water. External surface areas of montmorillonite, micas, kaolinite and halloysite average out at 50, 130, 30 and 37 $m^2\ g^{-1}$ respectively, but only montmorillonite can contribute an internal surface area, and that to the extent of 750 $m^2\ g^{-1}$. Thus, the combination of both an internal and external surface-active facility and the very large total surface area renders the montmorillonites particularly swell-prone.

Swelling pressures can be studied by measuring the one-dimensional swelling *in vacuo* of an undisturbed sample of clay under the action of de-aired distilled water and then consolidating the material back to its original volume in a standard oedometer (see Figures 3.39, 3.40 for some typical swelling curves). Swelling and subsequent consolidation take place in the same fixed-ring consolidation cell.

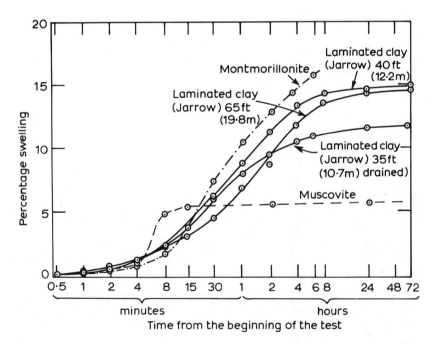

Figure 3.39 Some typical swelling curves for different materials (*from* Philpott, 1970).

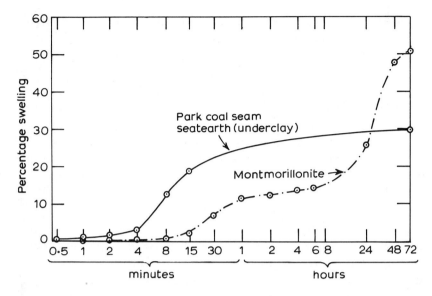

Figure 3.40 Two swelling curves (*from* Philpott, 1970).

Philpott (1970) found that the laminated clay in swelling to between 9 and 15 per cent created a pressure of between 1 and 2 kgf cm^{-2} (98.1 and 196.2 kN m^{-2}). The Coal Measures seat earth expanded about 33 per cent with a swelling pressure of 45 kgf cm^{-2} (4.4 MN m^{-2}) and the practical significance of such a result can be appreciated when it is realized that during coal extraction operations, very heavy machinery is required to traverse this material under conditions of continuous water-spray action for dust suppression purposes. Comparisons between the relatively inert muscovite and expandable montmorillonite produced respective figures of 6 per cent (0.5 kgf cm^{-2} or 49 kN m^{-2}) and 53 per cent (2 kgf cm^{-2} or 2.8 MN m^{-2}).

Depending upon the tortuosity and degree of continuity of the void channels in a porous material, air may be trapped and pressurized within a rock when the external void passages are sealed by water under pressure. Such a situation could arise along the banks of a river, lake or canal as the water level rises and falls periodically with seasonal flooding or diurnally under pumped storage conditions or even with the wave action from the passage of a ship. Cyclic immersion-drying tests performed *in vacuo* have shown that breakdown can indeed be arrested by the removal of air (Badger *et al*, 1956; Berkovitch *et al*, 1959; Taylor and Spears, 1970; Attewell and Taylor, 1971).

The mechanics of air-breakdown in clay shales and stiff clays can be explained in much the same way as was suggested by Terzaghi and Peck (1967) for the slaking of soils. During dry periods, evaporation from the surfaces of rock fragments promotes high suctions which in turn create an enhanced shearing resistance of the individual elements through the action of high contact pressures. When extremely desiccated, the bulk of the voids will be filled with air which, on rapid immersion in water, become pressurized by the capillary pressures developed in the outer pores. Consequential failure of the mineral skeleton along the weakest plane follows, so exposing an increased surface area to a further sequence of pressurization-breakdown events. Since, in theory, capillary pressure is inversely proportional to pore radius but is directly proportional to surface tension,* the finer-grained porous solids are most vulnerable but the breakdown facility is reduced if organic contaminants (and electrolytes also, if in sufficient quantities – Badger *et al*, 1956) are present in the water.†

The process of swelling and possibly mineral (especially clay) re-orientation, the latter upon a change of stress field particularly when exposed at depth in the earth, could be interpreted in terms of cation polarization in the presence of a suitable chemical environment. A high anion exchange facility under new pressure/chemical conditions could well promote the process of ionic dispersion (interlayer repulsion) so contributing further to breakdown. However, Taylor and Spears (1970) have suggested, on the basis of their work, that evidence pointing to the possible influence of ionic dispersion could rather more firmly be interpreted in terms of the overall physical assemblage of the constituent clay minerals (size, degree of floccu-

---

* $$p^* = \frac{-2T}{\gamma} \cos \alpha$$

where p^* is the pressure deficiency in the soil water with respect to the pressure in the soil air,

T is the surface tension of water at an air-water interface,

γ is the mean effective radius of curvature of the water meniscus

and α is the contact angle between water and soil (this is usually taken as zero although for shales Horsley (1951–52) has indicated that a 10 degree angle may be applicable.

†Nakano (1967), on the other hand, showed that Japanese mudstones broke down more easily in water–soluble organic liquids and his subsequent geochemical considerations strongly favoured the phenomenon of hydrogen bonding between the adsorbed water with the respective organic molecules (Taylor and Spears, 1970).

lation, tortuosity of voids etc). If such dispersion does take place then, together with chemical dissolution processes due to Nakano's hydrogen bonding, it will only become important over extended periods of time and after initial physical breakdown has generated a greatly increased internal surface area.

3.13 Chemical and mineralogical analyses of clays

(a) *Procedure*

Geochemical major element analyses are most easily performed on an automatic sequential analyser X-ray fluorescence machine preferably incorporating an automatic loading facility and a punched card (or tape) output for batch data processing. XRF analyses are referred to known, wet-chemically analysed standards which accompany the unknown materials through the system. Determinations on 11 major elements can then be processed using standard programmes with a suitable computer which will then list the elemental compositions in the combined oxide state (apart from sulphur). Listings will be in percentages of SiO_2, Al_2O_3, total iron oxide, MgO, CaO, Na_2O, K_2O, TiO_2, MnO, S, and P_2O_5. If required, 'wet' analyses using a carbon train will have to be undertaken for carbon and carbon dioxide determinations, but if it is known that these are present in only very low concentrations, then this time-consuming process can conveniently be omitted. Suitable computer programmes will also provide listings of the correlation coefficients for the major elements together with the ratios of the elements to Al_2O_3 and the correlation matrix of these ratios. The argument for expressing the oxides as a ratio of Al_2O_3 (alumina having been attributed to the clay minerals as alumino-silicates) has been outlined by Spears *et al.* (1971) as a means of revealing compositional trends which can be assigned to the clay minerals.

Mineralogical identifications, particularly for quartz and the clay minerals, are usually achieved semi-qualitatively by X-ray diffraction supplemented by thin section observations under the microscope. Both quartz and clay minerals can be determined quantitatively from smear mounts using an internal standard such as boehmite to 'spike up' the sample (Griffin, 1954) followed by area measurements under diffraction peaks (Section 1.5). Confirmatory XRD runs for cross-calibration would also be performed on 'spiked' and 'unspiked' standard minerals (for example, Morris illite or an Oligocene illite from Le Puy-en-Velay) although there can be problems in that such minerals may not always be of compatible crystallinity with the same natural minerals in a test sample. XRD quartz determinations can also be compared with the results from the more precise but protracted method of Trostel and Wynne (1940). Area measurements under XRD peaks can be made using a polar planimeter and 'shape factors' which would highlight, for example, a leading (higher dÅ spacing) tail to a 10 Å clay mineral peak (denoting a potentially expandable interlayer content*) are formulated by deriving the ratio of peak width at half-peak height to peak height. This same ratio, boehmite 6.18 Å to kaolinite 7 Å, is also used by Griffin (1954) as a crystallinity factor (S.F.) and by comparing lattice plane diffractions. Hinkley (1963) has also obtained a crystallinity factor for the 4.48 Å–3.584 Å peak broadening area of the kaolinite diffraction chart (Brown, 1961, p. 67).

Finally, geochemical alumina ratios, mineralogical data and geotechnical data (liquid limit and plastic limit) can all be cross-correlated for the determination of statistical significance levels.

(b) *Use of X-ray fluorescence chemical analyses and X-ray diffraction mineralogical analyses*

Several simple guide rules that have been used in or developed from the work of Attewell and Taylor (1971, 1973) may be of some assistance during evaluations of clays and clay shales:

(i) Since the free *quartz* percentage exerts a major control on shear strength, determine the percentage by XRD using a boehmite standard. Then subtract this percentage from the *total silica* determined by XRF. The difference can be assigned to the lattices of the clay minerals in the material.

(ii) Ratio any relevant oxides in the material to *alumina* (Al_2O_3), which is representative of the clay mineral content. These

---

*Expandable clay minerals of the illitic variety can be identified by XRD methods after treating the material with ethylene glycol solution and noting a translation of the untreated 10 Å diffraction peak to a lower Bragg angle (2θ) location on the diffractometer counter chart output record (see Section 1.5).

ratios can reveal trends within the clay mineralogy if combined with bound-quartz percentages. Any high positive correlation between silica and alumina (even taking total silica) reflects their combination in the sheet silicate minerals.

(iii) If the *total iron oxide* (ferrous and ferric) correlates positively with sulphur, this would suggest that most of the iron present is in the reduced ferrous state combined with sulphur in the mineral pyrite. However, there could be ferric sulphates present, so there must be a cross-check with XRD results. There must also be checks for other sulphates such as jarosite.

(iv) If *calcium and magnesium* oxides correlate positively, then this would indicate the presence of dolomite.

(v) If there is a high negative correlation between *sodium oxide* and *potassium oxide*, it is likely that they are mutually exclusive with respect to the interlayer cationic sites of the sheet silicates.

(vi) *Titanium oxide* is usually present as rutile needles.

(vii) *Phosphorous pentoxide* combines with calcium to give the mineral apatite, but phosphorous can also denote the presence of organic matter.

(viii) An unusually high concentration of *calcium* could suggest fossil remains in a clay (such as, for example, in the Fort Union Series of North American clay shales).

(ix) If *montmorillonite* is shown to be present by XRD glycolation, and it correlates positively (highly significant) with the sodium/alumina ratio but negatively with the corresponding potassium ratio, then this suggests that sodium-montmorillonite dominates.

(x) Check the percentage of *mixed-layer illite-montmorillonite* by XRD. If this percentage has a high negative correlation with montmorillonite but does not correlate at all to illite then this could suggest that its presence is due to the weathering of montmorillonite.

(xi) If illite has a high negative correlation with sodium oxide to alumina ratios, and illite and montmorillonite also correlate negatively, then this would suggest that the sodium is dominantly associated with montmorillonite.

(xii) *Kaolinite* often seems to show a high positive correlation with *illite* (for example, in most North American clay shales) and it is worth checking to confirm that this normal condition is satisfied. Since K^+ is a major lattice element of illite and mixed-layer clay, any reduction in the K_2O/Al_2O_3 ratio at any zone in the clay would tend to indicate that the element is being leached from the clay mineral lattice to create the more stable clay mineral kaolinite. The possible development of kaolinite would be checked out by comparing the actual combined silica (as a combined SiO_2/Al_2O_3 ratio) under the changed conditions with the theoretical SiO_2/Al_2O_3 ratios for monomineralic kaolinite and illite representative of the parent material.

(xiii) Check for *secondary gypsum* (it is present, for example, in the English Oxford and Ampthill clays). Such gypsum could be precipitated from ground waters at depth and it is even present in the Upper Lias Clay at depths which exceed the depth of weathering (Chandler, 1972a). This implies that it could be transported in advance of weathering and could mask the geotechnical relationships which are a function of primary minerals. If sulphur does not correlate well with iron, then it is probably present as gypsum. On the other hand, if it does correlate with iron, then it is probably present as pyrite. An XRF increase in CaO might indicate an increase in secondary gypsum, but this must be confirmed by sulphur determinations.

(c) *Geochemistry/mineralogy of some stiff clays/clay shales*

The following conclusions are drawn from a series of analyses (Attewell and Taylor, 1973) on 66 geochemical samples* (46 N. American, 20 British) and 83 mineralogical samples (29 N. American, 21 British: untreated XRD; 28 N. American, 5 British: glycolated). The conclusions at the time of writing are thought, therefore, to be reasonably representative of these clays and clay shales in the two countries.

N. American material

(i) Quartz is the major massive mineral present (varying from zero to 43 per cent) and may be accompanied by non-detrital

---

*Sources, with sample numbers were Dawson (9), Claggett (9), Pierre (8), Bearpaw (8), Lias Clay (5), Oxford Clay (5), Ampthill Clay (4), Kimmeridge Clay, (2), Gault Clay (1), Speeton Clay (3), Colorado (6), Fort Union (6).

(ii) carbonate minerals (dolomite and siderite) or pyrite both of which are present only in minor quantities.
(ii) Total silica content (56 per cent to 72 per cent), even allowing for the free quartz influence, correlates highly significantly and positively with alumina (12 per cent to 20 per cent) reflecting their combination in the sheet silicate minerals.
(iii) Total clay mineral content (74 per cent average, although two samples contained 100 per cent and another seven more than 92 per cent) comprises mainly montmorillonite (average 36 per cent) and dominantly Na-montmorillonite, and mixed layer illite–montmorillonite (23 per cent average). The high negative correlation of this latter mineral with montmorillonite (>99.9 per cent significance) together with the lack of significant correlation to illite may well suggest, after Spears (1971), that the mixed layer mineral originates from the weathering of montmorillonite at some stage in the latter's history. In three of the clay shales (Claggett, Pierre and Bearpaw), illite does not exceed 9 per cent but in the Colorado and Fort Union Series it replaces mixed-layer clay as the most abundant clay mineral. The overall mean of 12 per cent correlates with negative significance (> -95 per cent) to the sodium oxide-to-alumina ratio, so confirming the presence of Na-montmorillonite. Kaolinite averages only 39 per cent, correlating positively and highly significantly with illite.
(iv) Rutile (<1 per cent), manganese oxide (<0.05 per cent) and P_2O_5 (+ calcium = apatite at ⩽0.22 per cent) are the other minerals present.

British Material
(i) Quartz is almost always the dominant massive mineral (3.5 per cent to 26 per cent; average 14 per cent) but calcite may be equally important in some cases (up to 10.5 per cent) and attributable to fossil remains rather than being a non-detrital mineral.
(ii) Siderite was detected only in the Lias Clay and then at less than 3 per cent concentration.
(iii) The Oxford and Ampthill clays only, of all the British clays examined, contained gypsum at a maximum concentration of 8.5 per cent. This secondary mineral can be precipitated from groundwater at depth.

(iv)   Pyrite is present as an accessory mineral (14 out of 20 samples) reaching a maximum of 11 per cent in the Kimmeridge Clay, a high value even for marine shales.
(v)   The potassium oxide to alumina ratio shows a highly significant negative correlation with total clay mineral content. It seems, therefore, that potassic-rich clays occur only when the total clay mineral content is low. This argument is further supported by a highly significant positive correlation between sodium to potassium ratios and total clay. Thus, it would seem that when the clay mineral content is high, sodium is more common in the interlayer positions of the clay sheet silicate structures.
(vii) Illite first and then kaolinite make up the bulk of the clay mineral content, with expandable clay being rare. There is mixed layer clay present and it suggests that non-expandable mixed layer clay may be a characteristic of British Jurassic and Cretaceous sediments. Illite shows highly significant negative correlations with both magnesium oxide and total iron oxide ratios to alumina, and this is partly explained by a strong correlation between magnesium and iron. Magnesium also correlates strongly with iron and vermiculite (Mg-rich clay mineral, often with Fe substitution) in the Lias material and it would therefore appear from this evidence that in these British Jurassic sediments, the 14 Å minerals exert a major influence on the geochemical interrelationships (this could be checked by plotting the ratio: mixed layer clay/vermiculite versus ratio Fe/Mg and versus total SiO_2). It should also be noted that chloritic minerals (Fe-Mg types) can also be components in a mixed-layer clay (Brown, 1961, p. 280).
(viii) Rutile (1 per cent) and P_2O_5, mainly as apatite (0.06 per cent to 0.25 per cent), are the other important minerals present.

(d) *General comparisons between N. American and British clay shale geochemistry and mineralogy*
(i)   Quartz is the dominant massive mineral in both areas (mean 23 per cent in N. America; 14 per cent in Britain). From the comparative total silica values, it can be concluded that the same amounts of combined silica for both areas are being used in the silicate structure of the clay minerals.
(ii)   Differences in alumina content (15 per cent N. America; 20 per

cent Britain), differences in SiO_2/Al_2O_3 ratio (2.5 N. America; 1.9 Britain), and 75 per cent total clay content in both cases imply considerable differences in clay mineralogy, as was confirmed by the mineralogical investigation proper. Whereas in the N. American group, montmorillonite appears to be the dominant clay mineral, illite (including relatively non-expandable mixed-layer clay) predominates in the British group.

(iii) Kaolinite is the second major clay mineral species in the British group; it is much more concentrated and widespread than in the N. American clay shales. It tends to replace illite in the British material but the two clay minerals co-exist in N. American clay shales.

The basic differences in the mineralogy of the two groups of clays tend to stem from the occurrence and importance of different clay minerals in each group. As an additional exercise, Attewell and Taylor (1973) have compared the chemistry of the Panamanian Cucaracha clay shale (see Section 9.4) with that of both the N. American and British groups. Cucaracha source materials were volcanic in origin (for example, Thompson, 1947) and although early work (for example, MacDonald, 1915) implied that the clay minerals of the deposit were dominated by chlorite it is now clear that montmorillonite is the major clay mineral present. Although chlorite is present in the clay, the second most important clay mineral appears to be kaolinite; chlorite grains appear to act as nuclei for numerous spherules of siderite ($FeCO_3$).*

From Table 3.7, there would seem to be a greater difference between the chemistry of the Cucaracha and N. American Upper Cretaceous/Tertiary clay shales than between the Cucaracha and the older Jurassic/Cretaceous clays of Britain. Cucaracha-British 'affinities' are greater with respect to total SiO_2 and although Na_2O in the two groups is not significantly different, the clay mineral types in which it is found are different (montmorillonite in the Cucaracha and micaceous minerals in the latter).

It seems reasonable, on the basis of Kenney's (1967) work, to assume that variations in chemistry and mineralogy can result in behavioural differences between clay shales in response to geotechnical constraints.

*Petrographic observation by Taylor in Attewell and Taylor (1973).

Table 3.7 Comparison of chemical means (student's 't' test) of Cucaracha clay shale with North American and British clay shales (data normalized to 100%) *(from Attewell and Taylor, 1973)*

	Means			Means		
	Cucaracha	N. America	Significance	Cucaracha	British	Significance
S_iO_2	60.26	67.00	Different 99.9%	60.26	56.39	Different 99%
Al_2O_3	20.02	17.60	Different 99%	20.02	22.11	No significant difference
Fe_2O_3	11.03	5.55	Different 99.9%	11.03	6.13	Different 99.9%
MgO	1.81	2.35	No significant difference	1.81	1.93	No significant difference
CaO	4.27	1.49	Different 99%	4.27	6.37	No significant difference
Na_2O	0.27	1.41	Different 99%	0.27	0.41	No significant difference
K_2O	0.61	2.73	Different 99.9%	0.61	3.65	Different 99.9%
TiO_2	1.28	0.80	Different 99.9%	1.28	0.99	Different 99%
MnO	0.19	0.11	No significant difference	0.19	0.08	No significant difference
S	0.08	0.83	Different 99.9%	0.08	1.80	Different 99.9%
P_2O_5	0.20	0.30	No significant difference	0.20	0.15	No significant difference

Cucaracha = 6 samples {3 Upper Cucaracha samples from Cerro Escobar Hill
            {3 Lower Cucaracha samples from East Culebra Extension Slide } see Figures 9.21, 9.22.
N. America = 46 samples
Britain = 20 samples

3.14 Relationship between mineralogy, geochemistry and geotechnical properties of clays and clay shales

The mineralogy of clay shales can most simply be divided into the inequant clay minerals and the more equant non-clay minerals such as quartz, carbonates and feldspar. Other minor minerals such as rutile, pyrite, calcium, dolomite and apatite may also be present.

There is little doubt that the type of minerals present influences the geotechnical properties of the material, a very effective mineralogical parameter being the ratio of clay mineral to quartz content. As might be expected, the *liquid limit* increases with increase in the ratio, as shown for example in Figures 3.41 a and b from Attewell and Taylor (1973). This same trend, albeit with data point scatter, is confirmed in Figure 3.42 for different N. American clay shales. Grim (1962) pointed out that pure montmorillonites give the highest liquid limit values and Kenney's (1967) data, with montmorillonite L.L. percentages ranging above 620, add support to this. The actual exchangeable cation in the mineral also seems to be important. Lithium and sodium montmorillonites have greater liquid limits than have calcium or magnesium montmorillonites. Introduction of a second mineral to a montmorillonite system in a

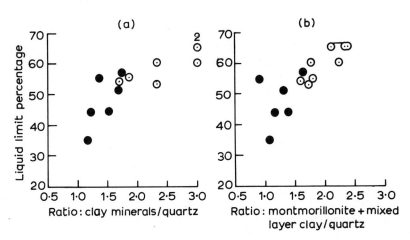

○ Data from U.S. Corps of Engineers Design Memorandum PC-24, Vol. 2, Plate 189, 1968
● Tests and analyses carried out in the Engineering Geology Laboratories, University of Durham, England

Figure 3.41 Relationship between liquid limit and mineralogy in certain Dawson shale samples taken from the site of Chatfield Dam, Littleton, Colorado, U.S.A. (*from* Attewell and Taylor, 1973).

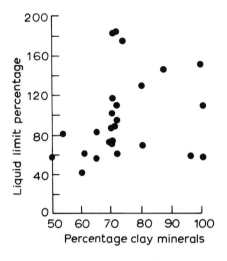

Figure 3.42 Influence of clay mineral percentage upon the liquid limit of some North American clay shales (*from* Attewell and Taylor, 1973).

concentration of 25 per cent (Grim, 1962) appears greatly to reduce the liquid limit. Conversely, small amounts of montmorillonite added to a natural soil significantly raise its liquid limit. Figure 3.43 plots the data from Tourtelot (1962) and Kenney (1967) and confirms the general increase in liquid limit with increasing montmorillonite content for different N. American material.

Mineralogy also influences the *activity ratio* of a soil (Section 2.2). This is the ratio of the plasticity index to the clay fraction (expressed as percentage weight of −0.002 mm particles) and it characterizes the slaking tendency of a shale. This tendency increases as the activity

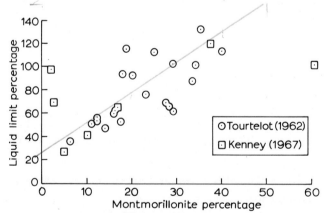

Figure 3.43 Influence of montmorillonite percentage upon the liquid limit in a number of North American clay shales (*from* Attewell and Taylor, 1973).

ratio increases. Typical activity ratios are:

Very fine sands	0.05
Clays very rich in kaolinite	0.10 to 0.20
Illitic clays	approaching 1.0
Bentonites (Kenney, 1967)	5.8 to 12.0

In the latter case, the exchangeable cation is a factor, with sodium montmorillonite providing the highest ratios.

Brooker and Ireland (1965) concluded that disintegration on submersion was dependent on absorbed strain energy which itself was dependent on clay mineralogy. Above about 10 per cent of the total clay mineral content, montmorillonite absorbs a significant amount of strain energy and it also absorbs large amounts of water leading to considerable mineral expansion upon submersion.

The question of mineralogical and physical controls on short-term and long-term breakdown is raised earlier in the chapter (see Taylor and Spears, 1970; also Attewell and Taylor, 1971) but mainly with respect to more indurated Carboniferous shales. It was concluded that an important control on immediate breakdown was the frequency of small-scale sedimentary structures such as bedding planes and laminations. Other major controls comprised air breakdown caused by capillary pressures developed on saturation and desiccation and the content of expandable clay minerals. On a larger scale, valley bulging in the Ampthill Clay in North Yorkshire, England has created some superficial folding with associated bedding plane separation and sub-vertical discontinuity planes which condition the material for more ready breakdown. In general, the mineralogy of the products of such physical breakdown tends to mirror that of the parent material and a feature of the breakdown of shales and mudstones is their disintegration through a brittle stage into a silty clay (Grice, 1969). This, for example, is the typical appearance of the superficial weathering of the Cucaracha clay shale forming the unstable part of the east bank of the Panama Canal.

In terms of direct mineralogical influences on breakdown, Kennard et al (1967) have concluded that the solution of calcite cement was the main cause of shale disintegration at the Balderhead earth-fill dam in the north of England. However, recalculations by Taylor (Attewell and Taylor, 1973) on their Ca^{2+} values as calcite suggested that only about 2.4 per cent was present in the unweathered material and only 0.4 per cent in their slightly weathered shales. Check-out slaking tests performed by the latter authors produced no systematic relationship between calcite content and degree of breakdown but there was a direct correlation between mixed-layer clay and breakdown. Work on weathered Jurassic Ampthill Clay has indicated an absence of calcite and pyrite, and it has been suggested (Attewell and Taylor, 1973) that in this case there has been a reaction between the soluble carbonate calcite and the sulphuric acid produced by low temperature oxidation of pyrite. It can be inferred that such an oxidation and solution process could be an early manifestation of a generally longer-term chemical process which is not to be found in the older, indurated shales of the British Coal Measures.

Mineralogy also influences the *residual shear strength* of natural clays. Kenney (1967) examined a range of monomineralic and polymineralic materials together with a range of natural soils (including weakly-bonded shales) to obtain ϕ'_r values from 29° to 35° for the massive minerals. The general trend downwards from these minerals is illite, or perhaps, hydro-muscovite ($\phi'_r = 16°$ to 26°) to kaolinite ($\phi'_r = 15°$) to the montmorillonites ($\phi'_r < 15°$). These trends mirror those derived for the variation of liquid limit and activity ratio with mineralogy but the ϕ'_r values exceed those considered later.

The type of montmorillonite can have interesting geotechnical ramifications. From his experimental work, Kenney (1967) has shown by implication that the presence of sodium-montmorillonite has a great influence on the strength of three weakly-bonded clay shales: the Bearpaw and Pierre (North America) and the Cucaracha (Panama). Chemical analyses reported by Attewell and Taylor (1973) on several North American clay shales indicated that montmorillonite was the dominant clay mineral (mean value 36.24 per cent), correlating significantly with the sodium-to alumina ratios but correlating equally significantly in a negative sense with the corresponding potassium ratio. This indicates that the mineral species is dominantly Na—montmorillonite and that some partial stabilisation might be achieved by admitting calcium to the clay and mobilizing its cation exchange capacity (C.E.C.) Such a procedure involving lime distribution has been adopted along the west bank of the Panama Canal where the Upper Cucaracha is exposed. However, chemical analyses on samples of Lower Cucaracha (north-west of East Culebra extension slide) and Upper Cucaracha (Cerro Escobar Hill), sampling areas which were unlikely to have been limed, have suggested that the Cucaracha contains a dominance of Ca-montmorillonite with a subsidiary Na-rich interstratification (Attewell and Taylor, 1973). Several tentative deductions of more general interest could be drawn from these observations: first, the Na-saturated montmorillonite could expand in water to the extent that it would dissociate into platelets the same order of thickness (10 Å) as the unit cell (see Gillott, 1968); second, if (not a strong possibility) the sampling areas had been limed, then the exchangeable Na would already have been replaced by Ca and this latter analysed; third, if the untreated Cucaracha is indeed rich in Ca-montmorillonite then liming procedures will do little or nothing to assist in ground stabilisation.

Kenney's general conclusions concerning the reduction in ϕ'_r with clay mineral content are confirmed by the plots in Figure 3.44 related to some North American clay shales (note that the clay shale of the Fort Union Series seems to be outside the general trend) and by the plots in Figures 3.45 and 3.46 from some stiff British clays. Figure 3.47 combining the data points also confirms the lower potential residual shear strengths associated with the presence of expandable clay minerals.

Several conclusions may be drawn with respect to the mineralogical influences in clay shales. First, both ϕ'_r and liquid limit are affected by the actual clay mineral content. However, although the

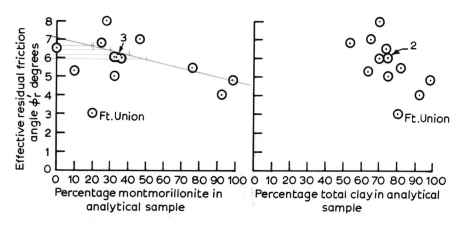

Figure 3.44 Influence of active clay mineral content in some North American clay shales upon the long-term residual shear strength (*Note:* the Fort Union Series clay shale is of Palaeocene age; the others are of Upper Cretaceous age and were loaded by Pleistocene continental ice sheets) (*from* Attewell and Taylor, 1973).

180 *Principles of Engineering Geology*

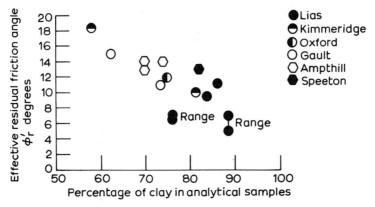

Figure 3.45 Influence of the percentage of clay minerals in some British stiff clays upon the effective residual friction angle (*from* Attewell and Taylor, 1973).

residual friction angle is designated as being 'effective', some of the values are so low as to suggest that there may have been incomplete pore pressure dissipation along the shear planes in the specimens. This would inevitably lead to an over-conservative design. A true effective residual angle in such material can only be achieved under full drainage with considerable shear displacement at very low strain rates. Second, as the proportion of montmorillonite increases, so does the liquid limit also increase to the extent of about 20 per cent per 10 per cent montmorillonite (North American material). Third,

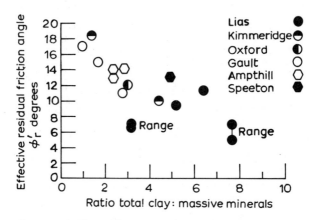

Figure 3.46 Influence of total clay-to-massive minerals ratio upon the effective residual friction angle for some British stiff clays (*from* Attewell and Taylor, 1973).

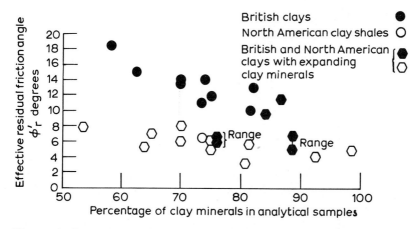

Figure 3.47 Showing the general decrease in the effective residual friction angle with increasing clay mineral content in British and North American stiff clays and clay shales (*from* Attewell and Taylor, 1973).

highly expandable clay minerals are not common in British stiff clays-to-clay shales; the expandable mineral that has been detected (Attewell and Taylor, 1973), vermiculite,* is unlike the North American expandable montmorillonite. Fourth, there is evidence that the liquid limit is not raised solely by an increase in the *total* clay mineral content nor by any expandable vermiculite in British clays, but due to the presence of montmorillonite, the liquid limits of North American clay shales are generally higher than those of British clays.

A final point concerns the direct influence of short-term and long-term geochemical changes upon discrete slip surfaces which pre-condition slopes to planar or rotational shear failure. Early and Skempton (1972) have noted the evidence of chemical reduction, signified by a lower Fe_2O_3 content in shear zone material, at the Walton's Wood slide in Staffordshire, England. There is, therefore, some support for the argument that during the period since late Glacial times, ancient slip surfaces have been subjected to reducing solutions in percolating groundwater. Such chemical changes acting initially along tectonically-created discontinuous surfaces and subsequent shear surfaces could be the *primary* root causes of shear strength reduction in post-glacial clays.

---

*Vermiculite and expandable smectites (montmorillonites) have been recorded in some British Jurassic and Cretaceous clay shales (see collation by Perrin, 1971).

4 ~ Rock as a Material

Rock mechanics is defined (see Judd, 1964) by the U.S. National Academy of Sciences Committee on Rock Mechanics as 'the theoretical and applied science of the mechanical behaviour of rock; being that branch of mechanics concerned with the response of rock to the force fields of its physical environment', and more simply, if less elegantly, by Coates (1965) as 'the study of the effects of forces on rocks'.

The basic objective of rock mechanics is to provide data which the engineer can use to design structures in or on naturally occurring rocks. Such data must therefore reflect two aspects of the rock's reaction to applied forces:

(a) the mechanical behaviour of the *intact* rock material, usually determined as a result of small scale laboratory tests on cored specimens; the term 'rock mechanics' is sometimes reserved only for this type of investigation.
(b) the mechanical behaviour of the *massive* rock modified by the presence of joints, fissures, bedding planes and other minor or major structural discontinuities and other environmental factors such as the presence of water; investigations into this type of behaviour and consequential design analyses are often referred to under the heading of 'rock engineering' or 'rock engineering mechanics' in order to differentiate what some people would regard as the only really relevant rock mechanics from the largely irrelevant intact material rock physics.

It is the fact that rock occurs in its natural state as a flawed, inhomogeneous anisotropic and discontinuous material, capable of only minor geotechnical modification, that separates rock mechanics from the general field of materials science. Design in rock as in all engineering materials requires some initial knowledge of the mechanical properties of the intact rock (see for example Clark,

1966; Farmer, 1968) and an analysis of the loads or forces acting on a future structure (see Coates, 1965; Obert and Duvall, 1967). However, it also requires, particularly in the case of slope design, a detailed knowledge of the presence and effect of discontinuities in the massive rock. This has been the area of emphasis of much recent research and is the subject of the later Chapter 6. The present chapter is concerned principally with the properties of rock as a material.

The basic components of any statement of rock properties required to define the material mechanically may be summarized as:

(i) The *strength* of the intact rock material, usually defined as the maximum force or combination of forces per unit area (that is, the active stresses) which the material can support. This information, usually derived in terms of an arbitrary laboratory experiment to determine uniaxial compressive strength, forms the basis of many rock property classifications and allows the intact rock to be designated between strong and weak limits.

(ii) The *short-term deformation* behaviour of the intact rock material, or the way in which it will alter in shape when subjected to sub-failure, or in some cases post-failure, stresses. When short term stresses are applied to the rock it may deform elastically, viscously or plastically (Figure 4.1) depending on the rock structure and the rate and magnitude of the stressing.

(iii) The *long-term deformation* of the intact material when subjected to maintained or cyclic sub-failure stresses.

Also of general interest is the response of the rock material when subjected to dynamic or impulsive shock loads and the effect of moisture on rock properties. A direct approach to the former is not considered relevant in terms of rock mechanics but some of the more important concepts are introduced in a later section (4.9). The concept of effective stresses in saturated granular materials is considered in Chapter 2. Its relevance to rock materials is limited since the rock 'frame' is very stiff (see for instance Skempton, 1961b), and until the rock is close to failure a very much smaller proportion of the total stress is transferred to the porewater. The concept is, however, highly relevant in the case of jointed rock masses and is considered in detail in Chapter 10.

In the present chapter, the phenomena of rock strength and deformation are considered initially in a simple way in order to provide a basis for rock classification and description. Later, some of

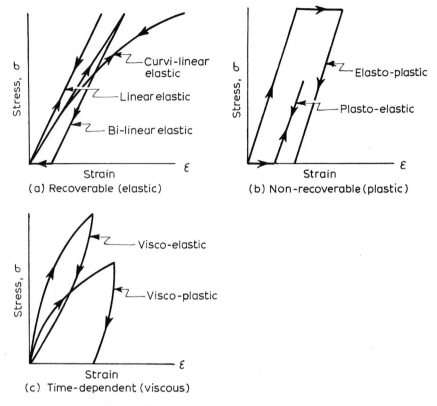

Figure 4.1 Some idealized deformation regimes.

the fundamental mechanisms of tensile failure on an intra-crystalline basis and which are now recognized as imposing a primary control on rock strength and deformation are introduced.

4.1 Uniaxial strength

The simplest index which can be used to describe the mechanical properties of rock is the uniaxial strength (or maximum stress acceptance level) of an unconfined cylindrical test specimen. Depending upon the direction of the applied force, strength may be measured in axial *compression* or *tension* (direct or indirect) or in *shear* (Figure 4.2). The test methods are described in detail by Hawkes and Mellor (1970) and typical test values are quoted in Table 4.1. Because of the difficulties associated with mounting and preparing specimens for the direct tensile test, and problems of interpretation in the case of the indirect tensile and shear tests, the

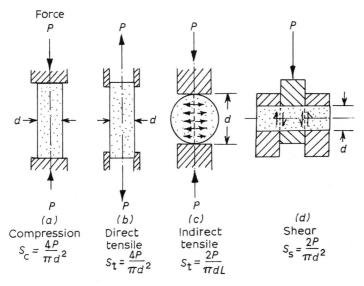

Figure 4.2 Force directions for uniaxial strength (S) measurement (in (c) L is the specimen length).

most widely quoted index property is *uniaxial compressive strength*.

The range of uniaxial compressive strengths in rocks usually falls between 5 to 400 N mm^{-2} (× 10^{-1} kgf cm^{-2}), although the lower limit is obviously discretionary. Within this range various arbitrary classifications have been suggested by Coates (1964), Deere (1966), Stapledon (1968), Coates and Parsons (1968). None is entirely

Table 4.1 Typical rock strengths, N mm^{-2}

	Compressive	Tensile	Shear
Granite	100–250	7–25	14–50
Diorite	150–300	15–30	—
Dolerite	100–350	15–35	25–60
Gabbro	150–300	15–30	—
Basalt	150–300	10–30	20–60
Sandstone	20–170	4–25	8–40
Shale	5–100	2–10	3–30
Limestone	30–250	5–25	10–50
Dolomite	30–250	15–25	—
Coal	5–50	2–5	—
Quartzite	150–300	10–30	20–60
Gneiss	50–200	5–20	—
Marble	100–250	7–20	—
Slate	100–200	7–20	15–30

satisfactory but they may be summarized roughly as:

5 to 20 N mm^{-2}	very weak	— weathered and weakly-compacted sedimentary rocks,
20 to 40 N mm^{-2}	weak	— weakly-cemented sedimentary rocks; schists,
40 to 80 N mm^{-2}	medium strength	— competent sedimentary rocks; some low density, coarse igneous rocks,
80 to 160 N mm^{-2}	strong	— competent igneous, metamorphic rocks and some fine-grained sandstones,
160 to 320 N mm^{-2}	very strong	— quartzites; dense fine-grained igneous rocks.

This basic classification is extended in Section 10.2 to include massive rock features.

It is particularly difficult, as Table 4.1 shows, to allocate specific rock *types* into any strength classification since many factors can affect the strength of a rock specimen. These are not the same factors as the mainly mineralogical and petrographic ones which are used in *geological* classifications (see Section 1.1), although as a rule competent igneous rocks are stronger than competent sedimentary rocks and dense, well compacted and cemented sedimentary rocks are more competent than low density weathered or poorly-consolidated sediments.

The *intrinsic* characteristics of rock as a material, which affect the strength of a test specimen, may be summarized (not necessarily in order of importance) as:

(a) *Porosity and density*. Although it is difficult to obtain a generalized relationship between porosity, bulk density and strength, many workers have related both the former properties to strength for a similar rock type such as sandstone or limestone (see for example, Morgenstern and Phukan, 1966). Porosity and

Table 4.2 Typical porosity and bulk densities of rock materials

	Bulk density Mg m^{-3}	Porosity n%
Granite	2.6–2.9	0.5–1.5
Dolerite	2.7–3.05	0.1–0.5
Rhyolite	2.4–2.6	4–6
Andesite	2.2–2.3	10–15
Gabbro	2.8–3.1	0.1–0.2
Basalt	2.8–2.9	0.1–1.0
Sandstone	2.0–2.6	5–25
Shale	2.0–2.4	10–30
Limestone	2.2–2.6	5–20
Dolomite	2.5–2.6	1–5
Gneiss	2.8–3.0	0.5–1.5
Marble	2.6–2.7	0.5–2
Quartzite	2.6–2.7	0.1–0.5
Slate	2.6–2.7	0.1–0.5

bulk density (Table 4.2) are themselves interrelated since most silicate mineral constituents of common rocks have a similar specific gravity. It is not surprising that porosity should affect strength, since in any mechanically (or molecularly) bonded aggregate (in this case, of mineral crystals) the overall magnitude of interlocking bond forces will depend on the total area of contact between individual particles. This will be inversely related to the amount of porespace. The absence of a generalized approach may reflect either the difficulty of measuring porespace accurately (although this can be overcome by air or mercury porosimeter saturation techniques) or the variable nature of the mechanical bond between different minerals. Experimental methods for determining intergranular adhesion are described by Savanick and Johnson (1974). Figure 4.3, based on results summarized by Judd and Huber (1962), illustrates the approximate relationship between strength and density.

The effect of porespace is most obvious in the comparison of strengths of fine-grained igneous or metamorphic rocks with coarse-grained, weakly-cemented sandstones; the former has negligible porosity, while the latter porosities are as high as 30 per cent.

(b) *Grain size and shape* (see also Section 2.2). Just as the intergranular contact area is increased with decreasing porosity,

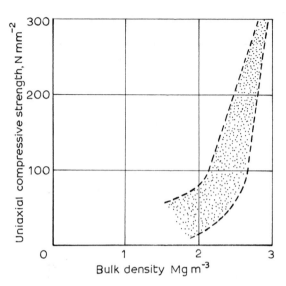

Figure 4.3   Approximate relationship between strength and bulk density.

so it is increased with decreasing average grain size and decreasing angularity. Thus, the interlocking forces resisting failure should be greater in a fine-grained rock than in a coarse, angular grained rock, and the strength should be consequently higher.

The size of grains is of some interest also in that grain boundaries may represent potential points of weakness where cracks may exist or be initiated during loading. It is known, for example, that edge dislocations within a crystal structure may pile up at a grain boundary so generating and wedging open a crack. It would appear that the longer the boundary, the more rapidly will any cracks spread and the more likely will be the chance of failure at low stress. This point is discussed later and it may be one of the reasons why mineralogically similar igneous rocks of similar high density but varying grain sizes may fail at different strengths.

The tendency to premature failure may also be exacerbated if the grain size (particularly in weakly-bonded aggregates) is large in comparison to the specimen size.

(c) *Anisotropy.* Most rocks, and particularly sedimentary rocks, tend to possess a preferred orientation of the constituent grains, a feature which can affect the apparent strength of rocks cored at different orientations. Although far less important in engineering terms than the effect of mass anisotropy in massive rocks, it will be shown subsequently in Chapter 5, that certain rocks exhibit specific directional properties on a small scale, and that these properties are more likely to be related to weakness planes or identifiable laminations, themselves being conditioned by the orientations that certain minerals acquire in a sedimentational or metamorphic environment.

(d) *Mineralogy.* Merriam et al (1970) show that tensile strength is inversely proportional to quartz content in granitic rocks, attributing this to increased crystal interlocking in low quartz rocks. Price (1960) shows that compressive strength is proportional to quartz content in Coal Measures sandstones and siltstones, and attributes this in part to the relative compaction of the calcareous or clay matrix. The only reasonable conclusion to draw is that it is dangerous to attempt to postulate a general relationship between mineralogy or petrology and strength.

(e) *Moisture content.* Specimens are usually tested and strengths quoted in an oven-dried state (100–110°C), the presence of water or electrolyte in significant quantities in the porespaces of a rock specimen leading to a reduction in strength. This may be high in shale containing active or swelling clay minerals. Various explanations ranging from particle surface energy reduction (see Rehbinder et al, 1948; Pugh, 1967) and interparticle bond modification to the existence of porewater pressures in a poorly-drained specimen have been suggested.

Equally important in determining specimen strengths are the size of specimen, aspect ratio and method of preparation and testing, in other words the *test specimen characteristics* as distinct from intrinsic rock properties. Cylindrical test specimens are normally cored from, or taken direct from, a diamond drill core sample. Common core *diameters* are:

DIAMOND DRILL	EX	⅞ in	22 mm
	AX	1⅛ in	28 mm
	BX	1⅝ in	41 mm
	NX	2⅛ in	54 mm
		4 in	102 mm
LABORATORY DRILL		1 in	25 mm
		1.5 in	38 mm
		2.5 in	63 mm

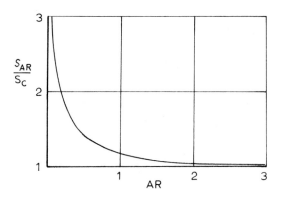

Figure 4.4 Effect of aspect ratio (AR) on compressive strength.

The specimen *aspect (length/diameter) ratio* (AR) (length/diameter) can have a significant effect on compressive strength. Originally, unconfined uniaxial tests were carried out on cores with an aspect ratio of 1, the uniaxial compressive strength S_c of other lengths being related to this standard by the U.S. Bureau of Mines formula (Obert et al, 1946):

$$S_c = S_1 \left(0.8 + \frac{0.2}{AR} \right) \qquad (4.1)$$

where S_1 is the compressive strength of a specimen having an aspect ratio of 1. These days it is generally agreed that provided AR > 2.5 (or a minimum of 2) a reasonably constant specimen compressive strength is obtained (Figure 4.4). The effect of AR on specimen strength at low aspect ratios is a function of the degree of lateral (or radial) restraint exerted by the relatively high elastic modulus of steel loading platens on the test specimen. The difference in modulus inhibits lateral expansion at the base, alters the test boundary conditions, and induces high tensile stress concentrations σ_t at the edges of the specimen in contact with the platen together with zones of compression at the centre of the platen. Figure 4.5 illustrates this using McLintock and Walsh's (1962) failure criterion applied to Balla's (1960) analysis of axial and radial stresses in an unconfined end-restrained cylinder of AR = 2. Failure will occur in the specimen when the tensile stress at any point in the rock exceeds its local tensile strength. The implications of this are discussed later, but it is evident that in the case of a low aspect ratio

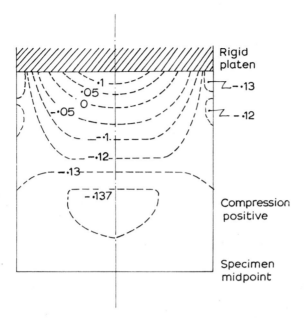

Figure 4.5 Contours of σ_t/σ_z for a uniaxially loaded cylinder subjected to a compressive stress σ_z, where:

$$\sigma_t = \frac{\sigma_a}{4}[(\mu^2+1)^{1/2} - \mu] - \frac{\sigma_r}{4}[(\mu^2+1)^{1/2} + \mu]$$

where σ_a, σ_r are axial and radial stresses in the cylinder, and $\mu = \tan\phi = 0.7$.

specimen, failure will tend to occur through a spread of outer spalling and mid-point collapse.

The effect of *specimen size* is less easily quantifiable. Various 'weakest link' theories exist (see Jaeger and Cook, 1969; Hawkes and Mellor, 1970), predicting lower strengths in increased size specimens due to the effect of flaws or discontinuities (see Section 6.6) which may be absent or less numerous in smaller specimens. These theories are based largely on Weibull's (1939) statistical approach. But in most cases, however, little correlation exists between specimen volume and apparent strength.

Specimen *preparation* and selection are important, particularly where a limited sample may lead to wide statistical variations in the measured specimen strength. Ideal test requirements comprise:

(i) a rational and representative sample of the drill core or rock sample, stored under good conditions. In addition, the specimen *history* should be known, particularly where its properties may have been modified by relief of geostatic stresses, weakening during coring or blasting, weathering, contamination, change in moisture content, or possibly by changes in temperature.

(ii) A smooth core surface (to avoid obvious discontinuities where failure may be initiated, particularly important in tensile or fatigue tests) and flat, smooth, lapped ends to ensure even contact over the loaded area. The core must also comprise a right circular cylinder so that the loading is applied parallel to the core axis.

The *method* of testing is also critical. The boundary conditions for a uniaxial test demand a uniform static stress acting on a plane at right angles to the specimen axis with zero shear stress at the platen/specimen interface and a stress-free exposed surface. Conventional use of a rigid platen means that an uneven stress is introduced almost invariably to the specimen during a compression test, and although it is more convenient to use a rigid rather than a flexible or hydrostatic platen (not necessarily so in a tensile test) the implications of so doing should be understood. The use of *stiff* as against *soft* testing systems can also affect failure levels by controlling (machine) energy release up to and after failure and so allowing a monitoring of the complete, post-failure stress-strain relationship. This aspect of rock testing is considered in detail in a later section (see Section 4.4). The rate of strain also becomes important at very rapid or very slow rates (Section 4.5). The boundary conditions for a conventional uniaxial compressive strength test obviously demand a low rate of strain. The ASTM specification (C 170–50 Compressive strength of building stone), the nearest approach to a standard, specifies rates of loading of 0.7 N mm^{-2} s^{-1} (stress controlled) and 1 mm min^{-1} (deformation controlled). Usually, any loading rate sufficiently low to avoid adiabatic effects and sufficiently high to avoid significant creep would be acceptable, and this is indeed the basis for a parallel ASTM specification (E 111–65 Determination of Young's Modulus at room temperature) for deformation testing.

A much simpler *field test*, which gives a result that can be related quite closely to compressive strength, is the point load test. This test has been described by Franklin *et al* (1971) and comprises a small hydraulic ram mounted in a loading frame (Figure 4.6). The ram

Figure 4.6 Imperial College point load test. (*Photograph by courtesy of* Robertson Research International Ltd.)

compresses a core of rock across its diameter between two standard pointed platens. The test results are expressed in terms of a point load strength index I_s, where $I_s = P/D^2$, P being the force required to cause failure, computed directly from the ram pressure and dimensions, and D is the distance between the points of the platens at the moment of failure, read from a graduated scale incorporated in the loading frame.

D'Andrea *et al* (1965) and Franklin *et al* (1971) have shown that it is possible to relate (Figure 4.7) the point load strength index to the uniaxial compressive strength with reasonable accuracy. The specimen, in fact, fails in tension at relatively minor loads in the point load test, and even irregular specimens can be tested with reasonable confidence. I_s is in fact quite similar in many cases to the rock tensile strength obtained by indirect methods (Figure 4.2c).

The principle behind the test is not particularly new, and, in the case of a regular core, the failure mechanism and strength are quite similar to those associated with tensile failure in the laboratory

194   *Principles of Engineering Geology*

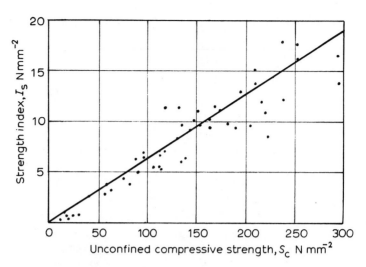

Figure 4.7   Relationship between point load strength index and unconfined compressive strength (*after* Franklin *et al.* 1971; D'Andrea *et al.* 1965).

indirect strength test. In the case of irregular specimens, Hobbs (1962) and Hiramatsu and Oka (1966) among others, have developed the concept of the P/D^2 strength index and related this to compressive and tensile strength. Other field tests yielding data which can be related to rock strengths include various types of penetrometer and rebound hammers, such as the Schmidt hammer (see, for instance, Franklin, 1974; Rankilor, 1974). These instruments have some history of use in rock and concrete testing and are generally regarded as unreliable – relating, as they do, the result of an arbitrary field experiment to the result of an equally arbitrary laboratory experiment. Where, however, a rapidly reproducible indicator of spot strengths in a rock core or mass (where stability controls are dominated by the rock structure) is required, they have some attraction.

4.2 Uniaxial short-term deformation

Uniaxial strength is almost certainly too simple a criterion to define the reaction of intact rock to external forces. A more adequate description of the reaction of intact rock materials to uniaxial compression can be obtained by considering their axial or volumetric deformation characteristics, both *short term* (that is, their 'elastic'

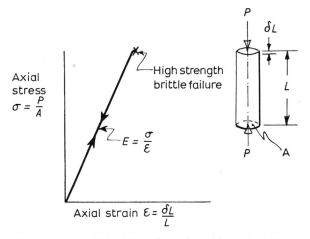

Figure 4.8 Uniaxial deformation characteristic — elastic material.

properties) and *long term* (that is, their *creep* and possibly fatigue properties).

Some rocks, for example the finer-grained stronger rocks, act virtually as elastic materials in *axial* compression — see Figure 4.8. In other words, they obey Hooke's Law and when subjected to compression (or tension), their stress ($\sigma = P/A$) versus axial strain ($\epsilon = \delta L/L$) relationship is linear and the strain — or more accurately the strain energy input — is *recoverable*:

$$\frac{\sigma}{\epsilon} = E \qquad (4.2)$$

where E is the *modulus of elasticity*. The basic deformation characteristics of an elastic material have been summarized by Jaeger (1969).

The weaker rocks, of lower strength, tend to project non-linear stress-axial strain characteristics with an initial concavity due to closing of excess porespace or micro-cracks, and a terminal plastic zone approaching a level of failure strain (Figure 4.9). In both these zones, some strain energy is obviously *irrecoverable*. In between these zones there is a linear zone, where the rock behaves in an approximately elastic manner. E values quoted in rock mechanics usually refer to the slope of this linear zone. Other values which may have importance under specified static or dynamic loading conditions are the *tangent* modulus (the slope of a line drawn tangential to the

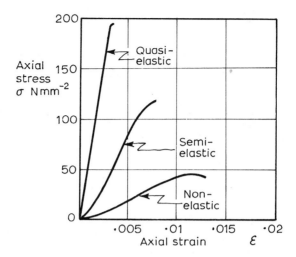

Figure 4.9 Typical rock deformation regimes.

curve) or the *secant* modulus (the slope of a line drawn from the origin to a point on the curve, usually at failure or 50 per cent failure, see Figure 2.24) at a particular stress level. One value which is often quoted is an E value obtained by surface seismic methods (see Section 4.10). This is effectively the initial *tangent* modulus (see Figure 2.24) and if derived from field measurements on rock masses represents a conservative design deformation modulus. It is possible as in Table 4.3 to propose a rock classification using the deformation modulus (in this case the *tangent* modulus of the linear part of the curve) as a reflection of the form of the stress strain curve. It should also be noted that *Poisson's ratio* is roughly proportional to E rising from very low (or negative) values in soft rocks to about 0.25 in stiff rocks (see U.S. Bureau of Reclamation, 1953).

The divisions in Table 4.3, together with strength criteria, form the basis for most rock material classifications. Judd and Huber (1962), in a statistical analysis of U.S. Bureau of Mines test results,

Table 4.3 Classification of intact rock on the basis of its stiffness

Quasi-elastic	— very stiff	$E > 8000$ N mm^{-2}
Semi-elastic	— stiff	$E = 4000–8000$ N mm^{-2}
	— medium stiff	$E = 2000–4000$ N mm^{-2}
Non-elastic	— soft	$E = 1000–2000$ N mm^{-2}
	— very soft	$E < 1000$ N mm^{-2}

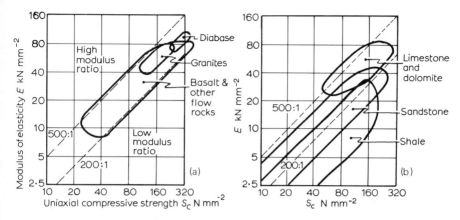

Figure 4.10 Engineering classifications based on modulus ratio (*after* Deere and Miller, 1966).

show that there is a confident relationship between E and compressive strength S_c in the form:

$$E = 350\, S_c \tag{4.3}$$

There is of course a certain amount of scatter, and Deere and Miller (1966) suggest that *modulus ratios* (E/S_c) combining deformation and strength be used to define rock properties. They suggest three divisions, defined by ratios of > 500, 500 to 200 and < 200 to indicate decreasing degrees of stiffness. This classification is reproduced in full in Figure 4.10 and in simple form in Figure 4.11. Modulus ratios for most igneous rocks (Figure 4.10a) fall in the range 200–500, but some sedimentary rocks (Figure 4.10b) have distinctly atypical modulus ratios. In particular some shales – as might be expected – have quite low modulus ratios indicating a distinct tendency to non-elastic deformation.

Classification of rocks on the basis of their axial deformation characteristics as well as strength is of considerable importance in rock design. For instance, many simple problems can be solved through elastic stress analysis based on the assumption that rock is a mechanical continuum, although the approach may differ slightly from conventional engineering limit state design.

In conventional engineering practice, a structure is designed so that the strengths of the materials used in construction are greater than the largest computed stresses acting on or in the structure. There may be some accommodation in the design for plastic

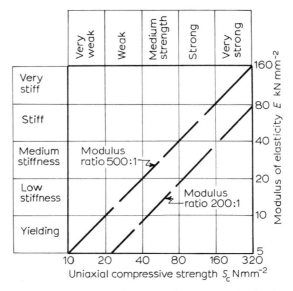

Figure 4.11 Classification of rocks on the basis of uniaxial strength and tangent modulus at 50% strength (*after* Deere and Miller, 1966).

deformation of the material (see Baker and Heyman, 1969). The purpose of design and of the critieria governing the selection of materials is to achieve these objectives safely and economically.

In the case of rock engineering, general design flexibility is limited, first by the discontinuous and hence partially failed character of the material and second by the structure itself which will usually take the form of an excavation or series of excavations in, and be wholly or partially surrounded by, the rock. A deep excavation in particular may impose on the rock a state of stress that exceeds the mass strength, and under these circumstances conventional limit state design, if based on elastic critieria, can at best only serve as a guide to design and at worst can be fatally misleading. It is much more important in these circumstances to appreciate the deformation characteristics of the rock material and the rock *en masse* up to 'failure', during 'failure' and after 'failure', the term 'failure' being used in this context to mean the greatest level of stress that can be supported by the rock.

In this respect, it is relevant to note the conclusion of Hudson and Fairhurst (1971) that 'rock material properties' — particularly those affecting and being affected by failure — should not be regarded as fundamental material properties but as experimental values. They point out that the process of rock material (or 'intrinsic') failure is extremely complex, starting virtually from the moment that a test specimen is first stressed and continuing throughout both an increasing and decreasing range of stress. It would follow that, for a rigorous design exercise, rather more valuable information can be derived from study of this breakdown process than by adopting an arbitrary selection of standardized tests to determine specific strength and deformation parameters. On the other hand, this argument does not rule out the use and value of fundamental properties as indices of rock quality to be adopted as the basis of classification or empirical design systems.

4.3 Deformation mechanisms in rock

When a solid crystalline elastic material deforms as a continuum, the deformation results from a change in interatomic spacing, the strain being equivalent to the magnitude of the average change of interatomic spacing in the direction of the applied force. This applied force is counteracted by a repulsive (compression) or, attractive (tension) resisting force tending to restore the original atomic structure of the crystalline lattice when the deforming force is removed. Rocks are agglomerates of crystals (often imperfect) which are mechanically bonded together where they contact each other. Any approach, therefore, to describing elastic deformation on the lines of classical materials science will obviously not work.

In fact, a more satisfactory description of rock deformation, based on the theories postulated by Griffith (1921) to explain the tensile deformation of glass, has been developed by Murrell (1958), Brace (1960) and McLintock and Walsh (1962) among others. These authors propose that progressive deformation and failure of rock, whilst possibly affected by classical crystal deformation (dislocation motion manifest as twin and translation gliding) and restorable adjustment of mechanically-bonded particles, is caused principally by extension of existing flaws and creation of new flaws in the rock fabric. The principal argument is that rock is a *flawed* material containing, on a microscale, porespaces and/or microfissures in the grains themselves at grain boundaries (see Table 4.4), and on a large scale joints, bedding planes and other macrofissures.

The requirement of any theoretical approach to deformation based on flaw or crack propagation is a mechanism to account for

Table 4.4 Classification of microcracks in rocks (*compiled after* Simmons and Richter, 1974)

Type	Symbol	Remarks
Generally grain boundary cracks (pressure temperature genesis)	dPdT	Development is a function of the different crystallographic orientation of two grains with respect to the grain boundary. Crack is produced when local linear strain near to the grain boundary exceeds the average linear strain of the rock. Refer to Simmons *et al* (1973 a, b). These cracks are normally found in intrusive igneous rocks.
Stress induced cracks	SIC	Chiefly a function of local principal stress differences in the rock and largely independent of relative crystallographic orientations of different minerals once the crack begins to grow. Several grains may be crossed. There may also be coincident grain boundary cracks produced by a mismatch between adjacent grains having different elastic properties (e.g. quartz and feldspar).
Radial (RD) cracks and concentric (CN) cracks about totally enclosed grains	RDC CNC	Created by a mismatch of the volumetric properties between a host grain and a totally enclosed grain under P–T changes. If volumetric strain of enclosed grain is less than volumetric strain of the host, CN cracks occur either within the host or along the grain boundary. If volumetric strain of enclosed grain is greater than that of the host by an amount sufficient to overcome the strength of the host, then radiating, non-coincident grain boundary cracks occur.
Tube cracks	TBC	These are 1–3 μm diameter, 10's to 100's of μm in length and are locally abundant in many terrestrial and lunar igneous rocks. Tubes may be hollow or filled with a solid phase. They may owe their genesis to one or more of the following processes: natural etching of dislocations by ground water or hot gases; solution by ground water or hot gases; healing of flat microcracks.
Thermal cycling cracks	TCC	Produced by slow heating or cooling of a rock. It is claimed that experimental and observational evidence clearly distinguishes this type of crack from others (Wang *et al*, 1971; Todd *et al*, 1972, 1973; Simmons *et al*, 1973 a, b).
Thermal gradient cracks	TGC	Caused by stresses which themselves originate from *high* thermal gradients. Temperatures below $350°C$–$450°C$ if changed slowly seem to produce few microcracks.
Shock-induced cracks	SHIC	Shock-induced crack porosity may be used to estimate peak shock intensity (Simmons *et al*, 1974; Siegfried *et al*, 1974). Detailed petrography of rocks affected by shock waves has been the subject of much study (for example, Ramez and Attewell, 1963; French and Short, 1968; Das Ries, 1969) and of particular interest is the correlation between microcrack (and other microstructural) petrography and shock pressures near underground nuclear explosions (Short, 1966).
Cleavage cracks	CLC	Cracks that are parallel to cleavage planes in minerals.
Thin section cracks	TSC	Cracks that are introduced during thin-section manufacture. The form of thin-section manufacture must either be such as to prevent their occurrence (Simmons and Richter, 1974) or they must be identified and removed by observation.
Cracks of unknown origin	UC	Simmons and Richter quote and show examples of cracks that cannot be genetically classified.

that propagation. This is the basis of the Griffith theories, which propose that: (i) crack propagation is initiated by very high tensile stresses at the tips of pre-existing ellipsoidal cracks or flaws in the rock; (ii) the cracks tend to propagate at right angles to the direction of maximum tensile stress. In uniaxial compression this will be approximately parallel to the axis of loading; (iii) the cracks will be initiated when the available stress is sufficiently high to satisfy the energy requirements for the newly created crack surfaces:

$$\sigma_t = \left(\frac{\rho E}{2\pi c}\right)^{1/2} \quad (4.4)$$

where σ_t is the tensile stress at the crack tip, ρ is the specific surface energy per unit length of crack and c is the crack half length.

Thus, when σ_t is sufficiently large, cracks will propagate along the axis of a uniaxially loaded specimen. In the case of a biaxial or triaxial (compressive) stress situation, then the cracks will tend to propagate in a direction parallel to the major principal stress.

Actual magnitudes of ρ, c, σ_t are difficult to determine (see Bieniawski, 1967). The overall effect of crack propagation can be demonstrated, however, in simple manner by examining the *volumetric* stress/strain characteristic of a uniaxially loaded cylindrical test specimen.

Consider, for instance, the elastic deformation of a right circular cylinder of length L and diameter D subjected to uniaxial compression as in Figure 4.12. If there is a change in length δL, and a

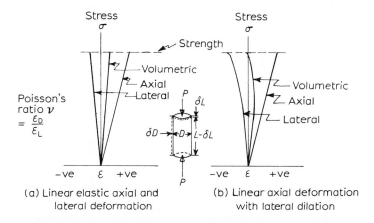

Figure 4.12 Deformation characteristics of a uniaxially loaded right circular cylinder.

change in diameter δD, then the change in volume δV will be given by (ignoring the direction of deformation):

$$\delta V = \frac{\pi}{4}[(D + \delta D)^2(L + \delta L) - D^2 L] \qquad (4.5)$$

In an *elastic* material the lateral and longitudinal strains will be related by a constant *Poisson's ratio*, v

$$v = \frac{-\epsilon_D}{\epsilon_L} = \frac{-\delta D/D}{\delta L/L} = \frac{-L\delta D}{D\delta L} \qquad (4.6)$$

(compression is taken as positive)

then substituting in equation 4.5 gives

$$\delta V = \frac{\pi}{4}\left[(D - v\frac{D}{L}\delta L)^2(L + \delta L) - D^2 L\right]$$

$$\simeq \frac{\pi}{4}\left[D^2\delta L - 2vD^2 \delta L\right] \qquad (4.7)$$

and $\dfrac{\delta V}{V} = \Delta = \dfrac{\delta L}{L}(1-2v) = \epsilon_L(1-2v) = \epsilon_L - 2\epsilon_D \qquad (4.8)$

If the material is elastically deformed and v is therefore constant, the axial lateral and volumetric stress-strain curves should be linear (Figure 4.12a), with axial and volumetric strain positive (compression) and lateral strain negative (tension). In fact, although the axial stress-strain curve may be linear, it can be shown by experiment that during short term deformation, neither the lateral nor volumetric stress-strain curves are actually linear (Figure 4.12b). The form of these curves presents certain very clear implications supporting the principal of crack propagation as a mechanism for progressive deformation.

The first conclusion is that there is a lateral *dilatant* effect, indicative of void space being created parallel to the core axis and apparently increasing the volume and the magnitude of Poisson's ratio, itself no longer a constant. The second is that there is a variable rate of crack propagation. Griffith's theory, based on a study of materials loaded in tension, in fact equates crack initiation with failure. Obviously in a compressive stress field, as is usually the case in rock mechanics, crack initiation involving initial tensile failure will

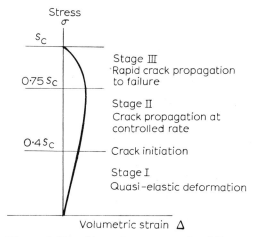

Figure 4.13 Stages in a progressive failure regime.

not lead to immediate structural collapse. Irwin (1957) was among the first to realize this and his modified Griffith theory allowed for controlled crack propagation to a critical level in a compressive stress field. Irwin's work was concerned with brittle steels and Bieniawski (1967) has publicized its applicability to rocks, showing how the approach can be used to postulate a three-stage progressive failure regime in terms of the stress-*volumetric* strain curve. These stages may be summarized in Figure 4.13 as:

Stage (I)   At stress levels up to about 0.4 S_c, quasi-elastic near-linear deformation prior to the initiation of new cracks or the mobilization of existing cracks;

Stage (II) Stable crack propagation at a controlled rate following crack initiation;

Stage (III) Rapid crack initiation leading to strength-failure and disintegration of the specimen.

A simple way in which these effects can be quantified is by considering the stress-volumetric strain relationship and also the energy balance for a loaded specimen.

Thus, the expression for volumetric strain (equation 4.8) may be written:

$$\frac{\delta V}{V} = \frac{\sigma}{E_0}[1 - 2(\nu_0 + \nu_c)] \qquad (4.9)$$

where E_0 and v_0 are the 'elastic' modulus and 'elastic' Poisson's ratio respectively.

A series of experiments on limestone, carried out by the authors, indicate that v_c, the variable crack dilation portion of v can be written:

$$v_c = \left[\frac{\sigma}{S_c} - 0.4\right]^n \qquad (4.10)$$

where $n = 3.4$ for limestone, and $\frac{\sigma}{S_c} > 0.4$

Then

$$\frac{\delta V}{V} = \frac{\sigma}{E_0}\left\{1 - 2\left[v_0 + \left(\frac{\sigma}{S_c} - 0.4\right)^n\right]\right\} \qquad (4.11)$$

This expression may be compared with a simple energy balance statement for deformation, namely that:

Energy input = elastic strain energy stored by the specimen + energy to satisfy surfaces of newly formed cracks

$$\sigma \delta V = \frac{\sigma^2}{E_0}(1 - 2v_0)V + \rho a V \qquad (4.12)$$

where ρ is the surface energy per unit area of crack and is a constant and a is the area of new cracks per unit volume of rock.

Dividing throughout by σV gives:

$$\frac{\delta V}{V} = \frac{\sigma}{E_0}(1 - 2v_0) + \frac{\rho a}{\sigma} \qquad (4.13)$$

and by equating 4.11 and 4.13, we have:

$$\rho a = \frac{-2\sigma^2}{E_0}\left(\frac{\sigma}{S_c} - 0.4\right)^n \qquad (4.14)$$

or

$$a = \frac{-2\sigma^2}{\rho E_0}\left(\frac{\sigma}{S_c} - 0.4\right)^n \qquad (4.15)$$

Thus, if ρ and n are known and a is taken as zero at $\sigma/S_c = 0.4$, then the amount of new crack surface can be estimated (Figure 4.14).

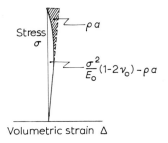

Figure 4.14 Proportions of strain energy expended in deformation and crack propagation.

Indications of crack growth also result from acoustic emission evidence. This emission, or microseismic activity as it is sometimes called, is caused by small, internal stick-slip events as cracks are extended and then halted, principally by grain boundary barriers. The sudden release of stored elastic strain energy from each local event generates a stress wave which travels from the point of origin within the material to an observation boundary. Support for this type of mechanism has been given by Gold (1960) in the studies of microseismic activity in ice.

In North America, acoustic emission studies related to geological materials were started in the late 1930's and early 1940's by Obert (1939, 1940, 1941), Obert and Duvall (1942, 1945, a,b), Hodgson (1943), Hodgson and Gibbs (1945) and others as part of research into mine design problems and rockburst prevention. A review of work in Russia was published by Antsyferov (1966) and an application of acoustic emission monitoring in Britain is described by Knill et al (1967). The emission from rock under stress is usually monitored by piezoelectric devices and at any given time, for a rock subjected to a known external stress, it may be expressed as a noise rate. This rate is the number of discrete emission events per unit time and takes no direct account of the acoustic emission amplitude which is the maximum amplitude of each recorded event in arbitrary units. Clearly, noise rate and the amplitude sum per unit time (the accumulated emission energy) are an expression of internal deformation. After a period of stress relief, such as might, for example, be induced by blasting in a deep mine, the noise rate decreases. It is therefore an expression of relative stability. Not only is a high noise rate indicative of the onset of failure in a brittle rock but it also is apparently associated with the closure of pre-existing pores and cracks at low stress levels. The intuitive genetic concept of the activity and the fact that it should be deformation-dependent is

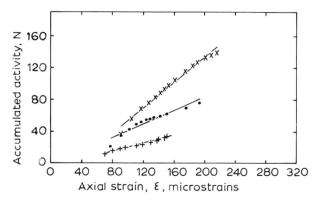

Figure 4.15 Typical curves showing the variation of accumulated activity with axial creep strain for Crab Orchard sandstone (*from* Hardy et al, 1970). Accumulated activity is defined as the total number of events observed during a specific period of time. Each curve relates to a different constant level of applied pressure.

confirmed by recorded increases in noise rate with strain under constant (creep) loading conditions (Figure 4.15).

An additional source of investigation arises from the frequency composition of a microseismic wave train. As in dynamic analysis generally, any transient signal may be regarded as the algebraic superposition of a large number of steady-state harmonic waves. It may therefore be resolved from the time domain into the frequency domain with the implication that the form of the frequency spectrum is in some way characteristic of the material and of the mode of deformation. This particular aspect of the microseismic story, and indeed, the detail of the total story itself, is somewhat controversial. For further reading, and for the ramifications of microseismic monitoring with respect to such civil engineering activities as slope and dam construction, reference may usefully be made to the work of Hardy (1969, 1971, 1973) for additional bibliography and to Goodman, 1963 (see Goodman and Blake, 1964, 1965).

4.4 Complete stress–strain characteristics of rock in uniaxial compression

Rapid release of residual elastic strain energy at peak stress levels will lead to disintegration and violent collapse of most *brittle* rock specimens subjected to uniaxial compression. This is not necessarily a fundamental property of the rock, nor a reflection of how it will react in practice, but is caused largely by the design of the testing machine.

Most hydraulic or mechanical testing machines are force-controlled 'soft' testing machines, where a constant force is applied to a specimen through a reservoir of oil or a screw feed acting against the machine columns. Coalescence of cracks in a specimen at peak strength results in a rapid release of elastic strain energy from the machine and from the specimen. This is the cause of the violent failure associated with 'brittle' rocks.

If the *stiffness* of the machine can be increased by reducing the oil reservoir to a minimum and by stiffening the testing frame, then it should be possible to reduce energy release when an imbalance develops with respect to the specimen. Alternatively, steel columns may be loaded in parallel with the specimen (Jaeger and Cook, 1969) or servo-controlled loading systems (Rummel and Fairhurst, 1970; Hudson *et al*, 1971) may be used to control deformation. Technically, the machine-specimen system is stable provided that the rate of energy release from the specimen — which is an indication of the reduced stress level that it can support — is accompanied by a greater reduction in the testing machine energy.

The principles of stiff system design are outlined by Rummel and Fairhurst (1970). The excess energy (δW) released by the testing machine over *any* unloading deformation increment δx may be defined in terms of the relative stiffnesses of the specimen (K_s) and of the unloading machine (K_m).

Then, the specimen stiffness (see Figure 4.16) is given by;

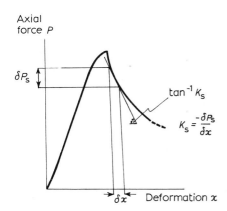

Figure 4.16 Specimen stiffness after failure.

$$K_s = -\frac{\delta P_s}{\delta x} \tag{4.16}$$

where δP_s is the change in force (stress × load-bearing area) which can be accepted by the specimen and the machine stiffness is given by:

$$K_m = -\frac{\delta P_m}{\delta x} \tag{4.17}$$

where δP_m is the change in force *applied* to the specimen by the loading platens.

In a soft system, $K_m > K_s$, and in a stiff system, $K_m < K_s$ (i.e. it has a larger negative magnitude), in which case:

$$\delta W = \frac{(K_m - K_s)}{2} \delta x^2 \leq 0 \tag{4.18}$$

In this latter case, a complete stress/strain curve may be derived for virtually all rocks except where δx is negative. Even then, it is possible to imagine an active system where K_m is positive (see Wawersik and Fairhurst, 1970; Wawersik and Brace, 1971). Where K_m has a large negative value, all except the steepest unloading curves can be monitored.

Typical stress-strain curves for rocks in uniaxial compression have been classified by Wawersik and Fairhurst into two types, designated Class 1, 2, on the basis of the form of the unloading leg of the curve. In both of these cases shown in Figure 4.17, the rising stress/strain curve is similar, with virtually no fabric change taking place until a stress level equal to about 70 to 80 per cent of peak stress has been

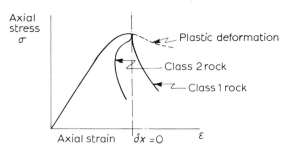

Figure 4.17 Typical complete stress-strain curves.

reached. This limit is approximately equivalent to the Irwin (1957) limit for the onset of rapid propagation of microcracks as proposed by Bieniawski (1967) on the basis of the volumetric stress-strain curve (Figure 4.13).

Propagation of microfractures increases towards the peak stress level, from which point the two modes of behaviour may be differentiated. Class 1 rocks, the weaker type of granite, limestone and marble used in the tests and having compressive strengths in the range 50 to 100 N mm^{-2}, are characterized by a stable post-peak-stress mode of deformation, requiring further energy input to further deform the specimen. Class 2 rocks, stronger granites, limestone and marble, with strengths ranging from 200 to 300 N mm^{-2}, are characterized by unstable or self-sustaining fracture development where, following peak stress, fracture continues whithout any further loading and indeed a negative unloading characteristic is required from the test machine. Fairhurst suggests a dividing line at $\delta x = 0$ to differentiate between those rocks where the stored elastic energy is sufficient to cause breakdown and other rocks where additional energy is required to cause final breakdown.

Possibly a *third* class (Class 3) might also be considered where the *plasticity* of the 'rock' is such that an eventual terminal, or 'residual' strength is obtained as in soils.

An interesting side effect of 'stiff' testing is the predominance of shear failure and the relative absence of typical 'cataclasis' type Griffith failure, particularly where no-friction end plates are used. Figure 4.18 sketches the form that the failure takes in most brittle rocks.

The implication of these arguments is that Griffith/Irwin failure mechanisms are subordinate to shear failure mechanisms once the

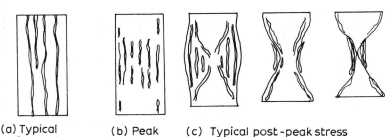

(a) Typical 'brittle' cataclasis type failure

(b) Peak stress

(c) Typical post-peak stress breakdown and collapse

Figure 4.18   Typical stiff testing specimen failure.

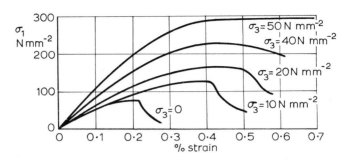

Figure 4.19 Confined tests on Tennessee Marble (*after* Wawersik and Fairhurst, 1970).

strength peaks and major crack coalescence commences. This contention is supported by the apparent tendency towards the third (plastic) class of deformation on the path to a post-peak residual stress during confined tests. The result of a series of triaxial tests on Tennessee Marble (Class 1) at confining pressures ranging from 3.5 to 50 N mm^{-2} described by Wawersik and Fairhurst (1970) are reproduced in Figure 4.19. Not only do the curves show a reduction in negative slope (following peak stress) with increasing confining pressure — and hence a reduction in brittle behaviour — but they also demonstrate an increase in the modulus of deformation during the loading process.

Wawersik and Fairhurst suggest a change from brittle to plastic behaviour in the region of 15 N mm^{-2} confining pressure, and at this pressure the form of the stress strain curve is indeed similar to those obtained during triaxial tests on soils. Since marble has a strong translation and twin gliding facility for plastic deformation within its constituent calcite crystals, this particular confining pressure level is low compared with that for a sandstone where the quartz crystals can only really deform cataclastically.

4.5 Effect of rate and duration of loading

Rate effects in the uniaxial unconfined loading of rock are usually considered to have a minor effect on observed strength and deformation characteristics at relatively *low* rates of loading and load amplitudes. Conversely loading rates are considered to be very important at high rates and high amplitudes. An important series of tests reported by Bieniawski (1970) and reproduced in Figure 4.20 in the form of complete stress/strain curves for sandstone (Class 1) puts

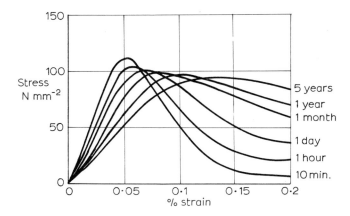

Figure 4.20 Stress-strain relationships for sandstone at various strain rates (*after* Bieniawski, 1970).

this into perspective. These test results show several interesting aspects of rate behaviour:

(a) Peak stress levels increase with increasing strain rate (Figure 4.21). Rinehart (1962) suggests that at impulsive rates (test duration 0 to 5 microseconds), compressive strengths may be as high as 10 times the low strain-rate strengths. This effect might be taken as a design constraint in certain instances since failure under impulsive loading will usually be imposed in tension by the change in sign of reflected compression pulses (see Section 4.11). Bieniawski (1970) suggests a specific relationship between σ and ϵ in an exponential form which would allow, from extrapolation, an estimate of strength at low strain rates.
(b) There is a similar reduction in modulus of elasticity with decreasing strain rate (Figure 4.21).
(c) The strain at strength failure, as might be expected, increases with decreasing strain rate and is accompanied by a reduction in slope of the stress/strain curve after failure.

This latter point, indicating as it does a transition from brittle (Class 1) behaviour at high rates of strain to plastic behaviour at very low rates of strain, is of fundamental importance in mine/pillar design. It implies that rock fractured at low rates of strain is more stable throughout its complete loading range than that subjected to high rates of strain.

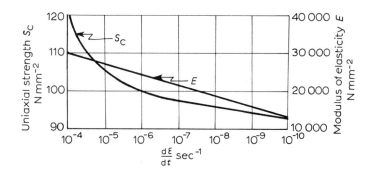

Figure 4.21 Influence of strain rate on uniaxial strength and modulus of elasticity of sandstone (*after* Bieniawski, 1970).

The results are also important in illustrating some aspects of *long term deformation* or *creep* in rocks. Time-dependent strain of a typical cylindrical rock specimen under a uniaxial maintained load is most simply described as primary creep with a logarithmic time function in the form (see Robertson, 1964)

$$\frac{d\epsilon}{dt} = A t^{-1} \qquad (4.19)$$

or

$$\epsilon_c = A \ln t, \; \epsilon = \epsilon_0 + \epsilon_c = \epsilon_0 + A \ln t \qquad (4.20)$$

where ϵ_c is the long term strain or creep
ϵ_0 is the short term or elastic deformation
and A is a constant, sometimes related to the stress in an approximate form $A = (\sigma/E)^n$, n being equal to 1 at low stresses and 2 at high stresses.

This relationship may be reproduced as a strain/time curve of the type shown in Figure 4.22, the general form of the equation having been derived empirically from experimental data. It describes a maximum creep strain that is much less than the initial elastic strain and which increases logarithmically with time. Creep is of practical significance only if ϵ_0 itself is large. The approach specifically excludes steady state or *secondary* creep ($\epsilon \propto t$) and increasing strain rate or *tertiary* creep ($\epsilon \propto t^n$), which are reported for ductile materials (Figure 4.23) but not for rocks except at high confining pressures and high temperatures. Possible exceptions to this rule are

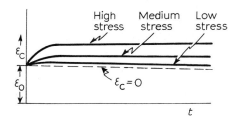

Figure 4.22 Typical creep curves in rock.

evaporites (see Dreyer, 1972) which exhibit significant ductility at high stresses.

The *mechanism* of creep is arguably similar to that of deformation under increasing stress. For instance, the volumetric strain/time characteristic associated with the creep strain characteristic has been shown to exhibit the same dilational features as the static volumetric strain/time curve (Figure 4.24). Thus after an initial minimum volume, microcrack propagation parallel to the maximum stress axis results in progressive lateral strain, and expansion in volume. This is only large enough to demand attention when the applied stresses are in the region of $0.7\,S_c$.

The Griffith crack deformation concept fits reasonably well as an explanation for creep deformation. In the case of cracks spreading under progressive load in a sample comprising part-randomly oriented voids, each crack movement will be accompanied by a stress

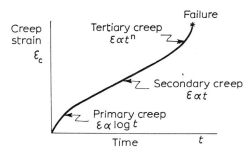

Figure 4.23 Complete creep curve in ductile materials.

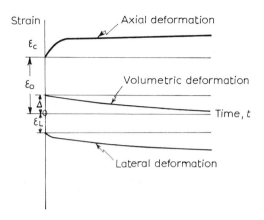

Figure 4.24 Effect of lateral dilation on creep strain.

relaxation at its tip, so tending to reconcentrate stress at the tip of an adjacent crack. Under increasing loading, local stress redistributions lead to an increasing density of crack propagation. In a *maintained* load situation, local stress redistribution leads to a reduction in the rate of increase of crack density with time which is reflected as a logarithmic or exponential strain/time relationship.

Since the rate of increase of *primary creep* strain does reduce with time, the strain history of a specimen undergoing increasing load must be influenced by the rate of loading. For instance, Figure 4.25 shows that if the elastic rock structure has an intrinsic deformation modulus E_0, an idealized no-creep stress/strain curve will take this same slope. A primary requirement in testing, therefore, is a

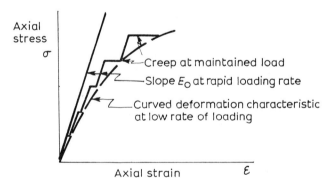

Figure 4.25 Effect of loading rate on deformation characteristic.

no-creep loading rate in a real rock, which rate must be arbitrarily fixed in any test for determining deformation/strength characteristics. If this rate is reduced, then time-dependent deformation will flatten the slope of the curve and lower the magnitude of the peak stress level since this latter must be related to the onset of uncontrollable crack coalescence, and through this to some *limiting strain*.

This latter point is controversial but logical, for if failure is due to the spread of cracks leading ultimately to breakdown of rock structure, it seems reasonable to propose that breakdown is strain, as well as stress-controlled. In fact, Bieniawski (1970) shows that the strain at failure increases by over 100 per cent with decreasing strain rate over the range of loading rates generating his complete stress-strain curves for sandstone (Figure 4.26). This graph implies the existence of some plasticity/ductility or other crack 'healing' mechanisms associated with creep. Bieniawski concludes that the long term stability of a rock in an intact or fractured state is determined by its previous loading history and particularly the rate of deformation to which it has been subjected.

An extreme example of rate effects during loading arises during *cyclic loading*. In the case of a rock loaded to a high stress level where some non-elastic creep strain might be expected, the unloading curve (short of failure) will not return to the origin but will exhibit some permanent deformation (ϵ_R). Repetition of the loading cycle (Figure 4.27a) will result in further deformation (ϵ_{R_1}, ϵ_{R_2} ... ϵ_{R_n}) until eventually strength failure will tend to occur at a peak stress level that is lower than the maximum peak stress sustainable at a higher strain-rate. The actual failure process is in some ways

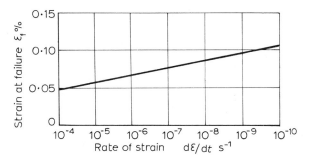

Figure 4.26 Influence of strain rate on strain at failure (*after* Bieniawski, 1970).

216 *Principles of Engineering Geology*

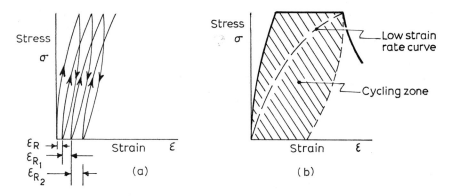

Figure 4.27 Deformation and failure under cycle loading.

analogous to low strain rate deformation and failure with reduced peak stress and extended strain (Figure 4.27b) as compared with high strain rate loading. The process is known as *fatigue failure* (see Burdine 1963; Hardy and Chugh, 1970; Attewell and Farmer, 1973) and can be defined in terms of the *number of cycles* to failure and the *peak or mean stress* for various materials.

A series of *S–N* curves (maximum stress at failure *S*/number of cycles *N*) can be obtained from tests on rock samples at various stress levels — the fatigue strength/compressive strength (S/S_c) ratio being related to the number of cycles to failure as shown in Figure 4.28.

Most rocks (and indeed concrete) have a fatigue limit in the region of 65 to 70 per cent of S_c, below which failure will not occur. This limit is reasonably satisfied by the deformation hypothesis

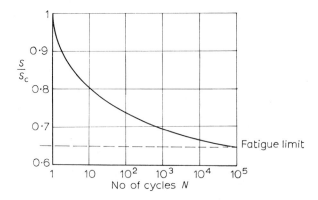

Figure 4.28 Typical *S/N* curve for rock subjected to cyclic loading.

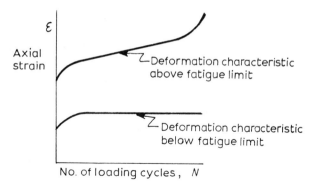

Figure 4.29 Strain-time curves during cyclic loading.

based on crack propagation, and the basic concept permits a comprehensive treatment of progressive creep and cyclic deformation.

Thus, the complete stress-strain curves for a rock, such as are shown in Figure 4.20 at various loading rates, will define (together with the applied stress) both the basic shape and slope of the loading curve and the magnitude of permanent (or recoverable) strain during the *first* cycle. Unless the recovery of the material is elastic and it takes place at the same rate at which the load is removed, deformation by the second and subsequent cycles will be added to the residual deformation from earlier cycles. This cumulative deformation in the case of a composite flawed rock will be affected by local stress amplitudes, stiffness and crack density distributions. In a situation with a high-degree of recovery, that is, *below* an arbitrary *fatigue limit*, successive strain increments will decrease logarithmically. In a situation where the peak stress is above the fatigue limit, it is evident that cumulative deformation will eventually lead to failure. It has been shown, as in Figure 4.29, that this failure path follows stage-by-stage a logarithmic to a constant strain rate to an increasing strain rate path prior to failure, so indicating a possible strain-hardening effect before accelerating to failure. This cyclic curve in the ϵ/N plane is directly analogous to a generalized 'static' creep curve and we may therefore term the generative process 'cyclic creep'. In an interesting contribution Haimson and Kim (1971) have confirmed that the strain at fatigue failure (when this occurs) is defined by the point of intersection of the upper limit of the cyclic stress and the unloading leg of the complete stress/strain

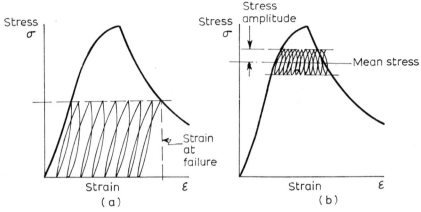

Figure 4.30 Intersection of cyclic loading regimes and complete stress-strain curves for a rock specimen.

curve (see Figure 4.30). This work was carried out at constant cycling rates of 1 to 2 Hz. The present authors (Attewell and Farmer, 1973) adopted a wider frequency range from 1 to 20 Hz for their cyclic studies on rock. Their results reinforce the creep analogy by showing that fatigue failure is related to the elapsed cycling time at a given peak/mean stress rather than to the number of stress reversals directly.

4.6 Deformation and failure of rocks in triaxial compression

So far, only uniaxial deformation and failure have been considered in detail, although some effects of confinement on the complete stress/strain curve have been noted in Figure 4.19. Although it is convenient to utilize uniaxial properties as an index of rock strength and deformation, rocks are only rarely subjected to true uniaxial stress situations in nature. In their natural environment, the state of stress acting on rocks *in situ* may usually be resolved in terms of three orthogonal principal stresses σ_1, σ_2, σ_3, or by any combination of three orthogonal normal stresses and three shear stresses acting on planes perpendicular to the normal stresses (see Section 2.5).

The stress situation in the earth's crust may most simply be represented by a vertical stress $\sigma_z = \gamma z$, where γ is the unit weight and z is the depth, and also by a homogeneous lateral confining pressure $\sigma_x = K\sigma_z$. A two-dimensional, plane stress situation simplifies analysis and analogous test procedures if the vertical and lateral normal stresses are assumed to be principal stresses.

Triaxial testing methods attempt to simulate and then restore the original confined stress conditions on cylindrical specimens by subjecting a jacketed specimen, immersed in an oil reservoir, to a lateral confining pressure. An axial (vertical) pressure is then applied through rigid platens at each end of the specimen in much the same way and at the same rates as in an unconfined compression test, whilst at the same time the deformation and ultimately the strength are recorded. The procedure follows much the same pattern as that adopted for soil testing (see Section 2.9). However, whereas soil specimens are first constrained by a hydrostatic confining pressure (representing confinement at depth) followed by the application of a deviator stress along the specimen axis, the standard type of higher pressure rock triaxial cell requires that the confinement be applied first of all around the curved surface of the rock specimen and that the first increment of independently-applied axial pressure to the flat end surfaces simply serves to bring the state of stress in the specimen up to hydrostatic. This 'difficulty' is offset to some extent by the use of a newer type of balanced-ram triaxial cell. Triaxial cell confining pressures are usually limited to the range 0 to 70 N mm^{-2}, sufficient for most engineering applications. Cell pressures up to 3000 N mm^{-2} (virtually up to Moho pressures) are available for small specimen 'structural petrological and petrographic' work. Cell facilities range in complexity from the simple Hoek (Hoek and Franklin, 1968) type shown in Figure 4.31 for undrained tests, the stiff membrane in the cell restricting the deformation after peak stress has been achieved, to more comprehensive types of cell for drained tests and controlled porewater pressure tests.

The principal effects of confinement are twofold: (a) to raise the peak strength of the rock and also its modulus of elasticity; (b) at higher total pressures to increase the ductility or plasticity of the specimen so that at a sufficiently high confining pressure rocks literally 'flow' at peak stress acceptance levels rather than fracturing (Figure 4.19). There is, for example, a brittle-ductile transition in a 20 per cent porosity sandstone at a confining pressure of about 1 K bar (100 N mm^{-2}). Ductility is manifest as a 'barrelling' of the specimen and the development of Luders lines on the exposed surfaces.

It is interesting to note that, whereas the first effect occurs only under confinement, increased ductility can also be obtained by reducing the rate of strain or by increasing the temperature

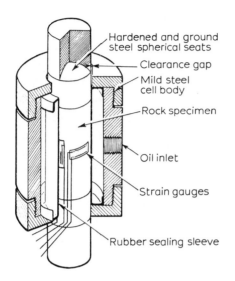 

Figure 4.31 Hoek—Franklin triaxial test cell (*Reproduced by courtesy of Robertson Research International Ltd.*)

(Figure 4.32), in the latter cases accompanied by decreasing strength and modulus of elasticity.* Significantly, the degree of confinement, the rate of strain and the ambient temperature are all conditioning factors in the long term orogenic processes resulting in large scale deformations in the earth's crust.

The borderline between brittle and ductile behaviour in rocks, sometimes known as the *brittle-ductile transition,* has been explored by many workers. Heard (1963), in his work in Solenhofen limestone, suggested the following limits for classification:

Brittle behaviour	— < 3% strain
Brittle-ductile transition i.e. semi-brittle	— 3%–5% strain
Ductile deformation	— > 5% strain

*The increasing ductility of rocks associated with increasing temperature can be described by the Arrhenius equation:

$$\dot{\epsilon}_c = A \exp -\left(\frac{Q}{RT}\right)$$

where A is a constant, Q is the activation energy in electron volts, R is the gas constant, T is the absolute temperature and $\dot{\epsilon}_c$ the creep strain rate.

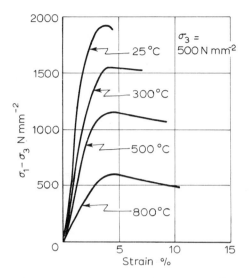

Figure 4.32 Stress-strain curves for granite at high temperature (*after* Griggs *et al*, 1960).

Materials vary tremendously in their response (Table 4.5), depending sometimes very finely upon the interaction between confining pressure, temperature, and any environmental fluids. For example, in Solenhofen limestone, a change in confining pressure from 75 000 kN m^{-2} to 105 000 kN m^{-2} at 150°C means a complete transition from brittle to ductile behaviour. From a surface foundation engineering point of view, it is unlikely that the pressures of confinement will be sufficient, together with any imposed loading, to induce a semi-brittle state upon any originally brittle rock. However, such a possible transition may have to be considered in connexion with tunnelling at depths greater than say 1800 to 3000 m (Table 4.6), and allowance made for possible 'flow deformation'.

Note that the assumption of brittle-to-ductile transitions at Table 4.6 pressures *hydrostatically* enforced may not always be correct. Considerable *differential stress* may also be required to induce 'flow'.

Price (1966) cites a typical example to show how temperature is necessary to induce ductility at depth. If a pressure increase of 22.6 kN m^{-2} per m of depth is assumed, then according to the experimental evidence of Griggs *et al* (1960), the igneous rock should remain brittle down to depths of 23 000 m *provided that the*

Table 4.5 Effect of confining pressure and temperature on rock deformation

Rock	Confining pressure σ_3 kN m^{-2}	$\sigma_1 - \sigma_3$ kN m^{-2}	Temperature °C	Strain %	Ref.
Coal	28 000	140 000	20° room	up to 4% elastic	Hobbs (1960)
Oil Creek Sandstone (well compacted)	200 000	1 100 000		3% elastic remains semi-brittle	Handin and Hager (1958)
Quartz	500 000	2 200 000	20–25°C	brittle	Griggs and Bell (1938)
Granite	500 000	200 000	25°C	5%	Griggs et al (1960)
Basalt	500 000	170 000	300°C	ductile	Griggs et al (1960)
Blair Dolomite	200 000			remains semi-brittle	Griggs et al (1960)
Solenhofen limestone	100 000			becomes semi-brittle	Griggs et al (1960)
Solenhofen limestone	75 000		150°C	brittle	Griggs et al (1960)
Solenhofen limestone	105 000		150°C	ductile	Griggs et al (1960)

temperature remains at approximately room. However, assuming a geothermal gradient of 1°C per 30 m, the temperature of 750°C (well in excess of Griggs' 350°C) will, together with the pressures associated with the depth, induce ductility.

At 350°C (equivalent to 11 000 m and 240 000 kN m^{-2}), from the work of Handin and Hager (1958) we would expect many *rocks* to behave in a *ductile manner* and the yield points to be generally depressed. In terms of an ultimate compressive strength, quartzitic igneous rocks have been found to exhibit flow and failure at confining pressures of 200 000 kN m^{-2} and temperatures in the range 200 to 400°C, the strength reduction being about 100 per cent.

Table 4.6 Brittle/Ductile transition points at 20°C (room temperature)

Rock	Confining pressure kN m^{-2}	Equivalent depth m
Igneous rocks (Bridgeman, 1952)	480 000	23 000 to 27 500
Compact sedimentary	205 000	9 000
Normal sedimentary	105 000	4 500
Porous sedimentary	10 500 to 52 000	450 to 1 800

Farmer (1968) has suggested the introduction of a brittle-to-ductile transition condition into a strength criterion, proposing a curvilinear relationship for stronger sedimentary rocks:

$$\sigma_1 \text{(at failure)} = S_c + \exp[m(\sigma_3 - \sigma_{3L})] \quad (4.21)$$

where m is a constant
S_c is the unconfined compressive strength
σ_3 is the confining pressure
σ_{3L} is the limiting confining pressure at room temperature for a brittle-to-ductile transition.

This equation can be extended to relate the ultimate compressive strength in rocks to depth of burial, taking into account geothermal gradient and making the following assumptions:—
A reduction of 10 to 15 per cent in strength for each 100°C increase in temperature and an equivalent 100 000 kN m^{-2} increase in confining pressure, or 3000 m increase in depth. Then

$$S_u = S_{RT} - \frac{15z}{3000} \quad (4.22)$$

where z = depth in m,
S_{RT} = ultimate compressive strength at room temperature, kN m^{-2},
S_u = ultimate compressive strength at any temperature, kN m^{-2}.

An increase in pore pressure affects the onset of ductility caused by increasing confining pressures. Griggs (1941) found that a 10 per cent solution of sodium carbonate at 400°C reduced the strength of quartz by 80 per cent. Also, even in water, a high degree of preferred orientation can be induced by shearing quartz at a temperature of 600°C and under an unjacketed porewater pressure of 100 MN m^{-2}.

It would seem that for porous rocks in the 'dry' state, permanent strain can develop at pressures equal to or greater than the pressures which were effective at their maximum depth of burial. In arenaceous rocks, this further compaction takes place by cataclastic flow (relative displacement of grain boundaries) with granulation and point fracturing. In limestones, there is also twinning on $e(01\bar{1}2)$ and $f(02\bar{2}1)$, and translation gliding on $c(0001)$ in the calcite.

The effect of solutions in conjunction with intense point stresses is to cause local plastic flow without elevated temperatures or high overall applied stresses.

Using evidence of undulatory extinction in quartz as an indication of local plastic deformation, Price (1966) concludes that in predominantly quartz-bearing rocks subjected to confining pressures of around 140 MN m^{-2} and temperatures around 250°C, plastic deformation can take place at depths greater than the 6000 m suggested by these environmental conditions. Compacted, non-porous limestones he suggests would deform plastically at shallower depths of around 2400 m to 3000 m.

4.7 Failure criteria for rocks

We have seen that rocks fracture or yield when they are deformed in *simple* compression, shear or tension. The purpose of a generalized failure criterion is to describe the stress (or strain) level at which failure occurs together with the orientation and location of any planes of failure in a rock body which is subjected to a more complicated stress system.

Failure criteria are summarized by Nadai (1950), Jaeger (1969), and Jaeger and Cook (1969) among others and they have been considered in relation to soils in Section 2.9. The simplest criterion is the maximum *shear stress criterion* which proposes that fracture occurs when the *maximum shear stress* developed in a material exceeds the shear strength of that material. The plane on which the maximum shear stress acts will be normal to the plane containing the major and minor principal stresses (σ_1 and σ_3 or in terms of the 'earth' convention, σ_z and $\sigma_x = \sigma_y$), and it can be shown (see Section 2.5 and Figure 2.7) that on this plane the *shear stress* will be equal to:

$$\tau = \frac{(\sigma_1 - \sigma_3)}{2} \sin 2\theta \qquad (4.23)$$

and the *normal stress* to:

$$\sigma_n = \frac{(\sigma_1 + \sigma_3)}{2} + \frac{(\sigma_1 - \sigma_3)}{2} \cos 2\theta \qquad (4.24)$$

It follows from the trigonometrical functions in the above equations that the *maximum* shear stress will occur on a plane at an angle of 45° to σ_1 and normal to the $\sigma_1 \sigma_3$ plane, and will be equal to $(\sigma_1 - \sigma_3)/2$.

The criterion is deficient in several respects, but the principal disadvantage is the implication that, independent of the confining pressures, the shear strength is always proportional to the stress difference and is not affected by the magnitude of the normal stress on the failure plane. This is implicitly a *yield* criterion and is considered as such in the next section.

It also implies that the uniaxial compressive strength and tensile strength are equal. These objections are eliminated in an adaptation of the *maximum shear stress* criterion which assumes that shear strength is related to normal stress across the plane, through a constant μ, which is analogous to a *coefficient of friction* along the plane (see Section 2.7). This criterion, usually known as the Coulomb–Mohr criterion, proposes, therefore, that fracture occurs on a plane when the shear stress along the plane is equal to:

$$\tau = c + \mu \sigma_n \tag{4.25}$$

where $\mu = \tan \phi$ is called the *coefficient of internal friction*, and c is the cohesion or shear strength of the rock when there is no normal pressure on the shear plane.

The Coulomb–Mohr criterion can be represented geometrically on a $\tau - \sigma_n$ plot as in Figure 2.27 using the Mohr method of representing plane stress as circle, having a centre $(\sigma_1 + \sigma_3)/2$ and radius $(\sigma_1 - \sigma_3)/2$. This limiting shear strength approach is considered in greater detail in Chapter 2.

The Coulomb–Mohr criterion works well for frictional materials, such as granular soils, and quite well for many intact rocks, giving a reasonable indication of the stresses required to cause shear failure and also the direction of that failure. Together with a general parametric relationship of the form $\tau = f(\sigma_n)$ (describing a curved envelope as in Figure 4.35) it operates as an empirical description of rock behaviour, and although it does not differentiate between failure through yielding and failure due to progressive brittle shear fracture, it does allow for different compressive, shear and tensile strengths. This relative versatility of the criterion can be illustrated by expanding the Coulomb–Mohr equation for a failure plane inclined at θ to the direction of σ_3:

$$c = \tau - \sigma_n \tan \phi$$
$$= \frac{\sigma_1 - \sigma_3}{2} \sin 2\theta - \left[\frac{\sigma_1 + \sigma_3}{2} + \frac{\sigma_1 - \sigma_3}{2} \cos 2\theta \right] \tan \phi$$

$$\tag{4.26}$$

or $2c = (\sigma_1 - \sigma_3)[\sin 2\theta - \cos 2\theta \tan \phi] - (\sigma_1 + \sigma_3) \tan \phi$

$$\tag{4.27}$$

and since $\tan 2\theta = -1/\tan \phi = \dfrac{-1}{\mu}$ fracture will occur when:

$$2c \leqslant (\sigma_1 - \sigma_3)(\mu^2 + 1)^{\frac{1}{2}} - (\sigma_1 + \sigma_3)\mu$$
$$\leqslant \sigma_1[-\mu + (\mu^2 + 1)^{\frac{1}{2}}] - \sigma_3[\mu + (\mu^2 + 1)^{\frac{1}{2}}] \qquad (4.28)$$

Thus, for failure in uniaxial compression where $\sigma_3 = 0$ and $\sigma_1 = S_c$ (compressive strength), and for tension where $\sigma_1 = 0$, $\sigma_3 = -S_t$ (tensile strength), we have the following relationships:

$$S_c = 2c/[(\mu^2 + 1)^{\frac{1}{2}} - \mu] \qquad (4.29)$$
$$S_t = 2c/[(\mu^2 + 1)^{\frac{1}{2}} + \mu] \qquad (4.30)$$

giving

$$\frac{S_c}{S_t} = \frac{(\mu^2 + 1)^{\frac{1}{2}} + \mu}{(\mu^2 + 1)^{\frac{1}{2}} - \mu} \qquad (4.31)$$

and

$$\frac{\sigma_1}{S_c} - \frac{\sigma_3}{S_t} = 1 \qquad (4.32)$$

or

$$\sigma_1 = S_c + \frac{S_c}{S_t}\sigma_3 \qquad (4.33)$$

These relationships seem to be confirmed in practice. A series of triaxial test observations collected by Hoek (1965) and plotted in Figure 4.33 suggests the general lines of a *triaxial rock classification* with stiff (brittle) rock having a high μ and high S_c/S_t ratio, and soft (yielding) rocks having a low S_c/S_t ratio.

A basic assumption in the above analysis is that the rock must fail in shear, although it can be shown by experiment that this in fact is only certain beyond the peak stress acceptance level, or the stress/strain level at which rocks disintegrate.

Griffith's theory of brittle fracture offers a more fundamental approach to the mechanics of rock failure. As outlined earlier (Section 4.3), this theory is based on the concept of progressive failure through microcrack extension in a direction normal to the

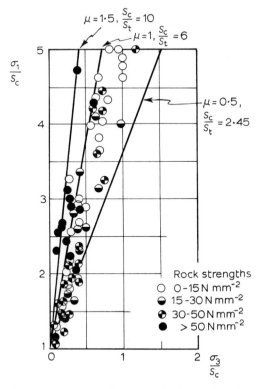

Figure 4.33 Collected triaxial test data (*after* Hoek, 1966).

axis of major principal stress. These cracks are propagated through a process of tensile failure at the crack tip even under conditions of all-round compression, and a solution for a *biaxial* state of stress may be derived (Jaeger, 1969) by considering the forces acting on a plane, randomly-orientated elliptical crack. (The case of directionally orientated cracks is considered in Section 5.11).

The general solution indicates that rock will fail if

$$\sigma_3 = S_t$$

or if (4.34)

$$S_t = \frac{(\sigma_1 - \sigma_3)^2}{8(\sigma_1 + \sigma_3)}$$

The general agreement with the Coulomb–Mohr derivative can be seen if $\sigma_3 = 0$, for then $\sigma_1 = S_c = 8 S_t$ or $S_c/S_t = 8$. The quadratic relationship implied in this equation can be expressed as a series of Mohr circles:

$$\left[\sigma_n - \frac{(\sigma_1 + \sigma_3)}{2}\right]^2 + \tau^2 = \frac{(\sigma_1 - \sigma_3)^2}{2} \qquad (4.35)$$

which by substitution and partial differentiation with respect to $(\sigma_1 + \sigma_3)/2$ produces a shear failure locus or envelope in the τ, σ_n plane having an equation (Murrell, 1963):

$$\tau^2 - 4S_t \sigma_n = 4S_t^2 \qquad (4.36)$$

This is a parabolic equation generating the curved envelope shown in Figure 4.34.

A similar criterion, which is probably more suited to a weaker, more compliant rock and where closure under normal compression of the ellipsoidal cracks leads to friction on crack surfaces, has been proposed by McLintock and Walsh (1962). Their analyses may be expressed as:

$$4S_t = [(\sigma_1 - \sigma_3)(1 + \mu^2)^{\frac{1}{2}}] - \mu(\sigma_1 + \sigma_3) \qquad (4.37)$$

or

$$\tau = \mu \sigma_n + 2S_t \qquad (4.38)$$

Thus, both the Griffith and Coulomb–Mohr approaches can be used to generate criteria for failure of rocks in terms of the imposed stress situation, and their simple rock properties.

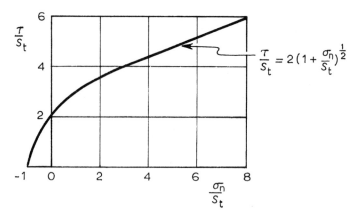

Figure 4.34   Representation of the Griffith failure criterion on shear stress–normal stress axes.

The Griffith criterion owes much of its attraction to the fact that it originates from a fundamental appreciation of the structural controls on failure. By way of contrast, the formulation of the Coulomb–Mohr criterion depends in no way upon the structural characteristics of the material. That the former criterion can be interpreted in terms of the latter, and that the latter criterion appears to describe the mechanical behaviour of rock in low to medium stress environments quite satisfactorily, does imply that any design problem in intact rock could be tackled with a reasonable degree of confidence.

4.8 Yield criteria

Failure is not a uniquely definable state. A *rock* confined at low hydrostatic pressure will fail catastrophically when a necessary and sufficient deviator stress is applied to it. Failure in this instance implies a complete loss of load-bearing capability. With an increasing non-deviatoric component of stress there will be a concomitant increase in the necessary deviator stress for failure, but the rate of increase will reduce as the level of all-round confinement is raised. Ultimately, the deviator stress will remain constant and quite independent of any further increases in the confining pressure. Having developed the Mohr envelope on a decreasing slope up to this point it thereafter continues as a horizontal line parallel to the normal stress axis. The rock has undergone a brittle-ductile transition of a progressive nature, the externally-applied pressures being accommodated internally by an increasing density of intracrystalline twin and translation gliding. Failure, under these conditions, could be specified only in terms of some *maximum acceptable strain* along the axis of deviator stress application.

The development of ductility, as described above for rock, is also shown in Figure 4.35. The symbols that are used to define the relationship between the curvi-linear, quasi-parabolic *failure* envelope and the linear *ultimate* envelope are just the symbols used by Bishop (1967) to define his *brittleness index* I_B for soils, where

$$I_B = \frac{\tau_f - \tau_r}{\tau_f} \quad (4.39)$$

In this equation, we may regard τ_f and τ_r as being the shear stress at failure and the *residual* shear strength respectively for a particular normal pressure on a potential and actual shear plane. Since I_B is

230   Principles of Engineering Geology

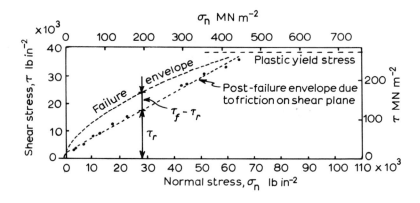

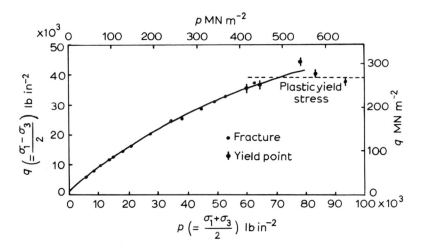

Figure 4.35   Failure envelopes for Darley Dale sandstone (generally *after* Murrell, 1966, but with some changes and additions).

rather intimately related to the normal pressure, the index can be regarded as a soil *behaviour* parameter rather than a *soil* property. Bishop (1972) discusses the influences of pore pressure conditions and time upon I_B of undisturbed and remoulded clays having different consolidation histories and sensitivities.

The greatest advances in determining the laws governing the failure of elastic behaviour in crystalline solids have been made with respect to metals, but it is suggested from time to time that the various criteria describing yield may also be applicable to polycrystalline rocks and soils. The *Tresca Law* states that plastic deformation is initiated in a polycrystalline material when the maximum shear stress attains a critical value, that is:

$$\sigma_1 - \sigma_3 = \sigma_0 \tag{4.40}$$

In this equation, σ_0 is a physical property of the material and depends on composition, structure, deformation history, temperature, and rate of straining. The Tresca law appears to provide a reasonable description of the behaviour of some undrained soils. For drained soils, Bishop (1966, 1972) suggests an *extended Tresca law* which is written in effective stress notation and which allows for some increase in yield resistance with normal stress resulting from the frictional characteristics of the material:

$$\sigma_1' - \sigma_3' = \alpha \left\{ \frac{\sigma_1' + \sigma_2' + \sigma_3'}{3} \right\} \tag{4.41}$$

where α is a variable parameter.

The Von Mises–Huber–Hencky law (see Hill, 1950) states that plastic deformation is initiated in polycrystalline materials when the distortional strain energy attains a critical value, that is, when

$$(\sigma_2 - \sigma_3)^2 + (\sigma_3 - \sigma_1)^2 + (\sigma_1 - \sigma_2)^2 = 2\sigma_0^2 \tag{4.42}$$

with its full derivation being given in Jaeger and Cook (1969). As before, this law provides a good approximation to the behaviour of metals, some rocks and undrained soils. For drained soils, an *extended* Von Mises law may be written:

$$(\sigma_2' - \sigma_3')^2 + (\sigma_3' - \sigma_1')^2 + (\sigma_1' - \sigma_2')^2 = \frac{2\alpha^2}{9} (\sigma_1' + \sigma_2' + \sigma_3') \tag{4.43}$$

It may readily be shown on a Mohr diagram that the original Von Mises law implies an initiation of plastic deformation when the maximum shear stress $(\sigma_1 - \sigma_3)/2$ attains a critical value which depends on the intermediate principal stress σ_2 and which lies between $\sigma_0/2$ (corresponding to the case when $\sigma_2 = \sigma_1$ or $\sigma_2 = \sigma_3$) and $\sigma_0/(3)^{\frac{1}{2}}$ (corresponding to $\sigma_2 = (\sigma_1 + \sigma_3)/2$) but which is otherwise independent of the hydrostatic stress. In the commonly assumed earth pressure situation (or the triaxial test simulation of earth pressure) where $\sigma_2 = \sigma_3$, $\sigma_2' = \sigma_3'$, the Tresca and Von Mises laws both reduce to the same form:

$$\sigma_1 - \sigma_3 = \sigma_0$$

$$\sigma_1' - \sigma_3' = \alpha \frac{(\sigma_1' + 2\sigma_3')}{3} \tag{4.44}$$

in the respective plane and extended cases. Triaxial tests in which σ'_2 is substantially different from σ'_1 and σ'_3 will therefore be required to differentiate between the Tresca and Von Mises criterion. An example is Wood's (1958) plane strain test.

The relative validities of the extended Tresca and Von Mises criteria compared with the Mohr–Coulomb criterion when applied to the behaviour of soils have been examined by Bishop (1966, 1972). These criteria can best be appraised by writing them in similar form:

Mohr–Coulomb: $\dfrac{\sigma'_1 - \sigma'_3}{\sigma'_1 + \sigma'_3} = \sin \phi'$ (4.45)

Tresca: $\dfrac{\sigma'_1 - \sigma'_3}{\sigma'_1 + \sigma'_3} = \left\{ \dfrac{1}{3} + \dfrac{2}{\alpha} - \dfrac{2}{3} b \right\}^{-1} = \sin \phi'$ (4.46)

Von Mises: $\dfrac{\sigma'_1 - \sigma'_3}{\sigma'_1 + \sigma'_3} = \left\{ \dfrac{1}{3} + \dfrac{2}{\alpha} (1 - b + b^2)^{1/2} - \dfrac{2}{3} b \right\}^{-1} = \sin \phi'$ (4.47)

where

$$b = \dfrac{\sigma'_2 - \sigma'_3}{\sigma'_1 - \sigma'_3}$$

and by plotting ϕ'_r against b in Figure 4.36 using a value of α computed for $\phi' = 39°$. The theoretical results for the yield criteria compare badly with actual test results quoted by Bishop (1972) on a sand (initial porosity $n = 39$ per cent) having an equivalent friction angle and using an independent stress control apparatus.

Except for a limited range of b values, it is evident that neither yield criterion describes actual test data as well as does the Coulomb–Mohr criterion. An alternative approach based on plastic theory has been considered by Jaeger and Cook (1969) and could eventually form a preferred basis for a rock/soil yield criterion. The reader will readily appreciate, however, that 'failure' might equally be defined on the basis of a limiting *strain* criterion, but developments in this direction have not progressed sufficiently to justify detailed consideration at this time.

4.9 Rock dynamics

With the exception of loading rate effects and cyclic loading, we have not considered any specific aspects of the dynamic properties of rocks. These are, however, of considerable importance in engineering

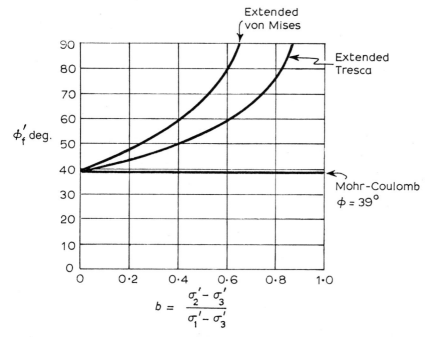

Figure 4.36 Predicted peak strengths using failure and yield criteria (*after* Bishop, 1972).

geology, having applications ranging from the crushing of aggregate through to the design of rock blasts and the control and analysis of ground wave propagation.

Probably the word *dynamic* is slightly confusing in this connection since what is invariably implied is a rapid or *impulsive* form of loading which results from impact or from explosive forces on or in rock. The result of this type of loading has been studied in detail by the authors (see Attewell, 1962; Attewell and Farmer, 1964a,b).

When an explosive is detonated in contact with a rock, either on the surface, or buried deep in a borehole, very high pressures are initiated at the explosive-rock interface. The magnitude and effect of these pressures will depend on the degree of confinement, the seismic velocity of the rock, the detonation velocity of the explosive, the quantity of explosive, and the intimacy of contact between the explosive and the rock. Given a reasonably good contact and velocity match, most of the explosive energy (or possibly impact energy from an external source) will pass into the rock as an unstable shock-wave front travelling at a greater speed than the sonic velocity of the rock.

234 *Principles of Engineering Geology*

Normally a true shock wave is only formed in rock at very high pressures when the rock behaves rather like a compressible fluid with zero shear strength. The shock wave develops a very steep leading edge and rapidly decays through an unstable form of plastic wave into a stable elastic or *seismic* wave. The shock wave and plastic wave zones are limited in areal extent and they can normally be identified as a zone of pulverized or fractured rock adjacent to a confined blast. The elastic wave, propagating at sonic velocity, transmits insufficient energy in compression to fracture the material in its path, but it may cause tensile fracturing upon reflection at a stress free interface (see Section 4.11).

4.10 Wave transmission through rocks

The laws governing transmission of seismic waves through rocks may be developed by considering rock as a deformable elastic body, but able to withstand a force of short duration. The simplest one-dimensional case is that of a rod, suspended in air and impacted at one end. The effect of this impact, which may be tensile or compressive, will be to extend or compress the element of material at the struck end of the rod and to impart a velocity to the particles in the rod. Temporarily, the rest of the rod will be at rest due to the relatively short period of the impact, but progressively, each adjacent element in turn will be affected by its neighbour causing a tensile or compressive stress wave to travel along the bar.

The length of bar subjected to compression or tension (that is, the wavelength λ of the disturbance) will be equal to the product of the velocity of the compressive wave front (c_p) and the duration of the impact t:

$$\lambda = c_p\, t \qquad (4.48)$$

The mechanics of the wave motion can be analysed relatively easily through a simple energy balance equating the input energy (W_i) to the stored (W_s) and kinetic (W_k) energy in the wave front:

$$W_i = W_s + W_k$$

or

$$\left(\frac{\sigma\lambda}{E}\right)\sigma = \frac{1}{2}\frac{\sigma^2}{E}\lambda + \frac{1}{2}\rho\lambda v^2 \qquad (4.49)$$

whence

$$v = \sigma(\rho E)^{-1/2} \qquad (4.50)$$

where ρ is the density of the rod material,
       v the amplitude of particle velocity of particles in the wave front,
       σ is the pressure amplitude of the wave,
and E is the modulus of elasticity of the rod material.

This simple relationship for a one-dimensional plane wave front can be expanded by considering the change in momentum of the wave front during time t, thus:

$$\sigma t = \rho \lambda v - 0 \tag{4.51}$$

and substituting for λ and v

$$\sigma t = \rho c_p t \sigma (\rho E)^{-\frac{1}{2}} \tag{4.52}$$

whence

$$c_p = (E/\rho)^{\frac{1}{2}} \tag{4.53}$$

Thus, one-dimensionally, tensile or compression wave velocity c_p is a function only of the modulus of elasticity and density of the rod. A combination of equations 4.50 and 4.53 can therefore be used to establish several general relationships. The strain in the rod, ϵ, for example, is related to v:

$$\epsilon = \frac{\sigma}{E} = v \left(\frac{\rho}{E}\right)^{\frac{1}{2}} = \frac{v}{c_p} \tag{4.54}$$

These relationships have considerable application in the measurement and interpretation of ground vibration. This aspect of rock dynamics is extended in Section 7.13.

The view of wave motion as a one-dimensional phenomenon is, of course, a little limited in practical applications of engineering geology interest. Any interpretation of wave transmission through a body of rock must also take into account the wave motion in a transverse direction to that of the wave travel and also at a boundary. So far, only a special case of compressional and tensile body wave or P-wave motion has been considered. There are in fact two basic types of elastic waves, *body* waves which travel through the interior of a rock mass and *surface* waves which can only travel through the immediate surface layers.

Body waves comprise two discrete components: compression or P-waves and shear or S-waves. P-waves induce longitudinal oscillatory particle motions in the direction of transmission and involve volumetric compression/dilation and shear. S-waves vibrate in a vertical plane normal to the axis of transmission and involve no compression or dilation. The velocity of the S-wave is always less than the P-wave (usually $c_s \simeq \frac{2}{3} c_p$) and of course the S-waves can only exist in a solid capable of resisting distortion.

The relationship between elastic constants and the P- and S-wave velocities in a solid medium can be determined from the equations of motion for an elastic solid, solved by Timoshenko and Goodier (1952) among others, in the form:

$$c_p = \left(\frac{K + 4G/3}{\rho}\right)^{\frac{1}{2}} = \left(\frac{E(1-v)}{\rho(1+v)(1-2v)}\right)^{\frac{1}{2}} \quad (4.55)$$

$$c_s = \left(\frac{G}{\rho}\right)^{\frac{1}{2}} = \left(\frac{E}{2\rho(1+v)}\right)^{\frac{1}{2}} \quad (4.56)$$

where K is the bulk modulus,
  G and E are the moduli of rigidity and elasticity respectively,
and v is Poisson's ratio.

The difference between equation 4.55 and the one-dimensional case of a thin bar derived in equation 4.53 is analogous to the difference between one- and three-dimensional static compression of an elastic body (see Section 2.8). The difference in this case is not great, and if v is assumed to take a value of 0.33, then equations 4.55 and 4.56 become

$$c_p = 1.2 \left(\frac{E}{\rho}\right)^{\frac{1}{2}} \quad (4.57)$$

$$c_s = 0.61 \left(\frac{E}{\rho}\right)^{\frac{1}{2}} \quad (4.58)$$

indicating a velocity ratio $c_s/c_p = 0.51$, which is rather lower than the actual case. Relatively stable relationships have been developed from a statistical analysis of measured values of c_p, ρ and E. Figure 4.37 derived by Farmer (1968) from the work of Talwani and Ewing (1960) and of Judd and Huber (1962) illustrates this point, and Table 4.7 gives a range of typical quoted values of c_p. A strong correlation between c_p and unconfined compressive strength (Elvery, 1973) also forms the basis for a laboratory and field test system using ultrasonic pulse transmissions to monitor rock and concrete strengths.

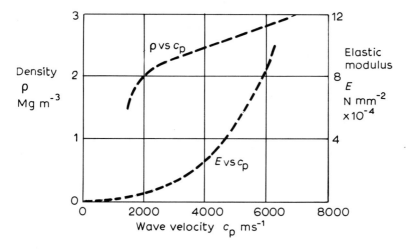

Figure 4.37 Relationship between ρ, E and c_p (*after* Farmer, 1968).

More accurate values of c_p, c_s for specific minerals, as quoted by Rzhevsky and Novik (1971), are given in Table 4.8.

Since the P-wave and S-wave velocities in rock can be measured with reasonable ease and accuracy, they provide a means of determining the modulus of elasticity and Poisson's ratio for the rock. Equations 4.55, 4.56 can be expressed in terms of the c_s/c_p ratio as:

$$E = \rho c_s^2 \left[\frac{3 - 4(c_s/c_p)^2}{1 - (c_s/c_p)^2} \right] \qquad (4.59)$$

Table 4.7 Typical ρ and c_p values for rocks

Rock	ρ Mg m^{-3}	c_p m s^{-1}
Granite	2.7	3000–5000
Basalt	2.85	4000–6000
Dolerite	2.9	4000–7000
Gabbro	2.9	4000–7000
Sandstone	2.3–2.7	1500–3500
Siltstone	2.6	1500–3000
Shale	2.3	1500–2500
Limestone	2.3–2.8	2500–5000
Clay	1.5–2.0	1200–2500
Sand	1.4–2.0	300–1200
Marble	2.7	3500–6000
Quartzite	2.65	5000–6500

Table 4.8 c_p, c_s values for minerals

Mineral	ρ Mg m^{-3}	E N mm^{-2} × 10^{-4}	ν	c_p km s^{-1}	c_s km s^{-1}	c_s/c_p
Albite	2.61	7.5	0.27	6.02	3.39	0.56
Augite	3.16	14.4	0.24	7.33	4.28	0.58
Biotite	3.10	7.0	0.28	5.36	3.00	0.56
Calcite	2.71	8.5	0.28	6.32	3.50	0.55
Corundum	4.03	43.9	0.22	11.15	6.67	0.58
Halite	2.35	3.7	0.25	4.32	2.50	0.58
Hematite	5.10	21.2	0.14	6.45	4.27	0.66
Hornblende	3.12	11.0	0.29	6.80	3.72	0.55
Magnetite	5.17	21.9	0.19	6.82	4.22	0.62
Muscovite	2.79	8.0	0.25	5.88	3.41	0.58
Orthoclase	2.54	6.3	0.29	5.68	3.09	0.54
Pyrite	5.03	26.3	0.18	7.30	4.98	0.68
Quartz	2.65	9.6	0.08	6.05	4.03	0.67

$$\nu = \frac{0.5 - (c_s/c_p)^2}{1 - (c_s/c_p)^2} \tag{4.60}$$

Similar expressions can be obtained for the other elastic constants (see Section 7.5).

Any wave velocity values measured in the field can be extremely variable since they depend on factors such as the structure of the rock and particularly the presence of open fissures or the presence of porewater. Water has a c_p value of 1500 m s^{-1} and its presence can significantly increase the measured wave velocity values for a rock mass.

At ground *surface*, where most measurements of wave arrivals and wave motion will be obtained, the interpretation is complicated by the presence of surface waves. The development of these waves is a rather complicated process (Ewing, Jardetzky and Press, 1957), but perhaps may be most simply illustrated by the sketch in Figure 4.38 (after Attewell and Farmer, 1973). P-waves and S-waves emanating from an explosive source will be reflected at the ground surface, subjecting the surface to minor disturbance. A much larger disturbance will result from the interaction of the on-coming S-wave and the reflected P-wave, and the head wave formed by this interaction will effectively travel below the surface, causing oscillations at the surface. Monitoring of wave arrivals at a surface station will therefore

Rock as a Material 239

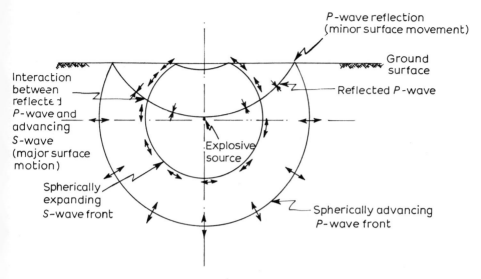

Figure 4.38 Wave propagation from a near surface explosive source.

detect initially a P-wave arrival of minor amplitude followed by an S-wave arrival and then by a large scale surface wave disturbance. In a fully-developed form, it has been claimed (Leet 1960) that surface waves can be partitioned into four modes of oscillation of which two — Rayleigh waves or R-waves and Love waves or Q-waves — are relatively easily detected and identified. Of these, the R-wave is dominant and carries the major part of the energy released by an explosion.

A rock particle undergoing an R-wave disturbance traces an elliptical orbit of retrograde motion (Figure 4.39) and the wave front itself is cylindrical rather than spherical as is the case with body waves. The Q-wave is the shear component of the body wave. The velocity of the R-wave is approximately half that of the P-wave and marginally slower than the S-wave ($0.9\ c_s$). A coda of waves is illustrated in Figure 4.40, where the P- and R-waves are detected in the vertical and forward (x,z) directions of the orthogonal detector array, and the S-waves in the transverse (y) direction.

4.11 Wave attenuation

With increasing distance from the source of the disturbance, the energy at any point in a seismic wave front becomes progressively reduced. This attenuation can be attributed to three main loss sources, geometrical, material and structure.

240 *Principles of Engineering Geology*

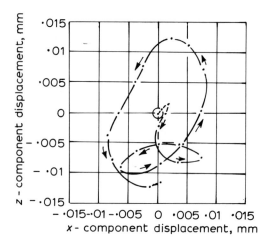

Figure 4.39 Minor R-wave orbit (*after* Davies et al, 1964).

Geometric attenuation is the most obvious, since even if the total energy in the wave remains constant, spherical or cylindrical divergence of the wave front means that the energy per unit area reduces in proportion to the increased area covered by the expanding disturbance. The actual energy per unit area of the wave front (W_r) is therefore related to the source energy (W_0) and to the distance r

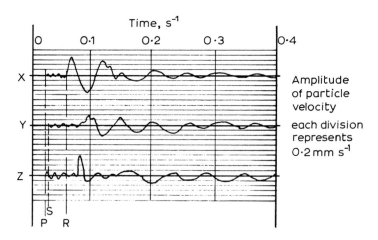

Figure 4.40 Wave profiles in X, Y, Z directions at ground surface 30 m from a small subsurface explosion in sandy shale (*after* Davies et al, 1964).

from the source. Geometrical attenuation is then written simply as:

$$W_r = \frac{W_0}{A} \tag{4.61}$$

where $A = f(r)$ is the wave front area and a point energy source is assumed.

For a cylindrical wave front A is equal to $2\pi rz$ where z is the depth of the wave from ground surface. For a spherically diverging wave A is equal to $4\pi r^2$. Thus, if $W_r = \frac{1}{2}\rho\lambda v^2$ (equation 4.49), then for a cylindrical wave of constant frequency,

$$v \propto \left(\frac{W_0}{r}\right)^{1/2} \tag{4.62}$$

and for a spherical wave:

$$v \propto \frac{W_0^{1/2}}{r} \tag{4.63}$$

This latter form of relationship has been shown (Attewell et al, 1964; Duvall et al, 1961) to be the basis of all wave propagation laws (see Section 7.13 for a more detailed treatment), and with various constants of proportionality can be applied to predict ground vibration levels from energy sources as diverse as piling or nuclear explosions.

Apparently, the properties of the rock itself have relatively little practical effect on attenuation. *Material* attenuation results from energy lost in permanently deforming the rock, internal friction and scattering mechanisms (Attewell and Farmer, 1964), and can be expressed, after Bornitz (1931), in a simple exponential form for a point source:

$$W'_r = W_r \exp(-\alpha^* r)$$

or in terms of particle velocity v' at distance r from a velocity v:

$$v' = v \exp(-\alpha r) \tag{4.64}$$

where α, the attenuation coefficient, has median values in the range 0.001 to 0.1 m^{-1} for most rocks and soils. Obviously, in the case of the lower values, the attenuation due to material damping is insignificant compared with the geometrical attenuation. Even in the case of organic clays, where α may be as high as 0.2 m^{-1}, material damping is only about one-fifth as severe as geometrical damping.

Energy losses caused by the presence of rock or soil *structures*, or generally by interfaces defining zones of different acoustic impedance, are less easily summarized in a general form. When a wave meets an interface or discontinuity of any type, part of the wave energy is transmitted through the interface and part is reflected back into the path of the incident wave. The magnitudes of the transmitted and reflected components are a function of the relative specific *acoustic impedance* of the interfacial rocks. Specific acoustic impedance is the product of density ρ and compression wave velocity c_p.

The equations for the stresses in the transmitted (σ_{tr}) and reflected (σ_r) waves, normalized in terms of the incident stress σ_i, may be written for a plane wave as:

$$\frac{\sigma_r}{\sigma_i} = \frac{\rho_2 c_{p2} - \rho_1 c_{p1}}{\rho_2 c_{p2} + \rho_1 c_{p1}} \qquad (4.65)$$

$$\frac{\sigma_{tr}}{\sigma_i} = \frac{2\rho_2 c_{p2}}{\rho_2 c_{p2} + \rho_1 c_{p1}} \qquad (4.66)$$

The wave is taken to be travelling from medium 1 to medium 2. Several examples illustrate the effect of wave travel across the boundary.

(a) $\rho_2 c_{p2} = \rho_1 c_{p1}$. In this case where the impedances are matched, there will be no energy loss at the interface and $\sigma_r = 0$, $\sigma_i = \sigma_{tr}$.
(b) $\rho_1 c_{p1} < \rho_2 c_{p2}$. In this case where the impedance of the receiving medium is greater than the transmitting medium, there will be a transmitted compression wave and a reflected compression wave.
(c) $\rho_1 c_{p1} > \rho_2 c_{p2}$. Where the impedance of the receiving medium is less than the transmitting medium, there will be a transmitted compression wave and a reflected tensile wave.
(d) $\rho_2 c_{p2} = 0$. This is, in effect, the case with a free air boundary, when $\sigma_{tr} = 0$ and $\sigma_r = -\sigma_i$. This simply means that there will be a fully reflected tensile wave.

Both cases (c) and (d), and most particularly (d), have important implications in engineering. Although compressive seismic waves remote from source do not carry a sufficiently high stress to damage a rock in shear or direct compression, the reflected tensile pulse can often be high enough to cause failure in tension. As we have seen earlier, the tensile strength of intact rock is low (almost an order of magnitude less than its compressional strength) and that of jointed rock may be non-existent. A rock that is being excavated is therefore able to support high input compressive stresses but will readily fail in tension when the wave reverses at, for example, a quarry face. Without the benefits of this peculiarly brittle phenomenon, explosive excavation in quarries and rock tunnels would probably cease to be an economic practicality. It should also be noted that although the dynamic stress vanishes at the free boundary, the particle velocity in the wave is required to double. The fractured ejecta acquire momentum according to their mass and the particle velocity in their portions of the wave front.

So far we have only considered travelling waves at *normal* incidence to an interface. Where a stress wave presents itself obliquely to an interface, shear waves are created and the reflection refraction pattern is altered. Rinehart and Pearson (1953) have considered the earlier important case (d) of a free face in air. Since this condition is the most widespread in an engineering geology context it may be shown that:

$$\frac{\sigma_r}{\sigma_i} = -\frac{\tan \beta \tan^2 2\beta - \tan i}{\tan \beta \tan^2 2\beta + \tan i} \qquad (4.67)$$

where

$$\sin \beta = \sin i \left(\frac{1 - 2\nu}{2(1 - \nu)}\right)^{\frac{1}{2}}$$

and where i is the angle of incidence (with respect to a normal to the plane). Further, a shear wave will be generated where:

$$\frac{\tau}{\sigma_i} = \left[\frac{\tan \beta \tan^2 2\beta - \tan i}{\tan \beta \tan^2 2\beta + \tan i} + 1\right] \cot 2\beta \qquad (4.68)$$

At an oblique rock/rock interface, it is found that the refracted, reflected and induced shear waves are dependent on the rock properties, but that the shear wave is of minor significance unless there is a serious impedance mismatch. The *angle* of refraction is, of course, given by Snell's law, with in this case:

$$\frac{\sin R}{\sin i} = \frac{c_{p2}}{c_{p1}} \qquad (4.69)$$

where R and i are respectively the angles of refraction and incidence.

4.12 Rock as a construction material

Discussion of the fundamental mechanical properties of rocks, as in the present chapter, often ignores some of their more commercially-relevant properties and particularly those affecting their use as a construction material. Specific uses include roadway subgrade aggregate and top dressing, rockfill, concrete crushed aggregate, rip–rap for dam facings and building stone. Apart from building stone (which is the least important since the stone now tends to be used only as cladding for prestige buildings) these uses involve rock in aggregate form, the properties of which are defined mainly by empirical tests designed to simulate specific construction and use conditions.

The fundamental properties of crushed aggregates and rockfill have been discussed in Chapter 2. These properties are specified principally in terms of *compressibility* and *shear resistance*. Compressibility is related to particle strength and size, aggregate void ratio and applied stress. Shear resistance is related to aggregate void ratio, particle size and frictional properties. Both of the two mechanical parameters are affected indirectly by particle shape.

The main commercial indices for aggregates in Britain are laid down in British Standard 812 for roadstones and British Standard 882 for concrete aggregates.

The requirements for concrete aggregates are relatively simple. Undoubtedly the most important of these is the stability of the aggregate, which should neither slake or swell when wet or shrink when dry. The most commonly–used natural aggregates, crushed limestone and siliceous gravels and flints tend to have high stability. Partial dolomitization of some limestones may lead both to a weakening of the aggregate due to the associated volume changes and also a possible predisposition for binder stripping when used as a tar or bitumen-coated aggregate (see Attewell, 1970). It has also been found that some natural aggregates which have traditionally been used in concrete do have a tendency to cause shrinkage. This is particularly the case in Scotland (see Edwards, 1970) where many sources of gravel are glacial in origin and contain a wide variety of material from the areas traversed by the glaciers. The worst type of aggregates from this point of view are those derived from greywacke, but all aggregates derived from porous, low density igneous rocks, particularly basalt and dolerite, tend to have a deleterious effect on concrete. This applies whether the aggregate is found as gravel or is crushed from intact rock. Although reasonably stable concrete can be made from most of these rocks, detailed trial-mix design is obviously required before final adoption.

Other requirements for concrete aggregates can be summarized in terms of the main functional requirements of concrete, usually including strength, durability, fire protection and thermal insulation. The manner in which various aggregates (including artificial aggregates) meet these requirements in summarized in Table 4.9 taken from the British Building Research Establishment Digest No. 150.

Strength is usually related to the crushing strength of the rock, all other factors such as cement and water content and aggregate grading being equal. *Durability* requires an inert aggregate and associated porewater, as well as a dense concrete, if it is subjected to frost action in a saturated state. *Thermal insulation* is inversely related to density.

In the case of fire protection, a good thermal expansion match between the cement and the aggregate in the concrete is required. This applies to virtually all artificial aggregates and to crushed limestone. Siliceous aggregates generally and all natural crushed rocks except limestone tend to have relatively high coefficients of thermal expansion and concrete made from them will have a tendency to spall if subjected to high temperatures.

Rocks used in *road-making* aggregates are classified by British Standard 812 under 11 group headings (Table 4.9). Apart from the final group, schists, most are suitable for roadway aggregates and the groups (apart from the artificial aggregates) are based on conventional geological classifications rather than engineering use. Properties can, of course, vary widely within a group and also within a rock type or even a local rock, and the values quoted in Table 4.9 are at best a rough guide to the properties of each group.

Sampling and testing techniques are described by Shergold (1960) and defined in British Standard 812. Sampling is particularly important if a test specimen truly representative of the bulk aggregate is to be obtained. The British Standard lays down that a

Table 4.9 Properties of various types of concrete (Crown copyright, from Building Research Est. Digest No. 150)

Aggregate	Typical range of dry density Aggregate kg m^{-3}	Concrete kg m^{-3}	Compressive strength N mm^{-2}	Drying shrinkage per cent	Thermal conductivity at 5% moisture content W/m °C	Main functional requirement
Clinker	700–1050	1050–1500	2–7	0.04–0.08	0.35–0.65	Class 1 fire-resistance / thermal insulation
Exfoliated vermiculite and expanded perlite	60–250	400–1100	0.5–7	0.20–0.35	0.15–0.39	
Pumice	500–900	650–1450	2–15	0.04–0.08	0.21–0.63	
Foamed slag	300–950	950–1500	2–7	0.03–0.07	0.30–0.65	
Expanded clay, shale or slate and sintered pulverised fuel ash	300–1050	700–1300	2–7	0.03–0.07	0.24–0.50	
Foamed slag	500–950	1700–2100	15–60	0.04–0.10	0.85–1.40	strength and durability / Class 1 fire-resistance
Expanded clay, shale or slate and sintered pulverised fuel ash	300–1050	1350–1800	15–60	0.02–0.12	0.55–0.95	
Crushed brick	1100–1350	1700–2150	15–30	—	0.85–1.50	Class 2 fire-resistance
Crushed limestone	1350–1600	2200–2400	20–80	—	1.6–2.0	
Flint gravel or crushed stone	1350–1600	2200–2500	20–80	—	1.6–2.2	

sample should comprise a number of small increments taken from widely-spaced points in a bin or stockpile or over a period of time in the case of conveyor sampling. The sample should then be thoroughly mixed and reduced by quartering or through a sample divider to the required size which will depend on the aggregate size.

The main tests are the aggregate crushing, impact and abrasion tests. The *aggregate crushing test* involves submitting a 4 in (100 mm) deep bed of ½ in (12.5 mm) single size aggregate contained in a 6 in (150 mm) diameter hardened steel mould to a uniaxially confined compressive load of 40 tonnes. The percentage of fines passing a 2.4 mm sieve is known as the *aggregate crushing value.* An aggregate crushing value below 10 indicates a very strong aggregate. A rock with a value greater than 35 would normally be considered too weak for road surfacing or any engineering use. Since with increasing degradation this test becomes progressively less sensitive, an alternative test using the same apparatus measures the load at which 10 per cent fines are created. Values for the 10 per cent fines test range from 40 tonnes for the strongest rocks (with an aggregate crushing value of 10) down to as little as 1 tonne for chalk or crushed brick.

The aggregate crushing test measures resistance to static loads. The *aggregate impact test* measures resistance to impact by subjecting a 1.1/8 in (28.6 mm) deep bed of 1/2 in (12.5 mm) aggregate in a 4 in (100 mm) diameter mould to 15 blows from a 30 lb (13 kg) hammer falling through 15 in (330 mm). The percentage of fines passing a 2.4 mm sieve is known as the *aggregate impact value.* It is numerically very similar to the aggregate crushing value except in the case of fine-grained siliceous aggregates (group 3) which are less resistant to impact, a fact which is evidenced by an impact value on average about 5 per cent higher than the crushing value.

Both impact and crushing values are also related quite closely to the results of compression tests on aggregates and through these to more conventional rock properties (see Chapter 2). A high crushing value is equivalent to a high strain under uniaxially confined loading.

Resistance to abrasion is obtained by the *aggregate abrasion test.* This test measures the percentage loss in weight of 35 No. 1/2 in (12.5 mm) particles mounted in a flat layer in pitch and subjected to abrasion by a standard sand on a standard grinding lap. The aggregate abrasion value ranges from below 1 per cent for flints to more than 15 per cent for softer rocks which would be considered unsuitable for road surfacing.

Table 4.10 Classification and properties of roadstone aggregates (*after* Shergold, 1955, 1960)

Group classification (B.S. 812)	Rocks	Specific gravity	Water absorption %	Aggregate crushing value	Aggregate impact value	Aggregate abrasion value	Polished stone coefficient
1. Artificial	Slags, Clinker, Burnt shale, Crushed brick	2.71	0.7	28	27	8.3	0.50
2. Basalt	Andesite, Basalt, Basic porphyrites, Diabase, Dolerite, Epidiorite, Hornblende schist, Lamprophyre, Quartz-dolerite, Spilite	2.80	1.1	14	15	6.1	0.56
3. Flint	Flint, Chert	2.54	1.0	18	23	1.1	0.35
4. Gabbro	Basic diorite, Basic gneiss, Gabbro, Hornblende rock, Norite, Picrite, Peridotite, Serpentine	2.86	0.3	12	12	3.9	—
5. Granite	Granite, Gneiss, Granodiorite, Granulite, Pegmatite, Quartz-diorite, Syenite	2.69	0.4	20	19	4.8	0.56
6. Gritstone	Agglomerate, Arkose, Breccia, Conglomerate, Greywacke, Grit, Tuff, Sandstone	2.69	0.6	17	19	7.0	0.69
7. Hornfels	All contact altered rocks except marble and quartzite	2.82	0.4	13	12	2.2	0.45
8. Limestone	Dolomite, Limestone, Marble	2.66	1.0	24	23	13.7	0.43
9. Porphyry	Aplite, Dacite, Felsite, Granophyre, Keratophyre, Microgranite, Porphyry, Quartz-porphyrite, Rhyolite, Trachyte	2.73	0.6	14	14	3.7	0.51
10. Quartzite	Ganister, Quartzitic sandstones, Re-crystallized quartzite	2.68	0.7	16	21	3.0	0.57
11. Schist	Phyllite, Schist, Slate. All severely sheared rocks	Not widely used as roadstone					

The *accelerated-polishing test* involves the use of a pneumatic tyre loaded with fine abrasives to polish 3/8 in (9.5 mm) particles mounted in a mortar matrix. At the end of a 6-hour run – the test is reputed (Shergold, 1960) to obtain a state of polish close to the ultimate reached by a road surface under heavy traffic. The state of polish is measured in terms of the coefficient of friction between the specimen surface and a pendulum-mounted rubber slider. The resulting *polished stone coefficient* gives an indication of skid resistance, ranging from 0.3 for a stone which will become highly polished under any traffic conditions to 0.8 for a stone which will remain rough under any conditions.

Apart from these specific tests, general aggregate tests such as specific gravity, water absorption, sieving, shape (including flakiness and angularity number), surface texture, strength and chemical analysis, are important in assessing compactability and durability. These tests are covered in the British Standard (BS812) and the equivalent ASTM standard (C131). An additional test to determine adhesion is important in surface-coated macadam dressings.

Interpretation of aggregate test results is rather difficult since there is no specified minimum quality and considerations of availability, cost and transport are often paramount in selection of aggregates in roadway construction. Ideally, materials with the best physical properties should be used and, in the case of surface dressings, a high polished stone coefficient is now considered essential. The main role of the tests is, however, comparative, using information of the type in Table 4.10 in order to evaluate available materials and to modify designs and performance predictions in the light of the evidence provided.

The accelerated pace of development in the middle eastern desert countries has highlighted the often poor availability and/or mechanical quality of their natural materials. Aggregates, both coarse and fine, tend to be carbonates, and if dolomitic there may be problems from alkali-carbonate reactions. Poorly-graded beach sands may contain organic matter with a high water absorption facility. With the very real possibility of high chloride and sulphate levels, there is the risk of concrete steel reinforcement attack from the former and concrete expansion from the latter. Then even if or when these problems can be overcome, the finished concrete structures must be further protected from salts in the ground.

5 ~ Preferred Orientation, Symmetry Concepts and Strength Anisotropy of Some Rocks and Clays

There is quite a close conceptual and analytical affinity between, on the one hand, the orientation distribution of microscopic planar and near-planar elements in a rock and clay, and on the other hand the distribution of larger-scale discontinuities in weaker rocks such as shales and mudstones and even in stiff clays such as the London Clay. The analysis of larger-scale 'field' discontinuities, together with the manner in which they impose a control on mass strength of the rock is considered in Chapter 6. In the present chapter, the intention is to introduce the idea of crystallographic preferred orientation in polymineralic aggregates and to consider the influence of such orientation density distributions upon intact strength. Intact strength anisotropy may be much more directly related to planar element distributions in a homogeneous material than it will be in a material that is also inhomogeneous.

Under the present treatment, strength anisotropy is considered first with respect to homogeneous and planar-inhomogeneous *rocks* and then with respect to homogeneous and planar-inhomogeneous *clays*. For a mathematical treatment of mechanical anisotropy, the reader would refer to Love (1892) and Hearmon (1961).

Although the present chapter is concerned with some rocks and clays that are fine-grained and obviously anisotropic, it should be noted that a great deal of research attention has been directed to the influence of microcracking upon the stiffness, strength and strength anisotropy of sedimentary and igneous rocks. Because of the generally much larger crystallographic and internal structure of these rocks, microcracks may be decorated by impregnation techniques and both their nature and their density distribution observed and measured directly under the petrological microscope. For further reading on this subject, reference may be made to Brace (1965), Walsh and Decker (1966), McWilliams (1966), Brace and Orange (1968), Thill *et al.* (1969), Willard and McWilliams (1969), Douglass and Voight (1969), Nur (1971), Peng and Johnson (1972), and Simmons and Richter (1974).

5.1 Studies of the orientation density distribution of clay minerals and other associated minerals

A key to the understanding of the anisotropic character and mechanical behaviour of clay mineral aggregates under consolidation and shear lies in the ability to quantify the degree of clay, and associated, mineral preferred orientation in these materials. As will be suggested later, there is also an orientation distribution compatibility between the clay minerals and the associated fields of intrinsic discontinuities which more directly condition the strength anisotropy of the intact material.

Experimental methods that have been used for clay mineral orientation determination have been summarized by Gillot (1968) and, with additions and comments, are listed below:

Technique	Advantages	Disadvantages
a) Petrographic microscope	Standard, basic equipment and relatively cheap to purchase. Can be adapted to make use of an integrated birefringent effect in which the intensity of light transmitted through a thin section is a function of the degree of preferred orientation (Mitchell, 1956; Wu, 1958; Pusch, 1964; Lafeber and Kurbanovic, 1965; Morgenstern and Tchalenko, 1967 a, b, c; Tchalenko, 1968; Ayyar 1969)	i) Resolving power of the microscope is inadequate for conventional petrographic observations. ii) Integrated birefringent technique is really only semi-quantitative, skill is required for information interpretation, and without considerable effort it does not lend itself to coverage of orientation distribution over the full sphere of projection.
b) Electron Microscope (particularly reflection scanning)	Adequate clay mineral resolution in the scanning mode. Visual evidence of orientation with photographic records (see Figures 3.6–3.13)	Expensive equipment. Specimen preparation problems and the further possibility of textural changes under electron bombardment. Only discrete, localized indications of clay mineral orientations.

Technique	Advantages	Disadvantages
c) X-ray diffraction i) 2θ (Bragg angle) scans on specimen surfaces cut at different but known orientations	Cheap using standard X-ray diffraction technique	Does not give full indication of orientation distributions. No visible or photographic evidence of actual minerals
ii) textural scans in reflection and/or transmission for specified Bragg angles over whole sphere of projection.	Relatively cheap modification to commercial goniometers. Output information needed in digital form for computer processing and direct print-out of sub-fabrics	Some specimen preparation problems. No visible or photographic evidence of actual minerals.
d) Optical data processing on holographic technique (e.g. Pincus, 1966; 1969 a, b).	Using spatial filtering methods, many features of a textural and fabric nature can be evaluated (for example, shapes of pore spaces and minerals in addition to orientation density distributions). In a suitable vibration-free and temperature-controlled environment, equipment costs need not be high.	Although the technique applied to mineral aggregates is potentially powerful, it is still in the development stage and will require much more work on it before it is suitable for serious 'production' analysis.

There are reasonable grounds for claiming that at the present time the optimum technique for detailed study of crystallite orientation distribution is that of X-ray texture goniometry provided that a high density of information is acquired, that the processing is fully automatic, and that the orientation distribution information is output directly from a computer in the form of a pole density or an axis density figure. This technique and the meaning of these latter terms can now be discussed in more detail.

5.2 X-ray texture goniometry

Introduction

In fine-grained geological materials of widely differing lithologies, hard rocks through to soft clays, it is the character of the distribution of the constituent crystallite orientations that determines the degree of intrinsic anisotropy. In the harder rocks, any associated small discontinuities also contribute to the strength anisotropy. It is to an investigation of the former and indirectly to an interpretation of the latter that our attention is now directed. To this end, it must be recognized immediately that direct optical observation is unsuitable for orientation determination of small size crystals and that there is a particular restriction on the necessary number of lattice planes that can be observed.

Measurement of X-ray diffraction from a sample offers the best means of studying crystallite orientation. In principle this can be evaluated by selecting one system of coordinate axes in a solid sample and another referred to the crystallographic unit cell and expressing the orientation distribution of one of these with respect to the other. With the solid sample frame of reference fixed, the crystallite orientation distribution takes the form of a pole figure diagram familiar to structural petrologists. Alternatively, orientation distribution of the reference axis with respect to a standard crystallographic axis delineates an axis distribution or inverse pole figure used to resolve textures of metals (Barrett, 1943; Jetter et al, 1956; Dunn, 1959; Dunn and Walter, 1959; Thurber and McHargue, 1960; McHargue and Jetter, 1960; Barrett and Massalski, 1966; Walter, 1967; Heckler et al, 1967), polymers (Krigbaum and Roe, 1964; Krigbaum and Balta, 1967), ceramics (Pentecost and Wright, 1963) and monomineralic quartz aggregates (Wenk et al, 1967; Baker et al, 1969). In general, inverse pole figure generation is restricted to axi-symmetric fabrics which are rare in the harder fine-grained metamorphic rocks. However, for more axi-symmetric fabrics, provided that the pole figures from a number of planes ($hkil$) can be recovered, then inverse pole figures can be generated using spherical harmonic analysis on the crystallite distribution function (Baker and Wenk, 1969; Baker et al, 1969). Unfortunately, limited resolving power between X-ray peaks tends to restrict the number of planes that can be examined in a monomineralic aggregate, so jeopardizing inverse pole figure development. In polymineralic aggregates, the intensity of possible overlap virtually precludes inverse pole figure construction unless some form of iterative peak elimination through computer processing from a basis of a few 'clean' low angle diffractions can be accomplished. But in practice, this would be excessively expensive even if theoretically possible and it is therefore proposed to limit the discussion to the technique of X-ray pole figure construction.

Mechanics of X-ray texture goniometry

The Schulz reflection technique (1949a) to be described subsequently forms the basis of the method widely used by metallurgists and petrologists but it is possible to trace a development from Decker *et al* (1948), Field and Merchant (1949) and Norton (1948). The Schulz reflection technique makes use of a flat specimen, but the spherical specimen method of Jetter and Borie (1953) and its use by Higgs *et al* (1960) and Chen (1966) deserves mention. The mechanical integration facility suggested by Suits (1951) should also be noted since it is incorporated in the modern Schulz-based equipment. For further information on historical developments in the technique, the reader is referred to Bradshaw and Phillips (1970).

The geometry of the Schulz reflection technique is described by Baker *et al* (1969) and its detailed implications are discussed by von Gehlen (1960). A flat-surface specimen, some 3 mm thick and 22 to 30 mm diameter, is prepared from an orientated block of the material to be studied and mounted on a quill within the latitude ring of the texture attachment to the goniometer (Figures 5.1, 5.2). A pre-adjustment of the incident X-ray beam with respect to the specimen surface ensures that those crystallites having lattice planes ($hkil$) inclined at half the Bragg angle to the incident beam and having a d-spacing appropriate to the Bragg angle setting (function of the type of radiation) will diffract incident energy to a counter to an extent dependent upon the number of diffractors at optimum orientation. If the specimen is slowly rotated in its own plane at the same time as it is progressively tilted about an axis lying within the plane containing the incident and diffracted beams (the Bragg angle setting being preserved throughout the motions) then the same lattice planes ($hkil$) lying at different orientations with respect to the specimen coordinate axes will be brought progressively into a diffracting status through an outward-directed spiral scan of the sphere of projection created by the combined motions in azimuth and latitude (see Figure 5.3). In commercial instruments there is a further reciprocating motion in the plane of the specimen, the mechanical integration serving to smooth the diffracted radiation from larger diffractors.

In operation, there is a fall-off in intensity due to defocusing and misalignment effects as the specimen is tilted through the latitudes at

Figure 5.1 Rock disc specimen in position on quill in Philips texture-goniometer showing input collimator, counter input slit and G.M. proportional counter. In position (a) the goniometer is set at a very low Bragg angle and at the beginning of the scan ($\eta = 0$). In position (b) the Bragg angle setting is high and the specimen is tilted well into the latitudes (η).

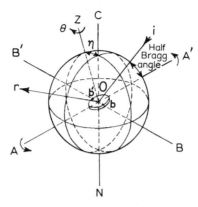

Figure 5.2 Geometry of the texture goniometer motions for the reflection condition. Incident (i) and reflected (r) X-ray beams at the chosen Bragg angle setting with respect to the specimen surface are within the plane ZNAA' and remain constant throughout the motions. The specimen is rotated clockwise (θ) about axis OZ which itself rotates (η) within the plane CB'NB about axis OA. At any time during rotation, the X-ray count rate is directly related to the density of lattice planes having poles parallel to OC, that is, planes themselves lying parallel to AB'A'B. Note that the effect of moving the specimen with respect to iOr (fixed) is to reverse the azimuthal motion on the plotting spiral with respect to the directional sense of θ. For η increasing counter-clockwise about OA (as shown), the spiral scan begins on the b side of $b'Ob$ specimen cartesian coordinate axis (which is also the axis of mechanical translation) and on the b' side for η increasing clockwise about OA. *After* Attewell and Sandford (1974b).

angles greater than about 70°. These effects were analysed mathematically by Feng (1965) and from their work on undeformed flint, Baker *et al* (1969) showed that the fall-off also increased with the Bragg angle setting. Up to this high latitude, Schulz (1949a) showed that changes in absorption with tilt are effectively compensated by changes in the scattering volume and, in practice, suitable input collimator slits together with a sufficiently wide counter slit (up to 4°) reduces fall-off in count-rate to a minimum but at the expense of lower resolution in Bragg angle during the spiral scan. Clearly, this lower resolution imposes constraints upon the actual lattice planes available for textural examination. These and other problems associated with the technique have been reviewed by Barrett and Massalski (1966).

There are several possible solutions to the fall-off problem. If a narrow counter slit is found to be essential for resolution of a narrow peak, then absorption losses through the latitudes, as determined from a randomly orientated crystallite aggregate, can be corrected by

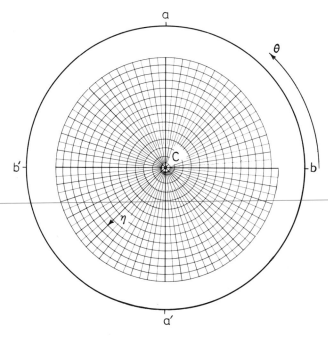

Figure 5.3 Outward-directed spiral scan of the sphere of protection in the *ab* plane.

fitting a polynomial of suitable degree to the fall-off curve (see, for example, Alty, 1968). This operation could be both difficult and time consuming to perform on texture intensity data that had been hand plotted from a strip chart record but is easily performed on computer-processed data. Another possibility is to supplement the reflection data, acquired up to a latitude of 70°, with transmission data (Schulz, 1949b) up to 90° latitude permitting about 5° of latitude overlap between the two sets of output data for intensity correlation purposes. Transmission specimens are only about 70–90 μm thick and are cut in the plane of the reflecting specimen surface, the incident X-ray beam passing directly through them to the counter. This technique has been used on flint by Baker *et al* (1969) and in the present authors' laboratory, but it is considered that the compatibility problems associated with the two modes of textural data acquisition militate against transmission. The most acceptable and economical method is to scan three orthogonal surfaces in the reflection mode each up to 70° latitude or thereabouts and then to combine all three for a complete pole figure.

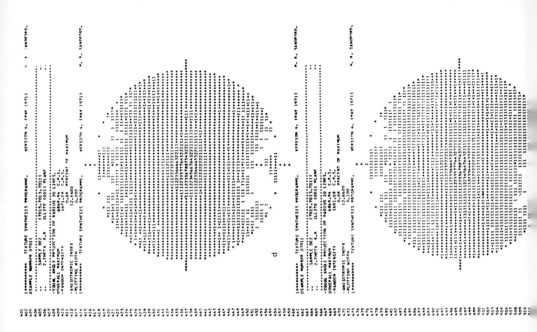
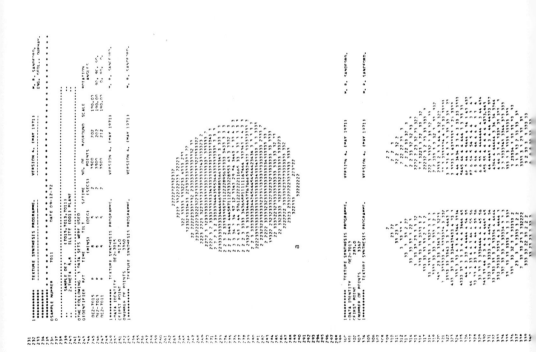

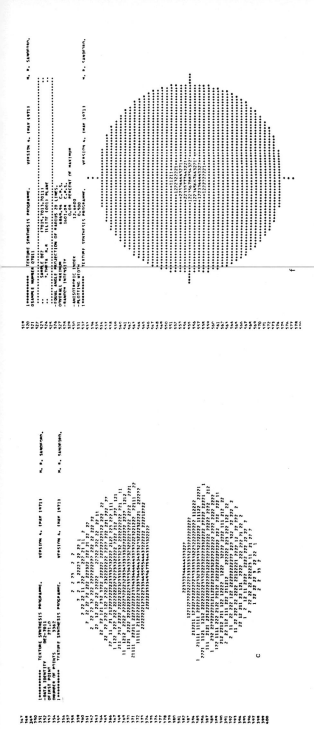

Figure 5.4 Computer synthesis of illite (001) sub-fabric in Penrhyn Slate. The final plane of projection is the primary cleavage, textures b and c being rotated into this plane (a). Numbers on the textures represent the diffracted X-ray count rate scaled to the maximum which is located by the computer search. Textures d, e, f incorporate the results of different scanning windows for the randomization operation, optimum definition of the randomized surface (+ symbol) being achieved in texture e with the 0.1 window. The anisotropic index, independent of window size, is 12.66. A total of 10,500 raw data entries of (hkl) orientation information is compounded into the final sub-fabric. Data were derived from a Philips X-ray texture-goniometer and diffractometer (PW 1050, 1078/10, 1310); CuKα Ni-filtered radiation. Data processed and direct sub-fabric print-out from IBM 360/67 computer. *Operator*: M. R. Sandford. (*From* Attewell and Sandford, 1974b.)

Usually, but not necessarily, one of the planes of projection will be designated as the plane into which the other two projections are to be rotated, and this will normally be performed by digital computer through suitable transformation matrices. X-ray intensity data will be continuously logged through the latitudes on punched paper tape, but since the raw information is acquired in a spiral configuration about the pole of projection, it is necessary not only to output in η (latitude) and θ (azimuth) coordinate form for direct fabric representation but also to interpolate and smooth the raw data for infilling of empty elements in the output arrays (for example, short range smoothing as per Baker *et al*, 1969). Further averaging and smoothing takes place between individual fabrics when superimposition procedures are called and when some additional numerical operations have been performed on the processed data (these operations are considered subsequently) the fabric can finally be printed out with point locations of numerical intensities, or these intensities can be directly contoured using a plotter on line with the computer. Examples of computer-processed fabrics appear later in the present chapter and an example of X-ray texture synthesis as outlined above, but which also incorporates features to be described subsequently, is given in Figure 5.4.

An alternative texture-goniometric method to the one described above makes use of photographic film rather than a counter for recording diffraction intensity. Microdensitometer scans on the film then quantify that intensity for computer processing. This technique, described by Starkey (1964), does not yet appear to be available commercially.

5.3 Symmetry concepts

A conceptual appreciation of symmetry is a useful aid towards understanding and describing the interaction between structural fabrics and the engineering process. It is a subject that cannot be considered here in the detail that it merits, but for further study and as a background to this section the interested reader is referred to the abundant crystallographic literature (for example, Hartshorne and Stuart, 1964) and also to the particularly appropriate paper by Patterson and Weiss (1961).

The symmetry presented by a body to an observer is an expression of its form when observed from different spatial coordinates. This form may be geometrical in character describing its outline shape. It may describe its internal stress and strain distribution, its elastic, electrical or optical properties. More pertinent to the present problems, it may describe the orientation density distribution of such fabric elements as defined lattice planes ($hkil$), internal microscopic cracks, and larger macroscopic discontinuities which can impose both general and directional strength limitations on a rock or clay mass.

It is usually necessary to describe the particular features of a body under investigation for symmetry with respect to defined coordinate axes. The symmetry of the property is then specified by the assemblage or group of all linear transformations of the coordinate axes which do not change the original description. This means that having performed the necessary transformations implied by the symmetry designation, the aspect of the property is the same to the observer as it was before the transformations were undertaken. Any such transformation from one set of rectangular coordinate axes to another for an apparent invariancy in the aspect of the property is termed a 'symmetry operation'.

Transformations of this type, which involve rotation only and contain no translational components, are termed 'point group operations' and are amenable to mathematical description. The nature of the point group symmetry can be described in terms of Schönflies symbols which comprise an alphabetic character subscripted by one or more additional characters. One of these latter characteristics, 'n', specifies the rotation angle about an axis through the relationship: angle = $2\pi/n$ ($n \geq 1$), but for a relatively high degree of symmetry n will be 2 for the 180° rotation. A great deal of apparent complexity will be removed from our discussion of fabric symmetry if we only consider the conditions of $n = 2$ and $n = \infty$. Five symmetry classes (Patterson and Weiss, 1961) are listed below:

a) *Spherical symmetry* $K_{\alpha h}$: This is the highest form of symmetry and is an expression of orientation randomness (or, more accurately, isotropy of orientation distribution). The fabric retains its distribution characteristics independent of the aspect of observation. Tectonite fabrics and sedimentological fabrics will rarely express total randomness since an otherwise inherited randomness of some particular fabric feature will be constrained by the symmetry of an imposed — or a sequence of imposed — stress fields.

b) *Axial symmetry* $D_{\alpha h}$: An infinity of planes of symmetry intersect a unique axis of symmetry which itself is normal to another plane of symmetry. One typical example of an axi-symmetric fabric is that of the orientation distribution of clay mineral basal planes that have been sedimented out of suspension and have been subjected to no subsequent pressures other than the consolidation pressure directed co-axially with the axis of settlement. On the other hand, subsequent tectonic pressures directed non-co-axially may be insufficiently high to impose a change in fabric symmetry (this is the situation with respect to clay mineral orientation distribution in mudstones and shales).

c) *Orthorhombic symmetry* D_{2h}: The projection fabrics have three orthogonal planes of symmetry and are typical reaction fabrics to a tectonic environment. For example, the clay mineral basal plane fabric in a low-grade metamorphic slate may be representative of an ellipsoidal distribution surface of oblate character in which the minor axis of the deformation ellipsoid (major axis of basal plane concentration since there is a reciprocal relationship between the two) reflects the axis of greatest tectonic pressure and is normal to the fold axial plane. The intermediate axis would be parallel to the axial plane of the local folding, while the axis of least shortening (or negative shortening) and lowest basal plane concentration would again be within the fold axial plane but would be directionally compatible with the line of greatest motion. These points are considered in greater detail elsewhere.

d) *Monoclinic symmetry* C_{2h}: The projection fabrics have a single plane of symmetry only. Such a symmetry is usually associated with linear or flow structures in sedimentary or igneous rocks and, kinematically, is associated with a single, dominant deformation vector.

e) *Triclinic symmetry* $S_2 = C_i$: The projection fabrics have no symmetry planes, and it must therefore follow that this lowest form of symmetry is the most generally common. Whereas, for example, a jointed rock mass may comprise orthogonal vertical and bedding plane joints to project an orthorhombic fabric symmetry, a shale may present a discontinuity fabric of triclinic symmetry. In this latter case, a field of systematic joints will be 'overlain' by a field of non-systematic, terminated discontinuities of non-random character. Note, that if this latter field is random, then the original D_{2h} symmetry is preserved. Non-randomness in this latter case is usually a reaction fabric feature to factors such as weathering and desiccation (proximity to an exposed surface), adjustment and accommodation (during early diagenesis), and morphology (as conditioned, for example, by slope profiles).

5.4 Deformation paths

Soils and rocks can acquire a final state of *deformation* which is compounded from quite different histories of incremental deformation. In a similar manner, and considering a two-dimensional soil mechanics problem, successive states of *stress* (the stress path) in a body can most conveniently be displayed in the $\frac{1}{2}(\sigma_1 + \sigma_3):\frac{1}{2}(\sigma_1 - \sigma_3)$ plane (the $p:q$ plane) and expressed via incremental K values (Lambe and Whitman, 1969, p. 112) where K is the ratio σ_3/σ_1 (see Section 2.5). A ratio of unity for isotropic compression (zero shear) would be expressed as K_1. A K_0 stress path (Figure 2.9) would reasonably approximate to the state of effective stress increase during the normal consolidation of a sediment. Similar concepts can be applied to biaxial strain and deformation paths.

For compression which is neither isotropic nor axi-symmetric with respect to the major axis of compression, an alternative method is required for representing the state of stress or strain or deformation in principal stress/strain/deformation space. Probably the most useful method is that in which the principal components of the relevant parameters are expressed in ratio form. By plotting, for example, σ_1/σ_2 against σ_2/σ_3 or ϵ_1/ϵ_2 against ϵ_2/ϵ_3 or, again, l_1/l_2 against l_2/l_3, the inter-relationships between the three parameters become immediately discernible on a two-dimensional plot. *Deformation* plots on these lines have been proposed by Flinn (1962, 1965) and used by Talbot (1970), both with respect to macroscopically-observable geological deformation under conditions of finite homogeneous strain.

5.5 Deformation ellipsoid

Ignore for a moment any more equant habit minerals in a clay, shale or slate and consider only the platy clay minerals. Then, within the scale of the material to be examined, the planar, platy elements can be regarded as having deformed homogeneously during sedimentation or metamorphism even though larger volumes of the constituent host material might have deformed inhomogeneously. On this basis, a single clay mineral can be taken as the fundamental deformation domain (using the terminology of Talbot, 1970) while still recognizing the existence of smaller deformation domains of an intracrystalline nature (typical examples in the more equant minerals are the lattice strains creating undulose extinction in quartz (Bailey *et al*, 1958) and polygonization in metals).

In terms of the development of preferred orientation created by the processes of *flow* of a matrix, Flinn (1965) has noted that discs, rods and plates embedded in a finer grained matrix take up a steady-state preferred orientation in a flowing matrix. This situation might be applicable, for example, to a state in which the micaceous minerals serving as X-ray diffractors repose in a ground mass of illitic mica which tends to a quasi-amorphous state. In homogeneous strain, linear and planar elements are assumed to suffer no distortion on the scale of the domain; they remain straight and parallel after responding to external forces if they were straight and parallel prior to deformation. As a consequence of this condition, such elements in a domain of homogeneous deformation are visualized as passing through the centre of a strain ellipsoid which serves conveniently to describe the state of strain on the scale of the domain.

A random orientation distribution of clay minerals, although a geological anomaly, is a necessary and fundamental concept at this stage. Randomness implies (although not strictly with accuracy) a statistically uniform density distribution of clay minerals in global space and it signifies that such a distribution can be represented by a unit sphere. The poles to the basal planes of the clay minerals can be regarded as intersecting a unit sphere of projection at uniform spacing, or alternatively the basal planes themselves of the minerals can be considered to lie tangentially at uniform thickness over the surface of the unit sphere.

Consider, now, the plots of Flinn (1965) and Talbot (1970) in Figure 5.5 which were conceived in a structural geology context but which have interpretative usefulness in the context of clay mineral fabrics. Referring to Flinn's plot, and making cross-reference to Talbot's it will be seen that for a $K' = \infty$ incremental deformation ellipsoid of *prolate* character, poles to disc and plate elements form a girdle about the longest axis (X) of the ellipsoid. For a $K' = 0$ incremental deformation ellipsoid, they form a concentration marking the axis of an *oblate* figure. For intermediate values of K', the configuration of the incremental deformation ellipsoid is intermediate between these two. Should there be asymmetry in the plane of the plates, their long axes will tend to be aligned parallel to the axis of a *prolate* incremental ellipsoid for $K' = \infty$ but for the

$K' = 0$ condition, the long axes of the plates would then tend to join a girdle about the shortest axis of an *oblate* incremental ellipsoid. Since an effective $D_{\alpha h}$ symmetry about their *c*-axes may be assumed for the clay minerals the discussion of clay mineral sub-fabrics in terms of deformation ellipsoids and their transformations is made rather more simple.

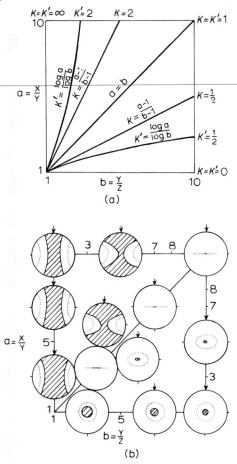

Figure 5.5 Representations of deformation plots due to (a) Flinn (1965) and (b) Talbot (1970). In Talbot's equal area projections, lines of no infinitesimal strain are shown dotted and the deformation field, in which planar elements are subjected to tension within their planes, is shown cross-hatched. Axial lengths $X \geqslant Y \geqslant Z$ in the deformation ellipsoids.

In the first place, and as determined by the character of most natural stress fields, deformation ellipsoids and, by implication, clay mineral basal plane sub-fabrics, will be either symmetric about the axis of greatest shortening or will be of oblate form. The former will be represented by a consolidation sub-fabric in the absence of any contemporary or subsequent tectonic overpressure non-concordant with the axis of uniaxial consolidation. The latter will be represented by the generality of metamorphic stress fields where the dominant reorientation mechanism is re-crystallization under elevated pressure/temperature conditions.

Consideration of the clay mineral sub-fabrics in slate (Figure 5.6) indicates that the orientation density distribution of the basal planes takes the form of an oblate spheroid in which there is an inverse correlation between the numerical density of the planes (function of η, θ) and length of the radius vector generating the deformation ellipsoid. This correlation is equally applicable to the more axisymmetric sub-fabrics characteristic of mudstones and shales (Figure 5.7). If N_3, N_2 and N_1 are taken to represent the numerical density of poles co-axial with the major, intermediate and minor axes respectively of the deformation ellipsoid, then the variables a and b in the deformation plots of Flinn (1965) and Talbot (1970) can be re-specified in terms of the fabric density parameters as a = N_3/N_2 and b = N_2/N_1.

5.6 Randomization

If a deformation ellipsoid can be transformed into a sphere of equivalent volume, then this operation provides a base from which

---

Figure 5.6 Equal area projections of poles to (a) illite (001) in Ballachulish Slate, Scotland and (b) chlorite (002) in Penrhyn Slate, North Wales. The projection plane ab is in each case parallel to the primary cleavage and the concentrations are expressed in percentages of the maximum X-ray counts per second (proportional to the number of basal plane diffractors). Axes aa' and bb' are parallel to cleavage full dip and strike respectively. Plus (+) characters (and broken line on fabric b) trace the intersection of the surface of random intensity with the fabric topography. Let axis X_3 be vertical through the centre of the projection, let axis X_2 be equivalent to the $b' - b$ axis on the projection, and let axis X_1 be equivalent to the $a - a'$ axis on the projection. The projected angle $\zeta_{X_2 X_3}$ ($\hat{=} 110°$) is the angle included by the randomized surface along the $b - b'$ axis and the projected angle $\zeta_{X_1 X_3}$ ($\hat{=} 60°$) is the angle included by the randomized surface along the $a - a'$ axis. Fabric data from Philips X-ray texture-goniometer and diffractometer (PW 1050, 1078/10, 1310); CuKα Ni-filtered radiation. Data processed and direct fabric print-out from IBM 360/67 computer. *Operator:* P. B. Attewell.

BALLACHULISH SLATE
Equal Area Cleavage Plane Projection
a a' – dip
b b' – strike
ILLITE (001)

PENRHYN SLATE (BLUE)
Equal Area Cleavage Plane Projection
a a' – dip
b b' – strike
CHLORITE (002)

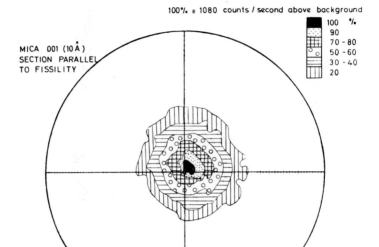

(a)

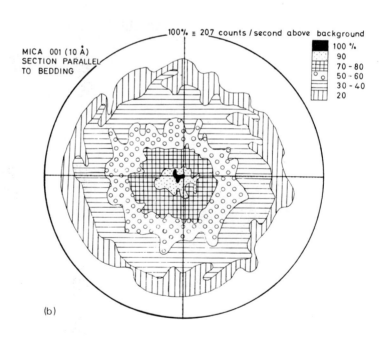

(b)

the intensities of preferred orientation of different sub-fabrics can be compared. The *anisotropic index* is the ratio of the maximum intensity of preferred orientation (or degree of ellipsoidal oblateness) to the randomness intensity, the mathematics of the operation being given in Figure 5.8. The trigonometrical sine function is taken inside the integral because the intensity maximizes at the origin (zero degrees) of the reference frame of axes and a cosine relationship is needed to achieve this.

Computationally, randomization is expressed on the sub-fabric by scanning the pole figure with a 'window' of pre-set width in order to locate the amplitude I_{Rand} and signify its presence with a chosen print character. In the sub-fabrics considered in the present section (Figure 5.6) the randomized surface is marked with with a series of plus (+) characters which also, of course, fix the line of intersection of the transformation sphere and the reciprocal deformation ellipsoid. This randomized surface, which is conical in shape (on a circular or elliptical base − in the present examples of metamorphic sub-fabrics, it is the latter) with an apex at the centre of the ellipsoid, is equivalent to a non-material surface of no finite longitudinal strain. The shape of this surface is most readily defined in terms of its apical angles in the two principal planes that it intersects, viz, two from $\zeta_{X_1 X_2}°, \zeta_{X_2 X_3}°, \zeta_{X_1 X_3}°$.

The narrower the 'window' ('plotting width' as expressed on the computer fabric print-out), the closer the resolution of the randomized surface on the fabric. In the three examples shown in Figure 5.4 the surface is much more clearly defined from a plotting width of 0.1

Figure 5.7 Axially symmetric equal area fabric diagrams of platy micaceous minerals in two dissimilar Coal Measures shales. Plane of projection is parallel to the bedding and the quality of preferred orientation decreases as the area covered by the counts per second contour lines (indicative of the volume of clay minerals possessing a particular orientation) increases. Fabric *a* shows excellent preferred orientation the silty quartz mineral content of the shale being very fine-grained and having little influence on clay mineral orientation. On the other hand, from thin section optical evidence, shale *b* contains large quartz grains, the clay particles being bent around the surfaces of the more equant quartz grains and thereby acquiring a more random orientation distribution. Shale *a* is much more fissile and tends to disintegrate in response to climatic changes much more readily than shale *b*. Fabric data from Philips X-ray texture-goniometer and diffractometer (PW 1050, 1078/10, 1310); CuKα Ni-filtered radiation. Data processed and direct fabric print out from IBM 360/67 computer. *Operator:* P. B. Attewell.

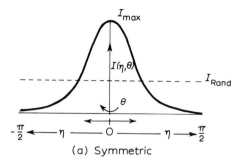

(a) Symmetric

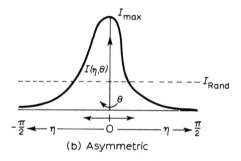

(b) Asymmetric

Area of projection hemisphere = $2\pi r^2 = 2\pi$ for unit radius hemisphere

$$I_{Rand} = \frac{1}{2\pi} \int_0^{2\pi} \int_0^{\pi/2} I(\eta, \theta) \sin \eta \, d\eta \, d\theta$$

$$\text{Anisotropic Index} = \frac{2\pi I_{max}}{\int_0^{2\pi} \int_0^{\pi/2} I(\eta, \theta) \sin \eta \, d\eta \, d\theta}$$

Check: If $I(\eta, \theta)$ is uniform, then

$$I_{Rand} = \frac{I(\eta, \theta)}{2\pi} \int_0^{2\pi} \int_0^{\pi/2} \sin \eta \, d\eta \, d\theta$$

that is, $I(\eta, \theta) = I_{Rand}$ and Anisotropic Index = unity

Figure 5.8 Intensity (density of crystallographic poles) variation as a function of latitude angle across projection hemisphere.

than from a plotting width of 0.2, but the actual amplitude of the anisotropic index remains independent of the 'window' at 12.66.

5.7 Symmetry elements and sub-fabrics

The symmetry of a regional fabric, and its implications with respect to both its formative stress fields and the stress fields to which it might be subjected in an engineering situation, must be considered in terms of the symmetry elements that are common to the several sub-fabrics from which the consensus fabric is compounded. For a fabric feature to have significance when several sub-fabrics are superimposed, the particular feature, or element, must be common to all the constitutive sub-fabrics even though it is accepted that the response strengths of individual sub-fabrics may differ significantly in terms of maximum pole concentrations to feature planes. An important corollary of this necessary compatibility condition is that a fabric cannot have a symmetry higher than that of any of its sub-fabrics. Fortunately, Patterson and Weiss have stated that, from their experience, very few of the possible sub-fabrics of a given fabric need to be considered in order to determine the symmetry of the fabric. We may now consider several sub-fabric requirements.

Basic sub-fabrics can be generated from the preferred orientation of crystallographic elements, preferably from dissimilar minerals. Typical planes that can be examined optically on the universal stage are [0001] axes in quartz,* (001) planes of micaceous minerals, and the three e (01$\bar{1}$2) planes parallel to visible twin lamellae in calcite. The only crystallographic planes that are technically available, in a polymineralic aggregate, for orientation analysis using X-ray texture-goniometric methods (discussed elsewhere) are the basal planes of the clay minerals illite, kaolinite and chlorite, although other planes of other minerals can be analysed and interpreted under certain conditions by invoking arguments of crystallographic symmetry and plane multiplicity (Attewell and Taylor, 1969).

The intrinsic symmetry of quartz axes and poles to micaceous mineral basal planes is $D_{\alpha h}$ and this implies that they will generate a usefully concentrated sub-fabric response in a non-hydrostatic stress field. Response sub-fabrics will usually be of $D_{\alpha h}$ or D_{2h} symmetry, the latter being the more general and the micaceous minerals reacting

---

*Square brackets [ ] represent a direction, in this case the optic axis direction. They are also used to represent an edge between two (*hkil*) planes.

more strongly in view of their platy geometry. Crystallographic sub-fabrics will then be compared with the distribution fabrics of structural features such as foliation planes, lineations, poles to fracture cleavages and so on. In general, if the necessary compatibility is achieved, then the total fabric can be specified in symmetry terms.

In view of the importance of the primary crystallographic sub-fabric symmetry, it will be useful in later sections to give one or two examples of how logical processes of interpretation must be adopted.

5.8 Crystallographic plane multiplicities and symmetry

There is often a tendency to allocate the orientation distribution of a particular lattice plane (*hkil*) within a crystal to the orientation distribution of the crystals themselves. This approach is somewhat limited because even if the orientation distribution of, say, the basal planes of a clay mineral can be determined, it is still not possible, without supplementary knowledge of the orientation distribution of other planes within the same crystals, to specify unequivocally the orientation distribution, and hence the symmetry, of the crystals in material space. Given the orientation distribution of several (*hkil*), and since the information cannot readily be assimilated in pole-figure form, the data are most usefully combined in the form of axis density, or inverse pole, figures (see for example, Barrett and Massalski, 1966).

Let us consider the implications of limited crystallographic orientation information. As an example, any clay mineral, as orientated by its basal plane sub-fabric, still possesses, according to the restricted information from that fabric, an infinite number of degrees of orientation freedom about an axis normal to the basal planes. Although the crystallographic symmetry is monoclinic, it can be approximated to hexagonal with three of the hexagon sides designated (110) and the other three designated (100). Suppose that (001) are concentrated at the centre of the projection, then for each crystal there would be six discrete concentrations (three of (110) and three of (100)) at the periphery of the projection. But due to these plane multiplicities and the relatively high crystallographic symmetry, there is little constraint on orientation within a plane normal to [001] and so the sub-fabric projections of (110) and (100) would take the form of *girdles* around the periphery of the net. The $D_{\alpha h}$ symmetry would be preserved.

This situation, as described, applies to a uniaxial stress situation such as might occur during the one-dimensional consolidation of a sediment. At an early stage in the consolidation process, the depositional environment conditions the quality of preferred orientation as evidenced by the orientation distribution of [001]. A high degree of concentration at the centre of the plane of projection axi-symmetric about the movement vector would indicate a dispersed sediment in a non-flocculating ground water environment. A more random clay mineral orientation distribution is reflected as a migration of [001] towards the periphery of the projection and is usually associated with edge-to-face floc and domain structure (Gillot, 1968) formation in an electrolytic environment. In the latter case, increasing consolidation pressure can generate clay mineral rotations, fracturing of bonds, and a rotation of [001] towards the axis of pressure, so concentrating the sub-fabric but retaining the $D_{\alpha h}$ symmetry. The degree of sub-fabric concentration may also be a function of the presence of more equant minerals in a polymineralic aggregate and the size of the minerals themselves. For the engineering significance of this point, the reader is referred to Attewell and Taylor (1970) and to Figure 5.7. Additional evidence of control on clay mineral orientation through the presence of more equant minerals in the same rock arises from the work of Attewell *et al* (1969) on Marl Slate (Permian) — (see Figure 5.9).

Dolomite in the Marl Slate can also be used to illustrate a further technique of symmetry interpretation. Reference to Figure 5.10 will show that when the plane of projection of the dolomite (1014) sub-fabric is parallel to the stratification, this sub-fabric again exhibits broad axial symmetry with a girdle of localized maxima at a latitude of about 40° encircling a central maximum. There are also subsidiary minima located between the girdle and the central concentration. The problem is to satisfy such a sub-fabric with the symmetry of {1014}.

The (1014) sub-fabric is compounded from six rhombohedral spacings, three direct and three reflection planes. If the *c*-axes lie normal to the stratification (as was indicated for some of the larger dolomite crystals seen in thin section) and if (conjecturally at this stage of the argument) all the dolomite crystals had acquired the same orientation within the plane of the bedding, the resultant sub-fabric would show, as in Figure 5.11, six point concentrations, each 60° displaced in azimuth from its neighbour and each displaced from the pole to (0001) by a latitude angle of 43°50'. If there is a lack of total constraint on orientation parallel to the stratification but at the same time all the c-axes remain normal to the bedding,

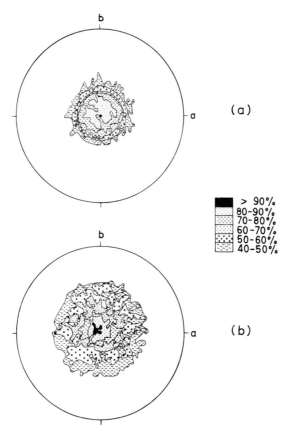

Figure 5.9 Axi-symmetric equal area projections of (a) illite (001) and (b) kaolinite (001) in Marl Slate. Plane of projection is parallel to the stratification and the orientation concentrations are expressed in percentages of the maximum X-ray counts per second. The shape of the lensoid carbonate-clay bands is controlled by the disposition of the dolomite rhomb faces in the slate and depending upon the genetic history of the dolomite some of the clay minerals lie sub-parallel to the stratification (optical evidence) up to maximum angles determined by the orientation of the rhombs. Fabric data from Philips X-ray texture-goniometer and diffractometer (PW 1050, 1078/10, 1310); CuKα radiation. Data processed and direct fabric print-out from IBM 360/67 computer. *Operator:* P. B. Attewell. *From* Attewell *et al.* (1969).

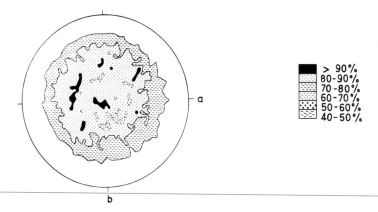

Figure 5.10 Equal area projection of poles to $(10\bar{1}4)$ dolomite in Marl Slate. The projection plane ab is parallel to the fissility and the concentrations are expressed in percentages of maximum X-ray counts per second. 70 per cent–80 per cent concentrations falling *within* the 80 per cent–90 per cent concentration are contoured using a broken line (N.B. The $(10\bar{1}4)$ are cleavage rhomb indices referred to the X-ray smallest cell; they correspond to the cleavage rhomb indices $(10\bar{1}1)$ for the cleavage rhomb pseudo-cell or true cell). Fabric data from Philips X-ray texture-goniometer and diffractometer (PW 1050, 1078/10, 1310); CuKα Ni-filtered radiation. Data processed and direct fabric print-out from IBM 360/67 computer. *Operator:* P. B. Attewell. *From* Attewell *et al.* (1969).

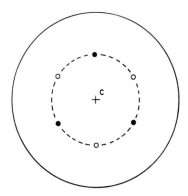

Figure 5.11 Point concentrations representing poles to rhomb faces in a dolomite crystal when the c crystallographic axis is parallel to the c fabric axis and both are normal to the stratification which itself is concordant with the plane of projection. Note that $0001 \wedge 10\bar{1}4 = 43° \, 50.5'$.

then the point concentrations merge to form a girdle of concentration axially about [0001]. Furthermore, any variation in the orientation of [0001] in excess of $43°50'$ about a mean central position on the projection can introduce a central sub-fabric concentration resulting from the overlap of the areas of dispersion of the discrete $(10\bar{1}4)$ reflections. Examination of the sub-fabric, therefore, leads to the conclusion that a preponderance of dolomite crystals possesses a good preferred orientation with their c-axes lying normal to the stratification. Through the $D_{\alpha h}$ sub-fabric symmetry these crystals contribute to the anisotropy of the slate. The sub-fabric also indicates a lower proportion of dolomite crystals orientated such that one of the rhomb faces of each crystal is concordant with the stratification and which, while not contributing in any way to girdle maxima, is responsible for the central concentration coincident with [0001] of the preferentially orientated crystals.

The same interpretation may be applied to the siderite $(10\bar{1}4)$ sub-fabric in Figure 5.12 but in this case there is no central sub-fabric concentration. This implies a much higher degree of preferred orientation within the $D_{\alpha h}$ sub-fabric symmetry and this has been attributed by Attewell *et al* (1969) to a characteristic flattening of siderite nodules parallel to the bedding in Coal Measures cyclothems. This flattening is in response to a rapidly accumulating detritus overload in the non-marine sediments which overlie the carbonate band in the Mansfield marine sequence. By way of contrast, the Marl Slate is essentially a precursor to a carbonate-evaporite sequence embodying a slower rate of sedimentation. Higher sedimentation rates in the Coal Measures seem therefore to promote higher degrees of preferred orientation through the earlier development of confining pressures and pressure differences necessary to stabilize the polymineralic aggregate.

So far, we have considered the situation in which a primary or compounded $D_{\alpha h}$ symmetry represents a response to a stress field that is essentially uniaxial during and after consolidation of a sediment. It is useful to consider now the rather more exacting pressure influences associated with subsequent metamorphism of sediments. In general, none of the axes of principal tectonic strain will be concordant with the original axes of consolidation which control the symmetry of the inherited sub-fabric. If, following metamorphism, there remains evidence of an inherited sub-fabric,

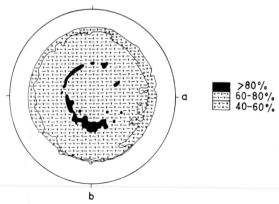

Figure 5.12 Equal area projection parallel to the bedding of poles to (10$\bar{1}$4) siderite in Mansfield carbonate band (Coal Measures). Concentrations are expressed in percentages of maximum X-ray counts per second. 0001 $\wedge$ 10$\bar{1}$4 = 43° 23' and the girdle maximum concentration conforms with this angle. Fabric data from Philips X-ray texture-goniometer and diffractometer (PW 1050, 1078/10, 1310); CuKα Ni-filtered radiation. Data processed and direct fabric print-out from IBM 360/67 computer. *Operator:* P. B. Attewell. *From* Attewell *et al.* (1969).

then it will be combined with an imposed sub-fabric with respect to the same lattice plane distribution to generate a symmetry of triclinic form. However, in three different slates that have been examined in textural detail in the authors' laboratories, there is every indication that an imposed sub-fabric completely overprints and obliterates any potential inherited sub-fabric.

It is useful to consider the implications of this latter situation. In a regional metamorphic stress field in which the amplitudes of the three principal stresses all differ from one another at a particular point, the original clay minerals are subjected to stress gradients in the presence of elevated temperatures and as a result they mechanically rotate and/or re-crystallize to a new orientation configuration. As a function of the symmetry of the stress field (D_{2h} for $\sigma_1 \neq \sigma_2 \neq \sigma_3$) eight discrete concentrations of [001], one in each octant of the *sphere* of projection, might be expected and these concentrations would reproduce on the chosen plane of projection as four double-amplitude concentrations, one in each quadrant.

Detailed X-ray texture-goniometric examination of the clay

mineral [001] distribution in a slate (Greenschist facies) does not reveal four such concentrations. Taking the projection plane parallel to the dominant cleavage the orientation distribution of poles to (001) illite and (002) chlorite take the form of a single concentration at the centre of the projection with a density fall-off that is more rapid along one of the orthogonal axes within the plane of projection than along the other (see for example Figure 5.6). Equal orientation density contours are generally of elliptical configuration in the plane of projection, one axis of the ellipse lying nearly parallel to line of dip of the cleavage and the other lying almost parallel to cleavage strike. In three dimensions, the intensity contours generate an ellipsoidal surface of revolution having an oblate character and it is the degree of oblateness created by the compressive deformation normal to the cleavage relative to the compressive deformation in the plane of the cleavage that ensures a restricted offset from the axis of major principal compression for the four quadrant concentrations. These four concentrations also lie in closer proximity to the axis of intermediate compression than to the axis of minimum compression. In effect, the overlaps of intensity contours from the four concentrations combine to produce a maximum at the centre of the projection which exceeds the maxima of the individual concentrations. Only theoretically under circumstances of reducing oblateness would the compounded concentration decompose back through its discrete component concentrations on the path towards $K_{\alpha h}$ symmetry.

Earlier mention has been made of technical limitations placed upon the choice of ($hkil$) that can be examined by X-ray methods for orientation distribution. These limitations arise through overlaps on diffraction peaks and resolution problems at the counter input slit, and unfortunately the difficulties increase in a polymineralic aggregate. However, even a texture record that is a composite of different ($hkil$) in different mineral crystals can yield useful information provided that all but one of the ($hkil$) composing the sub-fabric are high multiplicity planes in high symmetry, equant crystals, the planes thereby serving only to increase the quasi-random component of the sub-fabric within which the remaining low symmetry $\{hkil\}$ concentrates. Alternatively, the 'background' ($hkil$) may project a lower symmetry on the composite sub-fabric but may still be 'filtered out' by careful interpretation.

As an example of this latter point, consider the composite

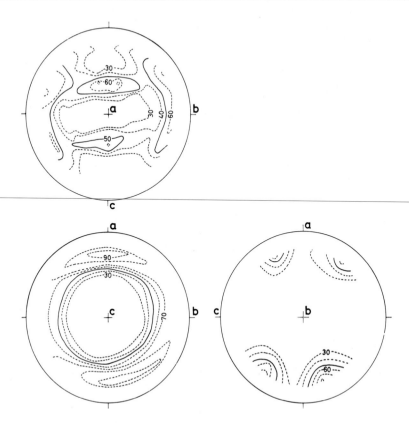

Figure 5.13 Composite equal area projection of poles to $(20\bar{1})$ albite and (112) muscovite in Penrhyn slate. *ab* is the plane of primary cleavage, *b* is the fold axis and *a* is the direction of fold movement within the axial plane. Concentrations are expressed in arbitrary units. Fabric data from Philips X-ray texture-goniometer and diffractometer (PW 1050, 1078/10, 1310); CuKα Ni-filtered radiation. *Operator:* P. B. Attewell. *From* Attewell and Taylor (1969).

sub-fabric for albite $(20\bar{1})$ and muscovite (112) in a sample of slate (Figure 5.13). The former planes diffract X-rays only at rather low intensities but the interference from the latter can be expected to be rather weak. By extrapolation from the known muscovite basal plane fabric, the (112) reflections can be expected to produce a low-intensity sub-fabric girdle about the *c* (crystallographic)-axis, the girdle being modified by the general orthorhombic symmetry of the stress field. In the case of the mineral albite (which would take triclinic symmetry in a low-grade metamorphic environment), the low multiplicity plane ($20\bar{1}$) in a mineral of nonequant habit should project a sub-fabric symmetry which should match the symmetry of this chosen family of planes (compatibility between the angular displacement of the fabric concentrations and the angular relationships between the planes (201) when adjusted for axial ratios) and thereby allow the albite crystals to be preferentially orientated, as it were, in material space.

Unfortunately, in the present example, this compatibility condition does not, on first examination, seem to hold and the earlier stated requirement for several constituent sub-fabrics to possess the same symmetry in order to define a confident fabric symmetry is not satisfied. But if some or all of the albite crystals are *twinned*, then the two concentrations shown on the *ab* cleavage plane projection are satisfied and the basic crystallographic compatibility condition is confirmed through a rather detailed analysis in Attewell and Taylor (1969). In order to facilitate examination of this sub-fabric and in order to demonstrate its basic D_{2h} symmetry as imposed by the external stress field, the three constituent orthogonal sub-fabrics are shown up to latitude angles of 70° on the projection.

The phenomenon of concentration overlap, discussed earlier for dolomite in Marl Slate, is applicable also to a fourth confirmatory sub-fabric for the slate being used as a vehicle for these symmetry discussions. The mineral this time is haematite, the planes (as before) are ($10\bar{1}4$) and there is a weak pyrite (200) overlap on the sub-fabric shown as Figure 5.14. The high symmetry of the cubic mineral and the plane multiplicity imply a quasi-$K_{\alpha h}$ symmetry for this constituent mineral fabric. For the haematite, each concentration along the *ac* axis represents a reflection overlap of three (1014) planes, the core of each concentration being compatible with a crystallographic *c*-axis. The crystals are therefore symmetrically-orientated within the *ac* plane of the fabric and examination of the constituent orthogonal fabrics will quickly indicate that the D_{2h} sub-fabric symmetry of the external stress field is again impressed.

The phrase 'control on orientation' can be used to imply that the crystals of a particular species are constrained to acquire a particular preferred orientation in a three dimensional stress field and generally one in which none of the principal stresses are equal to one another. It has been shown that symmetry considerations are vital to any interpretation. Shape factors and lattice compliances control the quality of preferred orientation and it is useful to consider these matters at this stage.

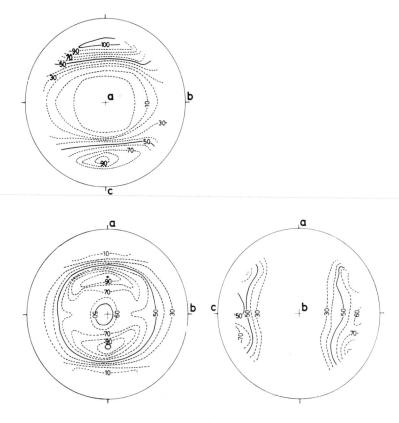

Figure 5.14 Composite equal projection of poles to (10$\bar{1}$4) haematite and (200) pyrite in Penrhyn slate. *ab* is the plane of primary cleavage, *b* is the fold axis and *a* is the direction of fold movement within the axial plane. Concentrations are expressed in arbitrary units. Fabric data from Philips X-ray texture-goniometer and diffractometer (PW 1050, 1078/10, 1310); CuKα Ni-filtered radiation. *Operator:* P. B. Attewell. *From* Attewell and Taylor (1969).

Whereas in the case of the platy clay minerals, the shape factor would appear to predominate over the lattice compliance factor as far as preferred orientation is concerned, the converse is probably true for the more complicated, more equant mineral situation. If, for these minerals, there is a high degree of differential lattice compliance within a crystal and there is a multiplicity of axes of maximum compliance, then it is likely that the axis of maximum linear compressibility within the crystal will be aligned parallel to the direction of maximum principal stress that is applied to the crystal. The axis of maximum linear compressibility is, in effect, the axis of maximum resolved compliance in the crystal and is equi-angular with all the most compliant axes as defined by the poles to planes of a particular form. As an example, and using the texture analyses on the slate, the c (crystallographic)-axis in albite (Figure 5.13) would appear to be the axis of maximum linear compressibility although it may not necessarily be the axis of maximum compliance.

Quartz is a major constituent mineral of many rocks which are of direct interest to the engineering geologist and it is worth noting that the question of a thermodynamically stable preferred orientation for the mineral in a non-hydrostatic stress field has been a matter of theoretical (Kamb, 1959; MacDonald, 1960; Brace, 1960) and experimental (Griggs and Bell, 1938; Crampton, 1958; Brace, 1965; Green, 1966, 1967; Wenk et al, 1967; Wenk and Kolodny, 1968; Baker et al, 1969) interest to structural petrologists for some years. Crampton (1958) has suggested that quartz re-orientates relatively easily through re-crystallization at low stresses and the work of Green (1966, 1967) and Wenk et al. (1967), subsequently confirmed by Wenk and Kolodny (1968), has shown that quartz acquires an extremely high degree of c-axis preferred orientation parallel to an axis of major compression following annealing. The mineral therefore contributes to the structural and strength anisotropy of a poly-mineralic aggregate in a rather less obvious manner than do the platy minerals.

Quartz optic axes are available for direct examination under the optical microscope equipped with a universal stage. Smaller crystals are less amenable to optical examination and the c-axes cannot be studied texturally by X-ray techniques because of diffraction peak overlaps. An X-ray textural study of the orientation distribution of $(11\bar{2}2)$ (second order trigonal pyramid, ξ) planes in the quartz of Ballachulish slate from the Dalradian of the N.W. Highlands of Scotland reveals (Figure 5.15) that the poles to these planes achieve their highest density distribution within a girdle that is centred about an axis normal to the slaty cleavage. Half the girdle included angle is compatible with a c (crystallographic)-axis distribution normal to the cleavage ($c_v \wedge \xi = 47° \ 43.5'$) and therefore, on the basis of thermodynamic theory, designates the crystallographic axis as the axis of maximum linear compressibility in the quartz crystals. With the Cu radiation used to produce the $(11\bar{2}2)$ sub-fabric there is a peak overlap with quartz (0003), and the presence of a central concentration on the fabric to the same intensity level as in the $(11\bar{2}2)$ girdle can be attributed directly to this third order basal reflection

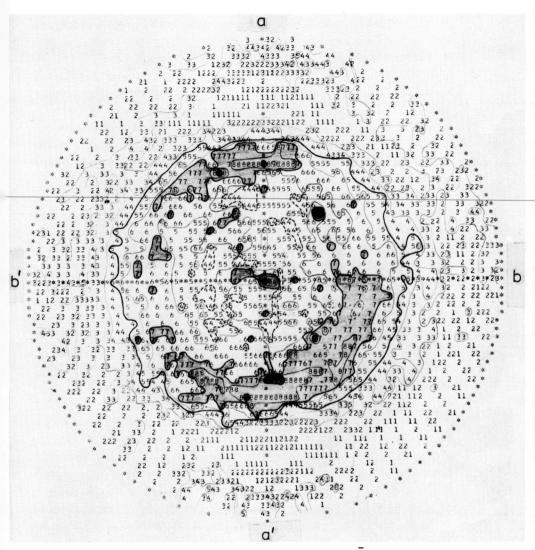

Figure 5.15 Composite equal area projection of quartz (11$\bar{2}$2) and (0003) in Ballachulish slate. Fabric axis aa' is the line of cleavage dip and bb' is the line of cleavage strike. The plane of projection ab is the primary cleavage plane. Concentrations are in percentages of maximum X-ray counts per second. 70 per cent and 80 per cent concentrations are shaded. 90 per cent is crosshatched. Fabric depressions $\leqslant 40\%$ are delimited by broken line for latitude angles $<$ girdle angle. Fabric data from Philips X-ray texture-goniometer and diffractometer (PW 1050, 1078/10, 1310); CuKα Ni-filtered radiation. Data processed and direct fabric print-out from IBM 360/67 computer. *Operator:* P. B. Attewell.

for re-confirmation of the crystallographic c-axis preferred orientation.

Several comments with respect to symmetry can be made on the fabric in Figure 5.15. The (1122) girdle is broken generally along the $b'b$ tectonic axis and the (0003) reflection is represented at a higher intensity than might be expected from comparison with the $(11\bar{2}2)$ level. These points, and the general computer fabric topography can be explained in terms of a greater crystallographic axis directional dispersion in the bc plane compared with the ac plane. Within the girdle, the pole intensities are reduced along the b axis within the bc plane and tend to be enhanced along the ac axis at the interfacial angle $c_v \wedge \xi$ of 47° 43.5'. The (0003) concentration is also elongate along the fabric b axis. These observations and interpretations point to a D_{2h} sub-fabric symmetry for this slate.

It is useful to continue the analysis a little further, but now on the basis of crystallographic compliance. A low quartz would be indicated from the metamorphic grade of the sample of Ballachulish slate examined in the texture-goniometer, in which case the three directions of maximum compliance form an angle of about 70° to the optic axis and are perpendicular to $f_3 (02\bar{2}1)$. The directions of minimum compliance (greatest stiffness) on the other hand are angled almost perpendicular to the first order r rhombohedron faces $\{10\bar{1}1\}$. Although the angle contained between the most compliant axis and the optic axis decreases slightly with increase in temperature, there is a concomitant decrease in the amplitude of the compliance. But since the effective girdle of maximum compliance describes an included angle, centred on the optic axis, greater than 90°, the optic axis might reasonably be designated *not* to be the axis of maximum linear compressibility but rather to lie at right angles to the direction of maximum compression. Axes $[hki0]$ would then be aligned parallel to the maximum compression. Considering now a β-quartz crystal and noting that the change in class from 32 (trigonal) to 622 (hexagonal) does not affect the issue, it is found that the most compliant direction is at about 30° to the optic axis, that is, less than 5 per cent angular offset from being normal to $(10\bar{1}2)$. Furthermore, calculations on the variation of elastic constants with temperature show that the most compliant direction more nearly achieves concordance with the optic axis as the temperature rises above 600°C, but this observation is, of course, irrelevant to the analysis of this low grade slate.

Based on thermodynamical reasoning, therefore, the fabric evidence is consistent with the existence of β-quartz in the slate, and since this would be an unusual occurrence the arguments must necessarily be taken a little further. It must be stated that even with a temperature enhancement from the aureole of the underlying Ballachulish granite, this particular slate was unlikely to have achieved the α–β quartz inversion temperature of 573°C, particularly when standard X-ray diffraction evidence is wholly consistent with the presence of low quartz and also when other mineralogical evidence, such as the presence in the slate of the mineral kaolinite, is taken into account. Furthermore, the higher polymorph is unstable and must rapidly revert to α-quartz when the rock cools below the inversion temperature. So, the analyst is left with a situation in which either these earlier theoretical concepts of crystallographic mechanical stability are incorrect or inadequate, or alternatively with one in which the inversion temperature at that particular metamorphic pressure was exceeded but in which the preferred orientation of the higher polymorph was retained even on the structural reversion to the lower polymorph.

In conclusion, and in spite of these interpretative problems in the example discussed above, it is useful, now that the basic arguments have been outlined, to state very briefly the concept behind the control on crystallographic orientation. On thermodynamical grounds it can be predicted that anisotropic mineral crystals comprising a rock aggregate should so orientate themselves in a non-hydrostatic stress field such that the strain energy of deformation is maximized at the expense of the potential energy of the external deforming forces plus the potential energy of deformation that is being minimized. Although the minimum potential energy theorem is applicable to plastically-deforming minerals, the theory is probably more applicable to the orientation of re-crystallized materials since the residual strain energy of the re-crystallized materials should generally be less than the induced strain energy of plastically-deformed grains and hence the former should offer a lower perturbing influence upon any subsequent metamorphically-applied stress field.

5.9 Engineering influence of intrinsic anisotropy

In a homogeneous, macroscopically-structureless *clay,* the mass strength anisotropy will be controlled almost totally by the

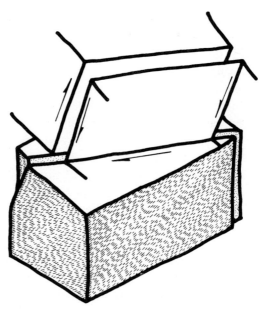

Figure 5.16 Step discontinuity shearing through a solid rock element at an angle to the intrinsic fissility.

orientation density distribution of the constituent clay minerals. In a regularly jointed *sedimentary or igneous rock* the mass strength anisotropy can be attributed almost exclusively to the joint orientations and spacing; any intrinsic anisotropy is ignorably small because of the dominance of more equant minerals. *Schistose rocks* (Akai *et al*, 1970; Masure, 1970; Pinto, 1970) owe their mass anisotropy in large measure to a crystallographic control which probably becomes more of an engineering issue as the *metamorphic grade* decreases (for example, the discussion on *slate* considered in the next section). In a *shale*, intrinsic strength anisotropy is low relative to that in a slate but, since the macroscopic discontinuities in the mass are irregular and terminated, its importance cannot be dismissed. This type of intrinsic anisotropy is derived from a crystallographic control unlike that in a *varved mudstone* which is more of a sedimentary origin.

It is possible to illustrate diagrammatically how a failure surface might be routed according to a macro- and micro-structural control in the type of rock which demands this form of consideration. Suppose that there is a shearing facility along a discontinuity but

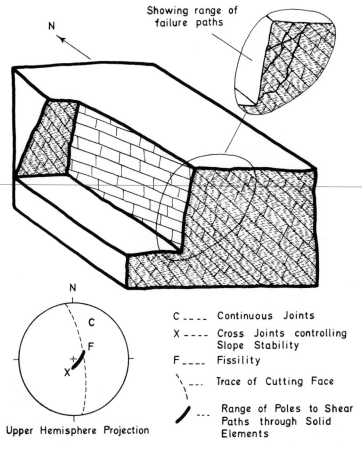

Figure 5.17 Cutting in jointed mudstone.

that if the shear fracture is to propagate further, it must then cut through an intact rock element at an angle to a plane of intrinsic anisotropy marked with broken lines in Figure 5.16. In the case of an over-steepened slope, Figure 5.17 shows how this through-cutting might develop in an idealized situation and Figure 5.18 illustrates the possible mechanics that might operate during cross-shear and the stress state that might apply in any analysis of intact shear strength. Of course, the shear strength of the mass would be compounded from the intact shear strength parameters (direction-dependent) and the shear strength of the joints, and as shown elsewhere the relative contribution of the two sets of parameters being a function of the shear path length within the (previously) intact and jointed rock. The

288   *Principles of Engineering Geology*

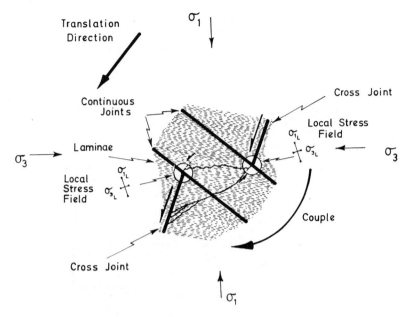

Figure 5.18   Possible secondary shear failure paths through solid elements of an intrinsically anisotropic rock.

fact that the shear strength parameters are direction-dependent in intrinsically anisotropic rock is confirmed by the curves in Figures 5.19 and 5.20.

5.10 Comparative degree of intrinsic anisotropy – mechanical evidence from rock experimentation

Whereas mechanical anisotropy in clays would be quantified through direct shear tests, it is necessary in general to perform uniaxial or triaxial compression tests on rocks. These latter tests involve a suite of specimens, each specimen being presented to the compression equipment with its plane of anisotropy – cleavage, fissility, stratification, inclined at different angles β to the major principal stress σ_1 direction. Subsequently, the angle λ formed by the shear failure plane and the σ_1 axis is noted. It is apparent that the greater the degree of anisotropy and the more dominant the primary feature plane in the specimen the more closely will λ be to β throughout the range of β. This means that for high and low β angles, the plane of final shear failure cannot be planar-concordant with a plane of maximum resolved shear stress.

Results from several rocks studied by Donath (1961), Akai *et al*

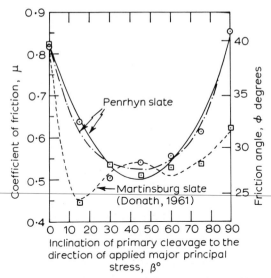

Figure 5.19 Variation of friction with inclination of an anisotropic plane in a stress field. A polynomial regression line links Donath's data points. The broken line linking the Penrhyn slate data points is a polynomial regression and the full line is fitted by the equation

$$\mu = \mu_1 + \mu_2 \cos 2(\lambda - \beta)$$

where λ is the angle formed by the shear failure plane and the major principal stress direction. *From* Attewell and Sandford (1974a).

(1970) and the present authors are plotted in Figure 5.21. Martinsburg slate follows the line $\lambda = \beta$ much more closely than does Penrhyn slate, and since both materials are homogeneous, albeit anisotropic, it can be concluded that the orientation density distribution of the clay mineral basal planes is much more preferred in the former compared with the latter. The two inhomogeneous mudstones are much less anisotropic with respect to this criterion compared with the slates and with the Longwood shale. The graphite schist is more homogeneous and can best be compared with Penrhyn slate.

5.11 Intrinsic strength anisotropy of brittle and semi-brittle rocks comprising a dominant clay mineral control

The mechanics of anisotropic shear failure have been considered in the papers of Jaeger (1960), Donath (1961) and Barron (1971 a,b,c)

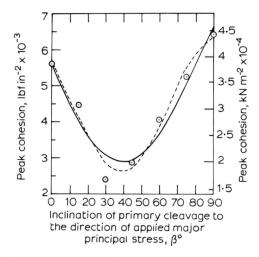

Figure 5.20 Variation of cohesion with inclination of an anisotropic plane in Penrhyn slate. Broken line represents a 4th degree polynomial curve fit. Full line is fitted according to the equation

$$c = c_1 - c_2 \cos 2(\lambda - \beta)$$

where λ is the angle formed by the shear failure plane and the major principal stress direction. *From* Attewell and Sandford (1974a).

and examined in detail in the book of Jaeger and Cook (1969). Genetic aspects of, and the controls on failure initiation have formed the basis of papers by McClintock and Walsh (1962), Walsh and Brace (1964), Hoek (1964), Weibols and Cook (1968), Brady (1969 a,b), Barron (1971 a,b,c), and Attewell and Sandford (1974 a,b,c).

Brittle and semi-brittle materials are replete with microscopic and sub-microscopic structural discontinuities (this has been treated in a simple manner in Chapter 4) around which are concentrated rather higher local stresses than exist in a 'smoothed' condition within the body of the material as a whole. These discontinuities, which are treated as 'Griffith cracks' (Griffith 1921, 1925) also serve as zones of weakness, since in the 'open' condition the elevated stresses at their tips facilitate extension and fracture and in the 'closed' condition more readily permit shear to develop along their surfaces. More equant voids, such as pore spaces, contribute less directly to a reduced strength condition and in the rocks such as slates and shales

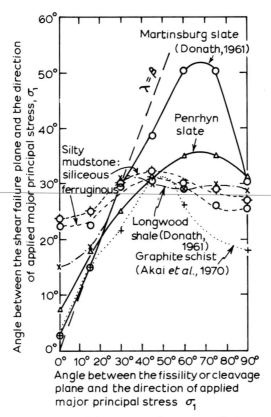

Figure 5.21 Inducement of an eventual plane of shear failure into a plane of natural weakness in certain rocks having varying degrees of intrinsic anisotropy.

that are intrinsically anisotropic to quite a high degree, such microscopic evidence as exists suggests that in a high proportion of the narrow cracks, their surfaces are in contact. A basic proposition is that the individual shears can be compounded into a final macroscopic shear surface, the failure condition of which can be analysed conventionally through the application of Coulomb–Mohr theory. Intrinsic strength failure can therefore be considered in terms of an energy criterion of failure.

Weibols and Cook (1968) considered the rather special case of a random distribution of discontinuities and proposed that the necessary and limiting strain energy was acquired through the summation of individual crack extensions which occurred when the available shear stress along a crack $|\tau_c|$ exceeded $\sigma_{nc} \tan \phi_c$ where

σ_{nc} is the normal stress on a crack surface and ϕ_c is the friction angle of that surface. If ϕ_c can be estimated for a consensus of cracks, then writing the direction cosines for the first quadrant, $|\tau_c|$ and σ_{nc} can be evaluated:

$$\sigma_{nc} = a_{11}^2 \sigma_1 + a_{12}^2 \sigma_2 + a_{13}^2 \sigma_3$$

and

$$\tau_c^2 = a_{11}^2 \sigma_1^2 + a_{12}^2 \sigma_2^2 + a_{13}^2 \sigma_3^2 - \sigma_{nc}^2$$

Rocks which comprise a predominance of more equant minerals such as quartz will possess a crack density distribution which is more nearly spherically-symmetric and which therefore approximates more satisfactorily to the Weibols and Cook situation. The intrinsic strength of these rocks is usually higher and so the practical controls on their mass stability are to be found in the larger-scale joint and bedding structural discontinuities.

Walsh and Brace (1964) and Hoek (1964) in their analyses specified fields of long cracks and short cracks, the former being preferred and conditioning the shear strength anisotropy and the latter being randomly distributed. The short cracks were really only introduced as a device into the model solely to inhibit the attainment of infinite strength. Attewell and Sandford (1974 b,c) have proposed that any arbitrary form of crack orientation density distribution can be accommodated in an analysis provided that the function describing this distribution can be derived from the distribution of the crystal faces of the minerals which comprise the solid rock constituents. This implies that each crystal face defines a small planar discontinuity along which shear can take place under a suitable combination of principal stresses, but several approximations must be made. First it is assumed that the higher symmetry crack orientations associated with the more equant minerals weaken the rock uniformly and do not contribute to the anisotropy. Second, it is assumed that there is a relative crack length influence to the extent that discontinuities associated with (001) of the clay minerals predominate with respect to strength anisotropy to the exclusion of such planes as (110) and (100), these shorter cracks, although creating girdles about (001), being assumed to generate a further isotropic field. Third, it is assumed that the *form* of the distribution is the same for each contributing mineral even if the *intensities* vary. The plots of intensity versus η (latitude angle) for two orthogonal sections in Penrhyn slate (Figure 5.22) confirm that this assumption, for that material at least, is quite valid.

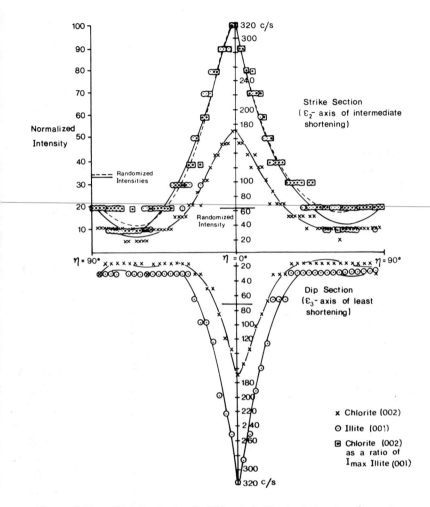

Figure 5.22 Distribution of diffracted X-ray intensity from two minerals in Penrhyn slate about two orthogonal axes between $\eta = -90°$ and $\eta = +90°$. Graphs indicate that; (a) there is a degree of asymmetry in the intensity distribution but the form of the distribution is generally symmetric along the scan axis about a normal to that axial plane; (b) the form of the intensity distribution is the same for both minerals scanned; (c) the rate of intensity fall-off through the latitudes is greater along the axis of least shortening than along the axis of intermediate shortening.

The first step in the process suggested by Attewell and Sandford (1971, 1974b) is to use a least squares curve-fitting procedure to approximate the experimental textural data to a series of associated Legendre polynomials which then represent an equivalent crack pole distribution function. It is this approximation that is used in theoretical strain energy computations.

This procedure is relatively simple for the high symmetry, axi-symmetric or orthorhombic, sub-fabrics of concern to the problem. Lower symmetry sub-fabrics are much more difficult to handle in this manner and it is worth noting that in the larger scale macroscopic discontinuity problem considered in Chapter 6, triclinic symmetry may be more usual. In order to illustrate the approach, the Penrhyn slate illite (001) sub-fabric synthesized earlier is shown with a polynomial fit in Figure 5.23. Similarly, for a mudstone, a processed kaolinite (001) sub-fabric is shown to the right of Figure 5.24 and the reconstituted version is shown on the left. The latter material is a Coal Measures mudstone, quite strongly laminated, with the plane of projection parallel to the laminations, and the symmetry of the sub-fabric appears to be slightly orthorhombic with the axis of least compression orientated north-south (top-bottom) on the projection. Some of the possible reasons for this symmetry have been considered by Cripps (1971). It will be noted that the reconstituted sub-fabric reproduces the original sub-fabric quite acceptably. This means that since the sub-fabric can be fully described in a mathematical sense, it is now available in a suitable form against which any triad of principal stresses can be tested.

Curves defining the strength anisotropy of the slate and the mudstone discussed earlier are shown in Figures 5.25 and 5.26. These are of the same form as the slate curves published by Donath (1961) and the shale curves of Chenevert (1965). For a more detailed discussion of the interpretation of such curves refer to Jaeger and Cook (1969) and for further experimental results on schistose and slate rocks see, for example, Rodrigues (1970) and Pinto (1970). Attempts by Attewell and Sandford (1974 a,b,c,) to satisfy quantitatively this anisotropic response by applying the theories of both Weibols and Cook (1968) and Brady (1969 a,b) to crack distribution functions formulated on the basis of a clay mineral basal plane pole figure met with no real success. Implicit in these attempts, however, was the notion that the crack *lengths remained constant* for all orientations but that their numerical densities decreased with

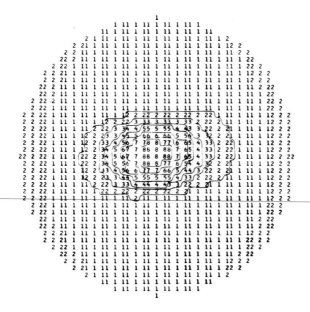

Figure 5.23 Associated Legendre Polynomial fit to pole figure data for illite (001) in Penrhyn slate. Function used:

$$I(\eta, \theta) = \sum_{l=0}^{\infty} \sum_{m=-l}^{\infty} [P_l^m(\eta) \sin(m\theta) A_{lm} + P_l^m(\eta) \cos(m\theta) B_{lm}]$$

where η is the dip angle of [001]
θ is the azimuth angle of [001]
l and m are quantum numbers

Polynomial fit accomplished with a limit value for l of 8 to give 45 coefficients. Processing by IBM 360/67 computer. *From* Attewell and Sandford (1974b).

increasing angular departure from the cleavage or fissility. The work of Walsh and Brace (1964) and of Hoek (1964) can, however, provide the starting point for the formulation of a new crack distribution function, one in which the *lengths decreased* with angular departure from the anisotropic plane, this length distribution being derived from the pole figure. Barron (1971 a,b,c) in fact, performed such a derivation for a sandstone.

Having obtained a form of crack length distribution, Griffith theory can be applied and predicted strength curves generated. From

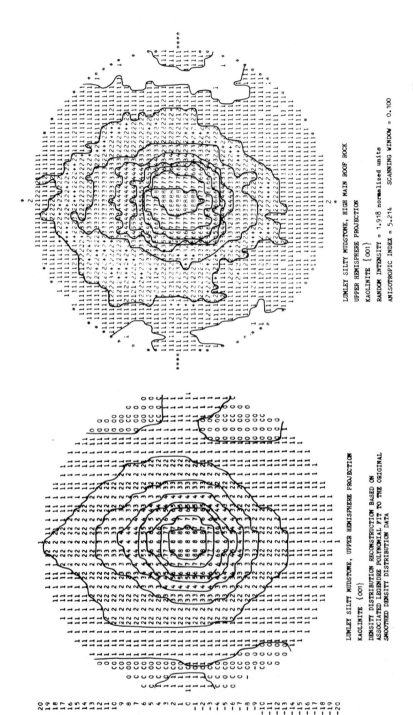

Figure 5.24 Kaolinite (001) sub-fabric in a laminated mudstone which forms the roof rock of the High Main coal seam in County Durham, England. Plane of projection is parallel to the laminations (bedding) and the projection is orientated NSEW. Intensities are expressed as a percentage of maximum intensity which, since it is highly concentrated at the centre of the projection, is smoothed out of the sub-fabric during the computer processing (eg $8 \equiv 80\%$). *From* Attewell and Sandford (1974b).

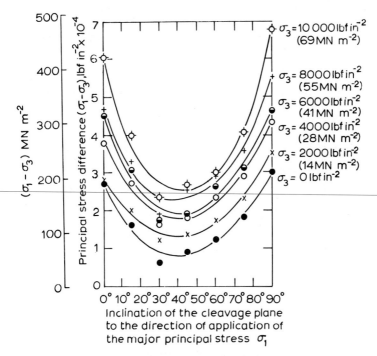

Figure 5.25 Intrinsic strength anisotropy of blue Penrhyn slate, North Wales. *From* Attewell and Sandford (1974a).

Griffith's theory, the stability criterion becomes

$$|\tau| - \mu_c \sigma_{nc} < 2 \left(\frac{K}{c}\right)^{1/2} \qquad (5.1)$$

where, as before, $|\tau|$ is the applied shear stress; μ_c and σ_{nc} are the coefficient of friction and normal pressure on a crack, respectively; K is a constant; c_c is the crack length.

Re-writing this equation, and taking c_c from the unmodified and normalized X-ray texture (Attewell and Sandford, 1974b), the criterion then becomes:

$$|\tau| - \mu_c \sigma_{nc} = \frac{B}{(I)^{1/2}} \qquad (5.2)$$

with I being the normalized diffracted X-ray intensity value. Constants μ_c and B have to be determined by mechanical experimentation.

If curves of $(|\tau| - \mu_c \sigma_{nc})$ as a function of the anisotropy angle

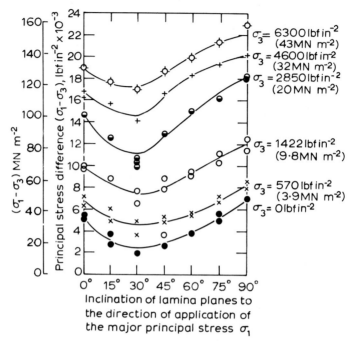

Figure 5.26 Intrinsic strength anisotropy of a laminated silty mudstone (Carboniferous), County Durham, England.

are constructed for various values of principal stress and for known values of μ_c (see, for example, Figure 5.27), then these curves may be used with a plot of $B/(c_c)^{1/2}$ versus β to determine strength levels, or, conversely, strengths may be used with the $(|\tau| - \mu_c \sigma_{nc})$ curves to generate $B/(c_c)^{1/2}$ values. If curves $B/(c_c)^{1/2}$ ($\equiv B/(I)^{1/2}$) versus β and $(|\tau| - \mu_c \sigma_{nc})$ are superimposed so that for a given σ_1 (or $\sigma_1 - \sigma_3$) the two curves just touch (Figure 5.28), then this defines a condition of potential shear failure. The angle through which the $(|\tau| - \mu_c \sigma_{nc})$ curve must be displaced along the β axis to achieve this result determines the shear failure condition.

The acceptability or otherwise of this concept of a smoothly decreasing effective crack length with orientation out of the cleavage rests upon the quality of fit between the experimental strength data points and the theoretical strength. Relationships for the four angles shown in Figure 5.29 suggest that there are some grounds for feeling that the form of the clay mineral distribution in a clay mineral-rich rock does indeed control the intrinsic strength anisotropy.

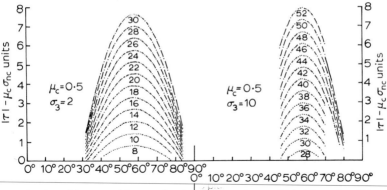

Figure 5.27 Families of $|\tau| - \mu_c \sigma_{nc}$ curves for different unit values of σ_3 in a slate. The value for σ_1 is marked against each curve. *From* Attewell and Sandford (1974c).

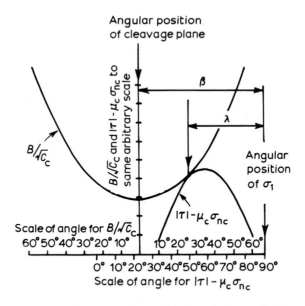

Figure 5.28 Showing graphical solution of the equation: $|\tau| - \mu_c \sigma_{nc} = B/(c_c)^{1/2}$. The two curves are moved relative to each other until they just touch. Displacements λ and β represent the orientation of σ_1 with respect to the failing crack and the cleavage plane respectively. *From* Attewell and Sandford (1974c).

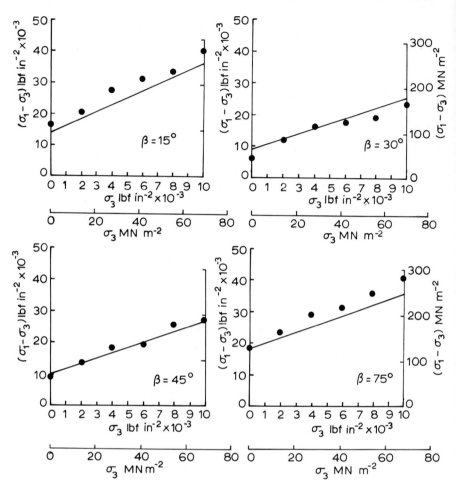

Figure 5.29 Theoretical and experimental principal stress relationships for Penrhyn slate. Continuous line represents theoretical fit with $B = 8.2$ and $\mu_c = 0.5$. Circles are experimental data points at failure. *From* Attewell and Sandford (1974c).

For further reading on the anisotropic behaviour of schistose and other rocks, reference may be made to Deklotz and Stemler (1966), Mendes (1966), Akai *et al* (1970), Pinto (1970), and Rodrigues (1970). The anisotropic response of coal has been studied by Pomeroy *et al* (1970).

5.12 Intrinsic anisotropy and sedimentation

Earlier in Chapter 3 we have considered the influences imposed on clay mineral sedimentation characteristics by the electrolytic content of the water through which they settle. The preferred orientation of these minerals is also conditioned by the nature of the more equant minerals that sediment out with them. Without delving further into this specialist subject, it will be appreciated that in a rock comprising closely-cyclic sedimentary structures, such as a laminated or varved mudstone, the degree of preferred orientation of the morphologically

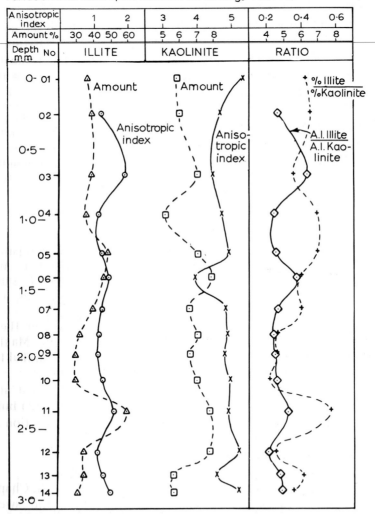

Figure 5.30 Variation of degree of anisotropy through the laminations in a sample of Coal Measures mudstone (*after* Cripps, 1970).

inequant minerals will tend to vary vertically through these structures. This problem has been studied by Cripps (1970) using X-ray texture-goniometric methods and the results discussed by him in a sedimentological context.

Figure 5.30 is a plot of a number of parameters related to clay minerals and their preferred orientation against vertical distance through the laminations in a mudstone. A cyclic variation with distance will be noted, and although it is somewhat irregular the general wavelength of the variation (~ 1.5 mm) coincides with the average lamination thickness. The illite percentage in the general sample is about five times that of kaolinite, each of these minerals rising and falling in phase through the laminations. It will be noted, however, that the anisotropic index values for the illite and kaolinite, that is, the degree of preferred orientation projected by their respective basal planes, have a tendency to vary half a wavelength out of phase, with the kaolinite being about three times as strongly preferred as the illite. This latter can probably be attributed to a size effect, kaolinite crystals being much larger than those of illite.

5.13 Anisotropy of clay shales

From the earlier evidence of strength anisotropy in mudstone, a similar response is to be expected from a more fissile shale. Wright and Duncan (1969) performed unconsolidated-undrained compression tests on specimens of Bearpaw shale and Pepper shale cut at various angles to the fissility. Figure 5.31 shows that although there are considerable variations in absolute strength between the two shales, the form of the anisotropy is similar in both. Maximum strength is mobilized when the compression is applied parallel to the 'horizontal fissuring'.

Directional strength differences are less clearly defined at low confining pressures in the Green River shale (Figure 5.32) but this material again demonstrates the manner in which the friction angle varies with specimen orientation.

5.14 Clay strength anisotropy

It will be apparent, from the discussion on *clay* fabrics in Chapter 3 and from Rowe's (1972) Rankine lecture, that the swelling, consolidation and shear behaviour of a clay in particular, and its engineering behaviour in general, will be to some extent direction-dependent. We might expect, for example, that a flocculated clay in

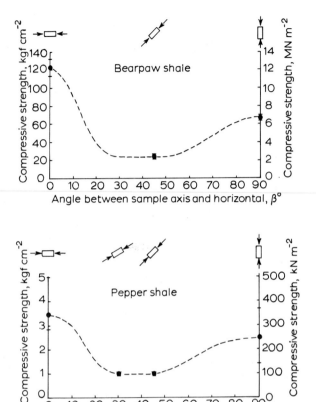

Figure 5.31 Strength anisotropy of two clay shales (*after* Wright and Duncan, 1969).

which the clay minerals are more randomly orientated would respond to pressure in a more isotropic manner. A similar argument would apply to a silty clay in which clay mineral stratification is locally disrupted by the presence of the more equant grains of such minerals as quartz, feldspars or carbonates. A dispersed sediment, on the other hand, comprising a very high clay mineral content and having a lower void ratio would exhibit, in theory at least, a stronger degree of anisotropy. The work of Lambe (1953), Mitchell (1956) and Rosenqvist (1959) on salt-flocculated clays, marine clays and lacustrine clay leads to a significant improvement in our understanding of the engineering significance of clay mineral structure, but whether structural anisotropy together with directional differences in

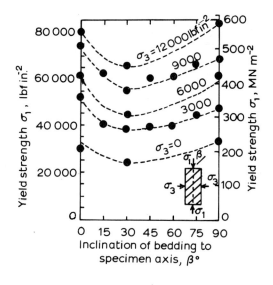
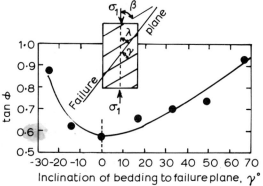

Figure 5.32 Variation of strength and friction angle with inclination of the bedding in Green River shale (*after* Chenevert and Gatlin, 1965).

shear strength and consolidation are sufficiently significant to become practical design issues is another matter. More recent work tends to suggest that these factors, together with an associated permeability anisotropy, will be allocated rather more consideration in future design work.

Duncan and Seed (1966) have defined two possible types of anisotropy with respect to undrained strength:

(i) Anisotropy with respect to the values of the shear strength parameters in terms of effective stresses;

(ii) Anisotropy of the soil with respect to the development of pore water pressure; application of the same change in total stress would result in pore water pressure changes dependent on the orientation of the applied stresses.

In practical soil slope stability terms, since the principal stresses are differently orientated along a slip circle, shear strength anisotropy, whether drained or undrained, should be taken into account during analysis. Furthermore, at any particular element of clay there will be a change of stress on that element with time and this will also result in a rotation of the principal stress axes with respect to a developing shear plane. The rotation problem in two-dimensions is drawn out in Figure 5.33. It will be noted that due to its fixed orientation within a clay mass, the anisotropic plane serves as a reference against which the changes in orientation of the principal stress axes may be evaluated from the Mohr diagram.

Just as the clay minerals in low-grade metamorphic rocks assume a generally perpendicular orientation to the direction of maximum compression, so do we find a similar orientation tendency in a clay during one-dimensional consolidation. Deformation of a clay mineral aggregate and changes in its void ratio involves the breakage and slipping of inter-plate bonds and also the bending of plates. Buessem and Nagy (1953) have described the translation and rotational effects of clay mineral platelets within a hydrostatic stress field, and more recently Geuze and Rebull (1966) have proposed a hypothetical model to describe local failure of a clay.

As an aid to understanding the mechanics of local, internal deformations, many researchers have used laboratory-sedimented suspensions of monomineralic kaolinite, finely-ground muscovite, illite and montmorillonite. Reference may be made, for example, to Olson (1962, 1963), Olson and Mitronovas (1962), Olson and Hardin (1963), Pusch (1966), Smart (1966), Slone and Kell (1966), Duncan and Seed (1966), Morgenstern and Tchalenko (1967 a,b,c), Smalley and Cabrera (1969), Mesri and Olson (1970), Foster and De (1971), Hambly (1972), Barden (1972), Kirkpatrick and Rennie (1972), Sridharan and Rao (1973), and Sankaran and Bhaskaran (1973). Similar work has also been performed in the authors' own laboratory (Over, 1969; Connelly, 1970).

Fundamental research with artificial monomineralic suspensions is of rather less practical relevance than is experimental work of an

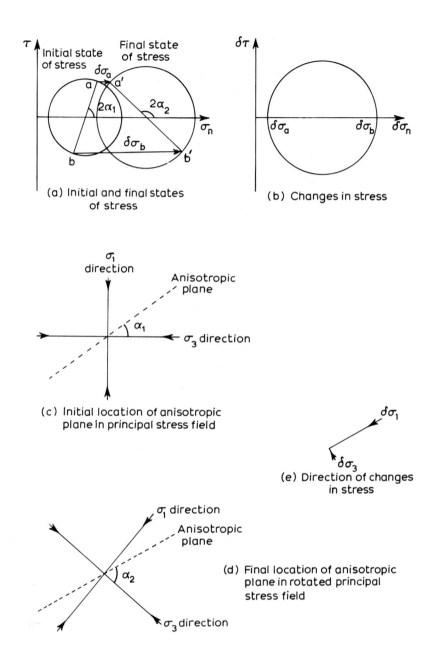

Figure 5.33 An example of the rotation of principal stress axes (generally *after* Duncan and Seed, 1966).

anisotropic character on natural polymineralic clays. Local deformation mechanisms are much more difficult to understand when there is a lack of physical control on the structure of the sediment and, for this reason, most quantitative and semi-quantitative investigations of the mechanical behaviour of polymineralic but *homogeneous* clays have been concerned solely with their overall shear strength and stiffness as a function of orientation with respect to a direction of maximum stress difference (see, for example, Agarwal, 1967; Pickering, 1970; Lutton and Banks, 1970; Bishop, 1966, 1972; Gibson, 1974). Permeability anisotropy has been considered by Poskitt (1970) and Smedley (1974), and the influence of anisotropy on slope stability by Lo (1965) and by Martin and Kayes (1971).

It is found, however, that most field problems where anisotropy becomes an issue relate to clays which are *inhomogeneous* on a scale of direct visual inspection. Over considerable areas of North America, Scandinavia and the U.S.S.R. there are, for example, clays which have a definite banded structure and which have resulted from cyclic changes in temperature and sedimentational environment. These cyclic changes have introduced to the clay structure two alternating layers, one of which is generally coarser in grain size and lighter in colour than the other. The term 'varved' is usually reserved for this type of clay, but alternative terms are 'lacustrine', 'lake', and 'laminated'. De Greer (1910), in first proposing the term 'varve', did in fact apply it to both marine and fresh-water deposits but it is now generally restricted to mean 'rhythmically banded sediments of a glacial lake' (Milligan *et al*, 1962).

Antevs' (1925) original hypothesis concerning annual deposition was examined by Burwash (1938), Legget and Bartley (1953), and Eden (1955). It is now accepted that rather than allocating silt and clay layers, light and dark layers, to summer and winter deposition, the control on the sedimentological style is much more one of flocculation caused by the presence of electrolytes. In these clays, therefore, the degree of flocculation is reflected as a distinct macrostructural anisotropy *and inhomogeneity*, the former being a direct consequence of the latter.

Because of the layered structure of a varved or laminated clay together with a marked difference in grain size and plasticity of the individual laminae comprising a varve, it is axiomatic that the horizontal permeability will tend to be greater than the vertical permeability. On the other hand, if there is a general lack of hydraulic continuity parallel to the varves due to depositional disturbances, perhaps in the form of bottom currents, then the material will not possess the ideal drainage characteristics that should, in theory, be its natural advantage. A number of foundation case histories for varved clays have been reviewed by Milligan *et al* (1962) and slope stability problems in laminated clays are described in Chapter 9. Consolidation characteristics of these clays have been considered by Rowe (1959, 1964) and the shear strengths of some 'quick', varved, anisotropic clays have been studied by Sangrey (1972).

Tests on the relatively homogeneous London Clay by Bishop (1971) have demonstrated its anisotropic character. Work up to 1966 on inhomogeneous clays has been summarized by Duncan and Seed (1966, Figure 3). More recently, Lo and Morin (1972) have studied the anisotropic behaviour of Canadian quick clays having 'cementation bonds' of the type discussed in Chapter 3 (Conlon, 1966; Kenney, 1967). Using drained and undrained triaxial tests with pore pressure measurement (Figure 5.34) they were able to show significant directional differences in both peak shear strength and drained tensile strength. On the other hand, the residual strength at large strains was found to be sensibly independent of specimen orientation.

Earlier in the present chapter it has been shown that the shear strength parameters for a slate vary with planar anisotropy angle in much the same manner as does the strength of the sample. Noting the early work of Casagrande and Carrillo (1944), this same behaviour may be assumed to apply to the clay in Figure 5.34. The form of c, ϕ variation with angle has been interpreted by Jaeger (1960), Donath (1961) and by Attewell and Sandford (1974a, Appendix 1).

The present authors have been concerned with the stability and surface settlement characteristics of tunnels in laminated clay and for that reason initiated a laboratory test programme on the clay (Figure 5.35, Table 5.1); the structure of the clay is also illustrated in Figure 11.10. The programme was directed specifically towards the consolidation − negative consolidation (swelling), and shear strength, anisotropy of the undisturbed clay. Although the results on a single laminated clay only are used in the text for illustration purposes, it can readily be shown (for example, Smedley, 1974) that laminated

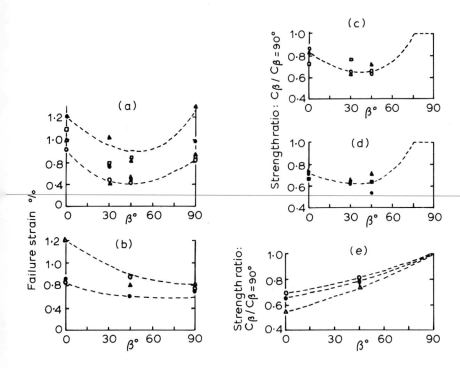

Figure 5.34 Anisotropy of two quick clays (*from* Lo and Morin, 1972) — St. Louis Clay (a, c, d) and St. Vallier Clay (b, e). Clays triaxially tested (unconsolidated drained and consolidated undrained with p.p. measurement) at confining pressures ranging from $1.5 - 10$ lbf in^{-2}. β is the angle that the stratification makes with the deviator stress direction.

Table 5.1 Geotechnical properties of laminated clay discussed in text

Natural Moisture content	28%
Liquid limit	53%
Plastic limit	23%
Plasticity index	30%
Specific gravity	2.61
Bulk unit weight	18.15 kN m^{-3}
Dry unit weight	14.22 kN m^{-3}
Percentage saturation	91%
Undrained shear strength (deviator stress applied normal to laminations)	$c_u = 73.2$ kN m^{-2} $\phi_u = 0°$
Drained shear strength (mean of nine sample orientations)	$c'_p = 21.7$ kN m^{-2} $\phi'_p = 19.5°$ $c'_r = 0$ $\phi'_r = 11.0°$

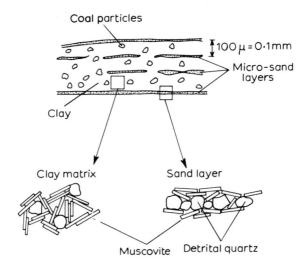

Figure 5.35 Schematic representation of laminated clay structure (*after* Leach, 1973).

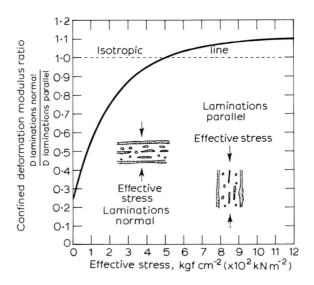

Figure 5.36 Influence of effective stress magnitude upon the confined modulus ratio of a laminated clay (*after* Leach, 1973).

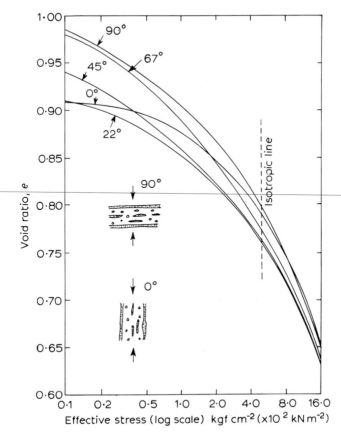

Figure 5.37 Consolidation curves for laminated clay tested at five different orientations (*after* Leach, 1973).

clays from different areas tend to respond both absolutely and anisotropically in a different geotechnical manner.

Derived from consolidation tests and the recording of stress-strain relationships, the curve in Figure 5.36 indicates that at low normal effective stresses quite large strains result from direct closure of the laminations whereas the material is much stiffer when the compression takes place parallel to the laminations. However, at higher normal effective stresses, with the stress-strain curves always concave-upwards, the stiffness anisotropy disappears.

Displaying a similar effect, the void ratio is plotted in Figure 5.37 against the normal effective stress for different attitudes of the laminations with respect to the direction of consolidation pressure.

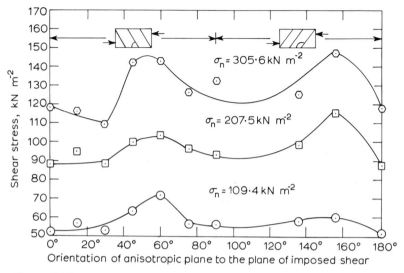

Figure 5.38 Variation of peak shear strength of an undisturbed sample of laminated clay with the orientation of the plane of anisotropy (*data from* Leach, 1973).

After initial stiffness anisotropy, the curves converge to become almost parallel at an effective pressure of about 5 kgf cm^{-2} (490 kN m^{-2}) as indicated in Figure 5.36. Although it is an academic point, the smaller radius of curvature associated with the more vertical laminations would make the determination of a pre—consolidation load rather more easy.

Although shear strength is most easily evaluated under conditions of direct shear in a shear box, definition of the orientation of an anisotropic plane with respect to an imposed shear plane requires a little care. There is a range of orientations (0° to 90°) for which the shearing takes place with the anisotropic plane and tends to rotate it into a configuration more parallel to the shear plane. But there is also an orientation range (90° to 180°) where the sense of rotation and accompanying dilation is towards a right-angled attitude, anisotropic plane to shear plane. The shear strength will tend to vary throughout this total range, as shown in Figure 5.38.

Contrary to what one might have expected, maximum shear strength does not occur when the anisotropic planes are orientated at 90° to the shear plane. Rather, the two shear strength peaks occur between angles 55° to 60° and at 155°, and are greatest overall at the latter angle. From the Mohr envelopes for each orientation, two

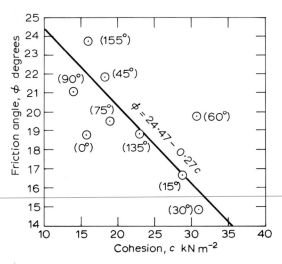

Figure 5.39 Correlation between friction angle and cohesion in a laminated clay for different inclinations of the laminations with respect to the imposed shear plane in direct shear tests (*data from* Leach, 1973).

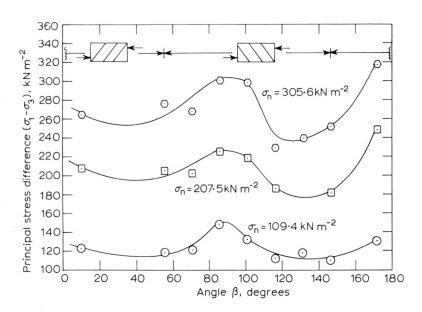

Figure 5.40 Variation of the principal stress difference with changes in the angle β for a laminated clay.

314 *Principles of Engineering Geology*

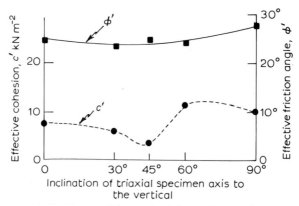

Figure 5.41 Influence of orientation upon the effective shear strength parameters in a boulder clay (*after* Courchée, 1970).

shear strength parameters, c_p, ϕ_p may be deduced. Figure 5.39 indicates that there is a reasonably linear inverse trend between the parameters.

If information on c and ϕ is readily available, it is then possible to transform the imposed shear and normal stress information on to principal stress axes, using a simple graphical construction on the Mohr failure envelope. The work load can be reduced if mean values of c and ϕ are taken into the analysis, resulting in a family of curves of the type shown in Figure 5.40. It will readily be noted that these curves follow exactly the same form as the equivalent curves for Penrhyn slate and laminated mudstone in Figures 5.25, 5.26 and for the quick clay in Figure 5.34.

It is difficult to classify boulder clay as either a homogeneous or inhomogeneous material with respect to the clay matrix. Many boulder clays are fissured, particularly through desiccation in the upper layers. These fissures have the effect of making the shear strength parameters direction-dependent, as shown, for example, in Figure 5.41.

6 ~ Rock Discontinuity Analysis

Chapter 4 has been devoted to a discussion of the mechanical properties of rock as a material. However, it will be apparent that the engineering behaviour of rock *en masse* is controlled more directly by the presence of discontinuities of a scale that can be physically measured. Sedimentary rocks are systematically jointed and it is this form of structure that probably imposes the greatest stability limitation in harder rock masses. Slope stability analyses based on the presence of continuously-jointed rock are contained in Chapter 10. But in weaker rocks, particularly, for example, in weaker argillaceous rocks and even in very stiff clays, we often find a more random field of terminated discontinuities superimposed on a primary joint system. The present chapter is devoted to a discussion of the various factors that should be considered when attempting to erect an analytical framework for an appraisal of the stability of these discontinuous materials *en masse*. In this context, therefore, stiff clays are treated as a rock.

Terminology tends to be a little confusing because to some extent the terms *discontinuity* and *fissure* are interchangeable. Fookes and Denness (1969) have noted that geologists usually look upon small-scale discontinuities as fissures (American Geological Institute, 1962). Engineers, on the other hand, do tend to regard large discontinuities, both open and closed, as cracks and joints respectively, and small discontinuities as fissures (British Standards Institution, 1957). The term 'joint' is probably best retained for a systematic 'continuous' discontinuity (large size) and the term 'fissure' for a small discontinuity.

6.1 The engineering interest in discontinuities

It will be obvious that there is a direct strength reduction effect due to the presence of discontinuities within the mass. Skempton and La Rochelle (1965) found, for example, that the peak strength in

London Clay would be reduced by up to 30 per cent through the presence of fissures and Muller (1968) has suggested a reduction factor of 1/300 on the intrinsic strength of rock due to the presence of weakness planes. Such reductions are due in part to the fact that a discontinuity is unable to support a tensile stress directed normal to its surface and also to the fact that the shear strength of fissure surfaces is usually reduced by weathering films and clay mineral alignments (Skempton, 1964) which render the surfaces more prone to sliding than the solid rock or stiff clay matrix. A further point, not always appreciated, arises from the stress concentration capacity of a fissure, or slickenside. At the termination tips of these discontinuities, the local strength of the intact rock can be exceeded – and its shear strength quite rapidly reduced to a residual state – if the shear strength along the discontinuity is low.

In order to assess the stability of discontinuous rock masses – and acknowledge that the stability status is going to be strongly influenced by the orientation and spatial density distributions of discontinuities – it is necessary to measure, record, and hopefully classify the orientations, lengths, spacings and any other characteristics in a planned manner. Terzaghi (1936) was probably the first person to recognize the engineering significance of small, non-systematic joints (or fissures) and, in a slope stability context, he pointed out that the water ingress facility that they offered also imposed a reduction on mass strength (see also Cooling, 1940). This strength reduction arises through both an increase in porewater pressure and a softening of the discontinuity surfaces. This effect will tend to increase both with proximity to an excavated surface (due to an enhanced fissure density associated with stress relaxation) and with time as the degradation caused by weathering increases. It should also be noted that discontinuities which are appropriately linked and suitably orientated towards a free surface offer a drainage facility for the rock mass.

Since Terzaghi's work, and up to the early 1960s, relatively little attention seems to have been paid in the literature to the engineering implications of systematic jointing. From 1962 the situation has been reversed to the extent that a wealth of data now exists on the effects of systematic jointing in slope stability studies both in hard rock (see, for example, Da Silveira *et al*, 1966; Bray, 1967; John, 1968; Jaeger, 1970; Brown, 1970; Hoek, 1970; Jennings, 1970; Hoek *et al*, 1973) and in soft rock (Fookes, 1965; Fookes and Wilson,

1966; Esu, 1966; Skempton and Petley, 1967; Marsland and Butler, 1967; Skempton *et al*, 1969; Fookes and Denness, 1969). Particularly in the case of smaller, terminated discontinuities and even in situations of continuous jointing it is recognized that shear failure of a rock mass can develop partly along a preponderance of discontinuous surfaces and partly through the intervening solid rock. Any resultant analysis must therefore incorporate strength parameters with respect both to the discontinuities and to the solid rock, their relative significance depending upon the area of the final shear surface that exploits or is entrained into the discontinuities compared with the area that cuts through the solid rock. With some reservations Skempton and La Rochelle (1965) did in fact attempt to determine the average area of a potential failure plane that would pass through open fissures, closed fissures and intact rock.

The cohesion and friction shear strength parameters of a discontinuous surface might be expected to approximate to the residual strength of the intact material. Marsland and Butler (1967) found from laboratory tests on stiff, fissured Barton Clay that the shear strength mobilized along a closed fissure was indeed barely more than a residual value under drained or undrained conditions. They have also noted that if a failure plane passes through discontinuities and thereby accelerates the drainage process, then there is a case for using drained shear strength parameters in a stability analysis even on a short-term basis. In the absence of specific laboratory evidence on discontinuity shear strength, trial solutions for potential failure in such discontinuous rock masses could initially be based on what would be regarded as conservative shear strength parameters as derived from laboratory tests on intact, undisturbed samples — perhaps residual values based on large shear displacements under drained conditions.

6.2 Genesis and modification of fissures and slickensides

The existence of fissures and jointing in clays received early recognition in 1849 by Martin, in 1875 by Kinahan and in 1882 by Gilbert. Crosby (1893) suggested that joints could develop in sedimentary rocks before consolidation processes had been completed and Hodgson (1961) collected and evaluated evidence to support this. Casagrande (1947) attributed early formation to electro-osmotic effects—transport of porewaters, the electric streaming potential for which can originate from the natural consolidation of the clay during deposition. Terzaghi and Peck (1967) attribute slickenside formation to shrinkages produced by chemical changes or by deformations resulting from gravitational or tectonic forces and it is interesting to record that Attewell and Taylor (1973) have proposed an origin from early shrinkage cracks for the slickensides in the unstable Cucaracha clay shale lining the banks of the Panama Canal. In the latter case, an early sedimentological origin for the slicks is deduced from the concentration of spherulitic siderite bodies along them and from the accepted fact that siderite is an early diagenetic carbonate. From comparative observations of discontinuity densities in brown and blue London Clay, Schuster (1965) implies, and the observations of Ward *et al* (1965) seem to substantiate the fact, that weathering agencies can create discontinuities. Processes of desiccation and syneresis can also produce discontinuities in a clay structure (Le Conte, 1882; Kindle, 1923, 1926; Jüngst, 1934; Berger and Gnaedinger, 1949; Twenhofel, 1950; Skempton and Northey, 1952; Skempton, 1953; Rosenqvist, 1955; White, 1961). Tectonic origins for right angled joint systems were mentioned in 1882 by Crosby, and seismic disturbances could also be responsible for discontinuity development in weaker rocks at an early stage during lithification. Changes in the physical conditions of deposition create bedding plane weaknesses (Otto, 1938; Pettijohn and Potter, 1964; Okeson, 1964) and in the cases of rocks having a laminar structure, a high density of discontinuities can be shown to be compatible with the horizontal or sub-horizontal bedding (for example, London Clay, Skempton *et al*, 1969). A high discontinuity density can be regarded as the only way of dissipating regional residual stresses (de Sitter, 1956) either through shear or through a dominant tensile origin caused by uplift and denudation (Price, 1966). Joint concentration seems to be inversely proportional to bed thickness (Price, 1959), or varies with it according to a hyperbolic law (Forcardi *et al*, 1970), and can be related to frictional forces between beds. Also, where competent and incompetent units are interbedded (Denness, 1969, quotes the Lower Greensand-Gault-Upper Greensand sequence in the south of England) joints in weaker interbeds are probably induced by the influence of the more competent material. There is also the strong possibility of joints being inherited through mechanisms of basement control (for example, Attewell and Taylor, 1971a, with respect to the Lower Lias in Robin Hood's Bay, Northern England).

Possible modes of formation and modification of fissures and slickensides may be summarized after Denness (1969):

(a) formed at the time of deposition or soon after by syneresis and/or changes in salt chemistry of the depositional environment;
(b) formed or modified some time later than (a) by *in situ* physico-chemical changes through agencies such as groundwater, weathering, ion exchange;
(c) formed or modified by tectonic or earthquake stresses during folding, or shearing of the beds;
(d) formed or modified by non-diastrophic processes such as hill creep, rebound on unloading, or stress release during erosion;
(e) inherited from underlying rocks.

6.3 Controls on fissuring and fissure patterns

Discontinuities can be placed into two main categories: those that are *inherited* and those that are *imposed*. The former category can develop at a relatively early post-diagenetic stage when a sediment is sufficiently lithified to respond in a semi-brittle manner to any other shear stresses imposed by regional or local earth movements. Such discontinuities can also result from later tectonism, and discontinuities which are dominantly tensile and tend to be orientated with the bedding are most probably associated with the progressive removal of overburden through geological denudation processes. Also, in general, the older the material and the more brittle it is, the more extended will be individual discontinuities. Other discontinuities are inevitably imposed on the much earlier inherited field through the dilatant relaxation of normal pressure associated with excavation and also through the creation of shear stresses in excess of the intact shear strength, again through excavation. Spatial densities of discontinuities will tend to increase with time adjacent to an excavation as a result of progressive stress relaxation and weathering. In some hot climates also, the breakdown of clay shales upon exposure is so rapid that the side walls of excavations require immediate protection. As an example, excavations in the Dawson formation at Littleton near Denver, Colorado for the Chatfield dam (see Figure 6.39 for location) must be bitumen-sprayed within two hours of exposure or concrete-covered within 48 hours (Figure 6.1). The subject of breakdown is considered generally in Chapter 3.

320 *Principles of Engineering Geology*

Figure 6.1 Protection of excavation against breakdown at Chatfield damsite, Littleton, Colorado. Photograph: P. B. Attewell.

Discontinuity density increases due to weathering are relatively superficial features in temperate climates, a weathered skin of little more than a third of a metre thick forming a protective seal to inhibit the drying out of the more deep-seated material. On the other hand, weathering agents other than simple desiccation may create an increase in fissure intensity; for example, there is a greater intensity of fissuring in the higher level brown than the blue London Clay (Ward *et al*, 1965).

Discontinuity weathering may be rather more insidious in that softer weathered zones may develop within the intact material either side of a discontinuous surface, and the discontinuity may be more difficult to trace. Delimitation of the gouge material will be equally difficult but it will be necessary to evaluate its liquid and plastic limits, its natural moisture content and its shear strength.

6.4 Classification of discontinuities

There have been several suggestions for discontinuity classification based on such features as genesis (Pettijohn, 1957; Braybrooke, 1966), size and distance between fissures (Fookes and Wilson, 1966), surface characteristics (Fookes and Denness, 1969) and fabric.

Discontinuity structures in sedimentary rocks can be classified broadly in a genetic sense into syngenetic and epigenetic forms. Syngenetic forms (those developed contemporaneously with the sediment) comprise bedding plane demarcations and associated features (such as mud cracks and ripple marks), and laminations. Epigenetic forms comprise such features as concretionary structures, corrosion zones, faults, joints and slickensides. Syngenetic forms have stratification characteristics in sedimentary rocks; this implies quasi-parallel surfaces defining layered structures. Bedding and lamination stratification can be differentiated on a thickness basis. *Bedding* is a layer greater than 10 mm in thickness visibly separated from adjacent parallel units by a discrete change in lithology (generally after Pettijohn, 1957). *Laminations* comprise similar discrete units but are less than 10 mm in thickness. When dealing with discontinuities, the engineering geologist is rather more concerned with certain of the epigenetic forms than with bedding and stratification, the surfaces of which can usually be regarded as providing a rather stronger type of weakness plane due to a higher cohesive strength.

Of the epigenetic forms, faults, joints, fissures, and slickensides can be termed 'fractures'. *Faults* are fracture planes or zones along which there has been displacement of the two sides relative to one another and parallel to the feature plane. *Joints* are fractures along which there has been little or no movement parallel to the feature plane (Price, 1966). *Fissures*, according to Fookes (1965), are small-scale discontinuities that do not cross the boundary of the bed or horizons within the bed in which they occur (that is, they are terminated 'cracks' within the scale of the bedding). *Slickensides* are smooth, grooved, polished features produced through frictional shear, often along fault planes, although slickensides are also found as volumetric, compressional shear features in such materials as the seat-earths (underclays) of coal seams; if they are representative of large shear displacements, then their strength is low.

These descriptive, non-quantitative terms often appear, with others, in technical descriptions of stiff, fissured, heavily-over-consolidated clays which are particularly problematical in engineering design. For example, Ward *et al* (1959) identified laminations, fissures, and 'backs' in the London Clay. Skempton *et al* (1969) have classified the London Clay in terms of five types of discontinuity; bedding, joints, sheeting (low angle joints), fissures, and faults. In general, classifications of discontinuities for engineering purposes

Table 6.1 Area classification of fissures (*after* Fookes and Denness, 1969)

Code	Type	Area
VL	Very large	$\geq 100 \text{ m}^2$
L	Large	$1-100 \text{ m}^2$
N	Normal	$0.01-1 \text{ m}^2$
S	Small	$1-100 \text{ cm}^2$
VS	Very small	$\leq 1 \text{ cm}^2$

should attempt to quantify significant morphological features of which size (surface area), surface geometry, surface markings, spatial density distribution, and orientation density distribution are the most important. Additional engineering classification features comprise: width of discontinuities (which together with area, spatial density and continuity controls the secondary permeability facility); nature of any infilling (weathering of the host rock or imported material), most conveniently quantified through its frictional properties; and chemical composition of any standing water.

From their analyses of fissure patterns in British Cretaceous sediments, Fookes and Denness (1969) have proposed area and spatial density classifications, as shown in Tables 6.1 and 6.2, and to which have been added a coding system to facilitate a shorthand designation of fissure size in an engineering report. Area density per unit volume and spatial density distribution tend to reflect the level of strain energy release from the rock, but unfortunately it is difficult to calculate fissure areas and therefore it may be necessary to rely on estimates. Measurements of surface geometry are especially tedious and classifications on this basis are less obviously useful unless the geometrical parameters can be related in some way to the shear

Table 6.2 Area Intensity classification of fissures (*after* Fookes and Denness, 1969)

Code	Type	Area per unit volume m^2/m^3	Average size of intact blocks
vl	Very low	≤ 3	$\geq 1 \text{ m}^3$
l	Low	$3-10$	$0.027-1 \text{ m}^3$
m	Moderate	$10-30$	$0.001-0.027 \text{ m}^3$
h	High	$30-100$	$27-1000 \text{ cm}^3$
vh	Very high	$100-300$	$1-27 \text{ cm}^3$
e	Excessive	≥ 300	$\leq 1 \text{ cm}^3$

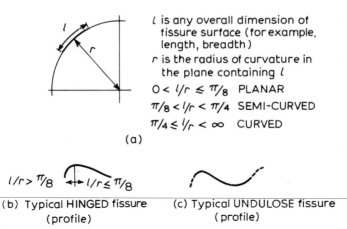

Figure 6.2 Fookes and Denness (1969) surface geometry classification of fissures.

strength characteristics of the discontinuities (undulatory characteristics would also have to be considered in the context of the normal pressures to which they are subjected since any shearing motion would either be over the protrusions or through them; see for example, Patton, 1966). Nevertheless, the Fookes and Denness (1969) proposals based on degree of curvature are given in Figure 6.2. Application of such a classification would involve measuring some length dimension l of the discontinuity and expressing that in terms of the characteristic radius of curvature r in the plane of l. Fookes and Denness (1969) have also produced a surface marking classification but in engineering terms (in contrast to any genetic overtones) roughness should be classified in terms of the shear strength parameters, probably by taking ranges of c and ϕ. Orientation fabrics are best classified in terms of their symmetry, concepts of which are considered in Chapter 5.

As an example, not of an engineering classification, but a classification that was developed for a specific engineering project, it

Table 6.3 U.S. Army Corps of Engineers classification of slickensides at Chatfield dam (DM PC-24, 1968)

Slickenside category	Definition
Ia	Small, wavy, discontinuous, poorly polished, randomly oriented, mostly curved surfaces. In this sub-group, *less than 10 slickensides per 6 inch length of core*.
Ib	The slicks have the same characteristics as in Ia, but the concentration is greater, that is, *exceeding 10 slicks per 6 inch (0.152 m) length of core* or an area of approximately equal size: 6 in^3 (0.0035 m^3) in an open excavation.
II	Slickensides become more continuous than in category I. They are still of the irregular wavy type, although more distinct with deeper striations. Some slickensided surfaces may extend beyond the width of the core. In open pits or trenches, a particular *slickenside surface may be traced for as much as 2 or 3 feet (1 m)*. Some few pieces may be slickensided on both sides.
III	These slicks are similar to the type seen in other shale formations. They are well polished, have roughly parallel orientation; tabular or elongated pieces within the zone are slickensided on all sides; crushed shale or gouge may be present. The planes are so continuous that they cut cleanly through the core. In a pit or trench, particular slickensides or zones of slickensides could be traced for some distance greater than a mere few feet. These represent the greatest movement and could be part of a joint system.

is useful to outline the U.S. Army Corps of Engineers classification of slickensides in the Dawson formation at the site of Chatfield dam, Littleton, Colorado, U.S.A. (U.S. Army Corps of Engineers, 1968). This classification is given in Table 6.3 and its designations are reproduced on the graph in Figure 6.3 with respect to residual shear strengths of the several slickenside categories. From the graph, it would seem that there is no consistent trend in shear strength with respect to category. It may also be mentioned that the Dawson (about 300 m thick formation) comprises a range of lithologies from clay shale to sandstone and although the slickensides occur more frequently in the 'pure' clay shale than in the more silty or sandy material, there is no obvious consistency in the directions of the category I and II slickensides. As an example of distribution of slickensides within the dam foundation, 60 per cent of the shale in the outlet works is slickensided, the remaining 40 per cent being free

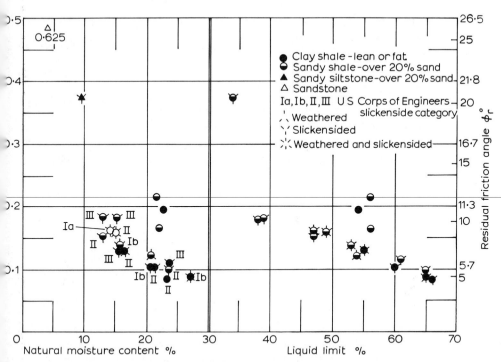

Figure 6.3 Direct shear residual strengths of Dawson foundation material at Chatfield damsite, Littleton, Colorado, U.S.A. (*Plots after* U.S. Corps of Engineers Design Memorandum C-24 Embankment and Excavation, vol. II, 1968).

from slicks. It should also be mentioned that in addition to compiling his own detailed classification of discontinuities into slickensides and fractures, the resident geologist at Chatfield engaged upon a very detailed and valuable exercise of identification, mapping, flagging and photographing of very soft horizons varying in thickness from 'paper thin' to more complex zones over 150 mm in thickness. Within the Dawson clay shale and its more silty variations, though not in the dense weakly-cemented sands ('sandstone'), soft seams could occur at any horizon and individual seams could be traced over distances of several metres. There is thus in this example a strong argument for allocating at least the same degree of concern to these weaker bands as is given to the slickenside distributions.

Further reference to Chatfield dam, discontinuity orientations and preliminary foundation stability analysis, is made in Section 6.21.

Discontinuity spacings, and particularly with respect to *joint* discontinuities, clearly exert a fundamental control on rock mass strength and they can therefore be embodied in a mass strength classification (for example, John, 1962). Such classifications are considered in Chapter 10.

6.5 Character of discontinuities

The character of discontinuities can sometimes be more important in a stability context than are orientation and spacing. Points to be noted and recorded include wall spacing, roughness and type of infilling between the wall surfaces. These matters are central to the tunnel support classification work of Barton *et al* (1974) and have been mentioned briefly in Section 10.2. Brekke and Howard (1972) have noted the following:

(i) Joints, seams and minor faults may be 'healed' through the precipitation from solution of quartz or calcite. 'Welds' of this nature may have become broken subsequent to crystallization but, if intact, the discontinuity shear strength could exceed that of the host material.
(ii) Clean discontinuities, those for which the wall rock is unaltered and which have never received any infilling material, should not be confused with those from which the filling material has been leached and washed away due to surface weathering.
(iii) Calcite and gypsum infillings may dissolve away during the lifetime of a structure.
(iv) Coatings or infillings of chlorite, talc and graphite give very 'slippery' low strength joints, seams or faults, particularly when wet.
(v) Inactive clay mineral in seams and faults may be washed out or squeezed out under construction stresses and changes in flow patterns.
(vi) Swelling clay, montmorillonite or inter-layered micaceous minerals, may create serious problems through unconstrained swelling and consequent loss of strength, or through high swelling pressure when confined by a temporary or permanent support.

6.6 Test specimen size–strength relationships

The greater the spatial density of discontinuities, the lower will be the overall mass strength. In terms of laboratory testing, the larger the sample the more discontinuities will it contain and therefore the

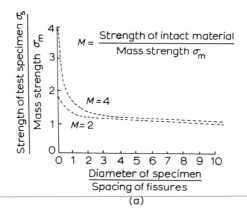

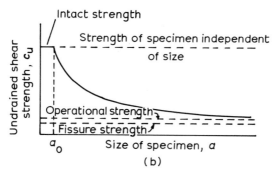

Figure 6.4 (a) Relationship between strength and specimen size (*after* Bishop, 1966b). (b) Influence of fissures upon specimen size-strength relationship (*after* Lo, 1970)

closer will it approach a 'lower-bound field strength' (Skempton and Henkel, 1957; Bishop and Little, 1967; Lo, 1970). Peterson *et al* (1960) and Bishop (1966b) found that the peak strength of large samples could be as much as 80 per cent lower than the peak strength of small samples due to the greater number of fissures. In a similar manner, the greater the orientation density distribution of the discontinuities along a plane of maximum shear stress the more likely is the rock mass to succumb to shear failure under shear stress amplitudes that tend towards the residual strength.

The strength influence of any fissures within a test specimen is probably most clearly illustrated by Bishop (1966b) and Lo (1970) — see Figure 6.4. From a small size of specimen a_0 which, it is assumed, contains no macroscopic fissures, there is a monotonical

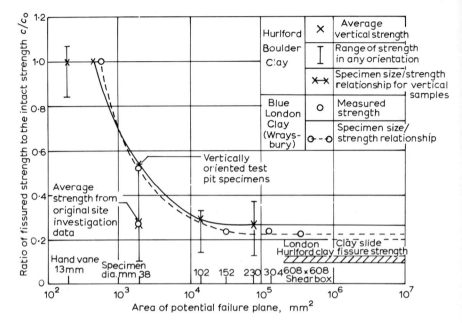

Figure 6.5 Size strength relationship for blue London Clay and Hurlford boulder clay (*from* McGown *et al*, 1974).

reduction in undrained shear strength as the specimen size a increases. Lo assumes that the fissures in the clay are randomly orientated and that failure in the soil mass will not take place preferentially along a continuous plane of weakness. A corollary of this assumption is, therefore, that the operational strength of the mass will be rather higher than the shear strength of an individual fissure surface.

Analyses by McGown *et al* (1974) on London Clay and on a boulder clay tend to reproduce in Figure 6.5 the general form of Lo's curve. Since, in most cases, discontinuity orientations are not random but are conditioned by previous tectonism and unloading, sample strengths are strongly dependent on sample orientation and will tend to minimize at the shear strength of the discontinuities, particularly under drained conditions.

6.7 Stereographic representation of discontinuity data

It is recommended that for basic information on the use of the stereographic projection in a structural geology context the reader refer to the book by Phillips (1971). He will then be able to acquire his discontinuity information, plot it directly on a stereographic projection, and if presented in an equal area manner he will then be able to contour his orientation concentrations for a qualitative appraisal of their possible influence upon the engineering stability of a rock mass. The present discussion proceeds on the assumption that the reader possesses this necessary background information.

In Figure 6.6, the basic features of two types of projection are presented. Since the engineering geologist will be concerned with grouped data, quantitative assessments on the equal angle projection will be inaccurate due to area distortions along any radius from the centre to the periphery of the projection. The Schmidt–Lambert equal area projection, although creating angular distortions, preserves a uniformity of any incremental area independent of latitude angle and thereby permits reliable quantitative assessments of discontinuity density distributions. Since the mathematics behind the construction of these nets is not immediately obvious and is not usually available in publications, it was considered worthwhile to develop the analysis in Figure 6.7 so that the reader, having confidence in equation 2, Figure 6.6, could construct his own equal area nets or develop his own computer programs for fabric analysis and direct print-out.

At the time of measurement, it is a useful practical procedure to record, not the dips and strikes of the discontinuities, but rather the extent of full dip and the direction of full dip. This recommendation is re-emphasized in Chapter 10 on rock slope stability, appropriate examples being given. Poles to discontinuity planes should also be plotted continuously on an equal area projection in order that developments in concentration trends can be discerned at the measurement stage. Use of the *polar* form of equal area net will considerably ease the plotting process, and the data points may be *directly* transferred to a Schmidt–Lambert equatorial *overlay* net at a later date without any further manipulation whatsoever.

6.8 Direct and inverse transformations from polar to equatorial angles

In Figure 6.8, a system of orthogonal references axes OX_1, OX_2, OX_3 is created in space and, for ease of correlation, it is assumed that these are parallel to north–south, east–west, and vertical respectively. Polar angles β and θ are the direct field observations of the dip of the discontinuity from a horizontal plane (which is the same as the dip of its pole from the vertical) and the azimuth angle of the discontinuity pole respectively. Equatorial angles δ, γ refer respectively to the corresponding angular displacements along the small (latitude) and the great (meridian) circles on a Schmidt equal area or Wulff equal angle projection. It should also be noted that

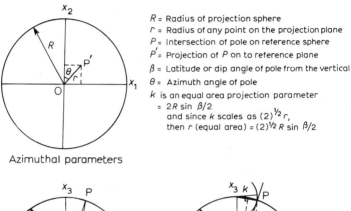

Azimuthal parameters

R = Radius of projection sphere
r = Radius of any point on the projection plane
P = Intersection of pole on reference sphere
P' = Projection of P on to reference plane
β = Latitude or dip angle of pole from the vertical
θ = Azimuth angle of pole
k is an equal area projection parameter
   $= 2R \sin \beta/2$
   and since k scales as $(2)^{1/2} r$,
   then r (equal area) $= (2)^{1/2} R \sin \beta/2$

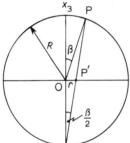
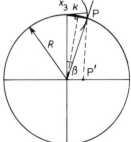

Wulff equal angle projection   Schmidt equal area projection

For cartesian coordinates in X_1 X_2 plane:

$$\left. \begin{array}{l} x_1 = R \tan \dfrac{\beta}{2} \sin \theta \\[6pt] x_2 = R \tan \dfrac{\beta}{2} \cos \theta \end{array} \right\} \text{Wulff equal angle} \qquad (1)$$

$$\left. \begin{array}{l} x_1 = (2)^{1/2} R \sin \dfrac{\beta}{2} \sin \theta \\[6pt] x_2 = (2)^{1/2} R \sin \dfrac{\beta}{2} \cos \theta \end{array} \right\} \text{Schmidt equal area} \qquad (2)$$

Figure 6.6 Representation of constructional procedures for equal angle and equal area projections.

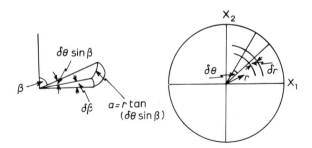

(Note: that when projected on to a plane, distance a remains constant but the radius of the segment decreases, i.e, the azimuthal angle $\delta\theta$ increases)

Area of element of sphere, radius R
$\delta A = R^2 \delta\theta\delta\beta \sin\beta$

Area of element of polar diagram
$\delta A = r\delta\theta\delta r$

Thus, for equal areas: $R^2 \delta\theta\delta\beta \sin\beta = r\delta\theta\delta r$ or $r\,\delta r = R^2 \delta\beta \sin\beta$.
Integrating: $r^2/2 = -R^2 \cos\beta$ or, $r/R = -(2\cos\beta)^{\frac{1}{2}}$. When $\beta = \pi$, ie, a hemisphere projecting on to a plane, $r = (2)^{\frac{1}{2}}R$ and the cartesian coordinate equations in Figure 6.6 are satisfied. This can be checked in the following way:

Area of circular projection $= \pi R^2 = A_n$
Area of surface of hemisphere $= 2\pi R^2 = A_s$
Therefore, $A_n = \frac{1}{2}A_s$

$$A_n = \int_r\int_\theta r d\theta\, dr = \int_r\int_\theta (2)^{\frac{1}{2}} R \sin\frac{\beta}{2}\, d\theta\, dr$$

(assuming at this stage that Equations 2, Figure 6.6 hold).

Now, if $r = (2)^{\frac{1}{2}} R \sin\beta/2$, then by differentiating,
$dr = R/(2)^{\frac{1}{2}}(\cos\beta/2)d\beta$
Hence, by substituting for dr

$$A_n = \int_\beta\int_\theta R^2 \sin\frac{\beta}{2} \cos\frac{\beta}{2}\, d\theta\, d\beta$$

Therefore $A_n = \frac{1}{2}R^2 \int_\beta \int_\theta \sin\beta d\theta d\beta$.
Now, an elemental area on the surface of the sphere
$= R\delta\theta \cdot R\delta\beta \cdot \sin\beta = R^2 \sin\beta\delta\theta\delta\beta$
Therefore $A_s = R^2 \int_\beta\int_\theta \sin\beta d\theta\, d\beta = 2A_n$ and Equation 2, Figure 6.6 is re-confirmed

Figure 6.7 Analysis of equal area projection.

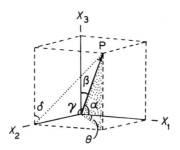

Figure 6.8 Spatial transformation parameters for a pole P to a discontinuity.

α, γ, β are direction cosine angles with respect to OX_1, OX_2, OX_3 respectively, a condition of their interrelationship being that $\cos^2 \alpha + \cos^2 \beta + \cos^2 \gamma = 1$ (in the i, j notation used later for transformation of axes this is equivalent to $a_{11}^2 + a_{12}^2 + a_{13}^2 = 1$).

Examination of Figure 6.8 indicates that the following transformation relations pertain:

(a) *Transformation from dip/strike (polar) to equatorial (stereographic)*
$\delta = \tan^{-1} (\cos \theta \tan \beta)$
$\gamma = \tan^{-1} (\cos \delta \sin \theta \tan \beta)^{-1}$

(b) *Inverse transformation from equatorial to polar*
$\theta = \tan^{-1} (\operatorname{cosec} \delta \cot \gamma)$
$\beta = \tan^{-1} (\cos \delta \tan \gamma \sin \theta)^{-1}$

The engineering geologist refers his discontinuity orientation data to global axes (NSEW and vertical) when he measures up with his compass-inclinometer (it is accepted that he will adjust his azimuthal data from magnetic north either subsequently or actually at the compass prior to taking any readings). Any stress system that acts naturally on the rock or is imposed externally can be resolved into three orthogonal principal stresses the axes of which will generally be incompatible with the global axes. In order to evaluate the stability status of any measured discontinuity, one set of rectangular axes must be rotated into the other and it is usually more convenient to rotate stress into global space rather than vice versa. Experienced structural geologists perform such rotations quite happily on the stereographic projection (see Philips, 1971). Since a number of authors process their discontinuity data for stability, perform such rotations and print-out stability fabrics directly by computer, it is relevant to indicate some of the logic behind the rotation operation.

6.9 Linear orthogonal transformations

Assume, as in Figure 6.9, that a rectangular cartesian system of coordinates is erected in three-dimensional space and that OX_1, OX_2, OX_3 are orthogonal axes. Coordinates x_1, x_2, x_3 now define any pole P to a discontinuity in space. If OX'_1, OX'_2, OX'_3 is another system of rectangular cartesian coordinates having the same origin at O, all poles P can equally well be described in terms of this alternative X'-system. Axis OX'_1 can be defined with respect to the X-system through the cosines of the angles that it forms with each of the OX axes. These direction cosines will be defined as a_{11}, a_{12}, a_{13}. In a similar manner OX'_2 and OX'_3 will create their own direction cosines with each of the OX axes and will be written respectively as a_{21}, a_{22}, a_{23} and a_{31}, a_{32}, a_{33}. The general orientation relationship between the OX and the OX' systems can most conveniently be listed as:

	X_1	X_2	X_3
X'_1	a_{11}	a_{12}	a_{13}
X'_2	a_{21}	a_{22}	a_{23}
X'_3	a_{31}	a_{32}	a_{33}

(6.1)

As the coefficients are specified in expression 6.1, the columns give the direction cosines of the X-axes with respect to the X' axes and

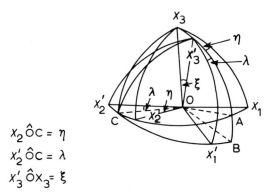

$x_2 \hat{O} C = \eta$
$x'_2 \hat{O} C = \lambda$
$x'_3 \hat{O} X_3 = \xi$

Figure 6.9 Linear transformation of orthogonal axes.

the rows reverse this. In any stability fabric computer print out produced by the present writers in the text, the direction cosines of orthogonal principal stress axes with respect to global axes are listed as above.

It is now apparent that if x_1, x_2, x_3 are the coordinates of P in the X-system and if x'_1, x'_2, x'_3 are the coordinates of P with respect to the X'-system, then the following fundamental relationships hold:

$$\begin{aligned} x'_1 &= a_{11}x_1 + a_{12}x_2 + a_{13}x_3 \\ x'_2 &= a_{21}x_1 + a_{22}x_2 + a_{23}x_3 \\ x'_3 &= a_{31}x_1 + a_{32}x_2 + a_{33}x_3 \end{aligned} \qquad (6.2)$$

It is more concise to express the equations in 6.2 as

$$x'_i = \sum_{j=1}^{3} a_{ij}x_j \qquad (i, j = 1, 2, 3) \qquad (6.3)$$

or even more simply

$$x'_i = a_{ij}x_j \qquad (i, j = 1, 2, 3) \qquad (6.4)$$

So far, we have considered the equations for the transformation of the coordinates x_1, x_2, x_3 of P from the X-rectangular cartesian system to the X'-system. This is the operation that would be performed, for example, on numerous poles to discontinuities if they were to be rotated into a principal stress space reference system. On the other hand if, for example, it was necessary to rotate some stress vectors into global space, the transformation inverse to equation 6.2 is derived by changing rows into columns and columns into rows within the scheme of the coefficients as given in expression 6.1 and interchanging x with x'. The inverse transformation equations are written:

$$x_i = a_{ji}x'_j \qquad (i,j = 1,2,3) \qquad (6.5)$$

Now each row in expression 6.1 gives the direction cosines of one of the axes in the X'-system (which is, of course, a straight line by definition) with respect to rectangular axes OX_1, OX_2, OX_3. It follows that:

$$\begin{aligned} a_{11}^2 + a_{12}^2 + a_{13}^2 &= 1 \\ a_{21}^2 + a_{22}^2 + a_{23}^2 &= 1 \\ a_{31}^2 + a_{32}^2 + a_{33}^2 &= 1 \end{aligned} \qquad (6.6)$$

Also, since the axes in the X'-system are mutually perpendicular, it will be quickly appreciated that the direction cosines of any pair of them can be related via:

$$a_{21}a_{31} + a_{22}a_{32} + a_{23}a_{33} = 0$$
$$a_{31}a_{11} + a_{32}a_{12} + a_{33}a_{13} = 0 \quad (6.7)$$
$$a_{11}a_{21} + a_{12}a_{22} + a_{13}a_{23} = 0$$

Thus, noting the character of the coefficient subscripts, the relationships between a_{ij} in equations 6.6 and 6.7 can be written in a condensed form as:

$$a_{ik}a_{jk} = \begin{cases} 1 & \text{if } i = j \\ 0 & \text{if } i \neq j \end{cases} \quad (6.8)$$
$$= \delta_{ij} \quad (6.9)$$

where δ_{ij}, the Kronecker delta, is equal to 1 or 0 if $i = j$ or $i \neq j$ respectively.

Inversion of columns and rows to change the frame of reference from one system to the other is simply specified by

$$a_{ki}a_{kj} = \delta_{ij} \quad (6.10)$$

The relationships described in equations 6.9 and 6.10 are termed orthogonality relationships. Transformations having coefficients which satisfy the orthogonality relationships are termed linear orthogonal transformations. More than one linear orthogonal transformation of coordinates can be applied successively, the resulting transformations at each stage being linear orthogonal. Suppose that two transformations are

$$x'_i = a_{ij}x_j \quad \text{and} \quad x''_i = b_{ij}x'_j \quad (6.11)$$

and it is required to determine expressions for the coefficients c_{ij} of the transformation

$$x''_i = c_{ij}x_j \quad (6.12)$$

in terms of a_{ij} and b_{ij}. Substitution for x'_j in equation 6.11 and comparison with equation 6.12 gives

$$c_{ij} = b_{ik}a_{kj} \quad (6.13)$$

and this result can be extended to the product of any number of transformations.

6.10 Eulerian angles

Since the nine coefficients a_{ij} of a linear orthogonal transformation are related by six orthogonality relationships, a consequence of this is that coefficients a_{ij} can be expressed in terms of three independent parameters to be known as the Eulerian angles.

In Figure 6.9 let OC, OA and OB represent the lines of intersection of the $X_1 X_2 : X'_1 X'_2$ planes, $X_1 X_2 : X_3 X'_3$ planes, and $X'_1 X'_2 : X_3 X'_3$ respectively. The Eulerian angles, η, λ, ξ, which completely specify the orientation of one set of axes with respect to another, are shown in Figure 6.9. It is worth noting that, sequentially, in the first rotation by η, the spherical triangle $X_3 X_1 X_2$ swings to X_3 AC about a fixed axis OX_3. In the second rotation by ξ, the spherical triangle X_3 AC swings to X'_3 BC about a fixed axis OC. In the final rotation by λ, the spherical triangle X'_3 BC swings to $X'_3 X'_1 X'_2$ about a fixed axis OX'_3.

In order to derive expressions for a_{ij} in terms of Eulerian angles η, λ, ξ and thereby to complete the transformation specification in terms of tractable trigonometrical relationships, we proceed systematically by close reference to Figure 6.9. Projection of OX_1 on to OX'_1 gives the cosine of the angle between these two axes which in the present direction cosine notation is a_{11}. Resolving this, it can be seen that OX_1 projects $\cos \eta$ along OA and $-\sin \eta$ along OC. Length $\cos \eta$ on OA projects $\cos \eta \cos \xi$ on OB and this in turn projects $\cos \eta \cos \xi \cos \lambda$ on OX'_1. The subtractive length $-\sin \eta$ on OC projects $-\sin \eta \sin \lambda$ along OX'_1. It follows from this step-by-step logic that $a_{11} = \cos \eta \cos \xi \cos \lambda - \sin \eta \sin \lambda$. Accepting that only those readers having the strongest willpower will derive the whole of the matrix in this manner, the transformation matrix for all a_{ij} is given in Table 6.4 below.

6.11 Discontinuity survey techniques

For studies of rock slope stability, a number of different practical techniques have been used to survey the discontinuity characteristics,

Table 6.4 Transformation matrix for Figure 6.9 in terms of Eulerian angles

	X_1	X_2	X_3
X'_1	$\cos \eta \cos \xi \cos \lambda - \sin \eta \sin \lambda$	$\sin \eta \cos \xi \cos \lambda + \cos \eta \sin \lambda$	$-\sin \xi \cos \lambda$
X'_2	$-\cos \eta \cos \xi \cos \lambda - \sin \eta \cos \lambda$	$-\sin \eta \cos \xi \cos \lambda + \cos \eta \cos \lambda$	$\sin \xi \sin \lambda$
X'_3	$\cos \eta \sin \xi$	$\sin \eta \sin \xi$	$\cos \xi$

their degree of continuity and their distribution in both orientation and space. Weaver and Call (1965) and Halstead et al (1968) used fracture set mapping, detailed line mapping, orientated core and borehole camera methods. Da Silveira et al (1966) mapped all the joints exposed on the faces of a rectangular tunnel at intervals along the tunnel. Hoek and Pentz (1968) working at Seville Spain pit used, in addition to conventional mapping techniques, an NX continuously-photographing borehole camera and adopted a photogrammetric method of studying the rock face using stereographs. Rosengren (1968) relied almost exclusively on the orientation features in drill cores for his information. Fookes (1965), on the other hand, used conventional joint sampling techniques on a fissure orientation study for the Mangla Dam project in Upper Siwalik Clay, West Pakistan, and later Fookes et al (1969) and McGown et al (1974) used a 'cavity technique' whereby a block of clay is excavated from a face and measurements are taken in the cavity that is created. This latter technique ensured that only fresh, unweathered discontinuities were sampled. Other surveys reported in the literature include brief descriptions of fissures in the Barton Clay at Fawley (Marsland and Butler, 1967)* and joints in unweathered Pliocene clays in Italy (Esu, 1966). More recently, Skempton et al (1969) made measurements on joints and fissures in the London Clay in borrow pits at Wraysbury and Edgeware, the observations including data on dip and strike, height, and where possible the length of each joint, and they supplemented the field data by excavating a large orientated block of clay for laboratory study.

In spite of advances in the use of photogrammetric methods for discontinuity orientation mapping (Ross-Brown and Atkinson, 1972; Ross-Brown et al, 1973; Moore, 1974) it would seem that for a number of years to come, most of the joint/discontinuity orientation and length data will need to be acquired from direct field measurements at exposures supplemented by measurements taken from drill cores. It is now proposed to consider the techniques of discontinuity orientation and length data acquisition together with the consequential adjustments for error. A supplementary check list of useful ancillary data is included in Table 6.5.

If the area of engineering interest is quite small and the

---

*For the very interesting story behind the practical reasons for examining these fissures, it is recommended that reference be made to the article in the New Civil Engineer (Institution of Civil Engineers) entitled 'Esso's giant oil tanks, a question of more haste, less speed' dated 28 February, 1974 (No. 81).

Table 6.5 Observer's check list on discontinuities

In addition to the formal measuring procedures that are outlined in Section 6.11, it is recommended that the engineering geologist be aware of the following:

Item	Notes
Geometrical state of discontinuities	Planar or curved, see Fookes and Denness (1969) for possible surface geometry if required.
Degree of wall roughness Degree of wall alteration Extent and type of any in-filling	These features to be recorded for each discontinuity, or if too time-consuming, ranges between discontinuities could be noted.
Relative ages	All discontinuities may not be penecontemporaneous, in which case it may be possible to age-date discontinuity suites by noting from any intersection pattern which discontinuity intersects a pre-existing one (Fookes, 1965) and interpreting with respect to a historical loading pattern.
Horizontal concentrations	Concentrations of horizontal fissures in lithified rocks might imply a stress relief origin (for example, Price, 1966) or, in the case of stiff clay like the London Clay, may be the visible reflection of an overconsolidated character.
Striations	Striations on opposing faces of discontinuities may indicate both a shear genesis and may be interpreted for relative directional movement (Fookes, 1965). If the extent of any shear movement can be estimated, this could be useful in any stability classification (for example Barton *et al*, 1974). However, if a rock mass quality parameter is to be specified in the manner of Barton *et al*, then it will be necessary to categorise the discontinuities and their surface characteristics far more rigorously.

discontinuities are not too numerous, it may be possible to achieve 100 per cent recording. But, more usually, the actual discontinuities that are exposed can only be *sampled* and it is necessary that the measurement programme and subsequent analysis is planned to ensure that the sampled data is representative of field conditions.

The errors inherent in data collection are of two forms: errors in the actual sample selection and errors directly or indirectly involved in the discontinuity measurement. Errors in selection are numerous; there is a tendency to disregard small discontinuities, large surfaces may be measured more than once, fractures parallel or near parallel to the bedding may be overlooked, and the direction of sampling may result in a sampling bias. Typical direct measurement errors arise from the presence of metal near a magnetic compass, non-conversion from magnetic north to true north and a misreading of dip direction for planes dipping at high angles. In a similar vein, the errors associated with the reading of strike directions increase as the plane tends to horizontal. Re-exposure of slopes has shown that dip and dip-direction angles have errors of the order of 5 and 10 degrees respectively.

Statistically, the most satisfying method of surveying discontinuities involves the use of a series of line scans, broadly as proposed by Jennings (1970) and extended by Piteau (1970). The technique used extensively by the present authors at surface exposures involves extending a metric tape along the face of the exposure, levelling it, and then securing it to the face at a number of points so that it cannot be disturbed by falling rock or strong gusts of wind. Two other scanlines would then be set out as nearly at right angles to the first as possible; for example, one line would be in a near-vertical plane parallel to the face of the slope and the other would run horizontal and orthogonal to the first if a suitable abutment is available. Additional or alternative scanlines could be run at an angle to the first line and still parallel to the face. Information from multiple scanlines tends to reduce the measurement errors outlined earlier. A possible measurement scheme is shown in Figure 6.10 and the results from two typical traverses are shown in Figure 6.11.

For every discontinuity that intersects the tape, the following are measured and recorded:

(a) Distance along the tape, from some datum, to the point of intersection of the tape with each discontinuity.
(b) Direction of the *pole* to each discontinuity expressed as an azimuthal bearing; this will require the addition or subtraction of 90° from the planar strike reading and it is recommended that these angles are expressed 0–360° clockwise from true north. The result is an expression of dip direction.
(c) Dip of the *pole* from the vertical, which is equivalent to the dip of the *plane* from the horizontal; readings (b) and (c) are taken quite easily with compass/inclinometer when the discontinuity *surface* is exposed, but if the discontinuity is simply indicated as

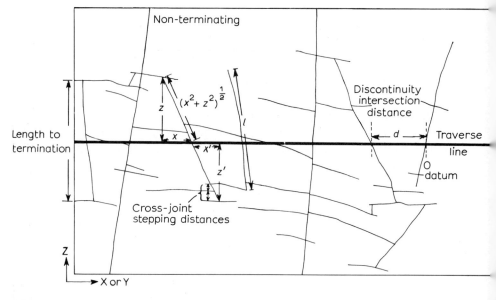

Figure 6.10 Typical discontinuity distance measurements taken at an exposed cutting in Carboniferous mudstone.

a line intersecting the plane of the exposure then the discontinuity surface can effectively be extended by inserting a non-ferrous plate into it and measuring on that.

(d) Whether or not a discontinuity which intersects the tape actually terminates; this, in effect, distinguishes between 'joints' *sensu stricto* and 'discontinuities' and is supplemented with the information below.

(e) Distance to termination of the discontinuity along its length; alternatively this information can be re-expressed as horizontal and vertical distances, x, z coordinates where joint length, l, is equal to $(x^2 + z^2)^{1/2} + (x'^2 + z'^2)^{1/2}$.

(f) Whether or not the discontinuity is open and by how much.

(g) Whether or not gouge material is present in an open joint, an estimate of its degree of infilling along the joint surface, and an analysis of it (moisture content, index properties, c', ϕ' from the shear box, undisturbed if possible and taken to residual); if the gouge material is difficult to extract, then a hand vane test should be performed *in situ*.

(h) Waviness of the surface; the usefulness of these estimates, expressed in terms of asperity amplitude and wavelength,

Rock Discontinuity Analysis   341

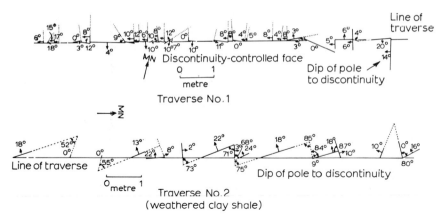

Figure 6.11 Two discontinuity-measurement traverses in the Kimmeridge Clay, North Yorkshire, England (*after* Little, 1972).

depends upon the openness, surface character and length of the fracture together with the degree of surface exposure.

If the rock is weak, there may tend to be fewer continuous joints and rather more smaller shear discontinuities and slickensides, the orientation and spatial density distribution of which will significantly influence and often control the orientation and location of any major shear failure. In order to record the interaction of such discontinuities, a further characteristic may usefully be recorded:

(i) The number of discontinuities intersecting a particular discontinuity within a distance of 250 mm along its length or within its total length if this is less than 250 mm; for completeness this information can be partitioned into the number of supplementary discontinuities intersecting the reference discontinuity on the left and on the right if the former terminate at the latter.

Even after arriving on site and trimming any loose debris from the exposure to be studied, it becomes apparent that the discontinuity data acquisition can be a very time-consuming process incorporating a possible 13 measurements on each discontinuity. It may be economically necessary to limit the number of discontinuities totally studied and to restrict a much larger population to the less-demanding dip-strike measurements. The newer photographic techniques mentioned earlier may eventually relieve the tedium of taking direct field measurements and permit a much higher measurement density to be achieved.

Tunnel face discontinuity survey technique

An alternative method of discontinuity scanning, which is suitable for restricted areas where a high density of information is required, has been used by Priest (1974) during tunnel stability work in Lower Chalk. The presence of a discontinuity intersecting the plane of a tunnel face will tend to create a very localized zone of reduced strength, and it may be argued that as the numerical density of these discontinuities increases per unit area of face, the strength of the exposed face will vary in some way as a function of that density. By gridding the face with a series of actual horizontal and vertical lines in the form of a scaled overlay, contour lines can then be drawn through equal numbers of discontinuity intersection with each grid element. The nature of the intersection density topography at the face can then be evaluated in the light, for example, of tunnelling machine pick performance. A typical contoured plot of discontinuity spatial density distributions is shown in Figure 6.12 as derived from the basic information in Figure 6.13.

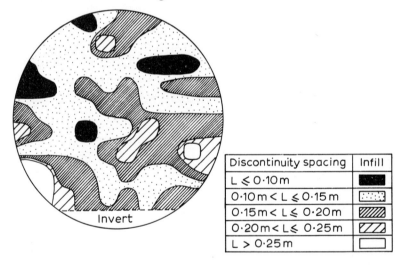

Figure 6.12 Discontinuity spatial density distribution at one tunnel face location in the Lower Chalk, Oxfordshire, England. Distribution is compounded from the evidence of nine equally-spaced vertical and nine equally-spaced horizontal scan lines (*measurements by* Priest, 1974).

A further element, that of time, imposes a constraint on discontinuity mapping at a tunnel face. Particularly with a machine-cut face, access time to the face for the purpose of measurement will always be limited. Almost certainly, horizontal and vertical scanline suites comprising nine lines in each mode would never be accommo-

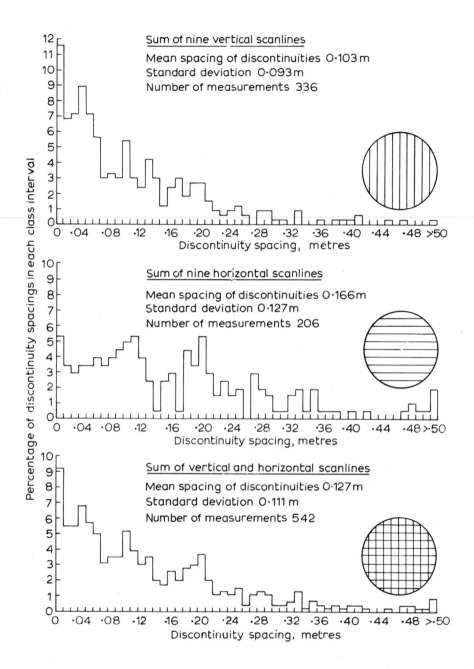

Figure 6.13 Discontinuity spacing histograms derived from measurements made at the face of a tunnel in Lower Chalk, Oxfordshire, England (*measurements by* Priest, 1974).

dated in a working tunnel. Priest was able to show by statistical analysis that within an available measurement period of 90 minutes, a quite adequate numerical appraisal of discontinuity orientations and spacings could be made using a combination of 6 scanlines comprising a pair of vertical and horizontal diameters combined with two pairs of vertical and horizontal scan lines spaced 1 m from the horizontal and vertical tunnel axes. He found that scanlines of this configuration produced a mean spacing that was within 14 per cent of the mean spacing, determined using the 18 scan line technique, at the 95 per cent confidence level.

6.12 Analysis of discontinuity data

It was earlier stated that the discontinuity measurements were subject to indirect errors but these errors have not yet been defined in any detail. It is now proposed to illustrate certain data correction procedures by taking as an example a discontinuity survey in two cuttings in the Kimmeridge Clay (Jurassic), Yorkshire, England (see Figures 6.11, 6.14). At one location (Golden Hill) the clay, strictly a

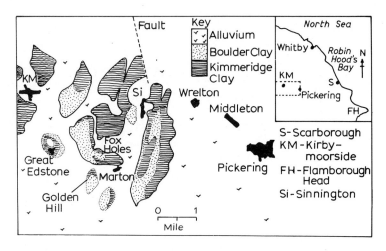

Figure 6.14 Geology of the Golden Hill and Fox Holes area of North Yorkshire, England.

clay-shale, varies from a stiff to a very-stiff consistency (classification of Terzaghi and Peck, 1967, based on unconfined compressive strength) olive-grey, fine-grained mudstone to a brittle, pale-yellow-brown, finely-laminated iron-stained mudstone. At the other location (Fox Holes) the material is mainly soft-to-brittle, medium and dark grey, calcareous, often carbonaceous, in places fissile, mudstone with very hard ironstone bands, and weathering to small iron-stained, disaggregated, brittle platelets.

The data will be considered first in its 'raw' state and then when it has been standardized.

(i) *Pre-standardization*

Orientation

It is recommended that all orientation readings from each traverse be considered on the basis of an *upper* hemisphere projection. This is contrary to general structural geology procedure but conforms to a trend in engineering geology presentation. It has an advantage in that the pole to a discontinuity is projected in the same direction as the discontinuity dip.

Each traverse can then be grouped according to the strike of the pole for each discontinuity. This can either be done quickly by plotting the pole azimuths on a separate 'rose' diagram for each traverse and specifying azimuthal boundaries for each group by inspection, or it can be achieved by an iterative statistical technique based on a lowest standard deviation concept. Then, for each group, the mean strike and dip are computed together with the standard deviation and percentage coefficient of variation for that set of discontinuities. Finally, this procedure is repeated for all traverses together, in this example 3 per location, 6 in total. These operations indicate the extent to which a particular traverse deviates from, or is representative of, the general discontinuity orientation trend.

Poles to discontinuities from 3 of the traverses in the Kimmeridge Clay are plotted on a strike-density-distribution basis in Figure 6.15 together with the total distribution. The statistical groupings from all 6 traverses are tabulated below in Table 6.6.

There are therefore 3 discrete joint sets represented by 5 discontinuity groupings (two groupings 180° apart having steep dips represent a single joint set where there is visible evidence of strong continuity), the groups being displaced in orientation by a mean angle of 92°. The joints themselves dip consistently at an average angle of 82° with a standard deviation of 1.3°. Such objective evidence substantiates immediate visual evidence of joint verticality (see Figure 6.16).

(ii) *Post-standardization*

Observation of Figure 6.15 will indicate immediately the basic differences between the pre- and post-standardization rose diagrams.

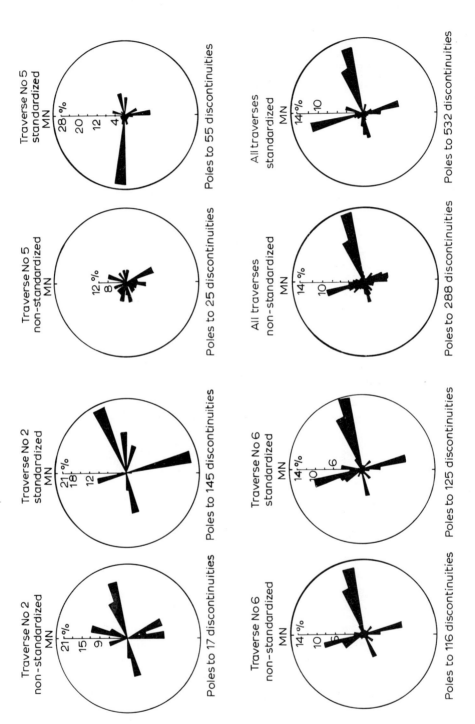

Figure 6.15 'Rose' diagram of dip directions of discontinuities in the Kimmeridge clay, North Yorkshire, England.

Table 6.6 Dominant discontinuity orientations in the Kimmeridge Clay at Golden Hill and Fox-Holes, Yorkshire, England

Group	Mean Dip Direction	Mean Pole Dip
1	16°	83°
2	71°	82°
3	163°	84°
4	255°	80°
5	337°	83°

These differences arise from the fact that the data from all the traverses must be normalized with respect to one or more of the

Figure 6.16 Showing near-vertical jointing in the Kimmeridge Clay, north Yorkshire, England. Photograph: J. A. Little.

Table 6.7 Traverse lengths in Kimmeridge Clay at Golden Hill and Fox-Holes

Traverse No.	Length (metres)
1	10.42
2	11.50
3	6.90
4	10.50
5	3.50
6	11.30

following criteria:
Length
Although every effort should be made to keep traverses to the same length, physical obstacles often lead to length discrepancies. The traverse lengths used in the present example are given in Table 6.7.

A length standardization procedure facilitates an unbiased statistical comparison of discontinuity data from traverses of differing lengths, the normalization usually being based on the length of the longest traverse. This latter requirement should be modified if necessary, however, as in the present instance where, although traverse 2 is the longest traverse, it only encompassed the measurement of 17 discontinuities due to the high degree of weathering in the clay shale and in spite of persistent attempts to clean the face prior to measuring. In this particular instance, traverse 6 in relatively fresh clay shale, and the next longest traverse, is used as the standard length.

If L is the length standardization multiplying factor, L_s is the standard traverse length and L_n (n = 1,2. . . .) is the length of the actual traverse to be normalized, then

$$L = \frac{L_s}{L_n} \qquad (6.14)$$

and is applied to the mean discontinuity of each group in traverses L_1 to L_n.

Degree of weathering
Although traverse 2 exceeded traverse 6 in length by 0.2 m, it sampled only 15 per cent of the number of discontinuities measured in traverse 6. This low percentage sample was a direct result of the

weathered, wet and sticky clay along the former traverse line and to allow for this anomaly a weathering standardization multiplying factor W is applied. This factor is really a discontinuity density index and is an expression of the discontinuity population per metre run of traverse length. Degree of weathering is only one of a number of factors which serve to constrain W. The index W is expressed as

$$W = \frac{n_{ds}/L_s}{n_{dw}/L_w} \times 100 \qquad (6.15)$$

where n_{ds} and n_{dw} are the number of discontinuities measured in the standard and weathered traverses respectively and L_w is the length of the weathered traverse.

Discontinuity orientation

It has earlier been stated that discontinuities more nearly parallel to a line of traverse tend to receive a negative sampling bias and that discontinuities aligned more nearly normal to the traverse line are more favourably chosen for measurement. Corrections for this bias can be applied in the manner outlined by Robertson (1970) and as indicated in Figure 6.17 or by a simplified modification of that procedure which lends itself more to hand processing rather than demanding computer facilities. The core of Robertson's argument can be expressed in the following way:

Since discontinuity planes normal to the standard traverse line should receive 100 per cent sampling attention, all other sub-normal discontinuity planes should be numerically upgraded according to the inverse of the cosines of the azimuth and dip angles necessary to rotate them into normality with the standard line. Any necessary rotation will vary from 0–90° for planes normal and parallel to the traverse line respectively.

These same arguments and corrections must also be applied to the interpretation of discontinuity intersections with an exploratory drill core (Figure 6.18 and see Ruth Terzaghi, 1965).

Clearly, as the discontinuity plane or mean plane of a group of discontinuities becomes more nearly parallel to a scanline, the cosines of the angular differences tend to zero and the correction factor tends to infinity. Robertson maintains that the standardizing factor should be limited to 5.

If discontinuity traverses are disposed orthogonally at the exposure, a simplified but less technically satisfying procedure can be

350 *Principles of Engineering Geology*

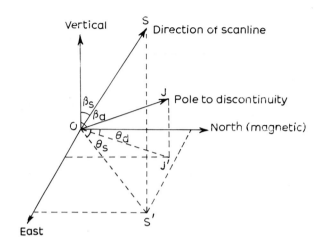

θ_d, θ_s are respectively the azimuth angles of the poles to a discontinuity or group of discontinuities and of the scaline,
β_d, β_s are the respective dip angles.
If $N(\theta_d, \beta_d)$ discontinuities fall within a prescribed group having a mean bearing $\bar{\theta}_d, \bar{\beta}_d$, then the corrected number of discontinuities $N'(\bar{\theta}_d, \bar{\beta}_d)$ in the group $\bar{\theta}_d, \bar{\beta}_d$ will be:

$$N'(\bar{\theta}_d, \bar{\beta}_d) = \frac{N(\bar{\theta}_d, \bar{\beta}_d)}{\cos(\theta_s - \bar{\theta}_d)\cos(\bar{\beta}_d - \beta_s)}$$

Figure 6.17 Numerical correction for measured discontinuities as determined by orientation differences between poles to discontinuities or groups of discontinuities and scanline direction.

adopted. All discontinuities normal (±20°) to the scanline are summed and expressed as a percentage of the total discontinuity sample for the traverse. This is repeated for those discontinuities having the same strike range as on the previous traverse, but this time on a scanline directed at 90° to the former standard line. The discontinuities in this second case will therefore be parallel and sub-parallel to the traverse line. A ratio is then taken of the percentage of discontinuities in the first scanline to the percentage of those in the second, the ratio being used as a *right-angle standardizing index*.

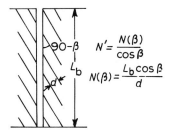

Figure 6.18 Vertical section through a borehole in discontinuous rock (*after* Terzaghi, 1965).

$$N' = \frac{N(\beta)}{\cos \beta}$$

$$N(\beta) = \frac{L_b \cos \beta}{d}$$

Effects of standardization

The total number of discontinuities sampled at the two specified locations was 288 and this increased after standardization to 532, an increase of nearly 85 per cent. Numerical differences on a traverse basis are summarized in Table 6.8 below.

In addition to the numerical changes, standardization procedures also affect the orientation distribution of discontinuities. Where these changes are significant in the Kimmeridge Clay, rose diagrams have been constructed for comparison purposes (Figure 6.15).

Some unease must be expressed at the proposal for 'manufacturing' data in this manner. If the operator can be satisfied that he has carefully measured *every* discontinuity intersection on his scanline however acutely-angled that intersection, then it is recommended that this form of numerical upgrading be ignored.

(iii) *Discontinuity orientation controls on stability*

Individual discontinuity planes dipping into a slope only impose a direct mechanical instability on the slope when they are of the same

Table 6.8 Effect of joint standardization on joint population per traverse

Traverse No.	Number of non-standardized discontinuities	Number of standardized discontinuities	% difference
1	50	58	16
2	17	145	752*
3	14	76	444
4	66	73	11
5	25	55	120
6	116	125	7

*Exceeds Robertson's (1970) tolerable correction factor of 5.

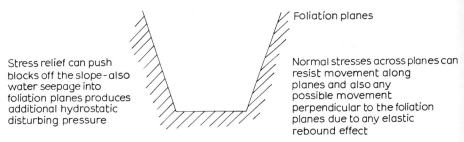

Figure 6.19 Diagrammatic cross-section of 30 m deep cuts in a group of highly foliated, folded and moderately-to-strongly weathered phyllites. Cuttings for re-location of 40A highway and Western Pacific Railroad around Oroville Dam and Reservoir along the edge of Sacramento Valley and the foothills of the Sierra Nevada, California, U.S.A. 8½ miles of re-location pass through these metamorphics of the Calaveras group (*after* O'Neill, 1963).

scale as the slope. They will, however, encourage the transmission of water into the body of the rock (Figure 6.19 and also see Section 8.2) and thereby promote a pore pressure increase in the vicinity of a failure zone. Discontinuities dipping outwards towards a slope adversely *affect* stability the more closely they parallel the surface of the slope. The slope is affected to the extent that its *form* is controlled by discontinuities so orientated. Its total *stability* status is a function of both orientation and spatial density distribution. Reference may be made to Wittke (1970) and Londe (1973b).

A stability analysis based on limiting shear stress along each discontinuity requires rather specific knowledge of the local stresses acting at points within the rock mass. The general approach to this problem is considered in Section 10.4, the magnitudes of the several principal stress ratios ideally being derived from the results of a finite element analysis of the slope being studied (the principal stress magnitudes and trajectories would then be computer-plotted on the slope section for visual examination in addition to being output in printed format). The general aim in the discontinuity stability evaluation is to sense the sensitivities of each of the stability parameters by re-cycling the analysis for each incremental change of each variable.

6.13 Influence of gouge material and the surface roughness characteristics of discontinuities

The shear strength of brittle rock joints in particular is influenced by the character and consequential shear strength parameters of any infilling gouge material. This influence is of great importance when the joint is continuous and it should be recognized that the resistance to sliding along a joint plane can be either decreased or increased depending upon the nature and thickness of the gouge and the character of the joint walls. If the line of shear is directed parallel to a joint infilled with a thick gouge layer then the shear strength parameters of the gouge exert an almost total control on mass stability.

If a gouge clay is at or near to its liquid limit, it may have little shear strength whereas at lower moisture contents it may have considerable strength and bearing capacity. A 10 per cent decrease in moisture content may create a 5-fold increase in the shear strength of the clay (Hough, 1957) but one way of assessing this effect is to study the relative strengths of gouge and host rock when they are both the same material. Tests on gouge material in the Kimmeridge Clay indicated a decrease in unconfined compressive strength of 820 per cent for a corresponding 24 per cent increase in moisture content. There would also appear to be a slight increase in the moisture content of gouge material with width of joint opening.

Undulations on surfaces of discontinuities

If the undulations on the discontinuity surfaces are systematic, the direction of their axial planes should be noted with respect to directions of possible movement along them. Movement parallel to the axial planes of the undulations will be unaffected by them. Movement normal to the axial planes could involve shearing through them if the normal pressure and shear stress across them are of sufficient magnitude. It is, however, more likely that any relative motion will take place by riding over the undulations in which case the friction angle ϕ_L referred to a flat surface will become $\phi_L + i$ (see Figure 6.20). Shear motion sub-parallel to the axial planes will be

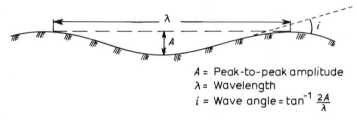

A = Peak-to-peak amplitude
λ = Wavelength
i = Wave angle = $\tan^{-1} \frac{2A}{\lambda}$

Figure 6.20 Specification of joint undulations.

directionally modified by the undulations. This question is considered in greater detail in Sections 10.5 and 10.6.

The reader will be aware from Figure 2.13 that there is a standard technique, used in a quality control metallurgical laboratory, of assessing in a quantitative manner the roughness of a metal surface. The device is known as a 'Talysurf'* and the degree of surface roughness is expressed as a centreline-average (CLA). If L is the surface length over which the roughness is to be assessed, and A is the undulation, up or down, from some average surface level, then

$$\text{CLA} = \frac{1}{L} \int_0^L |A| \, dL \tag{6.16}$$

The Talysurf instrument has been used by Welham (1969) to relate the friction parameter of cut rock and concrete surfaces to surface quality but more geological engineering application would be found in a larger scale profilometer instrument based on the same general principles. For further reading on joint surface morphology, reference may be made to Mogilevskaya (1974).

Cohesion and friction angle along discontinuity surfaces

Robertson (1970) has suggested the following method for estimating the shear strength of a discontinuity through the cohesion and friction angle parameters

(i) Determine c_{dc}, ϕ_{dc} for a clean joint having no gouge infilling; also determine c_{dg}, ϕ_{dg} for a joint with sufficient thickness of gouge that it controls its shear behaviour.

(ii) Knowing or estimating the normal stresses on the plane of failure, calculate the approximate average strength that either of these joints would possess and let these strengths be τ_{dc} and τ_{dg}.

(iii) If $\tau_{dc} > \tau_{dg}$, determine c_d and ϕ_d thus:
  (a) if percentage of discontinuities containing gouge is >30, then $c_d = c_{dg}$ and $\phi_d = \phi_{dg}$
  (b) if percentage (x) of discontinuities containing gouge is <30, then

$$c_d = c_{dc} + (c_{dg} - c_{dc}) \frac{x}{30}$$

$$\phi_d = \phi_{dc} + (\phi_{dg} - \phi_{dc}) \frac{x}{30}$$

---

*The 'Talysurf 4' is manufactured by the Rank Taylor Hobson Division of the Rank Organisation in England.

(iv) If $\tau_{dc} < \tau_{dg}$, determine c_d and ϕ_d thus:
   (a) if percentage of clean discontinuities is >70, then $c_d = c_{dc}$ and $\phi_d = \phi_{dc}$
   (b) if percentage (x) of clean discontinuities is <70 then

$$c_d = c_{dc} + (c_{dg} - c_{dc})\frac{70-x}{70}$$

$$\phi_d = \phi_{dc} + (\phi_{dg} - \phi_{dc})\frac{70-x}{70}$$

As an example of the use of these arguments, the cohesive parameter is calculated for the Kimmeridge Clay thus:—

Percentage of joints containing gouge in the three traverses —

Traverse 1: 9/50 = 18.0%
Traverse 2: 5/17 = 29.4%
Traverse 3: 2/14 = 14.3%

so, suggesting an average of 20.5 per cent for all three traverses. Also $\tau_{dc} > \tau_{dg}$. Using

$$c_d = c_{dc} + (c_{dg} - c_{dc})\frac{x}{30}$$

and if

$$q_u = 1.7 \text{ kgf cm}^{-2} \quad \text{and} \quad c_{dc} = 0.16\, q_u$$

where q_u is the unconfined compressive strength,

$$c_d = 0.272 + (0.034 - 0.272) \times \frac{20.5}{30} \text{ kgf cm}^{-2}$$

thus, $c_d = 0.1094$ kgf cm^{-2} (224 lbf ft^{-2} or 10.73 kN m^{-2})

6.14 Distributions

Analysis of the statistical distribution of a sampled variable is a rather specialist topic and a student in engineering geology would reasonably expect to refer to an appropriate text book such as, for example, the one by Krumbein and Graybill (1965). Nor are distributions the only statistical interest in engineering geology. A knowledge of linear and polynomial least squares regression analysis is an essential element in the statistical 'tool kit' of the serious student, as is correlation, goodness of fit and variance analysis. It would also be difficult to engage upon a meaningful exercise in geotechnical mapping without an appreciation of the qualities of trend surface analysis.

There are two aspects to the analysis of discontinuity distributions in rocks and stiff, fissured clays: these are the orientation and spatial density distributions. Orientation distributions are considered in Section 6.15 and a rather simplified appraisal of discontinuity lengths and their possible geotechnical ramifications is the subject of Section 6.20. It is appropriate, however, at this stage just to mention the possible forms of distribution that might characterize the *spatial* densities of discontinuities *en masse*. At the same time, it may be noted that similar distribution functions might be applied to orientation densities.

Probably the most important density function relating to a continuous geological population is the so-called normal, or normal probability distribution. Suppose that d is a spacing parameter, perhaps the minimum distance between two terminating, non-intersecting discontinuities or, more simply, the actual distance along a scanline between the points of intersection with the scanline of two adjacent discontinuities. Then if such a spacing distribution follows a normal probability curve, we may write

$$f(d,\bar{d},i) = \frac{1}{(2\pi)^{½} i} \exp\left[-\frac{1}{2i^2}(d-\bar{d})^2\right] \qquad (6.17)$$

Of these parameters, $\bar{d}$ is the arithmetic mean of all d and i is the standard deviation. A normal probability curve is 'bell-shaped', with a smaller radius of curvature at the crestal convexity than at the basal concavity, but the symmetry about a mean ordinate (density) axis will be preserved. A change in mean value displaces the axis of symmetry of the distribution along the abscissa (d axis) whereas a different standard deviation changes the shape of the curve. In general, for two different data sets on the same material, the two normal probability curves will be of different shape and will be displaced relative to one another horizontally along the graphical d-spacing axis. Other data may follow a log-normal probability curve.

There are examples of normal probability curves in Chapter 7. This type of curve also acquires some physical significance in that it fits the shape of surface settlement profile which results from tunnelling in clay (Litwiniszyn, 1956; Schmidt, 1969; Peck *et al*, 1969; Attewell and Farmer, 1974a, b). Statistical information having a distribution tending towards a normal probability form may also be

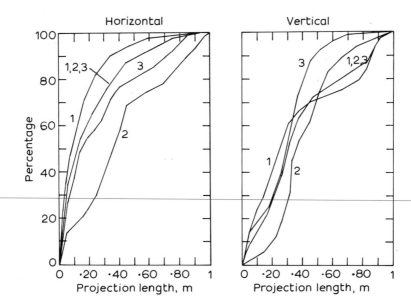

Figure 6.21 Cumulative continuity curves (three traverses) for discontinuity length projections on to horizontal and vertical axes — Kimmeridge Clay, Yorkshire, England.

presented in a cumulative manner. Figure 6.21 may be taken as an example of this, although some of these curves (particularly those representing horizontal projections of discontinuities on to scanlines) are not sigmoidal in shape. A further useful operation is to plot spatial data on normal or log probability graph paper. Cumulative probability sigmoidal curves may then be *linearized* and a best fit least squares regression line drawn through the data points.

Examination of discontinuity *length* histograms in Figure 6.22 for the stiff clay, clay shale, and chalk sites marked in Figures 3.26 and 3.27 and listed in Table 10.13 suggests that the tendency at some of the sites is towards a log-normal and bi-modal type of distribution. By way of contrast, the equivalent discontinuity *spacing* histograms in Figure 6.23 and those in Figure 6.13 are of a different general form. In these instances it would seem that the probability of recording a discontinuity spacing of a certain value d decreases in some exponential manner as d increases. A negative exponential distribution (a form of Poisson distribution) may be expressed as:

$$P(d) = \frac{1}{\bar{d}} \exp(-d/\bar{d}) \qquad (6.18)$$

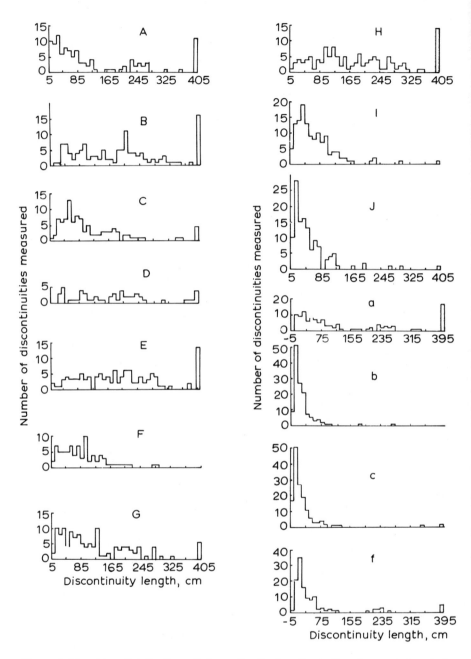

Figure 6.22 Size distribution of discontinuity lengths in the clays and clay shales listed in Table 10.13.

where $\bar{d}$ is the average of all d and is also the standard deviation in this distribution. The variance of the distribution is $\bar{d}^2$. Just as cumulative normal probability data may be linearized, so may the above equation:

$$\ln P(d) = \ln 1/\bar{d} - d/\bar{d} \qquad (6.19)$$

Spacing data may thus be plotted with respect to the following axes: logarithmic ordinate axis denoting percentage of discontinuity

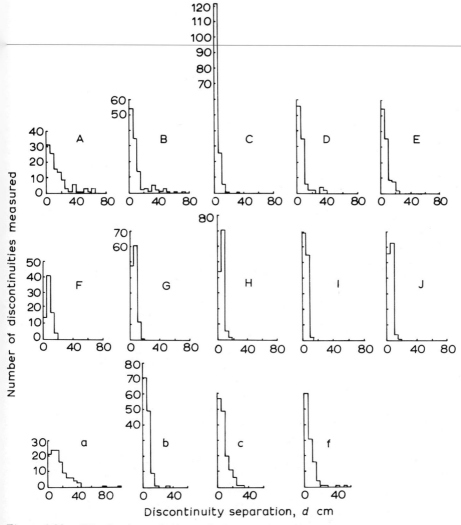

Figure 6.23 Distribution of discontinuity separations in the clays and clay shales listed in Table 10.13.

360   *Principles of Engineering Geology*

spacing values in each chosen interval and linear abscissa axis denoting discontinuity spacing. If the data can then be satisfactorily represented by a straight line, its slope would take the value $-1/\bar{d}$ and the ordinate intercept would be equivalent to the natural logarithm of $1/\bar{d}$.

Excluding the two chalk sites listed in Table 10.13, discontinuity spacings in the stiff clays and clay shales have been plotted in the manner of Figure 6.24 in order to see if they conform to a negative exponential law. It may be argued that the class interval of 0.05 m used for the plotting is rather too coarse in view of both of the

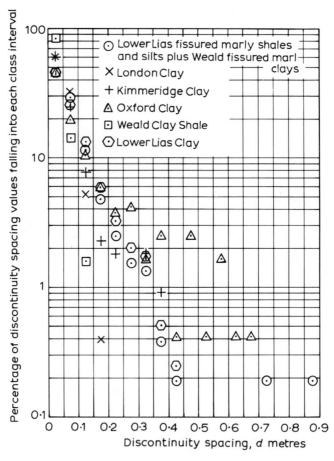

Figure 6.24   Spatial distribution of discontinuities in fissured clays and clay shales, plotted in class intervals of 0.05 m.

closeness of the discontinuity spacings in these materials and of the 0.1 m RQD characteristic spacing (discussed subsequently and in Chapter 7). The choice may, however, be justified on the basis of speed, and a finer class interval could be used at a later time if it were felt to be necessary.

Examination of the individual suites of points indicates that certain of the clays, such as the Lower Lias, conform quite well to a negative exponential law over their complete spectrum of spacings. Others, such as the Oxford Clay, fail to satisfy the law with respect to their larger discontinuity spacings. A composite plot of all the clay data in Figure 6.25 serves to confirm this latter shortcoming. For these particular materials, therefore, the case for a negative exponential law is not yet proven and there would be little merit in attempting to evaluate constants and slopes from data showing that degree of scatter.

There are analogies between a scanline that is intersected by discontinuities and a borehole core that is similarly intersected (Deere et al, 1967). In the latter case, since the spatial distribution of discontinuities, and hence the competence of the rock mass, is expressed in terms of an RQD value (see Section 7.7) there would seem to be scope for considering RQD in terms, for example, of a negative exponential distribution for those materials which reasonably satisfy that law. This operation has been developed by Priest and Hudson (1976) for different rocks in the manner described below.

In terms of RQD, let n be the number of whole lengths of core l_i ($i = 1, \ldots n$) $\geqslant 0.1$ metre (4 in). If L is the total length cored, then

$$\text{RQD} = 100 \sum_{i=1}^{n} l_i/L \qquad (6.20)$$

Note that the reference length is that of the borehole. The percentage recovery independent of l_i is also an indicator of ground quality.

In the case of a scanline, the probability of measuring discontinuity intersections with the line between distances d and $d + \delta d$ is given by $P(d)dd$. With L now the scanline length, the number of mean-spaced discontinuities would be $L/\bar{d}$. The summed length of the intact lengths will then be $Ld/\bar{d}. P(d)dd$. Expressing a lower limit of acceptable spacing as d^*, the length of scanline L^* for spacing

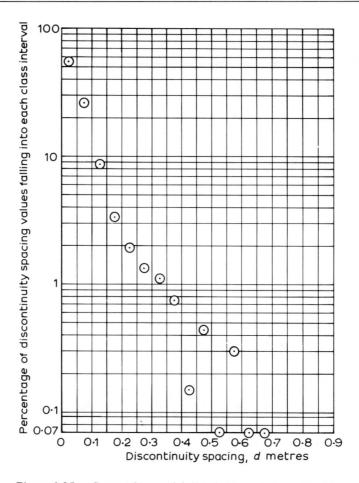

Figure 6.25 Composite spatial distribution of discontinuities in fissured clays and clay shales, plotted in class intervals of 0.05 m.

values above d^* must then be:

$$L^* = L/\bar{d} \int_{d^*}^{\infty} dP(d)\mathrm{d}d = \frac{L}{\bar{d}^2} \int_{d^*}^{\infty} d \exp(-d/\bar{d})\mathrm{d}d$$

$$= L/\bar{d}^2 \left[-d\bar{d} \exp(-d/\bar{d}) \Big|_{d^*}^{\infty} + \bar{d} \int_{d^*}^{\infty} \exp(-d/\bar{d})\mathrm{d}d \right]$$

$$= L \quad [1 + d^*/\bar{d}] \exp(-d^*/\bar{d}) \qquad (6.21)$$

Expressing L^* as a percentage of L, the above equation may be rewritten as:

$$\text{RQD}_{d^*}' = 100 \left[1 + \frac{d^*}{\bar{d}}\right] \exp\left(-\frac{d^*}{\bar{d}}\right) \quad (6.22)$$

Applying the standard RQD intact length of 100 mm ($= d^*$):

$$\text{RQD}_{(d^* = 0.1)}' = 100 \, [1 + (0.1/\bar{d})] \, \exp(-0.1/\bar{d}) \quad (6.23)$$

Equations 6.22, 6.23 therefore offer a means of estimating rock quality simply by *counting* the number of discontinuity intersections along a known length of scan-line. But although this would speed up the operation and eliminate the tedium of measurement, the value of the operation is dependent upon the discontinuity spacing values d following a distribution of negative exponential form. It is also sensitive to the choice of L.

Since the Lower Lias Clay discontinuity spacings appear to follow the negative exponential law quite well, the *measured* RQD values (equation 6.20) for nine scanlines at sites a,b,c (Figure 3.26: Table 10.13) are plotted against the equivalent mean discontinuity frequencies of $\bar{d}^{-1}$ ($= n/L$) in Figure 6.26. Also marked is the curve representing equation 6.23, although it would be quite easy to construct a family of RQD' curves for different d^* (equation 6.22) in order to gain an idea of the percentage of core that comprises lengths $\geqslant d^*$.

At higher mean discontinuity frequencies, $\text{RQD}_{(d^* = 0.1)}'$ exceeds RQD. This implies that the discontinuities tend to be more uniformly spaced than would be suggested by a negative exponential distribution. Priest and Hudson (1976) have noted that such an effect would be greatest in a rock showing a strong planar anisotropy. On the other hand, weaker rocks are more structurally prone to contain a more random field of terminated discontinuities superimposed on a more regular joint system. Since randomness is a characteristic producing a negative exponential distribution we may tentatively conclude that if a rock is materially weak then ideally it should also be intrinsically isotropic for it to satisfy a negative exponential distribution with respect to its discontinuities.

It is useful to record that Robertson (1970) also suggested the use of a negative exponential function to describe his joint trace *length* data. But we are by no means restricted to the forms of distribution

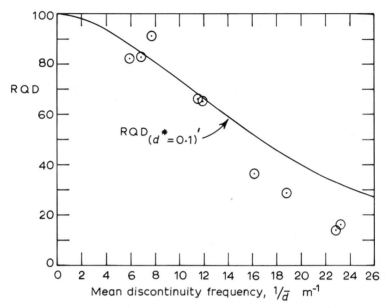

Figure 6.26 Comparison of measured and theoretical RQD values derived from discontinuity scans in the Lower Lias Clay, south coast of England.

so far specified. Others which may be considered include the binomial, Poisson, and geometric distributions. Two types of distribution which may find increasing use in geology in the future are the gamma density and circular density distributions (see Krumbein and Graybill, 1965). For further reading on distributions with particular reference to civil engineering problems, see the book by Benjamin and Cornell (1970).

6.15 Orientation density distribution of discontinuities

Provided that the discontinuity poles have been plotted on an equal area basis, they can be contoured to provide an expression of orientation-concentration. There are different methods of hand-contouring (see Phillips, 1971; Turner and Weiss, 1963; Fairbairn, 1949) and Stauffer (1966) concluded, from a study of their relative merits, that no one technique was suitable for all circumstances. The interested reader may care to refer to the Schmidt (1925) free-counting method (see also Billings, 1954), the variable-circle method of Mellis (1942), the variable-ellipse method of Straud (1945), the squared-grid method of Stauffer (1966) and the free counter method of Turner and Weiss (1963), and he will immediately appreciate the problems inherent in preserving as far as possible statistical accuracy with respect to discontinuity density through the latitudes. Basically, the preservation of sampling area uniformity on an equal area projection really demands a sampling cell of continuously varying shape. This point was made by Monahan (1963) and at the peripheral limit of the projection Attewell and Woodman (1971) have shown that there can be a fundamental inaccuracy associated with the necessary compounding of densities across a diameter. To these problems, which are essentially rooted in the *geometry* of the projection, must be added those *statistical* problems which are a function of the total number of data points and the nature of its distribution. Certain processing techniques may be viable for large numbers of data points whereas a different approach may be more valid for few data points. In a similar manner, the degree of randomness may act as a constraint upon the choice of processing technique. The problem can be viewed essentially in two parts; first an objective processing of the data, and second, estimates of the significance of orientation concentrations. As before, the interested reader is advised, for further study, to consult Winchell (1937), Chayes (1949), Larsson (1952), Flinn (1958), Kamb (1959) and Stauffer (1966). He should be ever cautious in his concentration interpretations and appreciate, for example, that randomness of orientation is equivalent to an isotropic distribution only when the number of data points is large and, conversely, a girdle or isolated concentration projected by a limited number of points may not indicate a statistical departure from randomness.

Probability concepts

It is useful to consider the application of probability theory in both the processing and interpretative modes. In the latter case, for example, the Poisson distribution (a mathematical model of randomly occurring events which defines a probability level of occurrence) can be used to evaluate point concentrations for petrofabric analysis (see Spencer, 1959 and Friedman, 1963 and note the Maximum Cell Value test and the Poisson O-Cell test suggested by Stauffer, 1966). The efforts that have been directed towards assessing the level of concentration significance and particularly whether weak concentrations, which appear as small perturbations on a quasi-spherically-symmetric background, have any significance at all have their origins in structural geology but the results of those efforts have applications in the field of mechanical stability.

It is now proposed to consider the mathematical concepts behind the processing of discontinuity data, as outlined by Attewell and Woodman (1971) for computer application. Fabric generation is a tedious and time-consuming exercise which is prone to human error, and if subsequent operations, such as the progressive application of incremented stress systems, are to be performed on the fabrics then computer processing and fabric presentation is the only acceptable answer. As backing references for some of the subsequent arguments, the reader may care to refer to Mellis (1942), Flinn (1958), Spencer and Clabaugh (1967) and Lam (1969), but it happens that the approach outlined below differs somewhat from that of Spencer and Clabaugh.

First of all it is necessary to acknowledge the fact that the distribution of elements comprising a fabric can only be specified accurately if the total number of these elements (in the present instance, poles to discontinuities) N' can be observed. Almost inevitably — and as considered earlier — the whole population N' will not be observed and so for any finite number of sampled elements $N < N'$, the probability distribution $P(\beta,\theta)$ for this subset of the population is a sum of Dirac delta functions. Noting that β and θ are the polar coordinates (latitude and azimuth respectively) of the poles to the discontinuities, then

$$P(\beta,\theta) = \frac{1}{N} \sum_{i=1}^{N} \delta(\beta_i,\theta_i) \qquad (6.24)$$

where, for integration over part ($\mathscr{S}$) of the sphere of projection,

$$\int_{\mathscr{S}} \delta(\beta_i,\theta_i) \sin\beta \, d\beta \, d\theta \;=\; 1 \text{ if the range of integration includes } (\beta_i,\theta_i) \qquad (6.25)$$
$$= 0 \text{ otherwise}$$

P clearly satisfies the normalization condition:

$$\int_{\text{sphere}} P(\beta,\theta) \sin\beta \, d\beta \, d\theta \;=\; 1 \text{ where the integration is carried out over the whole sphere of projection} \qquad (6.26)$$

Equation 6.24 gives the correct probability density in the cases where N samples represent the whole discontinuity population and are not merely part of it. But in the general case, in order to obtain a better representation of the total population, the delta functions must be smoothed out in some way. This can be done by allocating to each data point a small circular area on the *sphere* of the projection, the area being centred on the projection of the point. If A is this chosen area, then the locus $C(\hat{r})$ of the circle which delimits this area can be calculated from

$$\hat{r} \cdot \hat{f}_i = \cos \beta_a \qquad (6.27)$$

where $\hat{f}_i$ is the unit radius vector of the projected data point and β_a is given by

$$A = 2\pi a = 2\pi(1 - \cos \beta_a) \qquad (6.28)$$

The next step is to define a function S for each data point such that

$$S(\hat{r} \cdot \hat{f}_i) = \frac{w}{2\pi} \text{ if } \hat{r} \cdot \hat{f}_i \geq \cos \beta_a \qquad (6.29)$$

or,

$$= 0 \text{ if } \hat{r} \cdot \hat{f}_i < \cos \beta_a$$

It follows that

$$\left. \begin{array}{l} \int_{\mathscr{S}} S \, d\mathscr{S} = wa \text{ if } \mathscr{S} \text{ wholly includes } A \\ \qquad = wfa \text{ with } 0 < f < 1 \text{ if } \mathscr{S} \text{ partly includes } A \\ \qquad = 0 \text{ if } \mathscr{S} \text{ excludes } A \end{array} \right\} \quad (6.30)$$

If the $\dfrac{\delta_i}{N}$ in equation 6.24 are now replaced by S_i, then

$$P(\beta,\theta) = \sum_{i=1}^{N} S(\beta,\theta;\beta_i,\theta_i) \qquad (6.31)$$

It follows that

$$\int_{\text{sphere}} P(\beta,\theta) \sin\beta \, d\beta \, d\theta = \sum_{i=1}^{N} \int S(\beta,\theta;\beta_i,\theta_i) \sin\beta \, d\beta \, d\theta \qquad (6.32)$$

$$= wNa$$

It becomes apparent that there are two ways in which $\int P \, d\mathscr{S}$ can be normalized. Either the area window a can be set equal to $1/N$ with $w = 1$ or, alternatively, a can be fixed and then S weighted by the factor $1/Na$. In the latter case, a is usually fixed by convention at

0.01 (1% area) and although, computationally, any value of a can be specified, in practice higher values only serve to smooth out fabric concentrations which might be significant and smaller values create extraneous concentrations which could confuse meaningful interpretation.

The type of normalization chosen can exert a considerable influence on the character of the representation for small values of N. In general, the former area normalization, in which the window is adjusted inversely as N, produces rather better results for a near uniform data point distribution (near-spherical fabric symmetry) than for a highly preferred distribution, and vice versa.

One obvious fact, which has been mentioned earlier, is that $C(\hat{r})$ does not project on to an equal area net as a circle although its area a is preserved. This is a direct confirmation of the fact that the conventional, hand-operation method of plotting circles on the equal area projection itself, which also forms the basis of Spencer and Clabaugh's computer plotting program, leads to asymmetrical smoothing of data away from the centre of the projection (see also Mellis, 1942 and Flinn, 1958).

In the computer program based on the above analysis and used by the authors, the calculations are performed using a matrix the elements of which are matched, on a one-to-one basis, to the desired output format. The matrix element nearest to a data point is used to initiate a systematic outwardly-directed search for elements which satisfy the condition:

$$\hat{r} \cdot \hat{r}_i \geq \cos \beta_a \quad \text{(see equation 6.27)}$$

For each such element, the weighting constant w is added to the value already stored in that location. When no more elements are found to satisfy equation 6.27 (modified for inequality) the computer search is terminated and the search procedure recommenced on the next data point. This search procedure has the advantage of requiring only a small fraction of the matrix to be checked for each point. When all the S_i functions have been superimposed on the matrix, a corresponding output matrix comprising fabric character information can then be generated and printed out. Examples of output matrices appear later and elsewhere in the text.

As used by the authors, and in addition to the normalization options, the processing can be for an upper hemisphere, lower hemisphere, or whole sphere projection, the physical size of the output projection can be specified and any print characters can be called to represent the density distribution information. But at this stage of the processing, the fabric information is no more than visually qualitative, and although it is of interpretative value to the experienced engineering geologist it is obviously desirable to test its stability implications in an objective manner by subjecting it to a three-dimensional stress field and seeing if it will fail in shear against the inherent strength of the discontinuities. This operation is discussed in Section 6.16.

6.16 Discontinuity shear stability in a polyaxial stress field

In this section of the analysis it is intended to investigate the stability status of a single discontinuity having shear strength parameters c, ϕ and direction cosines a_{11}, a_{12}, a_{13} with respect to principal stresses $\sigma_1, \sigma_2, \sigma_3$ acting along axes OX_1, OX_2, OX_3 respectively in Figure 6.8. Tacitly, it is assumed, therefore, that global space and principal stress space are co-axially compatible for the purposes of this analysis although rotation procedures could be invoked were this not the case. It will also be appreciated that, for equilibrium, there are three additional quadrants to be balanced for stress in the reference hemisphere and four quadrants in the opposing hemisphere but these quadrants can be satisfied from the subsequent analysis by symmetry. The first row only of the direction cosine matrix need therefore be considered and in this example a discontinuity pole is substituted for one of the axes in the transformation field of Figure 6.9.

Geometrical analysis on the Mohr diagram

In Figure 6.27a are marked the three direction cosine orbits which mutually intersect in principal stress and global space at the point where the pole to the discontinuity itself intersects the hemisphere of projection. It is immediately assumed, in order to develop the analysis, that this discontinuity is just prone to shear under the influence of principal stresses $\sigma_1, \sigma_2, \sigma_3$ directed along OX_1, OX_2, OX_3 respectively.

In Figure 6.27b the stability condition is presented in the $|\tau|, \sigma_n$ plane and it is proposed to investigate the problem in a geometrical

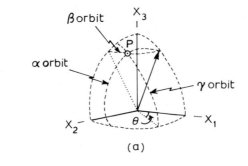

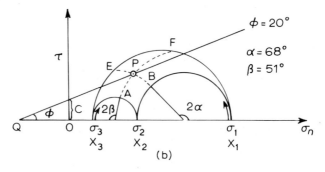

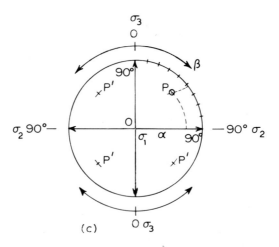

Figure 6.27 Discontinuity at critical equilibrium for shear.

manner. The reader requiring a rigorous mathematical justification for the operations that follow is directed to Jaeger and Cook (1969).

Construct normal stress (σ_n) and shear stress (τ) axes, both to the same, suitable, linear scale. Along the σ_n-axis strike off the amplitudes $\sigma_1, \sigma_2, \sigma_3$ and construct principal stress semi-circles with centres $\frac{1}{2}(\sigma_1 + \sigma_3)$, $\frac{1}{2}(\sigma_1 + \sigma_2)$, $\frac{1}{2}(\sigma_2 + \sigma_3)$ and radii $\frac{1}{2}(\sigma_1 - \sigma_3)$, $\frac{1}{2}(\sigma_1 - \sigma_2)$, $\frac{1}{2}(\sigma_2 - \sigma_3)$ respectively. To this picture of the external deforming stress field add the strength characteristics of the discontinuity by striking off its cohesion amplitude on the τ axis and completing its assumed linear envelope by projecting the friction angle ϕ. Quite arbitrarily at this stage, a ϕ-angle of 20° (which is reasonably representative for a clay shale) has been assumed so that the major principal stress circle is intersected (in this example, for c at its given amplitude and $\phi > 33°$, there would be no intersection and no shear failure possibility for any combination of α, β, γ). Following construction of angles 2α and 2β inside principal stress circles $\sigma_1 : \sigma_2$ and $\sigma_2 : \sigma_3$ respectively, arcs AF and BE are constructed from the centres of these circles; that is, from $\frac{1}{2}(\sigma_1 + \sigma_2)$ and $\frac{1}{2}(\sigma_2 + \sigma_3)$ respectively. For limiting equilibrium, the intersection point P of these arcs must lie on the c, ϕ envelope for the discontinuity, the shear stress τ_d and normal stress σ_{nd} along and across the discontinuity being defined by the coordinates of P in the $|\tau|, \sigma_n$ plane.

Figure 6.27c shows how the location of P is specified on the projection in which the axis $O\sigma_1$ (OX_1) is arbitrarily fixed at the centre. Angle α is measured from $O(\sigma_1)$ along the $O\sigma_2$ axis and then constructed for all β in the $\sigma_2 : \sigma_3$ plane using compasses at centre O. Angle β is measured along the periphery of the projection from σ_3 to σ_2 and its orbit generated along the computed small circle. Intersection point P fixes the discontinuity on the projection and at critical equilibrium marks the first step in the process of delimiting safe and unsafe orientation regimes in principal stress space.

Perusal of Figure 6.27b quickly indicates that there is an infinity of α, β, γ combinations for which a discontinuity pole P traverses along the c, ϕ envelope and from which the safe:unsafe orientation delimination can be completed (the reader will note, intuitively, that a construction for angle γ is unnecessary on Figure 6.27b since, from the direction cosine relationships discussed earlier, $\gamma = \cos^{-1}[1 - (\cos^2 \alpha + \cos^2 \beta)]^{1/2}$. Furthermore, if, for example, angle α is held constant, it can be seen that there is a possible range of β from 90° down to a low value determined from construction radius $\frac{1}{2}(\sigma_1 + \sigma_2)$: E, centre $\frac{1}{2}(\sigma_1 + \sigma_2)$, and its arc intercept on the $\sigma_2\sigma_3$ semicircle. In a similar manner, direction cosine angle β can be held constant and all α from 90° to the F-arc intercept on the $\sigma_1\sigma_2$ semicircle investigated. Of course, discontinuity orientations generated through arcs PE and PF are inherently unstable with respect to shear.

Suppose, now, that the shear strength of a discontinuity field tends towards a residual state with ϕ_r about 20° and $c = 0$. Under the same principal stress conditions as before, the discontinuity envelope now cuts through all three principal stress circles, as shown in Figure 6.28. Orientation areas $X_3 ab$, $X_2 cd$, and $X_1 ef$ represent shear stability (shaded areas) while the dotted area within the major principal stress circle $\sigma_1\sigma_3$ and above *abcdef* represents shear instability. The problem again is to determine $\alpha{:}\beta$ combinations along the envelope a to b, c to d and e to f. The manual procedure using a pair of compasses is exactly as before, assuming that pole P lies successively on $a,b;c,d;e,f$ and taking several intermediate points along these lines to facilitate a reasonably accurate delimitation on the projection. Points b,c,d,e, are immediately definable simply by inspection: at P(b), $\beta = \beta_1$ and $\alpha = 90°$; at P(c), $\beta = \beta_2$ and $\alpha = 90°$; at P(d), $\beta = 90°$ and $\alpha = \alpha_2$; at P(e), $\beta = 90°$ and $\alpha = \alpha_1$. Intermediate points become difficult to specify accurately, particularly along ab, due to the steepness of the limbs of the principal stress circles as they converge at X_3, X_2, X_1. It will readily be apparent that a high-speed computational approach will not only produce greater accuracy in these difficult regions but it will also permit point P to be incremented many thousands of times along the envelope for a

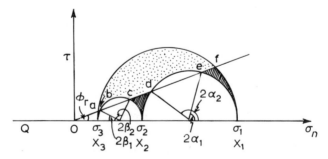

Figure 6.28 Intersection of a discontinuity Mohr envelope with all three principal stress circles.

definitive representation of shear stability under known input conditions of $\sigma_1, \sigma_2, \sigma_3, c, \phi$. At its simplest, analytical development for computer analysis might take the form outlined below, although in fairness to the reader it must be pointed out that additional programming complexities associated with, for example, discontinuity data incorporation and processing, rotations, and the negative side of the normal stress axis would have to be incorporated.

Simple analysis of stability

In Figure 6.27b let

$$\sigma_1^* = \sigma_1 + OQ \text{ and } \sigma_3^* = \sigma_3 + OQ \qquad (6.33)$$

then,

$$\sigma_m^* = \frac{\sigma_1^* + \sigma_3^*}{2} \text{ and } \tau_m^* = \frac{\sigma_1^* - \sigma_3^*}{2} (= \tau_m) \qquad (6.34)$$

Letting

$$F = \frac{\tau_m}{\sigma_m^*} = \frac{\sigma_1^* - \sigma_3^*}{\sigma_1^* + \sigma_3^*} \qquad (6.35)$$

then

$$\sigma_1^* = \left(\frac{1+F}{1-F}\right)\sigma_3^* = K_1 \sigma_3^* \qquad (6.36)$$

where

$$K_1 = \left(\frac{1+F}{1-F}\right) \qquad (6.37)$$

F and K are dimensionless parameters.

Now, σ_n^* (the normal pressure + OQ) can vary between σ_3^* and σ_1^*; therefore, $\sigma_n^* = \sigma_3^* + M(\sigma_1^* - \sigma_3^*)$ for $0 < M < 1$

$$= \sigma_3^* \left[1 + M\left(\frac{\sigma_1^* - \sigma_3^*}{\sigma_3^*}\right)\right] \qquad (6.38)$$

where M is a constant.

It follows that from equations 6.36 and 6.38,

$$\sigma_n^* = \sigma_3^* \left[1 + M\left(\frac{2F}{1-F}\right)\right] = K\sigma_3^* \qquad (6.39)$$

where
$$K = \left[1 + M\left(\frac{2F}{1-F}\right)\right] \quad (6.40)$$

and is a further dimensionless parameter.

In a similar manner, σ_2^* can take values between σ_1^* and σ_3^* and would take the form:

$$\sigma_2^* = \sigma_3^*\left[1 + N\left(\frac{\sigma_1^* - \sigma_3^*}{\sigma_3^*}\right)\right] = K_2 \sigma_3^* \quad (6.41)$$

for $0 < \underline{N} < 1$ where

$$K_2 = \left[1 + N\left(\frac{2F}{1-F}\right)\right] \quad (6.42)$$

and is another dimensionless parameter.

Now the triad of fundamental equations linking the direction cosines and the principal stresses is:

$$\left.\begin{array}{l}\sigma_n = a_{1\,1}^2 \sigma_1 + a_{1\,2}^2 \sigma_2 + a_{1\,3}^2 \sigma_3 \\ \tau^2 = a_{1\,1}^2 \sigma_1^2 + a_{1\,2}^2 \sigma_2^2 + a_{1\,3}^2 \sigma_3^2 - \sigma_n^2 \\ a_{1\,1}^2 + a_{1\,2}^2 + a_{1\,3}^2 = 1 \end{array}\right\} \quad (6.43)$$

(for the generation of the normal stress and shear stress expressions refer, for example, to Jaeger and Cook (1969); the third expression is given earlier). From equations 6.43, direction cosines a_{11} and a_{13} can be derived as

$$a_{1\,1}^2 = \frac{(\sigma_2^* - \sigma_n^*)(\sigma_3^* - \sigma_n^*) + \tau^2}{(\sigma_2^* - \sigma_1^*)(\sigma_3^* - \sigma_1^*)} \quad (6.44)$$

$$a_{1\,3}^2 = \frac{(\sigma_1^* - \sigma_n^*)(\sigma_2^* - \sigma_n^*) + \tau^2}{(\sigma_1^* - \sigma_3^*)(\sigma_2^* - \sigma_3^*)} \quad (6.45)$$

Also, for limiting equilibrium on a planar discontinuity,

$$\tau = \sigma_n^* \tan \phi \quad (6.46)$$

(note that under effective stress conditions, σ_n^* would not only incorporate the OQ component but would also accommodate the subtractive component of pore pressure u; also, the friction angle ϕ would be primed). Thus, substituting in equations 6.44 and 6.45, and cancelling for σ_3^*:

$$a_{1\,1}^2 = \frac{(K_2 - K)(1 - K) + K^2 \tan^2 \phi}{(K_2 - K_1)(1 - K_1)} \quad (6.47)$$

and
$$a_{13}^2 = \frac{(K_1 - K)(K_2 - K) + K^2 \tan^2 \phi}{(K_1 - 1)(K_2 - 1)} \quad (6.48)$$

From these expressions, a_{11} and a_{13} ($\cos \alpha$ and $\cos \beta$ respectively) can now be computed for various combinations of F and ϕ using values of M and N between zero and 1. It only remains to check the $\phi = 0$ case for which the same procedure is followed but with the introduction of a new dimensionless parameter G where

$$G = \frac{2c}{\sigma_1 - \sigma_3} \quad (6.49)$$

and c is the cohesion intercept. Since

$$\sigma_3 = \sigma_1 - \frac{2c}{G},$$

$$\sigma_n = \sigma_3 \left[1 + M\left(\frac{\sigma_1}{\sigma_3} - 1\right) \right]$$

$$= \sigma_1 + \frac{2c}{G}(M - 1) \quad (6.50)$$

and

$$\sigma_2 = \sigma_1 + \frac{2c}{G}(N - 1) \quad (6.51)$$

For $\phi = 0$, putting $\tau = c$, then

$$a_{11}^2 = \frac{(\sigma_2 - \sigma_n)(\sigma_3 - \sigma_n) + c^2}{(\sigma_2 - \sigma_1)(\sigma_3 - \sigma_1)}$$

which, after substitution and manipulation of the parameters, reduces to

$$a_{11}^2 = \frac{M(M - N) + \frac{G^2}{4}}{1 - N} \quad (6.52)$$

In a similar manner,

$$a_{13}^2 = \frac{(\sigma_1 - \sigma_n)(\sigma_2 - \sigma_n) + c^2}{(\sigma_1 - \sigma_3)(\sigma_2 - \sigma_3)}$$

which becomes

$$a_{13}^2 = \frac{(M-1)(M-N) + \dfrac{G^2}{4}}{N} \qquad (6.53)$$

Normalization to σ_1

Throughout the discussions in this section, stress has been considered as a parameter of absolute magnitude. It will be appreciated, however, that the hydrostatic component of the applied stress field can in no way contribute to the shear deformation of a discontinuity suite; it simply serves to change the volume (density) of the rock mass and it is the deviatoric component of the stress tensor that is responsible for the applied shear. For problems in which shear behaviour is under investigation it is possible, therefore, to express the stress components in a dimensionless form, most conveniently by taking their ratio to σ_1. On the Mohr diagram, this means that a suitable scale would be chosen for setting σ_1 at unity and σ_2, σ_3 and c would be calculated as fractions of σ_1.

6.17 Shear strain energy concepts

So far we have considered the criterion for specifying which orientations could be critical with respect to shear movement given a balance between discontinuity strength (c_d, ϕ_d) and the principal stress field $(\sigma_1, \sigma_2, \sigma_3)$. It is possible, in fact, to map on the projection the orientation traces for critical equilibrium delimiting, in effect, 'safe' and 'fail' in principal stress space, and these operations are quite independent of the actual presence of any discontinuity. The concept of an unstable *regime* in principal stress space does, however, embody the notion of potential discontinuity shear movements of varying amplitudes over a *range* of orientations; the relative amplitudes of τ_d and σ_{nd} vary as a function of α, β, γ with τ_d/σ_{nd} greatest towards the centre of an unstable area on the projection and reducing towards the projected line of critical equilibrium. It follows, quite logically, that when a discontinuity fabric is superimposed on the projection there will be a range of degrees of failure for those discontinuities that fall within the unstable area, as expressed by their proximity to the equilibrium line and the magnitude of τ_d/σ_{nd}. While the Coulomb–Mohr criterion tells us something about an ultimate failure plane direction, it might also be argued that any mass instability is a function more of an integrated movement over a range of discontinuity orientations than of a simple limiting τ, σ_n. There are some restrictions in this concept in that historical evidence clearly shows that discrete failure surfaces do develop in conformity with a critical resolved shear stress criterion of failure but it is proposed that such phenomena in discontinuous rocks are compounded from numerous discontinuity shears and are a final expression of such integrated internal movement. It is now proposed to examine this concept in terms of the critical shear strain energy acquired by the rock mass.

A basic assumption is that failure occurs in a stressed material when the total shear strain energy generated by sliding between opposing surfaces of closed discontinuities exceeds some characteristic value. This same criterion was used by Wiebols and Cook (1968) to examine 'intact' rock failure due to shear movement along Griffith cracks having a random orientation density distribution. In the present case, the criterion is applied to 'large cracks' having a non-random distribution.

The criterion for sliding is then

$$|\tau| - (\sigma_{n\,d} \tan \phi_d + c_d) > 0 \qquad (6.54)$$

Parameters $[|\tau| - (\sigma_{n\,d} \tan \phi_d + c_d)]$ on the left hand side of the inequality comprise the shear stress available for the promotion of sliding, τ'_d. When $\tau'_d > 0$ for a particular discontinuity, shear movement occurs to induce a corresponding shear strain energy

$$W_d = sl^3\, \tau'^2_d \qquad (6.55)$$

where s is a function of the material, the shape of the discontinuity and its position and orientation relative to other discontinuities, and l is a dimension of the discontinuity.

It follows that the total shear strain energy due to movement along N' discontinuity surfaces is

$$W = N' \int_{\tau' > 0} Psl^3\, \tau'^2 \, d\mathscr{S}\, ds\, dl \qquad (6.56)$$

where, as before, $d\mathscr{S}$ denotes the surface element on the sphere of projection.

For the N' discontinuities uniformly distributed in size and spatially but not necessarily directionally, equation 6.56 becomes:

$$W = \bar{k} N' \int_{\tau' > 0} P\tau'^2 d\mathscr{S} \qquad (6.57)$$

where $\bar{k}$ is some mean factor.

Substitution for P from equation 6.24 gives

$$W = \bar{k}\frac{N'}{N} \sum_{\substack{i=1 \\ \tau'_i > 0}}^{N} \tau'^2_i \qquad (6.58)$$

where $\tau'_i = \tau'$ with respect to the ith data point.

Substitution from equation 6.31 leads to:

$$W = \bar{k}N' \sum_{i=1}^{N} \int_{\tau' > 0} S\tau'^2 \, d\mathscr{S} \qquad (6.59)$$

$$= w\bar{k}N' \sum_{i=1}^{N} \int_{\substack{A_i \\ \tau' > 0}} \tau'^2 \, d\mathscr{S} \qquad (6.60)$$

the integral being taken over the area A_i (see earlier analysis) representing the ith data point. The shear strain energy W compounded from all the discontinuity potential movements can be computed from equation 6.60 and expressed in arbitrary units if the factors outside the integral are ignored (these factors would be extremely difficult to quantify with any degree of reliability). For comparative purposes, the shear strain energy W can also be computed for an equivalent isotropic discontinuity distribution (a probability density of unity) in which the potential fail areas on the projection contain a uniform density distribution of discontinuities. The expression for the isotropic shear strain energy magnitude is

$$W_I = \bar{k}N' \int_{\tau' > 0} \tau'^2 \, d\mathscr{S} \qquad (6.61)$$

and its importance arises from the fact that its amplitude is dependent solely on the magnitude of the critical deforming stress-to-resisting stress imbalance and is independent of discontinuity fabric. It therefore provides a base from which, for equivalent stress and strength conditions, different discontinuity fabrics can be compared for shear instability in a semi-quantitative manner. This can be done by taking the ratio of the fabric shear strain energy (anisotropic shear strain energy W_A) to the isotropic shear strain energy W_I, or vice versa, to produce a discontinuity anisotropic index. W_A/W_I can be regarded as a 'danger index' and W_I/W_A as a 'safety index', and whereas both W_A and W_I must be plotted on a logarithmic scale to accommodate their wide numerical range, their ratio, if plotted as a 'danger index', can comfortably be presented graphically in a linear-linear manner against the principal stress ratio. In the case of the 'safety index' W_I/W_A, it is important *not* to apply the conventional concept of limiting equilibrium (factor of safety equal to one) when $W_I = W_A$, since potential shear failure is implied when W_A is greater than *zero*.

In practical terms, stability trends are best discerned by plotting W_A and one of the index values against a suitable principal stress ratio for different values of c and ϕ. This can be best illustrated by taking the example of a cutting in the Kimmeridge Clay of Yorkshire, England. This clay is replete with discontinuities and in several areas these discontinuities have been mapped for an evaluation of their orientation and spatial probability density distributions. The latter problem is considered earlier but it is useful at this point to consider the end result of a series of processed orientation fabrics.

The stability fabrics in Figure 6.29 are all orientated with north at the top and east at the right and, in fact, the only variables in the suite are the normalized minor principal stress (R), which acts along the east–west axis of the projection, and the intermediate principal stress (Q) which acts horizontally and parallel to the slope face and which is dependent on a plane strain condition applying. Stress parameters are calculated from the slope height and the density of the clay. Discontinuity shear strengths used in this example are the effective strength parameters $c'_d = 9.35$ lbf in^{-2} (64.47 kN m^{-2}) and $\phi'_d = 22°$ derived from laboratory tests. Principal stress ratios, direction cosine matrix a_{ij} ($i,j = 1,2,3$), tan ϕ_d, c_d/σ_1, W_A and W_I are listed under each fabric and safe/fail areas are determined by the + (positive) characters. Reference to Figure 6.28 and its associated analysis will indicate that all the principal stress axes must be located in *stable* space (there can be no shear stress acting normal to a principal stress axis and therefore discontinuity poles coaxial with a principal stress are always stable) and this permits stable space to be differentiated from unstable space on more difficult fabrics. The numerical fabric print-out records directly from 1 ... 9,O (alphabetic O representing 10) per cent per 1 per cent area the probability density distributions of the discontinuity orientations. Alphabetic characters A to K represent densities of from 11 per cent to 21 per cent per 1 per cent area, and it is clear from this analysis that the poles to major joint sets in the Kimmeridge Clay at this location are orientated approximately N75°E and N167°E, the joint sets themselves lying vertically. At the former orientation, the maximum

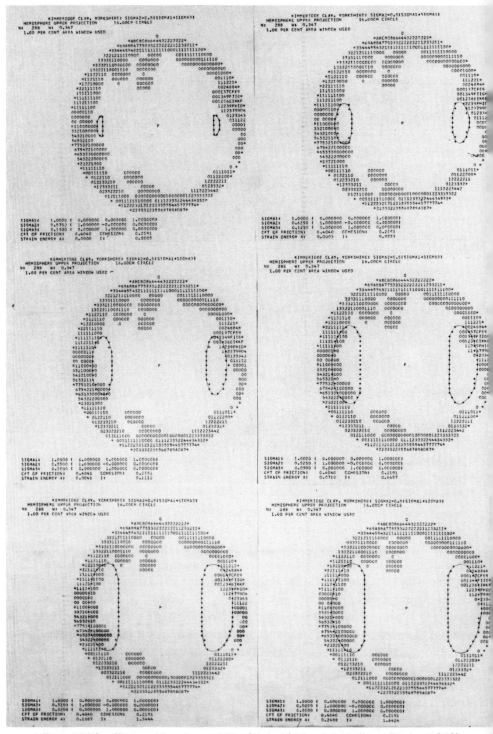

Figure 6.29 Kimmeridge clay stability fabrics. (*After* Attewell and Taylor, 1973).

discontinuity concentrations are 21 per cent per 1 per cent area while at the latter orientation they just achieve 12 per cent per 1 per cent area.

It is probably most useful, when inspecting the stability fabrics in Figure 6.29 and as an aid to interpretation, to regard the projected 'failure' window as a step function operating upon the fabric having certain numerical characteristics depending upon location within the window. There should be regular, monotonical changes in the *isotropic* parameter as a function of the shear stress-shear strength balance, and this is demonstrated in Figure 6.30. A range of c, ϕ shear strength parameters has been assessed, the present parameters $c'_d = 9.35$ lbf in^{-2}, $\phi'_d = 22°$ being defined in terms of the isotropic shear strain energy by an inverted solid triangle. Local, but intense,

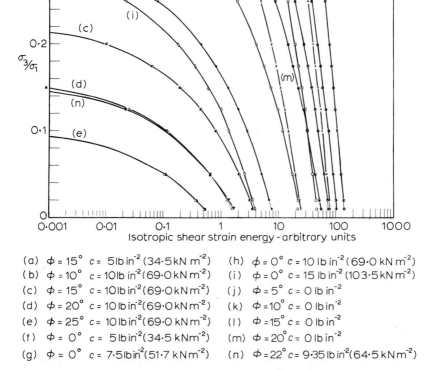

(a) $\phi = 15°$ $c = 5$ lb in^{-2} (34.5 kN m^{-2})
(b) $\phi = 10°$ $c = 10$ lb in^{-2} (69.0 kN m^{-2})
(c) $\phi = 15°$ $c = 10$ lb in^{-2} (69.0 kN m^{-2})
(d) $\phi = 20°$ $c = 10$ lb in^{-2} (69.0 kN m^{-2})
(e) $\phi = 25°$ $c = 10$ lb in^{-2} (69.0 kN m^{-2})
(f) $\phi = 0°$ $c = 5$ lb in^{-2} (34.5 kN m^{-2})
(g) $\phi = 0°$ $c = 7.5$ lb in^{-2} (51.7 kN m^{-2})
(h) $\phi = 0°$ $c = 10$ lb in^{-2} (69.0 kN m^{-2})
(i) $\phi = 0°$ $c = 15$ lb in^{-2} (103.5 kN m^{-2})
(j) $\phi = 5°$ $c = 0$ lb in^{-2}
(k) $\phi = 10°$ $c = 0$ lb in^{-2}
(l) $\phi = 15°$ $c = 0$ lb in^{-2}
(m) $\phi = 20°$ $c = 0$ lb in^{-2}
(n) $\phi = 22°$ $c = 9.35$ lb in^{-2} (64.5 kN m^{-2})

Figure 6.30 Analysis of Yorkshire Kimmeridge Clay stability using a plane strain condition such that $\sigma_2 = 0.5 (\sigma_1 + \sigma_3)$; 15 m slope; Isotropic Shear Strain Energy Parameter. (*After* Attewell and Taylor, 1973).

discontinuity orientation concentrations create temporary surges in the *anisotropic* parameter as the isotropic window expands or contracts through them. However, since the window has a smoothing capacity with respect to any non-isotropic fabric (noting, of course, that this capacity is a function of the stress-strength balance) the anisotropic parameter will usually plot quite smoothly against the principal stress ratio. Perturbations in this smoothness, if increasing the anisotropic shear strain energy, reflect orientation zones of steep fabric topography which could exert a particular control on structural stability with respect to shear.

Reference to Figure 6.31 will indicate that such a condition arises under plane strain conditions for a principal stress ratio σ_3/σ_1 of 0.125 and that the magnitude of the effect is conditioned by the c,ϕ

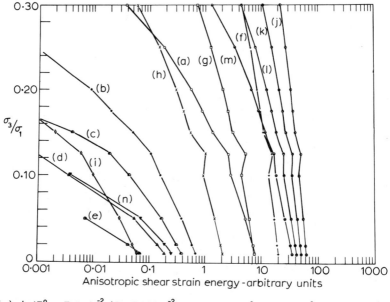

(a) $\phi=15°$ $c=5$ lb in^{-2} (34·5 kN m^{-2})
(b) $\phi=10°$ $c=10$ lb in^{-2} (69·0 kN m^{-2})
(c) $\phi=15°$ $c=10$ lb in^{-2} (69·0 kN m^{-2})
(d) $\phi=20°$ $c=10$ lb in^{-2} (69·0 kN m^{-2})
(e) $\phi=25°$ $c=10$ lb in^{-2} (69·0 kN m^{-2})
(f) $\phi=0°$ $c=5$ lb in^{-2} (34·5 kN m^{-2})
(g) $\phi=0°$ $c=7·5$ lb in^{-2} (51·7 kN m^{-2})
(h) $\phi=0°$ $c=10$ lb in^{-2} (69·0 kN m^{-2})
(i) $\phi=0°$ $c=15$ lb in^{-2} (103·5 kN m^{-2})
(j) $\phi=5°$ $c=0$ lb in^{-2}
(k) $\phi=10°$ $c=0$ lb in^{-2}
(l) $\phi=15°$ $c=0$ lb in^{-2}
(m) $\phi=20°$ $c=0$ lb in^{-2}
(n) $\phi=22°$ $c=9·35$ lb in^{-2} (64·5 kN m^{-2})

Figure 6.31 Analysis of Yorkshire Kimmeridge Clay stability using a plane strain condition such that $\sigma_2 = 0.5\,(\sigma_1 + \sigma_3)$; 15 m slope; Anisotropic Shear Strain Energy Parameter. (*After* Attewell and Taylor, 1973).

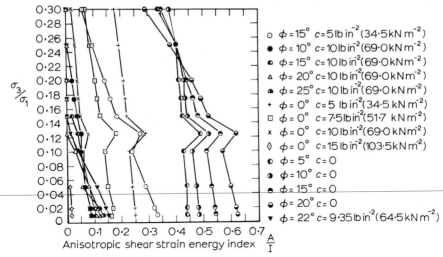

Figure 6.32 Kimmeridge Clay – Fox Holes, Yorkshire, England. Influence of magnitude of principal stress ratio upon the anisotropic shear strain energy index for a 15 m high slope and for various c, ϕ combinations across the discontinuities (computed for plane strain conditions throughout).

shear strength parameters of the discontinuities (note that there is a visible distortion caused by the log scale). The significance of this particular ratio is also, of course, a consequence of the principal stress directions as applied to the discontinuity fabric; had the slope face (and hence the principal stress orientations for the same magnitudes) been orientated differently then, depending again on the steepness of the fabric topography in other zones of the projection, anisotropic shear strain energy criticality could develop at a different σ_3/σ_1 ratio.

In Figure 6.32 the anisotropic shear strain energy index W_A/W_I is shown, for the Yorkshire Kimmeridge Clay fabric, to increase with decreasing σ_3/σ_1. The fact, for example, that W_A/W_I increases through an *increasing* range of discontinuity shear strength is an expression of increasing discontinuity orientation concentration as the failure window (effective shear stress $\tau' > 0$) condenses solely under c_d, ϕ control towards a situation of total discontinuity stability. In all such stability problems, it is the influence of the stress balance, increasing $c_d/\sigma_1, \phi$ and reducing $\sigma_2/\sigma_1, \sigma_3/\sigma_1$, or reducing $c_d/\sigma_1, \phi$ and increasing $\sigma_2/\sigma_1, \sigma_3/\sigma_1$, that can generate a specific anisotropic and isotropic parameter.

Rather than estimating principal stress amplitudes and directions

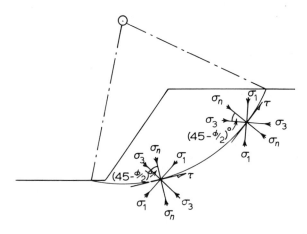

Figure 6.33 Principal stress trajectories along shear arc. Angle θ is measured between σ_n and σ_1.

(and in spite of any computational facility for rapidly incrementing these variables) they can be identified more precisely in two ways. One method involves the use of finite element analysis, an example of which is shown in Chapter 10 but not discussed in detail. Alternatively a soil mechanics circular failure type of analysis may be applied to the problem, the principal stresses and their directions being resolved from the shear stress acting at different positions along the slip surface (see Figure 6.33). A plane strain condition would normally be assumed to exist, and so with an arbitrary Poisson's ratio value of 0.5, the amplitude of the intermediate principal stress σ_2 normal to the plane of section would be equal to $0.5\,(\sigma_1 + \sigma_2)$. If τ and σ_n are known, then the major and minor principal stresses may be derived from the equations (see Section 2.5):

$$\left. \begin{array}{c} \sigma_n = \dfrac{\sigma_1 + \sigma_3}{2} - \dfrac{\sigma_1 - \sigma_3}{2} \sin \phi \\[1em] \text{and} \\[1em] \tau = \dfrac{\sigma_1 - \sigma_3}{2} \cos \phi \end{array} \right\} \quad (6.62)$$

where ϕ is the friction angle for the material.

Computer studies on discontinuity orientation density distribution and their stability in triaxial stress fields

It is clear that the analyses proposed earlier can only be performed satisfactorily given the use of a large digital computer with a fast central processing facility and adequate core storage capacity. Speed is required, since for any fabric that has been processed from raw discontinuity dip-strike input data, numerous stress computations and rotations must be performed. This is because none of the stress parameters, the relative amplitudes of the principal stresses and their orientations with respect to global space together with the cohesion and friction angle for the discontinuities, can be quantified with any confidence. Indeed, there will inevitably be a range of cohesion and friction for the discontinuities within any rock mass and so it will be necessary to increment all these variables in the problem step-by-step in a systematic manner in order to *sense the sensitivity* of each one of them on the fundamental fabric. The influence of stress orientation with respect to a discontinuity field has been at the core of the analysis in this section and it is quite obvious that it will considerably influence the rock mass strength (Braybrooke, 1966). A failure path can also be entrained by lower strength fissures (Hooper and Butler, 1966) and this is why it is important to analyse for a wide range of strength. The strong dependency of the mass strength properties on the type of discontinuity distribution and on the relative values of the principal stresses has also been demonstrated by Attewell and Woodman (1971).

6.18 Preliminary consideration of certain types of discontinuity structure in two-dimensions

The spatial density distribution problem is a particularly intractable one but it is necessary to attempt to specify in some way the spatial distributions of the discontinuities along which a potential failure plane could pass. But in so doing it must also, in the most general case, pass through solid elements of rock material. This problem is universally acknowledged (for example, Skempton and La Rochelle, 1965) and has been studied by, among others, Jennings (1970). Since ground engineering design should hinge on both orientation and spatial relationships, this latter problem is worthy of some preliminary, albeit incomplete, specification.

In Figure 6.34a, let each discontinuity be defined by its centre point, c_j, length, l_j and direction, θ_j. The form of each discontinuity is assumed to be linear. Initially, assume that discontinuities are non-intersecting and that the c_j form a lattice, the morphology of which, in a topological sense, is constant and continuous; that is, the c_j are at the corners of triangles, or quadrilaterals or pentagons, and so on, but are not mixtures of these. Each basic set of n-discontinuities forms an n-gon and unless the discontinuities are arranged in a

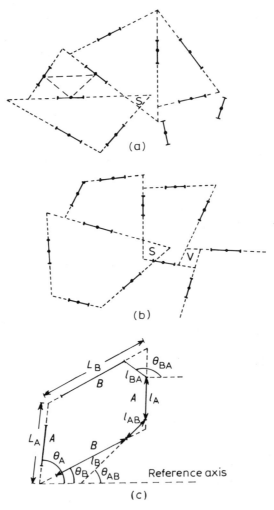

Figure 6.34 Two-dimensional discontinuity structures.

particular manner so that the vertices of the *n*-gons are common, the ensemble of *n*-gons does not correspond exactly to the space described by the discontinuities and solid elements in between. Some areas (in two-dimensions) are covered twice by *n*-gons (S in Figure 6.34b) and others not at all (V in Figure 6.34b). Generally these secondary areas are *n*-gons of different order.

Recognition of these secondary areas is important because the analytical aim is to characterize a rock mass structure by its basic *n*-gon unit; these latter are known as primary areas.

Failure of the mass will involve destruction of the primary units either by cracking along the edge, that is, discontinuity extension, or through the body, that is, initiating a new discontinuity direction. The existence of S secondaries will reduce the distance that the fracture must travel before reaching the next unit on the line of failure. Conversely, V secondaries will increase that distance. The tacit assumption here, of course, is that the fractures will be transmitted in the plane of the inherited discontinuities.

The effect of secondaries will be ignored at the present time. Suppose, now, that the basic element is a parallelogram. This implies that 4 mean directions and 6 mean lengths suffice to describe the structure, $\theta_A, \theta_B, \theta_{AB}, \theta_{BA}, l_A, l_B, l_{AB}, l_{BA}, L_B, L_A$. These are shown in Figure 6.34c in which AB and BA are directions through the body of the parallelogram. Let J be a discontinuity and C be a crack. J_A will be written for discontinuity A and so on. C_{AB} will define the crack from A to B or B to A within the included angle when going from A to B in an anticlockwise direction. C_{BA} is as above but the angle is described from B to A; that is, the crack is in an adjacent corner. It should be noted that $l_{BA}, \theta_{BA}, l_{AB}, \theta_{AB}$ cannot be determined geometrically from $l_A, L_A, l_B, L_B, \theta_A, \theta_B$ even if the mean positions of discontinuity centres are known also.

A fracture will comprise $J_A, J_B, C_{AB}, C_{BA}, C_{AA}, C_{BB}$, the relative frequencies depending on the geometry of the basic unit and the direction θ_F of the fracture. Vectorially,

$$\mathbf{F} = N_A \mathbf{J}_A + N_B \mathbf{J}_B + N_{AB} \mathbf{C}_{AB} + N_{BA} \mathbf{C}_{BA} + N_{AA} \mathbf{C}_{AA} + N_{BB} \mathbf{C}_{BB}$$

(6.63)

Therefore,

$$\mathbf{F} = (N_A + N_B + N_{AB} + N_{BA} + N_{AA} + N_{BB})[P_A \mathbf{J}_A + P_B \mathbf{J}_B$$
$$+ P_{AB} \mathbf{C}_{AB} + P_{BA} \mathbf{C}_{BA} + P_{AA} \mathbf{C}_{AA} + P_{BB} \mathbf{C}_{BB}]$$

(6.64)

where P_χ is the probability of event χ happening given θ_F.

Also $\quad P_A + P_B + P_{AB} + P_{BA} + P_{AA} + P_{BB} = 1$

P_χ are overall probabilities and must be distinguished from event probabilities p_χ. These latter arise in the detailed propagation of the fracture. Suppose that the fracture has reached discontinuity A. It may then continue along three routes to the next A or to one of the the two B's on each side with respective probabilities

$$p_{\overrightarrow{AA}}, p_{\overrightarrow{AB}}, p_{\overleftarrow{BA}}$$

Obviously, $p_{\overrightarrow{AA}} + p_{\overrightarrow{AB}} + p_{\overleftarrow{BA}} = 1$.

Let

$$N = N_A + N_B + N_{AB} + N_{BA} + N_{AA} + N_{BB} \tag{6.65}$$

Total length of the discontinuity in fracture is:

$$L_J = N(P_A l_A + P_B l_B) \tag{6.66}$$

Total length of the crack in fracture is:

$$L_C = N(P_{AB} l_{AB} + P_{BA} l_{BA} + P_{AA} l_{AA} + P_{BB} l_{BB}) \tag{6.67}$$

Note that $|\mathbf{F}| \leqslant L_J + L_C$.

Qualitatively, if θ_F is close to θ_χ, then P_χ is high.

The application of the above general procedure to more complex primary units is relatively straightforward. When the problem of discontinuity intersection arises, mixtures of n-gons are obtained and it becomes necessary to weight the influence of triangles, quadrilaterals, pentagons and so on according to their relative frequency.

This approach may prove to be fruitful in the long-term and it is certainly more relevant to the more general field of fracture mechanics. However, for more immediate engineering purposes some rather imprecise approximations must form the basis of much less complex analyses. One possible method of attack is considered in the next section.

6.19 Statistics of scanlines through discontinuity distributions

The methods of discontinuity analysis that have been discussed in this present chapter embody the assumption that the measured and recorded discontinuities possess orientation and spatial density distributions that are both representative of the mass and independent of the axes of measurement. This assumption has been tested on

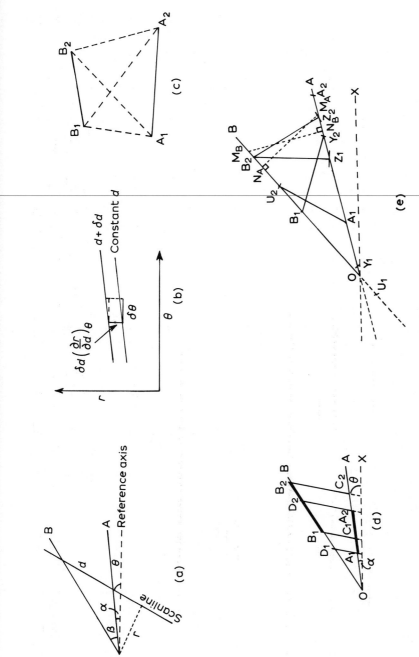

Figure 6.35 Scanline statistics through discontinuity distributions.

several occasions using different scanlines when field sites have been re-visited. As a theoretical background to these analyses, it seems appropriate just to introduce the statistical basis upon which the field procedures may be justified. It may be noted that the subsequent arguments could have equal relevance to the measurement of microcracks or other microstructural features in a thin section of rock under the petrographic microscope using a scanline method (see also Hilliard, 1962; Philofsky and Hilliard, 1969; Simmons *et al*, 1974).

Consider in Figure 6.35a two lines, OA and OB at angles α, β to an axis. The problem is to find the probability that a scanline intersecting at an angle θ with the axis will cut OA and OB at two points distance d apart. OA and OB may be regarded as intersection traces of discontinuity planes with the plane of measurement.

The equation of the scanline is:

$$y = px + q, \text{ where } p = \tan \theta \qquad (6.68)$$

The probability density associated with the scanline is $P(\theta, r)$ and with respect to these variables it is a constant, N say. The probability that the intersection length lies between d and $d + \delta d$ is given by $P(d)\delta d$. For each r and θ there is a result d and so curves may be plotted as in Figure 6.35b.

It follows that

$$P(d)\delta d = \delta d \int_{\Xi(d)} P(r,\theta) \left(\frac{\partial r}{\partial d}\right)_\theta d\theta \qquad (6.69)$$

the integral being taken over the projection of the curve $d = $ constant on to the θ axis.

Therefore,

$$P(d) = N \int_{\Xi(d)} \left(\frac{\partial r}{\partial d}\right)_\theta d\theta \qquad (6.70)$$

On transformation of variables from r and θ to r and p

$$P(r,p)dp = P(r,\theta)d\theta \qquad (6.71)$$

so that

$$P(r,p) = P(r,\theta) \frac{d\theta}{dp} = \frac{N}{1+p^2} \qquad (6.72)$$

and

$$P(d) = N \int_{\Lambda(d)} \frac{dp}{1+p^2} \left(\frac{\partial r}{\partial d}\right)_p \qquad (6.73)$$

where $\Lambda(d)$ is the domain on the p axis corresponding to the one described above on the θ axis.

The equation of OA is

$$y = ax,$$

the equation of OB is

$$y = bx,$$

and the equation of the scanline is

$$y = px + q.$$

There is an intersection with OA at

$$\left(\frac{q}{a-p}, \frac{aq}{a-p}\right)$$

and there is an intersection with OB at

$$\left(\frac{q}{b-p}, \frac{bq}{b-p}\right)$$

Thus,

$$d = \left[\left(\frac{q}{b-p} - \frac{q}{a-p}\right)^2 + \left(\frac{bq}{b-p} - \frac{aq}{a-p}\right)^2\right]^{1/2}$$

$$= \left|\frac{q}{(b-p)(a-p)}\right| [(a-b)^2 + \{b(a-p) - a(b-p)\}^2]^{1/2}$$

$$= \left|\frac{(a-b)q}{(b-p)(a-p)}\right| [1+p^2]^{1/2} \qquad (6.74)$$

We also have that

$$q = -r(1+p^2)^{1/2}$$

Therefore,

$$d = \left|\frac{(b-a)(1+p^2)}{(b-p)(a-p)}\right| r \qquad (6.75)$$

and

$$r = \left| \frac{(b-p)(a-p)}{(b-a)(1+p^2)} \right| d \qquad (6.76)$$

Thus,

$$\left(\frac{\partial r}{\partial d} \right)_p = \left| \frac{(b-p)(a-p)}{(b-a)(1+p^2)} \right| \qquad (6.77)$$

As expected, $(\partial r/\partial d)_p$ is independent of d for straight lines OA, OB. However, this is not true when scanlines are drawn between curved lines.

Finally,

$$P(d) = \frac{N}{|b-a|} \int_{\Lambda(d)} dp \frac{|(b-p)(a-p)|}{(1+p^2)^2} \qquad (6.78)$$

$$= \frac{N}{2|b-a|} \left[\left| \frac{(ab-1)p + a + b}{1+p^2} + (ab+1)\tan^{-1} p \right| \right]_{\Lambda(d)}$$

$$(6.79)$$

Notwithstanding the comment on the behaviour of $(\partial r/\partial d)_p$, $P(d)$ is not independent of d since the limits of the domain $\overline{\Lambda}(d)$ are dependent on d. Adjustment of the range of integration takes care of the case when intersections with only part of the lines OA and OB are required. Extreme scanlines are drawn in, as shown in Figure 6.35c. Extension to intersections with a primary unit (Section 6.18) poses no conceptual difficulty for it is necessary simply to consider each pair of discontinuities in turn. It is also possible to evaluate the probability that a scanline passes through a unit but misses the discontinuities altogether.

It is useful to consider the extent of the domain $\Xi(d)$ and to do so the acute angle case is taken first. Thus, as drawn in Figure 6.35a $0 < (\beta - \alpha) < \pi/2$. Further, using the notation in Figure 6.35d the two line or discontinuity segments $A_1 A_2$, $B_1 B_2$ are located on OA and OB respectively such that $OA_2 > OA_1$ and $OB_2 > OB_1$. Construction on Figure 6.35e takes the following form:

$N_B M_B$ is a perpendicular line from OA to OB having length d
$N_A M_A$ is a perpendicular line from OB to OA having length d
U_1, U_2 are points of intersection of circle, centre A_1, radius d, with OB

V_1, V_2 are points of intersection of circle, centre A_2, radius d, with OB

Y_1, Y_2 are points of intersection of circle, centre B_1, radius d, with OA

Z_1, Z_2 are points of intersection of circle, centre B_2, radius d, with OA

Intersection points $U_1, U_2 \ldots$ are labelled so that $\overrightarrow{U_1 U_2}$ takes the same sense as $\overrightarrow{B_1 B_2}$. In the example sketched in Figure 6.35e, $OA_2 > OM_A$ and V_1, V_2 do not exist. Let $m = OM_A = OM_B$.

The range of the angle θ for which it is possible to measure length d along a scanline between discontinuity intersection segments $A_1 A_2$ and $B_1 B_2$ is given by:

$$\mathrm{MAX}(\theta_{Y_1 B_1} - \alpha, \theta_{A_1 U_2} - \alpha)$$
$$+ \alpha \leq \theta \leq \mathrm{MIN}(\theta_{Y_2 B_1} - \alpha, \theta_{A_1 U_1} - \alpha) + \alpha$$

excluding

$$\theta_{Z_1 B_2} < \theta < \theta_{Z_2 B_2} \quad \text{when} \quad m > OB_2,$$

and

$$\theta_{A_2 V_2} < \theta < \theta_{A_2 V_1} \quad \text{when} \quad m > OA_2.$$

MAX and MIN denote the maximum and minimum values of their respective string arguments. In the system as shown, the angles are measured counterclockwise (taken as +ve) from OX in the range 0 to 2π. $\theta_{A_1 U_2}$, denotes the angle that $A_1 U_2$ makes with OX, and so on.

The obtuse angle case is as above except that

$$\frac{\pi}{2} \leq (\beta - \alpha) < \pi.$$

θ is restricted to the range

$$\mathrm{MAX}(\theta_{A_1 U_2} - \alpha, \theta_{Z_2 B_2} - \alpha)$$
$$+ \alpha \leq \theta \leq \mathrm{MIN}(\theta_{Y_2 B_1} - \alpha, \theta_{A_2 V_2} - \alpha) + \alpha$$

It should be noted that in the acute angled case only, A_1 may lie on the further side of O from A_2 without altering the range of θ already delimited for that case. In this situation, the sense of $\overrightarrow{A_1 A_2}$ is unaltered from that considered before and it is the direction of this vector which determines whether an acute angle is made with $\overrightarrow{B_1 B_2}$.

The discussion has basically been directed towards the behaviour of $\Xi(d)$ (ref. equation 6.69). However, the variation of $\Lambda(d)$ is not very different, given that the transformation $p = \tan\theta$ is fairly straightforward. Of course, if $\Xi(d)$ includes $\pi/2$, then $\Lambda(d)$ would have to be divided appropriately, vis:

$$\Xi(d): 0 < \theta_a \leq \theta \leq \theta_b < \pi \quad (\theta_a < \pi/2, \theta_b > \pi/2)$$
$$\Lambda(d): \tan\theta_a < p < \infty, \; -\infty < p \leq \tan\theta_b$$

By integrating with respect to θ or p, a lot of information is lost. The function $P(d,p)$ (the integrand) is much more useful than $P(d)$.

In Figure 6.35d, $D_1 D_2$ is the projection of $A_1 A_2$ on to OB in the direction θ and $C_1 C_2$ is the projection of $B_1 B_2$ on to OA in the direction θ.

We use the notation

$$d_{A_1 D_1} = \text{length of } A_1 D_1 \text{ and so on}$$

and specify

d_2 as the third highest value of $d_{A_1 D_1}, d_{A_2 D_2}, d_{B_1 C_1}, d_{B_2 C_2}$,

d_3 as the second highest value of $d_{A_1 D_1}, d_{A_2 D_2}, d_{B_1 C_1}, d_{B_2 C_2}$

(d_2 and d_3 are the two middle values of $d_{A_1 D_1}, d_{A_2 D_2}, d_{B_1 C_1}, d_{B_2 C_2}$).

Then, $P(d,p) = 0 \qquad \text{if } d < d_2$

$$= \left| \frac{N}{b-a} \frac{(b-p)(a-p)}{(1+p^2)^2} \right| \quad \text{if } d_2 \leq d \leq d_3 \qquad (6.80)$$

$= 0 \qquad \text{if } d > d_3$

Using this approach, $P(d,p)$ may be built up for more complicated shapes generated by the discontinuities.

6.20 Continuity

Although this problem has been discussed by Jennings (1970), Piteau (1970), and Robertson (1970), there still seems to be no really satisfactory method of assessing the degree of continuity of fracturing along a potential failure plane in a manner that would be useful in general stability analysis. The usefulness would arise from the allocation of shear strength parameters for the plane.

Continuity is a complex property depending upon discontinuity spacing, length and orientation with respect to a prescribed failure plane, analysis probably proceeding on the general lines of Section 6.18. However, suppose that the importance of a scanline

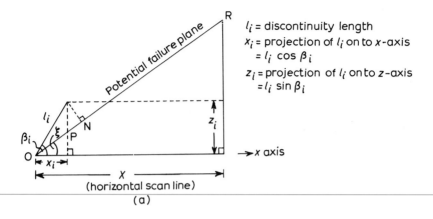

l_i = discontinuity length
x_i = projection of l_i onto x-axis
   = $l_i \cos \beta_i$
z_i = projection of l_i onto z-axis
   = $l_i \sin \beta_i$

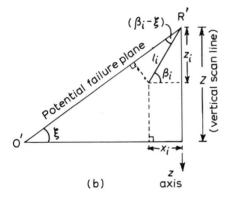

Figure 6.36 Simple construction for a possible estimate of slide plane continuity.

discontinuity spacing parameter d is acknowledged but ignored and that only the discontinuity length l_i and its orientation with respect to the potential failure plane angle are taken into an approximate two-dimensional analysis. We may then proceed on the lines of Figure 6.36a assuming that the number of discontinuities encountered per unit length of scanline is independent of scanline direction (although in general this will clearly not be so).

In Figure 6.36a, all l_i, x_i, z_i are measured as positive without regard to β_i, and without loss of generality β_i and ξ may be restricted to $0 \leqslant \beta_i < \pi$, $0 \leqslant \xi < \pi$.

$$ON_i = l_i |\cos(\beta_i - \xi)|$$
$$= |l_i \cos \beta_i \cos \xi + l_i \sin \beta_i \sin \xi|$$

If R_d is the summed length of discontinuity projections on to OR, then

$$R_d = \left\{ \left| \sum_{0 \leqslant \beta_i < \pi/2} x_i \cos \xi + \sum_{0 \leqslant \beta_i < \pi/2} z_i \sin \xi \right| \right.$$
$$\left. + \left| \sum_{\pi/2 \leqslant \beta_i < \pi} x_i \cos \xi - \sum_{\pi/2 \leqslant \beta_i < \pi} z_i \sin \xi \right| \right\} \frac{\text{OR}}{X} \qquad (6.81)$$

SPC $f(x,\xi) = R_d / \text{OR}$

= slide plane continuity with respect to the x-axis scan

$$= \left| \frac{1}{X} \left\{ \sum_{0 \leqslant \beta_i < \pi/2} x_i \cos \xi + \sum_{0 \leqslant \beta_i < \pi/2} z_i \sin \xi \right\} \right|$$
$$+ \left| \frac{1}{X} \left\{ \sum_{\pi/2 \leqslant \beta_i < \pi} x_i \cos \xi - \sum_{\pi/2 \leqslant \beta_i < \pi} z_i \sin \xi \right\} \right| \qquad (6.82)$$

Given the earlier assumptions, it would be argued that for $0 < \text{SPC} \leqslant 1$, the fraction SPC might be allocated shear strength parameters relevant to the discontinuities. The fraction $(1 - \text{SPC})$ might take c, ϕ relevant to the solid rock. One possible expression is

$$\left. \begin{array}{l} c = c_s + (c_d - c_s) \text{ SPC} \\ \text{and } \tan \phi = \tan \phi_s + (\tan \phi_d - \tan \phi_s) \text{ SPC} \end{array} \right\} \; 0 < \text{SPC} \leqslant 1 \qquad (6.83)$$

Suppose that the above operation is repeated for a vertical scanline of length Z and for the same failure plane angle ξ. The equation for SPC $f(z,\xi)$ will take the same form as that for SPC $f(x,\xi)$ in Equation 6.36b above with the exception that the controlling fraction outside both sets of brackets is this time $1/Z$. An average SPC could then be calculated:

$$\text{SPC } f(x,z,\xi) = \tfrac{1}{2} \{ \text{SPC } f(x,\xi) + \text{SPC } f(z,\xi) \} \qquad (6.84)$$

Several points need to be made. First, in many weaker rocks SPC will greatly exceed unity, so implying a considerable overlap of discontinuity projections in the potential slide plane. This overlap is a result of simple mathematical convenience and has no real physical meaning in the sense of implying total absence of any shear of solid rock. Nevertheless, shear strength parameters c_d, ϕ_d for the plane given SPC > 1 would reflect the highly discontinuous character of the mass. A second point concerns the measurement of l_i. This is often difficult to achieve with any degree of accuracy on a wet, smeared face, and in any case, the magnitude of l_i may not adequately represent the area extent and hence the physical importance of the discontinuity. Thirdly, if the discontinuity spatial density is not independent of scanline direction (the usual case, of course), the R_d and SPC should really be weighted appropriately according to direction OR. Finally, in view of interlocking effects on the slide plane, SPC should perhaps be additionally weighted so that the 'cut-off' point occurs at a value in excess of unity.

These concepts have been developed three-dimensionally and in the context of a multi-scanline situation, but it is not appropriate to present them at the present general level of discussion.

In order to characterize the 'blockiness' of the rock within a discontinuous mass, Reeves (1973) proposed the application of a 'granularity test' leading to the formulation of a 'Granularity Value' (G.V.). The steps are as follows:

(i) Plot the scanline directions on a stereonet and measure the angles between them by rotating each pair of axes in turn on to the same great circle (more experience of manipulation on the stereonet may be acquired by following the rock slope stability analyses in Chapter 10).
(ii) Re-draw the scanlines as axes and mark off along each axis the respective discontinuity intersection distances (DID).
(iii) A three-dimensional parallelogram (for example) having been constructed, the surface area of the block may now be calculated to give the G.V.

G.V.s for all the sites marked in Figures 3.26 and 3.27 and listed in Table 10.13 have been calculated by both Reeves (1973) and Gavshon (1974) who have also drawn projections of mean block dimensions. Since the Chalk site near White Nothe has been considered in some detail in Section 10.9, the average block dimensions at this site are shown in Figure 6.37 for illustration purposes. In view of the wide range of DID observed and measured at any exposure it will be appreciated quite readily that very little technical merit can be assigned to such a value. But although it would never play a part in any problem-solving exercise it probably serves as a quite useful indicator of the extent to which a mass of discontinuous rock may be analysed for stability as a particulate material approximating to a soil.

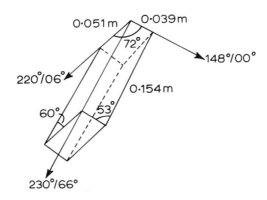

Figure 6.37 Average dimensions and angles for blocks of strongly-fractured Lower Chalk at White Nothe, East of Weymouth, England (Site f, Figure 3.26). Granularity value is 0.024 m^2.

6.21 Preliminary shear stability analysis of discontinuities at the foundation interface of an earth or rock-fill dam

A geologist charged with investigating the structural features of an earth or rock-fill dam foundation in a weak, discontinuous medium will often, as part of his investigation, produce a stereographic projection(s) depicting the orientations of as many discontinuities as he can measure at the exposures within the time period at his disposal. This information will form part of the earthworks design report and will be appraised in a qualitative, and often quite arbitrary manner, by the engineering designer or design team. It has been suggested earlier that, given certain information with respect to foundation material strength and the final stress field that is imposed on the discontinuity fabric, the engineering geologist could offer a preliminary and more meaningful assessment of foundation competence to the engineer for his further consideration. There would be no question of the engineering geologist acting as an analyst or designer: he would simply be presenting his geological information at a slightly higher level of assessment which would be taken note of, or be dismissed, by the engineer as the case may be. But the acceptability of his efforts will be in direct correlation with the quality and representability of both his geological data and his stress data. As might be expected, it is the latter which often presents a major difficulty.

Most basic analysis is directed to the stability of the embankment proper, ensuring that there is no possibility of rotational shear failure and that the dissipation of excess pore pressures set up during the construction stage is rate-matched to the embankment loading (see Section 11.3). Although there usually is a quite detailed geological and geotechnical assessment of the foundation quality during the initial investigation stage, only rarely in the case of an earth or rock-fill embankment is the stability of the foundation investigated directly (note that finite element analyses of foundation stresses have been undertaken by Clough and Woodward, 1967). It is usually assumed that the strength of the foundation at any point in the material will not be exceeded by the applied shear stress and it is further tacitly assumed that a soil foundation remains within an elastic behavioural regime during deformation. On this latter point, Clough and Woodward (1967) have deduced that the *degree* of deformation does not affect the applied vertical pressure, and that although horizontal pressures reduce with increasing foundation flexibility, they do not increase near the toe of the embankment as a result of foundation deformations. If any localized shear failure should occur, perhaps at points of stress concentration associated with discontinuities, then the excessive strain in such zones will raise a 'plastic' condition with some load-shedding to other adjacent areas that were originally less highly stressed. In theory, therefore, the 'plastic' zones within which limiting shear strength had been achieved could be specified by an iterative numerical technique, and those zones remaining elastically deformed could be determined by difference.

Unfortunately, there seems to be no practically acceptable method of plastic analysis applied to this problem, although numerical techniques in continuum mechanics can be used to investigate possible conditions of limiting equilibrium into the plastic state. On the understanding that earth/rock fill gravity dam foundations are unlikely to be overstressed, the foundation mechanics can usually be investigated as an elastic problem.

Sherard *et al* (1963) have traced some of the earlier attempts at elastic analysis and have pointed out the deficiencies in the approaches. All the analyses have approximated an earth/rock-fill gravity dam to a triangular elastic wedge on a horizontal foundation, and this was the configuration (with 3:1 side slopes) adopted by Bishop (1952) and analysed by relaxation methods.* Figure 6.38

---

*Refer also to the analyses of Middlebrooks (1936) and Mirata (1969).

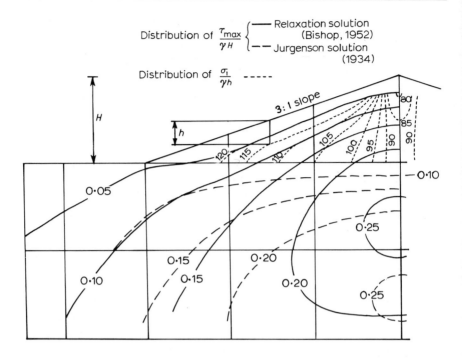

Figure 6.38 Possible distributions of major principal stress and shear stress at an earth-dam foundation (note that σ_1 is normalized to the overburden pressure at a particular point in the dam and not to overburden pressure at crest height).

reproduces this solution in terms of the distribution of major principal stress and maximum shear stress in the dam itself and in both dam and foundation respectively. The distribution of major principal stress, plotted in terms of the percentage of the overlying embankment weight, indicates that there should be little significant error involved in assuming that the major principal stress acts vertically and is equal in magnitude to the embankment weight, any errors only increasing directly under the crest of the dam and at its toe (see also Mirata, 1969). Thus, although there are inherent inaccuracies when accepting this latter assumption, in order to avoid computational complexity it seems reasonable to approximate the two-dimensional principal stress situation acting on any geological discontinuity in the foundation as σ_1 vertical and σ_3 acting in a horizontal plane normal to the axis of the dam. In a three-dimensional analysis of discontinuity stability, σ_1 would still be taken as acting vertically and its magnitude would be the product of the embankment height and the density of the placed fill. At the foundation contact with the placed fill, the minor principal stress σ_3 could be approximated to the impounded water pressure and the intermediate principal stress σ_2 acting along the axis of the dam could be approximated in magnitude by the summation of the other two principal stresses modified by some constant multiplying factor in order to preserve a plane strain condition within the σ_1, σ_3 plane (Clough and Woodward, 1967). A more rigorous analysis would not only accommodate variations in magnitude and direction of the principal stresses within the area of the embankment foundation but it would also take account of incremental stress balances created by construction works during progress towards the final equilibrium situation and by seepage forces (see Sections 8.2, 8.3).

A condition of shear stability, or instability, is not directly dependent upon the *absolute* magnitude of the principal stresses, but is, of course, a function of principal stress ratio (Section 6.16). If a cohesion shear strength parameter is to be used in an analysis, the absolute value of cohesion must be expressed as a ratio of the absolute magnitude of the major principal stress. With respect to this latter, the problem then arises as to what value of density to take into the γH product. On the upstream side of the dam, a buoyant density should be used for a height up to top water-level with the addition of a bulk (dry) density component from top water-level to the crest of the dam. On the downstream side of the core, a saturated density could be taken from top water-level plus again a dry density to crest. This latter might be considered an intermediate condition; if the core is quite leaky, the amplitudes and directions of the seepage forces should be entered into the analysis, but if seepage through the core is substantially inhibited, then an extreme dry density case might be considered. In both cases, for equilibrium, a maximum water pressure can be assumed to act hydrostatically at foundation level and thereby to provide a minor principal pressure component. For earth and rock-fill, the multiplying factor against the summation of major and minor principal stress (to satisfy the plane strain condition, and referred to earlier) can be taken as 0.5.

As an example of such an analysis, reference can be made to Chatfield Dam in Colorado, U.S.A. Discontinuity measurements, upon which the subsequent stability fabrics are based, were taken by one of the present authors and Dr R. K. Taylor at the spillway and weir (see Figure 6.1), a general location of the dam being given in

Figure 6.39. Major concentrations of discontinuities (without differentiating between 'fractures' and slickensides) appear to dip at about 50° in a direction slightly east of north-east and also vertically along north-south, east-west axes. These latter two concentrations are compatible with the two joint sets specified in the U.S. Corps of Engineers Design Memorandum No. PC-24. From this Memorandum, the following information has been accessed for computer input:

(a) *Dam alignment:* N108°E

(b) *Maximum height of embankment:* 142 ft (43.3 m) over South Platte River Channel. See also Figure 12.23.

(c) *Head of Water:* fully impounded 115 ft (35.0 m) allowing 27 ft (8.23 m) for a full pool below the crest of the dam

(d) *Density of fill:* from Plate 200, Corps D.M. No. PC-24.

Type	In situ weight lb ft^{-3}	In situ weight Mg m^{-3}	saturated weight lb ft^{-3}	saturated weight Mg m^{-3}	submerged weight lb ft^{-3}	submerged weight Mg m^{-3}
Impervious	120	1.92	126	2.02	63.6	1.02
Random	120	1.92	126	2.02	63.6	1.02
Pervious	130	2.08	136	2.18	73.6	1.18
Dawson	120	1.92	126	2.02	63.6	1.02

(e) *Foundation strength:* from Plate 187/8, Corps D.M.No.PC-24 (I to VI) additional data (VII to IX)

Description	ϕ	c lbf ft^{-2}	kN m^{-2}
(i) In lieu of Dawson 'S' strength	15°p	0	0
(ii) Minimum for clay shale	19.8°p	0	0
(iii) Boundary, clay shale-silty shale and siltstone	27°p	0	0
(iv) Boundary, silty shale and siltstone – sandy shale and siltstone	37.7°p	0	0
(v) Maximum, sandy shale and sandstone	42°p	0	0
(vi) Weathered sandstone	21°	0	0
(vii) Slicked lean or fat clay shale (Banks, 1971)	4.5°r	0	0
(viii) Dawson clay seam, Durham University direct shear tests	8.5°	393	18.8
(ix) Dawson silty clay shale from weir, Durham University direct shear tests.	24°$_{ult}$	185	8.9

Results of a series of preliminary stability analyses together with figure references to certain stability fabrics are listed in Table 6.10.

In the stability fabrics, X at the north and Y at the west define axes north-south and east-west respectively (the third Z axis is overprinted by character P). P,Q,R, define axes of major, intermediate and minor principal stress respectively. Plus (+) characters delimit stable and potentially-unstable discontinuity orientation regimes in gobal and principal stress space, stable space being differentiated by the requirement that it must incorporate the axes of principal stress (no shear stress across a plane normal to a principal stress axis). Also listed beneath each fabric is the direction cosine matrix which refers principal stress to global axes, normalized input stresses, normalized cohesion and tan ϕ. The anisotropic (A) and

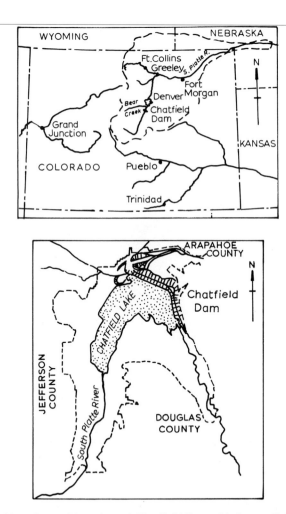

Figure 6.39 General location of Chatfield Dam, Littleton, Colorado, U.S.A.

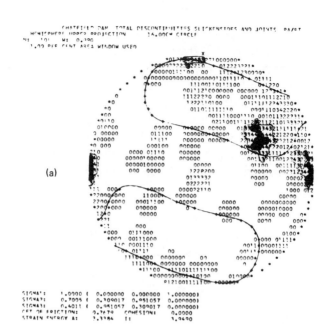

(a)

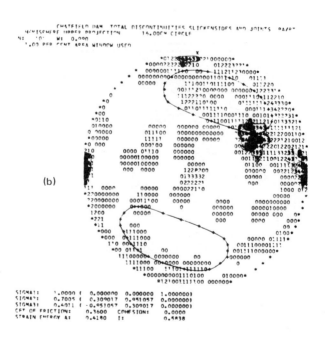

(b)

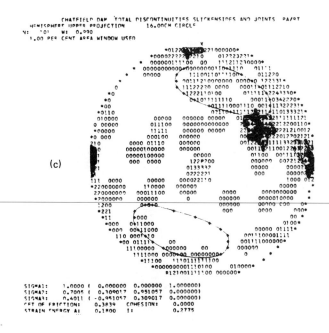

(e)

(f)

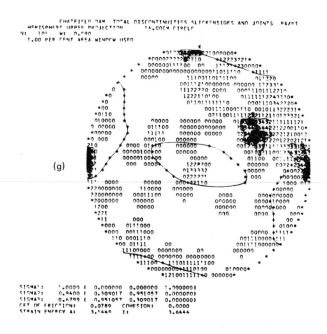

(a) Upstream face; $\phi'_p = 15°$, $c_p = 0$ (b) Upstream face; $\phi_p = 19.8°$, $c = 0$
(c) Upstream face; $\phi'_p = 21°$, $c = 0$ (d) Upstream face; $\phi'_r = 4.5°$, $c'_r = 0$
(e) Upstream face; $\phi = 8.5°$, $c = 393$ lbf ft^{-2} (18.82 kN m^{-2})
(f) Upstream face; $\phi_{ult} = 24°$, $c = 185$ lbf ft^{-2} (8.86 kN m^{-2})
(g) Downstream face; $\phi'_r = 4.5°$, $c = 0$

Figure 6.40 Discontinuity stability fabrics for Chatfield Dam, Littleton, Colorado, U.S.A.

isotropic (I) shear strain energy unit values (Attewell and Woodman, 1971) have been discussed in Section 6.17.

It should be re-emphasized that when discussing stability fabrics, the existence of an unsafe regime on the projection does *not* mean that failure will necessarily occur. It simply implies, through the relationship between those areas and the probability density distribution of the discontinuity orientations within them, that some shear movement could take place along such discontinuities given that the principal stress ratios and their directions of action as applied to the problem are valid and that the compounded shear can only be accommodated through free boundary displacement, such a displacement facility being available.

Table 6.10 Stability analysis on measured discontinuities in the Dawson formation at the site of Chatfield Dam, Littleton, Colorado

Position	No.	Shear Strength Parameters			Figure Number	Shear Strain Energy Parameter		Shear Strain Energy Index I/A
		ϕ degrees	c lbf ft^{-2}	c kN m^{-2}		Aniso-tropic	Iso-tropic	
Upstream	1.e.I	15p	0	0	a	3.3384	3.9490	1.1829
Upstream	1.e.II	19.8p	0	0	b	0.4180	0.5838	1.3966
Upstream	1.e.III	27p	0	0	—	0	0	0
Upstream	1.e.IV	37.7p	0	0	—	0	0	0
Upstream	1.e.V	42p	0	0	—	0	0	0
Upstream	1.e.VI	21	0	0	c	0.1800	0.2775	1.5417
Upstream	1.e.VII	4.5r	0	0	d	32.0298	36.2529	1.1318
Upstream	1.e.VIII	8.5	393	18.8	e	11.1563	12.6203	1.1312
Upstream	1.e.IX	24$_{ult}$	185	8.9	f	0.0001	0.0003	3.0000
Downstream	2.e.I	15p	0	0	—	0	0	0
Downstream	2.e.II	19.8p	0	0	—	0	0	0
Downstream	2.e.III	27p	0	0	—	0	0	0
Downstream	2.e.IV	37.7p	0	0	—	0	0	0
Downstream	2.e.V	42p	0	0	—	0	0	0
Downstream	2.e.VI	21	0	0	—	0	0	0
Downstream	2.e.VII	4.5r	0	0	g	3.1449	3.6444	1.1588
Downstream	2.e.VIII	8.5	393	18.8	—	0	0	0
Downstream	2.e.IX	24$_{ult}$	185	8.9	—	0	0	0

Examination of stability fabrics Figures 6.40a,b,c, suggests that, in the absence of any cohesion mobilization, any shear failure becomes increasingly less likely with respect to an *upstream* stress condition as the friction angle exceeds 21°. However, clay shale and weathered sandstone horizons could be suspect if the analytical stress-input conditions are realistic. Inevitably, the 4.5°ϕ_r condition (Figure 6.40d) could be problematical, and although the clay seam shear strength parameters (Figure 6.40d) are insufficiently high to preclude potential shear movement it can also be seen that there could be potential instability with respect to any silty clay horizon (Figure 6.40f).

From Table 6.10, only the 4.5°ϕ_r condition could create any comment on the basis of the simplified downstream stress levels (Figure 6.40g), but in this case the A parameter is an order of magnitude down on the equivalent parameter for the upstream face. Since, in this example, the dam foundation comprises a whole suite of rock types, it would obviously be totally unrealistic to select, quite arbitrarily, this single lithologic type as a source for any concern unless it could be demonstrated unambiguously that its presence was widespread over the whole foundation area.

Although the results of the fabric analysis imply that the upstream side of the core could impose rather more searching stress differences upon any foundation discontinuities, the fact that equilibrium must be maintained across the dam as an entity means that the stability of any structural foundation is only really as strong as its weakest part.

6.22 Stability of jointed rock in the foundation of an arch dam

It is now proposed to consider an example of rock-structure interaction and one which is of particular interest to engineering geologists. As in any investigation and design problem of this nature a number of assumptions have to be made and most of these will be stated in the text. A basic aim of this analysis is to suggest how similar foundation problems might be considered, and although the problem that is covered is one of continuous jointing, it is felt that it can most usefully and appropriately be included in this present section.

The example to be considered concerns the Monar dam in Scotland, and the basic information is taken from Henkel *et al* (1964). Monar dam, which was completed in 1963, was built across the head water of the River Farrar, 30 miles (48.3 km) west of Inverness in Inverness-shire (Figure 6.41) to form an impounding reservoir for electric power generation. The cupola double-curvature arch with a chord/height ratio of about 3 is about 134 m long

Figure 6.41 Location of Monar Dam in Northern Scotland.

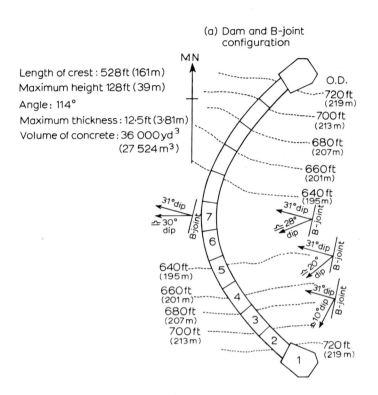

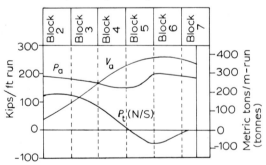

Figure 6.42 Monar Dam River Farrar, Inverness-shire, Scotland (*after* Henkel et al, 1964).

(c) Monar Dam Blocks 1 to 6 under construction. Photograph: J. L. Knill.

between abutment blocks, 161 m overall length and 35 m high above the river bed.* The arch comprises eleven blocks separated by contraction joints and is sketched diagrammatically in Figure 6.42.

The dam is founded upon psammitic granulites of the Moine Series of Northern Scotland, the rocks having been strongly folded during at least two periods of deformation. The rocks as a whole are massive, unweathered, and contain thin pelitic or semi-pelitic layers, pegmatites and occasional quartz veins. They tend to be rather heavily jointed, the bulk properties of the rock being largely controlled by the jointing. From the projection in Figure 6.43 three main joint sets were identified and designated by Henkel *et al* (1964): A-joints striking along the valley, B-joints dipping at about 30° upstream, and C-joints dipping at about 40° downstream. A further family of D-joints dipping gently upstream was exposed on the south bank at river-bed level during excavation. The most prominent A-joints dip very steeply (~83°) towards the centre of the valley on both banks,

*The site was one of the few in Britain suitable for a cupola or arch dam and the cost according to Walters (1971) was 10 per cent lower than the cost would have been for an equivalent gravity dam.

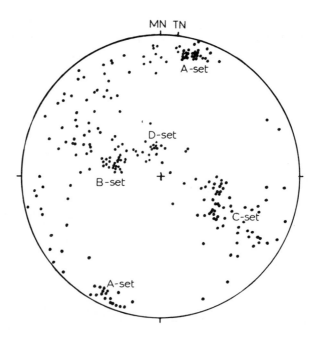

Figure 6.43 Orientation distribution of joints in Monar dam foundation (upper hemisphere projection). Photograph: J. L. Knill.

individual joints being traced for at least 4.5 to 6 m with, in places, the average spacing being as little as 30–45 cm. Smoothish joint surfaces were found to be weathered close to ground surface and filled with silty or sandy debris. In their stability analyses, Henkel *et al* (1964) allocate an assumed vertical dip to these joints, and this can also be assumed here. The B-joints were found to be well-developed on both banks but to be more scattered in orientation and generally more steeply inclined on the north bank. On the south bank, the B-joints dipping at 31° (N284°) were seen to outcrop over distances of at least 9 m and to occur at spacings of 0.6–1.2 m. The actual joint faces seemed to be irregular, with no clay infilling. Under a full, impounded head of reservoir water, fissure water flow was envisaged through the joint network, with the B-joints exerting the critical control on stability. The C-joints dipping at 40° (N 110°) were the most obvious on the north bank where, in association with the complementary B-joints, they form ridges trending up the side of the valley. D-joints occurring at a frequency of 60–90 cm were observed mainly at low levels on the south bank dipping at a shallow (~15°) angle, generally upstream. The fifth structural feature in the foundation area comprised a lamproschist dyke dipping at 35.5° at N86°.

Stability calculations

In order to assess the loads imposed by the dam structure on the foundation, Henkel *et al* assumed that the rock behaved elastically prior to any sliding movement occurring along the joints. From a superficial arch-cantilever type of analysis, they derived values for horizontal and vertical thrusts, shear forces and bending moments. They then resolved the thrusts and shears vertically and horizontally, parallel and normal to the axis of symmetry of the dam, as given in Figure 6.42b, and calculated factors of safety with respect to shear stability along the B-joints from these resolved forces. For further reading on the analysis of encastré arch structures, reference might be made to Bourgin (1953).*

For the purposes of the present analysis, the forces of Henkel *et al* have been resolved into principal forces V_a (vertical), P_r (radial) and

---

*See also Figure 12.12 showing abutment and foundation thrusts from a cupola dam. The stress amplitudes and orientations in the rock will be rather more complex than the ones considered in the present analysis.

414 *Principles of Engineering Geology*

P_t (tangential) for specific points on the dam of calculation interest. Their locations are: centre block 7, contact blocks 5/6, block 3 and abutment block 1 (see Figure 6.42a). Also to be taken into the analyses are forces W_c, W_r, U_w, U_b. These are respectively the weight of the concrete foundation block, the weights of the two rock wedges bounding the foundation concrete, the uplift water force along the B-joints, and the direct water force on the face of the wedge (see Figure 6.44). Concrete and rock densities are assumed to be 2.3 Mg m^{-3} and 2.5 Mg m^{-3} respectively. For the water pressure calculations, the full reservoir pressure has been taken to act down to the level at which the B-joint plane passing through the lowest point of the foundation concrete intersects the vertical through the upstream face of the dam. The uplift components are assumed to vary linearly along the plane from this maximum to zero at the downstream surface. Insertion of downstream drainage boreholes implies that the head of water at the end of the boreholes corresponds to the level at which the boreholes intersect the downstream slope, the nett effect being to modify the undrained triangular pressure distribution to two triangular distributions of reduced total area. A further point concerns the variation in apparent dip of the B-joints within the plane of section which in each case is taken to include V_a and P_r. These apparent dips, which affect the analyses, are derived stereographically in Figure 6.45 for the sections to the south of the vertical axis of symmetry of the dam.

In their analyses, Henkel *et al* have assumed that the mass of rock involved in the potential shear movement is bounded on either side by the A-joints. If, for the present three-dimensional assessments, it is assumed that there is B-joint planar continuity outside the plane of section and therefore through the A-joints, then the influence of U_w on P_t can be ignored; otherwise, these strong negative force components acting against P_t would greatly affect the relative amplitudes of the principal forces.

Henkel *et al* quote a range of friction angle results from their tests on rock joint surfaces. These range from a minimum peak friction angle, ϕ_p for the unweathered granulite of 48° and a minimum residual, ϕ_r, of 40°. The influence of weathering on a micaceous, clayey joint surface is to reduce ϕ_p to 25° and ϕ_r to 22°. The stability influence of these friction angle ranges can quite easily be checked.

In order to illustrate the general approach, attention has been

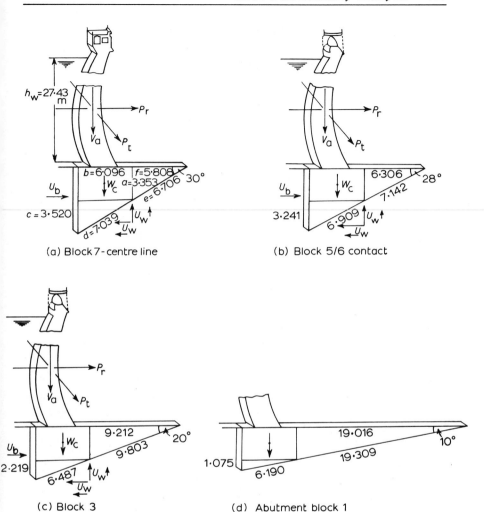

Figure 6.44 Sections of Monar dam foundation. (a) Block 7 – Centre line (b) Block 5/6 – Contact (c) Block 3 (d) Block 1 – Abutment.

directed only to the specific blocks to the south of the axis of symmetry of the dam. For the same block sections to the north of the axis, a further suite of calculations would be necessary due to the changes in apparent dip with the azimuthal differences in P_r and P_t. Although the blocks are investigated for irrotational instability at the foundation, it will be realized from the location of the force vectors that there are elements of rotation in the basic mechanics of the system. However, this latter problem is not considered.

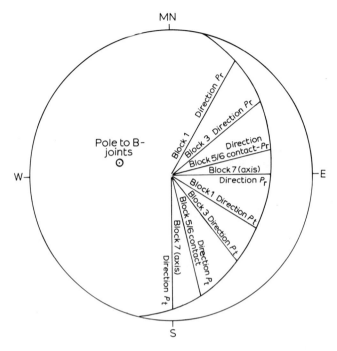

To determine the apparent dip for any section containing P_r, V_a or P_t, V_a, draw a radius for any P_r or P_t azimuth and read off the required dip arc by revolving a tracing of the projection (as above) on top of an equal area net until the appropriate arc lies above a principal diameter, ie, until a radius P_r or P_t lies along a diameter and then read off angular distance from P_r or P_t intersection with projection trace of joint plane to the periphery of the net.

Figure 6.45 Construction for determining apparent dips to B-joints, Monar dam, Blocks 1 to 7.

Using the symbols shown in Figure 6.43a, the basic method of calculation used in this simple approach would be as in Table 6.11.

Altogether 28 stability fabrics can be produced to cover the range of four friction angles quoted earlier and both the drained and undrained conditions. A 'drained' condition only is analysed for block 1.

It is possible to show only a few of these fabrics in Figure 6.46 but to draw general conclusions concerning the shear stability of the B-joints from the complete run of analyses. Principal axes of force P, Q, R, if not directly discernible on the fabrics, may be identified directionally from the information given in Table 6.11.

Parameter	Block 7	Block 5/6	Block 3	Block 1
V_a	Greatest	Greatest	Intermediate	Least
P_r	Intermediate	Intermediate	Greatest	Greatest
P_t	Least	Least	Least	Intermediate
Wt. of concrete foundation	$W_c = a \cdot b \cdot \gamma_c$			
Wt. of rock wedges	$W_r = \gamma_r/2[(b \cdot c) + (a \cdot f)]$			
Foundation water forces uplifting along 'B'-joint	U_w (undrained) $= \dfrac{\gamma_w}{2}[(h_w + a + c)(d + e)]$		—	—
	U_w^* (drained) $= \dfrac{\gamma_w}{2}[(a \cdot e) + (h_w + 2a + c)d]$			
Lateral water forces	$U_b = [h_w + \tfrac{1}{2}(a + c)](a + c)\gamma_w$			
Total forces without drainage	$V(\text{Vertical}) = V_a + W_c + W_r - U_w \cos\beta$	These ratios are specific to this problem and are applicable to all blocks		
	$R(\text{Radial}) = P_r + U_b - U_w \sin\beta$			
	$T(\text{Tangential}) = P_t$			
Principal ratios (Drained and undrained)	$F_1 = R/R$			
	$F_2 = V/R$			
	$F_3 = T/R$			
Total forces with drainage	$V(\text{Vertical}) = V_a + W_c + W_r - U_w^* \cos\beta$			
	$R(\text{Radial}) = P_r + U_b - U_w^* \sin\beta$			
	$T(\text{Tangential}) = P_t$			
F_1 direction	N90°/90° dip	N111°/90° dip	N128°/90° dip	N149°/90° dip
F_2 direction	N0°/0° dip	N0°/0° dip	N0°/0° dip	N0°/0° dip
F_3 direction	N180°/90° dip	N201°/90° dip	N218°/90° dip	N239°/90° dip

Note:
(a) $\gamma_c, \gamma_r, \gamma_w$ are the densities of concrete, rock and water respectively.
(b) h_w for blocks 7 and 5/6 is assumed to be the same at 27.43 m. h_w for block 3 is assumed to be 15.24 m lower than at block 5/6 contact (see Figure 1, Henkel et al, 1964).
(c) Because of the classification of the abutment block, it is assumed that the weight components of the system only are active and that water pressures can be ignored.
(d) Dip magnitudes of F_1, F_2, F_3 are expressed in angular measure with respect to the vertical.
(e) 1 tonne = 1 megagramme (Mg).
(f) β is the apparent dip angle of the B-joints.

(a)

(b)

(c)

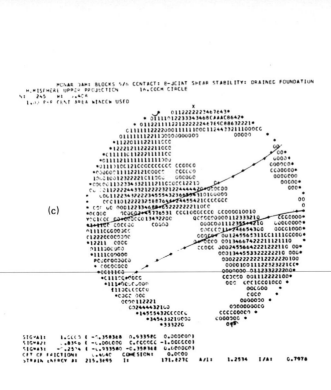

(d)

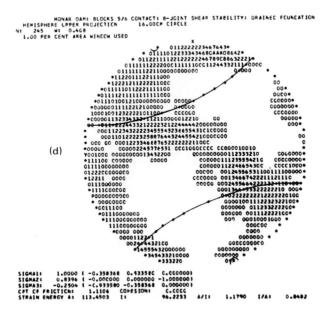

419

(e)
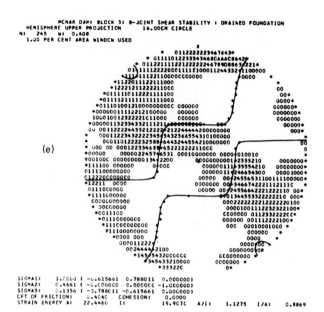

(f)

420

(g)

(h)

421

MONAR DAM: BLOCK 3: B-JOINT SHEAR STABILITY : UNDRAINED FOUNDATION
HEMISPHERE UPPER PROJECTION   16.00CM CIRCLE
N: 245   W: 0.408
1.00 PER CENT AREA WINDOW USED

(i)

```
SIGMA1:     1.0000 ( -0.615661   0.788011   0.000000)
SIGMA2:     0.3457 ( -0.000000   0.000000  -1.000000)
SIGMA3:     0.1428 ( -0.788011  -0.615661   0.000000)
CFT OF FRICTION:    0.8391    COHESION:    0.0000
STRAIN ENERGY A:    0.6696    I:    0.6783    A/I:    0.9872    I/A:    1.0129
```

MONAR DAM: BLOCK 3: B-JOINT SHEAR STABILITY : UNDRAINED FOUNDATION
HEMISPHERE UPPER PROJECTION   16.00CM CIRCLE
N: 245   W: 0.408
1.00 PER CENT AREA WINDOW USED

(j)

```
SIGMA1:     1.0000 ( -0.615661   0.788011   0.000000)
SIGMA2:     0.3457 ( -0.000000   0.000000  -1.000000)
SIGMA3:     0.1428 ( -0.788011  -0.615661   0.000000)
CFT OF FRICTION:    1.1106    COHESION:    0.0000
STRAIN ENERGY A:    0.0001    I:    0.0002    A/I:    0.5294    I/A:    1.8889
```

The following points may be noted:

(a) Beneath block 7, the B-joints are stable even if undrained and with a low residual friction angle of 22°.
(b) Beneath the blocks 5/6 contact, a marginal shear instability condition is raised only for a continuous, weathered, clayey joint surface at residual shear strength. The delimiting line on the projection intersects the joint concentration and does not totally encompass it. An unweathered granulite peak friction angle creates a situation of total stability with respect to B-joints.
(c) The same remarks apply to the foundation under block 3 as are made for blocks 5/6 contact. Stability fabrics for all four friction angles related to the undrained condition are shown and, as for blocks 5/6 contact, the fabrics for the two extreme friction angles, 48° and 22°, are also given for the drained foundation condition.
(d) Stability calculations on block 1 foundation show it to be stable under the lowest 22° friction angle and therefore no fabrics are shown.

Noting the earlier comments on the nature of the lowest friction angle used in the computations and the fact that the possibility of only marginal shear instability could be associated with the mobilization of that degree of shear strength, we may conclude that the B-joints do not constitute a foundation problem. An appraisal of the assumptions entered into the analysis reinforces this conclusion. There is, however, evidence of a potential shear facility along the other joint sets but those hazards may be discounted by virtue of the external constraints on any possible shear movement.

It is interesting to note that a system incorporating four inverted pendulums installed in the foundations and anchored below the dyke at a depth of 120 ft (36.58 m) from the surface was used to measure the total movement of the dam relative to stable rock below the

---

(a) Block 7 undrained $\phi_r = 22°$ (b) Block 7 undrained $\phi_p = 48°$ (c) Blocks 5/6 contact, drained $\phi_r = 22°$ (d) Blocks 5/6 contact, drained $\phi_p = 48°$ (e) Block 3 drained $\phi = 22°$ (f) Block 3 drained $\phi = 48°$ (g) Block 3 undrained $\phi = 22°$ (h) Block 3 undrained $\phi = 25°$ (i) Block 3 undrained $\phi = 40°$ (j) Block 3 undrained $\phi = 48°$.

Figure 6.46 Some foundation stability fabrics for Monar Dam in Scotland.

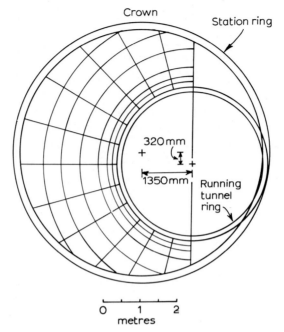

Figure 6.47 Running tunnel opened out into a station tunnel in London Clay showing the exposed clay face partitioned into discontinuity sampling segments.
Running tunnel: ring internal diameter = 3.850 m
outside diameter = 4.070 m
Station tunnel: ring internal diameter = 6.500 m
outside diameter = 6.900 m

lamproschist dyke. A deviation of 15 mm was recorded in the winter of 1964 when impounding was almost complete. When the dam was full, the movement then appeared to vary with water temperature.

The importance of the geological investigation in works of this nature was stressed by the dam engineers (Walters, 1971): 'The experience gained from this site has shown that an arch dam should be chosen only where the nature of the foundation has been explored to a degree not usual for other types of dam. It seems desirable to excavate the foundations before designing the concrete

Figure 6.48 (a) Anisotropic shear strain energy distribution (arbitrary units) around a 4 m diameter tunnel in London Clay. Soffit is at the top of the tunnel opening and invert at the bottom. Shear strength parameters (Skempton's fissure peak): $\phi'_p = 18.5°$; $c'_p = 7\text{kN m}^{-2}$ (b) Orientation density distribution of discontinuities (upper hemisphere projection; max. 6 per cent per 1 per cent area) as measured in the tunnel (stability fabric for 2 radii into the clay along the tunnel axis at 270°).

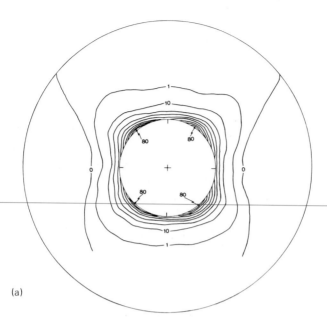

(a)

GREEN PARK FLEET LINE LONDON CLAY DISCONTINUITY STABILITY -0 DEG. 3RD QUAD
HEMISPHERE UPPER PROJECTION   16.00CM CIRCLE
N: 1598   W: 0.063
1.00 PER CENT AREA WINDOW USED

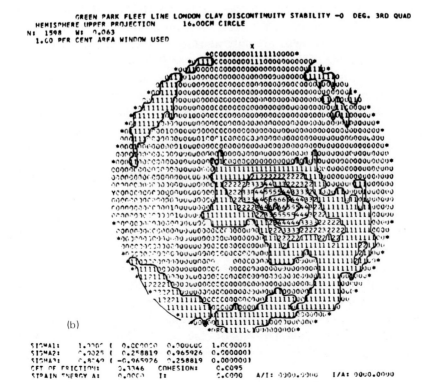

(b)

```
SIGMA1:    1.0000  (  0.000000   0.000000   1.000000)
SIGMA2:    0.0025  (  0.258819   0.965926   0.000000)
SIGMA3:    0.8040  ( -0.965926   0.258819   0.000000)
CFT OF FRICTION:    0.3346    COHESION:    0.0095
STRAIN ENERGY A:    0.0000    I:           0.0000    A/I: 0000.0000    I/A: 0000.0000
```

and deciding on the extent of rock fortification required'. Geological constraints upon the choice of the dam type are considered in Chapter 12.

6.23 Stability of a discontinuous clay surrounding an unlined tunnel

Another example of a three-dimensional stability analysis arises when a circular cross-section tunnel is driven through a stiff, fissured overconsolidated clay or shale. We may instance a theoretical investigation into the shear stability of London Clay, the results of a companion study of ground deformations in the vicinity of a tunnel driven in the clay having been described by Attewell and Farmer (1974a, b); see also Section 7.11.

Almost 1600 discontinuities were measured (Priest, 1974) when a running tunnel was opened-out into a station tunnel. Each measurement was related to its radial position above and below the tunnel axis and to its position into the clay from the original cut surface (Figure 6.47). For a rapid overall appraisal, however, the discontinuity parameters may be assumed to be independent of location. Two further complications arise. First, if a plane strain condition is assumed and an elastic analysis attempted, it will also be necessary to incorporate into the solution a continuous variation of the coefficient of earth pressure at rest (K_0) with depth (Skempton, 1961; Bishop *et al*, 1965). Second, the well-studied London Clay has been allocated a range of undrained and drained shear strength parameters by different research workers. These different parameters therefore represent a further variable in the problem.

An example of the anisotropic shear strain energy distribution around the 4 m diameter tunnel at a depth to axis level of approximately 30 m in London Clay is shown in Figure 6.48 together with the contoured orientation density distribution of the discontinuities. The rather asymmetric fabric, incorporating a subhorizontal bedding plane discontinuity suite and two near-vertical joint sets, appears not to exert a strong asymmetric control on the anisotropic shear strain energy distribution. Discontinuity orientation density distribution is not, therefore, a sensitive parameter in this particular problem. The very rapid reduction in shear strain energy with distance into the clay will be noted but it will be appreciated that no high degree of reliance can be placed on the energy levels in those elements of clay close to the cut surface of the tunnel where the shear stress differences are at their most intense.

7 ~ Site Investigation

It is relatively easy to discuss, in a general qualitative manner, the properties of rocks and soils on a small or large scale. It is less easy to obtain specific and reliable information for a particular site in practice, especially when the investigation is costed on a limited budget. Nevertheless, the ultimate aim of engineering geology is to provide information on the mechanical properties of a zone of rock or soil in order to enable 'an adequate and economical design to be prepared'. This last phrase is taken from the original British Standard Code of Practice for Site Investigation (1957), and whatever amendments may eventually be made, the four main objectives of site investigation stated in this Code remain sensible. They are in full:

(a) to assess the general suitability of a site for proposed engineering works,
(b) to enable preparation of an adequate and economic design,
(c) to foresee and provide against geotechnical problems during and after construction, and
(d) to investigate any subsequent changes in conditions, or any failures during construction.

Underlying all these *desirable* objectives is usually the *undesirable* fact that the process must be carried out at a cost which is often little more than 1 per cent of the total cost of the works. This cost fraction is usually consistent for most construction works (with the possible exception of very sensitive structures) since the combined laws of risk and dimension will normally demand increased investigation with increasing cost of the structure. The limitations imposed by the low cost allocation mean that most investigations follow a formal path, accepted by logic and experience as being the most economical approach to site investigation. This approach may be characterized as a four-stage process.

STAGE 1   PRELIMINARY INVESTIGATION

A desk study of the geology, geomorphology, history and relevant case histories aimed at isolating likely problems and enabling accurate planning and estimating of field work.

STAGE 2   FIELD INVESTIGATION

Drilling boreholes and digging trial pits; logging cores and examining pits; *in situ* testing; taking disturbed and undisturbed samples for laboratory testing. Evaluation of groundwater regime.

STAGE 3   LABORATORY TESTING

Re-examination and testing of samples, assisted by information on the geological structure obtained during the field investigation.

STAGE 4   FIELD MEASUREMENT

Monitoring of pre-construction trials such as pile tests, where desirable, and the behaviour of critical areas of the structure and surrounding ground (settlement, consolidation) in order to estimate the correctness of design, to control construction and to provide feedback information.

Obviously, in a cheap, well-documented structure, such as low-rise housing development, a detailed investigation would be unnecessary and Stage 1 accompanied by a limited field investigation, possibly involving site inspection and trial pits, would be adequate in most cases. A more extensive structure, such as a medium rise building or a highway, would require Stages 1, 2 and 3 and possibly some limited field measurement. A complicated or potentially dangerous structure, such as an atomic reactor or dam, will require all stages of investigation including a detailed field measurement study, often continuing for many years after construction.

In the present chapter, each aspect of site investigation is considered in some detail (although laboratory testing and joint surveying has been covered briefly in previous chapters), and in addition some alternative methods of field investigation, such as geophysical exploration techniques, are introduced. One aspect which is deliberately omitted is a detailed study of superficial and simple structural geology including problem maps. Adequate coverage of these subject areas can be found in many introductory texts.

In the case of *preliminary investigations*, the simplest first approach is to list the main sources to which an investigation would be directed. This has been comprehensively covered by Dumbleton and West (1971, 1972) who suggest three main areas for research in Great Britain: (i) Printed sources, (ii) Air photographs, (iii) Site features. These are considered initially.

7.1 Preliminary investigation

Most of the important preliminary information required for a site investigation can be obtained from published or private sources or records. These take two main forms: *maps* or *site investigation reports,* which are in many ways mutually complementary, although recent emphasis on geotechnical maps, aiming to combine both sources in a single detailed map, may lead to a reappraisal of the situation.

In Great Britain, probably the most readily available initial sources are the 1:25 000 (approximately 2½ in: mile) colour-printed Ordance Survey *topographical* maps. These each cover an area 20 km by 10 km and have a contour interval of 25 ft (7.6 m). They include most of the detail available on the larger scale 6 in : 1 mile (1:10 560) maps such as field boundaries, rights of way, woods, major mineral workings, springs, streams and areas of water, and marshy ground. New maps are 1:10 000 with 5 m contours.

From these maps, a general indication of the overall geological structure, slope angles, ground water level and type of soil/rock underlying the site can be obtained, as well as a certain amount of important preliminary information on access and possibly ancient monument sites. In areas where there may have been significant mineral working, a study of earlier editions of topographical maps can be useful. Dumbleton and West (1971) summarize the following features which can affect site conditions and which may be revealed by a study of older maps: (i) Location of concealed shafts, adits and wells, (ii) Location of concealed foundations and cellars of demolished buildings, (iii) Location of abandoned sewage farms, (iv) Location of *filled* ponds, clay pits, sand and gravel workings and quarries, (v) Original surface topography and drainage beneath filled areas, (vi) Change in surface level and resultant changes in drainage and stream and river courses, (vii) Changes in coastline due to erosion or deposition, (viii) Location of landslips and rate of movement.

This information may, of course, be supplemented by maps (and records) from local sources such as mine or quarry plans (the National Coal Board or the Department of Trade and Industry in the United Kingdom) or, in the case of coastal, harbour or estuary works, by Admiralty charts. The most important source of information will, however, be *geological* maps. In Great Britain, these are published by the Institute of Geological Sciences, principally as the 1 in/mile (1:63 360 now being replaced by 1:50 000) series, but many 6 in/mile (1:10 560) maps are available for more detailed study.

Geological maps are available in *solid* or *drift* editions. The former shows the major rocks which would outcrop at the surface if all recent surface deposits (alluvial, glacial, organic) were absent. The latter shows the actual *surface* deposits (usually with indications of thickness) and indicates the 'solid geology' only where drift-free outcrops actually occur. The principal object of a geological map is to indicate the major stratigraphical boundaries in the district covered by the map, rather than lithological differences, although obviously in the case of a drift map, and also in a large scale map, the characteristic lithology of strata will be an important factor.

This difference can and has caused confusion. For instance, the terms Keuper Marl or London Clay describe fairly widespread strata. The lithology of both, and particularly the former, can, however, vary widely with geographical position and depth due to weathering or modified depositional conditions. Since mechanical properties are affected by lithology far more than by stratigraphy, the stratigraphical description, whilst useful historically, will have minor engineering significance (unless modified by some lithological terminology).

It is likely, however, that much of the confusion can be eliminated by a careful study of large (1:10560) scale maps and accompanying memoirs. These usually show the sites of boreholes and give a description of depth, thickness and lithology of the deposits encountered in each borehole, so forming a basis for an adequate subsurface section from which the field investigation can be designed. Other information which may be obtained from the 1:10560 geological maps includes the dip of the strata, the presence of faults and igneous intrusions, and the location of buried channels, river terraces, solution cavities, escarpments and unstable zones such as landslips or disturbed strata.

This geological, geomorphological and topographic picture may be supplemented by groundwater information from various sources such as the Regional Water Authorities and the Institute of Geological Sciences in Great Britain, and by data from *previous site investigations.* This latter information has the disadvantage of being scattered and is of variable completeness and quality. In Great Britain, sources include the Institute of Geological Sciences, local authorities and road construction units, consulting engineers and contractors. Moves are at present afoot under the aegis of the British Construction Industry Research and Information Association to form a National data bank. Data would be stored in computer files, the information being available for immediate print out in its original form by calling appropriate identifying co-ordinates.

One aspect of this system, at present being studied by the authors and considered in a later section, is the possibility of applying probability theory to data collected from densely explored urban areas. An estimate of probability distributions for major geotechnical parameters collected from various sources might provide in some ways even more reliable data than the conventional site investigation method. Thus, whilst a numerical approach would never replace actual site investigation, it could certainly improve the design and interpretation as well as the cost-efficiency of conventional investigation procedures.

Study of topographical and geological maps, together with available water supply and site investigations data, is one approach to preliminary site investigation. An alternative approach is through *geotechnical maps* for regional and planning purposes, and more detailed geotechnical plans for specific civil engineering projects. These do not really exist in Britain at the time of writing in contrast to Europe, U.S.A. and U.S.S.R., but a fairly strong case has been made for them by a Working Party of the Geological Society of London (1972) and by Fookes (1969) and Dearman and Fookes (1974) among others. The basic idea is that the large-scale geological maps (1:10560) should be supplemented to include additional useful geotechnical data. These maps might include: (a) Detailed lithological descriptions of both bedrock and superficial deposits, ideally in the map margins. (b) Notes on the extent and type of any weathering or alteration. Six classifications were suggested by the Working Party ranging from *freshly-exposed* rock (WI) through slightly (WII) moderately (WIII) highly (WIV) and completely (WV) weathered rock or soil to *residual soil* (WVI). (c) Clear differentiation between solid and drift boundaries. (d) Notes on shear surface and shear zones. (e) Notes on joints and other structural discontinuities including information on frequency, direction, inclination, weathering, together with surface and infill characteristics. (f) Details of geomorphological features and particularly landslides and natural

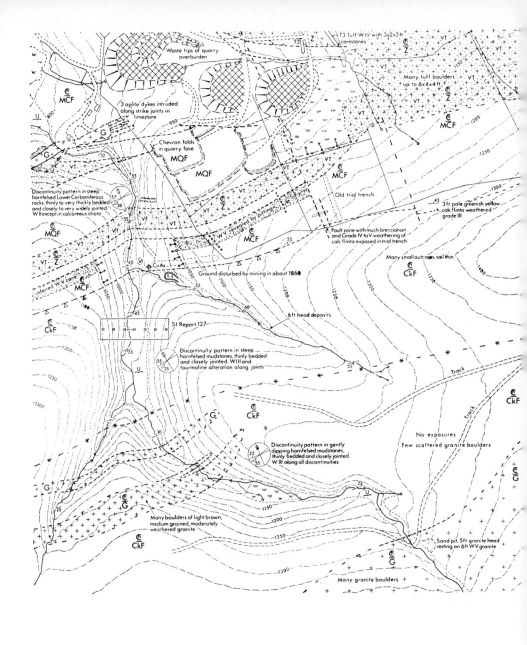

Figure 7.1 Part of a 1:10560 map supplemented with descriptive information in engineering geological terms with an extract from the accompanying explanatory legend (Geological Society of London, Working Party Report, 1972; reproduced by permission of the Geological Society of London)

EXPLANATION

SUPERFICIAL DEPOSITS (DRIFT)
RECENT AND PLEISTOCENE

Thickness in ft.

ALLUVIUM up to 8

On the granite, alluvium is a brownish-yellow, loose, sub-angular, coarse gravelly SAND with some peat and rounded boulders of moderately weathered granite up to 3 ft, and pebbles of quartz. Downstream, alluvium is a silty gravelly SAND with rounded granite boulders up to 3 ft and sub-angular cobbles and boulders of the solid rocks. The deposits are moderately to highly permeable. Locally much disturbed by streaming for tin.

RIVER TERRACES (UNDIFFERENTIATED) up to 40

Dark yellowish-brown, loose but locally weakly to strongly cemented in horizontal layers by manganiferous or ferruginous material, sub-angular to rounded, sandy GRAVEL with rounded to sub-angular cobbles and boulders of local rocks. Boulders occasionally up to 3 ft. The deposits are highly permeable except where cemented. Locally much disturbed by streaming for tin.

HEAD 6-10 locally >40

Almost everywhere present and largely obscures the solid formations. Represents solifluxion debris and grades downslope into alluvium and terrace deposits.

Within the outcrop of the granite, head comprises yellowish-brown, loose, layered, sandy GRAVEL with some clay, and gravelly silty SAND with cobbles and boulders of moderately weathered granite; grades down into moderately to highly weathered granite in situ. On the Upper Carboniferous outcrop next to the granite, head is typically reddish-brown, loose to compact, homogeneous, clayey gravelly SAND with many sub-angular cobbles; on steep slopes fines may be absent and head is then loose, clean COBBLES of the local rocks beneath 6 - 12 in. of humic soil.

On the Lower Carboniferous rocks, head is reddish-brown, loose to compact, homogeneous silty clayey SAND with some cobbles and boulders of local rocks; it may be layered with an upper gray horizon separated by a black cemented layer typically 3 in. thick from reddish-brown head down to bedrock.

SOLID FORMATIONS
CARBONIFEROUS
Upper Carboniferous (Namurian)

CkF **CRACKINGTON FORMATION** ?

Dark to very dark grey, very fine grained, thinly bedded to thinly laminated, very closely jointed, slightly to moderately weathered, poorly cleaved SHALE, weak, impermeable except along open joints. Interbedded with very subordinate grey to dark greenish grey, fine-grained, very thinly bedded, thinly laminated and cross-laminated, closely jointed, slightly to moderately weathered SILTSTONE, moderately strong, and dark greenish grey medium grained, very thinly to medium bedded, with closely to widely spaced joints slightly to moderately weathered, SANDSTONE, strong.

The shale slakes on exposure and is suitable for brick making.

S **SANDSTONE**

It has been possible to map groups of beds in which SANDSTONE predominates. Beds are usually less than 12 in. thick and are separated by very thin beds of siltstone and shale.

Sandstones are suitable for aggregate production.

Within the contact metamorphic aureole of the granite, dark grey, very pale orange to dusky yellowish brown, fine to medium grained, thinly bedded, closely jointed, slightly to moderately weathered, hornfelsed SHALE and SANDSTONE, strong, impervious except along open joints. Locally with fine grained black tourmaline developed as selvedges up to 1 in. wide along discontinuities and with irregular quartz veins up to 2 in. wide.

Lower Carboniferous (Dinantian)

MELDON CHERT FORMATION 240

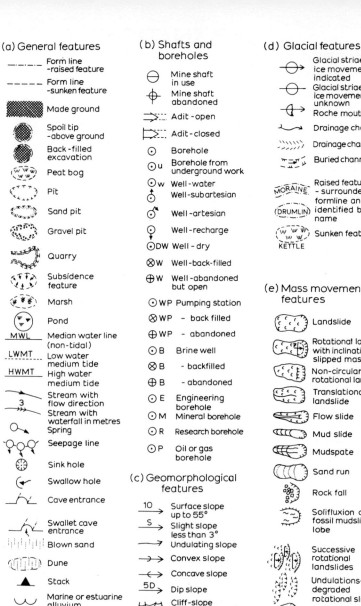

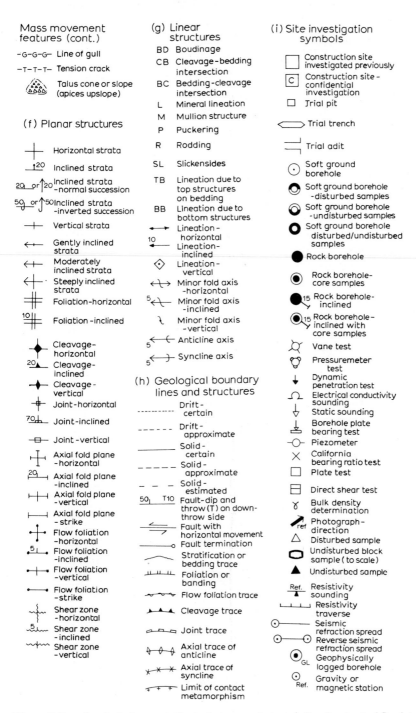

Figure 7.2 Symbols for geotechnical maps and plans (*after* Geological Society of London, 1972).

openings. (g) Hydrogeology, including notes on groundwater basins, groundwater movement, ground permeability together with groundwater salinity and chemistry. In particular, springs and periodic flows, together with notes on water-bearing properties of deposits are important. (h) Details of any recorded seismicity. (i) Notes on existing exploration, water or site investigation boreholes together with borehole logs and test data where relevant or available. (j) Sites of quarries, tips, mines, subsidence zones and disturbed ground.

Obviously, this is a great deal of information to include in a single map, and some idea of the complexity of the cartography, and the width of margins required, can be obtained from an excerpt of a map (1:10560) used to illustrate the Working Party's report (Figure 7.1).

Figure 7.2 illustrates a useful by-product of the Working Party report; namely, a standard list of symbols for use on geological and engineering maps and plans. Compiled from the lists of symbols produced by the Institute of Geological Sciences (1967), the International Geographic Union (1968), the British Standards Institution (1957) and the International Organisation for Standardization (ISO) the list, notwithstanding some controversy (see Dearman, 1974), will probably form the basis of an international standard for geotechnical cartography.

One possible solution to the complexity of geotechnical mapping suggested by the Working Party would be to produce in difficult areas a *series* of maps; say a geological map, a hydrogeological map, a geomorphological map, a mechanical property map. Alternatively, a detailed drift map with notes on the back or in the margins, updated frequently, would probably be a more elegant solution. A particularly good example of computer drawn mechanical property maps based on data from London Clay in the London and Hampshire Basins is illustrated by Barrett and Fookes (1974). Block diagrams show solid contours indicating changes in liquid limit, plasticity index, dry bulk density, undrained cohesion and clay mineral content.

When as much information as possible has been obtained from a study of printed sources, a *visual examination* of the site, possibly assisted by aerial photographs, is recommended. This is essentially an exercise in visual observation and classical field geology aimed at confirming or amplifying the information obtained in the desk study, and detecting any unusual features not covered by the desk study and requiring further investigation. A very detailed check list of features of particular interest to engineering geologists is provided by Dumbleton and West (1974). The scope of the examination will depend on the type of structure proposed. In the case of a highway, potential instabilities through old landslip areas, highly compressible materials (peat, organic clays) and signs of settlement or poor drainage will be important. In the case of rock or stiff clay foundations, emphasis may be placed on confirmation of or measurement of exposed fissure or joint orientations, together with simple mechanical property tests to estimate bearing capacity for preliminary design calculations. In the case of fill, some indication of the stability and degree of compaction may be required. In all cases, the spatial distributions of essential geotechnical properties, particularly any easily acquired index properties such as the Atterburg limits for a clay or simple strength parameters for a rock, will be expressed on a map of the area of interest and ideally each property will be contoured for easy visual reference. The ultimate aim in all cases will be to decide on the optimum layout of drill holes and pits during field exploration in order to provide the maximum amount of useful data on the strata succession and ground properties, for the minimum investment in site investigation.

Some of the more important items which an engineering geologist in particular should look for are listed in Table 7.1 as an engineering geologist's site reconnaissance check list.

7.2 Aerial photographs

However accurate a topographical, geological or even geotechnical map may be, it will always suffer from limitations imposed by scale, time and probably subjectivity. These are to a certain extent circumvented by a vertical air photograph, which will provide a detailed and definitive picture of the *topography* at the date the photograph was taken together with additional information on drainage and potential water supplies for remote urban populations, fence types, agricultural use, trees and vegetation capable of geotechnical interpretation.

Dumbleton and West (1970) and Norman (1969) among others have discussed the uses and geotechnical interpretation of aerial photographs in detail with specific reference to linear site investigations for roadways. This is one important use, the major use, of course, being in accurate map-making in unmapped areas. In these poorly-mapped areas, an aerial survey to establish a base map is an

Table 7.1   Engineering geologist's site reconnaissance check list

Item	Action
Lithology and structure	Anticipate lateral unmapped changes in basic geological features. Note the general attitude of joint sets and bedding with respect to possible instability and comment on the presence of water seepages and springs with respect to those structures. Layered deposits and structures may possess a strength anisotropy with respect to both shear and tension. Sedimentary laminated clay structures may possess low shear strength along the laminations. Gneissic and schistose rocks may be prone to spalling and flaking along fissile planes if exposed at depth in the ground.
Solifluction and periglacial disturbance	Look for the effect of these disturbances on the outcrops of softer Mesozoic and Tertiary formations (such as landslips, valley bulging, cryoturbation). *Landslips* are a feature of Pleistocene glacial retreat — the removal of support from oversteepened valleys and the rather gradual thawing-out of slopes that have been affected by deep permafrost (Higginbottom, 1971). Particular attention should be directed to possible old landslipping of Mesozoic and Tertiary clays near the base of escarpments of harder rocks. In Britain, for example, Higginbottom quotes the Inferior Oolite, sandstones of Wealden and Lower Greensand age, and the Chalk. Disturbance may at best have reduced the clay to a state of residual strength along a surface or zone of primary shear and at worst — with insufficiently rapid dissipation of pore water pressure — to a mudflow. Look for topographic evidence of such landslips (for example, hummocky ground) which may prove to be particularly troublesome if new road construction takes place through them. Note any evidence of springs and suggest locations for piezometric instrumentation. Since *valley bulging* can be the cause of abutment rock weakening at a potential damsite, be aware of possible oversteepening of valley sides and be prepared to specify an *ad hoc* programme of drilling and undisturbed sampling, pitting and trenching, and *in situ* permeability testing. Note that *cryoturbation* features are usually shallow and are often caused by localized thawing of the permafrost layer in Pleistocene times. These features are represented by the replacement of original material by a weaker one and may be more obvious as a repeated phenomenon when viewed on aerial photographs.
Stress release effects and frost wedging	These are again due to ice retreat and are manifest in stronger rocks. Typical phenomena are joint widening and infilling with weaker material, slabbing and cambering. Every attention should be directed towards these features for they are particularly important in slope stability and foundation engineering. Joint characteristics should be measured in the manner outlined in Chapter 6 and infill material should be

Table 7.1 (*continued*)

Item	Action
Stress release effects and frost wedging (*continued*)	sampled. Cambering may be particularly prevalent in stronger well-jointed rocks such as the Lincolnshire Limestone and the Millstone Grit in England. Note that the Millstone Grit cambering presented a problem during the construction of the Trans-Pennine motorway at Scammonden Dam (see Figure 10.59).
Frost shattering	Note that in rocks such as chalk the effects of frost shattering can be deep-seated and may lead to a general weakening of the material as a foundation or a slope. Pits may be required for observation and *in situ* plate bearing tests may be required for important foundations.
Solution cavities	In limestone country it is wise always to be on the lookout for possible solution cavities although these may not be readily apparent at ground surface. Higginbottom (1971) has pointed out that a concentrated inflow of near-freezing melt-water is a much more effective carbonate solvent than is water at normal temperature. He suggests that such cavities may be concentrated in melt-water channels or along the line of maximum ice advance. Hidden cavities likely to present engineering problems would have to be proved indirectly by geophysical methods supplemented by confirmatory boring. Newer detection methods, perhaps based on acoustic holography, may prove to be useful in the future.
Old mine workings	There must be an awareness of the several problems that can arise in mining areas. These problems, which are introduced later in the present chapter, comprise old bell pits, surface settlements, and the effects of tensile ground strain.

obvious preliminary to any site investigation. In well-mapped areas it is a useful additional tool which may reveal zones of actual or potential instability which are not easily detectable at ground level and which are not included in available maps.

The basic techniques of *photogeology* and *photogrammetry* have been described elsewhere (see for instance Allum, 1966; Lueder, 1959). They involve a careful study of stereographic pairs of scaled photographs, combined with surface exploration and sampling, in order to build up a geological map. Outcrop and drift areas can be detected and mapped through topographical features and changes in vegetation which are often detected by infra-red photography. Geological boundaries, faults and occasionally major joints can be traced through overburden cover, sometimes even through boulder

440 Principles of Engineering Geology

Figure 7.3 Aerial photograph of coastal landslipping at Black Venn, Axmouth near Lyme Regis in England (*reproduced by courtesy of* Aerofilms Ltd, Boreham Wood, Herts, England).

clay. Obviously, such interpretations depend for their accuracy on the skill and experience of the interpreter.

The main *geotechnical* features revealed by an aerial photograph (of which Figures 7.3 and 7.4 are examples) tend to be topographical, the most obvious being rotational slides (Section 9.2) in soils, or wedge failures (Section 10.8) in the case of rock slopes. These and other slip or flow features are far more easily identifiable from photographs than at ground level since the 60 per cent overlap between stereo pairs allows the ground to be seen in an exaggerated relief that is further highlighted if early morning or late afternoon photography is used. A useful preliminary exercise if aerial photographs are available during a site investigation would be to examine any slopes which are sufficiently steep to cause doubts as to their stability. This critical inclination can range from less than $5°$ in the case of poorly-drained mud or clay slopes to probably $45°$ in the case of rock slopes. The exaggerated vertical scale of the photographs should show any signs of movement, ranging from outright failures

Figure 7.4 Aerial photograph of cliff screes and rockfall at St. Albans Head near Swanage, England (*reproduced by courtesy of* Aerofilms Ltd, Boreham Wood, Herts, England).

and tension cracks to hummocks or steps associated with unstable solifluction deposits and hill creep.

Other important uses of aerial photography (discussed in greater detail in the following section) in site investigation lie in measuring topographic heights, computation of cut and fill volumes (see Horder, 1960), detecting subsidence zones due to old mine workings or solution cavities, monitoring the stability of completed structures, excavations or tips, and determining the drainage history of poorly-drained or swampy areas. In the case of subsidence cavities which are, of course, detectable at the surface, the exaggerated vertical scale of the photograph eases the recognition process, and the area that is covered is increased. Stability monitoring is essentially a follow-up to a linear investigation, where on-the-ground surveying may give insufficient coverage. The exercise would involve periodic photography of sensitive areas.

Detailed analyses on certain recognition features such as soil and rock slopes can be found in later chapters. One of the most important civil applications of aerial photography, in *terrain evaluation* for highway projects, merits further discussion, and is considered in the following section.

7.3 Terrain evaluation for highway projects

It is not often that the geology of an area imposes itself forcefully at the planning stage of a highway project. Exceptionally difficult geological conditions will usually generate modifications to the original plans but the major controls on alignment will usually be physical (in the form of topographic and water barriers), economic/industrial and social. A preliminary reconnaissance serves to bring the physical problems to the fore and this is usually followed by a detailed site investigation to provide necessary *design* information. But often the pre-design surveys are inadequate with the result that the roads and bridges are not sited in the best places and readily available construction materials for sub-grades, concrete and coated aggregates and fill are sometimes overlooked in favour of technically inferior rock, sands and gravels.

Terrain evaluation (or land classification) is a method of classifying ground into distinctive and recurring land types based on physiographic principles and using evidence derived both from field survey and from the interpretation of aerial photography (Dowling, 1968). Of use particularly in those developing countries where little basic geological mapping has been done, it involves the recognition of distinctive patterns of landscape, the form of the land surface, which go to make up land systems. The physical environment is compounded from numerous components such as rocks, soils, vegetation, topography, climate and it is axiomatic that when sub-sets of these components have the same character and interact in the same manner the physical form of the landscape should be the same independent of location. Dowling (1968) quotes the example of a particular rock type that has been subjected to the same weathering styles throughout the same geological period of time and notes that its surface form, including its associated soils and vegetation, would probably be of the same character wherever it occurred. Construction engineers are aware that certain ground features repeat themselves and in an area or country that they know well they would easily recognize associations between particular rock types, slope forms, soils and vegetation. When the various landscape features have been described it is then necessary to derive values for the physical characteristics of different parts of the landscape. In the present context, these characteristics are the geotechnical parameters upon which road design is based.

Ground can be sub-divided into distinct patterns of landscape, each pattern being compounded from several interrelated components which recur together to endow the pattern with its characteristic form (Dowling and Beaven, 1969). These patterns are readily discernible from aerial photographs of a suitable scale.

Land systems are landscapes which can be recognized through the use of aerial photography as having distinctive patterns at scales between 1:250 000 and 1:1 000 000 (Christian, 1958; Christian and Stewart, 1953).

Land facets are units of landscape having a reasonably high degree of homogeneity. They can readily be identified in aerial photography at scales between 1:10 000 and 1:80 000.

Land elements are the smallest units of landscape and are usually too small to be identified on aerial photographs at a scale of 1:20 000 or less (Brink et al, 1966).

Examples of land elements mentioned by Dowling (1968) are small rocky rises, a lateritic ironstone outcrop, incised water-courses and restricted areas of erosion. Land facets are formed from groups of genetically related land elements and it is the recurrence of combinations of linked land facets which characterizes a land system as a whole. Dowling has noted that adjacent land systems developed over the same bedrock material very often share common land facets and land elements. But, however, they may often be mutually distinguishable either by the presence of additional land facets in one of them or, where no new land facets exist, by differences in the total area covered by one or more of the component land facets. He points out that in one land system, a particular land facet may extend over only 20 per cent of the total area whereas in the neighbouring land system the same land facet may occupy 35 per cent. This could well be an important factor in the selection of a new highway route if, for example, the particular land facet was associated with low bearing capacity soils which demanded a heavier construction pavement and the necessary high volume of rolled fill and aggregate material was not readily available along that route. Expert examination of aerial photo-mosaics might even suggest that a high bore-hole sampling density during the site investigation would prove to be money wasted since the terrain environment and all its observable constraints seemed to vary very little over many miles.

Examples of the use of terrain evaluation are given in Millard (1967), Dowling (1968) and Dowling and Beaven (1969). It is quite obvious that the key to its successful operation lies in aerial photography and its successful interpretation by an expert (Dumbleton and West, 1970). There are several points that may be made or re-emphasized:

(a) Variations in the surface relief (topography) of an area often expresses differences in the type and qualities of rock beneath drift cover. The nature of a superficial soil covering is often a function of the relief of the parent rock and this also applies to the character of the vegetation. The extent to which these factors are interdependent depends on the geological history, the past and present climatic conditions, and the actual stage reached in the current erosional cycle. But there is always a basic interdependence and the landscape cannot be interpreted without a fundamental appreciation of the geology and geomorphology. Since information on slope contrasts is readily gleaned from examination of stereo-pairs it is obviously unwise to attempt to separate too rigidly the geological and geomorphological aspects of the picture.

(b) Terrain information obtained from aerial photographs can be broadly grouped into two divisions. First, there are the main relief features such as plains, escarpments and plateaux together with the more detailed features within the elements and facets of the landform, such as valleys, ridges, screes, slopes of different steepness, flood planes and levees. Second, there is information to be extracted on the lithology and structure of the underlying rocks. All these factors taken together allow an area to be divided into units of different shape and rock type and it is these units which are mapped in a terrain analysis.

(c) The drainage pattern of an area is important in terrain analysis, with the density of the pattern being related to rock type.

The more erosion-prone the rock, the more intricate the drainage pattern is likely to be. Thus, the pattern may suggest broad rock grouping but a particular pattern will not be exclusively characteristic of a particular lithology.

(d) Main breaks in slope will be mapped and a genetic reason for their location sought. Symbols will be allocated to bracketed slope angles so that areas having the same slope angles or patterns of slope can be grouped for comparative assessment on the basis of other incompatible parameters. Mapping of waterfalls and rapids with respect to slopes often serves to position the boundary between erosional cycles.

(e) Faulting is indicated on air photographs by the contact of different rock types, by a stepped topography or by a repetitive sequence of the same strata. Dykes at the surface can often be identified as long straight ridges.

(f) Air photographs create a vertical exaggeration and this has the effect of making low dips, across which roads are taken, more easily recognizable. Dipping sedimentary rocks are often characterized by differential weathering of strata sequences, particularly where strata dips are compatible with topographic dips. In areas where one dip differs from the other, and this will be the general case, the regional dip will be highlighted because the aerial photograph presents an extended view which shows the trend of numerous small exposures which would not be discerned singly.

(g) Of importance in certain parts of the world, the 'bird's eye' view shows quite distinct trend lines in basement gneisses and schists and other strongly-faulted and tectonically-disturbed metamorphics. These trends would not be picked out from investigations on the ground.

(h) In general, it is possible to extract more detail from aerial photographs taken over arid or semi-arid regions of the earth than from those taken over temperate or tropical regions. This is because in the latter cases deep and strongly leached soils often conceal the underlying rock and its structures. The associated and often lush vegetation also imposes a masking effect.

(i) Sedimentary rock areas are more informative than dominantly igneous rock areas and metamorphic rocks are least informative of all.

(j) The greater the differential resistance to erosion characteristic of adjacent strata, the greater is the volume of information that may be yielded by an area.

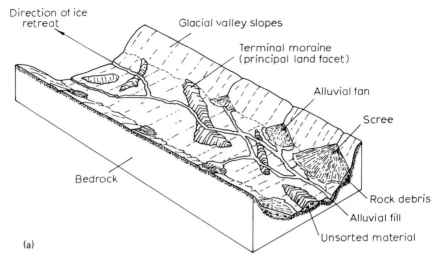

(a)

Recessional ablation moraines are formed throughout the length of the valley during halts in final ice recession. The valley may also contain buried fluvioglacial deposits of earlier glaciations which persist in face of subsequent overriding of the ice.

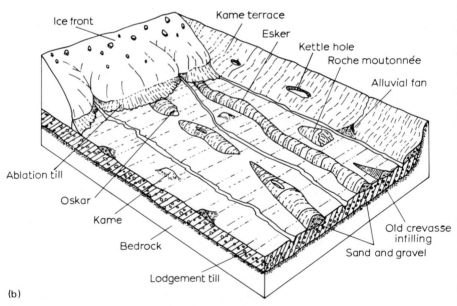

(b)

Fluvioglacial deposits are stratified and were laid down both on the ice surface and at its margin. Due to the rate of flow of the melt waters, these deposits contain little or no fines — sands and openwork gravel (permeabilities $\sim 10^{-4}$ cm s^{-1} to 1 cm s^{-1}). Also lenses of lake clay and silt (permeability $< 10^{-4}$ cm s^{-1}) may appear as kettle holes caused by melting of buried ice.

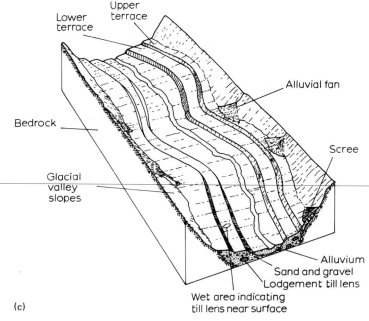

(c) Valley infilling with coarse flood material; erosional modification; lenses of open-textured gravel caused by deposition from fast-flowing water.

Figure 7.5 (a) Idealized glaciated valley land system (*from* Fookes *et al*, 1972). (b) Idealized fluvioglacial ice contact land system (*from* Fookes *et al*, 1972). (c) Idealized valley train land system (*from* Fookes *et al*, 1972).

Terrain evaluation is really only a part of the general project reconnaissance and feasibility study discussed earlier (see the paper by Edwards (1971) for the manner in which the constituent parts of the study are welded together). Edwards also produces an excellent example of map sequences for an area, topographic, geological, surface drainage distribution, hydrological, land facets, land elements and communications, and one can visualize such maps comprising an overlay suite.

Glaciated landforms are of particular concern in highway projects and they have been classified into three idealized systems by Fookes *et al* (1972). These are: the glaciated valley fluvioglacial ice contact deposits (including ablation moraines), and valley trains. A land facet would comprise two or more of these landforms. One facet, for example, may be represented by a till grading through ice contact deposits to an outwash plain, and as this pattern recurs it forms a land system. Fookes *et al* suggest another example of a glaciated valley with outwash deposits at its termination to form a land system comprising a glaciated region of many thousands of square miles. The scale of the system of reference must be such that the individual associations (such as eskers with kames, or outwash deposits with lacustrine deposits) can be evaluated in greater detail and it is the units in these associations that have to be identified in a site investigation and which pose some geotechnical problems.

These three idealized glaciated land systems incorporating the individual constituent units or elements are shown in the manner of Fookes *et al* (1972) in Figure 7.5 a,b,c.

7.4 Geophysical exploration techniques

The limitations of aerial photographs as an aid to site investigation and in the design of detailed site exploration is that they do not give any direct indication of what lies below the ground surface. This is an important omission in any engineering investigation being a vital element in the design of tunnels or extensive, heavy-loading or precise structures such as, for example, nuclear power stations. One obvious indirect investigation solution is the use of sub-surface *geophysical exploration* techniques. These have, of course, been used widely in exploration for minerals of economic importance and for oil, and, although in the former case the location problem is often eased by the high density and electrical and magnetic properties of the mineral, and in the latter case by the distinctive structures associated with the oil, there is, philosophically, no reason why geophysical techniques should not be used to detect important lithological differences in near-surface deposits as a preliminary to site exploration. These lithological and structural boundaries could subsequently be proved, as in the case of mineral/oil exploration, by a limited drilling programme designed to complement the geophysical exploration. This approach would seem to offer to the client a more attractive package than the somewhat arbitrary design of a direct sampling programme such as is usually adopted. Direct sampling methods are discussed in Section 7.7.

Although the arguments for geophysical methods as a preliminary or supplement to site investigation are sound, and although they have in fact been used extensively, the success rate has been less than good. The reasons for this can be seen to a certain extent in the

Table 7.2 Resistivities of some rocks and soils (*after* Jones, 1952)

Rock/Soil Type	Resistivity ohm – metres
Peat, clays, shales	0.5–15
Mudstones	10–200
Dry sand, sandstone	200–1000
Saturated sand, Sandstone	20–50
Igneous rocks	$100–10^7$

recognized limitations of the methods themselves. Thus, in the case of *gravity* methods, the object is to compare the theoretical gravitational force at any point on the earth's surface with the measured gravitational force, suitably corrected for elevation and topography. The difference is known as the Bouguer Anomaly, and the existence of large anomalies on an isogal map of zones of high and low gravitational force may indicate respectively the presence of denser rocks nearer to the surface or less dense rock or cavities near to the surface. The problem is that the anomalies are so minute that only a massive structure, such as an anticline or thrust block could be detected confidently. The same arguments apply to the analogous *magnetic* or *electromagnetic* methods, with the added emphasis that any non-magnetic structure would have to be very large indeed to distort the earth's magnetic field. These methods are described in detail in various geophysics texts such as Dobrin (1960), Griffiths and King (1965) and Parnasis (1973).

Electrical methods of surveying make use of three basic properties of rocks: resistivity (reciprocal of conductivity), electrochemical activity (self-potential methods), and electrical storage capacity (inductive prospecting methods). Only the former, the resistivity method will be considered.

Electrical *resistivity* logging is based on the principle that any change in the specific resistance (Table 7.2) of a rock or soil will change the flow of current through the material and thereby increase or decrease the electrical potential between two mutually-displaced measuring electrodes. Resistivity is a function of the electrolyte contained in the pore spaces of the material and is inversely proportional to the porosity (Deere *et al*, 1969). In massive but fractured rocks, therefore, the spatial distribution density of the fissures directly controls the resistivity (Griffiths and King, 1965).

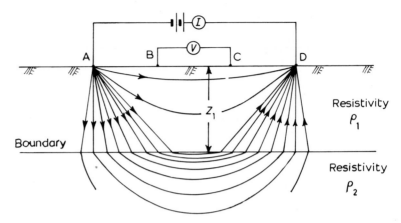

Figure 7.6 Wenner electrode configuration for electrical resistivity investigation (for condition $\rho_1 > \rho_2$).

One unfortunate result of this water-dominance feature is that the presence of a phreatic surface can completely obliterate information concerning vertical changes in rock type.

The Wenner electrode configuration sketched in Figure 7.6 is the one most commonly used. Current is passed into the ground through electrodes inserted at A and D and the associated potential gradient is measured by two secondary electrodes at B and C. The electrode spacing in this configuration is such that AB = BC = CD = $^1/_3$ AD.

When the electrode separation is small, very little of the induced current is able to penetrate to layer 2 and the apparent resistivity tends to ρ_1. This apparent resistivity can be expressed as

$$\rho = 2\pi a \frac{V}{I} \qquad (7.1)$$

where a is the inter-electrode spacing ($^1/_3$ AD), V is the measured voltage and I is the induced current. When the spacings are large compared with z_1 the apparent resistivity tends to ρ_2 because most of the induced current penetrates to the lower layer.

Interpretation of data for the two-layer case is conveniently performed through the use of Tagg's (1934) curves which graph relationships between ρ/ρ_1 and z_1/a, the relationships being expressed as different k values where k varies from ±0.1 to ±1.0 in increments of ±0.1 (k is taken +ve and −ve when $\rho_2 > \rho_1$ and $\rho_1 > \rho_2$ respectively). In the field, an expanding electrode technique would be used, with an increase from a very small value to a large one. When a is very small, $\rho \to \rho_1$ and therefore all ρ/ρ_1 can be divided by ρ_1 (for a small) to define a series of k-values ($k = f(z_1/a)$) by multiplying all 10 values of z/a by a. The k-values would then be plotted against z_1 on a separate graph to produce a discrete curve for each a. There should be a common, or nearly common, intersection point for the several $k:z_1$ curves representative of the several a spacings, this point defining the depth z_1 to the interface and, indirectly, the resistivity ρ_2 of layer 2 on the assumption that ρ_1 has been determined satisfactorily by the method of diminished a described above.

A less analytical method of data interpretation involves the comparison of observed apparent resistivity curves with theoretical master curves. Such a technique lends itself to a multi-layer interpretation ($z_{1 \text{ to } n}$), the actual matching being best performed by digital computer. The present authors have, in fact, used this computer-matching approach, based on the published curves of Mooney and Wetzel (1956), for the mapping of economic deposits.

As has already been stated, the resistivity is not a consistent rock property. Rzhevsky and Novik (1971) quote some very wide variations, and it is particularly affected by density and degree of saturation. There is, however, a sufficiently wide difference in resistivity between, say, a clayey drift deposit and an underlying bedrock or sand layer even if saturated for this to show up in a resistivity survey such as might be used as a preliminary for a site investigation. The difficulty, of course, lies in the layout of the survey which, for accurate analysis of deep zones, must cover a reasonably long traverse. This means effectively that, during the survey, a considerable length of layered strata beneath the surface traverse is sampled, so giving rise to obvious anomalies unless the strata are *even* and *flat*. The result is that resistivity surveying is really useful only in near-horizontally layered strata of sharply differing electrical conductivity.

Nevertheless, resistivity surveys carefully designed and analysed can be carried out successfully and economically in reasonably flat strata. An example described in the literature (see McDowell, 1970) concerns a linear site investigation for the experimental tracked hovercraft. Owing to the slightly artificial conditions of the tests (subsequently recognized in the controversial United Kingdom government decision not to proceed with further development) only a very low degree of deviation from the horizontal could be

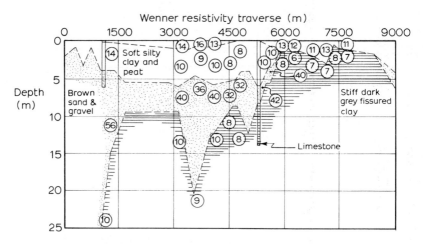

Figure 7.7 Section based upon resistivity results. Ringed figures are resistivities in Ω-m (*after* McDowell, 1970).

accepted in the finished track, necessitating an accurate knowledge of subsurface strata in order to design track foundations. This information was obtained (Figure 7.7) by sinking two boreholes and carrying out a resistivity survey. Conditions were assisted by the very high resistivity contrast between the overlying peat and soft clay ($\rho = 8 - 14$ Ω-m), the underlying stiff clay ($\rho = 6 - 10$ Ω-m) and the intermediate sand and gravel ($\rho = 170 - 400$ Ω-m above watertable; $30 - 60$ Ω-m below watertable).

The survey was very successful in determining the thickness of the peat layer, but less successful in obtaining the thickness of the gravel layer and the location of the top of the stiff clay. This latter was attributed partly to the differences in drainage at this level but mainly to the relative irregularity of the stiff clay/gravel interface. Accurate resolution of this interface would have entailed numerous surveys normal to the main traverse line, with a consequent increase in the complexity of analysis and escalating costs.

Seismic methods, like resistivity methods, depend on a specific property of the sub-surface strata, in this case the velocity of elastic waves in the rock. In the case of longitudinal P-waves (see Section 4.10) these velocities vary from 300–500 ms^{-1} in loose sands and poorly consolidated clays through 2000 ms^{-1} in stiff clays and weak rocks to 5000–6000 ms^{-1} in competent igneous rocks (see Rzhevsky and Novik, 1971).

The basic problem is to determine the thicknesses of a layer or layers of strata beneath the surface by measuring the arrival times of waves initiated by an impulsive or oscillating source over various distances. The technique used most widely in engineering site investigation is seismic refraction surveying. Its relative importance demands a rather fuller treatment of the method.

7.5 Seismic refraction surveying

This technique for indirectly locating changes in strata through the characteristic velocity of waves through them will be adequately described in any applied geophysics textbooks such as, for example, by Dobrin (1960). The assumption is that waves emanating from a point source S (Figure 7.8) will travel at a characteristic speed c_1 in medium 1 and at a faster speed c_2 when they pass into the lower, more dense medium 2. Wave fronts in medium 2 are therefore accelerating with respect to those directly transmitted in medium 1, and at a point A (when the tangent to a particular wavefront in medium 2 is perpendicular to the boundary) the boundary wave at velocity c_2 is projected back upwards in medium 1 at velocity c_1. The ray vector SA thereby strikes the boundary at an angle i_c of critical incidence. From the construction at point B, a wave spreading from that point will travel a distance BE in medium 2 in the same time that a wave will travel a distance BC in medium 1. The wavefront in medium 1 will therefore take the configuration of line CE at the same angle i_c as before. It will be seen from Figure 7.8 that

$$i_c = \sin^{-1}\frac{BC}{BE} = \frac{c_1 t}{c_2 t} = \frac{c_1}{c_2} \tag{7.2}$$

For distances y from the source greater than a critical distance y_c, the wave that reaches D first will be the one that meets the 1–2 boundary at critical incidence i_c, travels along the boundary at speed c_2 and returns to the surface at critical angle i_c at speed c_1. At distances less than y_c, the first arrival will be the wave directly transmitted through medium 1 at speed c_1. These two discrete velocities are shown in the travel-time graph of Figure 7.8.

Either a hammer or explosive source can be used depending upon the depth of penetration required. A single detector can be moved at increasing distances y from the source for repeated impacts or an array of geophones, each geophone spaced at different distances, can monitor the arrival times for a single source impulse. It is also

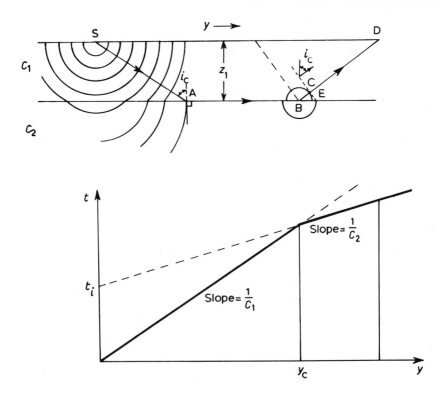

Figure 7.8 Refraction of waves – double layer case.

fortunate that in geotechnical exploration it is often information on the depth to rockhead through superficial drift cover that is required and in this case the condition of a high–speed marker bed overlain by a lower-speed formation is satisfied.

The characteristic two-layer equations, as derived for example by Milton Dobrin, are:

Arrival time

$$t = \frac{y}{c_2} + 2z \frac{(c_2^2 - c_1^2)^{1/2}}{c_1 c_2} \tag{7.3}$$

Ordinate intercept time

$$t_i = 2z \frac{(c_2^2 - c_1^2)^{1/2}}{c_1 c_2} \tag{7.4}$$

Depth of first layer

$$z_1 = \frac{t_i}{2} \frac{c_1 c_2}{(c_2^2 - c_1^2)^{1/2}} \qquad (7.5)$$

Critical distance

$$y_c = \frac{2z_1(c_2 + c_1)}{c_2 - c_1} \left(\frac{c_2 - c_1}{c_2 + c_1}\right)^{1/2} \qquad (7.6)$$

Once a travel-time record has been constructed, the characteristic layer velocities can be read off from the graphical slopes and an estimate made of the Young's Modulus (mass stiffness) values for the upper layer. With the close-in detector positions, it may be possible to detect both a first (P-wave, c_{P1}) and a second (S-wave, c_{S1}) arrival for layer 1 from which the following geotechnical parameters can be calculated:

Poisson's ratio

$$\nu_1 = \frac{c_{P1}^2 - 2c_{S1}^2}{2(c_{P1}^2 - c_{S1}^2)} \qquad (7.7)$$

Rigidity modulus

$$G_1 = \rho c_{S1}^2 \qquad (7.8)$$

Young's modulus

$$E_1 = \rho c_{P1}^2 \frac{(1+\nu)(1-2\nu)}{1-\nu} \qquad (7.9)$$

Bulk modulus

$$K_1 = \rho c_{P1}^2 - \frac{4}{3} G_1 \qquad (7.10)$$

where ρ is the density of the rock mass comprising the first layer.

In good seismic refraction work, the traverses will always be reversed; with horizontal bedding this serves as a check, but if the beds are thought, or known, to be inclined then this is mandatory practice. From the situation sketched in Figure 7.9, the following

456 Principles of Engineering Geology

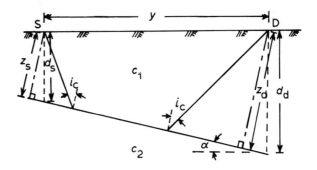

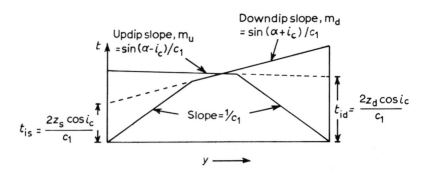

Figure 7.9 Reversal of seismic refraction profiling for evaluation of an interface dipping at α degrees.

critical relationships are established:

$$i_c = \tfrac{1}{2}(\sin^{-1}\overline{c_1 m_d} + \sin^{-1}\overline{c_1 m_u}) \tag{7.11}$$

$$\alpha = \tfrac{1}{2}(\sin^{-1}\overline{c_1 m_d} - \sin^{-1}\overline{c_1 m_u}) \tag{7.12}$$

$$c_2 = c_1/\sin i_c \tag{7.13}$$

$$z_s = \frac{c_1 t_{is}}{2 \cos i_c} \tag{7.14}$$

$$z_d = \frac{c_1 t_{id}}{2 \cos i_c} \tag{7.15}$$

$$d_s = z_s/\cos \alpha \tag{7.16}$$

$$d_d = z_d/\cos \alpha \tag{7.17}$$

These latter two parameters, d_s, d_d, together with the c_1 value are of basic geotechnical interest. Dynamic modulus values can be derived as suggested earlier.

If the bed separation is small and/or there is a strong penetration facility at source, multiple layer reflections are likely to be recorded and it may not always be easy to separate-out the different layers through discrete slopes on the travel-time graph. It should, however, be possible to fix a specific value for c_1.

Hammer seismographs usually incorporate their own data recording/processing facility but are suitable only for limited geotechnical surveys. More comprehensive systems using geophone arrays would use a direct-writing galvanometric recording system or, more usefully, a magnetic tape deck from which the information could be re-processed directly on-site or subsequently.

Geotechnical reliance would never be based exclusively upon such indirect seismic information. Rather, these refraction surveys would be used for information infill between exploratory boreholes put down to a comprehensive pattern.

One application where geophysical methods appear to have been very successful is the preliminary site investigation for the Deep Tunnel Project of the Metropolitan Sanitary District of Greater Chicago, described by Mossman *et al* (1973). Use was made of the Vibroseis oscillatory seismic system to map the top surface of two shallow rock strata in which extensive tunnelling was planned. The system was utilized in urban areas under conditions of quite intense interference and the results were shown by subsequent exploration to be relatively precise (±3 m, ±7 m). This type of area, with reasonably well-defined rockheads under glacial and lacustrine deposits, probably represents optimum conditions for the adoption of seismic exploration techniques.

7.6 Site exploration

There are no hard and fast rules for the *design* of site exploration programmes and for any consequent laboratory investigations of collected samples. The main requirement (previously quoted from the British Standard CP2001) is simply that the information collected should be sufficient to enable the preparation of an *adequate* and *economical* design. In practice, however, several accepted procedures have evolved to meet the analytical requirements of different types of structure.

In civil engineering, there are two main types of site investigation, each conditioned by the nature of the structure. These may be termed:

(a) *Foundation* (or compact) investigations where the structure is probably a building, dam, bridge or dock, and may exert a large force on a relatively limited area, and,
(b) *Linear* or extended investigations where the structure is probably a highway or tunnel and will exert a smaller force over an extensive area.

There are, in addition, some investigations of a remedial nature, such as coastal defence works, which may require a rather different approach, but the two quoted types are typical of the most common investigations and of the differing basic approaches demanded. Thus, in the case of a *foundation* investigation, a number of deep, closely spaced borings will be required to give an accurate and detailed picture of strata variations and properties to a depth at which the stresses exerted by the structure are minimal or until a stable, load-bearing stratum is reached.

This depth can be estimated beneath any regular foundation structure, such as a strip footing and a circular or rectangular pad from the Boussinesq equations (Boussinesq, 1885; Mayer, 1959) for the stresses mobilized by a force acting on an elastic halfspace. This topic is covered in most texts on elastic theory or on foundation engineering.

The assumption of elastic properties in compression is only a convenient approximation. For an irregular structure, the foundation stress can be obtained by the principle of superposition. This is based on the concept that the stress induced at a point beneath a structure is equal to the sum of the stresses induced at that same point by a number of forces acting at the centres of symmetical incremental units making up the structure. This procedure is often simplified through the use of influence charts (Newmark, 1942, 1947; Poulos, 1967; see also Capper and Cassie, 1969).

Since the stresses induced by a structure are greater near to the foundation/ground interface (Figure 7.10), it is vitally important that this area be carefully studied, and for *foundation* investigations it is good practice to take continuous borehole or trial pit (depending on depth) samples for testing and/or to carry out continuous *in situ* tests just below foundation level. If this level happens to coincide

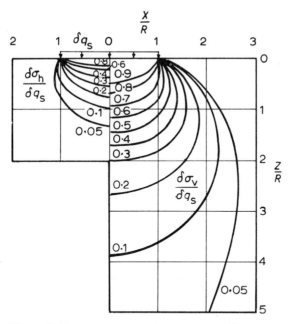

Figure 7.10 A sketch of horizontal ($\delta\sigma_h$) and vertical ($\delta\sigma_v$) stress increment contours induced by application of a uniform stress (δq_s) to a circular area (radius R) at the surface of an elastic half space ($\nu = 0.45$).

with rockhead, then rock samples should be taken. In addition, where feasible, frequent sampling at intervals less than 1 m should take place within the influence depth of the structure in some or all boreholes or pits. The actual depth of exploration boreholes is considered in detail by Smith (1970). Spacing of access holes will depend on the money available and on the complexity of the site, and except where symmetrical underground structures such as pillared mine workings exist, they should preferably follow some fixed pattern. Indeed, as has been noted by Knill (1974), much more thought and research needs to be devoted to the question of the controls on borehole location (geological or engineering) and spatial configuration (grids, straight lines, spirals, random and so on). The extent of the site investigation will be determined by the zone of influence of the structure and by the topography and the geological/ geotechnical properties of the surrounding ground. Foundation problems might also be expected from settlements in areas of old and

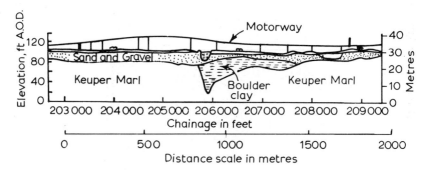

Figure 7.11 Section along the line of the London–Yorkshire M1 Motorway across the valley of the River Trent, showing a buried channel (*after* Rogers, 1970).

current mine workings and in areas of water (Davis *et al*, 1964), oil, or brine extraction.

In the case of *linear* investigations, closely spaced pits or borings and detailed test programmes will obviously be impracticable, and the site exploration must be designed principally to eliminate any weak points in the preliminary investigation. This means that the preliminary investigation must be meticulously carried out, probably utilizing aerial photography, geophysical techniques and detailed site inspection (these subjects are covered in more detail in the earlier sections of the present chapter). Borings and trial pits will then be required to confirm the proposed strata sequence, to obtain samples in sensitive areas and to identify accurately the ground water levels.

When new roads are taken across flood plains, the possible existence of buried channels must be investigated by special boring and geophysical site investigation programmes and excessive settlements of bridge foundations avoided. Rogers (1970) quotes the example (Figure 7.11) of a very deep boulder clay infilled channel which was located almost on the line of the present River Trent during the investigation for the M1 London–Leeds motorway.

A detailed examination of the site investigation procedures for *tunnelling* works in Great Britain is contained in a report by Attewell and Boden (1969). The importance of monitoring groundwater levels at the beginning and end of each drilling session and at each change of strata is emphasized. A very useful procedure on such linear investigations is for the drilling rigs to traverse the highway or tunnel route putting down primary holes and then to return along the route with a secondary drilling programme in order to resolve areas of stratigraphical or geotechnical difficulty. Investigation holes for tunnels should generally be taken to a depth of about one radius below the tunnel invert level and they should be offset alternately each side of the tunnel and at a distance from the tunnel centre line of about two radii in order to avoid water percolation into the tunnel from above or below during construction. Actual hole spacings will be very dependent on strata conditions, location and the importance of the tunnel. Reporting on an analysis of tunnelling contracts completed in Great Britain during the previous few years, Bevan (1972) has noted that these show an average borehole spacing of 150 m. In the case of bored tunnels beneath rivers, and in view of the cost element, the spacings will be greater, 500 m, for example, along the line of the River Tyne vehicular tunnel in Northern England. A very useful procedure in many instances is to seal a standpipe into selected holes when other tests have been completed. Changes in water level would be monitored for as long a period of time as possible, but before tunnelling operations begin, all holes should be sealed with concrete and capped at the top.

During tunnel construction, detailed mapping would be undertaken in the tunnel, a useful technique involving the production of numerous tunnel face cross-sections upon each of which are marked different strata horizons and the appearance of faults in the tunnel face. An excellent example of this type of presentation is shown in Figure 7.12.

The actual *methods* used in site exploration vary with the type of ground and the extent of the exploration. The options available for access to the ground for inspection include *trial pits, boring,* and in steep ground, *headings.* Trial pits tend to be used only for exploration at shallow depths in soils where they can give an accurate impression of the strata lithology, whilst boring is invariably used for a deeper and more extensive exploration in soils and for any subsurface exploration in rocks. Available methods of drilling ranging from manually operated post-hole augers for short holes in soft soils to diamond drills in hard rocks are described extensively in the literature (see for instance, McGregor, 1968). The most commonly encountered drills in engineering geology are the percussive shell and auger drill used for exploration and sampling in clays and sands (Figure 7.13) and the rotary mud-flush drill used for exploration in rocks, and in some soils where speed and accuracy are required (Figure 7.14). Samples in the former case can be obtained using

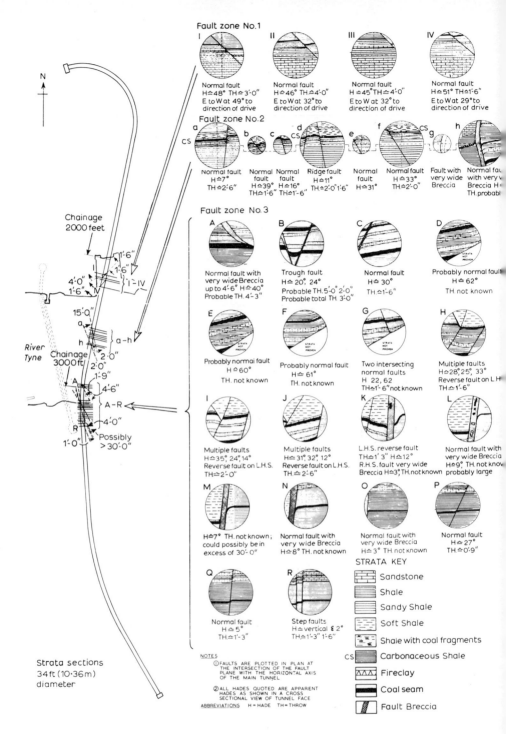

Figure 7.12 Presentation of strata and fault changes at the Tyne Tunnel face as observed during construction (*re-drawn and re-arranged from the original plan, by courtesy of* Messrs Mott, Hay and Anderson, Consulting Civil Engineers, *and of* the Tyne Tunnel Joint Committee).

Figure 7.13 Shell and auger rig (*photograph by courtesy of* Cementation Ground Engineering Ltd.— Soil Mechanics Dept).

various types of tube or shell sampler and in the latter case by the use of a core barrel.

A very desirable development in recent years has been the more extensive use of large diameter holes or shafts which permit direct strata inspection, together with sampling and *in situ* testing (see for example Knights, 1974). Such shafts are much more expensive to construct and protect and so they would be used only at critical points of uncertainty in the site investigation. A typical use of one to two m diameter inspection holes might arise during the investigation for a metropolitan rapid transit system in tunnel. In bad ground, a cased shaft would incorporate windows for inspection and sampling and larger windows for lateral access and testing. In the tunnel

Figure 7.14 Rotary drill (*photograph by courtesy of* Cementation Ground Engineering Ltd, Soil Mechanics Dept.).

investigation, two small but identical headings might be driven either side the access shaft at axis level. One type of support or ground treatment method might be monitored in one of the headings while a different support system or ground treatment might be tested in the other heading. It is also often thought to be good practice in tunnelling schemes to put down ventilation shafts (communication tunnels) or access shafts (sewage tunnels) early in order to probe the ground in advance of tunnelling. But these shafts represent a very early capital investment in the ground and they will not begin to play their part in earning revenue to set against interest charges until the scheme is completed.

The *end product* of the site exploration is twofold. First, through careful pit examination or borehole logging, we have the compilation of a series of *sections* or borehole logs. The second requirement is the collection of *samples* for laboratory testing and/or the compilation of results of *in situ tests* in order to present a mechanical property index alongside the lithological description.

7.7 Borehole logging

A major part of any site exploration, apart from sampling, is the building up of an accurate *borehole log* on which subsequent plans and sections will be based. A borehole log should provide an accurate and comprehensive record of the stratigraphy and lithology of the soils and rocks encountered in the borehole, together with any other relevant information obtained during drilling.

The type and accuracy of the log will depend upon the method of drilling and the type of material drilled. A typical borehole log should contain the following information (based on the Geological Society of London (1970) Working Party Report and British Standard CP2001):

(1) Borehole *identification*.
(2) Details of contract, site, grid reference, elevation and, if applicable, orientation and inclination.
(3) Method of *drilling* and sampling and details of equipment.
(4) Record of drilling *progress* with particular emphasis on the location of coring runs, samples, or *in situ* tests; the use, extent and diameter of casing; and detailed information on the rate of drilling interpreted as resistance to penetration or breakdown.
(5) Detailed, descriptions of *ground water* level, changes in standing water level, loss of water (in water flush drilling) or gain of water (in air flush drilling).
(6) Detailed *geological* description based on both a simplified description during drilling and on core or sample examination in the laboratory. This examination will include a systematic description of the soil or rock material (degree of weathering, micro-structure, colour, grain size, alteration) and correct nomenclature in the case of rocks or overconsolidated soils. There will be estimates of core fracturing, core recovery, RQD, discontinuity spacing and orientation in complete cores, and probably some simple field strength tests (see Rankilor, 1974).

Table 7.3 Rock Quality Designation

RQD (Percentage recovery > 100 mm lengths)	Description
0– 25%	Very poor
25%– 50%	Poor
50%– 75%	Fair
75%– 90%	Good
90%–100%	Excellent

The RQD or *Rock Quality Designation* system described by Deere et al (1967) is based on the premise that if the joint spacing in a rock mass is sufficiently wide, then the deformation modulus of the mass will tend to that of a small intact specimen. So, as RQD approaches 100 per cent, the field/laboratory modulus ratio approaches unity. The procedure is as follows:

(a) Having put down a cored borehole in rock and retrieved the core, sum the total length of core recovered by counting only the lengths which are 4 inches (100 mm) or longer and which are hard and sound. It is necessary to distinguish between those fractures caused by the drilling operation and poor handling and those that are true *in situ* fractures. The former type of fresh, fine, irregular breaks should be ignored and the pieces counted as intact lengths. On the other hand, imposed fractures along strong anisotropic planes (mica banding in foliated rocks, slaty cleavage, or mudstone fissility) would probably be counted in RQD if it is felt that a particular engineering operation would exploit them.
(b) Divide this sum by the total length of coring run to get the percentage core recovery.
(c) Describe the rock in the terms of Table 7.3.

It can readily be appreciated that the quite arbitrary choice of a 4-in (100 mm) standard could lead to problems where the joint spacing just happened to be an integral multiple of 4 in. There is, therefore, some merit in considering a length distribution, much on

Figure 7.15 Example of a borehole log in till, sandstones and mudstones. (Geological Society of London, Working Party Report, 1970; reproduced by permission of the Geological Society of London).

DRILLING METHOD	GROUND LEVEL	CO-ORDINATES	BOREHOLE NO.
Shell and auger to 4.80m Rotary core drilling, water flush to 25.00m	+43.63m O.D.	7268/5423	**30**

MACHINE	CORE BARREL AND BIT DESIGN	ORIENTATION	SITE
Pilcon '20' and B.B.S 10, truck mounted	F design barrel, diamond bit	Vertical	CASTLECARY DEVELOPMENT 'C', GLASGOW

SOIL SAMPLES DEPTH AND TYPE	DRILLING AND CASING PROGRESS	WATER RECOV. (%) & A.M. LEVEL	R.Q.D.	CORE RECOV % & SIZE	DESCRIPTION OF STRATA	O.D. LEVEL	SYMBOLIC LOG
0.50-0.96 U(10) 0.96 D 2.00-2.46 U(10) 2.46 D 3.50-3.96 U(10) 3.90 W 3.96 D 4.50 D		22 ▽			Stiff, becoming hard, brown silty CLAY with occasional cobbles and boulders (Till) 4.80	38.83	
PERMEABILITY cm/sec x 10⁻⁵ 10 20 30 40 13.2 3.5	22.3.67 22.3.67	23 ▼ 24 ▼		SF 9 HWF 12	Thick bedded pale grey and brown coarse strong SANDSTONE with fine pebbles and conglomerate bands. Steep clay lined joints 6.00 to 7.50m. Dark brown fine conglomerate 9.05 to 9.60m. Mudstone flakes at 10.15m. Very coarse at base (KIRKHILL SANDSTONE) 11.00	32.63	
				8	Thin bedded grey moderately weak MUDSTONE, sandy to 12.40m, with ironstone nodules throughout 14.00	29.63	
30.2	23.3.67			15	Medium bedded grey fine strong SANDSTONE becoming laminated 17.50 to 18.70m (Borehole continued to 25.00m)		

KEY: U(10) - 0.1m dia. undisturbed sample
D - disturbed sample ═══ Casing depth
W - water sample ─── Borehole depth
2 - day
▽ - ground-water depth first encountered
▼ - morning water level
② - rate of penetration (mm/min)

REMARKS
Borehole chiselled 1.05 to 1.90m, 4.0 to 4.45m.

LOGGED BY: M. Jones	SCALE 1/100		
BLOGGS BROS. INC.		CLIENT STRATHCLYDE CITY CORPORATION	REF. MJ/7964/30 FIG.I

468 *Principles of Engineering Geology*

the lines proposed for field inter-discontinuity measurements considered in Chapter 6. Indeed in Section 6.14 it is shown how spatial distributions of discontinuities as measured along scan lines in the field may be expressed in equivalent RQD terms.

Examples of borehole logs are illustrated in Figures 7.15–7.17. The format is usually tabular, with a section at the top for identification. The tabular description is usually divided into two parts, the left hand side giving details of drilling and casing progress, samples and tests, water levels and core recovery data, while the right hand side provides a detailed description of the *strata succession* and a symbolic illustrative log.

Since the log is essentially an observational description of the exploration process, the major part is the lithological/stratigraphical description. The symbolic log is drawn to scale using a standard petrographic (or soil) symbol to emphasize and represent the detailed description. These symbols are summarized in the Geological Society Working Party Report (1972) on the preparation of maps and plans in engineering geology and are reproduced in Figure 7.18. They can also be used in sketch maps and plans and in trial pit logs (Figure 7.19). They have the principal advantage of a certain simplicity and are clear and distinctive, whilst the symbols for soils and sediments can be combined for composite types.

The drilling information is mainly observational, although some simple test data, mostly of an *in situ* nature and usually involving penetration or permeability tests, can be included. The main purpose is to index, through core or sample locations, detailed analyses later in a completed report, and to provide information, only obtainable during drilling, from which a broad picture of the competence and permeability of the ground can be built, and from which relatively accurate information on groundwater conditions and local aquifer systems can be obtained.

A useful descriptive aid may take the form of a series of colour polaroid prints of split borehole cores or undisturbed samples. Additional palaeontological or micropalaeontological information may be required where engineering properties of specific horizons can be related to palaeontological data (see for instance, Hutchinson and Hughes, 1968).

Figure 7.16 Example of a borehole log in clay, sandstones and granites. (Geological Society of London, Working Party Report, 1970; reproduced by permission of the Geological Society of London).

DRILLING METHOD	GROUND LEVEL	CO-ORDINATES OR GRID REF.	BOREHOLE NO.
Rotary auger to 5.40m Rotary core drilling water flush to 17.60m	+401.80m O.D.	NL 6354/3482	**52**
MACHINE	CORE BARREL DESIGN AND BIT	ORIENTATION	SITE
BBS 10 (truck mounted)	F design barrel diamond bit	Vertical	OXBRIDGE DEVELOPMENT GREEN LANE, OXBRIDGE

WATER PRESSURE TEST cm/sec x 10^{-5} 1 10 100	WATER RETURN % & LEVEL 20 60	DRILLING PROGRESS	CASING	DISCONTINUITIES	FRACTURES per m 4 16	CORE SIZE AND RUNS	CORE RECOVERY % 20 60	DESCRIPTION OF STRATA	O.D. LEVEL	LOG
	13 ▽						1 2	Stiff dark yellowish brown (10YR 4/2) silty CLAY with occasional cobbles and boulders (Boulder Clay)		
		12.7.68					3 4 5 6 7			
								5.40	396.40	
					SF		8	Faintly weathered thick bedded yellowish brown (10YR 5/4) medium grained strong SANDSTONE		
	14 ▽ 8.1			Haematite stained rough tight small fissures	⊥ V		9			
		13.7.68		Fairly rough clay filled but open joint	∠			8.40	393.40	
0.7	15 ▼			Clean rough tight bedding plane fracture	∨	HwF	10	Slightly weathered thick bedded yellowish brown (10YR 5/4) medium grained moderately strong SANDSTONE with silty clay seams		
				Shattered zone 0.20m wide	∠		11	11.25	390.55	
	37			Fault zone (a)	∠			11.70 Highly weathered light grey (N6) coarse weak GRANITE	390.10	
				Many clean rough open joints	∠ ∠		12			
		14.7.68		Limonite stained slightly rough open prominent joint	V		13	Faintly weathered light grey (N6) coarse very strong biotite GRANITE		
8				Limonite stained slightly rough open prominent joint	∠		14			
				Shattered zone				15.50	386.30	
4.3				Clean slightly rough open prominent joint	V		15	Faintly weathered thick flow-banded light grey (N6) coarse extremely strong biotite GRANITE		
		15.7.68					16	17.60	384.15	
								Bottom of hole		

EXPLANATION:
- ☐ U4 sample
- • Disturbed sample
- ▮ Core sample
- W Water sample
- 22 Day
- ▽ Ground-water depth first encountered
- ▼ Morning water level
- 12.7.68 Depth of borehole
- ⊥ 80° – 90°
- ∨ 60° – 80° } Attitude of
- ∠ 30° – 60° } prominent fractures
- ∼ 0° – 30°
- --- Solid core recovery
- —— Total core recovery

REMARKS: Rock colours and colour index numbers (in brackets) are according to the "Rock Colour Chart" published by Geol.Soc. of Amer.

LOGGED BY: A. Smith   SCALE: 1/100

CONTRACTOR: JONES INTERNATIONAL   CLIENT: WESTSHIRE WATER BOARD   REF.NO. J1/498/52   FIG.3

SITE	SHEET 2 OF 4	BOREHOLE NO.
HADDLESHAM POWER STATION		17

INSTRUMENTATION	ROCK GRADE	DRILLING AND CASING PROGRESS	WATER RECOVERY MORNING LEVEL	CORE RECOVERY R.Q.D.		DESCRIPTION OF STRATA	O.D. LEVEL	SYMBOLIC LOG
Piezometer installed tip at 30.75m in pea gravel. Clay-bentonite grout from 29m.	III	18.9.66	▼ 19	52c 40r	HW 20.40	Highly weathered medium bedded red medium grained weak felspathic SANDSTONE	+5.10	
	II	18.9.66		58c 45r	MW	Moderately weathered medium bedded maroon medium grained moderately strong felspathic SANDSTONE. Occasional open joints with porous zone from 21.3m to 23.8m. Bedding at 5° to 10°		
				65c 50r	MW 26.20		-0.70	
	III			60c 54r	FW 28.80	Faintly weathered thick bedded maroon coarse-grained strong SANDSTONE	-3.30	
	IV			30c 0r	F 29.80	Thin laminated red weak MUDSTONE. Bedding 0°	-4.30	
	I			100c 100r	F 30.80	Thick bedded white strong ANHYDRITE	-5.30	
	IV			40c 22r	F 31.80	Laminated red very weak MUDSTONE	-6.30	
	II	19.9.66 20.9.66 20.9.66	▼ 20 ▼ 21 ▼ 22	80c 70r 70c 45r 79c 60r	MW 37.20	Moderately weathered medium to thick bedded maroon medium grained moderately strong SANDSTONE. Closely jointed from 29.4m to 30.2m. Bedding 4° to 12°. Occasional tight joints	-11.70	
	IV			60c 14r 40c 15r	FW	Faintly weathered thin bedded maroon weak MUDSTONE. Bedding 3°		
						(Borehole continued to 79m)		

REMARKS
c = % core recovery
r = % R.Q.D. (not plotted graphically)
▬▬▬ Casing depth
——— Borehole depth

Rock grade classification defined in text.

SITE EVALUATION LTD. NATIONAL ELECTRICITY COMPANY

REF. SE/5981/17
FIG. 4

A further approach to logging is known as well-logging or wire-line logging where various types of probe are lowered down the completed borehole to obtain information on groundwater or simple rock properties. This type of investigation can be useful, where core recovery is incomplete or unattempted, in establishing or confirming lithological boundaries. A summary of methods is included in a paper by Johnson (1968). The method is an established and expensive commercial process usually only justified in deep holes.

A typical example of this type of equipment is the seismic logger, described by Deere *et al* (1969) and designed for rapid borehole transit (10 to 30 m min^{-1}). The pulse is generated by means of a magnetostrictive transducer housed at one end of the instrument rod and is received by piezo-electric crystals at the other end of the rod after travelling through the rock adjacent to the borehole (Figure 7.20). En route from transmitter to receiver, the pulse (strictly train of pulses) undergoes several mode-conversions and, in fact, three waves, compressional, shear and boundary, are received for processing. Compressional and shear wave velocities cannot be computed directly from the arrival times, but for their solution additional information on the borehole diameter, rock density, and depth of energy penetration into the rock is required. Diameters are measured using a borehole caliper system and densities are assessed using gamma ray back—scattering methods. Once the true compressional and shear wave velocities have been determined, then the various rock modulus values at different positions along the borehole can be calculated from standard equations quoted by Deere *et al* (1969).

The system requires a fluid coupling through which to transmit to, and receive from, the rock. Obviously the hole cannot be lined nor can any drilling mud or grout be used for support. It follows that for the technique to be successful, the rock must be sufficiently strong and competent to stand without artificial support.

Another form of remote logging is through the use of still, ciné or television cameras designed to be lowered through an existing borehole that is dry or which contains clear water. Borehole cameras

---

Figure 7.17 Example of a borehole log in sandstones, mudstones and anhydrite (Geological Society of London, Working Party Report, 1970; reproduced by permission of the Geological Society of London).

(a) Sedimentary rocks

 Conglomerate

 Breccia

 Sandstone

 Siltstone

 Mudstone

 Shale

 Limestone

 Chalk

 Dolomite

 Chert

 Rock salt

 Gypsum

 Anhydrite

 Coal, Lignite

 Gravelly sandstone

 Silty sandstone

 Clayey sandstone

 Clayey siltstone

 Silty mudstone

 Sandy mudstone

 Oolitic limestone

 Dolomitic limestone

 Argillaceous limestone

 Cherty limestone

(b) Soils

 Gravel

 Sand

 Silt

 Clay

 Boulders, Cobbles

 Shells

 Peat

 Sandy gravel

 Boulder clay

 Silty clay

 Clayey silt

 Gravelly sand

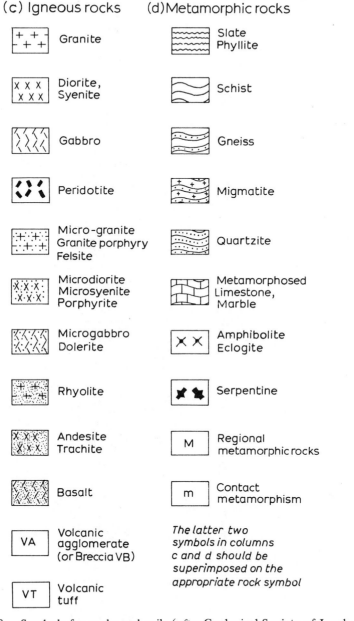

Figure 7.18 Symbols for rocks and soils (*after* Geological Society of London, 1972).

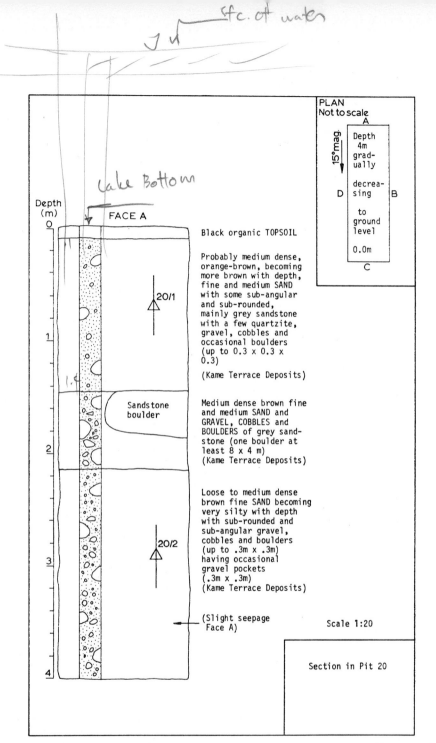

Figure 7.19 Example of a trial pit section in sands and gravels, (Geological Society of London, Working Party Report, 1970; reproduced by permission of the Geological Society of London).

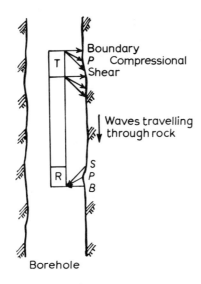

Figure 7.20   3-dimensional sonic logger (Deere *et al*, 1969 *after* Christensen, 1964).

can usually produce a 360° photograph in monochrome or colour of the interior of a borehole (see, for example, Burwell and Nesbitt, 1954) but there are problems associated with the collapse of soft rock into the unlined hole and also with the time delay required for the hole to be flushed with clean water. *Television cameras* can be used to transmit pictures of fracture orientation and spatial distribution in boreholes to the surface during a continuous traverse by means of closed circuit television, the picture being stored on video tape if required. An example of the use of such a camera is described by Dodd (1967), and Deere *et al* (1969) state that fractures less than 0.3 mm wide can be detected.

7.8 Sampling and testing

Sampling and testing in engineering geology are inextricably inter-related, for in a variable material that is defined through a largely empirical science, much information obtained through testing may be useless and some may be actually misleading unless the test sample is representative of the rock or soil as a whole. Accurate representation is plainly impossible in the case of many small scale (usually 4 in or 100 mm) core samples obtained for testing and this is the reason why a detailed description of the structure and fabric of a rock or soil is often more important in a final site investigation report than are isolated test data.

Some factors affecting borehole location and sample selection have been considered in previous sections. Although it would be desirable to base sample selection on rigorous statistical analysis (see Section 7.16), this is not always possible in a site investigation process involving elements of exploration. Dumbleton and West (1974) suggest several more flexible approaches to sampling which can be listed as:

(a) *Purposive* sampling, where the aim is to supply answers to specific questions, and sampling is concentrated on points where, in the judgement of the investigator, the most important information can be obtained.
(b) *Search* sampling, aimed at the location of features of specific interest, such as cavities or landslip areas isolated from a preliminary geological survey.
(c) *Sequential* sampling, in which a sampling plan is developed as information from preliminary investigations is processed.

A useful check list suggested by Dumbleton and West (1974) is reproduced in Table 7.4.

Samples obtained during boring or from trial pits are usually described as *disturbed* or *undisturbed* depending on how well the original structure of the material has been preserved during sampling. Rock core samples are considered to be undisturbed (although probably fractured by the torque exerted on the core during drilling in weaker rocks) whilst flushed drilling debris may be described as disturbed. For soils, undisturbed samples obtained in tubes from shell and auger soft-ground drilling are quite easily obtained in clays, but only with difficulty in relatively cohesionless soils. In particular, undisturbed samples of cohesionless soils from below the water table can only be obtained with difficulty.

Disturbed samples are useful only as a means of assisting in providing a lithological description of the strata, mineralogical and geochemical identification, and some simple index properties. In the case of rock these would be confined to irregular *particle strength* tests (see Hobbs, 1962, Franklin *et al*, 1971) and possibly durability. In soils the programme might include the *Atterberg limits*, particle *size* and shape analyses, *compaction* and other *aggregate* tests and

Table 7.4 Sampling in site investigation (*after* Dumbleton and West, 1974)

Sampling decision	Sampling method
1. Number of sample points	Determined by complexity of site, precision required and funds available
2. Location of sample points	Purposive, search or sequential sampling
3. Sequence of examination of sample points	Priority to points of major constructional expenditure and those crucial to geological interpretation
4. Location of material samples in boreholes or pits	At each change of strata and at specified depths (say 1 m) in each layer
5. Choice of sampling method	Drilling and coring method suited to materials and conditions
6. Selection of sample	Quartering or rifling where possible. Otherwise judgement
7. Selection of samples for laboratory testing	Base selection on geological formation or soil classification
8. Selection of test specimens from sample	Quartering or rifling where possible
9. Sequence of testing and selection of test operators	Random

possibly in some cases *remoulded strengths* where these can be related to undisturbed strengths.

Undisturbed samples, on the other hand, can be described as the raw material of soil or rock mechanics and form the basis for most of the standard tests mentioned in Chapters 2 and 4. The actual tests in all cases will depend on the type of soil or rock, its depth and the stress situation likely to be imposed by the resulting structure. Specific tests will usually concentrate, in the case of rock, on *bulk density, moisture content, porosity, permeability, strength characteristics* and *durability* (or weathering properties), based on something like the *slake durability* test. This latter is a very empirical test described by Franklin *et al* (1971) and originally developed by the British National Coal Board in its coal washery breakdown research programme. The test is designed to estimate the resistance of argillaceous rocks in particular to a combination of abrasion and moisture exposure (Badger *et al*, 1956).

In soils, emphasis will be on *moisture content, porosity/void ratio, particle size* and *plasticity, bulk density* and *specific gravity* and *strength* or *consolidation* characteristics under various stress conditions. *Permeability*, as in the case of rock, will also be important, but in soils having some degree of cohesion, an accurate value is unlikely to be acquired on scaling down to laboratory test size. *In-situ* tests must invariably be used to determine accurate levels of permeability. These tests will be considered in Sections 8.5 and 8.6.

Because of the major influence imposed by water on the geotechnical properties of soil and rock it is important when sampling to preserve the material at its natural moisture content during transport between site and laboratory. Preservation does not present a serious problem in the case of tube samples which are usually sealed at each end with wax and in turn are then protected by screw-on end caps. In the case of core samples, various methods recommended by the American Petroleum Institute include metal cans, plastic bags, metal foil, wax and dry ice for freezing. A useful survey by Stimpson *et al* (1970) suggests the use of polyurethane foam injected around continuous lengths of core in a pitch-fibre pipe. This is shown to serve the dual purpose of shock insulation and waterproofing.

In addition to moisture content, *chemical analysis* of the groundwater is important, particularly where salinity or the presence of corrosive effluents is suspected. Sulphates and acidic waters in significant quantities can have a deleterious effect on concrete. Acids, bacteria and oxidising agents will affect steel foundation structures.

Where the problem of scale or sampling proves insoluble and where a realistic assessment is paramount for an important structure, then *in-situ* tests can often prove effective. These tests can be divided approximately into two classes. The first comprises tests designed to provide substitute information for laboratory testing when sampling is not considered feasible in cohesionless soils or soft clays. The most common of these are *penetration* tests (cohesionless soils) and *vane* tests (soft clays). The second type of test consists of those designed to overcome problems of scale. These include *plate bearing* tests and *pile* tests designed to simulate or repeat actual foundation situations, some forms of *shear* test in hard rocks, and *pore pressure* measurements and permeability tests designed to investigate groundwater and groundwater flow conditions.

The most widely used *penetration test* is the *standard penetration test* in which a standard 2 in (50 mm) split-spoon sampler (Figure 7.21a) is driven into the ground by a series of blows from a 140 lb (64 kgf or 628 N) weight falling through a height of 30 in (0.76 m). In gravels and rocks, a 30° half angle cone is inserted into the sampler. During the test, the sampler is driven to its full depth of 18 in (0.46 m) and the number of blows required to penetrate the final 12 in (0.30 m) are recorded. The number of blows per foot, or the *N*-value, can be quite closely calibrated (see Peck *et al*, 1953)

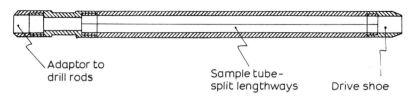

(a) Standard 2 in split tube sampler for Standard Penetration Test (SPT)

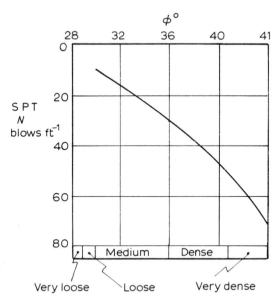

(b) ϕ-N relationship for Standard Penetration Test (after Peck *et al.* 1953)

Figure 7.21 Standard Penetration Test.

against the angle of shearing resistance ϕ (Figure 7.21b) and against Terzaghi's bearing capacity factors N_γ and N_q.* Penetration may be calibrated against *relative density* (see Section 2.2), and in exceptional cases against the unconfined compressive strengths of clays (see Table 7.5). In general, however, vane tests are considered to be a superior method of measuring the strength characteristics of clays where S.P.T. results may be misleading.

Another penetration test which is widely used in practice is the *static sounding* method typified by the *Dutch Cone test* (Figures 7.22, 7.23a), the equipment for which comprises a solid cone with a 30° half angle and a 0.001 m² cross sectional area. This is thrust into the ground by hydraulic pressure through a cylindrical casing for a distance of 100 mm at a rate of 1 m min⁻¹, and the resistance (usually up to a maximum of 10 tonnes with conventional thrust rigs) is recorded. Cone resistance measurements (the casing, which is pushed forward between each measurement, generates additional resistance which can be related to wall or skin friction) are normally obtained at 200 mm intervals and plotted as a depth/resistance log (Figure 7.23b). There is little relationship between this log and the SPT log. However, it is sometimes assumed that N is equal to 0.25 per cent of the static cone resistance expressed in kN m⁻², or one quarter of the static cone resistance expressed in bars. Cone resistance can be related directly to shear resistance and can also be used to determine foundation bearing capacities (see Tomlinson, 1969; de Beer, 1958; and Meyerhof, 1956). An approximate formula for shallow foundations is

$$q_a = C_R/40 \qquad (7.18)$$

---

*Terzaghi's bearing capacity factors are not covered in detail in the present book. They are derived from an empirically modified solution of an active-passive wedge failure model under a strip foundation giving the ultimate bearing capacity Q_{ult} for a foundation on cohesionless soil as:

$$\frac{Q_{ult}}{B} = \frac{\gamma B}{2} N_\gamma + q_s N_q$$

where B is the foundation width,
  q_s is the overburden pressure at the foundation base and
  N_γ, N_q are bearing capacity factors given approximately by:

$$N_\gamma = \tfrac{1}{2}(K_p^{5/2} - K_p^{1/2})$$
$$N_q = K_p^2$$

where K_p is the coefficient of passive earth pressure equal to $\dfrac{1 + \sin \phi}{1 - \sin \phi}$

The derivation and modifications for square, rectangular and circular foundations and for cohesive soils are given in virtually all soil mechanics texts.

Table 7.5 Relationship between standard penetration, density and strength (*after* Terzaghi and Peck, 1967)

Soil	Penetration Resistance N blows ft^{-1}	State	Relative Density percentage	Unconfined Compressive Strength kN m^{-2}
Sand	0– 4	Very loose	0– 15	
	4–10	Loose	15– 35	
	10–30	Medium	35– 65	
	30–50	Dense	65– 85	
	>50	Very dense	85–100	
Clay	< 2	Very soft		< 25
	2– 4	Soft		25– 50
	4– 8	Medium		50–100
	8–15	Stiff		100–200
	15–30	Very stiff		200–400
	>30	Hard		>400

where q_a is the allowable bearing pressure
and C_R is the cone resistance expressed as force per cone cross sectional area.

Static sounding is essentially a small-scale *plate test*. It works particularly well in normally– or lightly–overconsolidated soils where the soil fabric can affect its overall mechanical behaviour. In overconsolidated soils, where the soil macrostructure has increasing geotechnical importance, there is a stronger case for a larger-scale test. This is particularly important where bearing capacities are being determined since the term *allowable bearing capacity* in foundation design implies a limit on settlement as well as on load.

In this case a plate of equal area to the foundation and loaded to one and a half times the allowable bearing capacity to encompass a margin of error will be desirable in order to check design assumptions. Similarly, with deep foundations, a full-scale pile test is usually necessary. The test methods and the implications when plates in particular are scaled down are discussed by Tomlinson (1969).

Vane tests can be carried out in the laboratory but are of more value in the field. They are similar to static sounding or penetration tests, the uniform cruciform vane being driven into undisturbed soft clay at the base of a hole cased for support. The vane (usually 150 mm deep x 75 mm diameter – BS 1377) is rotated by a torque applied to the rods, so creating a shear surface in the soil at the vane

Figure 7.22 Static penetration test in progress (*photograph by courtesy of* Cementation Ground Engineering Ltd, Soil Mechanics Dept.).

perimeter. The torsional strain is applied at a uniform rate and the undrained shear resistance c_u is given by:

$$c_u = \frac{t}{\pi \left(\frac{d^2 h}{2} + \frac{d^3}{6} \right)} \qquad (7.19)$$

where t is the torque at failure,
       d is the vane diameter,
and h is the vane depth.

If the vane motion is continued after peak torque, then lower torque values may be related to a remoulded or possibly residual strength for the day. For further reading, refer to Aas (1965).

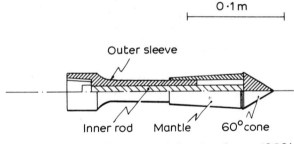

(a) Sketch of Dutch Cone (after Tomlinson, 1969)

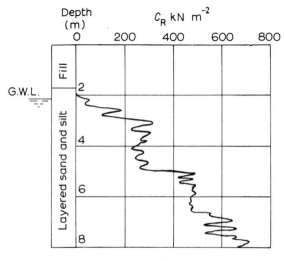

(b) Static sounding record

Figure 7.23 Static penetration test.

7.9 Site investigation reports

Site investigation reports should comprise a complete record of the investigation and all borehole logs, tests and hydrogeological information, prefaced by a precise introduction comprising notes on any special field or laboratory tests, a description of major topographical and geological features and a general description of soil and groundwater conditions. The report may vary at the instigation of the client, from a brief attempt to identify sensitive or anomalous areas, to a wide ranging introduction to structural design. The most important part of the report, however, will be the actual results clearly and objectively presented and these will usually include: (a) a

484 *Principles of Engineering Geology*

table summarizing groundwater levels; (b) a table summarizing chemical analyses of groundwater samples; (c) a table summarizing laboratory test results; (d) a detailed and complete record of each borehole with summary sections, type of strata, sample and *in situ* test positions and results; (e) detailed results of laboratory tests and particle size analyses where relevant; (f) plans and sections showing the positions of boreholes and any important geotechnical features.

7.10 Mechanical tests *in-situ*

Many of the mechanical tests conducted on site are very expensive items in the site investigation budget. They will therefore be reserved for critical situations where the capital cost of the civil engineering project is high. The most obvious example is that of an arch dam where the high abutment pressures must be resisted by competent rock with minimum deformation. Probably a less obvious example is that of a nuclear power station foundation. In this case, a strong rock foundation is required to carry the very high loads from the reactor cores and, due to the nature of the construction of the adjacent structures, the loads must be supported with minimal creep over the design life of the station (a maximum tolerable deformation of about 40 mm over a design life of, say, 30 years).

Within the site investigation budget, there is scope for only a few expensive tests and results from them must be correlated with the more numerous and much less expensive laboratory test results. This correlation is, in effect, the establishment of a relationship between rock or soil mass parameters and the equivalent intrinsic parameters for the rock or soil material, which is introduced in Chapter 6.

Plate jacking tests are probably the best-known of the larger-scale mechanical field tests on rock (see, for example Rosa *et al*, 1964). The idea is to load an area larger than the average spacing of the discontinuities, noting the displacements and calculating a modulus of deformation. Sometimes (Talobre, 1957) a quite small rigid cast iron or steel plate of about 10 in (0.25 m) diameter is used to transfer the load to the prepared surface of the rock. In these cases, a Boussinesq rigid (elastic) punch solution gives a mass deformation modulus as

$$E_m = \frac{P(1 - v^2)}{\delta d} \qquad (7.20)$$

where P is the total force applied to the plate, v is Poisson's ratio of the rock, and δ is the measured displacement of the plate, diameter d. Reference may be made to Figures 3.30 and 3.31.

The shearing stresses at the edge of the small diameter plate strongly influence the deformations and so larger diameter plates of up to 76 cm have been recommended (Waldorf et al, 1963) with 80 cm plates having been used in Russia (Sapegin and Shiryaev, 1966). Loads will usually be cycled at increasing stress levels and a secant modulus will usually be taken from the hysteresis curves at the stress level of design interest. The plates will usually incorporate a central hole through which the rock to be tested will have been originally cored. Such a hole will permit measurements of deformation to be taken at varying depths beneath the stressed rock surface and, in fact, the deformation modulus can be calculated as a function of depth from these deformations:

$$E_d = \frac{q}{\delta_z} \left\{ (1-v)z^2 \left[\frac{1}{(a_2^2 + z^2)^{\frac{1}{2}}} - \frac{1}{(a_1^2 + z^2)^{\frac{1}{2}}} \right] \right.$$
$$\left. + 2(1-v^2)[(a_2^2 + z^2)^{\frac{1}{2}} - (a_1^2 + z^2)^{\frac{1}{2}}] \right\} \quad (7.21)$$

In this equation, q is the pressure uniformly applied to the plate annulus, v is Poisson's ratio for the rock, a_1 is the radius of the central hole in the plate, a_2 is the radius of the plate, z is the depth of measurement from the directly loaded surface, and δ_z is the displacement at this depth due to the load.

In the equipment shown in Figure 7.24, the pressure is applied to the rock through the Freyssinet flat jacks with the adjustable screw tubes acting as reaction members. Dial gauges are used to monitor the rock movement. A slightly different type of plate jacking system used by the Laboratório Nacional de Engenharia Civil in Lisbon for testing in foundation galleries at arch dam sites is shown in Figure 7.25. Both these systems and the one used at the Karadj Dam in Iran (Waldorf et al, 1963) are designed for use in galleries where there is a natural reaction support by an opposite surface of rock. They could equally well be used in access shafts. For jacking at ground surface, very heavy kentledge (dead loading) would be required for the reaction in order to achieve the necessary deformations.

As the pressure is increased per cycle, the trend of the pressure: rock deflection envelopes will usually be concave downwards (for

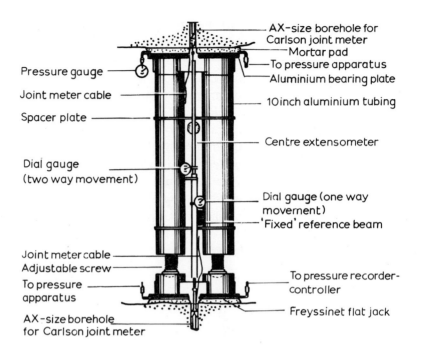

Figure 7.24 Plate jack test apparatus used at Dworshak Dam, U.S.A. (*after* Shannon and Wilson, 1964 and *after* Deere *et al*, 1969).

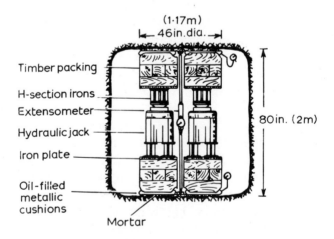

Figure 7.25 Plate jack test as used by the Laboratório Nacional de Engenharia Civil, Lisbon (*after* Rocha *et al*, 1955).

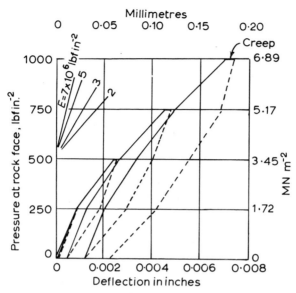

Figure 7.26 Concave – downwards pressure – deformation curve from a plate jacking test (*after* Shannon and Wilson, 1964; *after* Deere *et al*, 1969).

example Figure 7.26 from Deere *et al*, 1969) but sometimes there is a 'work–hardening' tendency as friction is rapidly mobilized across discontinuities in the rock and the mass-modulus tends towards that of the intact rock. The types of deformation are discussed by Lane (1966).

A useful procedure for a site investigation under a very heavy structure such as a nuclear power station is to excavate trial pits, mapping the geology and structure with depth and also conducting both horizontal and vertical plate loading tests for each increment of depth in order to deduce the deformation modulus anisotropy. Three loading cycles should be attempted in each case and the load-deformation curves condensed to those representative of a bi-linear elastic solid by taking two linear approximations to the loading leg of each cycle (see Figure 7.27). The break-point on each loading would be fixed at the design stress level to be imposed by the structure and the mass deformation modulus would therefore be the secant modulus (E_{SM}^{11}, E_{SM}^{21}, E_{SM}^{31}) represented by the first slope of each bi-linear curve.

An associated series of laboratory compression tests on rock cores will have produced a laboratory secant modulus E_{SL} (see Figure

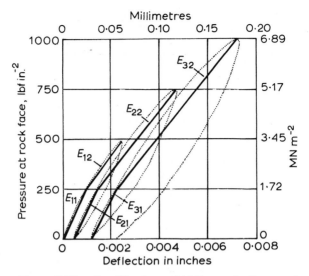

Figure 7.27 Specification of bi-linear elastic secant moduli with E_{11}, E_{21}, E_{31} defined for the 250 lbf in^{-2} (1.72 MN m^{-2}) point.

2.24) at the same stress level. The ratio E_{SL}/E_{SM}^{11} will be the 'mass factor'. As an example, tests on a Jurassic sandstone produced an E_{SL} (200 lb in^{-2} or 1.38 MN m^{-2}) range of 1.4 to 2.0 x 10^5 lb in^{-2} (965 – 1379 MN m^{-2}) and an E_{SM}^{11} (233 lb in^{-2} or 1.61 MN m^{-2}) range of 0.78 to 1.98 x 10^5 lb in^{-2} (538 – 1365 MN m^{-2}). In that case, the mass factor of 0.8 to 1.8 suggests only a marginal reduction in mass strength with respect to intact strength.

It is worth noting at this stage that probably the biggest 'plate loading' test yet staged, certainly in England, was the surface loading experiment performed on the Chalk in Norfolk and in which the loads were applied through a million-gallon tank filled with water. The idea was to test the response of the discontinuous rock to the loadings that might have been applied by the CERN 300-GeV proton accelerator had it been constructed in Britain. The proposed load intensities were on average up to 4 kgf cm^{-2} (0.392 MN m^{-2}) and the deflection requirements were that the differential settlement should be no more than about 0.15 mm over a length of 50 to 100 m.

The results of this test, described by Ward *et al* (1968), showed that the stiffness of the chalk mass increased with depth and that the deflections were recovered on removal of the load. This increase in stiffness was attributed to a reduction in the degree of fracturing

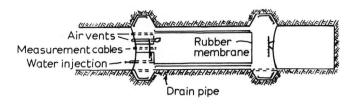

Figure 7.28 Pressure chamber installation (*after* Deere *et al*, 1969).

with depth and it was concluded that the experimental results could be extrapolated over the whole of the proposed accelerator site by simple measurements of the spacing and openness of the joints.

Even with a large diameter plate it can be claimed that the loading conditions do not even approximate to the loads that will be applied by the actual structure. The area of load application can, however, be increased by using an even more expensive test involving a pressure chamber. This test was first developed for determining the deformation modulus of rock around pressure tunnels in order to see if the rock would reliably support part of the internal pressure and so permit a reduction in lining thickness. It was later used to assess the strength characteristics of rocks (in which the joint spacing can be quite large) forming the abutments of arch dams.

The *pressure chamber* installation as described by Deere *et al* (1969) is shown in Figure 7.28. Instead of the two bulkheads as drawn, expense could be saved by dispensing with the innermost seal and using the end of the tunnel. Water would be used as the pressurizing medium and it is necessary to use a flexible lining to prevent seepage losses through joints in the surrounding rock. Deere *et al* have proposed using a very lightly reinforced concrete/bitumen mix for this purpose. It is also necessary to remove any protrusions of rock and to lightly infill any test section depressions before pressurizing.

Radial deformations at the stressed surface of the rock, and also into the rock, are measured at the central one-third of the length between bulk-heads in order to minimize end effects. Analysis of the results can be accomplished through the use of elastic, thick-walled cylinder theory even though the external radius of the (rock) cylinder is (theoretically) infinitely large. The basic equation is:

$$E_r = \frac{a^2 \sigma_i}{r \delta r} \qquad (7.22)$$

where E_r is the deformation modulus at radius r, σ_i is the internal water pressure, a is the internal radius of the pressurized test section, and δr is the deformation of the rock measured behind the skin of the chamber at a radius r. At the surface of the chamber, δr becomes δa and $r = a$.

This type of test also has obvious application where underground gas, oil or water storage is contemplated. In the latter context one thinks of pumped storage schemes (Figure 12.11 outlines one conventional scheme) but in this case where the lower reservoir is located underground — possibly in rock caverns excavated for this purpose in suitable rock — and the upper reservoir is at ground surface. Feasibility studies have demonstrated the viability of this method of power generation.

Information on rock or soil mass deformation modulus values is more directly expressive of a state of potential failure through excessive normal *displacement* although shearing both within the fabric of the intact rock and along discontinuities obviously contributes. But since, under compressive stresses, the most acceptable total failure criterion is expressed through the shear strength of the material, it is clearly useful to get some idea of the mass shear strength more directly. Shear jacking tests on rock discontinuities and *in situ* shear box tests for soils were developed for this purpose. They also allow the influence of progressive shear displacement to be assessed and expressed through ultimate (rock) or residual (soil) shear strength parameters.

Direct shear tests are not quite as expensive to arrange as plate jacking tests, but since they are often conducted at ground level or in a test pit, the normal loads must be applied either by jacking against kentledge or against a heavy site vehicle. Alternatively steel cables can be anchored deep below the test pit (Zienkiewicz and Stagg, 1966) and the jacking force exerted against the cables in tension. The direct shearing stresses are usually applied by jacking from a concrete reaction block cast at the bottom of the pit, or even, in the case of a soil, from the bucket of a site excavator (J.C.B., HyMac, or 'back-acter' in Britain). In rock tests, the shearing can be arranged to take place either through the solid rock of a block carefully isolated without the use of explosive or along a joint plane. A water pressure facility can also be incorporated (Figure 7.29). If shear strength parameters are required, then at least three tests must be performed, the required test conditions being most easily achieved by progres-

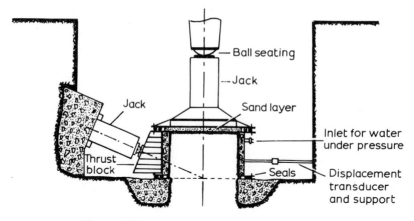

Figure 7.29   *In situ* shear test on block of rock.

sively deepening the pit after each test and re-constructing the concrete shear jack block.

It is preferable to provide a block of rock with a concrete facing particularly on the side withstanding the jacking thrust. Problems concerning the stability of the system arise if the shear displacement is continued on the path towards an ultimate state. It will also be noted from Figure 7.29 that the shearing thrust is vectored at an angle of about $(45 - \phi/2)$ degrees to the prescribed shear plane.

In situ shear box tests on undisturbed clay and soil are either in the form of single or double shear, the latter having merit with respect to the former. The box (or boxes in the case of double shear) is first placed over the material to be sheared by carefully excavating around the box and allowing the latter to fall slowly around the block. In operation, the direct shear force is usually applied parallel to the prescribed shear plane or planes and for a given normal load on the box, twice the horizontal shearing force will be required under conditions of double shear compared with the necessary capacity for single shear. A double shear facility is sketched in Figure 7.30.

It is not always necessary to indulge in expensive *in situ* shear jacking tests particularly if it is known that discontinuous rock at an engineering site will be subjected only to low normal loads. Whilst the engineering geologist may not participate directly in a full-scale test he himself can actually conduct some almost cost-free tests by jacking blocks of rock along joints or bedding planes. Hydraulic jacks can be pressurized using hand pumps and shear displacements can be

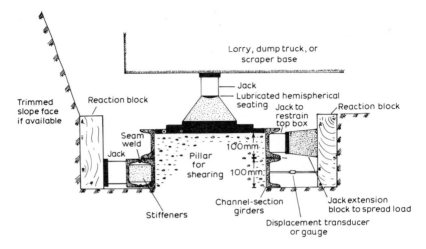

Figure 7.30 Double shear facility for clay or clay shale.

monitored by means of simple dial gauges. The block weight itself is taken as the normal force on a horizontal shear plane or if the plane is dipping downhill then the cosine of the dip angle would be used to modify the weight for the normal force. Similarly, there would be a sine of the dip angle addition to the downhill shear jacking force. Further dial gauges can quite easily be used to measure

Figure 7.31 Simple jacking system on a block of limestone using a Blackhawk hand-pump and jack. Photograph: A. C. Forbes.

Site Investigation 493

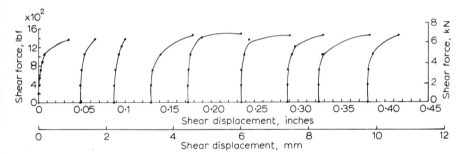

Figure 7.32 Repeated field jacking on single block of Great Limestone in Weardale, England. Weight of block = 3469 lbf (15.43 kN) and lying horizontally. Rough joint surface with loose limestone fragments (typical of general joint conditions) (*after* Forbes 1971).

normal dilation of a block as it tracks over asperities during shear. Block weights can be calculated from densities and volumes but metal weights can be added to increase normal loadings.

This very simple system, as used by Forbes (1971) in northern England, is illustrated in Figure 7.31. He examined numerous blocks of limestone and dolerite in this way during a slope stability investigation and produced primary shear force-displacement results of the type shown in Figures 7.32–7.34. The joint spacing in the dolerite is much closer than that in the limestone, although, as shown in Figure 10.7, there are close orientation compatibilities between the joints in the two rocks. For this reason, higher normal loads can be applied to the limestone shear planes. The results from these tests are plotted in Figures 7.35 and 7.36 and although, as might be expected, there is scatter in the data, it is possible to suggest friction angles for the joints. There is no obvious cohesive contribution to the

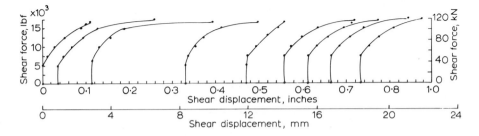

Figure 7.33 Repeated field jacking on single large block of Great Limestone in Weardale, England. Weight of block = 39,467 lbf (0.175 MN) and dipping at 4° in direction of jacking. Effective normal load = 39,372 lbf (0.175 MN). Rough joint surface with loose limestone fragments (typical of general joint conditions) (*after* Forbes, 1971).

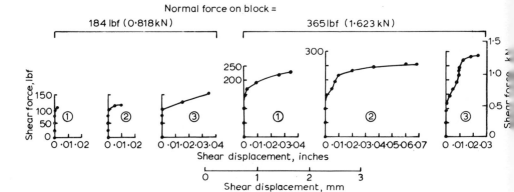

Figure 7.34 Repeated field jacking tests on two blocks of quartz-dolerite in the Whin Sill, Weardale, Northern England. Joint surfaces relatively smooth with both blocks dipping at 14° in the direction of the shear jacking (*after* Forbes, 1971).

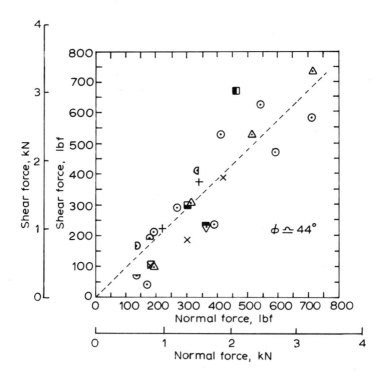

Figure 7.35 Simple *in situ* direct shear tests on blocks of quartz dolerite. Symbols denote suites of tests on different blocks.

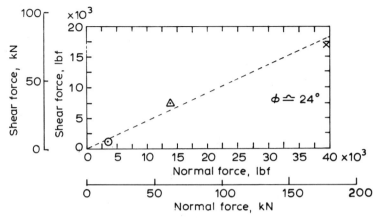

Figure 7.36 Simple *in situ* direct shear tests on blocks of limestone.

shear strength and this implies that the normal force is insufficiently high to cause the asperities to shear during the displacement. A clay gouge infills the limestone joints and contributes to the low *in situ* friction angle. Field test loadings are applied quite slowly and there is likely, therefore, to be a partial drainage facility in the clay. From laboratory tests performed on this clay (Figure 7.37) a ϕ'_p value of 14° is probably appropriate and is responsible for reducing a true limestone friction angle of about 40° to the field value of 24°.

Another joint shear tester, although not strictly speaking an *in-situ* test, has been developed by Hoek (1970). This is essentially a field shear box (Figure 7.38) designed to take plaster of paris mounted

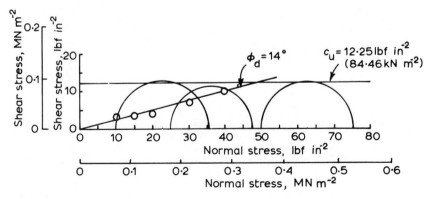

Figure 7.37 Undrained triaxial and slow, drained, direct shear data on the clay infilling joints in the Great Limestone, Weardale, England.

102 mm (4 in) core samples or block samples with a face area of 125 mm x 125 mm. The box, operating in the same manner as a laboratory shear box, comprises two halves, the lower half having a ram for shear loading and the upper half a ram for normal loading. A box of this type was used to derive the shear strength parameters for discontinuity surfaces in chalk, as detailed in Section 10.6.

Borehole devices comprise a family of *in situ* test and monitoring

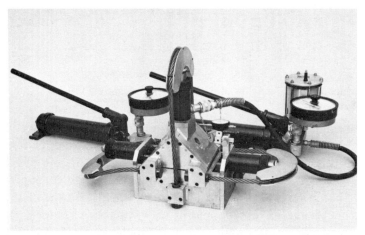

(a)

(b)

Figure 7.38 (a) Hoek joint shear box (*photograph by courtesy of Robertson Research International Ltd.*) (b) Partially mounted sample for the Hoek joint shear box.

instruments that have undergone progressive development since the early 1960s.

These instruments to be inserted into boreholes can be classified as either *active* or *passive* devices: they will either compress and strain the rock or soil, or they will themselves be compressed and strained. Of the latter type, there are some instruments designed specifically to monitor *deformation* and others to measure the *stress* pre-existing in the rock. Results from the deformation instruments are usually transformed to *in situ* stress values through the use of laboratory-determined moduli which, of course, can never be particularly satisfactory unless a laboratory/field modulus ratio is known.

Instruments of the former type, called dilatometers, have been designed by the Laboratório Nacional de Engenharia Civil in Portugal and described by Rocha *et al* (1966) — see also Rocha and Silvério (1969). Another instrument has been designed in Japan for the investigation of mudstone walls of an 86 m high arch dam and its use described by Takano and Shidomoto (1966). These instruments, used in rock strata, perform the same test function as the pressure chamber but, of course, on a much smaller scale and at a much lower cost.

The equivalent active pressure instrument for soils and which can also be used for soft rocks is the *Ménard Pressuremeter* (Ménard, 1966) sketched in Figure 7.39a. A similar device, for use in soft soils, is the Cambridge soil probe, developed and described by Wroth and Hughes (1973). The Ménard Pressuremeter performs a load test on the walls of a small diameter borehole, a direct measurement being made of the ultimate bearing capacity. An elastic modulus is calculated from the shape of the pressure-deformation curve (Figures 7.39b, 7.40a), the deformation being expressed as a function of volume change in the pressuremeter cell.

The borehole probe comprises three cylindrical metal cells with rubber membranes which can be inflated from the CO_2 and water cylinder to exert a radial pressure on the borehole wall. Whilst the central measuring cell is inflated by water pressure, the two end guard cells are pressurized with CO_2 gas (liquid nitrogen can be used when higher pressures are needed in rock) to the same pressure as in the measuring cell and they serve to reduce the end effects in the central measuring cell. In operation, the volumemeter is filled with water and then CO_2 gas is fed at a controlled pressure to the upper surface of the water, forcing it along plastic tubes to the probe measuring cell. At the same time, gas is admitted directly to the

498 *Principles of Engineering Geology*

guard packer cells at a separately-controllable pressure. On site, several pressure increments are applied to the borehole wall with the volume changes being read from the volumemeter for each pressure until such time as the volumemeter is empty. By this time, the failure point or limit pressure P_L will have been reached. In some instances near the end of the linear 'elastic' phase the pressure is reduced in

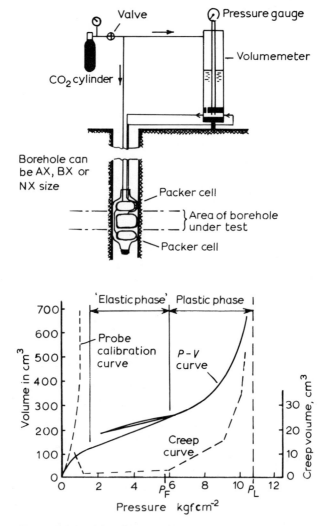

Figure 7.39 (a) Ménard Pressuremeter equipment (diagrammatic). (b) Typical pressure-volume curve from Ménard Pressuremeter.

stages, with volume readings being taken, before the system is re-pressurized again up to the limit pressure. Corrections have to be made to the *P-V* readings before plotting a graph as in Figure 7.39b.

The limit pressure provides a direct measure of the ultimate bearing capacity of the soil and the elastic modulus is related to the slope of the 'elastic' phase curve. The slope of the re-loading line gives a modulus which, when taken as a ratio of the elastic modulus, is a function of the type of soil.

The general creep curve shown in Figure 7.39b shows the way in which the soil deforms with time at any single pressure (see the example in Figure 7.40b). P_F marks the pressure at which the creep curve makes a definite upward break and it generally coincides with the end of the 'elastic' phase.

In granular soils, the borehole has to be lined with purpose-built casing containing six narrow longitudinal slots cut uniformly around its periphery. These slots, in effect, reduce the radial stiffness of the casing so that it can expand with the probe, the pressure necessary to expand the casing alone having previously been calibrated outside the hole.

Pressuremeter tests are usually made at about 4 ft (1.22 m) intervals of depth to any depth in a borehole, above or below the water table.

Dilatometer moduli should be compared with plate test moduli and with laboratory-derived moduli. It is preferable, and certainly necessary in bedded strata that might exhibit some mass anisotropy, to operate the dilatometer in horizontally-drilled holes for the comparisons to have meaning. A detailed site investigation would also specify a series of laboratory creep tests, ideally at confining pressures appropriate to the depths of each Ménard Pressuremeter test, against the significant parameters of which the borehole pressuremeter curves would be check-correlated for the determination of a mass factor in the manner of the plate bearing tests.

There is quite a long history in rock mechanics of the use of the *over—coring stress relief technique* in order to determine the actual state of stress existing in a rock mass. The method was probably first used by Nils Hast in 1958 in Sweden (see Hast, 1958) but the instrumentation side of the business was developed notably in England by the Sheffield and Newcastle groups (Potts 1956, 1964a; Roberts *et al*, 1964a,b, 1965; Roberts, 1965; Hawkes and Moxon, 1966; Roberts, 1968), in America by the U.S. Bureau of Mines (Panek, 1961; Wisecarver *et al*, 1964; Merrill *et al*, 1964), in South

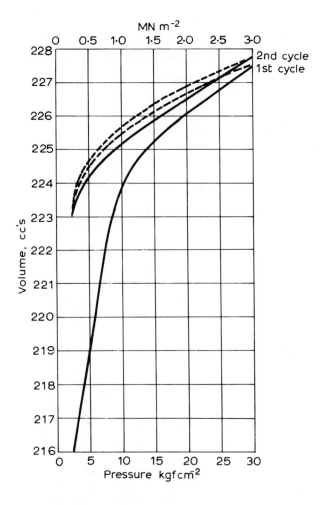
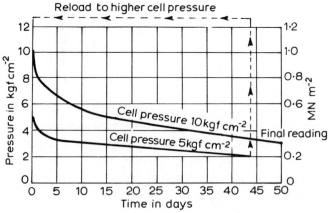

Figure 7.40 (a) $P-V$ curve for a sandstone. (b) Sandstone creep curves for two cell pressures.

Africa by the National Mechanical Engineering Research Institute (Leeman, 1964a,b,c; van Heerden, 1969) and in Canada by the Department of Mines and Technical Surveys (Barron, 1965). The work of Wilson (1961) of the British National Coal Board and of Hoskins (1967) should also be mentioned.

Since this is a specialist subject having stronger connections with deep mining problems than with the surface engineering province of the engineering geologist very little consideration will be given to it here. But basically, the procedure involves drilling a hole deep in the rock and installing a transducer or transducers (Figure 7.41). These

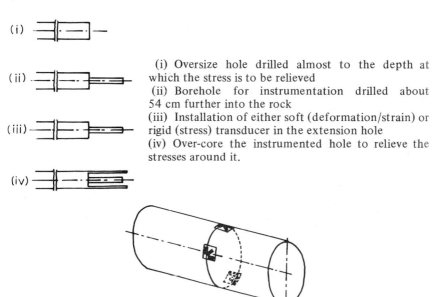

a) Overcoring for stress relief and resolution of secondary principal stresses in a plane normal to the plane of section.

(i) Oversize hole drilled almost to the depth at which the stress is to be relieved
(ii) Borehole for instrumentation drilled about 54 cm further into the rock
(iii) Installation of either soft (deformation/strain) or rigid (stress) transducer in the extension hole
(iv) Over-core the instrumented hole to relieve the stresses around it.

b) A system of 45° strain gauge rosettes cemented at 120° spacing around the walls of the borehole. The special pneumatic installation and cable device would also cement a rosette to the flattened back of the borehole (this latter is the South African C.S.I.R. 'doorstopper' element of the multi-component borehole strain cell). Both this C.S.I.R. instrument and a similar one developed by L.N.E.C. Lisbon allow the complete principal stress field in the rock to be resolved from a single borehole. Stress-measuring devices would be similarly installed but unlike a strain system they would resist inward borehole deformation.

Figure 7.41 Borehole transducer installation, and a versatile strain-measuring system. (*After* Leeman and Hayes, 1966.)

may take the form of diametral displacement gauges (inductance or differential transformer types) three of which may be spaced at 120° inside the hole, wire resistance strain gauge rosettes cemented around the walls and at the back of the borehole (special devices required for fixing in place), rigid inclusion devices the most popular of which have been the birefringent glass 'plugs', or most simply a pneumatic/ hydraulic pressure cell. When the chosen device has been fixed in place, the borehole is then overcored (or trepanned) using a larger diameter core barrel which penetrates for some distance beyond the back of the borehole. This de-couples and relaxes the annulus of the rock between the borehole and the circular slot, the magnitude of the absolute relaxation being dependent upon the magnitude and direction of the primitive stress pre-existing in a plane normal to the axis of the borehole. Provided that the stiffness (Young's modulus) of a rigid inclusion, cement-coupled tightly to the rock, is more than double the stiffness of the rock it will respond directly in terms of *stress* (for which it will have been calibrated) independently of the type of rock and will record either electrically or visually the directions and magnitudes of the two secondary principal stresses. Making the same observations in two other orthogonal boreholes produces two more sets of secondary principal stresses from which, by the analyses outlined by Obert and Duvall (1967) and Jaeger and Cook (1969), the magnitudes and directions of the three principal stresses pre-existing in the rock mass can be calculated. In the case of the strain or deformation devices, the secondary principal strains have first to be resolved from the electrical outputs and then transformed to secondary principal stresses through the application of Young's modulus.

This modulus is usually determined experimentally in the laboratory by testing the rock core from the borehole. Determination of the principal stress field then proceeds as before.

Deere *et al* (1969) have cross-correlated different *in situ* test moduli and in many cases obtained good agreement between different tests on the same rock. In a similar manner, their *in situ* static versus dynamic moduli correlate quite well, albeit with the inevitable scatter that must be expected when dealing with natural materials.

As discussed earlier in the context of pressuremeter and plate bearing tests, the ideal aim must be to obtain a laboratory-to-field correlation index which would allow the investigator to reduce the number of his expensive field tests and to derive correlated values from simpler laboratory tests. The simplest of laboratory tests on small samples are the *dynamic* ones which basically consist of either putting the specimen into *resonant vibration* (commercial equipment can be purchased for doing this, or see, for example Attewell and Brentnall, 1964) or of passing pulses through it at frequencies prescribed by the 'ringing' action of piezoelectric transducer elements. Deere *et al* have correlated velocity index values (field velocity/laboratory velocity) against modulus ratios to again produce acceptable trends which will give first approximations to field quality conditions by interpolation or extrapolation.

7.11 Field monitoring techniques

The assumptions involved in theoretical soil mechanics invariably mean that an element of speculation is involved in the design of many earth structures where sufficiently conservative safety factors may be unacceptable because of limitations of cost, time or space. This may occur in, for example, the case of construction on soft clays where rapid consolidation is required, deep retaining walls where anchors rather than timbers are used to create working space, or construction of low-level structures on backfill where deep consolidation would prove too costly.

In those instances where some element of risk may be accepted, some form of field monitoring/measurement, however cursory, is desirable in order to check the stability of the structure during and after construction. The resultant feedback may be used to control the construction progress or to assist in early incorporation of remedial measures. Specific commercial systems that are available are described by Hanna (1973) in some detail. They fall into four main categories depending on the physical quantities to be measured; namely *force* and *total stress, porewater pressure, strain* and *deformation.*

Of these, the most simple and commonly utilized are *deformation* measurements, usually in the form of surface horizontal *movements* or surface *settlements.* They can be determined by conventional levelling techniques in the case of settlements, and geodetic surveying techniques (see for instance, Bannister and Raymond, 1972), either triangulation or trilateration, in the case of horizontal movement. In particular, trilateration techniques have improved markedly with the recent development of electronic distance measuring instruments such as the Telurometer and the (British) National Physical Laboratory Mekometer. These techniques have been developed notably by the Building Research Establishment in England (see Penman, 1972; Penman and Charles, 1972a) and have been used to measure deformation associated with the construction of large dams, particularly the Llyn Brianne dam and Scammonden dams in Wales and England respectively.

Surface deformation measurements are of limited value when used in isolation. They can only provide an accurate interpretation of the behaviour of an earth structure or the ground affected by a structure when they are combined with observations of *subsurface* movement. Subsurface movements can be measured by various methods including:

(a) *inclinometer* and *sensing* devices (usually reed switches operated by magnetic fields which are induced by pre-set magnetic rings or plates) introduced through vertical or inclined boreholes,
(b) *hydraulic* systems buried during construction and designed to measure small changes in head caused by vertical settlement,
(c) tensioned and *anchored wires* or rods introduced through boreholes and attached to dial gauges at ground or underground surfaces,
(d) *laser* systems operating through tunnels or boreholes: these can, of course, also be used for surface movement measurements.

In all cases of subsurface movement, the relevant system must be referred to a surface datum or bench mark determined by accurate surveying or a point deep in the ground which is known to be movement-free. Specific examples of successful surveys are illustrated in Figures 7.42 and 7.43 which show vertical and horizontal strain contours obtained in two rockfill dams by Penman and Charles (1972a,b) using geodetic surveying to determine surface movements, and horizontal plate gauges to determine subsurface movement. These plates were connected through horizontal pipes containing an induction probe to measure horizontal movement (Figure 7.44). A hydraulic system was used to measure vertical movement. The movements, often greater than 1 m and expressed as strain (positive compression), demonstrate that the position and inclination of the clay cut-off core of the dam exert a significant effect on the magnitude of the horizontal strain. The much greater strains at Llyn Brianne are attributed to a major lateral thrust imposed on the downstream rockfill

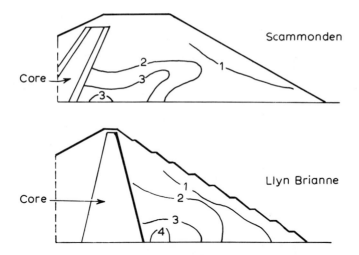

Figure 7.42 Contours of percentage constructional strain in a vertical direction (*after* Penman and Charles 1972a). Crown Copyright 1972: adapted from Building Research Establishment Current Paper CP 18/72.

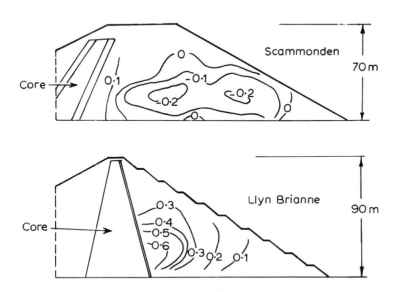

Figure 7.43 Contours of percentage constructional strain in earth dams in a horizontal direction (*after* Penman and Charles 1972a). Crown Copyright 1972: adapted from Building Research Establishment Current Paper CP 18/72.

506 *Principles of Engineering Geology*

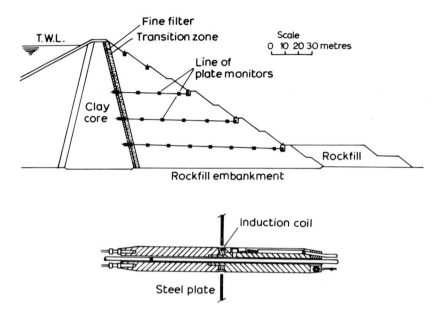

Figure 7.44 Location of displacement monitoring plates in the rockfill shoulders of Llyn Brianne dam in Central Wales. Site is a post-glacial gorge cut by the River Towy in slaty mudstones of Lower Palaeozoic period. Lower section is of the induction probe which is towed along interconnecting tubing between plates and records the presence of the plate (*after* Penman and Charles, 1972b). Adapted from Building Research Establishment Current Paper CP 19/72, Crown copyright 1972.

by the centrally-placed wet clay core, a situation which is ameliorated if the core is located on the upstream side as at Scammonden dam. We may note that this dam (Figure 10.59) was designed to incorporate a carriageway at the crest, a feature which may not always prove to be economical.

Figure 7.45 illustrates instrumentation layouts to determine ground movements above and around tunnels in London Clay and in laminated clay investigated by the present authors (Attewell and Farmer, 1974a,b).

Surface movements were determined by surveying and levelling. Subsurface movements were recorded through a reed switch/magnetic ring system (vertical) and an inclinometer tube (horizontal) located in a series of vertical boreholes positioned above and normal to the tunnel centre line. The results, summarized by Figure 7.46 and given in detail in Attewell and Farmer (1974a); Attewell *et al* (1975),

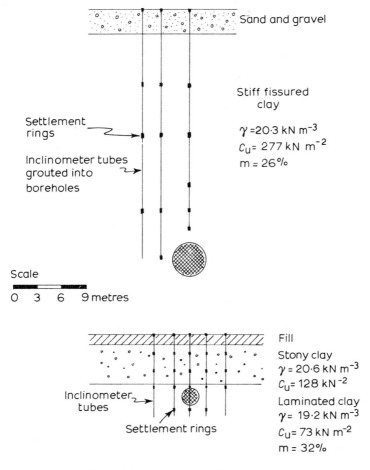

Figure 7.45 Typical instrumentation layouts transverse to tunnel line for vertical and horizontal deformation measurements around tunnels in clay.

indicate a significant and rapid clay yield into the tunnel face, into the gap around the protective tunnelling shield created by the leading edge overcutting 'bead' after excavation and into the annular gap after insertion of a rigid segmental lining but before setting of the contact grout (Figure 7.46). These ground losses in the tunnel are eventually reflected as a transverse surface settlement or 'subsidence' profile generally of an 'error curve' form associated with mining subsidence (see Farmer and Attewell, 1975 and Figure 7.60).

Typical of the anchor wire systems used to measure relative ground displacements at different depths along a borehole is that

508 *Principles of Engineering Geology*

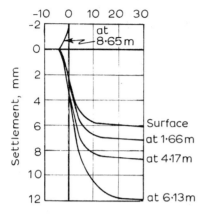
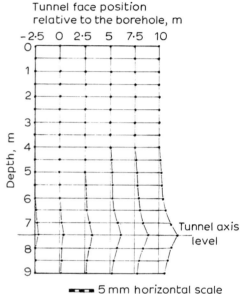

Figure 7.46 Ground deformation around a tunnel in laminated clay. Strata section in Figure 7.45. (*After* Attewell *et al.*, 1975).

illustrated in Figure 7.47. High tensile steel wires are anchored at different depths in the borehole and brought out to the surface of the hole through a measurement head which, in its simplest form, comprises a series of micrometer screws through which pre-set markings on each wire are re-positioned to zero. The micrometers can be replaced by inductance gauges, or linear variable differential transformer transducers, feeding to a multi-channel pen recorder or to a magnetic tape deck for continuous monitoring. Data logging systems such as this can readily be designed to incorporate high level limit alarms which give early warning of any unacceptable move-

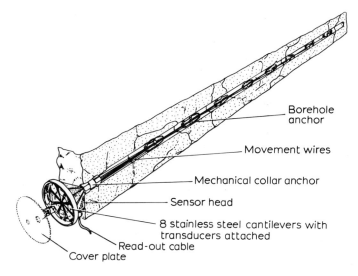

Figure 7.47 Terrametrics Multiple Position Borehole Extensometer, Model MPBX, 8PT, Type F.

ments. Up to about 8 wires can usually be accommodated in a single borehole and since, in its basic form, the system is designed to monitor displacements that occur parallel to the long axis of the hole, it is particularly suited to the measurement of differential vertical settlements that might develop over tunnels or mineworkings. The system can also be adapted for measuring shear movements in a plane normal to the axis of the borehole and as such would be more suited to the monitoring of slope stability. Potts (1964b) also describes the use of anchor wire systems for monitoring rock mass movements around underground mine workings.

A technique used for the monitoring of rock displacements at the surface of a dam abutment relative to the displacement at depth during and after impounding is based on an inverted pendulum principle. This system, sketched in Figure 7.48, was used for measurements at the Monar Dam in Scotland (see Section 6.22) and has also been mentioned by Londe (1973).

Most *strain* measurements in soils and soft rocks are, in effect, deformation measurements which may subsequently be interpreted in terms of strain. In some hard and competent rocks subjected to minor changes in stress there may be some justification for strain measurements directly. Devices capable of doing this would usually take the form of strain gauges cemented onto the sides or ends of

510 Principles of Engineering Geology

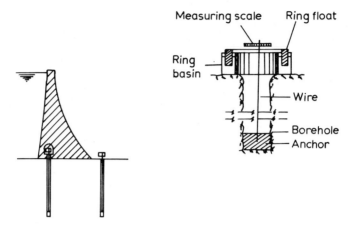

Figure 7.48   Inverted pendulum system for monitoring surface movements at dams (*from* Londe, 1973a).

boreholes or onto borehole insertions in order to monitor deformation of the borehole. These are rather specialist forms of instrumentation and have been outlined in the previous section.

Total stress measurements involve measurement of the absolute stress in the ground and any subsequent change in stress. They therefore differ from deformation measurements which are usually concerned only with changes after they commence, and for this reason total stress measurements are much more difficult to perform. One approach, as with the Cambridge soil probe (see Wroth and Hughes 1973), is to drill into a soil with as little disturbance as possible and then to wait for the ambient pressure to revert to its *in-situ* levels. A similar method used in harder soils or rocks is to insert some form of transducer or pressure cell into a borehole and to grout this into place, often at a grout pressure approaching the assumed ground pressure. Such systems are complicated by the non-hydrostatic earth pressures usually existing in the ground and also by the difficulty of accurately locating the pressure cell at a position remote from the installation point.

Even in the case of simple flat diaphragm *earth pressure cells* of the type located in earth structures or most successfully at soil-foundation interfaces – for example, the Glötzl cell (Figure 7.49) – some difficulty may be experienced in accurately determining the correct average stress acting on the surface rather than a modified stress caused by the insertion of the cell. This problem was

Site Investigation 511

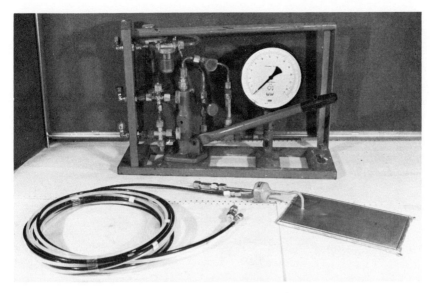

Figure 7.49 Glötzl earth pressure cell and pressure pump.

first considered by Taylor (1948) and his proposed correction to allow for the substitution of a soft soil 'cell' by a rigid inclusion is outlined by Hanna (1973) to which further reference should be made.

Pore pressure measurements are probably more important in field measurement than any of the others considered. In most pre-loading consolidation applications (see Section 11.3) or during construction on soft ground they are an integral part of construction. In *site investigation* they form an important but sometimes indirect part of the observation programme. The simplest type of porewater pressure measurement device is a *cased borehole* down to a known, water-bearing stratum. The rest-water level in the borehole provides, through its measured head, a direct indication of the porewater pressure in the layer at the foot of the casing.

For long-term observations, and for observations in unsaturated strata or strata of low permeability, a cased borehole would prove to be either uneconomical or unsatisfactory. In such cases, an alternative form of *piezometer* is required. Some of these devices are illustrated in Figure 7.50. They can also be used to measure permeability, and this soil/rock property is considered in Section 8.6. For porewater pressure measurement the main requirement is to maintain an open, porous path into the soil. Thus, in coarsely

Figure 7.50   Piezometer tips.

textured soils, an open standpipe tends to be the most satisfactory, with water levels being measured by some sort of dipping device containing an audible water-contact alarm. In clays or collapsing soils, a porous tip surrounded by coarse sand and sealed above the relevant strata by a clay plug will probably be successful. In fine-grained soils where a significant *time lag* will affect equalization of pressures, a closed circuit piezometer system, flushed through with de-aired water, can be used in place of an open standpipe. Various diaphragm or pneumatic/hydraulic automatically-recording piezometers are also available. These are described in detail by Hanna (1973).

Where both air and water are present (earth fills), a high-air entry piezometer is required. This means that the piezometer porous stone permeability must be sufficiently low ($\sim 10^{-8}$ ms^{-1} in most embankment applications) to prevent air under a pressure u_g flowing through the piezometer which will be subject to a hydraulic head equal to the porewater pressure u_w.

7.12 Use of field seismic techniques in engineering geology

When seismic waves are transmitted through fractured and/or weathered rock, their transmission characteristics are modified according to the reduction in quality of the medium along the transmission path. In general, there is a reduction in *wave velocity* with increased fracture density (Onodera, 1963; Deere *et al*, 1967;

Helfrich et al, 1970; Paolo and Armando, 1970; Lykoshin et al, 1971) and with degree of weathering (Iliev, 1970). One useful application of seismic velocity logs is in quarrying or tunnelling where a wave velocity can be used to define limits of rippability or cuttability. Generally, a wave velocity of about 6500 ft s^{-1} (2000 m s^{-1}) can be said to represent the upper limit of economically rippable rock.

Using what is probably a more sensitive parameter, *wave attenuation* can also be correlated with fracture intensity (Lykoshin et al, 1971). On the assumption that small laboratory test specimens can be regarded as intact and that they therefore transmit pulses at the maximum velocity for that material, Onodera (1963) and Deere et al (1967) have suggested using the field/laboratory velocity ratio as a quantitative index for the general character of the *in situ* mass and Knill (1969) has related this (fracture) index to the grout 'take' at dam sites in the United Kingdom. More usefully, a velocity index would be specified in terms of the *square* of the field:laboratory velocity ratio so that it would then be directly proportional to the respective (field:laboratory) dynamic moduli.

An emphasis on shear wave velocity propagation is probably advisable since these waves appear to be more affected by rock mass discontinuities than are longitudinal waves (Londe, 1973a) and indeed there would appear from the work of Roussel (1968) and Londe (1973) to be a correlation between the modulus of deformation determined from field jacking tests and the band-pass frequency of a discontinuous rock mass. Londe (1973a) has argued that the results from, for example, Figure 7.51 give an insight into the average spacing of discontinuities in the mass. The concept of a natural filtering mechanism is an attractive one and is rather more tenable in the case of a homogeneous, systematically jointed mass than with a layered, bedded sequence (where wave-guide mechanisms may operate) or with a weaker, more randomly discontinuous mass.

A fundamental problem arising from any attempt to correlate transmission characteristics in small laboratory specimens and in the field lies in the different frequency spectra associated with the two general types of exploratory wave forms. Velocity measurements on small specimens will usually be conducted at resonant frequencies (depending upon the specimen length) of between about 10 and 30 kHz or ultrasonically by pulse excitation of between about 100 kHz and about 5 MHz. Group frequencies of wave transmission in

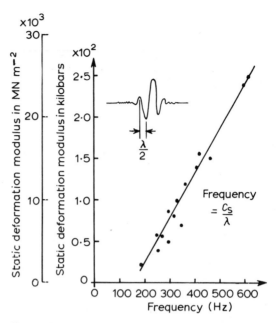

Figure 7.51 Relationship between the mass deformation modulus, derived from jacking tests, and the recorded group frequency of shear waves travelling at a velocity c_s and a wavelength λ (*from* Londe, 1973a).

the field will rarely exceed a few 100 Hz. Not only will there be a relative velocity dispersion effect but there will also be frequency-selective attenuation (over a very wide frequency range, there seems to be a direct proportionality between attenuation and frequency — Attewell and Ramana, 1966 — and this is compatible with the concept of a frequency-independent material friction parameter Q^{-1}). But such frequency-selective attenuation could be exploited in the field by composing frequency spectra for disturbed and undisturbed rocks (for example, Saul and Higson, 1971).

As a basis for a rock mass classification system, Deere *et al* (1969) have produced a number of correlation curves relating different index properties such as R.Q.D. and velocity index. Interested readers should refer to that publication for further information.

7.13 Analysis of ground vibrations

The importance of ground vibrations in site investigation is not confined to their utilization in seismic exploration processes or in detecting rock fracture zones. Of considerable concern is the effect of vibration, whether initiated by earthquakes, blasting, traffic, or industrial processes, on structures which lie in the path of wave transmission. Reference should be made to Steffens (1974).

As discussed in Section 12.15 on dam seismicity, ground accelerations vary very rapidly with time and although a peak acceleration may induce maximum inertia forces in a rigid structure, it will last for only a very short time and therefore the real damage capacity of the wave may not truly be expressed by an acceleration criterion. The response of a building to acceleration will be dictated by the properties and geometry of the structure itself with respect to the varying direction of the force and by the coupling of the structure to the foundation. Induced accelerations in the structure may be greater or less than the inducing ground accelerations and a situation could arise when integral lengths of sub-structures could resonate to a harmful pitch. Total structural response to applied forces will depend upon its composite resilience and its capacity to dissipate energy. The structure will, as it were, seek out and resonate with periods of the ground motion that are matched to its own. This means that individual structural elements each having a particular resilience will filter-out different components of the ground motion and will respond strongly to those components of the movement with which they can resonate. In general, the lower the capacity of a structure for dissipating energy, the greater will be its response.

In order to evaluate the characteristics of an incoming waveform it is usually necessary to analyse not only for amplitude but also for frequency composition. A Fourier transform of the signal must be generated and it is also useful to know the spectral distribution of power in the signal. These operations are best accomplished either by processing magnetic tape records on commercial equipment or alternatively by digitizing from light-sensitive paper records and analysing on a suitable digital computer.

Limiting criteria with respect to vibration amplitude have been proposed on the basis of several vibration parameters and, in the case of structural damage, some of these are listed in Table 7.6.

Probably the most common source of ground vibration arises from the movement of vehicular traffic (Walthall, 1970; Whiffin and Leonard, 1971; Knights, 1971; Leonard *et al*, 1974). Such vibration amplitudes are, however, very low. Typical values measured by the authors adjacent to the carriageway range from about 0.062 mm s^{-1}

Table 7.6 Some structural damage criteria that have been proposed for ground vibration

Mode	Criterion	
(a) *Acceleration* experienced by the structure (Thoenen and Windes, 1942)	(i) Accelerations <0.1 g	→ safe
	(ii) 0.1 g to 1.0 g	→ caution required
	(iii) >1.0 g	→ high damage probability
(b) *Energy* of the vibration. Crandell (1949) defined an *energy ratio* as the second power of the acceleration divided by the second power of the frequency, i.e. $$E_r = \frac{\text{acceleration}^2}{H_z^2}$$	(i) $E_r < 3$	→ safe
	(ii) $3 < E_r < 6$	→ caution required
	(iii) $6 < E_r$	→ dangerous condition
(c) *Ground velocity* (Langefors et al, 1958; Edwards and Northwood, 1960; Duvall and Fogelson, 1962)	(i) $\leqslant 71$ mm s^{-1}*	→ no structural damage, or minimal structural damage
	(ii) 71–109 mm s^{-1}	→ very slight structural damage
	(iii) 109–160 mm s^{-1}	→ moderate damage
	(iv) at 230 mm s^{-1}	→ serious damage
	*Note that a consensus of case history evidence suggests that ground vibration velocities less than 50 mm s^{-1} will cause no damage even to internal renderings and plasterwork of houses. These criteria can be re-expressed as: <50 mm s^{-1} → safe 50–100 mm s^{-1} → caution >100 mm s^{-1} → high damage probability	
(d) *Power*. Vibration intensity Δ (Alpan and Meidav, 1963) is proportional to the power transmitted to a unit mass during a quarter cycle	$$\Delta = \frac{\text{acceleration}^2}{\text{frequency}}$$ $$= E_r \times \text{frequency}$$	

peak for a motorway to about 0.62 mm s^{-1} peak for a light construction A-class road in England. Multi-wheel low-loaders generate higher ground vibration (2.8 mm s^{-1} peak recorded adjacent to an A-class road) but lightly-loaded lorries bouncing over settlement trenches in a road can generate even higher vibrations. There is also a vehicle speed effect in that up to about 60 m.p.h. (96 Km h^{-1}), the vibration velocity increases at about 0.00061 mm s^{-1} peak per m.p.h. increase (0.00038 mm s^{-1} pk/Km h^{-1}) in speed. On some roads, the figures may be much higher.

Interpretation of an incoming waveform is particularly complex. A very low frequency disturbance depresses the pavement in phase with the movement of the vehicle. A wave also emanates from each wheel acting as a point, but moving, source. Higher frequency components result from vehicle bounce on an uneven surface and the highest frequencies of all can be ascribed to out-of-balance engine vibration. Since the receiver is stationary and the source is moving there arises a doppler effect to complicate any analysis. But in view of the general vibration levels, no direct structural damage can be ascribed to them although they may well introduce long-term fatigue problems to old buildings.*

A more serious source of vibration arises from rock blasting works in urban areas (trenching and foundation excavation or rock tunnelling) or during quarrying operations. Quarry owners may sometimes find some of their potential reserves sterilized through the need to restrict the level of blast vibrations transmitted to adjacent property. In the past, a great deal of research aimed at establishing a vibration attenuation law has been performed in the U.S.A. particularly by the Bureau of Mines (Thoenen and Windes, 1938, 1942; Theonen *et al*, 1940; Duvall *et al*, 1961; Duvall and Fogelson, 1962; Duvall *et al*, 1963) and by Leet (1946, 1948, 1960), Crandell (1949) and Willis and Wilson (1960). There have also been developments in Britain (Fish and Hancock, 1949; Fish, 1951 a,b,c; Morris, 1950 a,b; Teichmann and Hancock, 1951; Teichmann and Westwater, 1957; Edwards and Northwood, 1960; Attewell and Farmer, 1964 a,b; Davies *et al*, 1964; Attewell *et al*, 1965) and in Sweden (Langefors *et al*, 1958; Langefors and Kihlström, 1963).

---

*Probably in a more indirect manner by disturbing dust and small particles which then fall into cracks, gradually wedging them open during each vibration cycle and over an extended period of time.

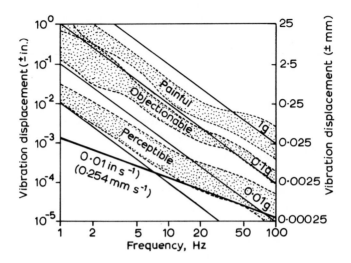

Figure 7.52 Effect of vibration on people (*from* Waller, 1969).

In blasting seismology, ground *velocity* criteria seem now to be universally accepted. Berger (1971) has given some useful statistics. He suggests, on the basis of the U.S. Bureau of Mines work, that 50 mm s^{-1} is a *conservative* limit. Research workers from the Bureau, Sweden and Canada have found that more than 97 per cent of the instances of actual damage were associated with ground velocities of 71 mm s^{-1} or more. Evidence also suggests that there is a 50 per cent chance of causing minor damage with a ground velocity of 137 mm s^{-1} and a 50 per cent probability of causing minor damage with a ground velocity of 193 mm s^{-1}. Awojobi and Sobayo (1974), on the other hand, have proposed a ground velocity limit of 25 mm s^{-1} with respect to structural damage and have introduced the depth of the disturbance source into their transmission equation.

Structural damage is not, however, the only consideration. Problems may also arise from personal environmental disturbance at vibration levels considerably below any damage threshold (Waller, 1969; see also Figures 7.52 and 7.53). Sensitive equipment may also suffer at quite low vibration levels, relays may trip, turbo-alternators may be harmfully excited and microscopic work may be difficult to perform. It is difficult to specify maximum tolerable vibration velocity levels without considerable on-site experimentation (for

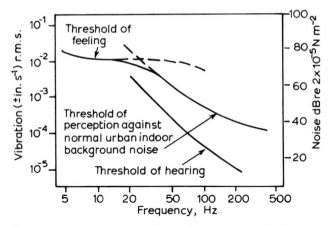

Figure 7.53 Sensitivity of people to noise and vibration (*from* Waller, 1969).

example, Skipp and Tayton, 1971) and even then any recommendations may be quite arbitrary.

Ground motion occurs in a three-dimensional sense, so that unless there are strong grounds for claiming that motion in one direction only is offensive, it will be necessary to record along three orthogonal axes and then accept a peak resultant displacement, velocity or acceleration for further analysis. Thus if $\hat{v}_x$, $\hat{v}_y$, $\hat{v}_z$ represent the peak ground velocities along axes x,y,z respectively, then the peak resultant ground velocity will be $(\hat{v}_x^2 + \hat{v}_y^2 + \hat{v}_z^2)^{1/2}$ and will exceed any one of the constituent orthogonal velocities. In practice, two of the measurement axes (say x and y) would be horizontal with one pointing towards the known source. The third axis would then be vertical.

Scaled distance law

In order to develop a predictive equation for future ground velocities on the basis of recorded vibration data and to present the information graphically for easy extraction, it is necessary to reduce the number of variables into dimensionless groups (Ambraseys and Hendron, 1968). We begin with the following:—

Dependent variables	*Dimension*
u = ground displacement	L
v = ground velocity $\left(\dfrac{\partial u}{\partial t}\right)$	LT^{-1}

a = ground acceleration $\left(\dfrac{\partial^2 u}{\partial t^2}\right)$ LT^{-2}

f = frequency T^{-1}

Most significant independent variables   *Dimension*

W = energy released by explosive (yield) FL

r = radial distance from source L

ρ = rock density $FL^{-4}T^2$

c = velocity of compressional seismic wave LT^{-1}

t = time T

Number of dependent variables   = 4
Number of independent variables   = 5
Number of fundamental parameters   = 3
Therefore, number of π-groups   = 9−3 = 6

Building up these π-groups, we have:

	x_n	y_n	z_n		x_n	y_n	z_n
π_n (n=1)	u	c	W	$v = L$	LT^{-1}	FL	LT^{-1}
π_n (n=2)	u	c	W	$a = L$	LT^{-1}	FL	LT^{-2}
π_n (n=3)	u	c	W	$f = L$	LT^{-1}	FL	T^{-1}
π_n (n=4)	u	c	W	$r = L$	LT^{-1}	FL	L
π_n (n=5)	u	c	W	$\rho = L$	LT^{-1}	FL	$FL^{-4}T^2$
π_n (n=6)	u	c	W	$t = L$	LT^{-1}	FL	T

Thus,

$$\pi_1 \begin{cases} F: z_1 = 0 \\ L: x_1 + y_1 + z_1 + 1 = 0 \\ T: -y_1 - 1 = 0 \end{cases} \begin{vmatrix} x_1 = 0 \\ z_1 = 0 \\ y_1 = -1 \end{vmatrix} \quad \pi_2 \begin{cases} F: z_2 = 0 \\ L: x_2 + y_2 + z_2 + 1 = 0 \\ T: -y_2 - 2 = 0 \end{cases} \begin{vmatrix} x_2 = 1 \\ y_2 = -2 \\ z_2 = 0 \end{vmatrix}$$

$$\pi_3 \begin{cases} F: z_3 = 0 \\ L: x_3 + y_3 + z_3 = 0 \\ T: -y_3 - 1 = 0 \end{cases} \begin{vmatrix} x_3 = 1 \\ y_3 = -1 \\ z_3 = 0 \end{vmatrix} \quad \pi_4 \begin{cases} F: z_4 = 0 \\ L: x_4 + y_4 + z_4 + 1 = 0 \\ T: -y_4 = 0 \end{cases} \begin{vmatrix} x_4 = -1 \\ y_4 = 0 \\ z_4 = 0 \end{vmatrix}$$

$$\pi_5 \begin{cases} F: z_5 + 1 = 0 \\ L: x_5 + y_5 + z_5 - 4 = 0 \\ T: -y_5 + 2 = 0 \end{cases} \begin{vmatrix} x_5 = 3 \\ y_5 = 2 \\ z_5 = -1 \end{vmatrix} \quad \pi_6 \begin{cases} F: z_6 = 0 \\ L: x_6 + y_6 + z_6 = 0 \\ T: -y_6 + 1 = 0 \end{cases} \begin{vmatrix} x_6 = -1 \\ y_6 = 1 \\ z_6 = 0 \end{vmatrix}$$

Hence,

$$\pi_1 = c^{-1} v \qquad = v/c$$

$$\pi_2 = uc^{-2} a \qquad = \frac{ua}{c^2}$$

$$\pi_3 = uc^{-1} f \qquad = \frac{uf}{c}$$

$$\pi_4 = u^{-1} r \qquad = r/u$$

$$\pi_5 = u^3 c^2 W^{-1} \rho = \frac{u^3 c^2 \rho}{W}$$

$$\pi_6 = u^{-1} ct \qquad = ct/u$$

We may now write

$$\frac{v}{c} = \Phi \left[\frac{ua}{c^2}, \frac{uf}{c}, \frac{r}{u}, \frac{u^3 c^2 \rho}{W}, \frac{ct}{u} \right]$$

Now, for explosions of different yields in the same rock medium, ρ and c should be constant. Taking π-groups 4 and 5, and noting that any of the groups may be multiplied by any of the other groups or their powers, we have that

$$r^3/u^3 \times \frac{u^3 c^2 \rho}{W} = \frac{r^3 c^2 \rho}{W} \qquad (7.23)$$

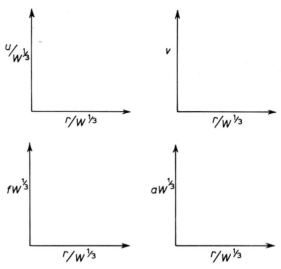

Figure 7.54 Some methods of plotting blast vibration parameters according to the results of dimensional analysis (*see also* Ambraseys and Hendron, 1968).

Thus, the following scaling laws apply:

r scales as $W^{1/3}$,

t scales as r and so also scales as $W^{1/3}$,

u is also proportional to $W^{1/3}$,

v should be independent of W,

a is inversely proportional to r at scaled times and ranges and therefore varies inversely as $W^{1/3}$,

f is inversely proportional to time and therefore scales inversely as $W^{1/3}$ at equal values of scaled time and range.

The most useful plots on a scaled basis are as shown in Figure 7.54. In practice rather than expressing W in energy terms, and since the energy released is taken as being proportional to the weight of explosive, the cube-root scaling can be in terms of *weight* of explosive. If, however, data are to be scaled from different types of explosive having different energy densities, then a common reference energy must be used.

The dimensional analysis and Figure 7.54 indicate that an equation of the form

$$v = k - m\left(\frac{r}{W^{1/3}}\right) \tag{7.24}$$

might be appropriate for fitting to measurement data and for estimating ground vibration velocities. In this assumed linear relationship, k and m are constants defining ordinate axis intercept and slope respectively. It is found, however, that a power law relationship is more applicable:

$$\log v = \log k - m \log \left(\frac{r}{W^{1/3}}\right) \qquad (7.25)$$

It might be expected that equation 7.25 would be used for the processing of blast vibration data and for the formulation of a 'site law' in which, in effect, k and m are 'site constants'. Blair and Duvall (1954) say that this cube-root scaling is acceptable for wave propagation from a spherical source. By analogy, the square-root of the charge weight might be used as the scaling factor for propagation on the surface from a long cylindrical source (Devine, 1962). Attewell and Farmer (1973), on the other hand, have argued on the basis of driven pile vibrations that geometrical attenuation should be related to the surface area of a wavefront and hence should be proportional to r^2 for spherical waves and r for cylindrical waves (see also Section 4.11). Where the predominant ground vibration amplitudes result from the transmission of surface waves which lag behind any preceding P and S body wave motion (and amplitude dominance of surface wave motion is the rule) then the case for a $W^{\frac{1}{2}}$ scaling is most convincing. Indeed, the published data of Duvall and his co-workers at the U.S. Bureau of Mines and of the present authors have always been scaled on a square root basis.

In Figures 7.55 and 7.56, some arbitrary ground velocity measurement data are plotted according to both a $W_T^{1/3}$ and a $W_T^{1/2}$ scaling and appropriate 'y on x' regression lines inserted. W_T is taken to be the *total charge* weight of explosive detonated. On the other hand, massive weights of explosive will rarely be detonated simultaneously in quarry blasting. For more efficient fragmentation, charges in different holes or groups of holes will be delayed in their detonation with respect to others so that, in fact, a single blasting operation may comprise a sequence of smaller discrete detonations, each delayed in time by about 17 or maybe 25 milliseconds with respect to its predecessor. A *close-in* geophone record would thereby depict a series of interrupted wave trains and it could reasonably be assumed that any structural response would be a function of the split charge

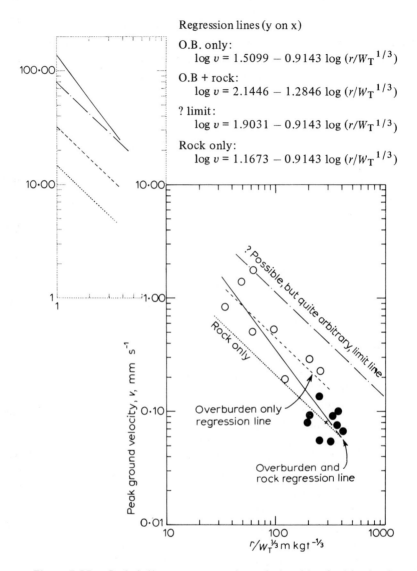

Figure 7.55 Scaled distance attenuation relationships for blasting in a rock. *Total charge* (W_T) and *cube root* ($W_T^{1/3}$) scaling. Open circles refer to overburden ground velocity readings. Infilled circles refer to rock velocity measurements.

rather than the total charge. However, as the transmission path lengthens, there is a frequency-selective attenuation and surface wave velocity dispersion, and as a result of these phenomena, it becomes increasingly difficult to resolve the individually-delayed blasts. For this reason, and for transmission distances of up to and beyond, for example, 1000 m to 1500 m, two attenuation laws, one with respect to W_D (delay blasts) and one with respect to W_T (total charge), should be formulated. The laws for a $W_D^{1/3}$ and $W_D^{1/2}$ scaling are derived from Figures 7.57 and 7.58 respectively.

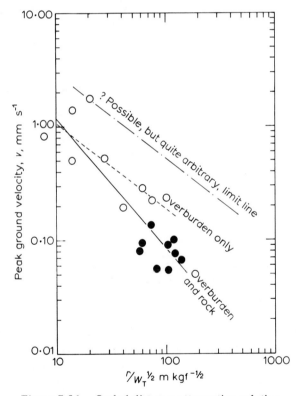

Regression lines (y on x)
O.B. only : log v = 0.7815 − 0.7533 log $(r/W_T^{1/2})$
O.B. + rock : log v = 1.2124 − 1.1316 log $(r/W_T^{1/2})$
? limit : log v = 1.2253 − 0.7533 log $(r/W_T^{1/2})$

Figure 7.56 Scaled distance attenuation relationships for blasting in a rock. *Total charge* (W_T) and *square root* ($W_T^{1/2}$) scaling. Open circles refer to overburden ground velocity readings. Infilled circles refer to rock velocity measurements.

Regression lines (y on x)
O.B. only : $\log v = 2.0612 - 1.0588 \log (r/W_D^{1/3})$
O.B. + rock : $\log v = 2.7816 - 1.4197 \log (r/W_D^{1/3})$
? limit : $\log v = 2.3856 - 1.0588 \log (r/W_D^{1/3})$
Rock only : $\log v = 1.7993 - 1.0588 \log (r/W_D^{1/3})$

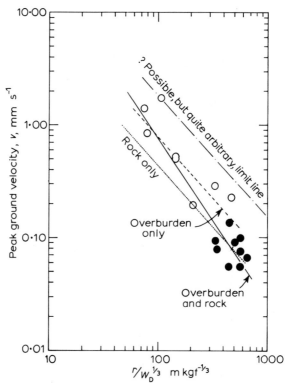

Figure 7.57 Scaled distance attenuation relationships for blasting in a rock. *Delay charge* (W_D) and *cube root* ($W_D^{1/3}$) scaling. Open circles refer to overburden ground velocity readings. Infilled circles refer to rock velocity measurements.

There are several matters of interest pertaining to Figures 7.55 through 7.58. Ground velocity measurements taken by geophone or integrated from accelerometer response will be a function of the ground conditions. Gauges should be buried beneath the surface of overburden or they must be rigidly attached to bedrock. But for the same input wave conditions, there will be a 2 to 3 dynamic magnification in the overburden compared with the velocity response in the rock. If the structure at risk is founded on drift material, then the

Regression lines (y on x)
O.B. only : $\log v = 1.4683 - 0.9773 \log (r/W_D^{1/2})$
O.B. + rock : $\log v = 2.0070 - 1.3225 \log (r/W_D^{1/2})$
? limit : $\log v = 1.8722 - 0.9773 \log (r/W_D^{1/2})$

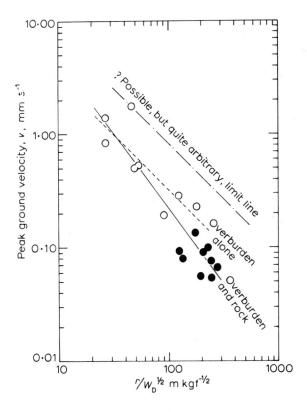

Figure 7.58 Scaled distance attenuation relationships for blasting in a rock. *Delay charge* (W_D) and *square root* ($W_D^{1/2}$) scaling. Open circles refer to overburden ground velocity readings. Infilled circles refer to rock velocity measurements.

ground velocity data with respect to the overburden alone would be used to formulate an attenuation law. If a structural foundation is intimately keyed into bedrock, examples being arch or concrete gravity dams, or nuclear power stations, then if possible, the attenuation law should be formulated from rock velocity measurements only and any internal vibration magnification assessed separately (see for example, Skipp and Tayton, 1971, on this latter point). However, it is not always feasible to measure on bedrock, and so a composite bedrock-overburden attenuation law may have to be formulated. The difference between the two laws may be quite significant.

Should the regression line equation(s) be taken directly for design purposes? It might possibly be argued (with an equally arguable degree of conviction!) that an attenuation line should take in the worst ground velocity measurement point and possibly run parallel to a regression line. One would then be using a statistical slope and a single maximum velocity-axis intercept to express an empirical law. Alternatively, the centroid of the rock data might be computed and a regression line drawn through that centroid and parallel to the overburden-only regression. When computing such a centroid, note should be taken of the logarithmic scales. These options are shown on the graphs.

A third explanation to be made for the less numerically-minded readers concerns the derivation of the constants in the attenuation law. The method is best illustrated by reference to Figure 7.55 from which it will be noted that the origin (0,0) of the axes corresponds to the point $v = r/W_T^{1/3} = 1$. The antilog of the constant in the regression line equation will then be marked off as shown (for example, 1.5099 units intercept on the v axis, O.B. only, corresponds to an actual ground velocity of 32.35 mm s^{-1}). It also follows that the slope of the regression line (tan^{-1} 0.9143 = 42.44° in the case of the O.B. only line) can be set off most easily from the v ordinate axis by taking the complement of the slope angle (47.56° for the O.B. only line).

Having derived a number of possible attenuation lines and then decided which of them is most appropriate for design purposes, it further remains to express that information in a manner that is more useful for blast design. There are two obvious ways of doing this. Acceptable charge levels can be calculated and plotted against stand-off distance for the introduction of specific levels of ground vibration at a remote point of interest. Alternatively this same information can be expressed as a series of quarry limit lines (equivalent to contour lines, or 'iso-vibes' as so termed by Attewell *et al*, 1965), each line defining a maximum encroachment distance towards property by the quarry given a specific weight of explosive charge and a specific vibration level at the property. Since the second method follows from the first, it is appropriate to show how charge limit graphs would be constructed.

Table 7.7 Calculation of charge weights from specified ground vibrations — $W^{1/2}$ scaling; overburden plus rock least squares regression

Charge	Ground vibration	Equation
Total	2.5 mm s^{-1}	$W_T = 2 \log r - 1.4395$ kgf
Total	4.0 mm s^{-1}	$W_T = 2 \log r - 1.0787$ kgf
Total	6.0 mm s^{-1}	$W_T = 2 \log r - 0.7675$ kgf
Delay	2.5 mm s^{-1}	$W_D = 2 \log r - 2.4334$ kgf
Delay	4.0 mm s^{-1}	$W_D = 2 \log r - 2.1247$ kgf
Delay	6.0 mm s^{-1}	$W_D = 2 \log r - 1.8584$ kgf

For the illustration, Figures 7.56 and 7.58 relating to the $W^{1/2}$ scaling will be used and the regression lines on the overburden-plus-rock may be adopted. The relevant equations are, then,

$$\log v = 1.2124 - 1.1316 \log (r/W_T^{1/2}) \qquad (7.26)$$

and

$$\log v = 2.0070 - 1.3225 \log (r/W_D^{1/2}) \qquad (7.27)$$

From these equations, we derive the data in Table 7.7.

These data are then transferred to Figure 7.59 for direct design of the charge weights. Up to the specified range, delayed charge weights would be prescribed, but beyond that distance, the total charge curves would be used even under delayed blasting conditions.

It must be realized that ground vibration is not usually the only control on blasting in the proximity of surface or underground structures that might be prone to damage. In certain instances, more severe controls may be imposed by the possibility of air blast overpressure waves should a large quantity of explosive in the quarry store be detonated accidentally. Potential damage levels from different overpressures may be estimated from data available in the literature. Similarly, the range of ejecta from unconfined charges may be estimated.

7.14 Marine geotechnical exploration

The discovery of submarine oil deposits, particularly those beneath the British Continental Shelf in the North Sea, emphasized the need for submarine soil exploration.

Submarine exploration methods for determining physical and

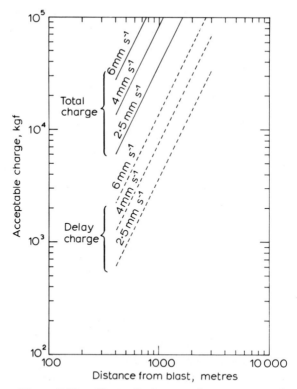

Figure 7.59 Charge limit curves for a square root scaling on the charge weight and for least squares regression parameters referred to measurements on both overburden and rock.

mechanical property data can be subdivided into four basic types:

(a) *Sea floor sampling* – using grabs for disturbed samples and piston samplers or gravity corers to obtain undisturbed samples for laboratory testing. These are well-proved traditional methods and operate at water depths up to 300 m. and for distances up to 10 m. below the sea floor.
(b) *Rotary drilling* to water depths up to 6000 m. and sub-sea floor distances greater than 1000 m. Sampling may be by conventional coring or wire line methods.
(c) *In situ tests* ranging from vane testing to measure shear strengths of clays to cone penetrometers and accelerometer monitored sampling devices (see Scott, 1970) to measure shear resistance in sands.

(d) Reflection seismic surveys to determine seismic (P) wave velocities in submarine sediments.

Since seismic surveys obviously cover a wider area per unit cost than do the other methods, and because large data banks of seismic information are in existence, there is a considerable incentive to utilize the well-known relationship between P-wave velocity data and other material properties as a means of estimating some mechanical and physical properties which cannot otherwise be measured remotely. Much research in this direction has related P-wave velocity and related variables to various properties measured directly or from samples. Typical of this is work by Buchan, McCann and Smith (1972) and McCann and Smith (1973) who propose, among others, statistically viable relationships between P-wave velocity and mean particle diameter, porosity, wet density and sand fraction; between acoustic impedance and wet density; between attenuation coefficient and mean particle diameter, sand fraction and clay fraction; and between dynamic compressibility and laboratory values for uniaxially confined compressibility. There is also some correlation between electrical resistivity and porosity.

However, in a variable non-elastic material correlation is not, and probably never will be, sufficiently good for geophysical techniques to replace conventional sampling and testing procedures as an exploration method for the heavy and expensive structures usually installed at sea for oil well exploitation. In general, seismic data can only be used to indicate major sediment boundaries and rock-sediment interfaces with any degree of confidence. The major application is, therefore, probably confined to the preparation of small-scale geotechnical maps which have very limited uses in foundation design.

Detailed site investigations for sea floor structures must utilize alternative methods of exploration and this usually means sea floor rotary drilling, together with conventional sampling and *in-situ* testing. Recent advances in each field include respectively; submarine drills for remote or diver operation (Tirey, 1972): remotely operated corers and piston samplers of extreme sensitivity (Noorany, 1972); and highly sophisticated penetrometers, vane tests and down-the-hole pressure cells (Richards *et al*, 1972). These and other developments are described in an ASTM symposium (D'Appolonia, 1972).

In the past 15 years a considerable amount of literature has evolved on the mass physical and mechanical properties of submarine sediments. The bulk of this literature is summarized in the comprehensive reviews of Bjerrum (1973), Noorany and Gizienski (1970), Keller (1969) and Richards (1967).

The major conclusions, which are supported by overwhelming evidence, are that near shore and continental shelf deposits (up to 200 m deep) comprise sands, silts and clay materials, the distribution of which is related to the local geology of the area, and whose physical and mechanical properties are controlled by the same basic mechanisms as control terrestrial sediments. Deep sea deposits of average submarine depth 4000 m but which are not the concern of the present book, are predominantly fine brown clays, derived partly from atmospheric dust and partly from sedimentation. These have some atypical mechanical properties. For instance, the compressibilities and strengths of some deep sea deposits appear to be controlled to a certain extent by poorly understood particle bonding mechanisms which do not affect shelf sediments.

The actual mechanical properties of shelf sediments vary with local geology. Typically they have high water content, low (dry) bulk and relative density, high porosity, and in the case of clay soils high plastic limit/liquid limit ratios. When subjected to force they have low shear strength and high compressibility. This is particularly true in the well researched area of Southern California (Inderbitzen, 1970) and the Gulf of Mexico (Bjerrum, 1973) and also for recent lacustrine deposits in Lake Marcaibo (Bjerrum, 1973). These last two are significant in that they are the major submarine oil fields worked up to 1974 and represent, together with information on harbour and coastal works, the major source of application of, and information on, marine geotechnics.

The properties of sediments on the British Continental Shelf, and particularly the North Sea, are radically different from most of those found elsewhere. The North Sea sediments typically comprise thick surface layers of 100 per cent relative density, wave-compacted glacial or post-glacial uniform fine sands and coarser materials underlain by layers of Pleistocene clays and sands. The clays are heavily overconsolidated having high strengths of the order 500 kN m^{-2} and low compressibility. As a foundation material they are in Bjerrum's words 'excellent', and are easily able to support the proposed heavy well head structures with a minimum of foundation preparation.

For this reason a variety of structures with weights up to 200 000 tonnes and designed to rest directly on the sea floor have been proposed and in some cases constructed. The foundation design requirements are summarized by Bjerrum:

(a) there must be a reasonable safety factor against sliding and shear failure when the structure is subjected to transient or repeated maximum wind and wave loading, and settlements and displacements under such conditions must be acceptable,
(b) the natural frequency of the structure must differ from that of the wave motion to avoid resonance,
(c) there must be good initial and maintained contact between the base of the structure and the sea floor,
(d) there must be protection against scour.

In the North Sea all of these requirements present interesting problems. Bearing capacity calculations for instance must allow for (i) high dynamic forces exerted by 100 year-type waves which can create undrained loading conditions in fine saturated sands, and (ii) accelerations up to 0.4 g exerted in the soil beneath the structure by repeated wave loadings. This latter point is particularly important since the uniform soil grading is typical of that of soils in Japan (Seed and Idris, 1967) liquefied by earthquake loading.

These aspects are considered by Bjerrum (1973) as also is the natural frequency of the structure, which depends on the compressibility of the soil and the resulting transient settlement under wave forces. The computed differential settlement of 300 mm for a typical structure under the action of a 100-year wave, compared with 200 mm construction settlement indicates the scope of the problem.

Not all submarine sediments are as stable as the North Sea sediments. Bostrom and Sherif (1974), for instance, show that soil and sediment accumulations flanking continents, and particularly deltaic sediments, can have negative strengths. Thus, application of a relatively small force can lead to the release of energy from the mass. In some cases the yield threshold may be so low and the potential energy for deformation so large that stabilization may not be economically feasible.

The effect of negative strengths can be seen where overloads due to ongoing sedimentation — such as the deltaic front of the Mississippi river in the Gulf of Mexico — can lead to slumping (over 100's of square kilometres in extreme cases), creep and faulting of sediments on the continental margin. Instability arises as fluid moves from the centre of the compressed sediment mass to the outer layers which comprise mainly clay minerals. As overburden pressures increase, water is released, through the operation of squeezing, consolidation and possibly potassium absorption, to the surface layers, which are already soft and unstable, and upon which construction must be based.

Bostrom and Sherif isolate deltaic aprons of major rivers and the flanks of tectonic trenches as particular problem areas and suggest concentration on *in-situ* pore pressure measurement in deep boreholes, and on depth-strength profiles (see also Section 3.5) as major exploration areas. They illustrate their argument with some depth-strength profiles and show how loss of strength due to sliding or sediment collapse can reduce strength over several hundred metres. They conclude that anchored, floating structures may represent the best construction option.

7.15 Mining subsidence

An important aspect of preliminary site investigation is the prediction of subsidence and resultant ground strains resulting from the working of minerals beneath a structure. A secondary problem is the estimation of the likely effect of construction on existing uncollapsed cavities.

The magnitude of subsidence will be related to the type of deposit mined and the method of work. It is most severe where caving methods of working and strata control are used. These methods are becoming increasingly common and involve total collapse of the worked area to relieve stresses on adjacent access tunnels and workings. The types of deposits worked can be considered under three basic headings:

(a) Near-vertical *veined* ore bodies, which include most metalliferous ores and which are formed mainly by deposition from hydrothermal solutions forced into pre-existing fissures or fractures.
(b) *Massive* orebodies which are usually formed in the same way as veined orebodies (or possibly by replacement of limestone) but which are irregular or lenticular in shape.
(c) *Layered* deposits, mainly sedimentary in origin and often near-horizontal, and including coal, evaporites and stratified ironstones.

Prediction of mining subsidence magnitudes above vein or massive deposits is difficult. Obviously, the subsidence will increase and extend (in the absence of support) with progressive deepening of the mineworkings. The only safe prediction is that subsidence will usually be severe and surface construction in the subsidence zone should not be contemplated. The extent of the zone may be roughly equated to that of an active pressure zone (Section 2.14) extending from the lowest point of the workings through the hangingwall zone at an angle of $45 - \phi/2°$ to the vertical.

In the case of tabular or layered deposit workings, subsidence is generally of a lower order and is more easily predicted. In this case, construction can be contemplated provided that the magnitude and extent of the subsidence zone can be estimated and its form predicted. The standard work in this type of exercise is the National Coal Board (1963) Subsidence Engineer's Handbook which not only collects together a large amount of case history data from subsidence observations in British Coal Mining practice but which also has much wider application.

Curves in this handbook express the relationship between the ratio of maximum subsidence and seam thickness and the ratio of width to depth for advancing longwall workings. With pillar and stall workings, the degree of subsidence will depend on the stability of the pillars. If they are sufficiently strong, then subsidence should be negligible except in near-surface workings.

The shape of the subsidence profiles in directions normal to and along the centre line of the direction of advance of a longwall face are similar in many ways to that of the subsidence profile above a tunnel in clay (see Section 7.11). Schmidt (1969), who has summarized data from many sources, shows that the transverse sigmoidal subsidence semi-profile takes on a near-Gaussian form provided that the width/depth ratio for the longwall excavation is less than unity. If the ratio is greater than unity, there is a tendency for the form of the 'stretched' profile to approximate to a harmonic function (see Farmer and Attewell, 1975). The traditional method of building up a profile normal to the centre line is to extend a 'draw' line from each side of the excavation, defining an active state wedge at an angle $45 - \phi/2°$ to the vertical, and to construct Gaussian or Harmonic curves on the basis of $x = 3i$ (see Figure 7.60) at the intersection of this line with the surface. The maximum subsidence (s_{max}) is then computed from tables or nomograms.

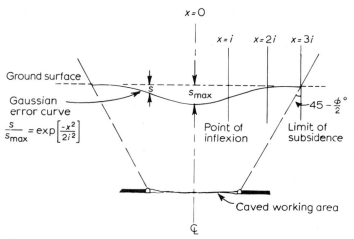

Figure 7.60 Representation of a transverse subsidence profile above a tubular mining zone.

From this construction, the points of maximum *curvature* can be identified and the *strain*, which is a critical factor in determining surface or underground structural damage, can be estimated, perhaps by using stochastic theory (Litwiniszyn, 1956). Alternatively, a more analytical procedure based on elastic or plastic analysis (see Voight and Pariseau, 1970) can be considered.

In some coal mining areas, sub-aqueous tunnels may be prone to hazard associated with residual ground strains. When a coal seam is extracted by the British longwall method, a zone of nett ground tension precedes and moves in phase with the coal face and a zone of compressive strain follows the face. Similar zones of tension and compression are developed transversely along the face. When a face stops, the tension and compression remain in the ground, the former to open up any joints that may be present in the over-burden and the latter to tighten up the joints. Should the tensile zones protrude, for example, beneath a river, and should it be necessary subsequently to tunnel through these zones, then depending upon the magnitude of the ground strain the tunnel is likely to encounter an increased and variable water make.

Considerable water problems requiring high air pressures for temporary ground support were encountered during construction of the River Tyne vehicular tunnel in England (see Prosser and Grant, 1968; Falkiner and Tough, 1968). Construction of a 3.23 m diameter sewage syphon tunnel beneath the same river commenced in 1974, but at the planning stage a detailed analysis of ground strain distribution from the old coal mine workings (Raw, 1940) on both banks of the river was undertaken by Boden (1969) on much the same lines as Knill's (1973) analysis for the Tyne vehicular tunnel. Superimposition of ground strains from workings in the same seam and in different seams permitted the construction of a tunnel line tensile strain profile from which the critical zones could be identified. Although the line of the tunnel was changed at a later stage, it is not a difficult matter to re-profile and reinterpret the strain distribution along any projected line once the strain contours have been computed and drawn in plan. Detailed consideration of the strain calculation technique, which is based on information provided by the British National Coal Board in its Subsidence Engineer's Handbook, is outside the scope of the present work. Contoured residual strains relative to the discussion above are shown in Figure 7.61 and the strain distribution along the tunnel line as originally planned is drawn in Figure 7.62.

The ground is unlikely to respond homogeneously to the distribution of applied tensile strain. Instead of each joint within a strained zone opening up a relatively small amount so that the integrated dilation satisfies the total zonal strain, a few joints may be more likely to open up to quite a large extent and thereby prove to be more hazardous when intercepted by a tunnel under construction.

Suppose, for example, that major vertical joints are spaced at about 1.5 m along the direction of tunnel advance and that over a length of 100 m the average ground strain across the joints is 0.3 per cent. On a homogeneous strain effect basis, the average opening of the individual joints would be only 0.45 mm whereas a single joint could accommodate the tensile strain by opening a third of a metre. In-tunnel observation of joint displacement trends and their correlation with predicted strain variations along the tunnel line may well indicate the manner in which the joint structure has accepted the imposed strain. By extrapolating this type of evidence to points ahead of the tunnel face it may be possible to locate quite limited zones of special hazard.

Residual strains associated with old mine workings are also considered (Woodland, 1968) to have contributed in a major way to the South Wales Aberfan spoil tip disaster on Friday 21st October, 1966. The mudflow of waste material was attributed to the presence of abnormal volumes of water within the tip, a situation which at

538 *Principles of Engineering Geology*

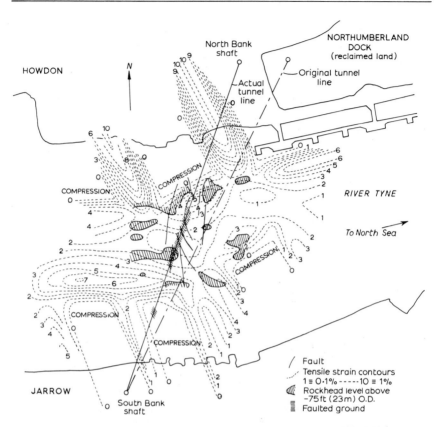

Figure 7.61 Line of the River Tyne syphon tunnel together with the distribution of residual tensile strains from the longwall coal mine workings in the underlying High Main and Maudlin seams (generally *after* Boden, 1969).

first sight conflicted with the potential drainage capacity of the well-jointed Brithdir sandstone which was underlying the tip. Water seeped downwards until its progress was impeded by an impermeable horizon from which it emerged in the form of springs and steady seepages. At Aberfan, a tongue of boulder clay sealed off a portion of the drainage path but there still appeared to be an adequate facility for drainage around the boulder clay. However, an analysis of residual strains created by mining subsidence (Figure 7.63) indicated a 'funnel' of tension generally below the boulder clay but flanked by zones of nett compression within which the groundwater transmissibility would have been very much reduced. Both the boulder clay and the ground compression barriers to seepage encouraged, in effect, a build-up of pore water pressure in the sandstone at the base

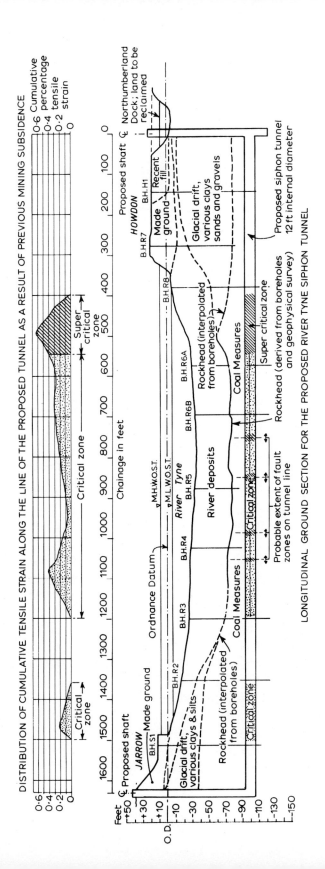

Figure 7.62 Distribution of tensile ground strain, from underlying old coal mine workings, along the originally proposed line of the River Tyne syphon tunnel in north-east England.

540 *Principles of Engineering Geology*

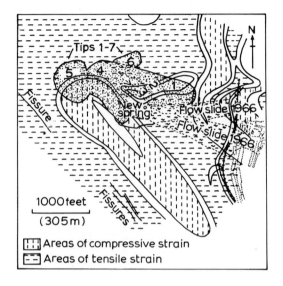

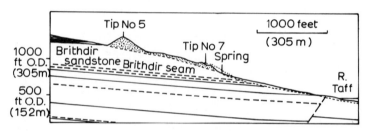

Figure 7.63 Geology and strain distribution in the vicinity of the tip complex at Aberfan, South Wales (*from* Woodland, 1968).

of the tip, promoting earlier evidence of 'back-sapping' and local instabilities to be followed, after two days of very heavy rain, by rupture of the boulder clay seal and the drainage of some 84 million litres of water within the mudflow.

With the benefit of hindsight, it is all too easy to say that the disaster should have been avoided. But with a more technical awareness of possible trigger mechanisms for mudflows and certainly of the importance of water pressure in the problem, engineers and geologists are better equipped to prevent such disasters in the future.

A further interesting example of subsidence engineering in the context of site investigation may be quoted. It occurred during an engineering geological survey of the Houghton Cut in County Durham, England. The survey, conducted by Turner (1967) several

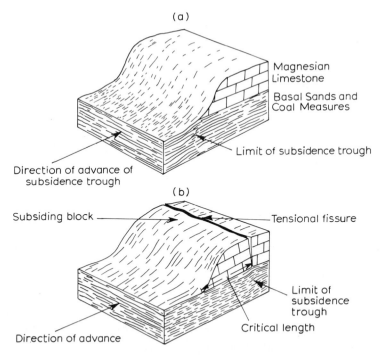

Figure 7.64 Diagrammatic representation of a subsidence trough advancing on the Magnesian Limestone ridge at Houghton-le-Spring, County Durham, England. (a) Edge of trough approaching the ridge (b) Edge of trough had advanced a critical distance beneath the ridge, causing a wedge-shaped block of limestone to break-off and subside *(after* Turner, 1967).

years before the Durham City-Sunderland road widening scheme was started, was concerned with the stability of a cutting in the Permian Magnesian Limestone scarp and also, as a by-product of the work, with the influence of underground coal mine workings on the presence of tension gashes running roughly parallel to outcrop in the Lower Magnesian Limestone.

The Magnesian Limestone rests on the Basal Permian Sands which themselves lie unconformably on the Coal Measures and it has been argued (Turner, 1967; Attewell and Taylor, 1970) that there is a critical advance distance at which the Magnesian Limestone cantilever beam, pinned at one end, actually fractures in tension along its upper fibre (Figures 7.64a and b). Calculations of the tensile strength of the rock for different configurations of beam section (Figure 7.65) based on the distribution of tension gashes were consistent

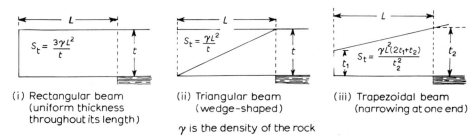

(i) Rectangular beam (uniform thickness throughout its length)

(ii) Triangular beam (wedge-shaped)

(iii) Trapezoidal beam (narrowing at one end)

γ is the density of the rock

Figure 7.65 Equations used in the calculation of the mass tensile strength of Magnesian Limestone (*after* Turner, 1967).

with the tensile strength of the rock determined experimentally. Subsequent *re-analysis* based on this strength then predicted the location of a further tension gash which had not been mapped thereto. When a field search was instituted, the predicted gash was found, infilled with mud and partially concealed. It was concluded that discontinuities in the Lower Magnesian Limestone and the underlying blanket of Basal Sands had not affected to any great extent the action of the cantilever beam when underground support had been removed near to outcrop.

In coal mining areas, pillar and stall or room/bord and pillar methods of working offered an early system of roof support but generally at the expense of poor percentage extraction. In Britain, pillar and room sizes have varied through the centuries and the system is used in modern times in, for example, Staffordshire under thick (9 m) seam conditions. Room sizes from about 1.8 m to 4.5 m and pillar sizes from about 3 m to 5.5 m might be found in Britain. The room and pillar system of coal mining is, of course, common in North America.

The question of long-term pillar collapse and instability under surface loading involves a relationship between the room width and pillar width, the height of the openings, and the spread of the loading. Documentary evidence of pillar widths may be suspect if illicit 'robbing' of pillars has taken place. Empirical evidence on room and pillar stability within the top 30 m. or so suggests that long-term, under ground pressure alone, collapse will not occur if the ratio of depth to pillar width is less than 0.1. Alternatively, using a simple, end-pinned, beam analogy and ignoring ground discontinuities such as joints, evidence suggests that the pillar width should be less than one-fifth of the depth of cover, or one-quarter for a competent sandstone, if the beam is to remain stable. By estimating a thickness of beam (t) and allocating a tensile strength (S_t) and density (γ) to the roof rock, a limiting beam span (L) may be calculated as $L = (2S_t t/\gamma)^{\frac{1}{2}}$ by equating bending moments.

Suppose that the actual L is too large and that shear failure of the beam occurs at the pillar edges. A void will migrate upwards in arch configuration until it either closes to become stable under the conditions then prevailing or it meets a higher bed of greater tensile strength (Figure 11.29). Bulking of de-coupled material in the room beneath may limit upward movement. At the limit, the void may break through to ground surface.

Various estimates have been made of the upward extent of the arch, but this is clearly dependent not only upon the factors mentioned above but also upon the discontinuous character of the rock and the presence or otherwise of water. One expression for the height of collapse (H_c) used by mining engineers and incorporating the seam thickness (h_t), the intact density of the roof rock (γ) and the bulk density (γ_k) of the collapsed roof rock is:

$$H_c = h_t \left\{ \frac{\gamma_k}{\gamma} \bigg/ \left(1 - \frac{\gamma_k}{\gamma}\right) \right\} \tag{7.28}$$

Provided that γ_k can be estimated for broken rock, some idea of H_c may be postulated. Measured H_c values seem to range from about $4h_t$ to $8h_t$.

When re-development is to take place over old pillar and stall workings it is necessary to appraise the strength of the pillars with respect to the imposed surface loading. Although the surface loading dissipates with depth, this factor is somewhat offset by the stress concentrations at the pillars. Consider, for example, the problem shown in two dimensions in Figure 7.66. The pillars are loaded both by the overburden and by the building. Symbol η_e is used to represent the degree of extraction and may vary from zero to unity.

Pressure imposed on the pillars by the strata

$$= \frac{\gamma z}{1 - \eta_e} = \frac{104}{0.34} = 306 \text{ kN m}^{-2}$$

Assume a surface loading (q) of 215 kN m^{-2} (2T ft^{-2}) on a rectangular foundation area of 6 m x 6 m. Using Fadum's chart in the Boussinesq example given by Capper et al (1966, p 71), we note that parameters m = 3/5 = 0.6, n = 3/5 = 0.6 and the influence factor

544 Principles of Engineering Geology

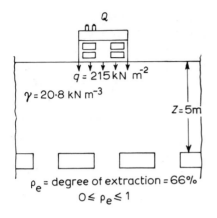

Figure 7.66 Problem on pillar loadings.

$I_f = 0.109$. Thus, vertical pressure from building

$$= 4 \times q \times I_f$$
$$= 4 \times 215 \times 0.109 = 94 \text{ kN m}^{-2}$$

which, when divided by 0.34 becomes 276 kN m^{-2}.
Thus, the total vertical pressure on the pillars is $306 + 276 \simeq 582$ kN m^{-2}.

Other surface loading configurations may be analysed by the methods discussed in Capper et al (1966).

As an alternative method of analysis at the site investigation stage, Terzaghi's (1943) analysis for tunnel loading may be employed. The vertical pressure σ_z acting on the tunnel is expressed as:

$$\sigma_z = \frac{B\left(\frac{\gamma}{2} - \frac{c}{B}\right)}{K \tan \phi} \left\{1 - \exp\left(-K \tan \phi \cdot \frac{2z}{B}\right)\right\}$$

$$+ q \exp\left(-K \tan \phi \cdot \frac{2z}{B}\right) \qquad (7.29)$$

In this equation, c and ϕ are the shear strength parameters (c may be approximated to zero), $K = \sigma_x/\sigma_z$, and B is the width of the pressure arch. Symbol z is used to denote depth and the remaining symbols are also as defined earlier. The first term in the equation refers to the self weight of the superincumbent strata and the second term defines the stress at depth z due to surcharge q.

Assume that $K = 1$ for the shallow depths considered and take $\phi = 45°$ for Coal Measures rocks. Then using the figures in the previous example we have

$$\sigma_z \simeq \frac{B \times 2080}{100 \times 2}\left\{1 - \exp\left(-\frac{10}{B}\right)\right\} + 215 \exp\left(-\frac{10}{B}\right)$$

where B = bord width *plus* $2h_t \tan(45° - \phi/2)$.

If, say, bord width = 4 m and h_t = 1 m, then B = 4.83 m and σ_z = 71 kN m^{-2}. Suppose, after Terzaghi, that H_c = 1.5 B. Arching will then occur up to ground surface in the above example. If $z > 1.5 B$, then 1.5 B replaces z in the equation and to q will be added a further rock pressure of $\gamma(z - H_c)$. Thus if, for example, z = 15 m, σ_z is then 66 kN m^{-2}. In practice, rock structure will affect H_c (Terzaghi, 1946).

Having estimated the stress applied to coal pillars it remains to estimate the actual strength of those pillars against compression collapse. Pillars in deeper workings (perhaps greater than 50 m) will probably have crushed out anyway (for example, at 50 m depth and 70 per cent extraction, the 'Boussinesq' *self weight* loading is in the region of 3.5 MN m^{-2} and this probably does not fall far short of the unconfined compressive strength of a weak coal). Thus, in the context of the redevelopment problem and surface surcharge loading, the real problems are likely to arise in the shallower workings at depths from about 30 m to 50 m which preclude the possibility of 'digging out'.

As is well known from laboratory tests on small specimens (Section 4.1), the compressive strength (S_c) of a cylinder or cube is conditioned by its height (h_t)-to-width (d) ratio for aspect ratios of less than about 2.5 to 1. The following empirical relationships have been proposed at various times for the strength of coal pillars:

(i) S_c (lb in^{-2}) = 1000 $d_{(ft)}^{0.16} \times h_{t(ft)}^{-0.55}$ \hfill (7.30)

\hfill (Bieniawski, 1968)

(ii) S_c (lb in^{-2}) = 1320 $d_{(ft)}^{0.46} \times h_{t(ft)}^{-0.66}$ \hfill (7.31)

\hfill (Salamon and Munro, 1967)

(iii) S_c (lb in^{-2}) = 1810 $d_{(ft)}^{0.16} \times h_{t(ft)}^{-0.49}$ \hfill (7.32)

\hfill (Jenkins and Szecki, 1964)

Although only expressions (i) and (iii) are based on the same coal type, estimates of pillar strength might usefully be drawn from all three expressions. For example, suppose that the pillar width is 12 ft (~ 3 m) and the seam thickness is 6 ft (~ 2 m). The appropriate pillar strengths S_c are then: (i) 555 lb in^{-2} (3.827 MN m^{-2}); (ii) 1269 lb in^{-2} (8.749 MN m^{-2}); (iii) 1120 lb in^{-2} (7.722 MN m^{-2}). Further problems in pillar design are described by Grobbelaar (1970).

Again, although these expressions are extrapolative to some extent from laboratory to full scale and must be treated with caution, they do at least act as some guide during the decision-making process. When investigating a field problem, the many parameters would be permutated one against the other in order to assess the sensitivity of each unknown in the problem. A factor of safety might then be expressed as the ratio of pillar strength to pillar stress, and the possibility of infilling (Section 11.8) considered.

Certainly in Britain, large-scale pillar failure at shallow depth is uncommon. Very old pillars may progressively oxidise and spall over a period of centuries and there may be local failures in the proximity of geological faults. Most pillars are of such a size that they will remain unaffected by surface loading. They may penetrate a yielding seat earth (underclay) floor and there may be some heave into adjacent rooms, but the overlying problem seems to be one of roof stability and surcharge loading design to offset the possibility of collapse. Flooding of the workings may also cause instability.

In Britain, the presence of medieval 'bell' pits and old, abandoned mine shafts pose special hazards in some redevelopment areas. They must first be located and decisions then made as to how they are to be treated. Bell pits used for the extraction of minerals such as ironstone, fireclay and coal will only be found where the drift cover is thin. They will be unlikely to exceed about 12 m depth and access to the mineral below will be through a shaft of only just over 1 m diameter. Using the characteristic self-support properties of an arch or dome, the base of the shaft will then have been excavated outwards until such time as the experience of the miner suggested that collapse was imminent. The pit would then be abandoned and another one started nearby. Adjacent pits were sometimes linked by narrow passageways both to improve the mineral yield and to enhance access and egress. Bell pits would sometimes be backfilled when no longer required, but more often they were left to collapse naturally. Most will have collapsed and may be identified by aerial photography as an irregular pattern of depressions in the ground. Sometimes suspect pits may be detected by resistivity or seismic shoot methods, but these are not always reliable. The best confirmatory technique is to make use of rotary drilling to locate the void. In a densely pitted area, ground treatment by infilling after positive location would be expensive and the best remedial measure, in view of their limited depth, would be to dig them out and recover the remaining mineral. Some shafts may have been partially backfilled and a wooden raft needled into the walls of the shaft one or two metres below ground surface. Wooden timbers will have decayed and if it is decided not to dig the shafts out, the original raft must be replaced by a reinforced concrete platform after restoring any original backfilling. New structures at ground surface would be located in firm ground and any suspect areas would be allowed to remain unloaded as open spaces.

7.16 Probability theory in site investigation

Terzaghi's logical observational approach to geotechnical problems has been outlined by Peck in both his State of the Art Review at the 7th International Soil Mechanics and Foundation Engineering Congress (1969a) and his Rankine lecture to the British Geotechnical Society (1969b). With a wealth of fundamental knowledge and experience, Terzaghi was able to make decisions and progressively update his advice as new geotechnical information came to light. With sufficient system flexibility being built into the earthworks design at the outset, it was thus possible to converge towards an optimum economic solution.

It is useful to mention just a little of the logic behind decision theory which would allow the engineering geologist to combine subjective judgements with limited objective data. This involves the concept of probability distributions of random variables, one starting point being Baye's theorem which is outlined by, for example, Cramér (1954). Rosenblueth (1969) also discusses the implication of Bayesian statistics in a geotechnical context.

Bayes theorem may be written as

$$P(A_i|B) = \frac{P(B|A_i)P(A_i)}{\sum_i P(B|A_i)P(A_i)} \quad (7.33)$$

In this equation, A_i are mutually exclusive events; for instance, if A_i represents the value of some physical property of the ground which varies (with location), then $P(A_i)$ describes the form of that variation. The quantity $P(B|A_i)$ may be termed the *conditional probability of B relative to the hypothesis that A_i has occurred* and when $P(B) \neq 0$ the quantity $P(A_i|B)$ will be the *conditional probability of A_i, relative to the hypothesis that B has occurred* (that is, it is the probability that A_i is the true value of A given that B is the result of the test).

As an application of Bayesian decision theory, consider a preliminary site investigation for a metropolitan rapid transit system in tunnel. A useful procedure is for the drilling rigs to traverse a particular length of tunnel line putting down holes at intervals that are pre-conditioned by the engineer's or geologist's general experience and by some prior information from other investigations or from geological/geotechnical maps of the area. At the end of its run, a rig is then available to return along the line inserting additional, probably large diameter, holes where more information is needed due to prior lack of consistent geological/geotechnical trends.

Suppose that due to an initial rather low density of holes, specified rather rigorously on the basis of a seemingly consistent map geology, there arises a particular length through which the tunnelling is expected to take place in a firm laminated clay but which could conceivably change rapidly to a saturated silt. On being pressed for an opinion, the geologist assigns a 90 per cent probability to the former and a 10 per cent probability to the latter. He and the engineer then happen to observe some structural distress in a nearby building, and from observation of the character of the problem they decide that its probability of having arisen through construction on firm laminated clay is only about 25 per cent but that it would have had a 95 per cent probability of developing with foundations on the saturated silt. They would then be able to calculate the posterior probability of there being a saturated silt over that particular tunnel length as

$$P(A_i|B) = \frac{0.95 \times 0.1}{(0.95 \times 0.1) + (0.25 \times 0.90)} = 0.297 \simeq 30\%$$
(7.34)

The argument then is that, on the basis of the posterior probability figure, the geologist can then make a rather more informed decision on whether to re-call the rigs and the engineer can decide whether to make appropriate provision for compressed air or ground treatment in the tunnel. This procedure can be iterated and progressively updated as more information becomes available.

Let us denote the *event* when a variable takes the *value* A by A, although strictly this is incorrect for it confuses the event with the value. Suppose that A_* is the prediction of the value A_*, which again is strictly incorrect for here the event of prediction is confused with the value of the prediction. However, if A is continuous, we may proceed as follows:

$$P(A \mid A_*) = \frac{P(A_* \mid A) P(A)}{\int_{-\infty}^{\infty} P(A_* \mid A) P(A) \, dA} \qquad (7.35)$$

Suppose that A_* is the value of A as predicted by a young geologist or engineer who is absolutely confident that he knows the answer. This then implies that

$$P(A_* \mid A) \simeq \delta(A - A_*) \qquad (7.36)$$

where δ denotes a Dirac delta function.
Thus,

$$\int_{-\infty}^{\infty} P(A_* \mid A) P(A) \, dA = \int_{-\infty}^{\infty} P(A) \delta(A - A_*) \, dA = P(A_*)$$

$$(7.37)$$

or

$$P(A \mid A_*) = \frac{\delta(A - A_*) P(A)}{P(A_*)} = \delta(A - A_*)$$

A 'blinkered' approach to the problem is therefore shown to preclude any possible improvement by updating.

On the other hand, if A_* is defined with an arbitrarily large range of width ΔA by an open-minded engineer, then $P(A_* \mid A)$ is approximately equal to a constant, that is,

$$P(A_* \mid A) \cong 1/\Delta A$$

Therefore, $P(A_*)$ is given by

$$\int_{-\infty}^{\infty} P(A) \frac{1}{\Delta A} \, dA = \frac{1}{\Delta A} \qquad (7.38)$$

and it follows that

$$P(A \mid A_*) = \frac{(1/\Delta A) P(A)}{1/\Delta A} = P(A) \qquad (7.39)$$

from which, as expected, the *a priori* judgement of the engineer has no effect.

The critical importance of an experienced decision is therefore proven. It is also clear that if the width of the *a priori* distribution is narrow, then it will never be economically viable to do more sampling. The economic aspects of site investigation may further be considered on a probability basis with some reference to Wu and Kraft (1967) and Folayan *et al* (1970). At the present level, the treatment cannot be exhaustive and is only of an outline nature.

When the main variables in a problem have been identified, the following logical sequence of action may be adopted for a general analysis:

(i) Establish an overall prior probability distribution of the random variable,
(ii) Formulate a so-called utility or loss function to be used in the overall decision-making process,
(iii) Determine the optimum number of soil or rock samples to be tested or other significant variables to be evaluated,
(iv) After testing, find the sampling probability distribution for combination with the prior probability distribution in order to formulate a new prior probability distribution and evaluate whether further sampling/testing would be beneficial,
(v) When additional information is not needed, then the combined prior and sampling distributions form a so-called posterior probability distribution upon which the value of the random variable may be based,
(vi) Come to an optimal decision by computing an expected loss or gain from the adopted value of the variable and then minimizing any loss,
(vii) Estimate the reliability of that decision.

Prior probability distributions of a random variable may well be fitted from a number of professional subjective judgements before acquiring any field or laboratory test data. These distributions are simply expressed as means and standard deviations, or they may be drawn as a number of normal probability curves as in Figure 7.67. The wider the spread of distribution, the more the uncertainty. Some geotechnical properties may well be described more accurately by a log normal distribution.

If the number of variables in the problem controlling the material

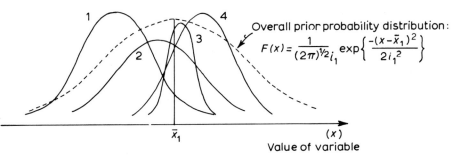

Figure 7.67 Some possible prior probability distributions of a random variable.

behaviour, and relating to each of N prior distributions, is expressed as n_a $(a = 1, 2, \ldots)$ (with mean $\bar{x}'_a$ and standard deviation i'_a), the mean $\bar{x}_1$ and standard deviation i_1 of the overall distribution will be quite simply:

$$\bar{x}_1 = \frac{\sum_{a=1}^{N}(\bar{x}'_a n_a)}{\sum_{a=1}^{N} n_a} \qquad (7.40)$$

$$i_1^2 = \frac{\sum_{a=1}^{N}(n_a i'^2_a) + \sum_{a=1}^{N} n_a (\bar{x}'_a - \bar{x}_1)^2}{\sum_{a=1}^{N} n_a} \qquad (7.41)$$

The overall mean can also be expressed in terms of the individual means inversely weighted by the variances, $i^2_{\bar{x}'_a}$ (Schlaifer, 1959):

$$\bar{x}_1 = i^2_{\bar{x}_1} \sum_{a=1}^{N} \frac{\bar{x}'_a}{i^2_{\bar{x}'_a}} \qquad (7.42)$$

where

$$i^2_{\bar{x}'_a} = i^2/n_a$$

and (7.43)

$$i^2_{\bar{x}_1} = i^2 \bigg/ \sum_{a=1}^{N} n_a$$

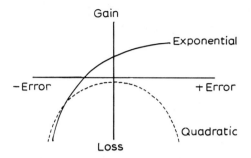

Figure 7.68a Possible loss functions.

The purpose of a utility, or loss function, is to relate any extra costs incurred to actual errors in the determination of the variable. It may take a variety of forms of which the exponential and quadratic are perhaps the most immediately useful. For example, should a high loss be incurred if the variable is under-estimated, but little be gained by an over-estimation, the function could then take the form of an exponential curve, as in Figure 7.68a. Alternatively, with a quadratic utility function, equal loss may be incurred irrespective of whether the error is positive or negative. Two functions as used by Attewell and Cripps (1975) are shown in Figure 7.68b.

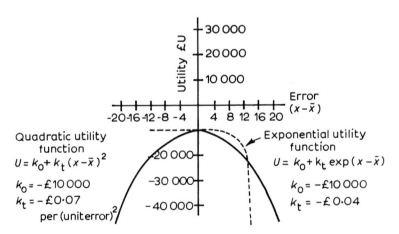

Figure 7.68b Quadratic and exponential utility functions as used by Attewell and Cripps (1975). k_0 is introduced in this case as a constant and is generally taken as the original cost. The form of this particular exponential curve is controlled by the values and signs of the constants k_0 and k_t.

If, for the purpose of this outline, it is assumed that the minimum loss, the so-called 'optimal act', coincides with the mean of a distribution, then the severity of the expected loss, using a simple *a priori* approach for illustrative purposes, may be expressed as:

$$I_{n1} \simeq \int_{-\infty}^{\infty} \frac{n_1^{\frac{1}{2}}}{(2\pi)^{\frac{1}{2}} i_1} \exp\left\{\frac{-(\bar{x} - \bar{x}_1)^2}{2i_1^2}\right\} k_t(\bar{x} - \bar{x}_1)^2 \, d\bar{x} \quad (n_1 \text{ samples})$$

$$I_{n2} \simeq \int_{-\infty}^{\infty} \frac{n_2^{\frac{1}{2}}}{(2\pi)^{\frac{1}{2}} i_1} \exp\left\{\frac{-(\bar{x} - \bar{x}_2)^2}{2i_1^2}\right\} k_t(\bar{x} - \bar{x}_2)^2 \, d\bar{x} \quad (n_2 \text{ samples})$$

$$I_\infty = \int_{-\infty}^{\infty} \delta(\bar{x} - \bar{x}_\infty) \underbrace{k_t(\bar{x} - \bar{x}_\infty)^2}_{\text{Loss function}} d\bar{x} \quad (\infty \text{ samples}) \qquad (7.44)$$

where k_t is a constant expressing the degree of loss due to error (cost per unit squared for a quadratic loss function). For example, adopting a quadratic loss function, k_t could be in £/(kN m^{-2})^2 (ie £/strength^2) or \$ mm^{-2} (ie \$/settlement^2).

Since it is obviously uneconomic to increase the sampling and testing cost $k_s n$ beyond the expected reduction in the cost of the works resulting from the increased knowledge so gained, it should be possible to consider an optimum number of samples and/or tests. We may expect the severity of the expected loss to decrease with the number of additional samples or tests and from Figure 7.69 it would be economic to take more samples and do more tests while $I_{n1} > I_{n2} + k_s(n_2 - n_1)$. Now if $I_{n1} - I_{n2} = V$, the difference in loss between two sampling densities, then the greatest potential saving should accrue when $V - k_s(n_2 - n_1)$ maximizes, that is, when

$$\frac{\partial V}{\partial n_2} = k_s; \quad \frac{\partial^2 V}{\partial n_2^2} < 0 \qquad (7.45)$$

It should be noted that with the function $k_s(n_2 - n_1)$ as drawn in Figure 7.69, saving is possible from the outset. There comes a point, however, when $I_{n_1} - I_{n_2}$ varies too slowly for any further saving to be possible.

In the case where absolute information provides just a single value for $\bar{x}$, the probability distribution takes the form of a δ-function

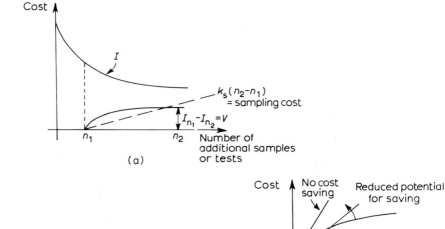

Figure 7.69 Sampling/testing economics.

centred on $\bar{x}_\infty$. Thus, assuming known variance and using equations 7.44,

$$I_\infty = \int_{-\infty}^{\infty} \delta(\bar{x} - \bar{x}_\infty) k_t (\bar{x} - \bar{x}_\infty)^2 \, d\bar{x} = 0 \qquad (7.46)$$

$$I_{n2} = \int_{-\infty}^{\infty} \frac{(n_2)^{\frac{1}{2}}}{(2\pi)^{\frac{1}{2}} i} \exp\left\{\frac{-(\bar{x} - \bar{x}_2)^2 n_2}{2i^2}\right\} k_t (\bar{x} - \bar{x}_2)^2 \, d\bar{x} \qquad (7.47)$$

and similarly for I_{n1}.
Writing x^* for $(\bar{x} - \bar{x}_2)$,

$$I_{n2} = 2 \int_0^\infty \frac{(n_2)^{\frac{1}{2}}}{(2\pi)^{\frac{1}{2}} i} \exp\left\{\frac{-x^{*2} n_2}{2i^2}\right\} k_t x^{*2} \, dx^* \qquad (7.48)$$

Now, since

$$2 \int_0^\infty e^{-ax^2} x^2 \, dx = 2 \left[-\frac{1}{2a} e^{-ax^2} x \right]_0^\infty + \frac{1}{a} \int_0^\infty e^{-ax^2} \, dx = \frac{1}{2a} \left(\frac{\pi}{a}\right)^{\frac{1}{2}},$$

then

$$I_{n2} = \frac{k_t i^2}{n_2} \qquad (7.49)$$

Therefore,
$$V = \frac{k_t i^2}{n_1} - \frac{k_t i^2}{n_2} \quad (7.50)$$

The range of n_2 values over which it is economical to perform tests is given by the solution of:

$$\frac{k_t i^2 n_2}{n_1} - k_t i^2 = k_s n_2^2 - k_s n_1 n_2$$

or,

$$k_s n_2^2 - \left\{\frac{k_s n_1^2 + k_t i^2}{n_1}\right\} n_2 + k_t i^2 = 0 \quad (7.51)$$

That is,

$$n_1 < n_2 < \frac{k_t i^2}{k_s n_1}$$

Now, the optimum sampling density arises when

$$\frac{\partial V}{\partial n_2} = k_s, \text{ that is, when } n_2 = \left(\frac{k_t i^2}{k_s}\right)^{1/2} \quad (7.52)$$

Thus,

$$\frac{\partial^2 V}{\partial n_2^2} = \frac{-2 k_t i^2}{n_2^3} < 0 \quad (7.53)$$

In general, however, 100 per cent information will not give a δ-function, but instead — and assuming a normal distribution:

$$P(x) = \frac{1}{(2\pi)^{1/2} i} \exp\left\{\frac{-(x - \bar{x})^2}{2 i^2}\right\} \quad (7.54)$$

The inferred distribution for unknowns $\bar{x}, i$, from n_1 tests will be:

$$P(\bar{x} \mid \bar{x}_1, i, n_1) = \frac{(n_1)^{1/2}}{(2\pi)^{1/2} i} \exp\left\{\frac{-(\bar{x}_1 - \bar{x})^2}{2 i^2} n_1\right\} \quad (7.55)$$

and

$$P(i \mid i_1, n_1) = \frac{2}{\Gamma\left(\frac{n_1 - 1}{2}\right) i} \left[\frac{(n_1 - 1) i_1^2}{2 i^2}\right]^{(n_1 - 1)/2} \exp\left\{\frac{-(n_1 - 1) i_1^2}{2 i^2}\right\}$$

$$(7.56)$$

We also have

$$I_\infty = \int k_t(x - \bar{x})^2 P(x) \, dx = k_t i^2$$

$$I_{n1} = \int k_t(x - x_1)^2 P(x) \, dx \, P(\bar{x}, i \mid \bar{x}_1, i_1, n_1) \, d\bar{x} \, di = k_t i_1^2 \frac{n_1^2 - 1}{n_1(n_1 - 3)}$$

$$I_{n2} = \int k_t i_2^2 \frac{n_2^2 - 1}{n_2(n_2 - 3)} P(i_2, n_2 \mid i) \, di_2$$

$$P(i \mid i_1, n_1) \, di = k_t i_1^2 \frac{n_2^2 - 1}{n_2(n_2 - 3)} \frac{n_1 - 1}{n_1 - 3} \quad (7.57)$$

where the pre-posterior analysis has been used for I_{n2} (Raiffa and Schlaifer, 1961). The value V of n_2 tests is $I_{n1} - I_{n2}$, and so

$$V = k_t i_1^2 \frac{n_1 - 1}{n_1 - 3} \left[\frac{n_1 + 1}{n_1} - \frac{n_2^2 - 1}{n_2(n_2 - 3)} \right] \quad (7.58)$$

It should be noted that apart from the completely *a priori* nature of the loss integral estimates, the above procedure may be faulted on the use made of probability statements about hypotheses, and it is probable that a new technique based on the method of support (Edwards, 1972) will need to be developed.

The reliability of the decision, which must be a reflection of the loss function adopted, is taken as one minus the probability of failure. Since 'failure' is not always amenable to absolute definition, it is valid to express the probability of the variable with respect to a certain range. For example, in the case of an exponential utility function, the reliability may be one minus the probability of the variable being 20 per cent higher than predicted, that is,

$$R_e = 1 - P(x > 1.2 x_a) \quad (7.59)$$

where R_e is the reliability of the estimate on an exponential utility function, and x_a is the predicted value.

Alternatively, for a quadratic utility function, failure may be defined as occurring should the variable exceed or undervalue the prediction by 20 per cent. Thus

$$R_q = 1 - P(x < 0.8 x_a \text{ and } x > 1.2 x_a) \quad (7.60)$$

where R_q is the reliability of the estimate on a quadratic utility function.

The overall reliability of being within or outside the allowable range can be computed for the upper and lower limits.

Suppose, for example, that in a site investigation comprising q boreholes, p determinations of a parameter x are made and the value adopted for design, a_p, is the least of the q values, $\bar{x}_p$. If the number of samples increases from $p.q$ to r so that a high sampling density tends towards a 'total' sampling of material in the foundation area (or area of design interest), the mean value so obtained tends towards the true site value $\bar{x}_r$. It follows from a general normal probability distribution for x that the form of distribution of $\bar{x}_r$ is

$$f(\bar{x}_r) = \frac{(r)^{\frac{1}{2}}}{(2\pi)^{\frac{1}{2}} i} \exp\left[-\frac{(\bar{x}_r - \bar{x})^2 r}{2i^2}\right] \quad (7.61)$$

and the distribution of a_p is

$$g(a_p) = q f(a_p) \left[1 - \int_{-\infty}^{a_p} f(\bar{x}_p) d\bar{x}_p\right]^{q-1} \quad (7.62)$$

The reliability of a_p is then

$$P(\bar{x}_r > a_p) = 1 - P(\bar{x}_r < a_p)$$

$$= 1 - \int_{-\infty}^{\infty} g(a_p) \int_{-\infty}^{a_p} f(\bar{x}_r) d\bar{x}_r \, da_p \quad (7.63)$$

This general approach to the analysis of site investigation data has been used by Cripps (1977), particularly with respect to foundation consolidation.

7.17 What is 'safety' in soil and rock mechanics?

Any geologist entering an engineering environment will very quickly be faced with the concept of a 'factor of safety'. The term itself represents an attempt to draw a strict line of demarcation between a state of incipient failure and a state of absolute safety. The factor of safety for a foundation, a deep excavation, or a slope will usually be defined as the ratio of the total force resisting failure to the total deforming force attempting to produce failure. In most cases, these forces will be shear forces and at critical equilibrium the factor of safety is unity. This implies that the resisting shear forces are just able to balance the deforming shear forces. As the factor of safety increases above unity, the rather unlikely concept of 'degree of safety' is invoked, but it is really only a statement of the extent to which the deforming forces would have to be increased with respect to the natural shear resistance of the geological material before failure would ensue.

Throughout the text, there is reference to, and calculations on factors of safety for various engineering situations, usually those related to slope stability. This approach must be adopted because it has gained universal acceptance and because it is most convenient for engineers to be able to hang their decisions upon a finite peg of reference. Nevertheless, it so happens that any natural system can never be 100 per cent safe and an alternative evaluation of stability would hinge upon the probability of failure as assessed using probabilistic methods of analysis. Unfortunately, the many interactive variables in any ground engineering problem render not only the solutions but also the problem specification difficult. In any case, a designer would much prefer to think in terms of a factor of safety of, say, between 1.5 and 2 than perhaps a probability of failure of 10^{-3} per cent.

Suppose that the factor of safety concept is considered to be the most appropriate for design purposes. There will be a number of variables, the magnitude of each variable having a range of uncertainty within the limits of the design problem. It may well be found that quite large variations in the magnitude of one of the variables will not influence the factor of safety significantly when all the other variables are held constant. On the other hand, small changes in one of the other variables may cause a dramatic swing in the factor of safety. A serious stability study will therefore require prior information on the amplitude distribution of each variable together with an iterative solution facility. In this manner, the sensitivity of each variable is expressed as the rate of change of the factor of safety with respect to that variable.

A question which persistently arises in ground stability design concerns the allocation of shear strength parameters, cohesion and friction, to the ground material and the calculation of a factor of safety based on those shear strength parameter values. The choice of these values may be statistically conditioned, or more usually it will be based on the experience of the engineer through a process of deductive reasoning and elimination.

One approach that may be used involves taking the ratio of both shear strength parameters with respect to calculated factors of safety.

The relationship between c and $\tan \phi$ for $F = 1$ will take the form of a straight line having a different negative slope for different values of the pore pressure ratio r_u (Bishop, 1955; Bishop and Morgenstern, 1960). Extreme values of c and $\tan \phi$ as $\tan \phi$ and c respectively tend to zero would usually be determined by extrapolation rather than by direct calculation due to a possible sensitivity in the relationship under these conditions. Having determined maximum values for both c and $\tan \phi$ it is then possible to enter these in the Coulomb equation in order to recompute a range of shear strengths for different input normal stresses. The whole operation can then be repeated for a different series of c, $\tan \phi$ values derived for the same material at different locations, and from a best fit line to the points plotted in the τ, σ_n plane, two final c, ϕ shear strength parameters may be derived.

Implicit in the above argument is the notion that the factor of safety of a slope with respect to shear is intimately controlled by the shear strength parameters. Obviously, this must be correct, but it is also apparent that the slope geometry and its possible interaction with geological structure exerts at least as strong a control on the eventual stability. It also happens to be the case that cohesion is a more sensitive parameter than the friction angle and more sensitive even in some instances than the pore water pressure. An increase in water table height (c', ϕ' constant) above a slip surface may reduce the factor of safety rather less than would the erosion of cohesive contact between the fixed and the moving mass. In the stability analysis of a jointed rock mass, there is always the temptation to ignore a small cohesive contribution to the total shear strength and to analyse solely on the basis of surface friction. This may lead to an excessively conservative estimate of potential hazard, as reflected by too low a factor of safety.

8 ~ Groundwater

Groundwater in engineering geology (as distinct from engineering hydrology) is mainly important because of the effect its presence can have on the stability and on the efficiency of ground engineering works. Thus, the presence of groundwater, by reducing effective stresses (see Section 2.6) in soils can cause instability in excavation sidewalls, and by uncontrollable inflow through fissures in pervious rocks can make working conditions in shafts and tunnels intolerable. In order to combat these and other effects, including piping caused by seepage beneath dams, the geologist must have a knowledge of simple hydrology and of some of the factors controlling groundwater flow and the seepage forces caused by groundwater flow. A discussion of these factors forms the initial part of this chapter. An opportunity is also taken to introduce the subject of water flow through earth dams. These early sections are followed by a consideration of remedial measures to reduce groundwater and some methods of measuring permeability.

Groundwater is also a valuable commodity in its own right with abstraction schemes playing a major part in any water resources management programme. Aquifer performance, the analysis of pumping test data, and simulation techniques therefore provide the main topics of discussion in the later sections of this chapter. The theme of water management is also carried over into Chapter 12 which is devoted to the question of surface water and reservoir storage.

8.1 Types of subsurface water

Following the precipitation phase of the hydrologic cycle, part of the precipitated water is retained in or on vegetation, part infiltrates the ground surface, and part runs off into streams or rivers. The actual amount penetrating the surface will vary, the proportion depending on the infiltration capacity of the surface soil, and probably more importantly on the infiltration capacity of subsurface layers. The water penetrating the ground will continue to flow towards major or minor discharge points, usually observable as springs, in a predominantly lateral motion, the actual flow paths and types of flow depending on the ground properties. Aquifer flow may be interrupted by cross-measures faulting or by dyke intrusion which may generate springs and marshy conditions at ground surface. Examples of some geological configurations that are conducive to spring formation are given in Bundred (1969). Three main types of flow are generally described:

(a) intermediate saturated flows above near-surface impervious layers. The upper surfaces of these flows are sometimes described as *perched* water tables.
(b) a major saturated flow zone usually defined at the top (that is, the water table) by a discharge source and at the bottom by an impermeable layer.
(c) an unsaturated flow zone between the surface and the water table through which water percolates or is held by capillary action.

A rough illustration of these flow zones is contained in Figure 8.1 and a more detailed classification suggested by Davis and De Wiest (1966) in Figure 8.2.

In this classification the term *soil* is used in the scientific or agricultural sense, rather than in the engineering sense, and *soil water* defines a near-surface zone (~ 10 m thick) where there are wide fluctuations in water content due to evaporation, minor precipitation through rainfall or condensation, and the discharge and intake of

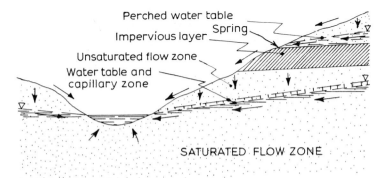

Figure 8.1 Typical groundwater flow regime — arrows represent direction of flow.

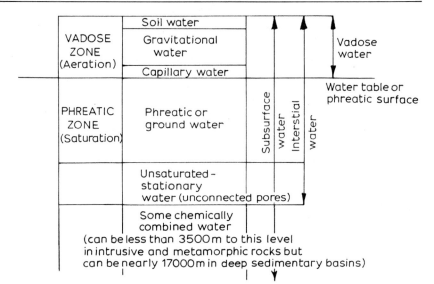

Figure 8.2 Subsurface water classification (*generally after* Davis and De Wiest, 1966).

water through the physiological functioning of vegetation. This zone is part of the intermediate, suspended water or *vadose* zone, in which water, not held by adsorption or capillary forces in the soil structure, flows downwards through unsaturated pore spaces under the influence of gravity to the water table. In moist environments and in arid regions the thickness of this zone may range from zero up to 350 m respectively.

The *water table* or *phreatic surface* is generally recognized in practice as the level to which water rises in a borehole or series of boreholes drilled into the saturated *ground water* or *phreatic* zone. This is not necessarily an exact definition, since the level of water in a borehole will be affected by the soil fabric* and by the drainage induced by the borehole, unless the water table is exactly horizontal. A more exact, although impracticable, method of determining the water table would be to measure porewater pressure at intervals from the surface. Then, provided the situation was unconfined, the porewater pressure would be zero (atmospheric) at the water table, and would increase linearly with depth into the groundwater zone. However above the water table an increasing negative porewater pressure would be recorded reaching a maximum at the upper limit of the *capillary* water zone.

*Rowe (1968), in particular, emphasises the effect of fabric on apparent piezometric levels in clay soils and the effect that misinterpretation of piezometric levels can have not only on groundwater prediction, but also on clay strength.

The upward flow of water from the groundwater zone into the suspended water zone is attributed to surface tension in the water at the water–air interface in the capillary passages of the soil. The extent (or height h_c) of capillary rise (see Section 2.6) depends on the capillary radius (a), the surface tension force (T_s) and the meniscus contact angle α – a complicated factor depending on the soil properties and the presence of impurities, such that

$$h_c = \frac{2T_s}{a\gamma_w} \cos \alpha \qquad (8.1)$$

for a capillary tube, which may be empirically stated for soils (after Terzaghi and Peck, 1967) as

$$h_c = \frac{C}{eD_{10}} \qquad (8.2)$$

where C is an empirical shape/impurity constant having a magnitude between 10 and 50 mm, e is the void ratio and D_{10} is the D_{10} particle size in mm. h_c can therefore have a magnitude of several metres in fine soils.

The groundwater or phreatic zones and the lower part of the capillary zone are both saturated, and water flow in both will be at a similar rate in a similar near-lateral direction. It is important, however, to differentiate between the two in *engineering* terminology, since the effect of negative (capillary) and positive (groundwater) porewater pressures will obviously have a widely different effect on soil or rock reactions. For this reason the definition of water table in terms of borehole or well (phreatic) levels (thus eliminating capillary effects) is important.

Formal presentation of hydrological data can be obtained through a hydrogeological map – possibly combined with a geotechnical map as discussed in Section 7.1. Simplest layouts would be a water-level (or depth-to-water) map representing the water table in contour form (Figure 8.3), although in some cases of complicated geological structures including confined and unconfined zones, piezometric maps may be more useful.

Provided the aquifer system is uniform and hydraulic gradients small then the water table contours are a reasonable representation of lines of equal and representative pressure (potentiometric or piezometric surfaces) and a flow net can be constructed relatively

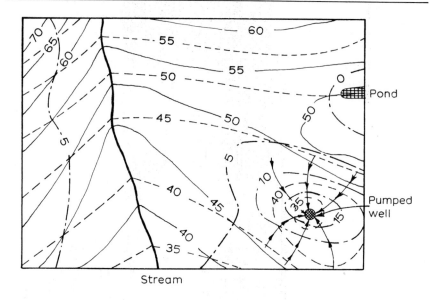

Figure 8.3 An example of a hydrological map illustrating ground (———) and water table (— — — —) contours, and contours of depth to water table (—·—·—). The arrowed lines indicate flow directions into the well. The map is not to scale and arbitrary contour units have been selected.

accurately (Figure 8.3). From a knowledge of permeabilities, this can be used to determine the quantities and rates of flow and areas of recharge (inflow) and discharge (outflow) from the aquifer.

The vertical extent of the groundwater varies with the type of rock, its geological structure, and mass porosity of the rock. The ultimate depth is about 30 km where increasing rock ductility will eliminate significant porespace. A more realistic figure is probably 10 to 15 km in sedimentary rocks, and 2 to 3 km in igneous and metamorphic rocks, where increasing compression and density will tend to eliminate the interconnecting pores essential for free water migration. Even within the phreatic zone, however, most rocks (as distinct from soils) contain relatively trivial quantities of groundwater. Those which contain economically significant quantities of water are called *aquifers*. These include uncompacted sediments, porous sandstones (for example, the Bunter Sandstones in Great Britain), cavernous limestones, and highly fractured zones in some igneous rocks. From an engineering point of view, aquifers are certain to create water flow as well as stability problems which will be considered later. Usually, where the yield is below economical

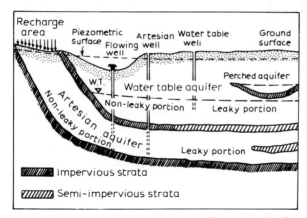

Figure 8.4 Types of aquifer (*from* Hantush, 1964)

levels, problems are also limited in rock engineering. In this case the terms *aquitard* or *aquiclude* can be used to describe rocks with decreasing transmission facilities. The term *aquifuge* is used to describe a rock which cannot absorb or transmit water.

In layered sedimentary rocks, the presence of aquifers and aquicludes in the succession can create a complicated progression of confined or semiconfined groundwater zones under *artesian* pressures beneath an upper perched water table. Where these are associated with synclinal structures in particular, the artesian pressure can often be higher at some parts of the structure than the hydraulic head represented by the cover to ground surface so creating conditions for artesian wells (Figure 8.4). Such a condition pertained originally in the London (England) Basin until overabstraction depleted the artesian head.

In engineering, the presence and recognition of confined groundwater zones is particularly important in shaft sinking and in tunnelling. This is not so much because the pressure of groundwater will critically affect the stress situation, as in uncompacted sediments generally located in unconfined groundwater zones, but because the water inflow encountered may be impossible to deal with by conventional means. In this case drainage, grouting or possibly even freezing may have to be considered in the light of flow estimates based on measured water table levels and ground permeability.

8.2 Groundwater flow

Some of the factors affecting groundwater flow have been considered (see Section 2.4) under the general heading of rock particle systems, or soils. This was a deliberate decision since to a great extent many of the theoretical treatments of groundwater flow are based on particle physics. In reasonably homogeneous porous rocks and incompressible soils, this is an acceptable assumption. In non-homogeneous or anisotropic rocks and soils (a common occurrence in sedimentary deposits) the simple approach may need modification.

The simplest way to represent groundwater flow is through a two-dimensional *flow net*. For a specific two-dimensional representation, this comprises *equipotential* lines which are lines joining points

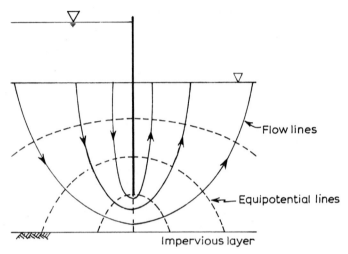

(a) Sheet pile coffer dam

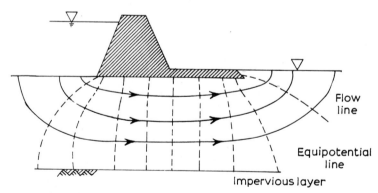

(b) Concrete gravity dam

Figure 8.5 Typical flow net sketches.

of equal porewater pressure and *flow* lines which are lines indicating the direction of water flow through the ground. Since the flow lines must intersect the equipotential lines at right angles in an isotropic medium, a flow net can be sketched for the boundary conditions existing for any particular case (see Figure 8.5a,b). Both of the cases illustrated are common engineering examples; the first represents confined flow underneath a sheetpile coffer dam and the second pictures the case of confined flow beneath a concrete gravity dam. In the former case, the boundary conditions are represented by equipotential lines at the ground surface on each side of the dam, and by flow lines along each interface of the buried pile (vertical) and along the impervious layer (horizontal). In the second case the extended dam length creates an extended confined zone of horizontal flow lines and vertical equipotential lines. In the case of an earth dam, the situation would be complicated by flow through part of the dam. Problems of this nature can be studied very easily in the laboratory using a two-dimensional model flow tank and/or a simple electrical analogue technique.

Using Darcy's law, and making the reasonable assumption that flow is laminar, this approach can be used to predict quite accurately the flow conditions for a given *two-dimensional* situation, using the graphical method of Forchheimer (see Muskat, 1947). In this method a flow net is constructed from equipotential lines having an equal pore pressure difference or *equipotential drop* δh, and from flow lines separated so that each intermediate *flow channel* has an equal discharge flow rate, δQ (Figure 8.6). Then, if N_f represents the number of flow lines and N_D the number of equipotential lines:

$$\delta h = \frac{h}{N_D} \qquad (8.3)$$

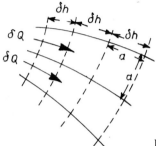

Figure 8.6 Element of a Forchheimer flow net.

$$\delta Q = \frac{Q}{N_f} \qquad (8.4)$$

where h is the total hydraulic head and Q is the total flow rate.

If N_f and N_D are chosen so that each element contained between equipotential and flow lines is approximately equal-sided, then for a given element of side length a, the discharge velocity $\delta v = \delta Q/a$, the hydraulic gradient $i = \delta h/a$, and from Darcy's law

$$\delta v = \frac{\delta Q}{a} = \frac{Q}{aN_f} = ki = \frac{k\delta h}{a} = \frac{kh}{aN_D} \qquad (8.5)$$

whence,

$$Q = kh \frac{N_f}{N_D} \qquad (8.6)$$

The selection of N_f and N_D is entirely arbitrary and depends on the way in which the sketched lines on the scale drawing work out from a trial and error process (see Cedergren, 1967 for a detailed review), possibly assisted in complicated cases by electrical or hydraulic models (see Karplus, 1953, and Johnson, 1963). From the scale distance on the flow net, individual hydraulic gradients and seepage flows can be computed.

Flow nets can also be developed theoretically or numerically, most simply by considering groundwater flow through a two-dimensional soil element having dimensions dx, dz (Figure 8.7). The total flow Q through the element will then be equal to the sum of the components of flow in the x and z directions:

$$Q = Q_x + Q_z \qquad (8.7)$$

which for flow into the element will be

$$Q_1 = v_z \, dx + v_x \, dz \qquad (8.8)$$

and for flow out of the element:

$$Q_2 = \left(v_z + \frac{\partial v_z}{\partial z} dz\right) dx + \left(v_x + \frac{\partial v_x}{\partial x} dx\right) dz \qquad (8.9)$$

where v_x, v_z are the flow velocities in the x and z directions respectively. Thus, if the volume of groundwater in the soil remains constant (that is, steady-state flow) and $Q_1 = Q_2$, then equating

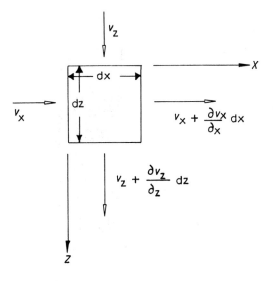

Figure 8.7 Groundwater flow through a two dimensional soil element.

equations 8.8 and 8.9 gives

$$\frac{\partial v_x}{\partial x} dxdz + \frac{\partial v_z}{\partial z} dxdz = 0$$

or

$$\frac{\partial v_x}{\partial x} + \frac{\partial v_z}{\partial z} = 0 \qquad (8.10)$$

Now Darcy's law states that Q = coefficient of permeability (k) × hydraulic gradient(i) × c/s/a of flow (A) or, flow velocity $(v) = ki$ Thus,

$$v_x = \frac{\partial h}{\partial x} k_x \text{ and } v_z = \frac{\partial h}{\partial z} k_z$$

and from equation 8.10

$$k_x \frac{\partial^2 h}{\partial x^2} + k_z \frac{\partial^2 h}{\partial z^2} = 0 \qquad (8.11)$$

where k_x, k_z are the respective coefficients of permeability (or hydraulic conductivities) in the x and z directions, and h is the total fluid head at the point in question.

If $k_x = k_z$ (an unlikely situation in layered strata) then equation 8.11 becomes

$$\frac{\partial^2 h}{\partial x^2} + \frac{\partial^2 h}{\partial z^2} = 0 = \nabla^2 h \qquad (8.12)$$

which is the 2-dimensional Laplace equation for flow in a resistive medium. It states that in an isotropic soil the sum of the changes in hydraulic gradient in the x and z directions is equal to a zero change in hydraulic gradient, or, in other words, flow lines and equipotential lines intersect at right angles. The equation can be solved manually for simple boundary conditions or numerically for more complex boundary conditions.

In the more likely case of an anisotropic soil, equation 8.11 may be re-written:

$$\frac{k_x}{k_z}\frac{\partial^2 h}{\partial x^2} + \frac{\partial^2 h}{\partial z^2} = 0 \qquad (8.13)$$

or

$$\frac{\partial^2 h}{\partial x_D^2} + \frac{\partial^2 h}{\partial z^2} = 0 \qquad (8.14)$$

where

$$x_D = x\left(\frac{k_z}{k_x}\right)^{1/2} \text{ or } x = x_D\left(\frac{k_x}{k_z}\right)^{1/2} \qquad (8.15)$$

This is again a Laplace equation which will give a transformed flow net for a soil with effective permeability k_e equal to $(k_x k_z)^{1/2}$. The seepage in an anisotropic soil can therefore be obtained graphically by sketching an orthogonal flow net as outlined previously, and substituting k_e in equation 8.6. The transformed flow net can be changed to a true or natural flow net by transforming x and z dimensions on the cross section using equation 8.15 (Figure 8.8). Then the cross section of the natural flow net will be composed of rectangles elongated in the direction of greatest permeability.

It should be noted that except in those instances where flow is parallel to a direction of permeability anisotropy (a condition which

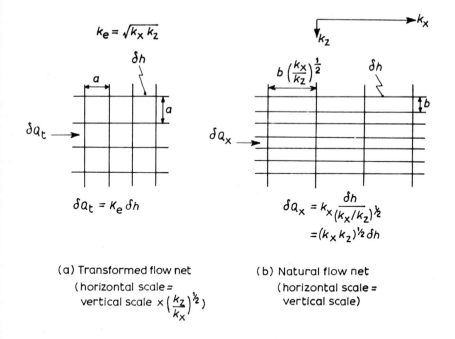

Figure 8.8 Flow net pattern in anisotropic soil.

is illustrated in Figure 8.8), the flow lines and equipotential lines of the natural flow net will not intersect at right angles, as is the case in an isotropic soil or on a transformed flow net. The natural flow net in an anisotropic soil will appear as a series of parallelograms.

Groundwater flow can also be affected by strongly directional open joint orientations in a rock mass of limited extent. This has been illustrated by Wittke and Louis (1966) who show that separation and shape of equipotential lines is affected by the joint orientations. They consider a specific case of two non-orthogonal joint sets having directional mass permeabilities k_1 and k_2 where $k_1 \gg k_2$, and the k_1 dip direction may be either out of or into a slope. In the former (Figure 8.9) case, the equipotential lines are well separated and inclined to the horizontal while the flow is nearly parallel to the slope. In the latter case the equipotential lines are steeper and closer together, whilst the flow is nearly horizontal. Seepage forces acting parallel to local hydraulic gradients could affect the stability of a slope which is near to critical equilibrium.

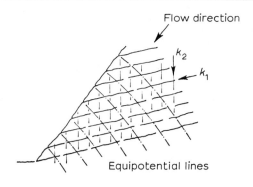

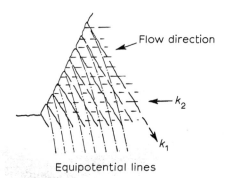

Figure 8.9 Effect of open joint orientations on flow into a slope, $k_1 \gg k_2$ (*after* Wittke and Louis, 1966).

Since flow lines and equipotential lines are only mutually perpendicular in an isotropic soil or for flow in a direction of maximum or minimum permeability in an anisotropic soil, the assumption of isotropy in many layered soils can lead to some errors in graphical flow net prediction, and it is essential that, as a preliminary to flow net design, an assessment of directional permeabilities be obtained either by measurement or by borehole core examination. Fortunately, as emphasized by Casagrande (1937), flow nets are an extremely powerful tool for groundwater assessment, and, provided that sensible allowances are made for anisotropy, even a crude net can give a relatively accurate estimate of flow, pore pressure magnitudes and gradients.

Groundwater 573

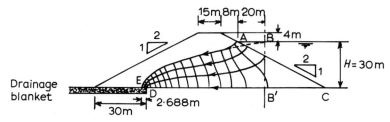

(a) Construction of free surface - earth dam with under-drainage

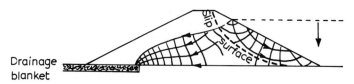

(b) Rotational failure on sudden drawdown

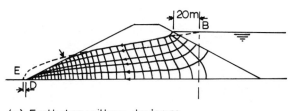

(c) Earth dam with no drainage

Figure 8.10 Seepage lines and equipotential lines through an earth dam.

A study of seepage flow is also an important factor in earth dam design. In particular, it is necessary to define the form of the free surface, which is the upper boundary of seepage flow and upon which the pressure is assumed to be zero (strictly, atmospheric). For an introduction to the problem it is necessary to consider only a homogeneous, coreless type of embankment but, of course, earth dams of the type described in Section 12.13 pose a much higher degree of analytical complexity.

In the example shown in Figure 8.10a, the fundamental boundary conditions are that the face of the dam AC is an equipotential line and that the flow lines must therefore meet it at right angles. The base of the dam CD is assumed to be impermeable and it therefore constitutes a flow line. Since an underdrain is present, the flow lines

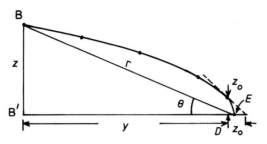

Figure 8.11 Parabolic construction.

must also intersect the drain at right angles along the seepage length ED as drawn.

The form of line AE is generally determined in the manner of Casagrande (1937); see also Jumikis (1962) and Capper *et al* (1966). Along the water surface, a distance AB equal to 0.3 B'C is marked off and a parabola, with focus D, is drawn through B. Figure 8.11 has been drawn with the same markings as in Figure 8.10a in order to illustrate how the parabolic line is constructed.

Within the cross-sectional area of most concern to the problem in hand, the phreatic surface (line) takes the general equation:

$$r(1 - \cos\theta) = \text{constant} \tag{8.16}$$

where r is the radius vector and θ is the angle formed between r and the axis of symmetry of the parabola. From Figure 8.11 we may write:

$$r = (z^2 + y^2)^{1/2} \quad \text{and} \quad y = r\cos\theta \tag{8.17}$$

If $\theta = \pi/2$ then:

$$\cos\theta = 0 \quad \text{and} \quad r = z_0 \, (=\text{constant from equation 8.16}) \tag{8.18}$$

Thus, from equation 8.16:

$$z_0 = (z^2 + y^2)^{1/2} - y \tag{8.19}$$

Points D and B are given and length z_0 is calculated from equation 8.19. Therefore, using equation 8.19 and for different values of y, the several values of z necessary for defining the parabolic surface may be derived. Also, from the properties of a parabola, the length z_0 is equal to twice the length from the focus D to the vertex E, that is $DE = z_0/2$. In Figure 8.10a it will be apparent, from the measurements shown, that y_0 = 5.377 m and so the focal distance DE to vertex E is 2.688 m.

The marking off of distance AB is simply a convenience for we know that the upper flow line must, in practice, take a sigmoidal form for intersection normal to AC. If the vertical height of the water (30 m) is then subdivided into a number of convenient equal intervals, the intersections of these grid lines with the upper flow line fix, in effect, the equipotential lines. Other seepage lines and the complete definition of the equipotential lines are best investigated using a finite difference digital procedure or, most simply, in an analogue manner using a physical flow tank model with an injected dye to trace several seepage paths or by using electrically-conductive teledeltos paper with a suitable potential dividing circuit.

In the example discussed, under-drainage has been assumed to exist on the down-stream side of the dam. It will be appreciated that different forms of drainage may be present, but in all instances the flow lines should intersect the drainage interface at right angles.

Suppose now, as in Figure 8.10c, that there is no under-drainage. The focus D of the parabola then moves to the base of the downstream slope of the dam and the procedure is as before except that the actual upper flow line diverges from the parabolic line to converge tangentially along the downstream sloping face of the dam. Length ED is calculated at 1.991 m and Figure 8.12 is used to calculate length a equivalent to the point at which the flow line converges on the sloping face of the dam. Length y is 111 m and so the ratio y/H is 3.7. The slope angle is 26.56° and so factor m is about 0.35. It follows from the inset equation in Figure 8.12 that distance a is equal to 23.48 m.

The example in Figure 8.10c may readily be adapted to the situation of a water table of unknown configuration in a soil slope. In the absence of quite extensive borehole evidence, it is not always possible to fix the position of a phreatic surface. Such knowledge is, however, necessary if the correct strip densities are to be entered into the limit equilibrium type of analyses outlined in Section 9.6. Provided that some limited information on water table height is available, perhaps remote from the slope, then a parabolic surface may justifiably be fitted by the method discussed earlier. Water table problems in rock slope analysis are discussed in Section 10.11.

Should a reservoir be drawn-down too rapidly, a lag on pore

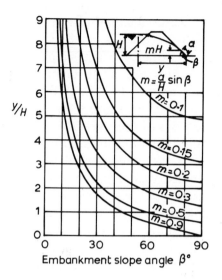

Figure 8.12 Determination of intersection point of upper flow line with downstream face of earth dam. (*from* Capper *et al*, 1966, *after* Gilboy, 1940).

pressure dissipation could lead to the development of rotational failure inside the upstream face of an earth dam. A generalized form of re-distributed seepage flow and equipotential lines is sketched in Figure 8.10b for the example considered earlier. The pore pressure distribution along the slip surface at the time of failure would be determined from the points of intersection of the original equipotential lines with the failure surface, the actual potential along each line being controlled by the hydrostatic head h equivalent to the point of intersection of the line with the upstream face of the dam.

Suppose that under normal operating conditions we are required to estimate the seepage loss through an earth dam. Let the coefficient of permeability be 30 m d^{-1}, and assume that the length of dam along an axis normal to the cross-section is 40 m. We note that in Figure 8.10a there are 3 flow lines (N_f) and 16 equipotential lines (N_D). Thus, from equation 8.6, we have:

$$Q = 30 \times 30 \times 3/16 \times 40 \text{ m}^3 \text{ d}^{-1}$$
$$= 6750 \text{ m}^3 \text{ d}^{-1}$$
$$= 6.75 \times 10^6 \text{ l d}^{-1}$$

In Figure 8.10c, N_f = 5 and N_D = 31. Thus, without the underdrainage, $Q = 5.81 \times 10^6$ l d^{-1}.

8.3 Seepage forces

When groundwater is stationary in the porespace or fissures of a rock or soil element, the porewater pressure u will be equal to the boundary water pressure applied to the element (usually multiple of head h and porewater unit weight):

$$u = \gamma_w h \tag{8.20}$$

Provided that the porewater is *confined*, then any changes in the boundary water pressure will lead to equivalent changes in porewater pressure. However if the porewater is not confined, then the changes in boundary water pressure will cause the water to *flow* and the flowing water will be subject to frictional drag at the water/flow channel interface. This drag, which will be proportional to channel size and flow rate, will be transmitted as a *seepage force* acting on the soil or rock in the direction of flow. The seepage force, equal to the multiple of the pressure head causing flow (h), the discharge area and the water unit weight, represents, compared with the static confined condition, a transfer of part of the porewater pressure, expressive of a loss in total head, as an effective stress acting in the direction of flow. Since the *drained* shear strength of a soil is directly related to the effective stress acting on the rock or soil fabric, the existence of seepage forces can have considerable engineering significance.

For instance, if the effective stress in a soil is reduced to zero through the existence of seepage forces acting against gravity, then a *quick* condition will be created in the soil as its effective strength and consequent bearing capacity are reduced to zero. This will occur when a *seepage force* acting vertically upwards is equal to the *submerged weight* of the soil, a situation which may occur at, for instance, the toe of the dam in Figure 8.5 — or virtually anywhere where flow lines are directed vertically upwards and the flow has significant velocity. The conditions for loss of strength are in that case:

$$\text{seepage force } hA\gamma_w = LA(\gamma - \gamma_w) \text{ submerged weight} \tag{8.21}$$

where γ and γ_w are respectively the soil bulk unit weight, and the water unit weight,
  L is the length of the vertical flow path,
  A is the area of force application,
and h is the groundwater head causing flow.

Thus the conditions may be simply stated in terms of a critical hydraulic gradient:

$$i_{crit} = \frac{h_{crit}}{L} = \frac{\gamma - \gamma_w}{\gamma_w} = \frac{\gamma_b}{\gamma_w} \qquad (8.22)$$

which is equal to the submerged unit weight of soil per unit weight of water, or about unity for a typical cohesionless soil. This critical gradient can be compared with the gradient at the 'upflowing' element on the flow net to check the possibility of quick conditions. The condition will be reduced if flow is sub-vertical or if the ground surface is inclined.

A quick condition can also be obtained in poorly-drained soils by the impulsive loading of a loose soil skeleton. This will result in a reduction of pore volume unaccompanied by instantaneous flow. There is a transfer of some pre-existing effective stress to an enhancement of porewater pressure.

A *quicksand* is simply a cohesionless soil in a quick condition and, as pointed out by Lambe and Whitman (1968), is approximately equivalent to a suspension having twice the density of water, and into which, contrary to popular belief, it would be difficult for humans or animals to sink, let alone be sucked! In fact, in the coalmining industry a process known as the Chance Cone comprising a conical bath (see Pryor, 1965) containing a fluidized suspension of sand and water in a quick condition was once widely used to separate coal (density 1.4 Mg m^{-3}) which floats, from rock (density 2.0 Mg m^{-3}) which sinks.

The effects of true quick conditions induced naturally in fine sands (and in some extrasensitive clays and clay/fine sand combinations) by changes in groundwater conditions, shocks, or earthquakes can be spectacular (several cases involving extensive slope failures are discussed by Terzaghi and Peck, 1967), but their occurrence is rare. In practice, most quick conditions are artificially induced as a result of high hydraulic gradients, possibly at the base of a deep excavation and more probably at the toe of a dam. In the latter case the situation may be exacerbated by *piping*. This occurs when soil near the toe in a quick condition is washed away to create a cavity which attracts further flow, at an increasing hydraulic gradient and over a decreasing seepage path, as the cavity is extended by erosion to form a large flowpath under or in the dam. This phenomenon may eventually lead to virtual collapse of the dam if untreated.

Seepage, and seepage forces, may be reduced and controlled by reducing permeability (k) or by reducing the hydraulic gradient (i) along the flow path. These objectives can be achieved in several ways. In natural soils or rocks, permeability can be reduced by filling pore or fissure space with a low-permeability filler such as cement, clay or various organic polymers, a process known as *grouting* (see Section 11.5) which tends to be expensive but which is effective if properly carried out.

Where the soil or rockfill is artificially constructed, permeability can be increased by careful grading and introduction of dispersed clay into the matrix. In the case of earth or rockfill dams, the objective can be achieved by incorporation of an impermeable clay core or impermeable membrane into the structure (see Figure 8.13). Various types of impermeable concrete, clay-concrete and pvc cores have also been described by Walters (1971). A flexible core is usually

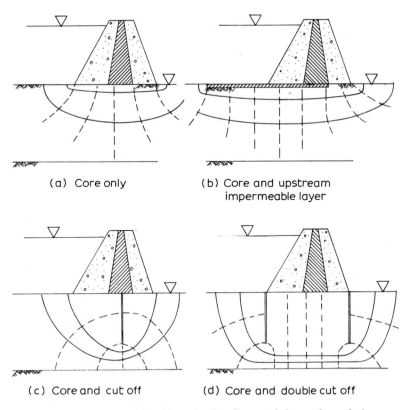

Figure 8.13 Methods of lengthening flow path beneath earth dams.

favoured due to the very high compressions often recorded during dam construction, particularly in rockfill. Earthfill dams are considered subsequently in Sections 12.13, 12.14.

In the case of dams sited on permeable strata where complete water-proofing is not feasible, hydraulic gradients can effectively be reduced by lengthening the flow path beneath the dam. Thus $i(=h/L)$ can be reduced if L is increased. This is commonly achieved by constructing a *cut-off* beneath the dam. A cut-off is an impervious barrier or sheet constructed beneath either the impervious core, centre or toe, or upstream (or both toe and upstream) ends of the dam. It usually comprises sheet piles or plastic (to improve flexibility in compressible soils) concrete or clay-sand filled diaphragm wall excavations. In addition, many cut-offs are grouted using a line of vertical injection holes along the cut-off line. The effect of the cut-off (Figure 8.13c,d) is to increase porewater pressures on the upstream side of the cut-off (to full hydrostatic heads in the case of a complete cut off to an impervious layer) and to reduce seepage or *uplift* forces on the downstream side of the cut-off. An alternative to a cut-off may be an upstream layer of impermeable material (Figure 8.13b).

8.4 Drainage and drain wells

Since the drained shear strength of a soil or fissured rock mass is related to the effective confining stresses acting on it, then it is possible to increase its strength by increasing the effective stresses. In a saturated soil with near-immobile groundwater, this can be done by reducing the porewater pressure, most effectively by replacing or reducing the groundwater. There are various ways in which this might be achieved, such as by grouting, freezing, or by the construction of diaphragm walls to isolate a particular section. But the simplest and most obvious method is to lower the groundwater level by pumping water from a well or wells bored vertically into the ground. Consequential problems arising from the settlement of adjacent structures due directly to de-watering may be offset by pumped recharge into boreholes a little more remote from an excavation.

In initiating the present treatment of drainage in the general context of a geotechnical operation, the reader will also appreciate the importance of groundwater as an economic commodity in its own right. Since wells are also sunk solely for the purposes of ground water exploitation as part of groundwater/surface water management

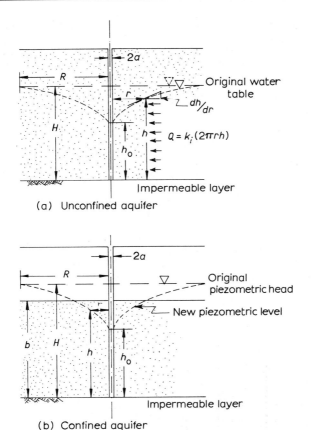

Figure 8.14 Drain well flow assumptions.

programmes, the subsequent analyses have a wider groundwater hydrology interest.

The simplest case (Figure 8.14a), and the one that is most relevant to ground de-watering in a foundation engineering context, is a *single well in an unconfined* aquifer, penetrating the full section of aquifer. Pumping will reduce the level of piezometric surface symmetrically around the well in the approximate two-dimensional form of a logarithmic spiral.

This flow condition was first analysed by Dupuit in 1863 (see also Jacob, 1950, for a full treatment) who assumed:

(a) that flow into the well was predominantly radial and horizontal, and
(b) that the flow velocity was uniform (steady state) over the depth of flow.

Then if the aquifer rock is homogeneous, isotropic and of infinite extent, the hydraulic gradient can be equated to the slope of the piezometric surface, equal to dh/dr at a radial distance r and height h above the well base, and the discharge area will be equal to $2\pi rh$, giving an equation for *steady state* flow based on Darcy's law in the form:

$$Q = kiA = k\frac{dh}{dr} \cdot 2\pi rh \qquad (8.23)$$

and for the boundary conditions, $r = a, h = h_0, r = R, h = H$:

$$\frac{Q}{\pi k}\int_{r=a}^{r=R}\frac{1}{r}\,dr = 2\int_{h=h_0}^{h=H} h\,dh \qquad (8.24)$$

where k is the permeability coefficient.

Equation 8.24 solves to give a flow rate:

$$Q = \pi k \frac{(H^2 - h_0^2)}{\ln(R/a)} \qquad (8.25)$$

which is generally agreed to be a reasonable approximation for steady state flow in layered strata, where the slope of the piezometric surface is less than $20°-30°$.

A similar analysis of flow into a well in a *confined* aquifer (Figure 8.14b) of thickness b gives:

$$Q = \frac{2\pi kb(H - h_0)}{\ln(R/a)} \qquad (8.26)$$

Equations 8.25 and 8.26 indicate that the well yield Q is directly proportional to k (the method can be used to determine permeability), so drainage will only be feasible, or at any rate rapid, in soils or rocks of reasonably high permeability. Since Q is inversely proportional to $\ln(R/a)$, errors in R estimation or differences in a will be relatively unimportant. (For instance, if $R/a = 1$, $\ln(R/a) = 0$; $R/a = 10$, $\ln(R/a) = 2.3$; $R/a = 100$, $\ln(R/a) = 4.6$; $R/a = 1000$, $\ln(R/a) = 6.9$ and so on.)

In design of pumping or drainage installations, therefore, the most important criteria are k and the depth of the hole H below the piezometric surface. H should be taken to the *base* of the relevant aquifer (where feasible) in order to maximize the discharge area. An actual or assumed complete aquifer penetration satisfies theoretical

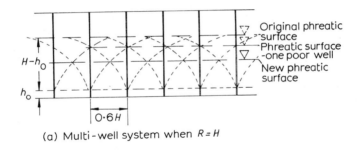

(a) Multi-well system when $R = H$

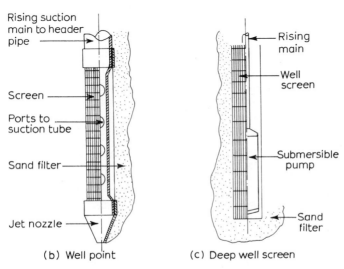

(b) Well point (c) Deep well screen

Figure 8.15 Groundwater lowering – layout and equipment.

conditions for flow parallel to bedrock and radially into the well. R is generally taken as equal to H, although in a multi-well system, closer holes may be required since one poor well, where k is not representative of the whole strata, could have a serious effect on the new piezometric surface (Figure 8.15a). The very considerable importance and sensitivity of k in designing drainage systems, particularly deep or extensive systems, justifies the largely empirical design approach to groundwater lowering, since analogue or numerical techniques are wholly dependent on a correct assessment of permeability.

Groundwater lowering as a stabilisation technique is used most frequently in clean sands and sandy gravels (see Cashman and Haws, 1970) to a depth of 4 to 5 m, using a multi-well point system connected through a range to a single rotary suction pump. Well points comprise a perforated metal tube, driven, jetted or placed in a drill hole, with an appropriate filter medium in finer sands (Figure 8.15b). In the case of deeper wells, perforated well screens are lowered into drilled holes and the annular cavity behind the screen is filled with a suitable filter medium. A submersible pump is then lowered into the hole. The only limitation on depth is the time and cost required to lower the piezometric surface sufficiently to drain the soil or, in the case of artesian pressures in confined aquifers, to reduce porewater pressures to manageable proportions. Eventually, an alternative method at depth, such as grouting, or possibly compressed air, will prove more economical.

Although groundwater lowering is probably the most important aspect of drainage requiring a knowledge of engineering geology, other types of drainage are of interest. These include surface drainage to prevent saturation or erosion of earth structures such as embankments or dams and subsurface drainage to reduce uplift* or hydrostatic pressures on buried or water retaining structures such as retaining walls, storage tanks or dams. The basis of good drainage is that water pressure is relieved by creating relief flow channels through which excess water can flow without eroding the foundation or structure. This invariably means the inclusion of a layer of filter material in the channel.

A filter material serves a dual purpose; it should be sufficiently *fine* to prevent the passage of eroded soil particles and yet be sufficiently *coarse* to offer less resistance to groundwater flow than the drained soil. The exact design may be obtained using the same principles as in grouting with particulate grouts (see Sections 8.12 and 11.5).

In practice, the US Corps of Engineers quoted by Lambe and Whitman (1969) or Sowers and Sowers (1971) suggest the following rule of thumb ratios between filter and soil sizes:

---

*It is important to note that uplift is affected by the contact area between water and the structure. This is high in the case of soils, lower in the case of fissured rock with low material permeability. If A is the area of the buried structure, the total uplift force F_u may be written

$$F_u = \Xi A u$$

where Ξ (see Section 11.7) is equal to 1 for soils and about 0.5 for fissured rock.

$$\frac{D_{15} \text{ FILTER}}{D_{85} \text{ SOIL}} < 5$$

$$4 < \frac{D_{15} \text{ FILTER}}{D_{15} \text{ SOIL}} < 20$$

$$\frac{D_{50} \text{ FILTER}}{D_{50} \text{ SOIL}} < 25$$

where D_{15}, D_{50}, D_{85} are the particle sizes at which respectively 15, 50 and 85 per cent by weight of the soil is finer.

8.5 Permeability tests — rock

Consideration of flow through rock is governed by the same factors as consideration of mechanical reactions — namely, that the properties of the rock *material* differ widely from the properties of the rock *mass*. In fact, the intrinsic porosity and permeability of rock as a material are often negligible ($k < 10^{-9}$ m s^{-1}), although anomalous materials, particularly sandstones (such as the Bunter Sandstone in Great Britain) can present some problems in engineering. On the other hand, water percolation through joints, bedding planes and other fissures, often comprising marked flow paths, can represent apparently high mass permeability.

Intrinsic permeability can be measured directly by injection of air or water* through a cylindrical specimen contained in an impermeable membrane. Daw's (1971) adaptation of the Hoek cell is an example of a simple apparatus (see Figure 4.31). Then, the coefficient of permeability is given by Darcy's law:

$$k = \frac{Q}{A} \frac{h}{L} \qquad (8.27)$$

where h is the applied head,
       Q is the volume flow rate,
and L, A are the specimen length and cross-sectional area.
Directional permeabilities can be obtained by specifying the specimen orientation.

---

*Although the results should be similar when adjusted for relative viscosities, air permeabilities are invariably greater. In practice it may be considered that if permeability is so low as to be unmeasurable using water, it has little engineering significance.

586 *Principles of Engineering Geology*

Not all rocks are sufficiently competent to allow the cutting of a suitable core for this type of permeability testing and not all samples can be obtained in a suitable state for coring. In this case, permeability may be estimated using the Kozeny equation (equation 2.10) from porosity n, and median particle or pore diameter measured by mercury injection of quite small and irregular specimens using some form of mercury porosimeter (see apparatus descriptions by Winslow and Shapiro, 1959 and Ulmer and Smothers, 1967).

The method represents a reasonable approach to rock material permeability, giving an acceptable value of k for most engineering purposes. As stated earlier, the main difficulty is that except for intrinsically porous or deep rocks, rock permeability is a mass rather than a material feature.

The fissures which control flow through a rock mass tend on a macro-scale to be evenly distributed both spatially and directionally so allowing many rocks to be regarded as *homogeneously fissured media*. Nonveiller (1968) states that a large proportion of geotechnically relevant rock may be treated as such, and this means that rock masses are directly analogous to the generally assumed homogeneously porous media of groundwater flow theory. The flow of water through fissured rock may therefore be treated as gravitational flow through a homogeneous porous medium subject to Darcy's Law. This is a useful assumption, since it means that well theory is actually valid both in rock masses and in soils, and that a simple pumping test can be used to determine mass permeability in a reasonably homogeneous aquifer.

Because the Dupuit assumptions do not describe accurately the drawdown conditions near the well (Figure 8.14 and p. 604), realistic permeability determination requires the monitoring of water levels h_1, h_2 above the aquifer base at two or more observation wells radially distant r_1, r_2 ($r_2 > r_1$) on the main surface of the seepage spiral (Figure 8.23). Then assuming steady state flow, equation 8.24 becomes, for the unconfined case:

$$\frac{Q}{\pi k} \int_{r=r_1}^{r=r_2} \frac{1}{r} \, dr = 2 \int_{h=h_1}^{h=h_2} h \, dh \tag{8.28}$$

The solution is

$$Q = \pi k \frac{(h_2^2 - h_1^2)}{\ln(r_2/r_1)} \tag{8.29}$$

and for permeability,

$$k = \frac{Q}{\pi(h_2^2 - h_1^2)} \ln(r_2/r_1) \qquad (8.30)$$

For the confined condition, the solution for permeability for the yield of the well is similar to that in equation 8.26:

$$k = \frac{Q}{2\pi b(h_2 - h_1)} \ln(r_2/r_1) \qquad (8.31)$$

Water levels can be determined by using direct measurement systems allied to various water sensing devices.

A commoner type of test is a packer permeability test (Figure 8.16) in which water is injected directly into the strata through a perforated tube located between two inflated packers or into the hole below a single packer. This has advantages over the pumping test since it enables the permeability of a selected section of rock to be determined. Packer permeability test lengths in discontinuous rock may be optimized by using the RQD arguments outlined in Section 6.13. The results of packer permeability tests are often expressed in *Lugeon* units. One lugeon is equal to a flow of 1 litre m^{-1} min^{-1} at a pressure gradient of 10 kgf cm^{-2} m^{-1} of hole, and is equivalent to a permeability coefficient of 10^{-7} m s^{-1}.

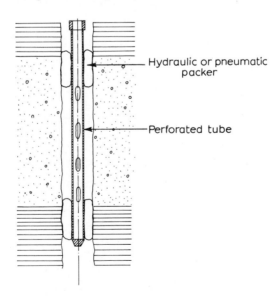

Figure 8.16 Packer permeability test.

Packer permeability tests may be carried out above or below the water table, but are most successful below the water table. The test length should be at least 10 times the hole radius, when an estimate of permeability can be obtained from equation 8.33 below by substituting L for R.

An actual measurement of mass permeability is unsatisfactory, particularly where the results are to be used to design ground treatments, unless it contains some information on the actual spacing of fissures and their thickness. Fissure spacing can usually be obtained by site investigations (see Chapter 7), and with this basic information, an estimate of fissure thickness is possible. Two similar approaches may be quoted. Both assume that the rock material is significantly less permeable than the mass.

Vaughan (1963) assumes N fissures of width δ normal to a borehole of radius (a) passing through impermeable rock. He then calculates the rate of flow (Q) of water under a typical flow test over a length (L) of the borehole from the standard form (see Sabarly, 1968):

$$Q = \frac{\pi N L P^* \delta^3}{6\eta \ln(R/a)} \qquad (8.32)$$

where P^* is the pressure drop over radial flow distance (R), and
η is the coefficient of viscosity of water.

This is then compared with an equivalent test section of non-fissured material having a radial permeability k_r, and zero permeability in a direction parallel to the borehole, whence:

$$Q = \frac{2\pi L k_r P^*}{\ln(R/a) \gamma_w} \qquad (8.33)$$

where γ_w is the density of water.

Combining equations 8.32 and 8.33 gives an expression for equivalent permeability in terms of fissure thickness and spacing in the form

$$\delta^3 = \frac{12 \eta k_r}{N \gamma_w} \qquad (8.34)$$

from which it can be seen (Table 8.1 and Figure 8.17) that for a particular equivalent permeability coefficient, the fissure width is only marginally sensitive to the fissure spacing. Thus an effective

Table 8.1 Estimates of fissure width from spacing at various k_r values

k_r m s^{-1}	δ mm			
	$1/N = 10$ m	$1/N = 1$ m	$1/N = 0.1$ m	$1/N = 10$ mm
10^{-1}	11.5	5.3	2.5	1.15
10^{-2}	5.3	2.5	1.15	0.53
10^{-3}	2.5	1.15	0.53	0.25
10^{-4}	1.15	0.53	0.25	0.115
10^{-5}	0.53	0.25	0.115	0.053
10^{-6}	0.25	0.115	0.053	0.025
10^{-7}	0.115	0.053	0.025	0.011
10^{-8}	0.053	0.025	0.011	0.005
	Massive igneous rocks	$\longrightarrow$	Laminated shales	

permeability of 10^{-5} m s^{-1} will represent a fissure width of between 100 to 300 μm over a reasonable range of sedimentary/igneous rock types, and quite approximate investigation data on spacing will allow accurate width prediction.

Snow (1968) derives an identical relationship to equation 8.34 on the basis of laminar flow discharge through a smooth, parallel-walled channel in a cubical fissure system, and similar flow through an equivalent permeable continuum. He adjusts the result by a factor of

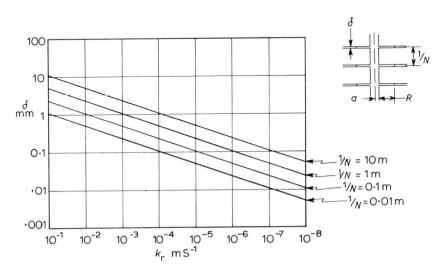

Figure 8.17 Fissure width — spacing — permeability relationship.

590   Principles of Engineering Geology

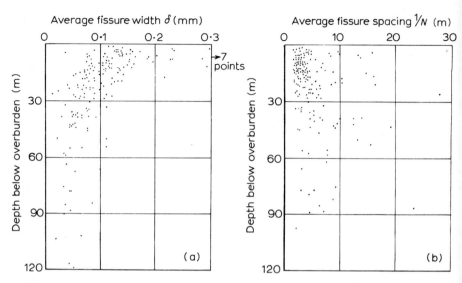

Figure 8.18   Effect of depth on fissure width and spacing at damsites in metavolcanics, quartzites, sandstones, shales, granites, gneiss, slates, phyllite, schists and other rocks (*after* Snow, 1968).

1.25 to allow for random orientation of drill holes to give:

$$\delta^3 = \frac{15 k_r \eta}{N \gamma_w} \tag{8.35}$$

Snow extends his analysis to include results of *in situ* permeability tests in a series of igneous and sedimentary rocks. He assumes that the number of fissures intercepted by a sample of random drill holes of equal length will be relatively small and will take the form of a Poisson distribution, having a percentage of zero discharges (no fissure present) which can be related to the average number of fissures per test length (L). It is therefore possible to obtain a value for N with a high degree of statistical probability from a large number of drill hole tests.

Data obtained by Snow from a large number of tests from dam cut-offs are shown in Figure 8.18. This shows a close correlation between fissure width and depth (Figure 8.18a), indicating widths between 75 and 400 μm in the upper 10 m which decrease to an average 50 μm at 60 m. There are also some indications that fissure spacings (Figure 8.18b) increase with depth, having a minimum separation of 1 m in competent rock. Snow states that neither spacings, openings, effective permeability or porosity are notably different in the different types of rock tested under similar conditions. There is however a general decrease in permeability with depth in the rocks tested which comprised the top 120 m of relatively sound rock in the sloping abutments of dams.

8.6 Permeability tests – soils

In soils and weakly cemented or porous rocks, groundwater is usually assumed to flow through a uniform homogeneous and isotropic porespace, and to be unaffected, or not significantly affected, by the soil or rock structure. Where this assumption can be defended, permeability can be measured through large scale *in situ* pumping tests (see Davis and De Wiest, 1966, or Todd, 1959), through multiple-well pumping tests (see Childs *et al*, 1952; Young and Smiles, 1963) or through packer permeability tests using the theoretical approach outlined in previous sections. Where the assumption is less defensible, or where the permeability is so small that large scale drainage is not feasible, smaller scale *in situ* tests of the type conventionally associated with soil permeability testing may be used. Three tests are usually considered:

(a) the auger hole method for shallow soils,
(b) the open-tube method,
(c) the piezometer method.

In these tests, depending on the type of soil being tested, the rate of flow associated with a *falling, constant* or *rising* head is measured, and related through an assumed flow path to the permeability of the soil. Typical assumed flow paths are illustrated in Figure 8.19 for each of the three tests. These suggest that the auger hole method of testing samples mainly *horizontal* permeability (this is similar to the well pumping test), the open tube samples *effective* permeability and the piezometer method samples *horizontal* permeability.

In all methods, assumptions regarding the flow paths surrounding the hole or the sampling point are critical to the success of the test, since for constant head conditions, the rate of flow (Q) will be given by the basic Darcy form:

$$Q = Akh \qquad (8.36)$$

592 *Principles of Engineering Geology*

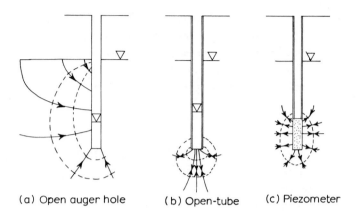

(a) Open auger hole   (b) Open-tube   (c) Piezometer

Figure 8.19 Notional flow nets for different permeability tests.

where k is the permeability coefficient (L/T dimensions),
   h is the hydraulic head (L) $\equiv s_t$ on p. 602,
and $A = f(L)$ is a *shape factor*, a function of the dimensions of the borehole and the boundaries of the flow medium.

Accurate determination of k therefore requires measurement of Q and h under a specified set of boundary conditions, and an accurate estimation of A. It is useful to refer to some specific methods of estimating the parameter A.

In the *auger hole* method (Figure 8.20a), water is pumped from a shallow open hole to a depth h below the water table, and the rate of inflow of water Q equivalent to rate of change in h is then measured immediately without waiting for the steady state conditions of the pumping test. Then:

$$k = \frac{Q}{Ah} = \frac{\pi a^2}{Ah}\frac{dh}{dt} = \frac{\pi a}{(A/a)h}\frac{dh}{dt} \qquad (8.37)$$

Values of the dimensionless shape factor A/a are quoted in the literature (see, for instance, Maasland and Haskew, 1957). For a borehole fully penetrating a saturated porous medium of infinite horizontal extent and terminating vertically in an impermeable layer Kirkham and von Bavel (1948) suggest:

$$\frac{A}{a} = \frac{16H}{\pi h}\frac{\sin(n\pi h/2H)}{n^2}\frac{K_i(n\pi a/2H)}{K_0(n\pi a/2H)} \qquad (8.38)$$

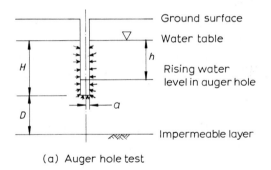

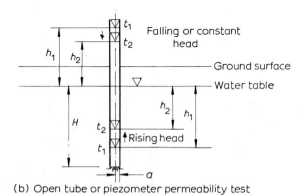

Figure 8.20 Open hole permeability tests.

where K_i, K_0 are Bessel functions,
  $n = 1, 3, 5 \ldots$
  D, the depth from the base of the hole to the base of the aquifer, is in this case equal to zero.

This equation assumes *horizontal* flow through a medium of uniform permeability. A simpler form is Ernst's (1950) formula for similar flow conditions, quoted by Luthin (1966):

$$\frac{A}{a} = 0.75 \left(\frac{H}{a} + 10 \right) \left(2 - \frac{h}{H} \right) \quad (8.39)$$

and for $D = \infty$, allowing for vertical flow:

$$\frac{A}{a} = 0.68 \left(\frac{H}{a} + 20 \right) \left(2 - \frac{h}{H} \right) \quad (8.40)$$

In fact, equation 8.40 holds, according to Kirkham (1955), for intermediate values of D down to $D/H = 1$, and for high H/a values. It is suggested that the two equations should cover most auger hole permeability tests. For more accurate computation of k, values of A/a calculated for intermediate D values are tabulated by Boast and Kirkham (1971).

In practice, the auger hole method is found to be a fairly accurate method of permeability measurement in shallow soils, although expansion of entrapped air may cause high initial inflow rates. In collapsing soils, a pervious lining may be used to support the borehole.

In deeper soils, *horizontal* permeability of a soil layer may be measured by using a *piezometer* sealed into a borehole. Various installation methods may be used, the best being to place the piezometer tip into a borehole so that it is in contact with the borehole sides and then to seal off the hole above the piezometer. Alternatively, the piezometer may be surrounded by sand, so effectively widening the diameter of the flow source. Average permeability may be measured by the *tube* method in which a tube is driven into the ground (and later augered out), or is placed in an open borehole and sealed in place if necessary.

The system (open tube or closed circuit in the case of some piezometers) is then allowed to register the piezometric head. This head is then increased and the rate of fall in water level measured (*falling head* test), or it is reduced and the rate of rise in water level measured (*rising head* test). Alternatively, the head can be raised to a specified level and the rate of flow measured at this level (*constant head* test). Then in the variable head case (Figure 8.20b), from equation 8.37 the permeability coefficient is given by:

$$k = \frac{\pi a^2}{Ah} \frac{dh}{dt} \qquad (8.41)$$

This differential equation in terms of h and t may be solved for the boundary conditions in Figure 8.20b to give:

$$k = \frac{\pi a^2}{A} \frac{\ln(h_1/h_2)}{(t_2 - t_1)} \qquad (8.42)$$

where a is the radius of the tube or piezometer source,

A is a shape factor with dimensions of length,

h_1 is the distance between the water level in the tube and the piezometric level at time t_1, and h_2 is the water level-piezometric level distance at time t_2.

For the constant head (h_1) case,

$$k = \frac{Q}{Ah_1} \tag{8.43}$$

where Q is the rate of volume flow of water into the soil.

Various workers have used empiricism, numerical analysis and electrical conductivity analogues to determine A for both tube and piezometer (see for instance Reeve and Kirkham, 1951; Hvorslev, 1951; Smiles and Young, 1967; Schmidt, 1967; Hanna, 1973). Of these, the most comprehensive and original work is by Hvorslev, who developed a series of empirical shape factors for various cases. Two important examples, a tube or piezometer in an infinite isotropic medium and in an anisotropic soil bounded by an impervious boundary are illustrated in Figure 8.21 and methods for calculating k_v and k_h summarized. Recent experimentation (Smith, 1973) has indicated that these shape factors give a reasonably accurate indication of relevant flow patterns.

Hvorslev also developed a relationship between permeability and *time lag* (T) in low permeability soils, time lag being the term used to describe the time between depression or increase of the water level and registration of *full* piezometric pressure. The time lag can be related to permeability using the same approach as for other flow situations, particularly the variable head case. For instance, if at any time t the head difference between the water level in the borehole and the piezometric level is h_t, then

$$Q = Akh_t \tag{8.44}$$

If it is assumed that Q is the average rate of flow from $t = 0$, $h_t = h_{t0}$ to full pressure equalization at $t = T$, $h_t = 0$, then we can substitute

$$Q = V/T = \pi a^2 h_t / T$$

in equation 8.44 giving

$$k = \frac{\pi a^2}{AT} \tag{8.45}$$

where a is the standpipe radius,
and V is the volume of water ($= \pi a^2 h_t$) required to bring the water level to the piezometric level.

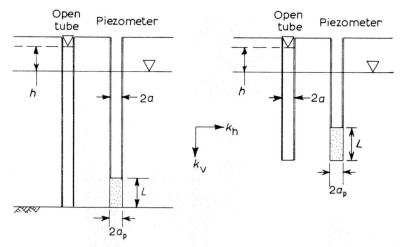

(a) anisotropic soil bounded by impervious layer — inlet at boundary

(b) infinite isotropic soil

constant head (h_1) — open tube (radius a)

(a) $k_e = \dfrac{Q}{4ah_1}$ (b) $k_e = \dfrac{Q}{5.5ah_1}$ where $k_e = (k_v k_h)^{1/2}$

variable head (h_1, h_2 at t_1, t_2) — open tube (radius a)

(a) $k_e = \dfrac{\pi a}{4(t_2 - t_1)} \ln \dfrac{h_1}{h_2}$ (b) $k_e = \dfrac{\pi a}{5.5(t_2 - t_1)} \ln \dfrac{h_1}{h_2}$

constant head (h_1) — piezometer (length L, radius a_p)

(a) $k_h = \dfrac{Q \ln\left[\dfrac{L}{a_p} + \left\{1 + \left(\dfrac{mL}{a_p}\right)^2\right\}^{1/2}\right]}{2\pi L h_1}$ where $m = (k_h/k_v)^{1/2}$

(b) $k_h = \dfrac{Q \ln\left[\dfrac{mL}{2a_p} + \left\{1 + \left(\dfrac{mL}{2a_p}\right)^2\right\}^{1/2}\right]}{2\pi L h_1}$

variable head (h_1, h_2 at t_1, t_2) — piezometer (length L, radius a_p)

(a) $k_h = \dfrac{a^2 \ln\left[\dfrac{mL}{a_p} + \left\{1 + \left(\dfrac{mL}{a_p}\right)^2\right\}^{1/2}\right]}{2L(t_2 - t_1)} \ln \dfrac{h_1}{h_2}$

(b) $k_h = \dfrac{a^2 \ln\left[\dfrac{mL}{2a_p} + \left\{1 + \left(\dfrac{mL}{2a_p}\right)^2\right\}^{1/2}\right]}{2L(t_2 - t_1)} \ln \dfrac{h_1}{h_2}$

Figure 8.21 Shape factor values (*after* Hvorslev, 1951).

Equation 8.45 is identical to equation 8.42 for the specific condition $\ln(h_1/h_2) = \ln(h_{t0}/h_t) = 1$ or $h_t/h_{t0} = 0.37$ and $T = T_0 = t_2 - t_1$. So, the permeability coefficient k can be calculated most simply by substituting in equation 8.45 for a *basic* time lag T_0 at $h_t = 0.37 h_{t0}$ for low permeability soils. In a field test this will involve plotting h_t/h_{t0} against time on a log-linear scale for a raised or depressed head and estimating T_0 from the graph (Cedergren, 1967).

In order to obtain a required degree of accuracy with any method of permeability measurement, some basic precautions are required. In the *tube* method, a clean tube-soil interface and an empty tube are essential. Penetration of soil into the tube, or creation of a cavity beneath the tube, will cause serious errors. In addition, to overcome the effects of inexact scaling or silting, rising head rather than falling or constant head tests are desirable. In the case of *piezometer* and auger hole tests, smearing (see Section 11.4) can affect permeability of the sidewalls, and augering rather than driven piezometers or driven mandrels is essential in both cases to avoid remoulding in clayey soils. The permeability of any porous lining or piezometer tip should be an order of magnitude greater than the soil or sand filter, and that of the filter an order of magnitude greater than the soil. In low permeability soils, periodic flushing of the system may be needed to remove trapped air or gas bubbles (see Penman, 1972). In the case of swelling clays, a constant head test may be required to isolate non-linearities caused by adsorption of water at increased porewater pressures. In all clay soils, an equalizing head of water during drilling is desirable to prevent changes in microstructure and water content in the clay as a result of induced suction pressures.

Actual soil permeabilities will vary widely with position and direction of water flow in most layered soils, and the theoretical approximation of a uniform, continuous, homogeneous and isotropic bed is unlikely to obtain in practice, except in the case of thick sand/gravel deposits and some homogeneously fissured rocks. Drift deposits containing isolated or interconnected inclusions of permeable gravels, sands or silts in a predominantly low permeability matrix probably represent the extreme case of inhomogeneity, and they present difficulties of fabric description as well as permeability measurement. Layered and fissured clays ranging from laminated clays containing silt partings to alternating clay, silt, sand deposits represent the extreme case of anisotropy and would normally have k_h/k_v ratios of between 5 and 10, a significant factor in drainage and consolidation. In extreme cases the permeability of the soil, as in the case of fissured rocks, may be controlled by the presence of widely separated, highly pervious layers which might easily be ignored when (or if) sampling for laboratory permeability or consolidation testing.

8.7 Economic exploitation of groundwater

In many parts of the world, the potential groundwater resources are vast. In Britain, for example, there is probably 50 times more water beneath the surface than lies on top, but the quality of that water is quite variable. The cost of extraction is low compared with the cost of surface reservoirs for impounding equivalent volumes but the efficient exploitation of that water demands a refined approach to the spacing of abstraction wells and an appreciation of consequential water movement.

Groundwater offers a relatively cheap supply of generally good quality water. Two basic types of aquifer have been mentioned earlier, but it is useful to define three major types of system (Figure 8.4). With a confined system, the storage is of a permanent character and there is no seasonal recharge of the aquifer. If water is abstracted, there will be a progressive depletion in the total volume of water stored. In an *unconfined or water table* system, storage is in part permanent and in part temporary as the aquifer is recharged seasonally. Water table aquifers offer a degree of flexibility in an overall water development or conjunctive use plan, but since they are in hydraulic continuity with nearby rivers, pumped abstraction from them leads to a depletion in river flow which may require supplementation through a water-transfer scheme. This aspect of hydrology is discussed in Sections 12.2, 12.3. The third system relates to a *semi-confined* situation in which the aquifer is confined, and so embodies some permanent storage, but in which there is recharge through a natural process of leakage from overlying and possibly underlying aquicludes. Semi-confined aquifers are more suitable for development in a water resources programme than are water table aquifers since they are not in direct hydraulic continuity with any surface water sources except under certain circumstances. The only requirement is that any pumping programme shall be so defined as to ensure that the rate of abstraction does not exceed the rate of natural recharge which itself may vary depending upon the precipitation characteristics and the infiltration capacity of the aquiclude.

Development of an aquifer may lead to long-term problems. In a confined system such as exists in the London Basin, there may well be a progressive depletion of storage, a lowering of the water table and ground settlement (Wilson and Grace, 1942). Abstractions from the aquifer increased from about 45.5 million $l\,d^{-1}$ in the years 1850–1875 to about 230 ml d^{-1} between 1925 and 1940. The fountains at Trafalgar Square were originally artesian but since 1900 there has been an approximate two-thirds of a metre per year decline in water head. In the Liverpool area of England, there has been a reduction in the quality of water abstracted with possible salt water intrusion (the same problem has arisen in the San Joaquin Valley California – see Prokopovich, 1974) and in some areas of the world, for example the Santa Clara and San Joachin valleys in California, U.S.A (Poland and Green, 1962), abstraction has led to surface subsidence. Subsidence of up to about 4 m was first noticed in the San Joachin Valley in the 1930's and it has been found that artesian head levels have been lowered by up to 80 m by about 1965 with a rate of lowering up to about 3 m a year. In the London Basin there seems to have been about a third of a metre subsidence related to about an 80 m decline in water head. In east Durham, England, groundwater abstraction from the Permian aquifer has resulted in a lowering of the groundwater level by an estimated 80 ft between 1935 and 1969, and Wood (1923) believed that some of the subsidence attributed to coal mine workings in the Carboniferous was, in fact, caused by water abstraction. The only satisfactory remedial measures involve a long rest period during which there is no abstraction (usually an impractical solution) together with a programme of recharge. When, for example, ground is de-watered to facilitate tunnelling operations, it may be found necessary to adopt borehole recharge methods in order to stabilise property adjacent to the tunnel. But probably the classical case of overpumping over an extended period of time arises in Venice (Ricceri and Butterfield, 1974) where until about 1948 sweet water under artesian pressure was enjoyed in the City. Braun (1973) has suggested that the average settlement of about 5 mm per annum varies according to the number of wells in the area and the amount of water abstracted. One suggested combative measure involves the use of a diaphragm wall cut-off 12.8 km long by 0.6 m thick penetrating 120 m deep. It would be installed in the sea bed along the perimeter of Venice so that it would enclose and isolate soil formations beneath the City. The proposal also foresees 10 to 15 large diameter recharging wells within the cut-off wall, each pumping in sweet water at a rate of $10\,l\,s^{-1}$

at an excess pressure of 1 kgf cm^{-2}. It has been calculated that it would then require a period of 3 to 5 years to restore the original groundwater pressure beneath the City to halt the sinking. Further pumping at 10 atmospheres pressure at a rate of 120 l s^{-1} would cause the City to rise again at a rate of 3 cm to 5 cm per year.

In the case of an unconfined aquifer, the major problem involves river flow depletion and its effect upon the dilution of effluent from down-stream industry, upon down—stream abstraction requirements, and upon fish life in the river. Progressive groundwater lowering can have a widespread effect upon agricultural crop yields, and near to the coast there is also the risk of saline water intrusion into the aquifer. Examples of all these problems are to be found in England. Partial remedial measures lie in the more widespread adoption of conjunctive use schemes.

8.8 Ownership of groundwater and permitted abstractions

Groundwater tends to enjoy a different legal status in different countries. Probably most widespread is the concept which accepts that although water is a public property it is available to be exploited, and the first person so to exploit in a beneficial manner thenceforth enjoys the firmest title to the water against claims by subsequent persons. Another system would permit an individual landowner to abstract water under his land without any restriction but obviously if many adjacent landowners engaged in heavy pumping operations then the aquifer would become grossly over-developed and possibly depleted. Such a system rules out the possibility of the managed development of water resources, but it operated under British Common Law until the passing by Parliament of the 1963 Water Resources Act. Under a modified system, an individual ownership and unrestrained groundwater development would be restricted with respect to the reasonableness of the abstraction. A further modification to the total abstraction rights of an individual arises from a requirement for correlation between several abstractors during times of drought. Finally, at the other extreme end of this ownership spectrum there is the situation when the rights to and control of all groundwater abstraction is vested in the state with no recognition of private water rights.

In England and Wales, the 1963 Water Resources Act (which was not affected by the 1973 Water Act) requires that all abstractions must be licensed on the basis of a protected right. Licences issued after the Act must be of such a form as not to derogate pre-existing licences. There are also controls on groundwater abstraction with respect to river flows; there must be no diminution of river flows below a minimum acceptable level.

8.9 Groundwater exploration

Exploration starts as a basic geological exercise of mapping surface exposures, faults, unconformities and intrusions together with interpolations and extrapolations from visible evidence. Structural features such as faults and igneous intrusions may serve to delimit an aquifer and their development at depth may be determined only by borehole evidence. Base topographic maps may be used, and because water heads and potential gradients are so important, it may be necessary to re-survey an area using a portable aneroid barometer reading to better than 1 metre. Aquifer information will also be derived from observation of stream flows and spring lines. Aerial photographic techniques, including infra red and ultra violet methods, will also play a part during an initial survey.

Of the geophysical methods that are available to the hydrogeologist, the gravity approach may be useful for delineating basins but the resistivity method will be most generally used for sensing aquifers up to depths of about 150 m (see Section 7.4). A simple hammer seismograph may be able to resolve a velocity contrast due to the presence of water at a depth of up to about 100 m.

The minimum acceptable diameter for direct exploratory drill holes is 150 to 200 mm (6 to 8 in). They will indicate the depth to and thickness of an aquifer and will provide access for determining the hydraulic properties of a water-bearing formation. Samples may also be taken for water quality determination. Well logging is a necessary but expensive exercise (see Section 7.7). It may take several forms: electrical logs (resistivity, which defines the conductivity of the formation, and probably self-potential); radioactive logs (gamma, gamma-gamma, neutron); mechanical (acoustic, caliper); temperature. Inflows will also be monitored at discrete points down the hole to measure the depth-conductivity relationships and to combine with caliper information for assessing the influence of borehole diameter on that inflow.

Each exploratory borehole, after being cleared-out, would be subjected to a programme of test pumping, probably using a 6 in pump in an 8 in diameter hole. The aim would be to examine an

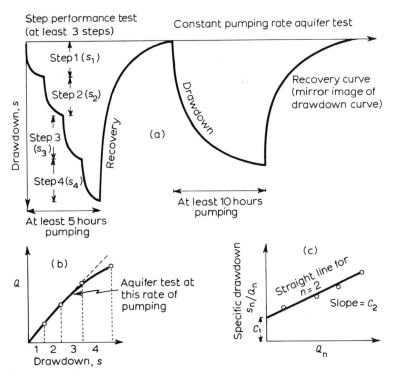

Figure 8.22 Step pumping tests at different rates followed by an aquifer test at an optimum constant pumping rate determined by the break-point on the Q/s curve.

optimum rate of abstraction with respect to the drawdown for the pump at a prescribed horizon in the well. From such tests, it is then possible to estimate the transmissivity of the aquifer with respect to that particular well.

Some empirical well-test curves are shown in Figure 8.22. The total drawdown s_t in a pumping well will comprise an aquifer loss (loss of water head due to frictional resistance as the water flows towards the sink) and a well loss (due to construction and development of the well). If flow into the well is assumed to be laminar, then the form of the yield-drawdown curve might be approximated to a quadratic function but since suction at the pump might create conditions of localized turbulent flow, the relationship might be expressed more generally as

$$s_t = C_1 Q + C_2 Q^n \quad (2 \leqslant n \leqslant 3.5) \tag{8.46}$$

where C_1 and C_2 are constants.

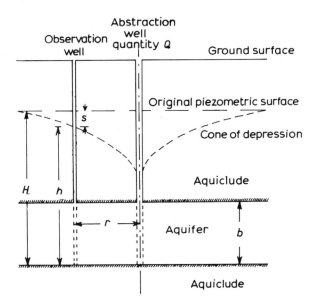

Figure 8.23 Non-leaky aquifer condition with a well fully penetrating into the aquifer.

Thus, for each step of the pumping test (Figure 8.22a,b), the ratio s_n/Q_n ($n = 1,2,\ldots t$) may be plotted against Q_n in the manner of Figure 8.22c and if a linear, least squares regression line can be drawn through the data points, then exponent $n = 2$ and the flow approximates to laminar.

It is useful to extend the general differential flow equation to the solution of several idealized aquifer and pumping well conditions. In all cases, aquifers are assumed to be homogeneous, isotropic and infinite in extent and the boundaries between layers are assumed to be regular. It is also assumed that a discharge or recharge well penetrates and receives water from the entire thickness of aquifer and that the transmissibility is constant at all times and at all places. Solutions to the several systems to be outlined arise from a substitution of specific boundary conditions into the basic differential equations of flow. It is also assumed that due allowance will be made for barometric pressure changes on the long-term water table level.

The theory of aquifer testing and evaluation of potential economic yields is based on either a steady or a non-steady state condition of flow within the aquifer. The *steady state flow* concept, providing the framework for analysis of well hydraulics, developed from the end of the 19th century with the Thiem formula for *confined* flow and the Dupuit-Forchheimer equation for *unconfined* flow. These conditions have been expressed earlier in equations 8.25 and 8.26. The Dupuit-Forchheimer equation can be used to predict discharge to within an error of about 5 per cent but Boulton (1951) has shown that it cannot be employed to determine the position of the free surface at a distance from the well greater than 1.5 times the undisturbed saturated aquifer thickness. The main factor producing divergence of the free surface from that predicted by Dupuit–Forchheimer is the existence of a seepage face in the well above the pumped water level. Boulton (1951) has shown by relaxation methods that the height of this seepage face may be determined from the well radius, the volume of water pumped and the hydraulic conductivity, and Hantush (1964) has proposed a method for more accurately determining the free surface position. Further developments are due to Zee *et al* (1957) and Herbert (1965).

The application of steady state conditions is an attractive proposition in view of a basic simplicity, but some of the limiting assumptions are rather too rigorous to allow more than an approximate solution. From basic definition, the supply of water under steady state conditions comes from lateral flow within the aquifer rather than by de-watering, and so an estimate of the storage coefficient cannot be derived.

One further term, that of Specific Capacity (SC) should be defined. It is usually expressed as the yield in gallons d^{-1} ft^{-1} of drawdown under steady state conditions. For a confined aquifer, the specific capacity is constant for drawdowns less than the artesian head. For an unconfined aquifer, SC is a function of the percentage drawdown of maximum, or the percentage de-watered. SC values have tended to be normalized to the value represented by a 50 per cent de-watering, and using a relationship derived from the graph of Johnson (1966) we have

$$SC_{50} = SC \left[\frac{0.75}{1 - \%dw/200} \right] \qquad (8.47)$$

where SC is the calculated specific capacity and %dw is the percentage of the aquifer de-watered.

Consider, now, the case of a non-leaky artesian aquifer subjected to *non-steady state* radial flow during pumped abstraction. There is no recharge source, and as pumping progresses, the rate of decrease in the water head itself decreases as the cone of depression expands outwards in the absence of recharge. Water is released from its storage through elastic contraction of the aquifer and adjacent beds and also by its own expansive properties. The differential equation describing the flow will usually be written in polar coordinate form as

$$\frac{\partial^2 h}{\partial r^2} + \frac{1}{r}\frac{\partial h}{\partial r} = \frac{\partial h}{\partial t} \cdot \frac{S}{T} \tag{8.48}$$

S is the dimensionless storage coefficient (defined by Jacob, 1950) = specific storage (L^{-1}) x b (Walton, 1970, p. 127) and T is the transmissivity (the coefficient of transmissibility) of the aquifer, or the hydraulic conductivity multiplied by the thickness of the hydraulic path. T is in units of area per unit time ($L^2 T^{-1}$). The other units are defined in Figure 8.23 which also indicates the use of observation wells to define the form of the drawdown surface. Lateral impermeable barriers (Figure 8.24) or stream recharge may affect this surface.

The solution of equation 8.48 (see Davis and De Wiest, 1966) leads to the so-called non-equilibrium, or Theis (1935) equation:

$$h = H - \frac{Q}{4\pi T} \int_{\frac{r^2 S}{4Tt}}^{\infty} \frac{e^{-x} dx}{x} \tag{8.49}$$

where x is the horizontal cartesian coordinate in the plane of the problem.

This equation is a function of the lower limit $\frac{r^2 S}{4Tt}$ which is replaced by a parameter u.

Re-writing,

$$h - H = -s = -\frac{Q}{4\pi T} \int_u^\infty \frac{e^{-u} du}{u} \tag{8.50}$$

$\int_u^\infty \frac{e^{-u} du}{u}$ is defined as the well function $W(u)$ – see Wenzel (1942). $W(u)$ is a series equal to

$$-0.577216 - \ln u + u - \frac{u^2}{2.2!} + \frac{u^3}{3.3!} - \frac{u^4}{4.4!} \cdots \tag{8.51}$$

Expressing equation 8.50 in terms of $W(u)$ we have

$$s = (H - h) = \frac{Q}{4\pi T} W(u)$$

which is equivalent to, say,

$$s = \frac{Q}{4\pi T} \frac{(\text{gallons min}^{-1})}{(\text{gallons d}^{-1} \text{ ft}^{-1})} W(u)$$

$$= \frac{Q}{4\pi T} \frac{\text{ft-d}}{\text{min}} W(u)$$

$$= \frac{Q}{4\pi T} \left(\frac{\text{ft-d}}{\text{min}} \times \frac{1440 \text{ min}}{\text{d}} \right) W(u)$$

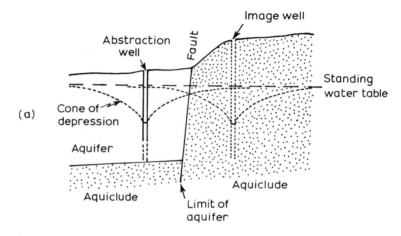

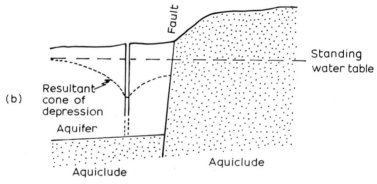

Figure 8.24 Use of the image well method to re-define the form of a cone of depression when drawdown is interrupted by an impervious barrier (in this case created by a geological fault). In (a) an imaginary discharging well is located at the same distance from the barrier as the real well in order to satisfy a condition of no flow across the barrier. In (b) the resultant cone of depression — flatter towards the barrier and steeper on the side away from the barrier — arises from the algebraic summation of overlapping real and image well cones. Using the same superimposition method, stream recharge will steepen the surface of the cone of depression.

Therefore, $s = 114.6 \dfrac{Q}{T} W(u)$ ft, (8.52a)

or

$$s(\text{m}) = \dfrac{Q}{4\pi T} \dfrac{\text{ls}^{-1}}{\text{ls}^{-1} \text{ m}^{-1}} W(u)$$

$$= 0.08 \dfrac{Q}{T} W(u) \text{ m} \tag{8.52b}$$

Since $u = r^2 S/4Tt$, taking transmissivity in U.S. gallons d^{-1} ft^{-1} and noting that 1 U.S. gallon = 0.13368 ft^3,

$$u = \dfrac{1.87 S}{T} \cdot \dfrac{r^2}{t} \tag{8.53}$$

Then re-writing equations 8.52 and 8.53 in logarithmic form:

$$\log s(\text{ft}) = \log W(u) + \log 114.6 \dfrac{Q}{T} \tag{8.54}$$

or

$$\log s(\text{m}) = \log W(u) + \log 0.08 \dfrac{Q}{T}$$

and

$$\log \left(\dfrac{r^2}{t}\right) = \log u + \log \left(\dfrac{T}{1.87 S}\right) \tag{8.55}$$

Wenzel tabulated the function W(u) for values of u from 10^{-15} to 9.9. A condensed version of the table (Table 8.2) is listed by Davis and De Wiest (1966) and it will be apparent that the relationship between W(u) and u may be represented by a family of curves contained within two boundary curves. The classical method of solution for S and T is first to plot the *type curve* of W(u) against u on log paper according to the tabulated data. Measured values of drawdown $s(=\overline{H} - h)$ derived from several observation boreholes are then plotted against equivalent values of r^2/t on the same size log paper but this time of the transparent variety. Provided that the discharge Q is constant for all s,r,t then, since W(u) is a function of u in the same manner that s is a function of r^2/t, it is possible to superimpose the *data curve* on the type curve maintaining the coordinate axes of the two curves parallel so that the best fit is achieved between the two curves. A common point is usually selected from the overlap portion of the curves to give mutual values

Table 8.2 Values of W(u) for different u (after Wenzel, 1942, condensed by Davis and De Wiest, 1966)

u	1.0	2.0	3.0	4.0	5.0	6.0	7.0	8.0	9.0
$\times 10^{-1}$	0.219	0.049	0.013	0.0038	0.0011	0.00036	0.00012	0.000038	0.0000124
$\times 10^{-2}$	1.82	1.22	0.91	0.70	0.56	0.45	0.37	0.31	0.26
$\times 10^{-3}$	4.04	3.35	2.96	2.68	2.47	2.30	2.15	2.03	1.92
$\times 10^{-4}$	6.33	5.64	5.23	4.95	4.73	4.54	4.39	4.26	4.14
$\times 10^{-5}$	8.63	7.94	7.53	7.25	7.02	6.84	6.69	6.55	6.44
$\times 10^{-6}$	10.94	10.24	9.84	9.55	9.33	9.14	8.99	8.86	8.74
$\times 10^{-7}$	13.24	12.55	12.14	11.85	11.63	11.45	11.29	11.16	11.04
$\times 10^{-8}$	15.54	14.85	14.44	14.15	13.93	13.75	13.60	13.46	13.34
$\times 10^{-9}$	17.84	17.15	16.74	16.46	16.23	16.05	15.90	15.76	15.65
$\times 10^{-10}$	20.15	19.45	19.05	18.76	18.54	18.35	18.20	18.07	17.95
$\times 10^{-11}$	22.45	21.76	21.35	21.06	20.84	20.66	20.50	20.37	20.25
$\times 10^{-12}$	24.75	24.06	23.65	23.36	23.14	22.96	22.81	22.67	22.55
$\times 10^{-13}$	27.05	26.36	25.96	25.67	25.44	25.26	25.11	24.97	24.86
$\times 10^{-14}$	29.36	28.66	28.26	27.97	27.75	27.56	27.41	27.28	27.16
$\times 10^{-15}$	31.66	30.97	30.56	30.27	30.05	29.87	29.71	29.58	29.46
	33.96	33.27	32.86	32.58	32.35	32.17	32.02	31.88	31.76

of s, $W(u)$, r^2/t and u which, if inserted in equations 8.54 and 8.55 produce values of T and S. The control expressions for a non-leaky confined aquifer are therefore:

$$T = \frac{Q}{4\pi s} W(u) \qquad (8.56)$$

and

$$S = \frac{4Tt}{r^2 \cdot 1/u} \qquad (8.57)$$

Davis and De Wiest (1966) provide a numerical example of the use of this technique. It may also be noted that Theis's non-equilibrium formula can also be used to derive S and T from a reverse analysis of a recovery condition.

Jacob (1950) — see also Cooper and Jacob (1964) — introduced an approximation method using a straight line fit and restricting the use of the earlier series in u to the first two terms. This method is discussed by Davis and De Wiest (1966) and a numerical example given.

Many aquifers fail to satisfy a condition of zero recharge. They are overlain by semi-pervious beds which permit downwards percolation into the pervious horizons or there may even be a recharge from below. A type-curve method for non-steady state drawdown was proposed by Hantush and Jacob (1954); see also Hantush (1956, 1964). The drawdown solution given by Hantush and Jacob is

$$s = \frac{Q}{4\pi T} W\left(u, \frac{r}{L}\right) \qquad (8.58)$$

where this time the well function may be expressed as

$$W\left(u, \frac{r}{L}\right) = \int_u^\infty \frac{1}{x} \exp\left(-x - \frac{r^2}{4L^2 x}\right) dx \qquad (8.59)$$

and, as before,

$$u = \frac{r^2 S}{4Tt}.$$

It will be noted that as L tends to infinity, the leakage parameter tends to zero and equation 8.59 reverts to equation 8.50. Tabulation of the function $W(u, r/L)$ should therefore lead to a Theis-type graphical superimposition solution. With t expressed in minutes, Q in U.S. gal. per minute, T in U.S. gal. d^{-1} ft^{-1}, s in ft, r (the radial distance from observation well to pumped well) in ft, the fundamental equations become (note: 1 U.S. gallon = 0.83267 Imp. gallon):

$$T = \frac{114.6Q}{s} W\left(u, \frac{r}{L}\right) \tag{8.60}$$

$$S = \frac{3.7Tt \times 10^{-4}}{r^2 \cdot 1/u} \tag{8.61}$$

$$\frac{r}{L} = \frac{r}{\sqrt{T/(k'/b')}} \tag{8.62}$$

In equation 8.62, the ratio k'/b' is known as the leakance, where k' and b' are respectively the coefficient of permeability (gallons per day per ft^2 in U.S. units or metres per second in SI units) and the saturated thickness (ft or metres) of the aquitard.

If equations 8.60 and 8.61 are re-expressed in logarithmic form it then becomes clear, by analogy with the Theis method, that log $W(u,r/L)$ may be plotted against $\log(1/u)$ for a range of values r/L and that, for a constant discharge Q, a graphical plot of measured values s against t may again be superimposed on the type curves. As before, a match point is selected anywhere on the overlapping section of the sheets of graph paper and its coordinates $W(u,r/L)$, $1/u$, s and t, together with the r/L value that generated the data-type curve fits, are then inserted into equations 8.60, 8.61 and 8.62. Walton (1960, 1962) has produced type curves for a leaky artesian aquifer (see also Davis and De Wiest, 1966, p. 228).

A further type-curve method, this time with respect to a water table aquifer, is due to Boulton (1963). In the very early stages of pumped drawdown, the relationship between s and t conforms to a Theis type curve, the behaviour being that of a confined aquifer with the water being drawn from elastic storage. But, as time goes by, extra water drains downwards into the aquifer and de-watering takes place. This continues until such time as any excess gravity drainage is completed from which time the falling water level keeps pace with the drainage front. A Theis solution seems to apply not only up to point a in Figure 8.25 but also to the water table condition beyond point b. There will therefore be two storativity coefficients, one relating to the confined aquifer at the beginning of pumping and the other being an ultimate value with respect to the water table aquifer:

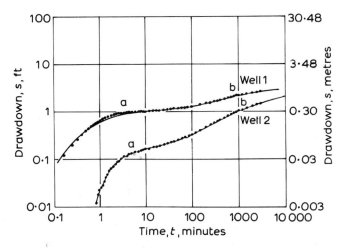

Figure 8.25 Boulton's (1963) theoretical drawdown curves matched to Walton's (1960) pumping test data from two wells near Dayton, Ohio, U.S.A. Test data are the discrete points. Letters 'a' and 'b' are added to the Figure taken from Boulton (1963).

$$T = \frac{Q}{4\pi s} W\left(u_{a,y} \frac{r}{B}\right) \tag{8.63}$$

$$S_a = \frac{4Tt}{r^2 \cdot 1/u_a} \tag{8.64}$$

$$S_y = \frac{4Tt}{r^2 \cdot 1/u_y} \tag{8.65}$$

Further equations may be added for the leaky water-table aquifer condition:

$$B = \sqrt{T/\alpha S_y} \tag{8.66}$$

$$\alpha = \frac{\left(\frac{r}{B}\right)^2 \frac{1}{u_y}}{4t} = \text{reciprocal of the delay index.} \tag{8.67}$$

and

$$\eta = 1 + \left(\frac{S_y}{S_a}\right) \tag{8.68}$$

S_y/S_a will tend to infinity in Boulton's equations, but if $S_y > 4S_a$, the technique still appears to be valid.

Boulton's non-steady state water-table type curves* are shown in Figure 8.26 with the two Theis-type curves bounding a family of curves having different r/B values. The upper-bound curve refers to the confined state at the beginning of pumping, while the lower-bound curve relates to the ultimate water table condition.

Having plotted measurement data s against t on the same transparent log-log paper as the type curves, the measurement data, as before, are superimposed on the type curves keeping the graphical axes parallel. At first, as much of the initial test data as possible should be matched to one of the curves at the 'a' end of the range, selecting a suitable match point at the intersection of major axes and writing down coordinate values s, $W(u_a)$, t, and $1/u_a$. These values would then be substituted into equations 8.63 and 8.64 and then solved for T and S_a (Q and r being known). The field data plot would then be moved horizontally to the right in order to match as much of the later water-table data as possible to one of the 'y' family of curves. The match type-curve 'y' should ideally have the same r/B value as the earlier match type-curve 'a'. As before, a second match point would again be chosen at the intersection of major axes and its coordinates s, $W(u_y)$, t, and $1/u_y$ substituted into equations 8.63 and 8.65 for the solution of T and S_y. Parameter η in equation 8.68 may now be calculated. As a result of differences in storativities, when $\eta > 100$ the line linking early and late field data points is essentially horizontal, but as η falls to lower values, the line ceases to be horizontal and the early transmissibility of the aquifer differs from its late transmissibility. Nevertheless, although T is strictly an aquifer parameter independent of the time-history of abstraction, this matching technique still appears to provide a reasonable, approximate value and provided that the difference between the two T values is not large, an average value may be taken.

From observation of the type curves in Figure 8.26, it is evident that as $1/u_y$ increases, a type 'y' curve for any particular value of r/B becomes indistinguishable from the Theis curve for water-table conditions. The value of $1/u_y$ at the point of conjunction is proportional to the time t_0 at which delayed yield ceases to influence the drawdown, that is, when gravity drainage ceases to be effective in the water transfer operation. It could therefore be used

---

*Boulton (1963) tabulates the numerical values used for constructing these curves in order to facilitate type-curve plotting to a larger scale for practical use.

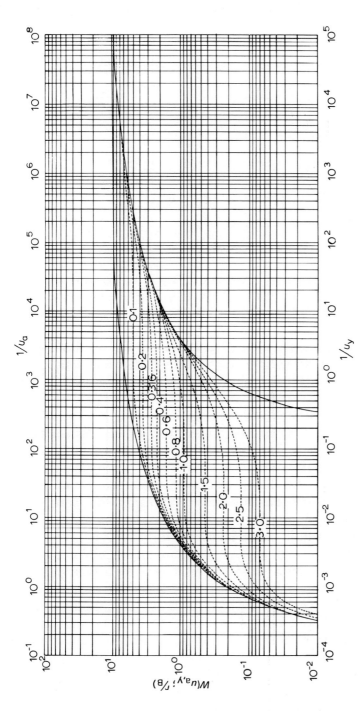

Figure 8.26 Non-steady state water-table type curves from Boulton (1963). The two bounding curves are Theis-type curves and the broken lines represent leaky-type curves for different r/B ratios as marked. Note that the general form of a leaky-type curve corresponds to the form of the curves in Figure 8.25.

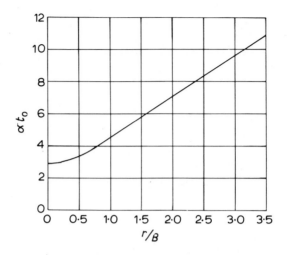

Figure 8.27 Boulton's (1963) curve for estimating t_0 when delayed yield ceases to affect drawdown.

to calculate for known or assumed values of T and S_y. However, it is more convenient to use Boulton's curve (Figure 8.27), given calculated values of r/B and α.

8.10 Regional investigations

Over an extended area, and with respect to a particular aquifer, the actual groundwater system may change from place to place depending upon the type and thickness of drift cover, upon the regional dip of the beds and upon the distribution of structural features such as faults, folds and unconformities. The Permian of the north-east of England (Figure 8.28) is taken to illustrate the distribution of ground water levels. In that area, shown in Figure 8.29, the beds dip gently to the east-south-east on the Carboniferous-Permian unconformity with the areas close to the Lower Magnesian Limestone escarpment (a prominent sinuous feature to the east of Durham City) constituting the major source of groundwater recharge since they are usually covered with only a thin veneer of drift. In addition, there are numerous quarries on or near to the scarp slope, and the exposed rock must greatly facilitate percolation. It was estimated by Burgess (1970) that there may be a recharge of some 45 million gallons (British) d^{-1} over the whole of the Permian shown in Figure 8.29, although the uncertainty due to mine water abstractions from the

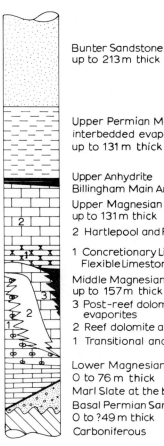

Figure 8.28 Generalized Permo-Trias succession in North East England (*after* Smith and Francis, 1967).

Carboniferous and river flows could lead to an error of 10 mg d^{-1} either way.

If the aquifer as shown were isotropic and of uniform transmissibility, with recharge along the western margin and discharge into the sea to the east, then the equipotential lines would be uniformly spaced. Piezometric maps compiled from borehole evidence of rest-water levels did, in fact, show a wider spacing of contours south east of Ferryhill (Rushyford-Aycliffe area) where the Middle Magnesian Limestone is present beneath the drift. This wider spacing corresponds to lower potential gradients.

If the 1969 and the pre-1969 (Anderson, 1941) groundwater

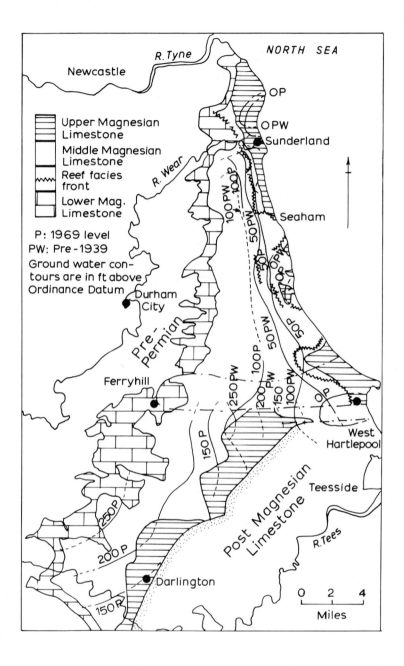

Figure 8.29 Geology of the Permian outcrop in North East England showing changes in groundwater level contours. Broken lines denote uncertainty in the interpretation of borehole information (compiled *after* Burgess, 1970).

contours are compared, some differences will be noted. In the Sunderland area, there has been a decrease in the number of private abstractors and a more readily available source of surface water supply with the construction of the Derwent Reservoir in 1967. This has resulted in a rise in the groundwater level. On the other hand, in the Hartlepools area to the south, increased abstraction has produced a fall in the groundwater level by over 6 m, bringing forward the ultimate future possibility of some sea water intrusion to the aquifer. Any further managed abstraction from the aquifer to the rise (west) must ensure that down-dip seepage is not substantially depleted.

In the northern part of the area, the system may be treated as unconfined from geological evidence and from yield-drawdown curves. Boreholes penetrate through the Lower Magnesian Limestone and Marl Slate to the Basal Permian Sands beneath the groundwater level. The Marl Slate does not act as a confining bed because the Permian is abundantly faulted and a throw of only $^2/_3$ m would be sufficient to bring the lower Magnesian Limestone and the Basal Sands into hydraulic continuity. South of a line from Hartlepools to Ferryhill, the main aquifer, the Middle Magnesian Limestone, is invariably confined by drift deposits or by marls of the Lower Evaporite Group. The underlying Lower Magnesian Limestone has, in comparison, a sufficiently low hydraulic continuity to be regarded as a basal aquiclude.

Burgess (1970) analysed pumping test (drawdown and recovery) data from both northern and southern areas using the several methods outlined for steady and non—steady state flow. From these analyses for the northern area he listed the hydraulic conductivities for the different geological formations as in Table 8.3. He also found a good linear relationship between the specific capacity and hydraulic conductivity in the form

$$SC_{50} = 84.4\, k,$$

the latter parameter being based on the Dupuit—Forchheimer relationship. For the southern area, transmissibilities calculated by non-equilibrium methods of analysis showed wide variations even around a single well, and it was felt that impermeable and semi-impermeable barriers created by faulting and basement 'highs' exerted a rather large influence upon the development of cones of depression. Pumped hole recovery transmissibilities varied from 4000 gpd ft^{-1} (6.9 x 10^{-4} m^2 s^{-1}) to 36,000 gpd ft^{-1} (62.3

Table 8.3 Hydraulic conductivities for the Permian in north-east England as derived by Burgess (1970)

Formation	\multicolumn{3}{c}{Hydraulic conductivity k}		
	gallons $d^{-1} ft^{-2}$	litres $d^{-1} m^{-2}$	$m\ s^{-1}$
Upper Magnesian Limestone (brecciated)	300	14680	1.70×10^{-4}
Middle Magnesian Limestone	300	14680	1.70×10^{-4}
Lower Magnesian Limestone (except basal unit)	110	5382	6.23×10^{-5}
Lower Magnesian Limestone (basal unit)	40	1957	2.26×10^{-5}
Basal Permian Sands	40–120	1957 to 5871	2.26×10^{-5} to 6.80×10^{-5}

$\times 10^{-4}$ m² s^{-1}) but averaged out at about 13410 gpd ft^{-1} (23.2 $\times 10^{-4}$ m² s^{-1}). On the basis of a Jacob recovery value, an average permeability of 215 gpd ft^{-2} (1.22 $\times 10^{-4}$ m s^{-1}) was calculated for the Middle Magnesian Limestone. In order to avoid de-watering an aquifer, pumping rates should be limited to prevent the drawdown exceeding the artesian head. This was a necessary condition in the Permian, the maximum safe yield from a well being expressed as

$$Q = h_a \, SC \qquad (8.69)$$

where h_a is the artesian head. It was found for this southern area that there was a similar trend as in the northern area between specific capacity and hydraulic conductivity in the form

$$SC_{50} = 65.5 \, k \qquad (8.70)$$

8.11 Simulation of groundwater regimes

The application of digital and analogue simulation methods to a study of groundwater movement gained impetus from the beginning of the 1950's. Simulation is a basically accurate predictive operation subject to the satisfactory choice of user-determined parameters and also subject to the approximations that always arise in the numerical methods.

The use of numerical methods for the study of field problems owes much to the work of Southwell (1946) on relaxation, but electrical analogues had been used in groundwater studies a decade earlier (Wyckoff and Reed, 1935). Electrical resistance-capacitance models have been described by Skibitzke (1963), Stallman (1963), Walton and Prickett (1963), Zee *et al.* (1957) and Herbert (1965). But with powerful digital computation facilities, problems may be solved directly (see, for example, Pinder and Bredehoeft, 1968) using the finite difference method in continuum mechanics. This digital method, applied three-dimensionally, has application in a wider field of engineering hydrology; it is being used, for example, in the authors' laboratory to study seepage flow in tunnels driven below the water table. But it will not totally replace the electrical analogue, which is simple, cheap, versatile and suitable for operation by inexperienced personnel.

A comprehensive explanation of the application of digital finite difference and electrical analogue simulations is outside the scope of the present work, but the interested reader will have no difficulty in obtaining further information from the literature and pursuing an understanding of the procedures. In the *digital method*, for a steady state solution the Laplace equation is simply written in finite difference form, which is an equation for the 5—point solution of potential and which is used in problems of heat, electricity and water flow. There are modifications for transmissibility anisotropy and a balancing term which accommodates recharge and discharge at the central node of the 5—point local potential field (see Figures 8.30 and 8.31). Boundary conditions are easily modelled, but since the finite difference method is, by its nature, an approximate method, an evaluation of errors is necessary (Noble, 1964), usually by expanding the equations which solve for the individual heads of water distributed about the finite difference mesh. At the point of actual drawdown into a well, the change in potential is so rapid at the singularity that it cannot adequately be represented by the Laplace equation. A method of replacing the transmissibility elements by 'effective' well elements for suitable finite difference modelling has been explained by Herbert and Rushton (1966). For a non-steady state situation, the finite difference approximation and solution by an alternating direction-implicit procedure has been described by Pinder and Bredehoeft (1968).

In general, the digital method of analysis would seem to have inherent advantages where a considerable number of programmable iterations must be executed. Burgess (1970) used the technique for

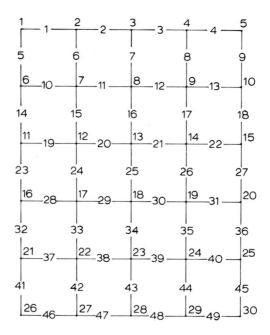

Figure 8.30 Typical overlay mesh for aquifer simulation. Nodes and transmissibility elements are numbered consecutively from top left to bottom right.

simulation work in a 'hybridized' style in conjunction with an electrical analogue model in order to carry out time-consuming and tedious tasks at the analogue model calibration stage.

Fundamental to the electrical analogue method are four scaling factors:

$$q = K_1 . Ch \qquad (8.71)$$
$$H = K_2 . v \qquad (8.72)$$
$$Q = K_3 . i \qquad (8.73)$$
$$td = K_4 . ts \qquad (8.74)$$

where the link constant

K_1 is in gallons (or litres) per coulomb,
K_2 is in feet (or metres) per volt,
K_3 is in gallons (or litres) per day per amp,
and K_4 is the time in days per model time in seconds.

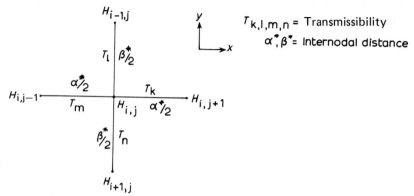

Nodal flow for above element is:

$$Q_{i,j} = \frac{\beta^*}{\alpha^*} T_k(H_{i,j+1} - H_{i,j}) + \frac{\alpha^*}{\beta^*} T_l(H_{i-1,j} - H_{i,j}) + \frac{\beta^*}{\alpha^*} T_m(H_{i,j-1} - H_{i,j})$$

$$+ \frac{\alpha^*}{\beta^*} T_n(H_{i+1,j} - H_{i,j}) \qquad (1)$$

With respect to the nodal mesh in Figure 8.30, equation (1) may be written:

$$Q_d = \frac{\beta^*}{\alpha^*} T_k H_e + \frac{\alpha^*}{\beta^*} T_l H_f + \frac{\beta^*}{\alpha^*} T_m H_g + \frac{\alpha^*}{\beta^*} T_n H_h$$

$$+ \left\{ -\frac{\beta^*}{\alpha^*} T_k - \frac{\alpha^*}{\beta^*} T_l - \frac{\beta^*}{\alpha^*} T_m - \frac{\alpha^*}{\beta^*} T_n \right\} H_d \qquad (2)$$

or, in general matrix form:

$$\{Q\} = [T_A] \cdot \{H\} \qquad (3)$$

For an impermeable boundary along $\overline{H_{i,j-1}; H_{i,j+1}}$

$$Q_d = 0 = \frac{\beta^*}{2\alpha^*} T_k H_e + \frac{\beta^*}{2\alpha^*} T_m H_g + \frac{\alpha^*}{\beta^*} T_n H_h$$

$$+ \left\{ -\frac{\beta^*}{2\alpha^*} T_k - \frac{\beta^*}{2\alpha^*} T_m - \frac{\alpha^*}{\beta^*} T_n \right\} H_d \qquad (4)$$

For a fixed flow boundary:

Replace zero flow in equation 4 by Q_d. But if Q_d is not known it is necessary initially to use a fixed boundary positioned outside the area of influence of abstraction wells.

For fixed potential boundary:

Make all off-diagonal elements of the row zero in the coefficient matrix $[T_A]$ and replace the diagonal element by an arbitrary number of the same magnitude of others in $[T_A]$. Q_d is then put equal to $T_{Ad,d} \times H_d$.

Solution of matrices:

In a rectangular model of l columns and m rows, the potential distribution is determined by the solution of (lm) nodal equations in (lm) unknowns. The Gauss-Jordon method of solution solves linear equations by a system of successive pivotal condensations but there are serious drawbacks. The Gauss-Siedel iterative method (McCracken, 1967) or the Choleski method are to be preferred.

Figure 8.31 Specifications for matrix solution of aquifer simulation.

From these parameters, Walton and Prickett (1963) have given the following identities:

$$\frac{K_3 K_4}{K_1} = 1 \qquad (8.75)$$

$$R = \frac{K_3}{K_2} \cdot \frac{1}{T} \qquad (8.76)$$

$$C = 7.48\alpha^{*2} S \frac{K_2}{K_1} \qquad (8.77)$$

where C is the electrical capacitance in farads and R is the electrical resistance in ohms,
 α^* is the mesh length in feet,
and S is the storage coefficient with T being the transmissivity.

Of these scaling factors, only K_2 can be fixed absolutely. But from the range of expected transmissibilities, likely recharge, and abstraction rates, a working value for K_3 can be chosen for establishing working values for currents and resisters.

A *steady state analogue model* requires only banks of resistors linking the nodal points shown in Figure 8.30. Boundary potentials may be applied from a bank of potential dividers with a large enough input voltage. But this method suffers from the fact that both the potential and the current depend on the load, which may vary during simulation studies. A preferred alternative is to estimate the recharge inflow and discharge over the area being studied, and then to apply these values as constant *currents* using parameter K_3. A problem then arises in that this pre-supposes a knowledge of recharge and discharge levels, which may not be available. This technique may be used with positive and negative constant current generators (Figure 8.32) based on a simple bias emitter resistor circuit. A simple voltage dropper circuit maintaining constant current over a predetermined range would be another option. Discharge areas, such as spring lines, may be simulated by means of a diode lead to a potential divider reference, equivalent to the actual topographic height at the location. When the nodal potential exceeds the reference potential, current drains from the model but when the model potential falls below reference during pumping, the diode becomes negatively biased and no current flows. For resolving nodal response to any model input conditions, a digital voltmeter with a very high input impedance and drawing negligible current is required.

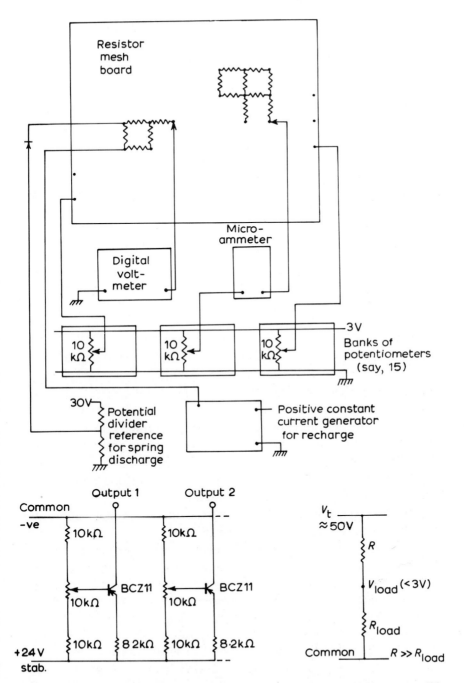

Figure 8.32 A steady state excitation – response network with two possible constant current circuits below (*after* Burgess, 1970).

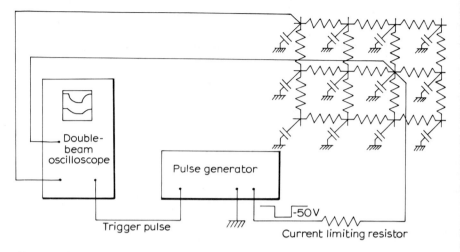

Figure 8.33 A non-steady state excitation response network for pilot study (*after* Burgess, 1970).

A readily available micro-ammeter will be suitable for measuring current flows.

Non-steady state flow, by definition, incorporates a time-variant which may be modelled by injecting transient voltages and measuring the circuit response. Depending upon circuit parameters, the pulse decay time may need to be monitored by oscilloscope (fast decay <1 second) or by galvanometric ultra-violet recorder (slow decay >1 second). Ideally, for excitation, a battery of pulse generators able to simulate abstraction and recharge at different rates and times is required, but for a pilot study a single pulse generator arrangement in the manner of Figure 8.33 may be suitable.

The first stage in any simulation is one of model calibration. Resistor values given in Figure 8.32 have been used to model the Permian aquifer steady-state over a mesh length of 3124 ft (952 m), 600 resistors in total being used. Areas requiring more detailed attention could be enlarged on separate panels. $10K\Omega$ resistors were equivalent to 10,000 gpd ft^{-1} (17.3×10^{-4} m^2 s^{-1}) transmissibility, the final scaling factors being:

$$K_1 = 10^{14} \text{ gallons C}^{-1}, \text{ or } 4.546 \times 10^{14} \text{ litres C}^{-1}$$
$$K_2 = 10^2 \text{ ft V}^{-1} \text{ or } 30.48 \text{ m V}^{-1}$$
$$K_3 = 5.58 \times 10^{10} \text{ gdp A}^{-1}, 25.37 \times 10^{10} \text{ lpdA}^{-1}$$
$$K_4 = 10^4 \text{ days s}^{-1}$$

An initial estimate of relative transmissibilities may be made from a piezometric contour map and the corresponding resistors plugged in using, in the present example, a working K_3 value of 10^{10} gpd A^{-1} (4.546×10^{10} 1pdA^{-1}). Boundary potentials are applied in a step-by-step process and the resistors similarly adjusted to obtain reasonable agreement with reference nodal potentials. Actual pumping test parameters may be simulated by imposing the known drawdown at the pumped well and altering the surrounding resistors until the drawdowns in the observation wells are reproduced on the model. If in some areas it is difficult to satisfy both the pumping and the piezometric conditions, this may be a function of local recharge, perhaps by stream flow, which may be modelled by appropriate inputs at selected nodes. Having completed the best possible calibration, the piezometric contours would be plotted out from the model and compared with actual piezometric data such as in Figure 8.29.

Such a calibrated model is then available for the application of different pumping scheme simulations, the results from which could be of use in the field. In the area outlined in Figure 8.29, for example, it would be useful to know what combinations of pumping rates at existing wells would optimize the total abstraction without substantially depleting the ground water availability at the Hartlepools 'end of the line'. Alternatively, it would be useful to know if the sinking and pumping of new wells would improve the situation. For the south-east Durham reference area considered for illustrative purposes, most of the results were obtained for steady state conditions because the Permian aquifer had a low storage coefficient. From model calibration, a flow through the area of some 16 mg d^{-1} (73×10^6 litres d^{-1}) was indicated, including 4 mg d^{-1} (18×10^6 litres d^{-1}) discharging in the Hartlepools area. This implied that a nett 12 mg d^{-1} (55×10^6 litres d^{-1}) was available for abstraction inland. Of this 12 mg d^{-1}, there appeared to be a maximum yield in any one area which was independent of the number of wells actually pumped, and it was possible also to propose a well – replacement scheme producing 11 mg d^{-1} (50×10^6 litres d^{-1}) while reducing the flow at the coast by less than 2 mg d^{-1} (9×10^6 litres d^{-1}).

With an electrical analogue, therefore, complex abstraction schemes may be studied by simulation, old schemes optimized and new schemes proposed. With increasing accumulation of field data, the model may be progressively updated to mirror more accurately the aquifer response.

8.12 Well losses

Equation 8.26 expressed the generalized formula for horizontal radial flow towards a well. Although, and as might be expected, this formula only approximates to the conditions in a real aquifer, it does appear to describe reasonably satisfactorily the flow conditions in the immediate vicinity of the well itself. From equations 8.25 and 8.26, it will be noted that for a particular s, k and b, the discharge Q is inversely proportional to the natural logarithm of the ratio of the radius of influence to the well radius. For actual wells, the hypothetic radius of influence generally has values that are effective in the range 150 to 600 m.

Suppose it is decided to increase the radius of a well from 0.10 m to 0.20 m. It can easily be calculated that, for a radius of influence of 150 m, the nett gain in discharge is only 9.5 per cent. The equivalent figure for a radius of influence of 600 m is 8.0 per cent. It follows from these figures that the gain is relatively insensitive to the radius of influence and also to the well diameter, since for most practical purposes a doubling of the diameter introduces a gain in the range of only 8 per cent to 14 per cent. There is little merit, therefore, in increasing pre-existing well diameters unless there are reasons other than yield for doing this.

In some wells, there may be an area near to the radius R where the velocities of flow could be high enough to create conditions of turbulence outside the laminar flow regime appropriate to Darcy's law. Losses due to turbulence are reflected in a necessarily increased drawdown for a given discharge. Since they will be proportional to Q^2 rather than Q it may be worth increasing the well diameter to avoid them.

In lined wells, pipe losses will only be a small fraction of the total drawdown, comprising a roughness factor and a turbulence factor caused by the lateral entry of the water. As a general policy, wells in fissured rock aquifers, such as chalk and volcanics, will not need lining unless the fissure density is so high as to introduce instability. In highly cohesive sedimentary rocks, a lining should not be required. But in unconsolidated sediments and porous, poorly cemented rock a lining screen with a suitable graded gravel filter will be required, the lining to prevent erosional ground collapse and the filter to prevent both movement of aquifer rock into the well during pumping and also to offer a zone of higher permeability around the well.

Different filter design criteria are suggested at different times by different people. Graded filters are most desirable, and the lower the

Uniformity Coefficient (D_{60}/D_{10}), the higher the permeability. The C.U. should therefore be as low as possible consistent with the requirement for a smooth grading and an ability to maintain the stability of the ground. An optimum C.U. value would probably be 2 to 3. Experimental work (Terzaghi, 1922; United States Bureau of Reclamation, 1947) has suggested that if D_{50} (ground material) = 1/5 to 8 times D_{50} (filter), C.U. (ground) = C.U. (filter), and if D_{50} (ground) relates to the average grading properties of the aquifer then the protective filter should perform its duties satisfactorily (also refer back to the end of Section 8.4). The filter material would be placed continuously to a thickness of about 6 in (~150 mm) and would consist ideally of clean, well-rounded siliceous gravel or coarse sharp sand.

Filter screen slot sizes must also be determined. An upper limit for the slot size would reasonably be D_{60} (ground) — Terzaghi and Peck (1967); lower limits tend to be a matter of choice but could reasonably be around 0.5 D_{85} (ground).

8.13 Improving aquifer yield

In fissured carbonate materials, it is sometimes useful to open up existing fissures, or even create new ones, in the zones surrounding the well where turbulent flow is possible. A strong acid may be injected into the well and left for several days before being pumped out. The acid must be doped to inhibit harmful metal attack and it must also be stabilized to prevent any dissolved carbonates being deposited. Unfortunately, the results of such an exercise are not always successful.

Small explosive charges are sometimes detonated down wells in a similar attempt to improve well performance or to release gravel and screen slots encrusted with previously dissolved deposits after many years of pumping. Explosive well development would usually be followed by acidising.

Clogged filters may also be relieved by overpumping and high pressure surging, but if the filter is correctly designed at the outset, mobile particles will be quickly lost and those particle fractions which are designed to be retained will be so retained.

8.14 Groundwater quality

This is a specialist subject which is rarely treated at length in hydrology textbooks and which will be only mentioned here. For further reading, reference may be made, for example, to Hem (1959) and Holden (1970).

Groundwater composition is very largely dependent upon the nature of the soil and rock encountered during downwards percolation and also upon the time taken to reach aquifer storage. Most groundwaters are complex dilute solutions, the principal ions present in them being calcium (Ca^{2+}), magnesium (Mg^{2+}), potassium (K^+), sodium (Na^+), chloride (Cl^-), sulphate (SO_4^{2-}), nitrate (NO_3^-) and bicarbonate (HCO_3^-). Ionic concentrations in a particular groundwater can only be determined by fundamental analysis but their relationships may be most conveniently expressed on trilinear diagrams in the manner of Figure 8.34. Its general character can, however, be appraised by deriving information on its hardness (mainly due to dissolved carbonates), hardness excluding carbonates, alkalinity, sulphate and chloride content, its pH and its electrical conductivity. This latter may be used to get an idea of the total dissolved solids in the groundwater (there will be an upper limit of about 500 mg l^{-1}). Of the cations (positive charge) and the anions (negative charge), it is generally found that most apart from the bicarbonate are precipitated with the rainwater but there may be added Ca and Mg acquired during infiltration. Water abstracted from chalky formations is usually notoriously hard with a high bicarbonate content.

Water at the gathering ground of an aquifer at an early stage in its downwards percolation often contains a high calcium bicarbonate dissolved content. As it moves through the aquifer, there tends to be on-going ion-exchange reactions resulting in the gradual replacement of the calcium and magnesium ions by sodium to create a softer water containing sodium and bicarbonate ions. A further progressive depth—dependent increase in the chloride content leads to an abstraction product that comprises a predominant sodium and chloride ion content. If there is a substantial bacteria content in the rock (for example, where a sewage disposal problem exists), then further reduction of sulphates and nitrates will also lead to changes in groundwater chemical composition (note that excessive concentrations of nitrates may cause blood disorders in children). Natural sodium in rocks may also be supplemented by the percolation of effluent and industrial waste materials. Sodium contents in deep groundwater may be as high as 10^5 mg l^{-1}. Sulphates will usually be derived from the oxidation of pyrite and may be present in sufficiently high quantities to form a very acidic water capable of neutralizing dissolved carbonates. The presence of chlorides is usually regarded as being indicative of fossil sea water, present at depth as connate water. Fluoride in drinking water is still an emotive public issue, but it is present as a natural constituent in water. Whether supplemented or not, it would appear that up to about 1 mg l^{-1} in potable water is effective in delaying dental decay but above about 1 mg l^{-1} it can be harmful.

In the generality of cases, groundwater needs no treatment upon abstraction. This tells us a great deal about the natural purifying capability of the ground. Sometimes the water needs chlorination and sometimes it needs softening (there is, for instance, a statistical relationship between water softness and the incidence of heart disease). Any bacteria may be removed by filtering the water through a bed of sand. Open refuse tipping and land-fill reclaimed areas can create problems of toxic drainage downwards into an aquifer (Walker, 1974; Gray *et al*, 1974) if the extent of the hazard is not anticipated at the outset. Tips should always be located and designed in the context of enlightened appreciation of aquifer hydrology. Ideally, each tip would have a puddle clay base serving as a cut-off with an extensive network of land-drains into the sewage system. An added option involves tip oxygenation, but in the presence of carbonaceous matter, this operation would clearly be hazardous.

An aquifer housing poor quality water may also be 'cleansed' by artificial recharge. There would be an attempt to raise the oxygen level in the water and to reduce the CO_2 content. Unfortunately, recharge water is not easily acquired except periodically in times of flood. Transferred river water may itself be bacteriologically unacceptable, but the cleansing effect of natural infiltration may remove this disadvantage. Natural infiltration may also be supplemented on the ground surface by rapid natural sand filtration to re-oxygenate the water, schemes of this type being operated in Holland and Germany.

Detailed groundwater chemical determinations would include not only the principal ions but also the trace metals. Water quality is most conveniently expressed in terms of the principal ionic content through groupings related to the degree of salinity. Five classifications, originally proposed by Palmer (1911) for natural waters, involved primary salinity, primary alkalinity, secondary salinity, secondary alkalinity, and tertiary salinity (or acidity). Since tertiary salinity is rarely found in underground aquifer water, the remaining

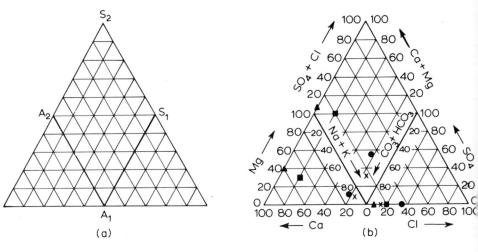

Figure 8.34 Trilinear diagrams for the graphical treatment of groundwater chemical analyses (a) *after* Hibbert (1965) and (b) *after* Piper (1953). Points on (b) relate to a borehole through Triassic sandstone — samples at 20 m (●), 60 m (x), 120 m (▲) depths and total sample (■). For the plotting technique and interpretation see, for example, Walton (1970, pp. 439–484).

four classifications were described by Hill (1941) as:

S1 Primary salinity: the maximum grouping of Na, K, Cl, SO_4, and NO_3 ions.

A1 Primary alkalinity: the minimum grouping of Na, K, HCO_3, and CO_3 ions.

S2 Secondary salinity: the minimum grouping of Ca, Mg, SO_4, Cl, and NO_3 ions.

A2 Secondary alkalinity: the maximum grouping of Ca, Mg, HCO_3, and CO_3 ions.

Of these groupings, A1 and S2 cannot mutually exist in any given solution. This means that only three, vis, S1, A1, A2 or S1, S2, A2, can exist together. Expressing these properties as percentages of the sum of the three, a point on a trilinear coordinate diagram adequately represents the quality of a single sample of water. An alternative form of trilinear presentation ignores the *groupings* of the ions and plots directly as a function of the *individual* ions (Piper, 1953). Both types of trilinear diagram are shown in Figure 8.34.

For further reading on this subject, reference may be made to Hibbert (1956) and Walton (1970).

The classical view of water supply was that, having located a contamination-free source of water, it should then be stored in a rather remotely-accessible area and be totally restricted to the general public. A rather more up-dated approach is directed towards purification of water that could originally have been impure. At one end of this activity spectrum, reservoirs, or lakes in N. American terminology, are treated as the foci of amenity areas for public enjoyment in the form of sailing and fishing. Minimal purification is thereby required. At the other end of the spectrum, there is the concept of direct re-use and closed-circle purification and regeneration. On a unit basis, the development of such a system could revolutionise present-day thinking on problems of sewage disposal. Re-use of water is quite common; water drawn from rivers like the Thames may be used several times over, having received only a dilutionary form of purification. One thing is certain: that is that of all the subjects discussed in this book, the subject of water and its management will probably respond to the greatest scientific development. This subject is further discussed in Chapter 12.

9 ~ Stability of Soil Slopes

Questions involving the stability of new cuttings in soil or clay or the stability of natural slopes arise in connexion with the construction of railways, highways, canals and deep basement excavations. There may even be problems with submarine landslides (Henkel, 1970). Similar stability problems can arise both during and after the construction of manmade embankments as waste tips or as earth dams. In the former examples, constructed slopes are invariably oversteepened with respect to an angle of long-term stability for the logical reason that the costs of acquiring greater ultimate stability rise very much out of proportion to the decrease in slope angle so acquired. But although the elements of cost and design life are present, a slope must be shown, by some accepted method of mechanical analysis, to be stable. This stability is usually expressed in terms of factor of safety F (see Section 7.17), the degree of potential stability increasing as the value of F increases above unity.

Observation of those slopes in which mass failure has occurred along deep-seated shear planes has suggested that at limiting equilibrium ($F = 1$) a suitable criterion of failure would be a balance between the shear stresses on a shear plane and the shear strength parameters of the material. The time variant in the problem can then most conveniently be accommodated by changing the shear strength parameters.

The subject of soil slope stability is a specialist one in the field of soil mechanics, and the reader is directed towards any appropriate text book (for example, Terzaghi, 1943; Terzaghi and Peck, 1967; Jumikis, 1962; Lambe and Whitman, 1969) or papers (refer, in particular to publications in the Journal of Soil Mechanics and Foundation (now Geotechnical) Engineering of the American Society of Civil Engineers and in Géotechnique of the British Institution of Civil Engineers). It does seem, however, to be desirable at least to introduce the subject in a limited manner in the hope that it will

Stability of Soil Slopes 633

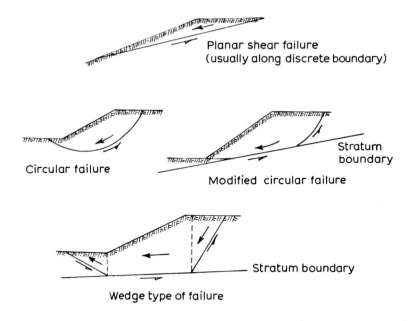

Figure 9.1 Different modes of sliding and analysis.

encourage engineering geologists to read more deeply elsewhere. The problem of seismic influences on the circular shear failure of embankments is discussed in Section 12.15 when it is considered in the context of earth dam stability.

Excluding long-term superficial degradational effects which specifically condition slope morphology (see Sections 9.10, 9.11), three styles of movement can be considered explicitly while acknowledging that many failures are compounded from all three. These are: *planar slides, rotational slides, mudflows*. There are variations on the two former types of slide which are amenable to different analytical approaches (Figure 9.1).

9.1 Planar slides

If groundwater conditions are known, an exact mathematical analysis of an infinite planar slide is possible. Consider an actual problem of a boulder clay slope adjacent to a river, a tall slender mast being founded in the boulder clay which itself is in planar contact with underlying laminated clay. Shear failure is considered likely along the contact interface between the two clays which dips at 15° towards the river and is the same as the general dip of the

634 Principles of Engineering Geology

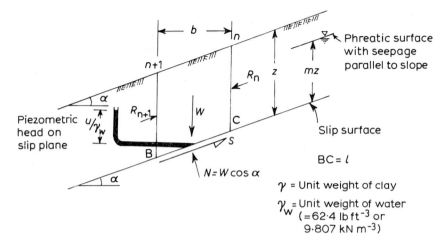

Figure 9.2 Forces in planar slide analysis (*after* Skempton and DeLory, 1957).

slope. A two-dimensional analysis of the safety of the mast is as shown in Figure 9.2.

In the problem, a single slice of width b is considered and vertical interslice forces should strictly be taken into account. However, ignoring the contribution of these forces to stability, and noting that the shear force acting up the slip plane is

$$S = \{(N - ul) \tan \phi' + c'l\} \quad (9.1)$$

then if the water table creates a hydrostatic component of pressure on the slip surface with flow out of the slope (Haefeli, 1948) then the factor of safety can be expressed as (Skempton and DeLory, 1957):

$$F = \frac{c' + (\gamma z \cos^2 \alpha - u) \tan \phi'}{\gamma z \sin \alpha \cos \alpha} \quad (9.2)$$

or

$$F_1 = \frac{c' + (\gamma \cos^2 \alpha - m\gamma_w)z \tan \phi'}{\gamma z \sin \alpha \cos \alpha} \quad (9.3)$$

If the water table is maintained at a constant depth below ground surface with a seepage pressure component of flow parallel to the slope then

$$u = \gamma_w mz \cos^2 \alpha \quad (9.4)$$

and
$$F_2 = \frac{c' + (\gamma - m\gamma_w)z \cos^2 \alpha \tan \phi'}{\gamma z \sin \alpha \cos \alpha} \tag{9.5}$$

In the example mentioned, the following conditions applied: $\phi'_r = 12°$; $c'_r = 250$ lb ft^{-2}; $\gamma = 100$ lb ft^{-3}; $\alpha = 15°$; $z = 20$ft; $m = 0.8$. Calculating on these values, $F_1 = 0.87$ and $F_2 = 0.9$ and the slope is prone to long-term failure. In both cases, if the water table can be lowered to just below the slip plane ($m = 0$), then $F_{1,2} = 1.3$, that is, the slope will be stable long-term. The answer therefore appears to involve a lowering of the water table by suitable drainage measures together with permanent prevention of further ingress of surface water.

It should be noted that when evaluating F for an infinite slope, the worst condition, the phreatic line at the slope surface, should generally be assumed unless there is unambiguous evidence to the contrary.

9.2 Circular failure surfaces

Deep seated shear failures in homogeneous soils that are relatively unaffected by bedding or other structural anomalies which might precondition slip in a planar manner tend, from visual inspection, to be rotational on circular slip surfaces. There are two general methods of analysis. Of these, the friction circle method is really a graphical trial and error procedure for finding the factor of safety for different assumed failure circles in a c, ϕ soil. Basically, the analysis is performed in two ways: to allocate a known cohesion c and determine the required friction angle ϕ_{req} to secure equilibrium, or to reverse this procedure. In the first case:

$$F_\phi \text{ (factor of safety with known } c) = \frac{\phi}{\phi_{req}} \tag{9.6}$$

and in the second case,

$$F_c \text{ (factor of safety with known } \phi) = \frac{c}{c_{req}} \tag{9.7}$$

Using trial circles, the aim is to find the most dangerous surface of rupture, that is, for a given ϕ-value, the surface that demands the greatest c-value for equilibrium, and vice versa. Reference to Figure 9.3 shows that the shear stress and normal effective stresses are

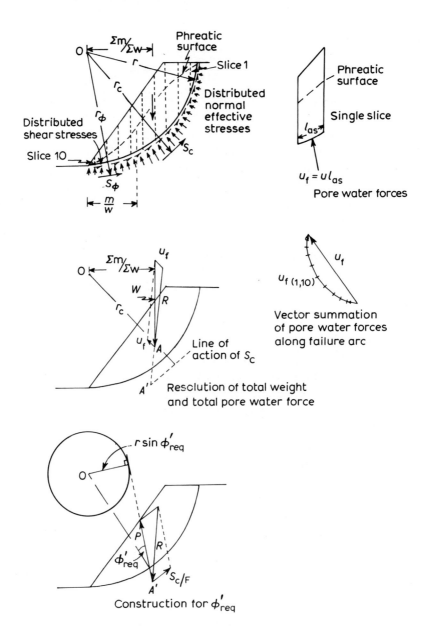

Figure 9.3 Friction circle analysis.

distributed along the failure surface. The analysis proceeds as follows:

(a) Divide the slope into several vertical slices. Find the weight, w, of each slice by multiplying unit weight by slice width by average height, and noting that the two extreme slices will be triangular in cross-section. $\Sigma w = W$

(b) Find the moment, m, of each slice. This will be the perpendicular distance from the centre of the slip circle to the line of action of each w. Note that towards the toe of the slope, the moment could be negative. $\Sigma m \doteq M$.

(c) Thus, the resultant of all w lies at a perpendicular distance $\Sigma m/\Sigma w$ from the centre of the circle, i.e. this fixes the line of action of W, the total weight of slide material.

(d) Multiply the pore water pressure (piezometric head) at the base of each slice by the length of each slice base to give the pore water force u_f. Since all u_f will act normal to the failure surface they can all be added vectorially to give a resultant u_f which must pass through the centre of the slip circle.

(e) Construct the line of intersection of W and u_f (say letter A). Then, from A, thicken in the amplitudes W and u_f. Construct the resultant R of W and u_f by closing the force triangle but then displace the line of action of R to pass through A.

(f) Next, determine the line of action of the shear resistance due to the cohesive force in the material, i.e. S_c. The moment of cohesive force about O is $cl_a r/F$ where l_a is the length of the failure arc. However, it will be seen from examination of the failure arc, that the resultant S_c is equal to $c'l/F$, where l is the length of the chord of the failure arc. This arises because components of c normal to the chord cancel out. It follows that

$$r_c S_c = r_c \frac{c'l}{F} = \frac{c'l_a r}{F} \qquad (9.8)$$

or

$$r_c = r \frac{l_a}{l} \qquad (9.9)$$

Thus, the cohesion radius arm can be established from the arc length to chord length ratio directly.

(g) Displace vector R along its own line to intersect circle radius r_c at A'. Mark off S_c/F ($= c'l_a/F$) tangent to this circle from A' and acting up the slip circle. Close the force triangle with resultant P and then displace P in a parallel manner to pass through A'. If it is then a prior assumption that $r_\phi = r$, the line of action of P must form an angle ϕ'_{req} with the radius through the intersection of P with the failure arc, where ϕ'_{req} is given by

$$\tan \phi'_{req} = \frac{\tan \phi'}{F} \qquad (9.10)$$

It follows that P must be tangential to a circle about O having a radius equal to $r \sin \phi'_{req}$; this circle is termed the 'friction circle'.

(h) An equilibrium condition is satisfied only when the $P, R, S_c/F$ force triangle closes, and a series of trial solutions involving several friction circles may be necessary to achieve this. For each circle that is used, two safety factors are derived, vis:

$$F_\phi = \frac{\tan \phi'}{\tan \phi'_{req}} \qquad (9.11)$$

$$F_c = \frac{c'l}{S_c} \qquad (9.12)$$

and the aim must be to find the circle giving the lowest factor of safety. In general, the condition $r = r_\phi$, in which it is assumed that the normal stress along the arc is concentrated at a single point, produces a lower bound solution to the problem.

An alternative method of analysis still considers the potentially unstable, or failed mass of clay above the shear arc to be divisible into slices, the forces at the base only of each slice being resolved as in Figure 9.4. The centre of the potentially most critical circle in an unfailed mass of clay originally had to be located by trial and error but nowadays a computer would perform the search.

If F_s is the factor of safety with respect to the base of each slice, then in total stress notation

$$F_s = \frac{\text{Restoring moment at the base}}{\text{Disturbing moment at the base}} = \frac{crl + Nr \tan \phi}{rT} \qquad (9.13)$$

or, for the whole circle, with α taking both positive and negative values

$$F = \frac{cr\Sigma l + r \tan \phi \; \Sigma N}{r\Sigma T} = \frac{cr\theta + \tan \phi \; \Sigma \overline{W \cos \alpha}}{\Sigma W \sin \alpha} \qquad (9.14)$$

Stability of Soil Slopes 639

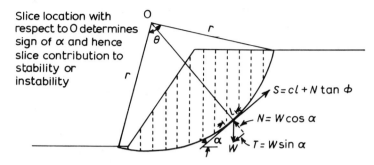

Figure 9.4 Basis of the Swedish Strip Method, or the Conventional Method of Slices, for the analysis of circular slope failures.

This Swedish Strip Method, otherwise known as the Conventional Method of Slices, was first proposed by Fellenius (1936) and was originally applied on a total stress basis. It is assumed, as with all these methods, that cohesion and friction are mobilized absolutely and uniformly along the shear surface and that the sliding develops at all points at the same time. That such an occurrence is unlikely will be considered later in the context of what shear strength parameters should be adopted in the analysis of clay slopes.

Probably the most widely used of the newer, developed versions is due to Bishop (1955) and is known as the *Simplified Method of Slices**. The main parameters involved in this analysis are shown in Figure 9.5 and it should be noted that both interslice normal forces (E_n, E_{n+1}) and interslice shear forces (t_n, t_{n+1}) are present in the problem. However, it is assumed that the interslice forces have zero resultant in the vertical direction and do not, therefore, affect the force balance. With this assumption holding, the method has been shown to be very accurate in most cases (Whitman and Bailey, 1967) but the results may be misleading with deep failure circles when $F < 1$ (Lambe and Whitman, 1969; see also the analysis and comments of Spencer, 1967).

Referring to Figure 9.5a and resolving up the slip circle, it can be

---

*Skempton and Hutchinson (1969) uphold the success of this method if applied to non-circular surfaces, especially if a substantial portion of the slip surface is predominantly planar. The accuracy of this method decreases with increase in the D/L ratio (where L is the toe-to-crest chord length and D is the distance from this chord to the failure plane). There will thus be a tendency to underestimate the normal effective forces acting on steeply inclined areas of the slip surface. There will also be an underestimate in F. Comparisons on circular landslides in natural slopes, between the conventional and Bishop Simplified methods, show that the former may well underestimate F by over 20 per cent.

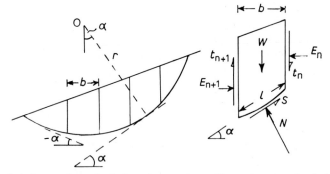

(a) General analysis for slopes above the ground water table

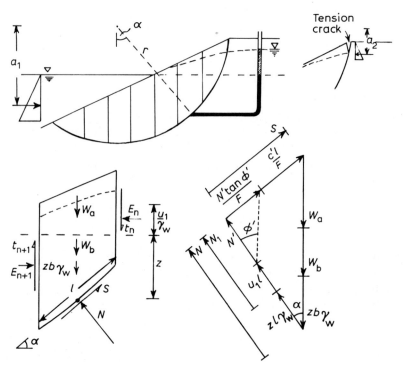

(b) General analysis for a partly submerged slope (for example, forming the banks of a river)

Figure 9.5 Analyses by the Method of Slices (*in the manner of* Terzaghi and Peck, 1967).

seen that since

$$S = \frac{\tau}{F} \frac{b}{\cos \alpha} \qquad (9.15)$$

where S is the shear force and τ is the shear stress, then,

$$F = \frac{\sum_{i=1}^{i=n} (\tau_i b_i / \cos \alpha_i)}{\sum_{i=1}^{i=n} W_i \sin \alpha_i}, \text{ or } F = \frac{\Sigma \tau b / \cos \alpha}{\Sigma W \sin \alpha} \qquad (9.16)$$

Substituting for τ and manipulating the parameters in the numerator, it can be confirmed in effective stress notation that

$$F = \frac{\Sigma \left(c' + \frac{W}{b} \tan \phi'\right) \frac{b}{m_\alpha}}{\Sigma W \sin \alpha} \qquad (9.17)$$

where, after Janbu et al (1956),

$$m_\alpha = \cos \alpha \left(1 + \tan \alpha \cdot \frac{\tan \phi'}{F}\right) \qquad (9.18)$$

Perhaps more conveniently, equation 9.17 can be re-expressed as

$$F = \frac{\Sigma \left(c' + \frac{W}{b} \tan \phi'\right) \frac{b}{n_\alpha}}{\Sigma W \tan \alpha} \qquad (9.19)$$

where

$$n_\alpha = \cos^2 \alpha \left(1 + \tan \alpha \cdot \frac{\tan \phi'}{F}\right), \qquad (9.20)$$

after Janbu (1957)*. Since the factor of safety F appears on both sides of equations 9.17 and 9.19, trial and error solutions are required for the necessary convergence. The functions m_α, n_α can be determined through the use of the charts in Figure 9.6. Trial input values of F would be used to evaluate m_α or n_α for each slice following which calculated values of F would be produced. If a slope is thought to be close to limiting equilibrium and prone to shear failure, an initial trial value of F of about 1.25 would be chosen.

*Janbu (1957) introduced a correction factor, f_0, in $F = f_0 \frac{\Sigma \ldots \ldots}{\Sigma \ldots \ldots}$
This factor depends upon the shear parameters and the form of the slip, and takes into account the influence on F of the vertical shear forces between the slices. $1.0 \leq f_0 \leq 1.13$.

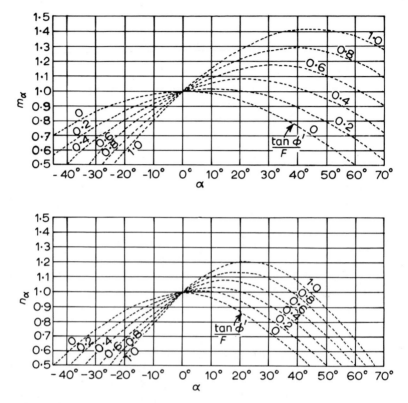

Figure 9.6 Graphical solutions for equations in m_α, n_α (generally *in the manner of* Terzaghi and Peck, 1967).

Terzaghi and Peck (1967) consider a typical problem of a partly submerged slope, an appropriate force polygon being drawn in Figure 9.5b. W_a is the total weight of the soil above the standing water level, W_b is the submerged weight of soil below standing water level and $W = W_a + W_b + zb\gamma_w$. If the whole of the slice under consideration is below standing water level, the weight of the water must be included in the $zb\gamma_w$ term. The porewater force acting hydrostatically on the base of the slice will be equal to $(u_1 l + zl\gamma_w)$. Thus,

$$F = \frac{\Sigma(c'l + N' \tan \phi')}{\Sigma(W_a + W_b) \sin \alpha} \qquad (9.21)$$

$$= \frac{\Sigma[c'b + (W_a + W_b - ub) \tan \phi'] \dfrac{1}{m_\alpha}}{\Sigma(W_a + W_b) \sin \alpha} \qquad (9.22)$$

or

$$F = \frac{\Sigma[c'b + (W_a + W_b)(1 - r_u)\tan\phi']\dfrac{1}{m_\alpha}}{\Sigma(W_a + W_b)\sin\alpha} \qquad (9.23)$$

where the pore pressure ratio

$$r_u = \frac{ub}{W*} \qquad (9.24)$$

(see Section 9.6) and

$$W* = (W_a + W_b) \qquad (9.26)$$

If a tension crack is discernible at the crest of the slope and it contains water, then this water, as a function of its depth, will create an additional disturbing force which can be allowed for by using the symbol Q as an additive parameter in the denominator of equation 9.23 and by adding a horizontal vector to the polygon in Figure 9.5b.

It may so happen that a potential slip circle may intercept a planar boundary to form composite shear surface. A planar slide analysis may not be considered satisfactory and an alternative approach might be to consider using a wedge analysis of the type to be discussed later. However, it might be possible to adapt the method of slices to this new situation, thereby more nearly preserving the form of the potential failure surface. Such a possibility is outlined in Figure 9.7 from which it will be noted that moments of the disturbing forces must now be taken explicitly. Selecting an arbitrary pole O then

$$F = \frac{\Sigma(c'l + N'\tan\phi')a}{\Sigma(W_a + W_b)x - \Sigma N_1 k} \qquad (9.26)$$

or

$$F = \frac{\Sigma[c'b + (W_a + W_b - ub)\tan\phi']\dfrac{a}{m_\alpha}}{\Sigma(W_a + W_b)x - \Sigma\left[W_a + W_b + (ub\tan\phi' - c'b)\dfrac{\tan\alpha}{F}\right]\dfrac{k}{m_\alpha}}$$

$$(9.27)$$

$$= \frac{\Sigma[c'b + (W_a + W_b)(1 - r_u)\tan \phi']\dfrac{a}{m_\alpha}}{\Sigma[(W_a + W_b)x] - \Sigma\left[(W_a + W_b)\left\{1 + \dfrac{r_u \tan \alpha \tan \phi'}{F}\right\} - \dfrac{c'b \tan \alpha}{F}\right]\dfrac{k}{m_\alpha}}$$

(9.28)

The Bishop Simplified Method is amenable to programming for a digital computer (for example, Little and Price, 1958) whereby a large number of slices can be input and rapid convergence for F assured. It is generally regarded as being a good method for non-circular slips. For circular slips, by comparison with much more sophisticated methods, the accuracy is to within 7 per cent and usually within 3 per cent to 4 per cent.

Computer programs for slope stability analysis usually calculate a factor of safety for known failure surface geometry, c', ϕ' parameters, densities and piezometric conditions for the different layered deposits through which the shear surface passes. Alternative programs calculate possible combinations of c', ϕ' and express them in matrix form for similar known geometries, geology, and piezometric heads under conditions of limiting equilibrium ($F = 1$). In Table 9.1, equations 9.14, 9.17 and 9.19 are re-expressed in effective stress terms and in a form that is more suitable for computer programming.

Table 9.1 Equations for three methods of soil slope stability analysis

Method	Equation
Fellenius (1936)	$F = \dfrac{1}{\Sigma W \sin \alpha} \Sigma[c'l + (W \cos \alpha - ul)\tan \phi']$
Bishop (1955)	$F = \dfrac{1}{\Sigma W \sin \alpha} \Sigma[c'l + W(1 - r_u)\tan \phi'] \dfrac{\sec \alpha}{1 + \tan \alpha \cdot \dfrac{\tan \phi'}{F}}$
Janbu (1954)	$F = f_0 \dfrac{1}{\Sigma W \tan \alpha} \Sigma[c' + (N - u)\tan \phi'] \dfrac{b \sec^2 \alpha}{1 + \tan \alpha \cdot \dfrac{\tan \phi'}{F}}$

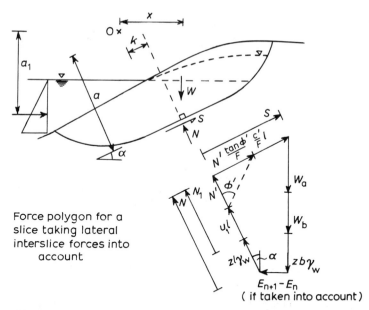

Figure 9.7 General analysis for a circular-planar composite slide (*in the manner of* Terzaghi and Peck, 1967).

9.3 Slope stability case histories

Results from a series of analyses on a problem slope in Durham City (Figures 9.8, 9.9) can be used to illustrate comparative factors of safety generated by different analytical approaches (Lynn, 1973). The surface of the slope, which dips in a general north-westerly direction towards the River Wear at Carrville, is of variable inclination (Figure 9.10) and is very humocky, the small irregular undulations giving the appearance of being localized shallow-seated slip circles. There is also evidence of some springs which have no definite trend in line or cross-section. The general geology is of glacial deposits of clays, silts, sands and gravels overlying sandstone. These deposits are thicker on the gentle upper slope and tend to thin out on the steeper portions. Near to the river, there are about 19.5 m of overlying deposits, while at the crest of the slope there are nearly 31 m. There is a history of embankment slips following road construction just below the crest of the slope (on overall movement of about 5 mm per annum) and this has necessitated temporary remedial work and the institution of a slope instrumentation programme comprising borehole piezometers, slip detectors, inclinometer and settlement gauging, together with surface surveying. Initial

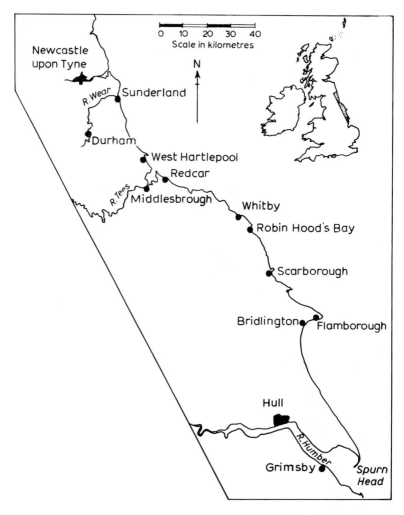

Figure 9.8 Location map for the soil slope stability areas.

considerations (Taylor, 1971) suggested that the low angle slips evidenced at the roadway were in a dominantly two-layer geological sequence of a more granular deposit (mixture of sands, silts, soft clays and water bearing laminated clays overlying dark brown, firm laminated plastic clays, and that pore pressures generated in the more granular horizon during wet spells were related to a longer-term strength decrease in the upper part of the laminated clays. Evidence of pore pressure re-distribution would categorize the slope as being in the long term stage (Skempton and Hutchinson, 1969), and from its

Stability of Soil Slopes 647

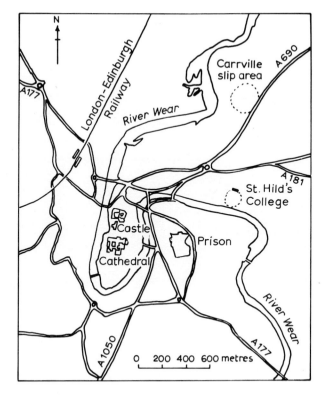

Figure 9.9 Location of two sites of instability in Durham City, England.

geology and geomorphology it can be concluded that it is a translational slide with accompanying circular failures in many places.

Considering the problem first in translation, in view of its long-term character, the groundwater flow can be taken parallel to the slope and equation 9.15 utilized. Analysis was directed towards two slope sections within which a slip plane depth and the level of water above it were known from the instrumentation records.

Using laboratory-determined residual shear strength parameters for the clay, the factor of safety variation with water table height is shown in Figure 9.11 for the two slope sections, the appropriate F values of 0.95 (Upper) and 1.07 (Lower) representing critical equilibrium.

Three slope sections (Figure 9.12) for which slip detection and water table information had been acquired were then subjected to

Figure 9.10 Slope instability in Durham City, England. (a) View south, with slope dipping to the right. Note evidence of remedial work to the carriageway. (b) View to the west down the slope. Photograph: C. B. McEleavey.

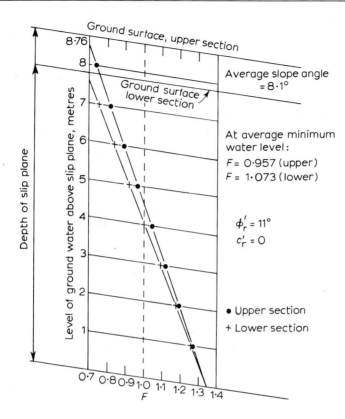

Figure 9.11 Infinite slope analysis – Durham City. Variation of F with level of phreatic surface above slip plane.

computer analyses, based on Fellenius (Conventional Method of Slices), Bishop (Simplified Method of Slices) and Janbu, in order to determine minimum factors of safety for the most critical slip circles for a range of c' and ϕ'. The results of the analyses are given in Table 9.2 together with some comments.

Noting the low factors of safety and assembling visual and other evidence, it was concluded that excavation of the claypit in the laminated clay at the foot of the slope served to re-activate earlier movement but in a different direction. Localized circular failures within the general translational slide facilitated the ingress of water to the latter, so encouraging the shear process. Using Skempton's (1964) 'residual factor' R (the proportion of the total slip surface in

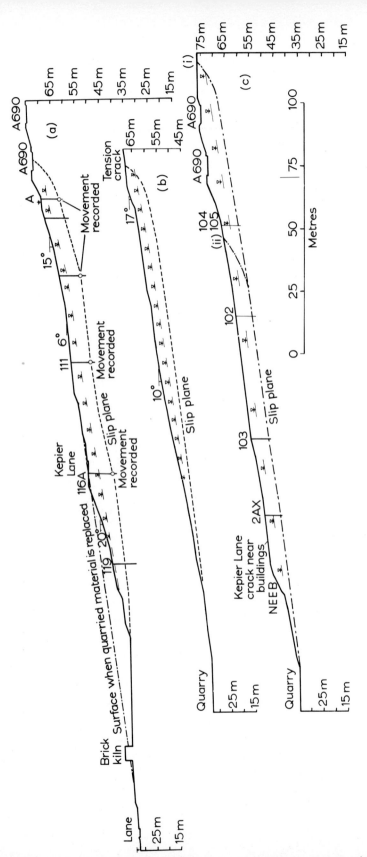

Figure 9.12 Slope instability of the transitional type in Durham City, England (showing positions of inclinometers, water table, and slip plane).

Table 9.2 Minimum factors of safety for critical slip circles at $\phi'_r = 11°$, $c'_r = 0$ in three sections of slope in a sand–gravel–silt–laminated clay complex in Durham City, England (*after* Lynn, 1973; Taylor, 1974).

Section Fig. No.	Factor of Safety			Remarks
	(A) Fellenius	(B) Bishop	(C) Janbu ($f_0 = 1$)	
9–12(a)	1.07	1.08	1.07	Over range $7° \leq \phi'_r \leq 13°$ and $0 \leq c'_r \leq 9$ kN m^2, there was very little difference between F(A) and F(C), as might be expected for a planar slide. However, F(B) is to within 0.01 of F(A) and F(C) over ϕ'_r, c_r range mentioned.
	Analysis as in diagram			
9–12(a)	1.25	1.26	1.25	Obvious improvement in F by loading the toe.
	'Replace' excavated material at foot of slope, leaving slip plane as before, but extrapolate water table, keeping at constant depth			
9–12(b)	1.09	1.11	1.07	For $\phi'_r = 11°$ $c'_r = 0$
	Analysis as on diagram. Tension crack defines upper boundary of slip			
9–12(b)	1.17	1.19	1.17	For $\phi'_r = 11°$ $c'_r = 0$
	Upper boundary (i)			
9–12(c)	1.15	1.15	1.14	For $\phi'_r = 11°$ $c'_r = 0$
	Upper boundary (ii)			

the clay along which its strength has fallen to residual) defined by:

$$R = \frac{\tau_p - \bar{\tau}}{\tau_p - \tau_r} \tag{9.29}$$

where $\bar{\tau}$ is the average shear strength and τ_p, τ_r are the peak and residual shear strengths respectively, then it is calculated that the general slope would possess an R value of, or close to, unity. This means that along the whole of the slip plane, the shear strength has fallen to residual.

Two immediately-obvious remedial measures that could be used are:

(a) Lower the ground-water-table by suitable drainage methods; counterfort vertical drains taken down to the depth of the slip plane augmented by horizontal drains comprising 23 cm diameter semi-porous concrete pipes with sand filter surround and extending continuously for 183 metres from the toe of the road embankment to the toe of the hillside. Horizontal drains could be driven at varying slopes and directions from a pit sunk at the head of each counterfort and would eventually discharge to an interceptor drain and thence to the river. But the costs of this scheme could be high, maintenance would be needed to prevent clogging and in the early stages of any movement to rectify any breakages, and in any case the analyses imply that there might only be marginal increases in F over a limited section of the hillside.

(b) Load the toe by infilling the pit. Although a great deal of material would be needed (such as coal mine waste which would be quite accessible) together with some drainage and landscaping, it would lead to a positive improvement of F and could be the preferred solution.

It is pertinent to mention also that a similar instability developed in about 1965 to 1966 after a winter of high rainfall a few hundred metres away from the case history just discussed but this time on the northern slope of the other arm of the loop in the River Wear. The caps of piles supporting an annex of one of the Durham Colleges (St. Hild's College) had become exposed, there was evidence of trees being dragged down-slope, and there were also some indications of superficial cracking inside the building. It seemed that construction of the wing had caused failure of the slope and that the piles, in

Stability of Soil Slopes 653

Figure 9.13 Christopher wing of St. Hild's College, Durham City, England with a southerly aspect over the River Wear. Photograph: P. B. Attewell.

stitching across a failure surface, had themselves become subjected to high shear stresses. The average slope of the ground towards the river was $20° - 22°$ and the same general lithologies applied as before; a dominantly 'granular' mass overlying laminated clays which dip towards the valley at about $5°$. Figure 9.13 shows the area after completion of the remedial works.

General evidence supported the diagnosis of a planar slide transition along the junction with the laminated clays but reverting to circular shear surfaces in the upper granular material (see Figure 9.14). It was also noted that in wet weather a spring line developed close to the toe at the junction with the clays.

The successful remedial measures involved a re-grading of the slope, a loading of the toe, and the incorporation of drains at the foot of the slope.

It is useful to note that the assumed shear strength parameters used in the back-figuring analysis for the Hild's slip were $c'_p = 12.4$ kN m^{-2}, $\phi'_p = 23.4°$; $c'_r = 0$, $\phi'_r = 11.7°$. These residual values are compatible with those entered into the previous Carrville slip analysis (Table 9.2). If back-analyses are performed on both these slips under

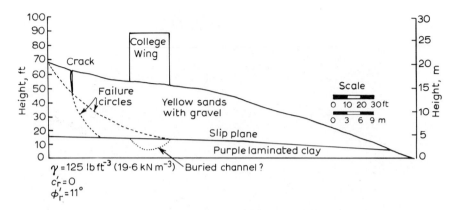

Figure 9.14 Analysis for St. Hild's slide, Durham City.

conditions of limiting equilibrium ($F = 1$) and for varying c', ϕ' inputs, it is found that in both cases the relationship between c and ϕ for no pore pressure along the slip surface produces a situation of $\phi' \simeq 11°$ when $c' = 0$. A similar situation arises with respect to two other slips (Browney and Crook Hall) in the same laminated clay material in Durham and which have been studied by Robinson (1969) and Clarkson (1974). This result indicates that the limit of stability of these slopes is at the residual shear strength of the laminated clay.

For the next example of slope stability, it is proposed to consider the boulder clay cliffs fringing the coast line at Robin Hood's Bay in North Yorkshire, England (see Figure 9.15). The boulder clay, which directly overlies the Lower Lias beach platform (Zwart, 1951; Attewell and Taylor, 1971), is thought to have been deposited by the melting of a single ice sheet during late Weichselian times (Penny *et al*, 1969; Francis, 1970) and radio-carbon dating gives its age as 18 000 years. Subsequent erosion has caused the whole cliff to become overconsolidated to a minimum of 285 kN m^{-2} (Courchée, 1970). From samples of the clay extracted at various points along the cliff section, it is geotechnically very consistent (see Figure 9.16) but there are considerable variations in the surface layers. It should be noted also that there is a wide range of boulders contained within the clay, comprising predominantly dark shales of Jurassic age and highly-weathered Carboniferous limestones and sandstones (see also Lamplugh's (1890) assessment of boulders in the cliffs at nearby Whitby).

Stability of Soil Slopes 655

Figure 9.15 Boulder clay cliffs, Robin Hood's Bay, North Yorkshire coast, England. Photograph: C. B. McEleavey.

Erosion of the cliffs must have begun when the glaciers retreated about 10 000 years ago. Magnitude and rate of erosion are dependent on many factors and, for example, from the work of Courchée (1970) seem to have varied over the last 100 years from about 0.073 m per year to 0.4 m per year depending upon location around the bay and height within the slope section. Taking the cliff as a whole, the rate of erosion of the top edge has been on average about 0.24 m per year since 1852*.

Other records of erosion along this coast, for example, from the Holderness area, Bridlington south to Spurn Head (Steers, 1964), where the boulder clay cliffs are only 5.25 m high, show it has taken place continuously and rapidly over the years. Holderness data carefully collected by Valentin (1954) and Robin Hood's Bay data acquired by Courchée (1970) are plotted in Figure 9.17 to indicate the manner in which recession rate varies with cliff height.

The following mechanisms have contributed to the boulder clay cliff erosion: (a) Surface weathering (b) Creep (c) Rotational sliding (d) Planar sliding (e) Mudflow.

*Plans for cliff protection from erosion were announced on 8 November 1973.

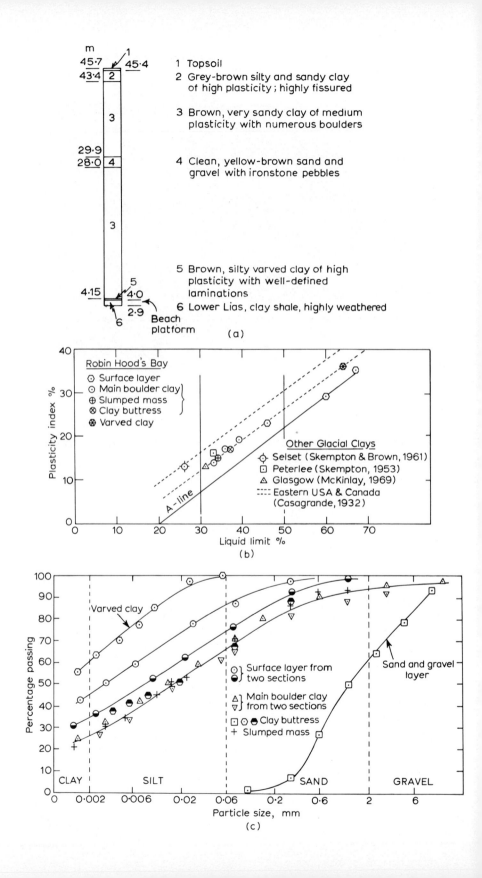

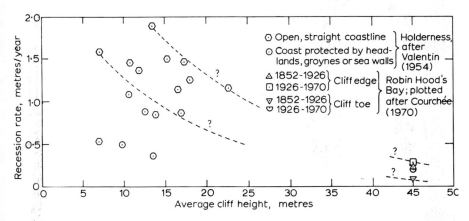

Figure 9.17 Boulder clay cliff erosion rates along the North Yorkshire coast, England.

Surface weathering, physical and chemical, caused by temperature changes, moisture variations, and frost action creates seasonal mantle creep. *Deeper seated creep* proceeds continuously through shearing mechanisms at greater depths due to gravity effects alone. Such a phenomenon can be investigated in the laboratory using creep shear equipment and the results used to predict creep rates in the field. If creep movement takes place parallel to the slope surface as in a planar slide, then by considering the stresses acting on a plane of movement, the effective normal stress is

$$\sigma'_n = z \cos^2\alpha\,(\gamma - m\gamma_w) \tag{9.30}$$

and the shear stress is:

$$\tau = \gamma z \cos\alpha \sin\alpha \tag{9.31}$$

Field measurements of groundwater levels suggested an average depth to the water table of 1 m parallel to the slope. In the laboratory, a normal stress of 65 kN m^{-2} was applied to undisturbed samples sheared at τ values of 3, 11, 21, and 41 kN m^{-2} and the rate of shear movement noted. σ_n of 65 kN m^{-2} is then equivalent to a depth of 5 m for a 1 m depth water table, and the ratio τ/σ_n can be

---

Figure 9.16 (a) Strata at typical cliff section, Robin Hood's Bay, North Yorkshire coast, England (*after* Courchée, 1970). (b) Plasticity chart for clay at Robin Hood's Bay (*after* Courchée, 1970). See also Fig. 3.17b. (c) Particle size distribution of cliff material at Robin Hood's Bay (*after* Courchée, 1970).

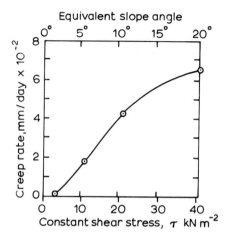

Figure 9.18 Relationship between creep rate and slope angle for the boulder clay cliffs at Robin Hood's Bay, N. Yorks coast, England (*after* Courchée, 1970).

related to an equivalent slope angle. This information is presented in Figure 9.18 in such a way that the equivalent slope for any value of shear stress can be read off for an expected rate of creep movement.

Evidence of rotational sliding was apparent in the upper third of the cliff section, with the base of the slides at the main sand layer (Figure 9.19). This layer can be shown to be weaker than the clay in shear at depths less than 8 m.

Four sections of cliff, two representing shallow areas of recent movement and two representing steeper sections (Figure 9.19) were analysed for critical slip circles using a Bishop Simplified computer program. No prior assumptions were made on the field evidence of tension cracks and rotational surfaces; the highest observed water table level of 0.65 m below surface was input to the analysis and it was assumed both that the surface layer had no strength and that critical shear failure took place tangential to the base of the sand layer (see Table 9.3 for shear strength parameters used in the analyses). For each section, 25 possible circles with their centres incrementing on a 5 x 5 (x,z) matrix were analyzed and the latter contoured with respect to computed F in order to locate the critical circle ($F_{m\,in}$) for each section. It is apparent from Figure 9.19 that in the less steep sections, the critical circles intersect the cliff top at greater distances from the edge. This feature proved to be entirely in accord with the field evidence, since major slips in recent times have occurred only at the northern end of the boulder clay cliffs (average slopes: northern end, 19.7°; southern end, 29.7°).

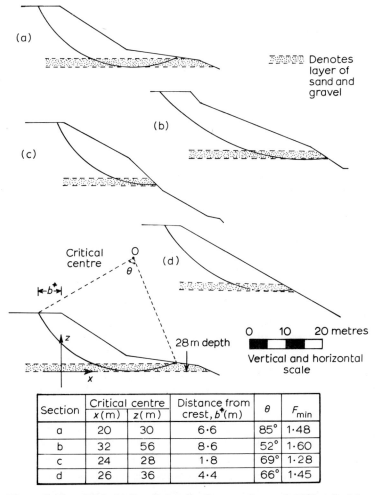

Figure 9.19 Critical slip circles for four sections of cliff at Robin Hood's Bay (*after* Courchée, 1970).

Section	Critical centre		Distance from crest, b^*(m)	θ	F_{min}
	x(m)	z(m)			
a	20	30	6·6	85°	1·48
b	32	56	8·6	52°	1·60
c	24	28	1·8	69°	1·28
d	26	36	4·4	66°	1·45

There is some evidence of *planar sliding* of virgin material or a renewal of movement of previously slipped clay. Recent sliding is confined to the northern end of the Bay.

Taking the Skempton and DeLory (1957) equation (9.3, 9.5) for flow parallel to the surface, and also taking the worst conditions when the water table is at its highest observable level (0.65 m below the surface), the factor m varies only slightly with variations in z, and F is reduced as z increases because of the declining influence of c'.

Table 9.3 Shear strength parameters derived from direct shear tests

Parameter	Main Boulder Clay	Sand Layer
c'	17.2 kN m^{-2}	0
c'_r	0	—
ϕ'	26.3°	34.0°
ϕ'_r	22.9°	—

For the limiting case of a deep slip in which

$$c' \ll z \cos^2 \alpha (\gamma - m\gamma_w) \tan \phi', \qquad (9.32)$$

then

$$F \simeq \frac{\cot \alpha (\gamma - m\gamma_w) \tan \phi'}{\gamma} \qquad (9.33)$$

and at limiting equilibrium, $\alpha = 14.3°$. It would appear to be unreasonable, therefore, to invoke the concept of a deep planar slide for this particular problem.

Shrinkage cracks can be observed within the surface layers and the possibility of shallow planar slides occurring can be examined. Analysing for a depth-to-slip plane z of 2 metres (the deepest observable shrinkage crack), Courchée (1970) was able to plot the variation of F with slope angle (see Figure 9.20) for $m = 0.7$ (highest observed water table level) and $m = 1$ (water table at the surface). Figure 9.20 indicates that all slopes mobilizing peak shear strength and inclined at less than about 40° to the horizontal should be safe, but that the influence of groundwater level on stability is quite marked. Differences between predicted and observed slope angles can be attributed to the variability of the cohesion parameter and the probability of lower-than-average c-values existing along some slip surfaces, and to the possibility of artesian pressures developing after heavy rain in the sandy layer and also leading to shear strength reductions. More generally, the analysis is useful in confirming that deep rotational slips are a little more likely to develop than shallow planar slips in clay underlain by a sandy stratum. It also demonstrates the significant lowering of the slope angle caused by a shear strength reduction from peak to residual. A slipped mass of clay can achieve very temporary stability through relief of pore water pressure, but if shrinkage cracking can develop as far down as the

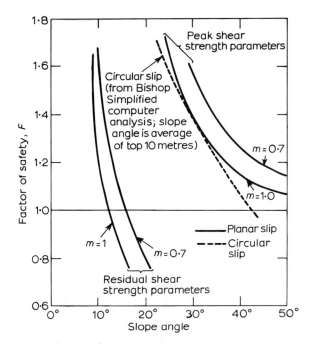

Figure 9.20 Shallow planar slide analyses (2m depth to slip plane) in boulder clay (*after* Courchée, 1970).

original slip plane, the plane is decomposed and entrained to areas having slope angles less than 20°. In fact, hydrostatic forces developed in shrinkage cracks within a locally sheared volume of clay can so break up a colluvial mass that it takes the form of a mudflow.

9.4 Simple wedge method of analysis

This type of analysis is particularly appropriate, although not restricted, to those slope situations where potential shear surfaces are constrained by the presence of fault and bedding plane structural features. It has been used, for example, as a means of back-figuring strengths of existing failure planes in the unstable Cucaracha clay shale which forms the banks of the Panama Canal (Figure 9.21, 9.22; see also Table 9.4 and 9.5) along the Gaillard Cut section (Lutton and Banks, 1970). Figure 9.23 is a reconstruction from borehole evidence of the original geology within a section across the East Culebra Slide in the Gaillard Cut and in Figure 9.24 are drawn two possible groupings of initial shear failure surfaces within the same section as analysed by Lutton and Banks. It will be noted from both

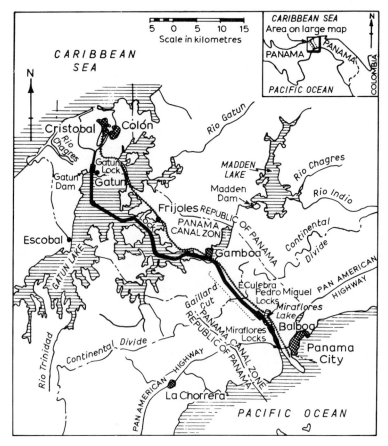

Figure 9.21 Map of the Panama Canal.

of these two sections that the main structural control on failure arises at the Culebra/Cucaracha boundary, direct evidence being obtained from borehole instrumentation such as inclinometers, slip indicators (plastic tubes) and microseismic (acoustic) monitors. Additional evidence for the configuration of the active and passive wedges arises from surface observations of the limit of deformation and from evidence of uplift on the passive wedge (there is 1915 photographic evidence of uplift at the Canal base, due to sliding of both East and West Culebra). It should be noted also that the transition zone between the two formations would be accepted as providing a strong basal control for any slip circle analysis.

As implied at the end of the last paragraph, in a wedge analysis the mobilized mass is sectioned into an active or driving wedge, a passive

Stability of Soil Slopes 663

Figure 9.22 View looking north along the Panama Canal from Contractors Hill on the West Bank. On the right foreground is the East Culebra extension slide area. Photograph: P. B. Attewell.

or resisting wedge, and a neutral block between the two wedges. The disposition of these wedges and the method of their analysis for the derivation of factors of safety are shown in Figure 9.25, the weights of the elements being calculated simply from the multiplication of their areas and the bulk density of the ground. It is recommended that during the preliminary use of this method, the force polygons be actually constructed and the factor of safety parameters ($W_{a,n,p}$; $\tan \gamma, \beta, \alpha$; $R_{a,n,p}$) determined by measurement rather than through immediate use of the equations. It will also be noted that the boundaries between the three elements comprising the sliding mass have been constructed vertically with earth pressure forces directed horizontally across them. This is an arbitrary, if obvious, construction leading to a possible loss of accuracy and factors of safety that are 10 to 20 per cent lower than for more complex methods of analysis (Sultan and Seed, 1967; Wright, 1969; Lutton and Banks, 1970).

There are two basic approaches in any type of stability analysis. In the first case, trial values of the two shear strength parameters, c and ϕ can be used to determine factors of safety for slopes currently stable but prone to instability. The absolute and relative sensitivities of the two parameters with respect to the factor of safety can then

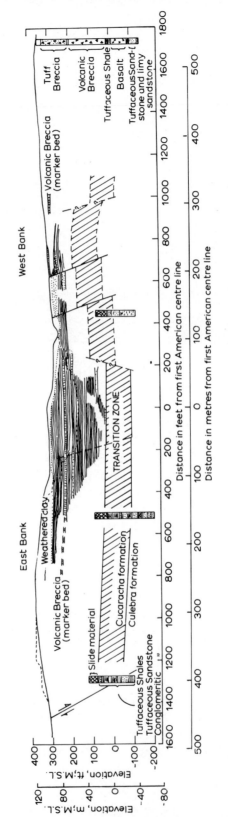

Figure 9.23 Original geological section across the Panama Canal at Station 1782 + 50 (extracted *from* Lutton and Banks, 1970).

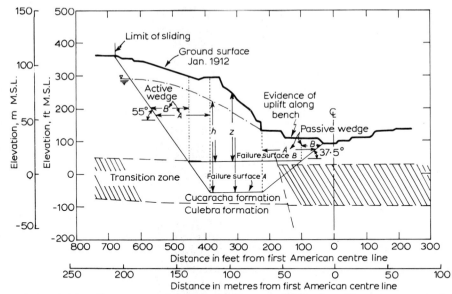

Figure 9.24 Two assumed failure surface conditions for effective stress wedge analyses (initial failure conditions) on East Culebra slide section, Station 1782 + 50, Panama Canal (*after* Lutton and Banks, 1970).

be examined by iterative methods. Alternatively, slopes that have failed for the first time can be assumed to have done so at limiting equilibrium under a factor of safety of unity. Since the geometry of the failed slope is known, a series of back-figuring exercises yields a range of possible c, ϕ combinations for further interpretation. Lutton and Banks (1970) have found that the simple wedge approach

Table 9.4 Some historical data on the Panama Canal

1883	French construction works begin
1899	French cease work on Canal due to construction difficulties with bank failures and health problems.
1903	United States treaty with Colombia giving to the Americans rights of construction, operation, and supervision of the Canal. Treaty rejected by Colombians. Subsequent revolt and secession of Republic of Panama from Colombia. New treaty signed in perpetuity with Panama.
1974 (Feb. 7)	Negotiations in Panama City between U.S. Secretary of State, Dr. Henry Kissinger and Panamanian Government lead to the United States surrendering its 'sovereignty in perpetuity' over the Canal.
1978 (Apr. 19)	President Carter's Panama Canal Treaty ceding ownership of the Canal passes through the U.S. Senate.
1978 (June 17)	President Carter signs the handing-over treaty in Panama.

Table 9.5 Some other Canal data

Original engineering options:

(a) Canal with locks: estimated at $m147 and 8 years to build;
(b) Sea level canal: estimated at $m250 and 12 to 15 years to build.

Option (a) was started in 1907 and finished in 1914 at an actual cost of $m380.

Canal is 51.2 miles (82.4 kilometres) long, deep water to deep water.

A total of 413.9 million yd³ (316 million m³) of material were excavated including 168.3 million yd³ (129 million m³) from Gaillard Cut.

Minimum width was originally 300 ft (91 m) through Gaillard Cut, later widened to 500 ft (152 m).

Minimum depth is 37 ft (11.3 m) in Balboa harbour at low tide.

There have been many Canal bank slides over the years, particularly during the construction period, in the Cucaracha shale. The last Canal closure was in August 1914. At one place, it has been estimated that the side slopes are moving in towards the Canal at a rate of 28 ft (8.5 m) per year.

---

produced factors of safety that were about 5 to 6 per cent low when more rigorous solutions indicated factors of safety near to unity. Thus, in the former case, the present type of analysis produces a conservative result. But under back-figuring conditions, the calculated strengths may be high by this same amount.

Back analyses are best performed in effective stress terms and the results ($F = 1$) presented as graphs of c' versus ϕ' for various pore pressure ratios. The pore pressure ratio, r_u, was defined by Bishop (1955) as

$$r_u = \frac{u}{\gamma z} \tag{9.34}$$

where u is the porewater pressure, γ is the total saturated unit weight of the soil and z is the vertical distance from ground surface to failure surface. With the three surfaces, ground, phreatic, and failure, in a generality of cases all varying with respect to one another (see, for example, Figure 9.24), and the probability of pore pressure changes during and following engineering excavation works as conditions equilibrate, it is advisable to assess the implications of an r_u range from zero to unity. But clearly it is useful to begin by assuming that the porewater pressures on the failure planes vary directly with the vertical distance between the phreatic surface and failure plane, that is:

$$u_{(z)} = \gamma_w h \tag{9.35}$$

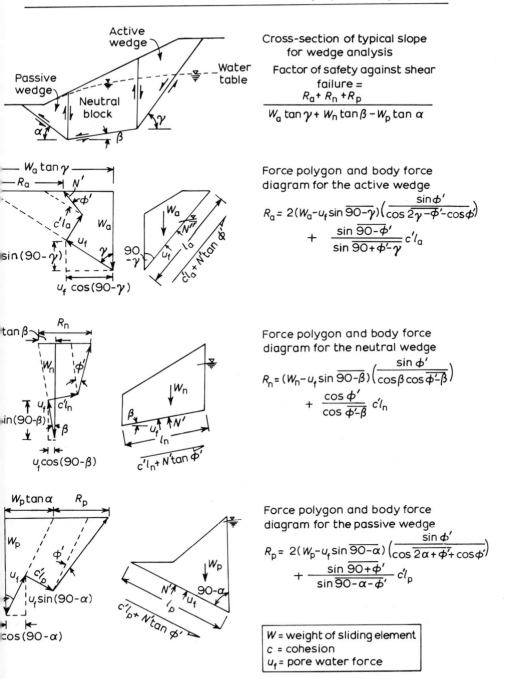

Figure 9.25 Typical use of wedge analysis for the solution of certain slope stability problems (generally *after* Lutton and Banks, 1970).

668 Principles of Engineering Geology

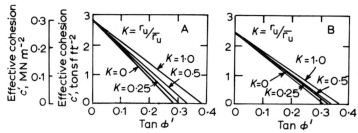

Figure 9.26 Relationship between effective shear strength parameters for two assumed failure surfaces in the Cucaracha clay shale (Station 1782 + 50, East Culebra, Panama Canal) and for a factor of safety of unity; average $\bar{r}_u = 0.38$ (critical failure surface A) and 0.36 (critical failure surface B); average shear strength of clay shale = (A) 3.03 tonf ft^{-2} (0.325 MN m^{-2}); = (B) 2.56 tonf ft^{-2} (0.275 MN m^{-2}) (*from* Lutton and Banks, 1970).

The c', ϕ' relationships for the two assumed slide conditions in Figure 9.24 are shown in Figure 9.26 and the assumed failure surfaces for the progressive slide situation in 1969 are drawn in Figure 9.27. For completeness, it is also useful to include Figure 9.28 from Lutton and Banks (1970) in order to show the character of the slide development in the Cucaracha clay shale at East Culebra, Panama Canal. On the Skempton and Hutchinson (1969) classification it would be entered as a 'multiple retrogressive translational slide'.

The Cucaracha formation as a whole consists of shale with

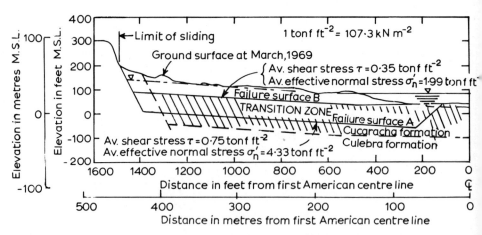

Figure 9.27 Analysis of Cucaracha clay shale slope (Station 1782 + 50, East Culebra, Panama Canal) for the configuration in March, 1969 (*from* Lutton and Banks, 1970).

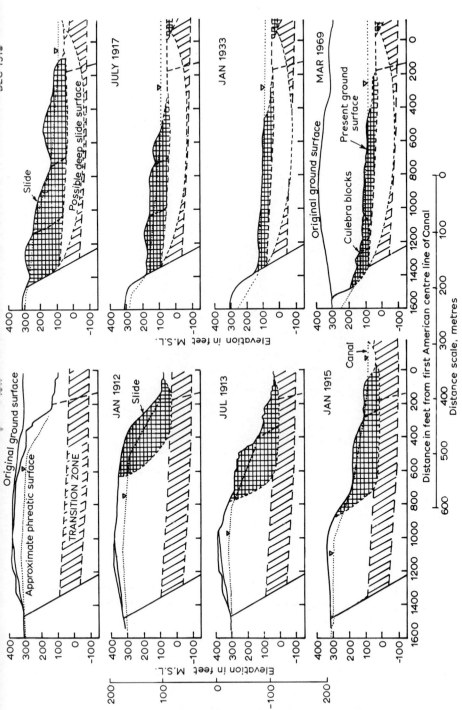

Figure 9.28 Progressive development of sliding in Cucaracha clay shale, East Culebra, Panama Canal, Station 1782 + 50 (*from* Lutton and Banks, 1970).

Figure 9.29 Corinth Canal, Greece, looking towards the Gulf of Corinth in the north-west. Photograph: P. B. Attewell.

subordinate layers of sandstone and conglomerate. Lower Cucaracha, of Lower Miocene age (23 to 25 m years), is mainly terrestrial in origin but the presence of some fossils suggests an intermittent marine contact. Piezometric levels in the material have been found to be generally below the water level in the Canal, so this implies that in the intervening years following construction the pore pressures in the banks of the Canal have not yet reached equilibrium. Until they have, there could be a continued decrease in the strength of the clay shale. Stability analyses by the U.S. Corps of Engineers on the East and West Culebra *initial* slides have produced field shear strength parameters of $\phi' = 19°$ and $c' \simeq 0$ but they have also indicated a considerable loss of strength with time ($\phi'_r = 7.5°$; $c'_r = 0$). As in the

Stability of Soil Slopes   671

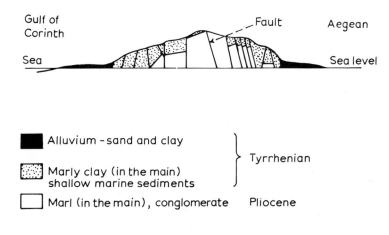

Figure 9.30   Geological section along the Corinth Canal.

earlier example, the use of horizontal drains to intercept water in cracks is recommended together with improved surface drainage in order to prevent run-off into open cracks.

The instability of the shallow banks of the Panama Canal may be contrasted with the apparent stability of the walls of the Corinth Canal in Greece which links the Gulf of Corinth and the Ionian Sea with the Aegean. Figure 9.29 presents a view of the Canal looking towards the Gulf of Corinth and from this view the steepness of the walls is immediately apparent. The maximum depth of the cutting is 79 m to water level, the width is about 40 m at water level and 21 to 22 m at the base of the Canal. The Canal is 6.3 km long and the water depth is 8 m. This latter depth contrasts with the maximum theoretical draft in the Panama Canal of about 13.7 m and a safe navigable depth in the Suez Canal of about 11 m. Some idea of the slope angles in Figure 9.29 may be derived from the fact that the rail bridge is 80 m long at a height (in that place) of 55 m.

A simplified section of the rocks in the cutting of the Corinth Canal is shown in Figure 9.30 (see also v. Freyberg, 1973). The Canal cuts through a normal, mid-Pleistocene horst, with Tyrrhenian sediments unconformably blanketing Pliocene sediments. Apparently, there is some occasional slight instability during periods of heavy rain or seismicity (return period: 9 to 11 yrs, $M \geqslant 5.5$; 23 to 27 yrs, $M \geqslant 6$; 55 to 70 yrs, $M \geqslant 6.5$; 140 to 165 yrs, $M \geqslant 7$) but nothing that could be termed deep-seated or 'on-going'. There are plans for widening and/or duplicating the Canal in the future.

9.5 Use of design curves

Instead of treating every slope stability problem as unique and requiring its own calculation for a factor of safety, the ideal approach is to draw up design curves in which the main parameters promoting instability — slope angle, slope height and pore pressure — are balanced against the shear strength parameters of the material at limiting equilibrium. The curves of D. W. Taylor (1948), although based on total and not effective stress conditions, are nevertheless useful in providing a first quick approximation for the factor of safety.

In Figure 9.31 it is assumed that fairly shallow slope has suffered circular base failure, the centre of the slip circle lying near to the vertical line passing through the centre of the slope. From a simple analysis, Taylor derived stability numbers $N = c/F\gamma H$ for different factors of safety F. His curves can also be used with a c, ϕ soil, with the factor of safety being applicable to both cohesion and friction and usually being taken as approximately linear with slope height.

As an example of the use of the chart in Figure 9.32 let us examine the factor of safety of a 50 ft (15.24 m) high slope at an angle of 40°, soil unit weight of 125 lb ft^{-3} (19.6 kN m^{-3}), cohesion and friction angle of 500 lb ft^{-2} (23.94 kN m^{-2}) and 24° respectively. Assume that minimum tolerable factors of safety are 1.25 for open ground and 1.75 for a built-up area and first try an intermediate F of 1.5, operating on the friction angle to produce an apparent ϕ of $24/1.5 = 16°$. Thus, $N \simeq 0.07$ from the curves and the relevant slope height would be equal to $500/(1.5 \times 0.07 \times 125)$ or about 39.5 ft

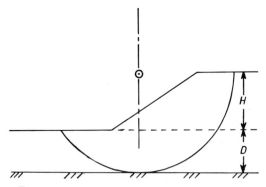

Figure 9.31 Definition of parameters for D. W. Taylor's curves. $D_f = (H+D)/H$.

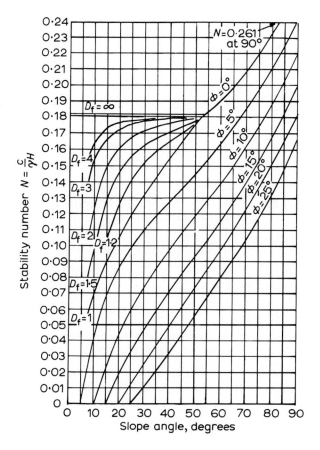

Figure 9.32 Curves of D. W. Taylor's stability numbers.

(12.0 m). However, $H = 50$ ft, so the factor of safety is too high. Trying, next, $F = 1.2$, the trial value of ϕ is 20° leading to $N = 0.05$. It follows that $H = 67$ ft (20.42 m) which this time exceeds the actual slope height. Perform this same operation for several trial values of F and by interpolation derive an actual F value of 1.36 for the actual 50 ft (15.24 m) high slope (Figure 9.33).

The curves of Bishop and Morgenstern (1960) are based directly on the Bishop Simplified Method of slope stability analysis and can be regarded as the effective stress equivalent of Taylor's curves. Their use again involves a convergence and interpolation for an actual factor of safety, and for the curves themselves together with details of operation the reader will refer to the above paper.

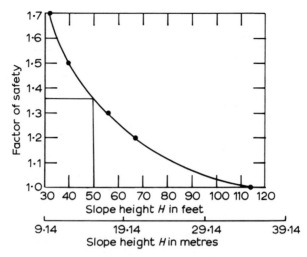

Figure 9.33 Interpolation for the factor of safety of the actual slope.

9.6 Pore pressure ratio

The influence of pore pressure in clays under conditions of total and partial saturation has been considered in Chapter 2 and in Section 2.11 of that chapter attention is directed specifically to the meaning of the pore pressure parameters. It will also have been noted earlier in this present chapter (Section 9.2) that the pore pressure ratio r_u has been expressed in the Bishop Simplified equation as $r_u = ub/W$ (equation 9.24) or more conveniently as $r_u = u/\gamma z$ (equation 9.34) where z is the depth to the element in question.

However since $\gamma z = \sigma_1$ as an approximation,

$$u = u_0 + \delta u = u_0 + \bar{B}\delta\sigma_1 \qquad (9.36)$$

Thus,

$$r_u = \frac{u_0}{\gamma z} + \frac{\bar{B}\delta(\gamma z)}{\gamma z} \qquad (9.37)$$

where u_0 is the pore pressure before a change in stress.

In the case of clays placed as fill and subject to compacting forces (see Section 11.1), then if the clay is on the dry-side of optimum, u_0 is negative. In that case r_u is very small and can be ignored for design purposes. When, however, fill is placed for an earth embankment, such as a dam, or is used for loading the toe of a slope, it is usually placed on the wet-side of optimum to improve workability and aid

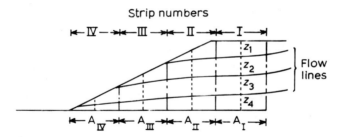

Figure 9.34 Determination of an average value for r_u

the compaction process. In this case, u_0 is small and r_u approximates to $\bar{B}$.

It will be appreciated that the actual value of r_u may vary within a slope and that some average value will be required for a Bishop Simplified analysis or for the use of the Bishop and Morgenstern curves. This can be observed by dividing the slope in a series of strips, as in Figure 9.34, and noting that with respect to area I (A_I) of the first strip,

$$\bar{r}_{uI} = \frac{z_1 r_{u1} + z_2 r_{u2} + z_3 r_{u3} + z_4 r_{u4}}{z_1 + z_2 + z_3 + z_4} \qquad (9.38)$$

This same procedure is adopted for the remaining strips. Each $\bar{r}_u$ is then multiplied by its area, the products are summed and then divided by the sum of the strip areas, vis:

Grand average $\bar{r}_u$ for the slope is

$$\bar{\bar{r}}_u = \frac{\Sigma[A_I r_{uI} + A_{II} r_{uII} + A_{III} r_{uIII} + A_{IV} r_{uIV}]}{\Sigma[A_I + A_{II} + A_{III} + A_{IV}]} \qquad (9.39)$$

The character of the upper flow line may not be completely defined by a drilling programme and it must therefore be approximated using a parabolic form of construction. The technique for doing this is outlined in Section 8.2.

9.7 Clay slopes and shear strength parameters

Most failures of excavated slopes in stiff, overconsolidated clays and clay-shales occur during the initial construction period if oversteepened. Cut-slopes will stand short-term at angles determined by the peak strength of the mass and failures may tend to take place under conditions of ϕ tending to zero. The undrained situation arises from the relatively low mass permeability (Terzaghi and Peck, 1967). Strength will also be conditioned by favourable or unfavourable bedding plane dips (Bhatia, 1967) and the nature and intensity of any jointing (Esu, 1966). Continuous bedding or jointing which dips into the mass admits water and quickly decreases the strength of the clay near to its surface. Dips out of the mass and into an excavation act as shear planes for early movement even though drainage might be improved. Initial failure is often presaged by the growth of tension gashes parallel to the excavated face.

If a slope remains stable initially, there will be a steady decrease in clay strength with time *towards a long-term residual condition* (Skempton, 1964; Chandler 1974) in which the basal planes of the clay minerals have become more nearly aligned parallel to an assumed continuous shear plane. During the intermediate stages, there will be softening and swelling effects due to stress relaxation and the ingress of water along discontinuity planes. There will also be slaking effects associated with the presence of any expandable clay minerals. Furthermore, a period of excessive rainfall may accelerate failure (Chandler *et al*, 1973). Slide failures often occur along toe circles mobilizing a relatively shallow body of clay since the shearing resistance will tend to increase quite significantly with distance from the exposed surface. Even in less-heavily pre-consolidated clays, the ratio of horizontal residual stress to terminal vertical stress may approach the coefficient (K_p) of passive pressure in the clay (Skempton, 1961), and the moisture ingress associated with the lateral decompressions continues to contribute to the mechanical disintegration of weaker clay shales adjacent to the slopes of river valleys or behind man-made cuts. Vaughan and Walbancke (1973) have noted that the failure of slopes in overconsolidated clay may be delayed by the rate at which pore pressure equilibriates after excavation.

In engineering terms, a strength decrease with time is usually related to progressive displacement along a shear plane.

If an undisturbed clay is caused to be sheared, the necessary shear stress may at first increase to a peak for quite little shear displacement but will then fall-off somewhat in an exponential manner with further displacement to a minimum residual level as the shear plane fully develops. This development of a continuous principal shear surface involves the linkage of numerous Riedel, thrust and displacement shear surfaces (Skempton, 1966), the grinding down and the rotation of individual particles. The extent of the fall-off in stress from peak to residual, characterized by Skempton (1964) as a 'residual factor', is greater for an over-consolidated than for a normally-consolidated clay. The form of the relationship between shear stress and displacement, as derived, for instance, from a normally consolidated laminated clay is shown in Figure 9.35 and the associated envelopes are shown in Figure 9.36. Furthermore since in a general *zone* of shear failure there is a distribution of normal pressure, shear failure can develop in the slope at different strains. Pruška and Thú (1974) have noted that the line connecting the graphical peak strengths as a function of normal pressure acting on the clay structure is a characteristic of the *rate* of strain at which discrete microshears develop in earth slopes.

When shear failure occurs *for the first time* in an unfissured clay mass, the undisturbed material lies very nearly at its peak strength and it is the effective peak friction angle ϕ'_p and the effective peak cohesion c'_p that would be used in the analysis. If the mass is fissured, then Skempton (1964) has shown that any first-time slides would be initiated at strengths well below peak. This reduced effective strength is probably related to moisture movement into the shear plane as a result of local dilations, readjustments, and rotations which create a softening of the clay and facilitate expansion of any expandable clay minerals that are present. The failure state would be conditioned by a necessary minimum shear displacement such that the material on or in proximity to a single or composite shear plane would be strained past peak strength. In natural slopes which have acquired their morphology through sliding, particularly landslipping that occurred in Pleistocene times, the shear strength of the clay will have fallen to something approaching a residual value. Indeed, if one takes the Skempton and DeLory planar slide equation, assumes a melt-water table to ground surface, and approximates the unit weight of water to one-half that of the clay, then it will be quickly proved that the slope angle at critical equilibrium ($F = 1$) is equal approximately to only $\phi'_r/2$ if a residual condition is accepted into the analysis. A corollary of these arguments is that some very low-angle slopes might be encountered in landslip areas and that old slides will be much more prone to reactivation during construction works in them (see Section 9.13). As an associated but further specific mechanism, Palladino (1972) in his back-analysis of 8 failures in the Lawton clay at Seattle, USA, has suggested that stress relief during glaciation caused horizontal slips along the bedding planes of the clay and so reduced their strength to residual at that time.

A question arises as to what is the approximate shear strength to

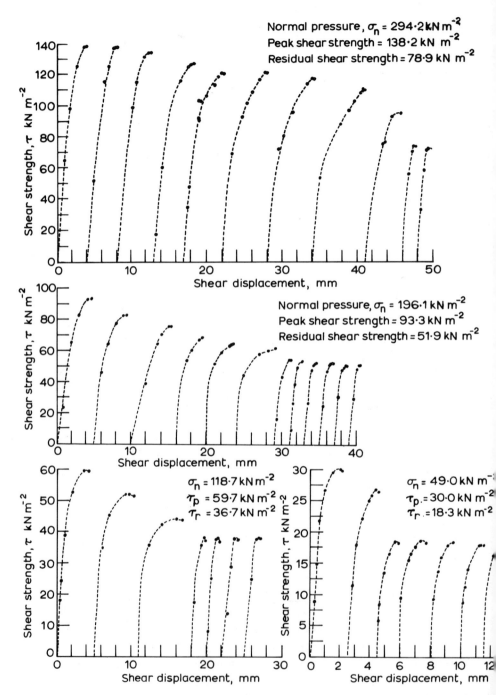

Figure 9.35 Variation in the shear strength with displacement and normal pressure laminated clay sheared parallel to the laminations.

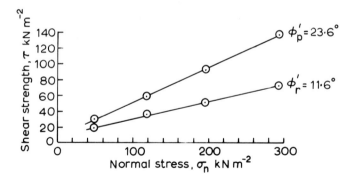

Figure 9.36 Peak and residual shear strength of laminated clay sheared parallel to the laminations.

take for design against first-time slides. Without specifically differentiating between first time slides and reactivation slides, Lane (1969) has proposed that a strength decrease in clay shales subsequent to excavation could perhaps be avoided by designing for a long-term strength substantially below peak strength but well above residual strength. He suggested that this strength should be the stress at which the deviator stress in a triaxial compression test departs from the initial linear portion of the stress-strain deformation curve. He argues that irreversible or progressive failure effects might thus be minimized or eliminated, thereby preventing continuous failure surfaces from ultimately developing but having first achieved peak strength in a progressive-type failure action. The concept of progressive strength reduction with time is borne out by Morgenstern's (1969) observation of a slope failure in London Clay that occurred after 30 to 40 years but, as has been noted by Johnson (1969), there is a lack of firm data on such reduction. Chandler (1974) has analysed the effective cohesion for the Lias Clay along twelve first-time slides and has shown that there is no obvious decrease in c' with time. Strength loss is more directly related to slow pore pressure increases.

Morgenstern (1969) has quoted data from stability studies in London Clay and has suggested that, for periods of about 100 years, stability could be estimated on the basis of $c' = 0$, $\phi' = \phi'$ peak from drained shear tests provided that residual conditions had not developed through appropriate displacements along pre-existing planes of weakness. Whether such an argument could be applied to clay shale slopes in general is rather uncertain. Probably the most relevant and obvious suggestion is the one calling for a time-dependent reduction factor (r) in conjunction with c'_p and ϕ'_p

(Skempton and Hutchinson, 1969). As a basis for short-term design, Skempton and Hutchinson (1969) advocate the application of a reduction factor (f) for fissures, plus a reduction factor (x) for the rate of testing, anisotropy and so on to c_u (peak strength parameter undrained). Iverson (1969) has proposed a rather more empirical approach to slope design based mainly on geological defects observed in test pits and other exposures in the area of interest. In the absence of obvious structural defects, the basis for slope design would be a function of the importance of the excavation and the strength would either be specified at peak modified by a factor of safety or some values between peak and residual. On the other hand, Hardy (1969) has objected to any suggestion that a residual strength concept be used in clay shale slope design unless, for example, the swelling behaviour is considered explicitly.

Skempton (1970) has shown that the strength for first time slides in London Clay cuttings tends towards but does not fall below the fully softened state. As defined by Roscoe et al (1958), a saturated clay is at its critical state, under conditions which permit drainage to take place, if any further increment in shear distortion causes no further change in the water content along the shear plane. A clay will tend to expand or contract during shearing with a drainage facility, and this will continue until the critical state is reached, at which point the clay then continues to deform under constant stress and at a constant volume. In particular, the critical state, or fully softened strength, is equivalent to the strength of a normally-consolidated clay and the water content of the clay at that stage of the shear is equal to the water content acquired by an overconsolidated clay as it dilates during shear. This means that c'_{crit} and ϕ'_{crit} correspond to c'_p and ϕ'_p of a normally consolidated clay — confirmed as per Skempton (1970) by the fact that $\phi'_{crit} = 22\frac{1}{2}°$ for London Clay (Schofield and Wroth, 1968) compares favourably with the normally consolidated peak ϕ'_p of $20°$. Experimentally, the critical state parameters for design purposes could be determined by the point of intersection of, on the one hand, an undisturbed shear stress-displacement curve and, on the other, a re-moulded curve from the same material.

In certain problems of a ground-structure interaction nature (for example, in the context of slopes, the design of a retaining wall, or in an entirely different field, the bearing pressures on a tunnel lining), it is necessary to consider the rigidity (the stiffness with respect to shear) of the clay. The higher the normal pressure applied to the developing shear plane, the steeper will be the initial rise of the shear stress-displacement curve, the higher will be the peak shear stress (and ϕ'_p) and the greater will be the critical shear stress point (and ϕ'_{crit}) on the downslope of an overconsolidated shear displacement curve. The stiffness of the soil increases both with pressure and overconsolidation ratio (Wroth, 1972).

Skempton (1970) cautions against accepting the fully-softened strength as a design parameter against first-time sliding in all overconsolidated clays. Drawing on slide analyses at Lodalen (Sevaldson, 1956), Selset (Skempton and Brown, 1961) and Kimola (Kankare, 1969) he states that although the clays at these sites are overconsolidated, the strength at failure was close to the undisturbed peak value. The clays at Lodalen and Kimola were unfissured and only lightly overconsolidated, and showed little tendency to dilation during shear. There would thus be little structurally-controlled tendency for water entry to the shear plane. By way of contrast, the clay at Selset is heavily overconsolidated and expands strongly during shear. Skempton argues that the higher failure strength in this case and the absence of appreciable softening is caused by the lack of a progressive post-peak failure strain facility, the most important aspect of this facility being a discontinuity presence.

Most clays have not been researched to the same extent as London Clay. There may, therefore, be a temptation to use residual shear strength parameters for design in undisturbed ground (with a probable consequential conservatism) in addition to their obvious use in old landslip areas. A true residual strength can be achieved only after very great shear displacements ($>$ 1 metre: Bishop *et al*, 1971) and only when, according to Bishop's (1972) ring shear evidence, the constituent minerals have been rotated and shear-degraded to their fundamental sizes. Residual is therefore a material characteristic which in theory will be approached asymptotically, and in order to eliminate the displacement uncertainty there may be merit in standardizing shear displacement to, say, 0.5 inch (13 mm) and expressing all shear strengths at that displacement* during testing.

It is usually accepted that cohesion tends to zero as the friction angle tends to residual. Experimental results may suggest, however, a

---

*The U.S. Corps of Engineers 'S' strength (direct shear) tests on Dawson clay shale gave the same ϕ'_r results at a constant rate of 0.5 inch in 24 hrs as at a rate of 0.5 inch in 12 days.

finite c'_r value. 'S' tests, for example, on the Kimmeridge Clay in a 12 inch (305 mm) shear box gave c'_r = 2.03 lb in^{-2} (14 kN m^{-2}) and ϕ'_r = 10° (linear regression analysis). Constraining c'_r to zero in a least squares analysis produced a ϕ'_r angle of 16° which should be regarded as an absolute maximum residual friction angle for long-term clay shale stability design. Alternatively, if a critical normal stress for a potential shear failure plane can be estimated, then ϕ'_r can be determined from the point of interception of this stress ordinate with the residual failure envelope projected linearly back to the origin.

In spite of the laboratory evidence of a finite value of c'_r for some clays – for example, the Kimmeridge Clay just quoted, brown (weathered) London Clay (Skempton and Petley, 1967; Bishop et al, 1971), and weathered Upper Lias Clay (Chandler et al, 1973) – there is often a tendency to dismiss this shear strength contribution during slope stability appraisal on the assumption that it will exert little influence upon the eventual slope design. However, as is very appropriately pointed out by Chandler and Skempton (1974), an assumption of c'_r = 0 for the design of cutting slopes for permanent and long-term stability in stiff fissured clays would incorporate a factor of safety greater than unity and would lead to slopes which were very much flatter than experience would suggest to be necessary. It would also lead to the conclusion that the limiting slope of a cut in a given clay would be independent of the depth of the cut – a conclusion that is not borne out by practical experience nor by the curves in the next section.

These arguments tend to re-emphasize the importance in 'clay mechanics' of a discontinuity presence. But it is also the form of the discontinuity field or fields that is important in more perceptive analyses of shear strength. There are 'inherited' discontinuities that are old and pre-date a particular shearing event. These discontinuities probably possess an orientation and spatial density distribution that does not change significantly from place to place within an area of immediate slope failure interest. The genesis of such discontinuities has been discussed in Section 6.2. Depending upon the distribution characteristics of the inherited field (the distances between adjacent discontinuities, their mutual orientations, and their orientations with respect to a developing shear surface or surfaces), there will in general be a further field of discontinuities necessarily 'imposed' through the shearing process and limited to areas adjacent to the

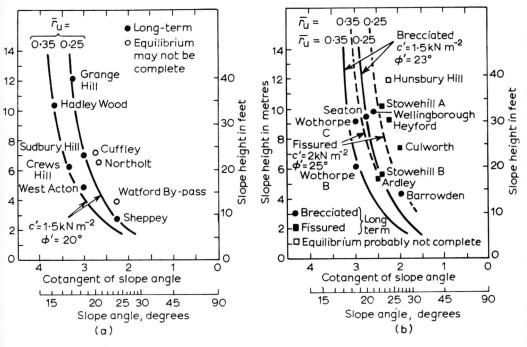

Figure 9.37 Slope height-slope angle curves based on circular arc analyses for cuts in two different clays compared with observed height-angle relationships (*from* Chandler and Skempton, 1974). $\bar{r}_u$ is the average pore pressure ratio (where $\bar{r}_u = \gamma_w h/\gamma z$, with h the local head of water on the slip surface and z the local depth of cover to the slip surface.)

shear plane. Studies of discontinuity field characteristics have, therefore, an important bearing on the selection of friction parameters for interpretation of past slope failures and design against new failures.

9.8 Slope angle measurements in clays and clay shales

Skempton (1969) has defined as the angle of ultimate stability the inclination at which a slope in nature finally becomes stable against any form of landsliding. The angle depends on the clay properties, the climate and the groundwater situation and, as he points out, it can be determined with accuracy only from field observations such as would be conducted as a natural course of events by geomorphologists.

Because clay shales weather, at least superficially, within a

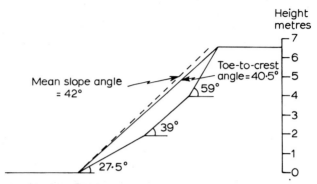

Figure 9.38 Slope profile after 18 years exposure — Kimmeridge Clay, Fox Holes, North Yorkshire, England.

relatively short period of time*, their failure tends to be progressive, the slope receding primarily by intermittent sliding towards an ultimate flatness. As the slope angle decreases, the average shearing stresses also decrease along potential surfaces of sliding. Ultimately, according to Terzaghi and Peck (1967), slopes are reduced to 1 vertical to 10 horizontal (6°), or even less.

The concern is often, therefore, less a matter of danger due to failure; it is more a question of whether the rate of superficial degradation and any combative measures are going to be tolerably expensive during the economic life of the excavated slope. Slope height – slope angle relationships for brown London Clay and weathered Upper Lias Clay have been published by Chandler and Skempton (1974), and are redrawn in Figure 9.37. The slope geometry parameters that are defined by these data points are related more to a discrete circular shear type of failure than to a state of progressive degradation.

Slope angles may often vary with height within a particular profile; if conclusions are to be drawn from numerous slope measurements, then these can either be considered individually, their average can be taken as characterizing the full slope height, or a direct toe-to-crest angle can be measured†. These options can be seen

---

*As an example, the specification for newly exposed cuttings in the Dawson at Chatfield dam site at Littleton, Colorado, U.S.A. is that they must be protected against weathering and slaking where concrete would subsequently be placed against them, and this is done by spray-sealing with a bitumen compound within 2 h after exposure (see Figure 6.1). Following this, a 4 in slab of concrete must be placed within 48 h. If concrete placement can be accomplished within 2 h of exposure, then the bitumen coat can be dispensed with.
†There are, of course, practical height limitations as the slope angle decreases. Heights are also more difficult to assess.

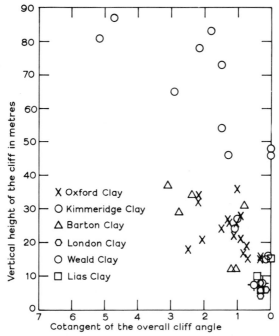

Figure 9.39 Height-angle relationships for some coastal cliff slopes in the south of England.

specified in Figure 9.38 and some relationships for coastal slopes and for short-term (≤ 18 years) cut slopes in the Kimmeridge Clay are shown respectively in Figures 9.39 and 9.40. Where natural phenomena of this nature are to be considered, considerable data should be amassed to offset the inevitable scatter and permit a reasonable statistical assessment to be made. The scarcity of information on Figures 9.39 and 9.40 prevents any satisfactory conclusions being drawn except to suggest that contrary to what might appear from the general form of Figure 9.40 data, the cotangent of the slope angle would be expected to increase with slope height up to the limiting values of cot ϕ'_r. This certainly seems to apply to the Chandler and Skempton data in Figure 9.37. However, these plots refer to relatively long-term slope failure situations in which pore pressures have equilibrated whereas it might be claimed that the slope data in Figure 9.40 are representative only of a much more recent intermediate stage of shear instability and degradation. Long-term also, the individual slopes within a profile should be planed smooth, that is, the mean slope curve and the toe-to-crest curve should converge at an ultimate angle determined by the residual friction state.

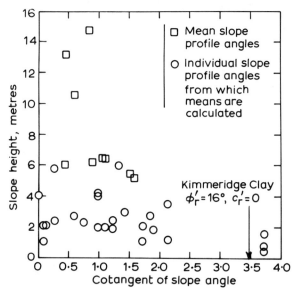

Figure 9.40 Slope height – slope angle relationships for the Kimmeridge clay of Yorkshire, England.

The influence of slope age is masked within any short period low profile curves, since over an extended period of geological time there must be a trend towards an increase in slope cotangent (decrease in angle) as the total slope height increases even though a state of residual friction might imply an independency with respect to slope height. Slope *design* curves would be based on criteria less conservative than a residual state of friction, which would condition the design only for the greatest slope heights envisaged. For slopes of reduced height, increasingly greater slope angles can be tolerated, as is indicated, for example, in MacDonald's (1915) design curves for the Panama Canal slopes (Lutton and Banks, 1970) and which were considered anyway on re-analysis by Hirschfield et al (1965) to be too conservative. For other work on slope angle-slope height relationships, reference should be made to Hoek (1970) and Banks (1972). The work of Shuk (1955), Kley and Lutton (1967) and Lutton (1970) provides a qualitative indication of typical slope height-slope angle relationships but the results are of rather limited value in the quantitative design of slopes. It may also be noted that Onodera *et al* (1974) have compiled relationships between slope heights, angles, widths and thicknesses. For their planar type of weathered surface layer movement in diorites, granites, tertiary

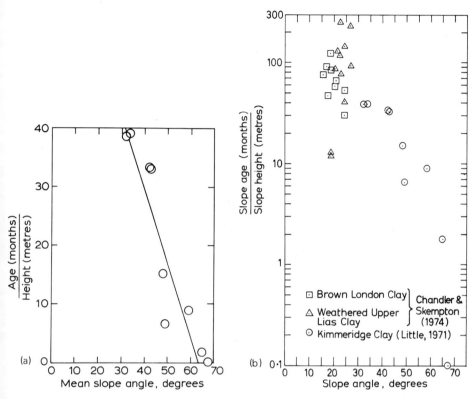

Figure 9.41 (a) Changes in the mean slope angle with variations in slope age/slope height ratio in the Kimmeridge Clay of Yorkshire, England. (b) Some slope angle–slope age relationships.

sedimentaries and surface deposits they deduced thicknesses and slope angles of most failed slopes to be 0.5 to 1.5 m and $35° - 45°$ respectively.

On the question of time controls on ultimate slope angles, Hutchinson (1967a) and Skempton and DeLory (1967) have analysed several naturally-occurring inland and coastal slopes in London Clay with respect to their age-angle relationships. Their results demonstrate a general tendency for inclination to decrease with increasing orders of magnitude of time, and they conclude that angles of ultimate stability can indeed be correlated with residual strength.

Analysis of slope angle decreases with time in the Kimmeridge Clay (Little, 1971) shows that slope angle differences are approximately proportional to constant differences in age over an 18 year period. Although in engineering stability terms, this time period is rather restricted, use of toe-to-crest data indicated that 75 per cent of all slope degradation processes over that time had taken place over 8 y (45 per cent of the time). Compilation of curves, such as that in Figure 9.41a, for particular clay shale horizons might perhaps, by extrapolation, be used to predict the age of slopes of known height. On the other hand, the Chandler and Skempton (1974) data tend to imply that, within their quite narrow band, slope angles for the London and Upper Lias Clays are generally independent of age. Data for these clays and for the Kimmeridge Clay are plotted on the same graph in Figure 9.41b in order to show that the two groups of clays respectively occupy different regions of the slope angle spectrum and thereby to imply that the comment made earlier concerning the different degradational conditions is probably quite reasonable.

9.9 Classification of gravitational mass movements in clay

Skempton (1953a, 1953b) has proposed three parameters (Figure 9.42) upon which a descriptive classification of clay slope mass movement can be based, but Skempton and Hutchinson (1969) have also pointed out that the form of the disturbed clay, or the movement picture, may be variable even when ratios of the slope morphology parameters are the same. Their proposed classifications are shown in Figure 9.43 and for detailed comments on each type of movement, see their Conference Report referenced above.

In general, *falls* promoted by a tension crack facility are short-term failures in new cuttings or river banks. *Rotational slides* are characteristic of slopes in more uniform clay or shale; they are usually quite deep-seated ($0.15 < D/L < 0.33$) and can be seen in actively eroding cliffs. *Circular rotational slips* occur in cut slopes of relatively uniform clays and in natural slopes of overconsolidated clay (for example, London Clay). *Non-circular slips* are a feature of overconsolidated clays that have weathered to provide a quasi-planar slide surface, or of structurally anisotropic unweathered strata. Circular and non-circular *shallow rotational slips* tend to develop on moderately inclined slopes in weathered or colluvial clays.

Compound and translational slides develop when rotation is inhibited by an underlying planar feature, the translational effect becoming more dominant, the shallower the slide. The control on the character of the slide may be a bedding plane or, as before, the base of a weathered boundary layer. Material subjected to part-rotational, part-translational sliding is quite severely distorted and broken up to

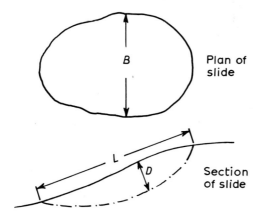

B is the maximum width of the slide
L is the maximum length of the slide up the slope
D is the maximum thickness of the slide

Figure 9.42 Fundamental dimensions of a slide (*after* Skempton and Hutchinson, 1969).

an extent which seems to be dependent on the degree of non-circularity of the failure. Translational slides are more superficial than compound slides, being conditioned by a more shallow inhomogeneity. *Block slides* are more characteristic of more lithified, jointed rock which tends to separate into blocks which are then prone to sliding as units on well-defined bedding, joint or fault planes. *Slab slides* are a type of translational failure, characteristic of more weathered, clayey slopes of low inclination and they involve the movement of material *en masse* with little internal distortion. Weathered mantle and colluvial materials are particularly prone to this mode of failure which rarely takes place with D/L ratios greater than 0.1.

Skempton and Hutchinson exclude from the category of *Flows* those extremely rapid flow movements 'which result from the sudden access of water to debris mantled slopes, as these seem to partake more of mass-transport than mass-movement'. Excluded are volcanic mudflows (average speed up to 25 ms^{-1}) and torrential mudflows (speeds of several ms^{-1}). Nevertheless, mudflows are often characterized by high speeds of movement commonly in the range 10 to 10^3 m min^{-1} and the ability to move over slopes of one degree or less. They include periglacial solifluction sheets and lobes. The term '*earthflow*' is limited to slow movements of softened, weathered

FALLS

ROTATIONAL SLIDES (SLIP, SLUMP)

Circular   Shallow   Non-circular

COMPOUND SLIDES   Graben

Competent sub-stratum

TRANSLATIONAL SLIDES

Block slide   Slab slide

FLOW

Lobate   Lobate or elongate

Earthflow   Mudflow

Sheet   Lobe

Solifluction sheet and lobe

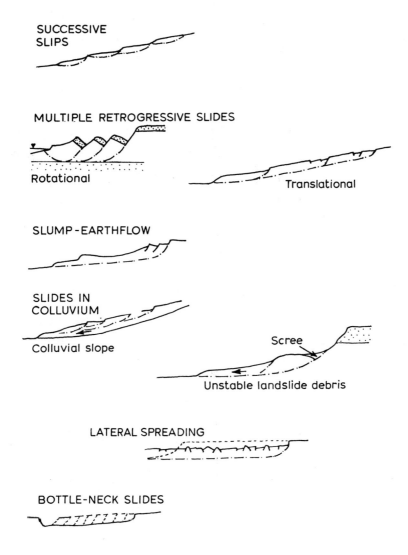

Figure 9.43 Some basic types of mass movement, multiple and complex landslides in clay slopes (*after* Skempton and Hutchinson, 1969).

debris such as forms the toe of a slide, and the style of movement is transitional between a slide and a mudflow. Skempton and Hutchinson (1969) say that earthflows accommodate a smaller degree of structural breakdown than is the case with mudflows (maintenance of the integrity of the mass implying a large degree of retention of the original vegetation cover). Carson and Kirkby (1972), on the other hand, see the majority of an earthflow movement taking place 'as a viscous liquid rather than the sliding downslope of a solid slab of earth' and the quick or extra-sensitive clays as being ideal vehicles for such flows. *Mudflows* (in addition to the references in Table 9.6, see Mullineaux and Crandell, 1962; Broscoe and Thomson, 1969; Hutchinson, 1970; Hutchinson and Bhandari, 1971) develop along discretely sheared boundaries in fissured clays and glacial varved or laminated deposits where there is a softening facility at the shear zone. Interdigitations or lensing of fine, water-bearing sands promote the necessary ingress of water. The movement mechanics involve development of forward thrust due to the undrained loading of the rear portion of the mudslide where the basal slip surface is inclined steeply downwards. According to Hutchinson (1970) this mechanism facilitates shearing movements on flat or even negatively-inclined surfaces. Morphologically, mudflows have a glacier-appearance and usually develop on inclinations of between about 5° and 15° (Skempton and Hutchinson, 1969). There are in fact quite close mechanical analogies between mudflows and the flow of glaciers and ice sheets (Nye, 1951), both being amenable to treatment as problems in plasticity (Nye, 1951; Brückl and Scheidegger, 1973). Elongate mudflows ($L/B \approx 10$) can develop their attenuated form when there is little or no erosion at the toe of the flow. Where the toe is eroded (such as on the coasts) a lobate type of mudflow has a much lower L/B ratio. *Solifluction lobes and sheets* apparently exhibit slickensided shear surfaces when fossil periglacial forms have been examined on slopes as low as 3° or 4° (Chandler, 1972), but such evidence of quasi-brittle discrete shearing seems not to be available in contemporary solifluction deposits. But it is quite apparent that flowslides of all types, whether imminent, contemporary, or fossil could present particular engineering difficulties both during any terrain evaluation stage and during a later construction stage of a highway project.

Skempton and Hutchinson (1969) also consider the question of multiple and complex landslides. They suggest that *successive slips* may develop upwards from the foot of a slope, the shallow rotational

Table 9.6 Listing of references to recorded slides and flows (as given by Skempton and Hutchinson, 1969)

Type of Movement	Clay and Location	Authors
Falls	Leda Clay (St. Lawrence Valley)	Bazett *et al.* (1961)
	Brown London Clay (Bradwell, Essex)	Skempton and La Rochelle (1965)
	Residual soil (Hong Kong) (decomposed granite)	Lumb (1962)
Rotational slides		
Circular	Slightly overconsolidated (Lodalen, Oslo)	Sevaldson (1956)
	Stiff and medium blue clay (Chicago)	Ireland (1954)
	Fen silt and soft plastic Fen clay (Eau Brink Cut, Norfolk)	Skempton (1945)
	Stiff, fissured, overconsolidated London Clay (Warden Point, Kent)	Hutchinson (1968)
	Intact clay till (Selset, Yorkshire)	Skempton and Brown (1961)
Non-circular	Jurassic clay (Barrage de Grosbois, France)	Collin (1864)
	Glaciolacustrine sand, silt and clay, glaciofluvial deposits, fluvial sand and gravel, alluvial fan deposits, windblown sand (nr. Fort Spokane, Washington)	Jones *et al.* (1961)
	Weald Clay (Sevenoaks, Kent)	Toms (1948)
Shallow	London Clay slopes	Hutchinson (1967a)
Compound slides	London Clay (Kent coast, England)	Hutchinson and Hughes (1968)
	Soft quick clay and weathered crust (Bekkelaget, Oslo)	Eide and Bjerrum (1954)
	Pliocene lacustrine clay, silt, gravels and capping of Pleistocene andesite tuff (Gradot, Yugoslavia)	Suklje and Vidmar (1961)
Translational slides	Unweathered, overconsolidated and jointed clays (Valdermo, Italy)	Esu (1966)
	London Clay low-angle slopes	Hutchinson (1967a)
	Residual soil (Canaleira, nr. Santos, Brazil)	Vargas and Pichler (1957)
	Quick clay (Furre, Norway)	Hutchinson (1961)

Table 9.6 Continued

Type of Movement	Clay and Location	Authors
Flows		
Rapid flows (mass transport)	Volcanic mud – 'Lahars' (Gunong Keloet, Java)	Scrivenor (1929)
	Volcanic mudflow (Mt. Bandai, Japan)	Iida (1938)
	Mudflow (Wrightwood, Southern California)	Sharp and Nobles (1953)
	Mudflows (Tenmile Range, Central Colorado)	Curry (1966)
Mudflow	Varved or laminated clays (Steep Rock Lake, Ontario)	Legget and Bartley (1953)
	Elongate Slumgullion mudflow (nr. Lake City, Colorado)	Crandell and Varnes (1961)
	Elongate mudflow (Stoss, St. Gall, Switzerland)	von Moos (1953)
	Elongate mudflow (Mt. Chausu, (Japan)	Fukuoka (1953)
Solifluction Lobes and Sheets	Hythe Beds (nr. Sevenoaks, Kent)	Skempton and Petley (1967); Weeks (1969)
	London Clay (Boughton Hill, Kent, England) and Wadhurst Clay	Weeks (1969)
Successive slips	Irregular slips in London Clay (Hadleigh Castle, Essex, England)	Hutchinson (1967a)
	Irregular slips in London Clay (High Halstow, Kent, England)	Hutchinson (1967b)
Multiple Retrogressive slips		
Rotational	Oligocene clays (Seagrove Bay, Isle of Wight)	Skempton (1946)
	Chalk and Gault Clay (Folkestone Warren, Kent, England)	Hutchinson (1969); Toms (1953)
	Heavily overconsolidated, stiff, fissured, lacustrine silts and clays (Meikle River, Alberta, Canada)	Nasmith (1964)
	Overconsolidated, stiff, fissured interstadial or interglacial clay (Sandnes, Norway)	Bjerrum (1967)
Translational	Overconsolidated clay (Jackfield, Shropshire, England)	Henkel and Skempton (1954)
	Miocene shale and bentonitic tuff, (Portuguese Bend, California)	Merriam (1960)
	Silty deposits (Vibstad, Norway)	Hutchinson (1965)

Table 9.6 Continued

Type of Movement	Clay and Location	Authors
Slump Earthflows	Material unspecified (nr. Berkeley, California)	Varnes (1958)
Slides in Colluvium	Weathered Carboniferous mudstone (Walton's Wood, Staffordshire, England)	Skempton and Petley (1967)
	Weathered colluvium-alluvium Weirton, West Virginia)	D'Appolonia et al. (1967)
	Clay and loose rock (Portland, Oregon)	Clarke (1904)
	Hamaishidake pebble strata (mudstone, tuff breccia, conglomerate) (Yui, Japan)	Taniguchi and Watari (1965)
Spreading Failures	Varved clay (Hudson River Valley)	Newland (1916)
	Quick/varved clay (Sköttorp, Sweden)	Odenstad (1951)
	Quick clay with sand lenses (Turnagain Heights, Alaska)	Seed and Wilson (1967); Seed (1968)
Bottleneck Slides	Late marine clays (Ullensaker, Norway)	Bjerrum (1954)
	Post-glacial marine clays (Nicolet, Eastern Canada)	Crawford and Eden (1957)

shears linking in a regular or irregular manner to form 'step' features. With sufficiently high density and overlap, the movement picture could be that of a *shallow translational retrogressive slide.* However, the higher the cohesive strength of the material, the greater the dimensions of the individual shearing units comprising the latter type of slide and the more, therefore, the translational form of the multiple retrogressive slide is compounded from unit slab slides. A feature of all multiple retrogressive slides is, of course, that the constituent failures combine to create a common basal shear surface, and it has been pointed out by Skempton and Hutchinson that the *rotational form of multiple retrogressive slides* occurs most frequently on actively eroding, high relief slopes in which a thick stratum of overconsolidated fissured clay or clay shale is overlain by a thick bed of more competent rock.

Slump-earthflows tend to develop from a long rotational slide in which the over-ridden fractured toe of the disturbed mass becomes softened by water ingress and develops into an earthflow or even into a mudflow.

The term 'colluvium' is reserved by Skempton and Hutchinson to denote weathered and degraded material in accumulation zones beneath cliffs, and *colluvium slides* involve movement of such inherently unstable material. They say that typical conditions conducive to the assemblage of such degraded material are, for example, severe melt-water erosion of the bases of cliffs during Pleistocene ice sheet retreat, followed by about 10 000 years of unconstrained degradation of the oversteepened clay or clay shale slopes. Colluvial slides can also involve the regeneration of older slide surfaces often in preference to the initiation of first-time slides in the total body of accumulated material.

Spreading failures are characterized by a sudden lateral spreading of a retrogressive translational type but on quite gentle slopes, the movement being completed within a few minutes to leave a profile of graben and horst structures. This type of failure is attributed to the presence of elevated porewater pressures in a more pervious basal layer, dissipation of these pressures (sometimes triggered by earthquake motion) being at the expense of a dynamic mobilization of the superadjacent clay mass. Quick clays and varved clays seem to be prone to such effects.

Bottle-neck slides are peculiar to quick clays and are a form of retrogressive, multiple rotational failure. They take place as an initial rotational failure of the stiffer weathered crust at the banks of an incised stream followed very rapidly by further failures into the weaker, less weathered material. Failed, rapidly remoulded clay is ejected through the opening in the bank and is washed away to be re-deposited elsewhere. The failure is more widespread in plan further away from the bank because of the relative weakness of the material, this type of slide taking its name from the constriction in the weathered crust at the stream bank.

References to recorded slides in different materials, as quoted by Skempton and Hutchinson (1969), are listed in Table 9.6. For interesting case histories of engineering slope stability problems, particularly in overconsolidated clays and clay shales, the reader is also referred to Wilson's 1969 Sixth Terzaghi lecture. For general reading on landslides, reference may be made to Zarubá and Mencl (1969).

9.10 Rock breakdown and landform development

Climatic conditions determine the relative importance of physical and chemical weathering in a rock but the discontinuous character of the rock itself, on a macro- and micro-scale, determines the nature of the weathered debris.

A jointed rock mass initially weathers into a talus (mantle) of loose cobble-sized material through the action, for example, of freeze-thaw and wet-dry (slake) cycles. The weathering is greatest at ground surface and probably decreases non-linearly towards the base of the mantle. Initially, the fragments are controlled in size by the spacing of the discontinuities in the parent, unweathered rock. Further breakdown takes place by the splitting of these fragments through suction pressure effects (air pressurization in tortuous capillaries by external immersion, leading to tensile microfissuring – see Chapter 3), by solution of the intergranular cement, and by the decay of weak minerals. Eventually, the rock is degraded into a colluvial mantle the constituent elements of which more nearly approximate to the fundamental minerals.

It is probable that this early basic pattern of breakdown of jointed rocks is valid for all climates although the actual degradational processes may differ. In the case of non-jointed rocks, or those in which the joints are very widely spaced, step-by-step size degradation processes are eliminated and the breakdown involves direct cement solution and the production of colluvium-grade material (for example, sand grains from a massive sandstone) *in situ* from the parent rock. The composition of the initial colluvium and its subsequent alteration depends on the prevailing climatic conditions (Jenny, 1941; Mohr and Van Paren, 1954). In humid areas, spheroidal weathering dominates, a mantle of corestones being formed in a matrix of residual soil through water percolation and solution down joint planes. Talus production in arctic areas is through frost thrust action in joint planes, so producing more angular blocks (Rapp, 1960). In the tropics, with an abundance of heat and precipitation at high cyclic frequencies, chemical weathering is much more of a live issue than in temperate climates.

The changing profile of a rock slope involves the removal of material from the crest through chemical and mechanical breakdown. If the depth-related weathering is in *equilibrium* with the landsliding process the production of new mantle material at the base of a weathered layer must be balanced by the sum of the chemical loss (as measured in terms of a chemical change in stream samples) and the mechanical removal of the surface layers of the mantle. Rates of chemical removal (independent of slope angle) of about 0.01 to 0.1 mm per year depth equivalent have been suggested for temperate climates although the evidence of Taylor and Spears (1970) points emphatically in the direction of negligible chemical effect below a skin of, at the very most, a metre thick.

During the processes of degradation, the material is undergoing a reduction in shear strength, primarily through losses in cohesion. Carson and Kirkby (1972) have quoted evidence to suggest that these changes may not be smooth with degree of degradation. Evidence to support this arises from soil tests on mixtures of different grain sizes which seem to show that as the coarse and fine content is gradually varied in composition, there can be a 'jump' change in the friction angle (Vucetik, 1958; Holtz, 1960). Once the debris has achieved a soil-sized aggregation of particles, weathering processes may continue to promote mineralogical changes in the more sensitive components such as the clay, iron, aluminium and calcium minerals. There is, therefore, a continuing facility for shear strength reduction of the *in situ* material.

Since they are a function of so many variables, *weathering rates* are themselves extremely variable (Kellogg, 1941). The thickness of a soil mantle, and any consequential engineering implications, is dependent not only upon rate but also duration and removal. If the transport processes are more rapid than the weathering, the weathered layer will be thin and the movement is said to be 'weathering-limited'. If the weathering rate exceeds the rate of removal of the weathered products, then the movement is termed 'transport-limited'. As pointed out by Carson and Kirkby (1972) these different controlling imbalances generate different sequences of slope development. At the weathering-limited end of the movement process spectrum there are rock falls, while at the transport-limited end we have soil creep under continuous cover conditions. Mass movements of soil in which discrete internal deformations are subordinate to larger scale planar or circular shears occupy an intermediate position in the movement spectrum.

Soil slope morphology as well as its engineering stability will be a function of the shear strength parameters of the material. There may be a temptation to ignore the cohesive contribution, particularly at low normal pressures, but stability analyses do tend to suggest that a computed factor of safety is quite sensitive to the magnitude of cohesion that is input to the analysis. Friction, however, tends to be the dominant parameter particularly in the control of more superficial transportation mechanisms. Shear strength is obviously a function of weathering, certainly for tropical residual soils (Lumb, 1962, 1965), the effective friction angle ϕ' itself being influenced by the degree of packing of the particles and by the mineralogy. Kenney (1967) has shown that at an advanced stage of weathering and breakdown, a change in mineralogy produces a much greater change in ϕ' than does any change in grain size. On the other hand, there is evidence to suggest that grain size can influence the friction angle.

ϕ' is greater for graded mixtures of coarse and fine particles than for mono-component soils, a phenomenon that can be explained at least in part in terms of their respective packing densities. Thus, in a detailed evaluation of what friction angles to use in any back-figuring of slope profiles the actual path or paths taken by a suite of particles undergoing degradational styles may be important. ϕ' may *rise* first with reducing gravel content, then it may fall to rise once again later in the weathering process when landsliding is no longer a real issue.

Shear strength data on taluvial (talus-colluvium), residual soils and some weathered rocks are given in Tables 9.7 and 9.8. Additional data will be found elsewhere in the text.

It is useful to record that, as a result of the arguments concerning ϕ' – particle size relationships, Carson and Kirkby (1972) have produced a general slope profile interpretation:

(a) Thick soil mantles (> 30 cm) of shale and 'shale grit' parent rocks: visible landslide scars and several slope angle distributions due to stepped ϕ' – particle size relationships and continuous breakdown;
(b) Thinner soil mantles of grits, sandstone and limestone parent rocks; visible terracettes and continuous slope angle distributions, the terracettes being confined to slopes on which the rock breaks down discontinuously.

After each large or small slide, the disturbed material comes to rest at a shallower angle, there to continue weathering until it again tends to instability. There is therefore the prospect of true multi-linear slope profiles. On the other hand there may be a continuous change in the stable slope angle. This could be simply a manifestation of a continuous sequence of discrete slope angles, which means that in

Table 9.7 Shear strengths of some talluvial materials (*partly after* Carson and Kirkby, 1972, but with additions)

Material	Gravel fraction		ϕ'	Test type cell or normal pressure ($kgf\ cm^{-2}$)	Reference
Alluvium	0.7		41°–44°	T 3.5–2.4	Lowe (1964)
Silty, sandy gravel	–		45°	T 0–7	Hall and Gordon (1963)
Glacial till	–		37°	T 0–7	Insley and Hillis (1965)
Penrhyn Slate (granular)	Zero	passing B.S. 72 sieve retained B.S. 100 sieve	$\phi'_p = 47°$ $\phi'_r = 40°$	D 1.4–8.5	Attewell and Sandford (1974a)
		passing B.S. 7 sieve retained B.S. 14 sieve	$\phi'_p = 49°$ $\phi'_r = 39°$	D 0.7–8.5	Attewell and Sandford (1974a)
Exmoor Slate	0.57		44°	D 0–0.2	Carson and Petley (1970)
Shale grit	0.4–0.6		43°	D 0–0.2	Carson and Petley (1970)
Fluvial, glacial, talus, etc.	0.6–0.8		37°	T 0–4	Pellegrino (1965)
Sandy clay (volcanic)	35% sand		38°	D –	Wallace (1973)
	45% sand		29°		

Note: T = Triaxial test, D = Direct shear test.

Table 9.8 Shear strengths of some residual soils and weathered rocks (*partly after* Deere and Patton, 1971 but with additions)

Parent rock	Degree of weathering	Strength parameters (cohesion in $kN\ m^{-2}$)	Test	Reference
Gneiss	Decomposed (fault zone)	$c_u = 150, \phi = 27°$	Direct shear on concrete rock interface	Evdokimov and Chiraev (1966)
	Very decomposed	$c_u = 400, \phi = 29°$		
	Partly decomposed	$c_u = 850, \phi = 35°$		
	Unweathered	$c_u = 1250, \phi = 60°$		

Rock	Description	Parameters	Test	Reference
Gneiss (micaceous)	Muram (Zone 1B) Decomposed rock	$c_u = 60, \phi = 27°$ $c_u = 30, \phi = 37°$	Direct shear	Gruner and Gruner (1953)
Schist	Partly weathered (Zone 1C)	$c' = 50, \phi = 15°, \phi' = 15°$ $c' = 70, \phi = 18°, \phi' = 21°$	50% sat. 100% sat. consol. undrained	Sowers (1963)
Granite	Weathered (Zone 11B) Partly weathered (11B) Sound (Zone 111)	$c = 0, \phi_r = 26-33°$ $c = 0, \phi_r = 27-31°$ $c = 0, \phi_r = 29-32°$	Direct shear	L.N.E.C. (1965)
Granite	Decomposed (fine-grained) Decomposed (coarse-grained) Decomposed (remoulded)	$c' = 0, \phi' = 25-34°$ $c' = 0, \phi' = 36-38°$ $c' = 0, \phi = 22-40°$	Drained triaxial	Lumb (1962)
Keuper Marl	Highly weathered Partly weathered Unweathered	$c' \leqslant 10, \phi' = 25-32°, \phi_r' = 18-24°$ $c' \leqslant 10, \phi' = 32-42°, \phi_r' = 22-29°$ $c' \leqslant 30, \phi' = 40°, \phi_r = 23-32°$	Drained and consol. triaxial	Chandler (1969)
London Clay	Unweathered	$c' = 7-33, \phi' = 19-23°, \phi_r' = 9.5°$ $\phi_r' = 14°$ $\phi_r' = 10.5°$	Direct shear Ring shear Direct shear	Bishop (1972) Marsland (1972b) Bishop et al. (1971)
	Weathered (brown)	$c' = 15, \phi' = 20°$		Skempton (1964)
Upper Lias Clay	Weathered	$c_r' = 1.5, \phi_r' = 9°$	Ring shear	Chandler et al (1973)
Siltstone Dark shale Seatearth	Weathered Weathered Weathered Weathered	$c_p' = 160, \phi_p' = 43.5°$ $c_p' = 0, \phi_p' = 26°$ $c_p' = 173, \phi_p' = 39°$ $c_p' = 24, \phi_p' = 31°, c_r' = 12, \phi_r' = 17.5°$	Triaxial Triaxial Triaxial Direct shear	Taylor and Spears (1972)

the field it should be possible under particular geomorphological circumstances to observe a range (albeit limited for a particular rock type) of straight slope angles, or it could truly reflect a spectrum of slope linearities, each a function of a specific ϕ' – particle grading and each so short that it is tending to zero.

The movement picture in slope profile development is therefore one both of continuity and discontinuity.

It is usually considered that in *soil creep* there is a continuous gradation in down-slope movements such that the rates of deformation range from zero at depth to a finite value nearer the surface. Discrete shear failure surfaces are not formed directly during creep.

Terzaghi (1950) differentiates between a highly seasonal mantle creep, which is conditioned by ambient temperature, vegetation, and water levels, and a more deep-seated mass creep. Skempton and Hutchinson (1969) suggest a range for the rate of mantle creep on clay slopes of from less than 0.1 cm per year up to a few cm per year, with movements extending to a depth of up to 1 m (see also Carson and Kirkby, 1972, p. 272 and 288). Mass creep, being controlled only by gravity, takes place at a more constant rate (Terzaghi, 1950) under stress differences that are only a fraction of the peak strength of the material (Bishop, 1966). Davison (1889) first proposed a freeze-thaw mechanism for mantle creep, a mechanism that required a soil expansion normal to the slope profile upon freezing and a retraction vector upon thawing that is angled intermediately between the vertical and the normal. Particle movements would thereby acquire a 'saw-tooth' trajectory for their down-slope motion, such motion being superimposed upon the deeper-seated creep. For a mathematical approach to the mechanics of soil creep, the reader is referred to Carson and Kirkby (1972, p 294–300).

The transformation from an essentially creep process to one of discrete shear can be considered by reference to Figure 9.44. Down-slope movement is relatively constant with time until some slide-producing process comes into operation, commonly as a result of a seasonal state of elevated porewater pressure. After a period of pre-failure particle acceleration, landsliding occurs, and it is clearly of engineering importance to be able to predict the onset of this acceleration phase which will be a manifestation of shear surface development. Overall movements develop more rapidly the steeper

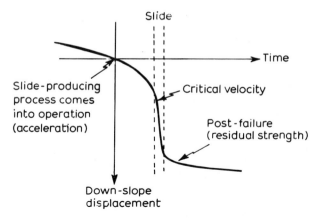

Figure 9.44 Transformation from creep into positive sliding.

the slopes and it is also necessary to try to determine whether the movement is superficial or more deep-seated.

After sliding, the clay will be at or near to a residual state of shear strength. It is therefore susceptible to further changes given suitable external agencies. In very broken and softened translational slides, post-failure movements can take the form of mudflows, seasonally-controlled, in which the displacements can be up to about 25 my^{-1}

Tell-tale observational posts can always be inserted to monitor creep development. On a more sophisticated level, inclinometer access tubes can be installed in boreholes and the deformation monitored as a function of depth. Discrete shear zones can be located using a system whereby weighted lines passing up and down a plastic tube in a borehole are intercepted at a point of tube shear. Micro-seismic monitors are also useful for fixing points of shear in a borehole through noise intensity enhancement (although problems can be encounterd through background disturbances in wet boreholes). But at the simplest surface observational level, the engineering geologist would take careful note of any telegraph posts, door posts, trees and so on; their inclinations can tell him a great deal about ground stability.

Solifluction is a more rapid version of seasonal creep produced in periglacial areas by annual cycles of soil freezing and thawing. During the thawing period, the more superficial layers of soil are mobilized and lubricated by the contained water, and under gravity on gradients as low as 2° to 5° they over-ride the still frozen ground in well defined lobes (see, for example, Figure 3.15).

9.11 Geomorphological description of slope profile development

Slope profiles are amenable to description on the bases of either physical (analogue) modelling or mathematical modelling (see, for example, Doornkamp and King, 1971; Carson and Kirkby, 1972). Such descriptions are of limited engineering interest and they are therefore not considered directly here. It is useful, however, to note the kinematic descriptions applied to various sub-sections along the idealized profile in Figure 9.45;

1. *Interfluve:* Pedogenetic processes associated with vertical sub-surface soil water movement.
2. *Seepage slope:* Mechanical and chemical eluviation by lateral sub-surface water movement.
3. *Convex creep slope:* Soil creep; terracette formation.
4. *Fall face:* Fall; slide; chemical and physical weathering.
5. *Transportational mid-slope:* Transportation of material by mass movement (flow, slide, slump, creep); terracette formation; surface and sub-surface water action.
6. *Colluvial footslope:* Re-disposition of material by mass movement and some surface wash; fan formation; transportation of material; creep; sub-surface water action.
7. *Alluvial toeslope:* Alluvial deposition; processes resulting from sub-surface water movement.

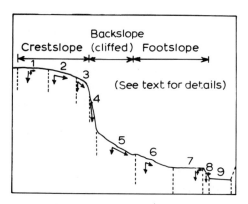

Figure 9.45   Hypothetical nine-unit land surface model (*from* Dalrymple *et al*, 1968).

Stability of Soil Slopes 705

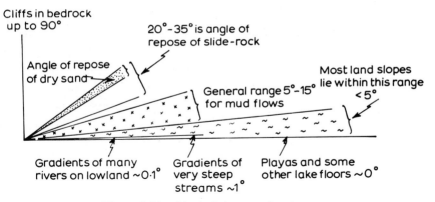

Figure 9.46 Typical slope angle ranges.

8. *Channel wall:* Corrasion*, slumping, fall.
9. *Channel bed:* Transportation of material down-valley by surface water action; periodic aggradation and corrasion.

In Figure 9.45, the vectors denote the direction and relative amplitude of movement of the weathered rock and soil by the major geomorphic processes. Units 7, 8, 9 experience down-valley movement normal to the plane of section. In terms of engineering stability, units 3 to 8 demand the greatest attention.

Some preliminary idea of the geomorphological range of slope angles can be obtained from Figure 9.46, and useful reference may be made to Brunsden and Jones (1972).

9.12 General methods of preventing slope failure

The techniques that have been adopted for preventing or alleviating the failure of slopes in soft ground are many and varied. Some of the techniques are described elsewhere in the book, but it is useful to mention one or two briefly at this stage.

The need for reliable borehole information as to the nature of the strata, its moisture content, and the standing water level is obvious. The presence of any particularly plastic layers along which shear could take place more easily will be noted. Piezometer tubes or piezometer tips will be installed in the ground and changes in water level recorded over an extended period of time.

---

*Corrasion is the term used to define the wearing-down of surfaces beneath, and due to the movement of scree slopes.

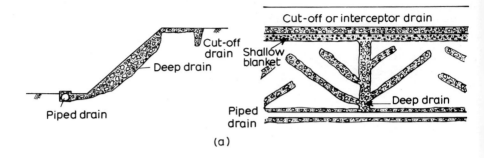

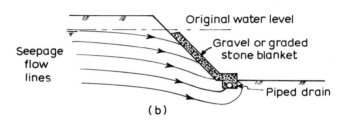

Figure 9.47 Typical drainage systems.

Infiltration to an exposed slope can usually be reduced by grassing it over or by covering it with sand. There are more exotic methods, such as a vacuum system to remove water from permeable strata or electro-osmosis to de-water silts. Hot, dry air can also be circulated from galleries. But these methods will usually prove to be too expensive and the engineer will usually look to standard drainage systems for reducing creep and enhancing stability. Typical systems are sketched in Figure 9.47. If there is a danger of sub-surface erosion resulting, for example, from piping, this can be offset to some extent by the use of an inverted filter loaded at the top.

Other methods involve: (a) regrading of the slope to a less steep angle, and total removal of material; (b) removal of material from the crest and its replacement (with appropriate drainage) at the toe; (c) re-establishment of original slope, after removal of crest material, through the use of imported 'rip-rap' material (construction of berms); and (d) the importation of material for toe-loading without any re-grading. In many instances, (a) may not be possible and (c) may be the answer (see, for example, Arrowsmith (1971b) and Figure 10.51b). In colliery districts, shale from waste heaps may be

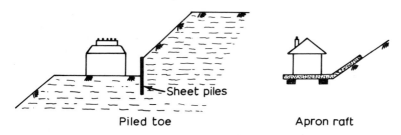

Figure 9.48 Other protection techniques at unstable slope locations.

suitable for (d). Earlier discussion has been directed towards a problem at one of the Durham Colleges (Figure 9.13) located on a slope forming the north bank of the east loop of the River Wear. In that case, a condition of progressive slope failure has seemingly been arrested by the adoption of measure (d).

Much more expensively, sheet piling techniques (Figure 9.48) have been adopted to hold back hillsides (there is at least one example in the Team Valley in north east England, but in this case the piles themselves have been rotated). Another possibility is to adopt an apron raft system (Figure 9.48) for the protection of building foundations, but this is usually restricted to the more intact rock masses. Yet another device for stabilizing excavations, usually on a temporary basis prior to permanent foundations being installed (deep-basement engineering), involves the use of well-pointing as described for example by Tomlinson (1963).

After, or during failure of a slope, there will be a facility for water seepage along the shear plane. If the ground water table is high, this seepage will accelerate the shear movement along the slip plane unless it is intercepted. At Herne Bay, for example, in the south of England, deep drains were laid behind the cliffs. In other cases, counterfort drains are used to break up the slip material, intersect the slip plane and provide deep drainage (Figure 9.49). Whichever method is adopted, it is always wise to re-turf over those areas where a slip plane is visible at the surface; the turf contributes a measure of cohesion and acts against direct water infiltration.

Post mortem analysis and back-figuring on failed slopes is of great importance. Together with the interpretation of laboratory test results and theoretical reasoning they have suggested that a fully-softened clay strength might be used in design. The factors of safety that are adopted in slope design analysis are always low (around 1.25

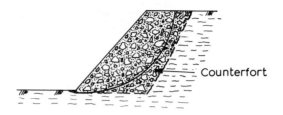

Figure 9.49 Counterfort aggregate drain.

for country areas and 1.5 for urban areas) compared with a factor of safety of 3 used in foundation design. This is mainly a reflection of the amount and cost of land needed to permit lower slope gradients.

9.13 Highway slopes

There have been published examples of slope instabilities in soils and clays requiring highway re-routings and ground treatment, both demanding expenditures over and above that originally budgeted for the project. Many of these problems can be traced to the re-activation of old Pleistocene landslips, the further disturbance of material the mass strength of which was probably close to a residual state. New highway routes are often taken along the sides of valleys in order to achieve tolerable gradients and also because the land in such locations can be purchased relatively cheaply. If these roads cross old landslips, the deep, steep cuts and heavy sub-grade treatments demanded by modern industrial traffic could well upset the pre-existing critical equilibrium.

What is probably now quoted as the classical example of re-activation of relict landslide debris is the Waltons Wood slide which occurred on the Staffordshire section of the M6 motorway. The road was taken along the side of a V-shaped valley the 12° side slopes of which had been cut by melt-water in Late Glacial times. Thin drift covered the bedrock of mudstone and shale with subordinate sandstone and coal of the Upper Coal Measures but unfortunately these materials weathered rapidly upon exposure with the mudstone reverting to a clay.

From the detailed geological surveys of Early, Glossop (1968) has outlined the stages leading to the natural development of and modification of the valley (Figures 9.50, 9.51) — see also Early and Skempton (1972). Basically, rapid-deepening of the valley (now a narrow alluvial plain underlain by a deep, sand-filled buried channel)

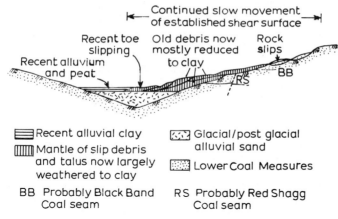

Figure 9.50 Landslide at Waltons Wood on the M6 (*from* Glossop, 1968 *after* Early).

had led to over-steepening and retrogressive rotational shearing of the valley sides with the old clay debris, along which the road was taken, at a very low state of shear strength ($\phi' = 10\frac{1}{2}° - 14°$). Before completion, the road showed distinct signs of landslip-distress, and Woodland (1968) quotes a figure of £600 000 as the cost of remedial works. These took the form of the emplacement of a massive berm of colliery shale at the toe of the embankment and thorough drainage of the valley side above the embankment by counterfort drains (Glossop, 1968). The slope was finally graded-off and re-afforested.

Another road embankment slope failure in England which has received some attention in the literature (Woodland, 1968; Symons and Booth, 1971) occurred during construction of the Sevenoaks By-pass in Kent. Instability developed during construction of earthworks along a 2½ kilometre length at the southern end of the by-pass (Figure 9.52) and the problem can again be traced to the re-activation of extensive solifluction-flows and landslips, this time in the Hythe Beds and the underlying Atherfield and Weald Clays. The Hythe Beds comprise alternating beds of hard sandy limestone and argillaceous calcareous sandstone, the Weald Clay is a heavy blue-to-brown (weathered) clay (LL 45 per cent, PL 20 per cent) and the Atherfield Clay is rather more sandy than Weald Clay (Symons and Booth, 1971). According to Woodland (1968), the low-strength slip zones at the bases of several solifluction lobes (Symons and Booth, 1971, Figure 2) were not detected in the original site investigations

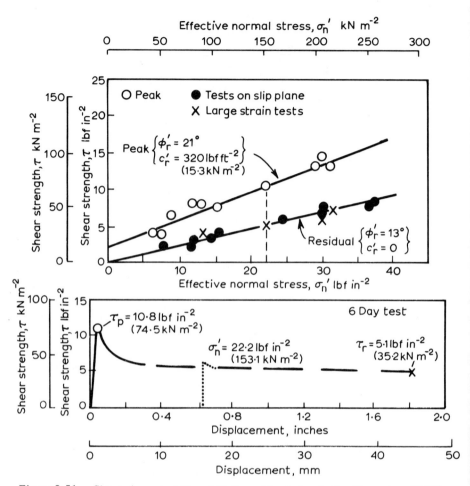

Figure 9.51 Shear characteristics of Waltons Wood clay (*after* Skempton, 1964).

although such slip planes have been observed down to a depth of about 20 ft. Indeed, he notes that in Kent it has been known for a long time that almost the whole length of the outcrop of the basal Hythe Beds is associated with cambering, springs, solifluction and landslipping. In the event, this section of the by-pass was abandoned and re-routed at considerable (> £1m) additional cost.

There is one point worthy of special mention. It has been shown earlier (Section 9.1, equation 9.5) that a factor of safety may be derived quite easily as per Skempton and DeLory under conditions of planar sliding and water seepage parallel to the slope.

Now if the problem concerning the re-activation of old slides is

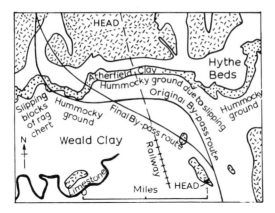

Figure 9.52 Southern section of Sevenoaks By-pass showing change of route necessitated by slipping ground (generally *after* Woodland, 1968).

considered it will be appreciated that many of these might originally have developed during early glacial retreat with the widespread presence of melt-water raising the water table to ground surface (m = 1). Since the unit weight of clay is about twice that of water, it has been noted earlier in Section 9.7 that, at limiting equilibrium, the critical slope angle will be about half the friction angle under the conditions envisaged above if any cohesion is ignored. Furthermore, if there has been significant shear displacement, then ϕ' will have tended to ϕ'_r and this suggests that when new highways are routed along slopes of Pleistocene clays the fact that a slope may be critical at half the residual friction angle should not escape the notice of the earthworks designer.

Two other factors, however, bear on this same point. During periods of intense storms, perched water tables may develop within a soil mantle, saturating ground completely up to the surface (Whipkey, 1965; Kirkby and Chorley, 1967) and thereby creating the same conditions as just discussed. On the other hand, very many slopes stand over the geological short-term at angles which are steeper than $\phi'/2$. In these cases, limiting equilibrium may be related to the critically softened strength (the point on the shear stress-displacement graph at which the remoulded curve intersects the undisturbed curve), cohesion being ignored at low normal pressures and ϕ' being taken at the point of critical softening.

Highway cuts in loess present rather locally-specialized* but nevertheless interesting problems to the designer and, for further detailed reading, reference should be made to Lutton (1969). That text represents one of the best expositions on the geotechnical properties and the *in situ* behaviour of this material.

Loess is a poorly-graded, porous silt with cohesive shear strength and is usually considered to be a wind-blown deposit (see Section 3.3). There is usually evidence of bedding and jointing together with vertical root tubes many of which are lined with calcite. It is claimed, with some justification, that these tubes add a dimensional reinforcement to the mass endowing it with the property of being able to stand, without slumping, in vertical road cuttings. On the other hand, the tubes also enhance the vertical permeability of the deposit, so it is likely that the latter effect militates to some extent against the former. Lutton (1969) suggests that the influence of both effects has probably been exaggerated in earlier literature.

Lutton's work was directed to the cuts made in the loess bordering the Mississippi River for a by-pass route of the U.S. Highway 61 around Vicksburg and for the Interstate Highway 20. There are also examples of steep cuts in the Vicksburg National Park. For the Highway projects, faces were cut at approximately 80° with 15 ft benches at 20 ft intervals and this resulted in total slope inclinations of 55°. Turnbull (1948) quotes canal slopes through loess in Nebraska remaining stable at 53° (80 ft or 24.4 m high), 63.5° (80 ft), 76° (55 ft or 16.8 m) and both Scheidig (1934) and Lutton (1969) quote the spectacular loess topography in China where U-shaped valleys with flat bottoms have vertical walls rising as high as 300 ft (91 m). Penetrating these valley sides are gashes as deep as 125 ft (38 m) and as narrow as 5 ft (1.52 m) across which facilitate tributary drainage, and, as Lutton notes, such remarkable geomorphology and apparent material competence have led some investigators to view loess more as a rock than as a soil. As with fissured clays, the borderline between the technologies of soil and rock mechanics does become blurred.

The main causes of loess slope instability seem to be related to post-construction mechanical disturbance and to blocked drainage at the toe. Rainwater drains vertically to the toe and tends to erode it.

*Although loess and associated soils cover about 10 per cent of the earth's land surface (Scheidig, 1934).

Stability of Soil Slopes   713

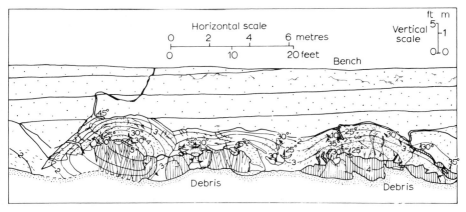

Figure 9.53 Undercutting and slumping at the toe of a highway embankment in loess, Vicksburg, Mississippi (*from* Lutton, 1969).

If there is considerable infiltration at the top of the slope the volume flows at the bottom will undercut the toe and eventually cause basal slumping (Figure 9.53). If drainage is inhibited and water ponds, the undercutting will deepen, diverting the drainage further into the slope and causing slab-failure of the overhanging loess. Failure debris is usually accommodated on a wide shoulder to the highway (see Figure 9.54).

Figure 9.54 Slumping of loess in a cutting, Interstate Highway 20, Vicksburg, Mississippi, U.S.A. Photograph: P. B. Attewell.

Embankments

Embankment problems usually take two forms: first, the stability of a natural slope surcharged at the crest and, second, the design of side slope angles, compaction and drainage of placed fill material. Total shear failure is not the only criterion; excessive settlements caused by a progressive creep-type of movement can be expensive since it is not possible to ballast-up and re-level as in the case of a railway line.

A natural embankment slope that has been correctly designed and drained, and has not been undercut at the toe should create no problems throughout its life. But it is essential that embankments constructed of fill should be compacted to optimum in layers, the moisture content necessary for this to be achieved being derived from laboratory compaction tests. It is not satisfactory to use end-tipped inhomogeneous rubbish. For instance, in England in a far-from-unique example of road widening of the A1 near to Gateshead, end-tipped material, which had originally been placed for the new slow lane of the north-bound carriageway, was removed and more suitable material, compacted and drained, used in its place.

9.14 Protection against coastal erosion

Along coast lines, cliffs are constantly under attack from wave action, particularly during winter months. Tidal wave theory has been discussed by Wood (1969) but of rather greater concern in the present context are those waves that are generated by wind action, particularly when the generating forces are complementary. The size of the waves generated in an open body of water is dependent on the 'fetch', that is, the distance over which the wind operates. The Stevenson formula

$$H = 1.5 \, (f)^{\frac{1}{2}} \qquad (9.40)$$

where H is the trough-to-crest wave height and f is the 'fetch' in nautical miles, has been used in design by British sea defence engineers (Thorn, 1960). Readers may also refer to the work of Wiegel (1964), but the most widely used tables for predicting H for deep water waves are those of Bretschneider (1952) – see also the U.S. Coastal Engineering Research Center (1966). Wind speed is obviously a major parameter controlling wave height, graphical relationships between wind speed and fetch for different wave heights and periods and for different wind durations being presented by Darbyshire and Draper (1963); see also Wood (1969). The Darbyshire and Draper curves provide data for waves in coastal waters, 100 to 150 ft (30–45 m) deep, based on observations around Britain.

In some areas, long swells can only reach the beach after much refraction or by direct oblique incidence. Allowance is usually made by taking the exposure of the beach relative to the dominant swell direction into account. As an example, the greatest 'fetch' on the Yorkshire coast is from due north, but this would imply very oblique incidence. The maximum effective 'fetch' on this coast is from the Danish coast 330 nautical miles away, the corresponding wave height being 26 ft (8.2 m). For Robin Hood's Bay on the North Yorkshire coast (see earlier slope stability problem) the British Admiralty Tide Tables quote a normal range of spring tides as -7.2 to 8.1 ft (-2.19 to 2.47 m) O.D. Wind speeds on 26 ft (7.9 m) wave heights and a 330 nautical mile 'fetch' would be about 35 knots according to the Darbyshire and Draper curves.

It seems likely that where the toes of cliffs are located in deep water, then the cliffs will be called upon to withstand waves of Stevenson height. Where there is a gently sloping foreshore, however, shoaling water will tend to cause the waves to decrease in amplitude to a critical depth, and then to increase considerably in height on the final run up the beach (Stoker, 1957). From his experimental work, Bagnold (1940) found that the mean swash height on steep shingle beaches measured from the lowest level of dry beach was $1.7H$, with H the deepwater wave amplitude. On such beaches, therefore, an amplification factor might be used partially to offset potential amplitude 'losses' due to incident obliqueness and/or refraction effects.

There are three main types of beach material: shingle, sand and mud. Each of these is not only an economic material which, as will be shown later, can play a part in a sea defence programme but they also impose a strong influence on the form of a beach. Most beaches comprise either sand or shingle, or a mixture of both. Globally, sand is much more ubiquitous than shingle and mud will only lie where a beach is protected from strong wave action.

The steepness of a beach increases with increasing coarseness of the deposit size, this being attributable to the relative permeabilities of coarse and fine material and a greater loss of energy with coarse material. Since a coarse material drains much more rapidly than a

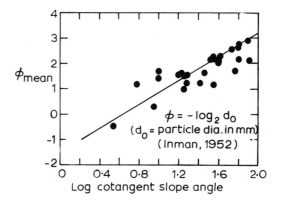

Figure 9.55 Relationship between steepness of beach and mean size of material *(from Doornkamp and King, 1971).*

fine one, the tidal backwash running down a beach is reduced in volume relative to the swash moving up the beach. As a result of this, the gradient on a coarser beach needs to be steeper to maintain equilibrium between the swash and backwash. The relationship between mean diameter of beach material and the slope of the beach shown in Figure 9.55 is taken directly from Doornkamp and King (1971).

A natural beach of sand or shingle is an economic means of absorbing the energy of breaking waves, and so protecting an erodable coastline (Wood, 1969). It is useful briefly to consider beach protection before turning to the problem of coastal slope protection.

Not only will a beach suffer a long-term deterioration, but any changes in a coast line caused by maritime works may change the depositional and erosional conditions at places along the coast quite remote from the source of the disturbance. An engineer has to decide whether the degree and direction of any littoral drift imposes a demand for remedial measures, and he also has to consider whether there is sufficient natural beach material for continuous coastal protection. His decisions, with respect to a particular stretch of coastline, will depend on whether there is an annual loss or gain of material and how this change varies over the years. He will also consider whether or not it is important to maintain a certain shoreline, or whether it may be allowed to change by increase or depletion in a controlled manner.

A dominant direction of littoral drift may be visibly obvious through morphological examination of a coast line, but the drift can only be quantified by using fluorescent tracers or radioactive tracers or artificial beach material made from crushed concrete containing a resin and fluorescent dye. In Britain, reference would also be made to Ordnance Sheets for information on breakwaters and other coastal protection works together with their dates of construction.

There are two fundamental methods of restoring a beach; by arresting littoral drift through the use, for example, of groynes or by artificially re-charging the beach. Groynes, which will usually require anchoring at the base in a suitable bedrock, should be built transverse to the mean direction of the breaking crest of storm waves, which usually means that they will be approximately at right angles to the general coast line with a possible seaward end slightly inclined downdrift (Wood, 1969). Wood has also discussed the spacing and lengths of groynes together with the special problems that they might introduce to the foreshore, particularly the possible development of scour downdrift of the terminal groyne. Beach replenishment usually makes use of existing beach material, feeding at the updrift end of the beach in depths of water that are not beyond the limit of littoral drift, and allowing natural drift increases in the downdrift direction. Intermediate feeding may be necessary to make good any obvious deficiencies that develop over a period of time. On a groyned foreshore, particular attention should be directed to the groyne compartments to ensure that they are filled and remain full.

Although supplementary coastal protection can be achieved by arresting longshore drift and encouraging the build-up and retention of shingle at a cliff toe by a groyne system, the final line of defence will usually take the form of a sea wall.

The sea wall will be designed on the basis of a maximum probable wave height and on the importance of the structure to be protected. A maximum probable wave concept is analogous to the maximum probable flood parameter used in dam design or a maximum probable earthquake. It represents, in effect, an event which statistically may be expected to occur once in so many (say 100) years. It is also important to realize that such a maximum wave might never occur over the design life of a sea wall, but on the other hand it may occur very soon after construction.

Concrete sea walls are one of the obvious protective devices against basal erosion of cliffs, but they can be expensive to construct

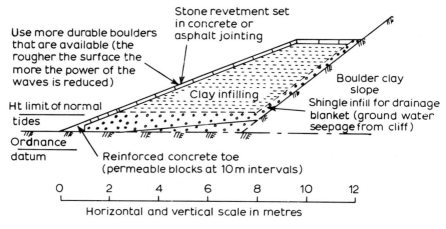

Figure 9.56 Possible sea wall for cliff toe protection (*from* Courchée, 1970).

and they become an exercise more directly in structural engineering (for dynamical analyses, refer again to Wood, 1969). In an engineering geology context, it is worth considering the principles involved in the use of readily available natural materials such as clay, sand and shingle.

Boulder clay, for example, is not normally the first choice for water-retaining structures but it has been used in those instances where it is easily obtainable (such as in the Cow Green dam, Co. Durham, England). When re-moulded and compacted its permeability is probably akin to that of marine and lacustrine clays because of the difficulty of re-working the latter in a short construction period (Courchée, 1970). Clay walls unprotected by revetment have been extensively used in Kent and elsewhere as flood protection barriers in estuaries, but when subjected to wave impact a revetment of stone or concrete blocks set in asphalt jointing is essential (Marsland, 1957; Thorn, 1960).

The design of a clay wall is usually a compromise between a steep front slope (say about 25°) to reduce the materials requirement and a flatter slope for increased stability, more effective dissipation of wave energy, and decreased scour at the toe of the wall due to backwash. Courchée (1970) suggested a design of clay sea wall for protection of the boulder clay cliffs at Robin Hood's Bay (see earlier). In order to achieve fully drained conditions in any clay wall, the clay should be emplaced in layers separated by thinner intercalations of sand and gravel. Using a factor of safety for the clay

Figure 9.57 Coastal protection works at Robin Hood's Bay, North Yorkshire coast, England. Photograph: C. B. McEleavey.

of 1.25 on a maximum angle of repose of 26°, a suitable slope for a seaward face would be 21°. For the shape shown in Figure 9.56, about 13 m^3 of shingle and 44 m^3 of clay would be needed per metre run. It will often be possible to obtain this amount of clay by re-grading the cliffs to be protected; otherwise the operation could be uneconomic.

There are numerous examples in different countries of more substantial concrete retaining and coastal protection walls. At the northern end of Robin Hood's Bay, there has also been long-term instability in the sandstones and micaceous shaly silt beds that stand proud of the beach platform. The progressive loss of land there has been particularly serious because of the proximity of buildings to the cliff edge. Figure 9.57 shows the progress of the remedial works in 1974.

10 ~ Rock Slope Stability

Engineering geologists and geomorphologists are interested in the stability of rock slopes for rather different reasons. The former are concerned with slope *design* and safety while the latter are concerned with the mechanics of natural slope development and change. Obviously there are areas of common interest and it is proposed first of all to consider some of the basic mechanics of natural slope development. At a later stage, some engineering design methods, with particular emphasis on the role of joints in slope stability developed from a consideration of the properties of massive rock, will be developed in some detail. It is recommended that Chapter 6 be read before the later sections of this present chapter are studied.

10.1 Geomorphological classification of rock slope instabilities

A geomorphological classification of the various instability processes as listed by Carson and Kirkby (1972) is given in Figure 10.1. These result mainly from weathering which reduces the surface strength of a rock slope and facilitates the gravitational removal of the debris to the foot of the slope. Joint planes can be rendered progressively less stable through processes of ice wedging and this results in the fall of rather larger blocks. The individual mechanical processes are explained briefly below:

(a) *Slab failure*

In order to appreciate the concepts behind this particular analysis the reader is referred to the chapters in any soil mechanics textbook which deal with retaining wall theory and the implications of Rankine active and passive states of stress. Particularly appropriate is Lambe and Whitman, 1969, pp. 337 to 347, and a brief outline of the theory is also given in Section 2.14 of the present text. It must be stressed that the slab failure model outlined below constitutes only one of several possible approaches to the problem of block

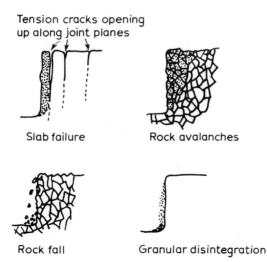

Figure 10.1 Development of rock slope profiles through instability processes (*from* Carson and Kirkby, 1972).

instability. It is introduced here to act as a link with earlier soil mechanics theory but the problem will be reconsidered later in the present chapter in the more specific context of rock engineering.

Consider that in an undisturbed state at depth z in the rock, the vertical pressure σ_v can be taken as γz and the horizontal pressure σ_h as $K_o \gamma z$, where K_o is the coefficient of earth pressure at rest and γ is the bulk unit weight of the rock. Imagine that the rock face is suddenly formed and that the outermost vertical slab acts as a retaining wall with no shear stress being mobilized along its contact face with the rest of the rock. Assume also a constant height of water table h at a relevant point, although any phreatic line will tend in reality to intersect the slope at its toe. This general picture is as shown in Figure 10.2.

Instability will develop under a Rankine active state of stress as the block moves outwards from the face. K_o will tend to a minimum value K_a with quite small displacements and at this stage,

$$\sigma_h = \sigma_3 = \frac{z\gamma}{N_\phi} - \frac{h\gamma_w}{N_\phi} - 2c\left(\frac{1}{N_\phi}\right)^{\frac{1}{2}} + h\gamma_w \qquad (10.1)$$

where c is the rock cohesion, γ_w is the unit weight of the water and $N_\phi = (1 + \sin \phi)/(1 - \sin \phi)$. One effect of the negative cohesive term is to place the rock at the crest of the slope in a state of lateral tension

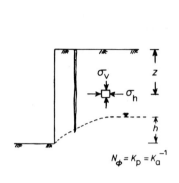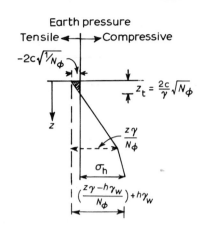

Figure 10.2 Analysis for slab-failure (*developed after* Carson and Kirkby, 1972).

and although this tension will be quite easily accepted by the intact rock between joints, there is a strong likelihood that vertical joints will be opened up progressively away from the free rock face. The depth of the joint face along which the lateral tensile stresses act against the cohesive strength of the rock is shown in Figure 10.2.

From this type of analysis, it can be inferred that the degree of potential instability is in part the relationship between the height of the rock face and the depth of the tensile zone. Nevertheless, stronger rocks that have been deeply gorged can still possess the ability to remain stable.

A number of examples of slab-spalling are quoted by Carson and Kirkby (1972). Typical potential instabilities are to be found at the sandstone walls of canyons, chalk coastal cliffs, and rock cliffs generally that have been oversteepened by glacial action. In the latter context, removal of lateral pressures imposed by ice sheets promotes stress relief and the creation of a distinct set of slabbing fractures.

A tension-relieved surface will be exploited by water in three fundamental ways. First, there will be a seepage path to facilitate further softening in the body of the rock and so reduce the effective strength parameters along any shear surface. Second, water in a tension crack will provide an additional overturning moment. Third, freezing and thawing cycles of water inside the crack will produce a wedging and overturning action, the depth-penetration of the wedging increasing with time. A particular water-filled tension crack problem will be considered later in Sections 10.7 and 10.8.

(b) Rock avalanches

If, as in the previous example, we consider a vertical slab of rock but this time refer to its intact unconfined compressive strength S_c then, theoretically, its critical height before self-weight failure in unconfined compression is simply the ratio S_c/γ. Insertion of any realistic strengths and densities into this ratio will quickly show that the necessary slab heights derived for intact failure greatly exceed actual observed rock slope heights (Terzaghi, 1962). The difference, of course, arises from the fact that a rock mass is replete with weakness planes which obviously preclude any consideration of intact strength. More direct consideration of the strength controls imposed by the presence of weakness planes is to be found later in the present chapter and also in Chapter 6.

As a typical example of a rock avalanche, the 1903 Frank slide on Turtle Mountain in Alberta, Canada may be quoted. The geological situation is sketched in Figure 10.3 and the slide is discussed by Terzaghi (1950) and Coates *et al.* (1965). Hoek and Bray (1974) have also shown how the failure mechanics of the slide may usefully be modelled.

The Frank slide involved about 30 million tons (0.3 million MN) of rock flowing almost as a fluidized mass for a distance of about 2½ miles (4 km) in less than 100 seconds. The maximum slope angle was about 53°, the height of the slope was 2120 ft (646 m) and the limestone dipped 49° into the slope. Day and night temperatures preceding the slide were respectively 70°F and 0°F with a result that there was a cyclic melting of snow during the day and melt-water freezing in fissures at night. It was concluded that the slide was

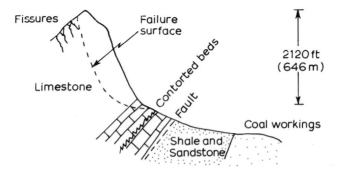

Figure 10.3 Diagrammatic cross-section of the Frank slide, Wyoming, U.S.A. (*after* Coates, 1965).

triggered not only by those factors but also by a combination of general water seepage into the rock, by the presence of compressible layers of shales and sandstones below the limestone at the toe of the slope, by underground coal workings at the toe of the slope area, and by earthquake activity during the preceding months.

Avalanches may also encompass block glides which consist essentially of a few blocks of relatively undeformed material moving at a moderate-to-extremely slow rate (say, less than 2 m per d). The mass moves as a unit along a weaker, planar surface which could be a bed of weak clay or shale beneath a more massive rock. At increasing speeds, the slide material disintegrates to form a rock flow of great speed and dimensions (refer also to Gerber and Scheidegger, 1974).

During the 1925 Gros Ventre slide in Wyoming, U.S.A. (Alden, 1928) an estimated 50 million cubic yards (38.2 million m^3) of material were involved, moving over 1½ miles (2.4 km) in less than 5 minutes. This slide took place along bedding planes inclined at 18° to 21° to the horizontal and we may conclude that major geological structures such as bedding planes, joint systems and fault planes (especially when slickensided), together with foliation planes, could give rise to such slides which are a terminal manifestation of the quasi-static stability analyses which are conducted subsequently in this present chapter.

The concept of a rock avalanche embodies the requirement of a discontinuous assemblage of rock units and the creation of a shear failure surface through rotational or planar instability. It also accommodates Terzaghi's concept of an 'effective joint area' along which shearing takes place so that an 'effective' cohesion c_e for the mass can be expressed as

$$c_e = c\frac{(A - A_j)}{(A)} \tag{10.2}$$

where c is the intrinsic cohesion of the intact units, A is the total area of shear plane and A_j is the total joint area within the shear plane. This simply formalizes the obvious fact that a mass cohesion (and a mass friction angle) will be controlled by the relative path length contributions of the discontinuities and the intact elements. An alternative approach to this discontinuity influence on shear strength is considered in Section 6.20. Lajtai (1969) has pointed out, however, that the maximum shear strength develops only if the 'solid' rock material strength and the joint strength are mobilized simultaneously.

Frictional angles for discontinuous rock seem to vary from about $43° - 45°$ in the loose aggregate state, through up to about $65°$ (Silvestri, 1961) in a densely-packed state, to a maximum of about $70°$ as determined from MacDonald's (1915) many measurements of stable rock slope angles in North and Central America.

(c) Rock fall

In contrast to the more deep-seated failures associated with rock avalanches, the processes under the heading of Rockfall encompass more superficial block shear and rotational failures. Such failure mechanics in a direct engineering stability context are considered in more detail subsequently.

Block movements are triggered by progressive long-term weathering, excessive saturation and water-ponding at a rear face (probably promoted in very wet weather by a build-up of imported clay gouge infilling joints), ice wedging, basal undercutting by stream or river flow*, disturbance by overhanging blocks, seismic disturbance, root wedging from adjacent vegetation, and thermal expansion and contraction.

(d) *Granular disintegration*

It is the more intrinsically weak rocks that break-up through, in part, the solution-weakening of inter-granular cement. Sedimentary rocks having an inter-connecting porous structure can rapidly become saturated through capillary entrainment, but it is only on the drying portion of a wet-dry slaking cycle that the tensile cohesive forces associated with capillarity break-down and fragmentation can take place. Frost action also assists disintegration on this scale. Reference may be made to Chapter 3 for a discussion on some of the processes of breakdown.

(e) *Mechanics of falling rock*

Protection against damage and possible loss of life from falling blocks of rock can take several forms — bolting back of discrete blocks, use of steel mesh along the faces of steep rock slopes (Figure 10.4a), during the design of highways through rock cuttings allocation of

---

*Particularly where a weaker rock, such as shale, underlies a more competent but jointed rock.

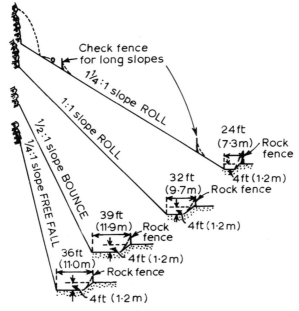

Figure 10.4b Rock fence location and protection from spalling rock on various slopes (*after* Ritchie, 1963).

sufficient 'shoulder' width to absorb debris, or the use of cable fencing at the base of slopes to decelerate rolling stones. A cable fence is spring-suspended to absorb the shock of falling rocks, and is supported by fence posts every 50 feet (15.2 m) or so. If trimming back an oversteep slope is impractical or undesirable, then the first option, combined with attention to the provision of suitable drainage, is probably most desirable, albeit expensive. The second option operates on the principle that if initial momentum can be inhibited by relatively low reaction forces, then unsafe blocks should remain just above the state of limiting equilibrium. For high slopes, it is unlikely that sufficient shoulder could be specified for safety, and it is the fourth option that is worthy of consideration. The design of rock fences has been studied by Ritchie (1963) and reference will be made to his approach.

It would be ideal to use a chain link fence only 6 ft (2 m) high to resist all rolling stones. Ritchie's (1963) idea is that the fence would

---

Figure 10.4a Use of steel mesh to resist movement of blocks of Cretaceous marly limestone down steep slopes which border the Athens to Patras motorway near the town of Megara, Greece. Photographs: P. B. Attewell.

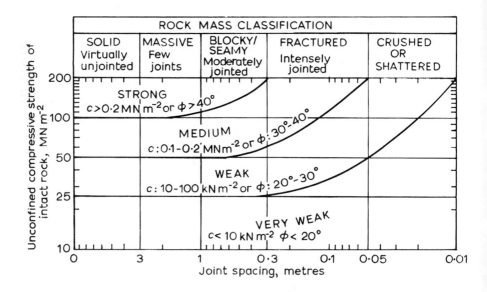

Figure 10.5a Strength classification of jointed rock masses (*after* Bieniawski, 1973)

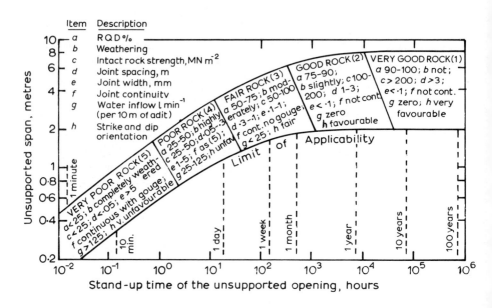

Figure 10.5b Rock mass classification for tunnelling (*after* Bieniawski, 1973)

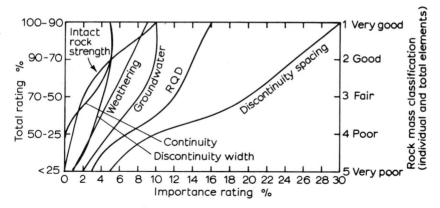

Figure 10.5c Individual importance ratings for discontinuity parameters plotted from Bieniawski's (1973) data. Evaluate importance rating for each element *a* through *h* (see Fig. 10.5b). Then allocate total rock mass classification from the total rating. Support system is then chosen from Table VII in Bieniawski (1973).

be suspended like a curtain wall from a cable, the cable in turn pulling on a compression spring to take up the force of block impact. It is necessary first to consider the mechanics of rock motion, acknowledging the common feature that they all start from a static position by rolling and then either continue to roll, or take several bounces, or take off into flight.

Assuming a spherical rock mass, Ritchie has suggested two cases:—

(a) For a smooth, uniform incline, most rocks remain at rest on the slope up to an inclination of about $1\frac{1}{3}:1$. On steeper slopes, rocks roll at an accelerated rate. If the slope is reasonably smooth, the rocks will continue to roll on the surface irrespective of inclination. Although this condition is never really satisfied in nature it is more nearly approximated the smaller the size of the particles that comprise the slope.

(b) A broken, or rough slope causes the rocks to take-off into trajectory, often after one or two short bounces. For slopes that are inclined at $\frac{1}{3}:1$ or steeper, the rocks seldom touch the slope again once they have left it.

Typical positions for the installation of rock fences are shown on Figure 10.4b.

10.2 Classification of rock masses

Two points will be immediately obvious to the reader: first, that a simple geological classification in terms of igneous, sedimentary, metamorphic, and second, that a strong dependence upon the intrinsic strength of constituent elements (Section 4.1), are factors that may well be irrelevant in a description of the engineering strength of rock masses. Such a description and classification should give due weight to the discontinuous character of the medium and perhaps also to the function that the rock mass will be called upon to perform. The opportunity has also been taken in this section to consider 'classification' beyond its specific relationship to slopes.

General classification

The classification by John (1962) was one of the first to acknowledge the engineering importance of discontinuities. Its merits are discussed in Deere *et al* (1969). Deere himself proposed standardized terminology for different discontinuity spacings in the mass (see Table 10.1) and this contrasts somewhat with the proposed classifications of Coates (1964), Burton (1965) and Coates and Parsons (1968) which make no specific attempt to quantify discontinuity spacing (Table 10.2). In fact, Coates' (1965) classification did rectify this latter point under the 'continuity' heading as in Table 10.3.

Classification with respect to tunnels and underground openings

Terzaghi's (1946) rock mass classification, as adapted and expanded by Deere *et al* (1969) is given in Table 10.4. It uses mining terminology related to discontinuity spacings and is still applicable

Table 10.1 Terminology for joint spacing and thickness of bedded units (*after* Deere et al, 1969).

Descriptive term (joints)	Joint spacing or thickness of beds	Descriptive term (bedding)
Very close	2 inches (50 mm)	Very thin
Close	2 inches – 1 foot (50 mm – 0.3 m)	Thin
Moderately close	1 foot – 3 feet (0.3 m – 0.9 m)	Medium
Wide	3 feet – 10 feet (0.9 m – 3 m)	Thick
Very wide	10 feet (3 m)	Very thick

Table 10.2 Rock mass classification *after* Coates (1964), Burton (1965), Coates and Parsons (1968)

Characteristic	Designation	Notes
Unconfined compressive strength of rock substance	Weak	<5000 lb in^{-2} (35 MN m^{-2})
	Medium strong	5000 lb in^{-2} – 10 000 lb in^{-2} (35 – 70 MN m^{-2})
	Strong	10 000 lb in^{-2} – 25 000 lb in^{-2} (70 – 170 MN m^{-2})
	Very strong	>25 000 lb in^{-2} (170 MN m^{-2})
Pre-failure deformation of the rock material	Elastic	–
	Viscous	At a stress of 50% of the unconfined compressive strength the strain rate >2 microstrains/hour
Failure characteristics of the rock material	Brittle	–
	Plastic	More than 25% of the total strain before failure is permanent (Coates & Parsons take quantitative values of ϵ, the ratio of irrecoverable–to–total strain)
Gross homogeneity	Homogeneous	–
	Inhomogeneous	–
	Inhomogeneous-welded	Layered igneous and metamorphic rocks but with strongly welded layers
Continuity of the rock substance	Intact	No planes of weakness
	Tabular	One group of weakness planes
	Columnar	Two groups of weakness planes
	Blocky	Three groups of weakness planes
	Fissured or seamy	Planes of weakness irregularly disposed – generally associated with faulting
	Crushed	In fragments that would pass a 3 inch sieve (75 mm)

Table 10.3 Quantification of continuity parameter (*after* Coates, 1965).

Continuity	Joint Spacing
Massive	>6 ft (1.8 m)
Blocky	1 ft – 6 ft (0.3 – 1.8 m)
Broken	3 in – 1 ft (75 mm – 0.3 m)
Very broken	<3 in (75 mm)

Table 10.4 Rock loads and classifications for tunnels (*after* Deere et al, 1969)

Fracture spacing cm	RQD			Rock load, H_p metres		Remarks		Rock load H_p metres	Remarks
				Initial	Final				
100 / 98 / 95 / 90 / 75 / 50 / 25 / 10 / 2		Hard Stratified or Schistose	1. Hard and intact	0	0	Lining only if spalling or popping	1 Stable	0–0.5	
			2.	0	0.25 B	Spalling Common	2 Nearly Stable	0.5–1	Few rock from loos with time
			3. Massive Moderately Jointed	0	0.5 B	Side pressure if strata is inclined, some spalling	3 Lightly Broken	1–2	Loosening time
			4. Moderately Blocky & Seamy	0	0.25 B to 0.35 C	Erratic load changes from point to point. Generally no side pressure.	4 Medium Broken	2–4	Immediat stable; bre after few
			5. Very Blocky Seamy & Shattered	0 to 0.6 C	0.25 C to 1.1 C	Little or no side pressure	5 Broken	4–10	Immediat fairly stab later rapid break-up
			6. Completely Crushed		1.1 C	Considerable side pressure. If seepage, continuous support required	6 Very Broken	10–15	Loosens o excavatio local roof
			7. Gravel and Sand	0.54 C to 1.2 C	0.62 C to 1.38 C	Dense			
				0.96 C to 1.2 C	1.08 C to 1.38 C	Side pressure $\sigma_h = 0.3 \gamma$ $(0.5 H_t + H_p)$ Loose			
		Weak and Coherent	8. Squeezing Moderate Depth		1.1 C to 2.1 C	Heavy side pressure Continuous support required	7 Lightly Squeezing	15–25	High pres
			9. Squeezing Great Depth		2.1 C to 4.5 C		8 Moderately Squeezing	25–40	
			10. Swelling		up to 80 m	Use circular support. In extreme cases use yielding support.	9 Heavy Squeezing	40–60	Very high pressures

TERZAGHI (1946)

Notes: 1) For rock classes 4, 5, 6, 7 when above ground water level, reduce loads by 50%
2) For sands (7), $H_{p\,min}$ is for small movements (−0.01 C to 0.02 C); $H_{p\,max}$ for large movements (−0.15 C)
3) B is tunnel width. $C = B + H_t$ = width + height of tunnel. For circular tunnel, $H_t = 0$

STINI (1950)

Note: Loads are for 5 metre wide tu
For L metre wide tunnel
$H_p = H_{p5m}(0.5 + 0.1 L)$

	Rock load H_p m Initial/Final	Side pressure m Initial/Final	Invert pressure m		
				A Stable	Sound
		Little Loosening		B Unstable after long time	
				C Unstable after short time	Sound stratified or schistose (some fissures?)
broken	0/3–4	0/0	0		
				D Broken	Strongly fissured
		Loosening with time		E Very Broken	Fully mechanically disturbed
·oken	3/11–13	0/1	1–2		
		Roof falls, loosening at time of excavation			
·ely	5–10/11–15	2–4/2–6	4		Gravel and sand
ng or ·te	10–13/15–25	4/4	6	F Squeezing	Pseudo-sound rock (properties change with time)
					Some squeezing (genuine rock pressures), small overburden
ng;	15–25/47–75	8/6	12	G Heavy Squeezing	Heavy squeezing. Large overburden
					Swelling
					Silt, clay
›AUMER (1913) and others ·NDEL (1948))				LAUFFER (1958) Note: This classification is correlated with stand-up time	RABCEWICZ (1957) Note: This classification has been used for evaluating feasibility of rock bolt types

to-day. However, Barton et al (1974) have concluded, after allocating an appropriate density to Terzaghi's rock load values, that support design predictions are somewhat conservative. Very useful case history data, relating the design and extent of unsupported openings in rocks of different types having different strengths and jointing characteristics, have been tabulated also by Obert and Rich (1971).

The work of Barton et al (1974a,b) constitutes an important advance in the field of rock classification and support practice for tunnels. The classification, which has wider ramifications in the field of engineering geology generally, is specifically related to Q, a Tunnelling Quality index, where

$$Q = \frac{RQD}{J_n} \times \frac{J_r}{J_a} \times \frac{J_w}{SRF} \qquad (10.3)$$

In this equation:

RQD is as defined in Section 7.7 and ranges from 0 (very poor) to 100 (excellent).

J_n is a joint structure number relating to the number of systematic joint sets and random discontinuities in the rock mass; its value ranges (with some qualification) from 0.5 for massive, almost jointless rock through increasing values for increased joint sets to 20 for crushed, earthlike rock.

J_r is a joint roughness number, being (with some possible modification) from 0.5 for slickensided, planar joints in contact to 4 for discontinuous joints.

J_a is a joint alteration number ranging from 0.75 for a tightly healed, hard, non-softening, impermeable filling, up to 20 when the surface is highly altered, with a thick, continuous zone of clay which precludes rock wall-to-wall contact even when the rock is sheared.

J_w is a joint water reduction factor, having values from 1.0 (relatively dry excavation, with water pressure $<$ 1 kgf cm^{-2} or 98 kN m^{-2}) to 0.05 (very high inflow at pressures $>$ 10 kgf cm^{-2} or 980 kN m^{-2}).

SRF is a stress reduction factor imposed by the capacity of the rock to influence the stability of the excavation as tunnelling proceeds. This factor could be as high as 20 where rock pressures are high and the rock is prone to 'flow' or it could be as low as 2.5 for competent rock near ground surface.

Numerical values of Q were derived by Barton et al for about 200 case histories (strongly biased towards igneous and metamorphic rocks) and were found to vary from 0.001 for very poor squeezing-ground conditions to 1000 for very competent, practically unjointed rock. This wide range of 6 orders of magnitude offers a potentially sensitive categorization of rock competence.

In order to match site evaluations of Q against the type of rock support used, Barton *et al* used a parameter D'_e. This parameter is defined as the ratio of an excavation dimension (span, diameter, or height) to a factor *ESR* (Excavation Support Ratio). *ESR* varies from about 3 to 5 for temporary openings down to around 0.87 for important underground installations such as power stations. By the use of *ESR*, the actual opening is either decreased or increased to a critical equivalent dimension, and it is this latter, D'_e, that is plotted against Q on double log paper. From such case history evidence, the investigators were able to conclude that when $D'_e < 2Q^{0.4}$, no artificial rock support would be needed. If $D'_e > 2Q^{0.4}$, then the actual relationship between the two parameters determines the type of support that might be used. Recommended support systems related to D'_e, Q combinations are listed by Barton *et al* on the basis of their site evidence.

Although the authors do consider support pressures adjacent to tunnels, neither tunnel depth nor the material strength of the rock are directly quantified in their design curves. These parameters may assume more significance when weak rock is excavated near ground surface.

A further rock mass classification and a similar 'rating system' approach to underground opening support requirement is shown in Figures 10.5a,b,c generally after Bieniawski (1973).

Classification with respect to slopes

A descriptive/pictorial classification proposed by Duncan and Goodman (1968) and St. John (1972) is given in Table 10.5. Reference can also usefully be made to the rock slope/rock quality classification listed by Knill and Jones (1965) for the Latiyan dam.

Classification with respect to dam-sites

Classification systems were developed by Knill and Jones (1965) in order to categorize the geology at the Rosieres (Sudan) and the Latiyan (Iran) dam sites. The classifications are based on the logging of borehole cores, *in situ* deformation tests and permeability measurements, geological mapping and seismic surveys. Although the classifications are restricted to three types of rock, gneisses

Table 10.5 Classification of rock masses for surface excavation – continuous (*a* to *e*) and discontinuous (*f* to *j*) rock masses (*after* Duncan and Goodman, 1968 *and* St. John, 1972.)

	Type	Description	Analysis	Figure
(a)	Strong Homogeneous Rock	As strong as concrete. Free from discontinuities	Elastic Continuum. Stress Analysis	
(b)	Weak Homogeneous Rock	Much weaker than concrete. Free from discontinuities	Soil Mechanics techniques. Plastic Analysis	
(c)	Terraced Rock	Horizontal beds of varying strength	Check behaviour of soft zones by Soil Mechanics techniques	
(d)	Ravelling Rock	Rock continuous but breaks upon weathering	Analyze as a cohesionless soil	
(e)	Slumping Rock	Altered or clay-rich rock with very low strength	Soil Mechanics techniques	
(f)	Sheeted Rock	Hard strong rock with planar weakness roughly parallel to natural slopes	2-dimensional analysis. Limiting Equilibrium Calculations. Check for toppling failure	
(g)	Slabby Rock	Hard rock with one strongly-developed set of discontinuities which control the strength	2-dimensional analysis. Finite element analysis may be used. Also Limiting Equilibrium	

Table 10.5 (*continued*)

	Type	Description	Analysis	Figure
(h)	Buttressed Rock	Hard rock with two sets of weaknesses which dominate the rock mass strength	3-dimensional analysis. Limiting Equilibrium by Vector/Stereographic Projection methods. Also Finite Elements	
(i)	Blocky Rock	Hard rock with three or more sets of weaknesses which dominate behaviour	Soil Mechanics if very heavily fractured. Search for dominant features and then as (h)	
(j)	Schistose Rock	Schistocity planes, usually steeply dipping, control mass behaviour	2-dimensional analysis by Soil Mechanics techniques. May also consider toppling failure	

(Rosieres), sandstone and quartzites (Latiyan), they are valuable on account of the inherent relationships between geology, structure, mechanical on-site operations (blasting and formation), and test data.

Knill and Jones (1965) together with many other workers have used an *in situ*/laboratory seismic velocity ratio to express the quality of intactness of the *in situ* rock, the argument being that the ratio of the velocities will tend to unity as the number of joints decreases and the rock quality improves. Questions of rock geometry (boundary conditions) with respect to divergence, and of the relative frequency composition of the pulses are ignored and the laboratory specimens are usually tested in a fully saturated state to maximize intergranular coupling and transmissibility. Deere *et al* (1969) define their *velocity index* as the *square* of the ratio *in situ*/intact compressional wave velocities; the merit in this procedure derives from the fact that the dynamic elastic modulus of a rock is proportional to the square of the compressional wave velocity through the equation:

$$E_{dynamic} = \rho c_p^2 \frac{(1+\nu)(1-2\nu)}{(1-\nu)} \qquad (10.4)$$

where ρ is the density of the rock,
$\quad c_p$ is the P-wave velocity,
and $\quad \nu$ is the Poisson's ratio.

Rock dynamics and wave transmission are considered in Sections 4.9–4.11.

10.3 Character of joints in rock masses

Slope failures in weaker rocks containing a high density of terminated discontinuities will be conditioned by the spatial and orientation density distribution of such discontinuities with respect to a planar or circular surface in the rock mass which just fails to satisfy one of the necessary conditions for limiting equilibrium. Shear failure must then be compounded from 'intact material' failure and shear along pre-existing discontinuities. A rather special problem of this type is considered by Jennings (1970) and the general handling of discontinuities is discussed in Chapter 6.

In harder, sedimentary, metamorphic and igneous rocks, failure is unlikely to occur unless a joint plane (*sensu stricto*) is orientated in such a manner that shear can take place along it with minimal fracturing of intact rock. This implies that the shear strength parameters entered into any stability analysis will be almost exclusively referred to the shear plane itself. Characteristic appraisals of shear plane topography and the significance of the shear process with respect to the shear strength parameters are reserved for a later section.

In the context of *sedimentary rocks*, true joints are continuous fractures through large volumes of rock and have experienced little or no shear displacement parallel to their planes. Any movement normal to the joint surface produces an open joint which may later become filled with degraded host rock material or with imported material. This material often serves to reduce the shear strength of the joint. A consensus of evidence tends to suggest that sedimentary rock joints have a tensile origin of an uplift character (Price, 1966) but the presence of shear markings on some joint surfaces raises one or two questions marks above this hypothesis.

Joints have been classified in a number of ways. *Genetically*, joints may be either systematic or non-systematic. *Systematic joints* occur in sets. Each set, in plan view, comprises parallel or sub-parallel joints which may or may not show the same mutual configuration in a vertical view. They possess planar or slightly curved surfaces and are about at right angles to the upper and lower surfaces of the rock units in which they occur. Most important, they *cut across other joints*. *Non-systematic* joints *do not cut across other joints*, they are often curved in plan and they frequently terminate at the bedding surfaces (Badgley, 1965).

On a *regional* basis, joints will usually be referenced to a dominant structural trend. This will often be determined by the folding, the fold axial plane being the pertinent reference feature. *Longitudinal joints* (or 'release' joints—Billings, 1954) are generally parallel to the fold axes and possess steep dips. *Cross joints* (or 'extension' joints—Billings, 1954) are generally perpendicular to fold axes and are also steeply-dipping. Cross joints generally seem to terminate against systematic joints (Hodgson, 1961) and have more irregular surfaces than systematic joints (Badgley, 1965). *Diagonal joints* occur in paired sets of reasonably symmetrical configuration with respect to the regional longitudinal and cross joints, forming an obtuse angle along the former joint trend and an acute angle about the latter joint trend. They are usually steeply-dipping. The interrelationships of these joints with a fold structure is shown in Figure 10.6 in a diagrammatically rather over-simplified manner.

Joints may further be classified with respect to the bedding, particularly when no dominant structural trend is discernible. *Strike joints* are parallel to the strike of the bedding planes in plan view but cut across the bedding in vertical view. *Dip joints* are perpendicular to the bedding in plan view and cut the bedding in vertical section. Strike joints are therefore analogous to longitudinal joints and dip joints have affinities to cross joints, although they may dip at any angle. As is self evident, *bedding joints* parallel the bedding in both plan and vertical views. Most joints, therefore, are either generally horizontal or generally vertical unless the rocks in which they have been formed have been subjected to subsequent tectonic upturning. The postulated nature of their formation does not permit an infinity of possible joint attitudes as in the manner, for example, of strike slip faulting. There will be no possibility of slope instability in the shear mode along horizontal joints. There will be little possibility of

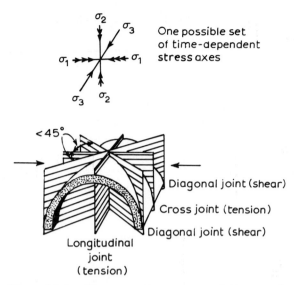

Figure 10.6 Diagrammatic representation of regional jointing with respect to a dominant fold system and to principal stress axes (*after* Badgley, 1965 and *after* Willis and Willis, 1934).

shear instability along vertical or nearly-vertical joints because the joints will barely intersect the free face of the slope (the necessary 'daylighting' condition for instability is discussed in a later section). Subsequent analyses will therefore assume that some condition for the tilting of the bedding has been satisfied.

The character of jointing in *igneous rocks* depends not only upon the type of rock and its mode of formation (intrusive or extrusive) but also upon its age with respect to any adjacent rocks. The late Upper Carboniferous quartz-dolerite Whin Sill in the English Northern Pennines seems to have acquired the same joint orientation density distribution as the Great Limestone of the Lower Carboniferous into which it was intruded (compare Figure 10.7a with Figure 10.7b although the fractures are more closely spaced in the former rock (Figures 10.8a,b)). However, jointing in igneous rocks is often related to shrinkage upon cooling, typical examples being found in volcanic flows and necks. As with desiccation fractures in muds, they are of tensional origin in which stress relief occurs around discrete centres and the final fracturing is hexagonal in plan and columnar in form (Beard, 1959). Probably the best-known locations for such jointing are the Giant's Causeway in Antrim, Northern

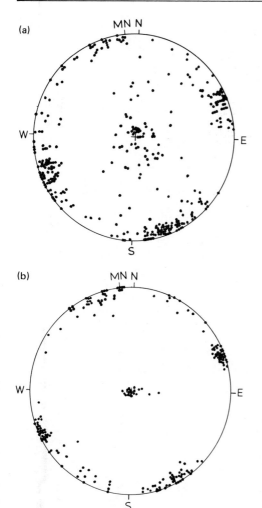

Figure 10.7 (a) Upper hemisphere equal area projection of poles to joint planes in the Whin Sill, English Northern Pennines. (b) Upper hemisphere equal area projection of poles to joint planes in the Great Limestone at Weardale, Northern Pennines, England. *After* Forbes (1971).

Ireland (basalt), Mount Rodeix in the Auvergne region of France (basalt), Devil's Tower in Eastern Wyoming, U.S.A. (phonolite), and Devil's Postpile of Eastern California, U.S.A. (basalt).

In the context of a specific engineering problem, Price and Knill (1967) have interpreted the relationship between different joint sets in the Edinburgh Castle rock, a plug of basalt emplaced in Lower Carboniferous rocks which were subsequently ice-scoured to leave the basalt exposed. Their interpretative classification (Figure 10.9) reflects the cooling and unloading events to which the basalt was subjected.

Figure 10.8 (a) Jointing in the Whin Sill dolerite, High Force on the River Tees, English Northern Pennines. (b) Jointing in the Great Limestone, Northern England. Photographs: (a) P. B. Attewell and (b) A. C. Forbes.

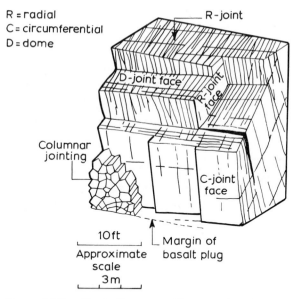

Figure 10.9 Block diagram to illustrate the relationship between various joint sets in the Edinburgh Castle basalt rock (*from* Price and Knill, 1967).

It is not possible to generalise on potential failure modes in igneous rocks. The problem with which Price and Knill were concerned related primarily to potential rotational instability of blocks of basalt from the cliff face. This type of problem and possible combative measures are considered later in the present chapter.

10.4 Engineering recognition of rock failure modes

The main types of rock slope failure requiring an engineering appreciation have been recognized by Hoek and Londe (1974) and by Hoek and Bray (1974) and are sketched in Figure 10.10. Stereoplot representation will become more familiar to the reader as he progresses through the text but it should be apparent at this stage that each of the stereographic projections embodies information on the orientation and dip of the slope face and discontinuity planes and on the poles to those planes. An essential feature, therefore, of any engineering investigation is the acquisition of fundamental structural data and the evaluation of that data in the context of the slope geometry. It is to an understanding of the techniques of evaluation that the subsequent text in this present chapter is devoted.

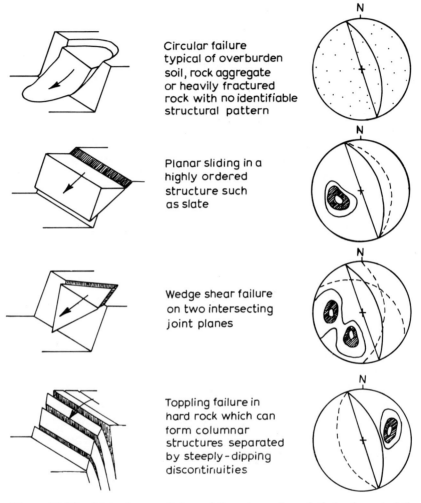

Figure 10.10 Main types of slope failure in rock and their characteristic stereoplots (upper hemisphere projection) (*after* Hoek and Londe, 1974).

Excavation in a soil or rock obviously creates a change of stress in the adjacent ground, the more 'plastic' the material, the more easily does it adjust to the stress imbalance by deformation along shallower but more attenuated strain gradients. Brittle materials develop tension cracks in these zones.

One approach to analysis of the stability of such slopes ultilises finite element methods. Such analyses are usually performed two-dimensionally as a plane strain problem. Some analyses assume

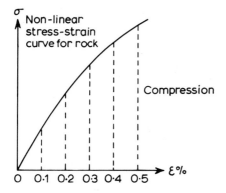

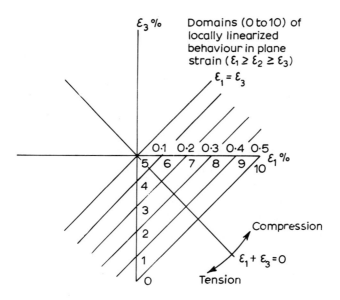

Figure 10.11 Specification for multi-linear finite element analysis.

that, when the pre-existing constraints on any element of rock are changed by the excavation process, the material response remains elastic independent of the magnitude of the stress release or the magnitude of induced strain. Other analyses have a built-in accommodation for stress-strain non-linearity through an element bi-linear or multi-linear facility (see Figure 10.11).

All finite element analyses will carry a pre-specification of regions of different material properties within the plane of the problem, but

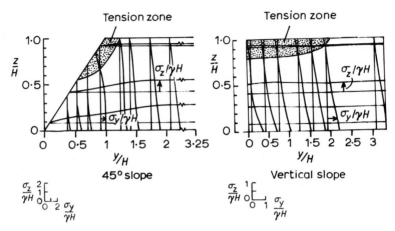

Figure 10.12 Designation of tension at the crest of cut slopes — by finite element analysis (*after* Reséndiz and Zonana, 1969).

within these regions it is usually necessary to assume a condition of homogeneity if not one of isotropy. On the basis that rock masses are replete with *discontinuities,* Heuzé et al (1971) have proposed incorporating a 'joint perturbation' or 'no tension' solution (Zienkiewicz et al, 1968) into the finite element analysis.

Further discussion of the finite element method in continuum mechanics and its detailed application would not be appropriate in this present work for, together with the finite difference method, it constitutes a special and powerful branch of stress analysis. For further specialist information on the technique and with reference to problems in rock and soil mechanics, the interested reader is referred to the books by Zienkiewicz and Cheung (1967) and Desai and Abel (1972). There are a number of papers in the rock and soil mechanics literature which apply the finite element method to soil and rock slope stability analysis (for example, Stacey, 1969) and which emphasize also the importance of correctly modelling the sequence of slope formation (Chowdhury, 1970).

Tension zones at the top of cuts have been studied by Reséndiz and Zonana (1969) using finite element methods. The same technique has been used by Stacey (1970) to study the distribution of horizontal, vertical and shear stress in cut slopes of different angles. Figure 10.12 after Reséndiz and Zonana is compatible with the existence of tension cracking in the ground at the point of intersection of a slip circle with ground surface. On the other hand, Figure 10.13 reproduced from a series of finite element analyses

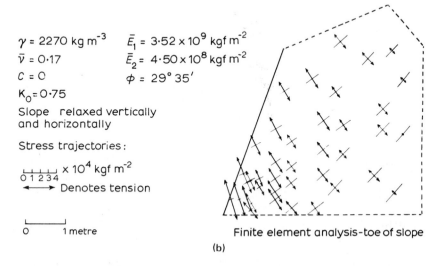

Figure 10.13 (a) Principal stress trajectories in a two-dimensional cross-section of a slope—multi-linear finite element analysis; plane strain condition. (b) Principal stress trajectories at the toe of a two-dimensional slope cross-section—multi-linear finite element analysis; plane strain condition.

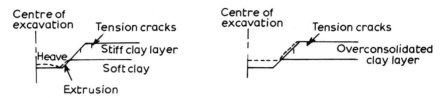

Figure 10.14 Tension crack development (*after* Reséndiz and Zonana, 1969; Wilson, 1970).

undertaken by the present authors shows that when the base of the excavation is accepted into the model, quite strong tensions are also generated at a sharp toe (tension is denoted by an arrow-head). From similar finite element modelling, Bhattacharyya and Boshkov (1970) found that the highest values of maximum shear stress were located near to the toe of the slope and in areas near to the slope face which had been relaxed most drastically. As with surface tension cracks, strong shears and tensions also facilitate moisture ingress and structural weakening which, in their case, can lead to undercutting of the toe and subsequent oversteepening of the slope. In the case of overconsolidated clay shales (such as the Bearpaw and Pierre in North America and Canada) which contain thin bentonite seams, excessive moisture ingress to these seams, and all that this entails in terms of expansion, can prove to be particularly hazardous.

Wilson (1970) also points out that a stiffer stratum is also predisposed to cracking in a tensile manner when it overlies a weaker one. Such cracking may be accompanied by extrusion and basal heave of an underlying softer clay (Figure 10.14). He has also noted vertical cracks as wide as 12 inches (0.3 m) in sandstones ending abruptly at an underlying shale interface (the plastically-deforming shale presumably dissipating the stress concentrations at the bases of the notches). Lateral relaxation of an overconsolidated clay or clay shale (typically reducing from the surface downwards due to increasing frictional restraints on displacement) can also create significant tensions at the surface (Figure 10.14). Hoek and Londe (1974) recommend a continuous monitoring of tension crack openings and possible correlations with seasonal rainfall.

Behavioural predictions must necessarily be based on experience compounded from observations of previous natural events and possibly from some physical or mathematical modelling technique. But although the latter method using the finite element approach is now in quite common use (see, for example, Mahtab and Goodman,

1970) it need hardly be stressed that the quality of the final result can be no better than the realism and quality of the constituent parts of the model.

Bearing in mind the discontinuous nature of rocks, a finite element approach based on limiting failure conditions associated with tensile zones is probably much less satisfactory than either a multi-block contact model using computer graphics as proposed by Cundall (1974) or a detailed analysis of shear failure along joint planes and discontinuities. This latter type of analysis presupposes a knowledge of joint orientation distributions and joint surface properties.

10.5 Surface roughness of joints

The specification of shear strength parameters for discontinuous surfaces is of fundamental concern to engineers who are charged with designing rock slopes or rock foundations for heavy structures (Jaeger, 1971). Where any shear failure would take place along a continuous joint surface any question relating to the shear strength of the intact rock is largely coincidental to the problem in hand. The safest field situation is where the discontinuities are stepped and the individual bounded blocks are locked together. Shear strengths of locked assemblages of regularly-shaped blocks can be assessed by model tests in shear boxes and triaxial cells, from which information on block rotations and mass dilations can also be derived (see, for example, Brown, 1970; Coulson, 1970; Rosenblad, 1971; Walker, 1971; Randell, 1972; Chappell, 1974).

The most dangerous situations for stability are obviously surfaces that are planar, smooth, possibly filled with clay gouge and without any interlocking. If, in addition to these conditions, the surfaces have been subjected in the past to large shear displacements, the stability situation is even more potentially hazardous. The shear strength characteristics for a discontinuity are mobilized in a very similar manner to those pertaining to soils. Reference to Figure 10.15 shows that at small displacements along a rough interface, a high shear strength is mobilized until such time as the asperities are planed down under a sufficiently high normal pressure and a sufficiently large shear displacement. This curve I is analogous to the shear stress-shear displacement characteristic exhibited by an over-consolidated clay (similarly, there will also be a dilational effect and a pore pressure reduction along the shear plane in the presence of

750 *Principles of Engineering Geology*

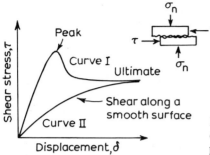

Figure 10.15 Changes in shear strength of a rock discontinuity with shear displacement (*after* Hoek and Londe, 1974).

water). The term 'ultimate' will be reserved for the shear strength condition at large displacements in contrast to the term 'residual' that is used for soils. Curve II is reflective of smooth discontinuity shear and takes a similar form to that of a normally consolidated clay which exhibits no cohesion or dilation.

Plotted in the τ/σ_n plane (Figure 10.16), the initial slope of the peak shear strength curve will be $\phi + i$, where ϕ is the friction angle of the general discontinuity material surface and i is the average angle of incidence of the surface undulations in the direction of shear displacement (Patton, 1966). There may be a small value of cohesion between the contacting surfaces. As the normal stress increases, shearing begins to take place through the asperities, dilation is inhibited and eventually the shear strength of the surface is controlled entirely by shearing through the material high spots. The discontinuity is then at its peak frictional strength ϕ_p. In general, the transition from $\phi + i$ to ϕ_p is smooth. The ultimate friction angle ϕ_{ult} is the angle appropriate to smoothed-down surfaces in contact and will usually be less than ϕ_p.

Figure 10.17 after Patton (1966) shows the trace of a bedding plane in limestone, the actual length of section being 5 ft (1.52 m). A series of second order, generally more steeply inclined, undulations is

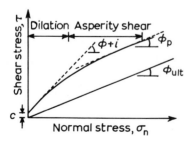

Figure 10.16 Influence of normal stress applied to a discontinuity upon the peak and ultimate shear strength (*after* Hoek and Londe, 1974).

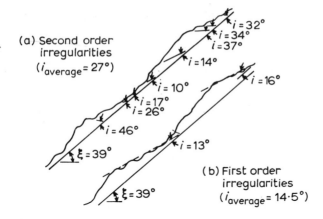

Figure 10.17 5 ft long bedding plane profile in limestone showing the nature of the undulations (*after* Patton, 1966).

superimposed on larger, more shallowly inclined first order undulations. Although all sizes of irregularities can influence the shear strength of the *in situ* rock mass, Patton and Deere (1970) have stated that field studies of natural slopes in sandstone and carbonate rocks have indicated that the shear strength of their discontinuities is more clearly related to the small *i*-values obtained from the first order irregularities. They draw the implication that some natural slope-forming processes such as creep, ice wedging, weathering and cumulative displacement resulting from repeated seismic forces have tended to cause failure of the steeper, smaller irregularities. Only the larger irregularities are left, therefore, to be overcome through larger shear displacements.

Shear box tests in the laboratory on small specimens will provide peak and ultimate friction angles which reflect the presence of the smaller irregularities having large *i*-angles. Even larger-scale *in situ* shear tests would tend to over-estimate the 'first-time' initial friction angle (refer to Figure 10.16) which would actually control large scale shear failure of a slope. The extent of this over-estimate is, of course, critical to the path of safe design but unless joint surfaces are sufficiently exposed in the field to lend themselves to profilograph examination, it is difficult to see how a reduction factor can be applied to a small-scale $\phi + i$ value. It is probably safest not to design on the basis of $\phi + i$, even for low normal pressures, but to enter a value of ϕ_p for all first-time slide analyses. On the other hand, due to pre-consolidation load effects, *in situ* normal pressures on discontinuity faces may be rather higher than might be estimated from overburden loadings and, as a result, rather greater shear stresses might be required to cause movement along those faces. The result of this phenomenon is that measured shear strengths may well be conservative due to inevitable sample disturbance.

Hoek and Londe (1974) argue strongly for the use of field shear box (see Figure 7.38) tests for determining shear strength parameters for discontinuous surfaces and they recommend against the adoption of *in situ* field shear tests 'except under very special circumstances'. This argument is based on the conclusion that the ultimate strength of a sheared surface is independent of the scale of the test since ϕ_{ult} is a dimensionless number, the value of which can be determined, in the absence of any cohesion, by tests on small samples (Londe, 1973). On the other hand, under first time slide conditions when the peak strength of a discontinuity has not been exceeded in the geological past and there is some cohesive contribution to the total shear strength along the discontinuity plane, there will be a scale effect. For realistic design, any cohesion however small should be entered into the stability balance equation unless the foundations of

Table 10.6 Some approximate friction angles and cohesion values for rocks (*from* Hoek, 1970; Hoek and Bray, 1974)

Rock	$\phi°$ (Intact rock)	$\phi°$ (Discontinuity)	$\phi°$ (Ultimate)	c (massive rock) kN m^{-2}
Andesite	45	31–35	28–30	
Basalt	48–50	47		
Chalk		35–41		
Diorite	53–55			
Granite	50–64		31–33	100–300
Greywacke	45–50			
Limestone	30–60		33–37	50–150
Monzonite	48–65		28–32	
Porphyry		40	30–34	100–300
Quartzite	64	44	26–34	
Sandstone	45–50	27–38	25–34	50–150
Schist	26–70			
Shale	45–64	37	27–32	25–100
Siltstone	50	43		
Slate	45–60		25–34	

Intact rock strengths are given in Table 4.1
Comparable soil ϕ values are given in Table 2.9.

Table 10.7 Approximate friction angles for possible joint infilling materials (*from* Hoek, 1970)

Material	$\phi°$ (approximate)
Remoulded clay gouge	10–20
Calcitic shear zone material	20–27
Shale fault material	14–22
Hard rock breccia	22–30
Compacted hard rock aggregate	40
Hard rock fill	38

very large structures, designed for a long life exceeding 100 years, are being analysed. In the latter case, Hoek and Londe (1974) recommend the adoption of an ultimate shear strength. For short-term stability, and based on back-figuring of slope failures, they suggest an envelopment of c, ϕ values which for cohesion ranges from about 200 kN m^{-2} (29 lbf in^{-2}) to zero and for friction ranges from about 15° to about 43°. Some ϕ values for discontinuous surfaces in different rocks are listed in Table 10.6 from Hoek (1970) and from Hoek and Bray (1974).

For infilling materials, Hoek (1970) suggests the friction angles listed in Table 10.7.

Although the shear strength parameters for joint infill material will usually be the ones to take for safe slope design, this need not always be so. Patton and Deere (1970), for example, have quoted results to show that the shear strength along a limestone-clay boundary is lower than the shear strength of the bentonite clay tested alone. Similar results were obtained, apparently, for illite- and montmorillonite-rich soils, the low strengths along the boundary contact being similar to, or rather lower than the residual strengths obtained along a pre-cut failure surface in the soil alone. They also state that the minimum strengths are mobilized with much smaller displacements than are needed in the soil alone and point the relevance of these laboratory test results to a natural contact situation between fault gouge and a slickensided fault surface.

10.6 Discontinuity roughness classification

In interpreting the non-linearity of shear strength envelopes related to rock discontinuities, and in using such an interpretation as the basis of a joint roughness classification, Barton (1973) has proposed an equation:

$$\frac{\tau}{\sigma'_n} = \tan\left[\left(JRC \cdot \log_{10}\left\{\frac{JCS}{\sigma'_n}\right\} + \phi_b\right)\right] \quad (10.5)$$

In this equation, ϕ_b is the 'basic friction angle' for the pre-formed discontinuous surface and it usually falls between 25° and 35°. Symbols τ and σ'_n are, as per convention, the shear stress and effective normal stress respectively along and across the surface. The 'joint roughness coefficient' (JRC) represents a sliding scale of roughness which varies from about 20 to zero from the roughest to the smoothest end of the spectrum. JCS, the 'effective joint wall compressive strength', is taken equal to the unconfined compressive strength S_c of the rock for an unweathered joint but may reduce, according to Barton, to a quarter of this value if the walls of the discontinuity are weathered. A quick indication of weathering potential may be derived in terms of a percentage *alteration index* which is expressed as a ratio of the weight of water absorbed by rock in a quick test to the dry weight of the rock (Hamrol, 1961).

Priest (1975) has been particularly concerned with the character of discontinuity spatial and orientation distributions and has performed a series of field shear box tests on discontinuous surfaces in the Lower Chalk near Oxford, England. The work has a special significance in the context of discontinuity influences on the processes of hard rock machine tunnelling and support in that material. 12 direct shear tests in total were performed on dry surfaces and the same number on wet surfaces, the results being summarized as:

Dry: c'_p mean = 0.07 MN m^{-2}; standard deviation = 0.07 MN m^{-2}
ϕ'_p mean = 34.2°; standard deviation = 8.7°
c'_{ult} mean = 0 MN m^{-2}; standard deviation = 0.01 MN m^{-2}
ϕ'_{ult} mean = 32.3°; standard deviation = 6.4°
Wet: c'_p mean = 0.04 MN m^{-2}; standard deviation = 0.04 MN m^{-2}
ϕ'_p mean = 26.8°; standard deviation = 6.6°
c'_{ult} mean = 0.01 MN m^{-2}; standard deviation = 0.01 MN m^{-2}
ϕ'_{ult} mean = 25.9°; standard deviation = 7.2°

One of the peak and ultimate shear strength envelopes for a dry discontinuity surface is drawn in Figure 10.18. From the results above, it is possible to conclude that there is a substantial difference between the average shear strengths of a dry discontinuity and those of a wet discontinuity and, further, that the peak-to-ultimate

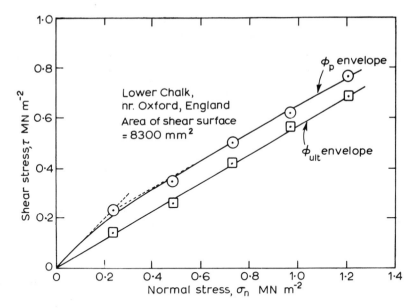

Figure 10.18 Peak and ultimate friction angles for a dry discontinuity surface in Lower Chalk (*after* Priest, 1975).

reduction in friction angle is a little more pronounced in the case of an average wet discontinuity.

Priest has also proposed an undulatory or roughness classification for his chalk discontinuity surfaces. In Table 10.8, A is the maximum deviation from a planar surface and is an expression of trough-to-crest height (see Figure 6.20), l is the length of a discontinuity, and r

Table 10.8 Possible discontinuity surface roughness classification (*after* Priest, 1975)

Trough-to-crest height A mm	λ_1 Undulations with $0 < \lambda < 50$ mm	λ_2 Undulations with 50 mm $< \lambda < \infty$	Curvature classification based on Fookes and Denness (1969) (for 100 mm diameter sample)
$0 < A < 3$	0	0	Planar $\left(0 < \dfrac{l}{r} \leq \dfrac{\pi}{8}\right)$
$3 < A < 6$	1	1	
$6 < A < 9$	2	2	Semi-curved $\left(\dfrac{\pi}{8} < \dfrac{l}{r} < \dfrac{\pi}{4}\right)$
$9 < A < 12$	3	3	
$12 < A < 15$	4	4	curved $\left(\dfrac{\pi}{4} < \dfrac{l}{r} < \infty\right)$
$15 < A < 18$	5	5	

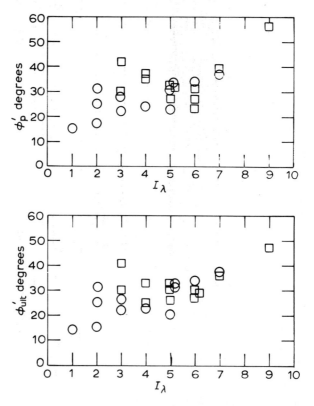

Figure 10.19 Relationship between the roughness index I_λ ($= \lambda_1 + \lambda_2$) and the peak and ultimate friction angles ϕ'_p and ϕ'_{ult} for discontinuity surfaces in chalk (*from* Priest, 1975). Circles represent wet sample data and squares represent dry sample data.

is a mean radius of curvature (see also Fookes and Denness, 1969). He enters two categories of undulatory wavelength, λ_1 and λ_2 where $0 < \lambda < 50$ mm and 50 mm $< \lambda < \infty$ respectively.

If I_λ, a roughness index, is the sum of λ_1 and λ_2, then Priest shows in Figure 10.19 that the friction angle correlates quite reasonably with this index on a rising trend. Chalk often incorporates a secondary mineral content which tends to slurry the rock in the presence of substantial quantities of water. Tunnelling below the water table and rock support in a jointed chalk formation that may also be replete with terminated discontinuities could therefore generate more problems than might usually be associated with tunnelling in rock. The plots in Figure 10.20 indicate that there is

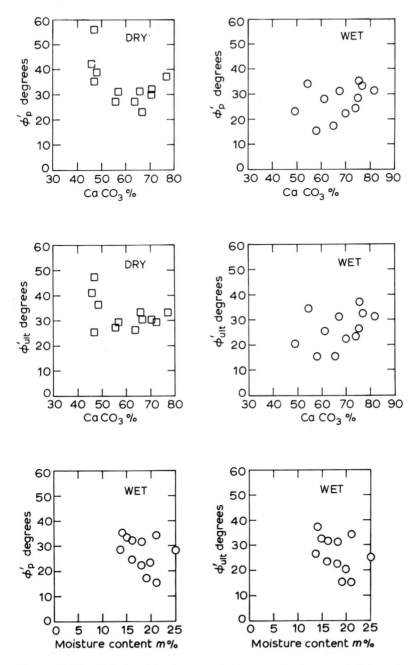

Figure 10.20 Relationships between friction angles (peak and ultimate), $CaCO_3$ and moisture content for wet and dry discontinuity surfaces in chalk (*from* Priest, 1975).

indeed a general reduction in shear strength as the $CaCO_3$ percentage decreases in the wet rock and also as the moisture content increases.

10.7 Planar sliding and the friction cone concept

For the purpose of the subsequent analyses, the simplest and least confusing method of expressing the orientation of a planar or line element is in terms of the clockwise azimuthal orientation of full dip and of the magnitude of full dip itself. The three examples given below in Table 10.9 should make this notation clear.

Table 10.9 Comparisons between two orientation specifications.

Example No.	Dip/strike convention for plane	Pole orientation	Method to be used in this chapter
1	N45°E/60°NW	N45°W/60°NW	315°/60° dip
2	N0°E/85°E	N90°E/85°E	90°/85° dip
3	N45°W/10°SW	N135°W/10°SW	225°/10° dip

Before proceeding further, it is assumed that the reader will have studied Phillips (1971) together with the appropriate sections in Chapter 6 dealing with the representation of planar orientation data on the stereonet.

Refer to Figure 10.21a in which a block of rock, weight W, is prone to slide down a continuous joint plane dipping at an angle ξ (in this example) of 35°. The plane is shown to be striking N80°E and, since the dip is southerly, the pole azimuth or direction of full dip ψ is 170° using the clockwise notation. In this simple example, it is assumed that there is a purely frictional restraint to shear motion down the slope, any cohesion that might exist along the joint plane not being mobilized. If P represents both the pole to the plane and the normal force component of the weight of the block, S is the shear force mobilized up the slope, and ϕ is the friction angle for the joint plane, then simply from the Coulomb equation,

$$\tan \phi = \frac{S}{P} \qquad (10.6)$$

Suppose that $\phi = 40°$, then since $P = W \cos 35°$, the condition for limiting equilibrium with respect to shear is written

$$W \sin 35° = W \cos 35° \tan 40°$$

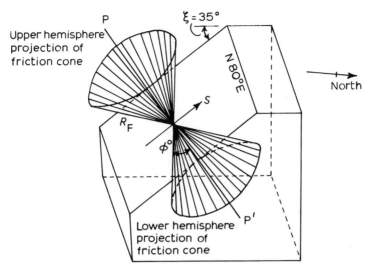

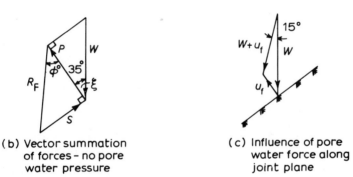

Figure 10.21 Analysis of forces contributing to potential shear failure along a joint plane in rock, with and without porewater uplift force.

or, more usually in terms of the factor of safety,

$$F = \frac{\tan 40°}{\tan 35°} = 1.2$$

This last equation, written in the standard symbols, is

$$F = \frac{\tan \phi}{\tan \xi} \qquad (10.7)$$

and will be referred to later.

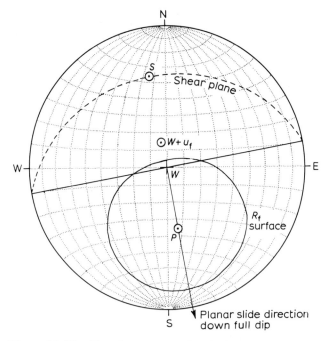

Figure 10.22 Equal area upper hemisphere projection of a stable situation against shear failure along a joint plane with no porewater pressure, but with the development of instability under a porewater force of $W/3$.

The reaction force to the imposed loading is the vector summation of the individual active forces. From the vector diagram in Figure 10.21b it can be seen that a combination of the only active forces S and W creates a reaction force R_F directed at an angle $\phi°$ with respect to the normal force P. The surface of the friction cone, apical angle $2\phi°$ and drawn in Figure 10.21a, defines all possible orientations of R_F for sliding. But in any actual loading situation on that slope for a specified friction angle $\phi°$, provided that the resultant driving vector R acts at an angle less than $\phi°$ to the normal then sliding will not take place in any direction.

In Figure 10.22, the 40° friction cone is projected from a 35° dip-slope on to the equatorial plane. As the slope angle increases and the cone intersection with the plane of projection tends towards the primitive circle, or periphery of the stereonet, the shape of the cone becomes progressively more distorted from a true circle. In all cases however, *the locus of critical equilibrium is generated by striking off*

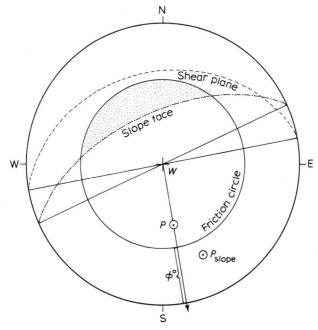

Figure 10.23 Equal area upper hemisphere projection of rock slope stability against shear failure along a joint plane.

the ϕ angles from P on the great circles passing through P as the overlay containing P is rotated through 360° on the base net.

Since the vector representing the weight of the block prone to sliding acts through the centre of the projection in Figure 10.22, it falls within the trace of the friction cone and so confirms the safe condition calculated in equation 10.7.

Suppose that the friction angle had been less than the slope angle. Obviously, the factor of safety would be less than unity and the friction cone in Figure 10.22 would not encircle the weight vector W. *Failure would take place, however, only if the dip angle of the slope face exceeded the dip angle of the shear plane.* In other words, the shear plane is required to 'daylight' in the slope.

This latter condition can be illustrated more fully by referring to Figure 10.23. Since from equation 10.7, and with respect to the shear plane alone, failure will only take place when the dip angle exceeds the friction angle, it is useful to construct on the projection a circle representing an infinity of planes all dipping at ϕ degrees. Provided that this circle intersects both great circles defining the trace of the *slope face* and the *slide plane*, and if the slope face dip > the shear plane dip, the shear instability is confirmed.

The new friction circle is centred on W and is displaced 40° from the periphery of the net. The slope face orientation is, quite arbitrarily, 155°/60° dip. It thereby satisfies the 'daylight' condition for instability. However, as shown in Figure 10.22, the shear plane would need to be steepened by 5°, or the friction angle reduced by 5°, for the second instability condition to be satisfied.

In order to make the problem a little more general, assume now that a porewater uplift pressure acts along the plane of potential shear failure to reduce the stability of the superincumbent block. This pressure is expressed as a force if multiplied by the joint area and then only the normal component of the force is considered, any seepage pressure component down-slope being ignored. The porewater force u_f, having a chosen magnitude of $W/3$, is added vectorially to the block weight (see Figure 10.21c) to produce a resultant $(W + u_f)$ offset angle of 15° from the weight vector. Force $(W + u_f)$ acts sub-normal to the slope, as shown in Figure 10.22, to promote the shear force driving the block down the joint plane and it therefore appears in the upper portion of the upper hemisphere projection outside the safe area of the friction cone (15° dip angle from W). The factor of safety with respect to sliding is now the ratio of the friction angle tangent to the tangent of the angle between the resultant driving vector and the normal to the slide plane, that is:

$$F = \frac{\tan 40°}{\tan (35° + 15°)} = 0.7 \qquad (10.8)$$

It has been pointed out in Chapter 9 that the location of tension cracks plays an important part in the interpretation and analysis of soil slope stability. Tension cracks may also be formed, usually by an opening-up of an existing joint structure, in rock slopes and if these cracks become filled with water there is then an additional driving force to promote shear instability. Hoek and Bray (1974) have considered two special cases of tension crack in planar slide analysis: a crack in the upper surface of a slope, and a tension crack in the slope face. These two possibilities, shown in Figure 10.24, may be analysed by resolving forces graphically and generally on lines similar to those in Chapter 9.

In these two examples of planar sliding, the same slope angle (35°)

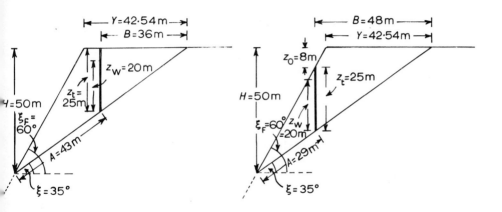

Weight of sliding wedge

$W = (\gamma/2)(HY - Bz_t)$
$\gamma = 22$ kN m^{-3}
$W = 1.34 \times 10^6$ kgf (13.14 MN)

Horizontal force of water in tension crack

$u_{ft} = \gamma_w/2 \cdot z_w^2 = 2 \times 10^5$ kgf (1.96 MN)

Uplift water force

$u_f = \gamma_w/2 \cdot z_w \cdot A = 4.3 \times 10^5$ kgf
(4.22 MN)

By measurement, A.c = 0.46×10^6 kgf
(4.51 MN) at limiting equilibrium ($F = 1$)

But since $A = 43$ m, the necessary cohesion for stability:

$c \simeq 10^4$ kgf m^{-2} (98 kN m^{-2})

Weight of sliding wedge

$W = (\gamma/2)\{HY - B(z_t + z_0) + z_0(B - Y)\}$
$\gamma = 22$ kN m^{-3}
$W = 0.64 \times 10^6$ kgf (6.33 MN)

Horizontal force of water in tension crack

$u_{ft} = \gamma_w/2 \cdot z_w^2 = 2 \times 10^5$ kgf (1.96 MN)

Uplift water force

$u_f = \gamma_w/2 \cdot z_w \cdot A = 2.9 \times 10^5$ kgf
(2.84 MN)

By measurement, A.c = 0.42×10^6 kgf
(4.12 MN) at limiting equilibrium ($F = 1$)

But since $A = 29$ m, the necessary cohesion for stability:

$c \simeq 1.45 \times 10^4$ kgf m^{-2} (142 kN m^{-2})

Factor of safety
$$F = \frac{f + A.c}{S}$$

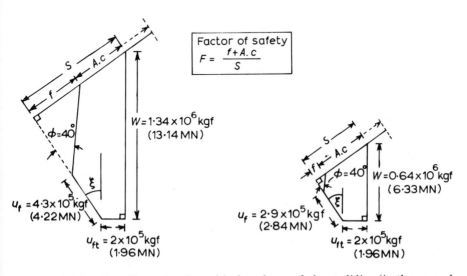

Figure 10.24 Two-dimensional graphical analyses of planar sliding (in the general manner of Hoek and Bray, 1974).

and the same friction angle (40°) have been taken as in the previous example. Quite arbitrarily, both cases are based on the same depth of tension crack and the same depth of water in the crack. In other respects, the slope geometry is the same and in order to quantify the problem, some actual dimensions have been added to the figures.

A factor of safety is derived from a balance of the shear resistance force (sum of the friction (f) and cohesion ($A.c$) contributions) and the downslope shear deforming force (S). The examples indicate that under the groundwater conditions envisaged, frictional resistance alone is insufficient to offset shear failure. In calculating the cohesive stress necessary to achieve balanced stability, the quite critical importance of this shear strength parameter is underlined for it will be noted from the first example that a cohesion of about 15 lbf in^{-2} (1.0 kgf cm^{-2} or 103.4 kN m^{-2}) contributes to approximately the same extent as does the friction along the shear plane. In the second example, the cohesion contribution to stability is relatively much greater. Any additional cohesion improves the factor of safety by extending the $A.c.$ vector further along the broken line. In the calculation of the uplift water force u_f which arises directly from the head of water in the tension crack it is assumed that there is a free drainage facility at the toe of the slope and that an average uplift force may be taken along area A. If drainage is inhibited at the toe, then u_f may increase from the present value up to a maximum of twice that value. The extent of the reduction of the factor of safety, or alternatively the requirement for an increased cohesive contribution, will be noted from the force diagrams.

Planar failures are probably less common in nature than those failures that are conditioned by rock wedges (considered in the next section). As in all rock slope failures, the geology and the structure of the rock mass exert an overwhelming control. Planar failures are probably less likely to occur along continuous joint surfaces in sedimentary rocks because of the near-verticality of major joint sets in regions that have received no significant tectonic disturbance. Planar failure would therefore involve the intersection of solid rock elements with a combination of block shearing, tensile failure and rotation. If the site investigation reveals the presence of a fault daylighting into the slope face, then this situation is probably the most conducive to planar shear failure particularly if there is an uninhibited facility for water ingress to the fault plane.

10.8 Instability on intersecting joint planes

It may already have been suggested from the simple planar slide example that the representational advantages of the stereographic projection with respect to force vector *orientation* might have to be supplemented by ancillary vector addition in the form of force polygons for more complex joint situations. But although force magnitudes cannot be depicted on the stereographic projection it will be shown that, as before, factors of safety can still be extracted by angular measurement even for the rather more demanding rock wedge problem.

In order to develop further the advantages of stereographic representation, consider the possibility of rock wedge sliding under gravity and against frictional restraint along two intersecting joint planes. Two such planes, p_α and p_β, are shown in Figure 10.25 and they are also drawn in Figure 10.26 with their respective poles P_α and P_β. Planes p_α and p_β are orientated 265°/70° dip and 140°/50° dip respectively. The slope face is also shown to be orientated 175°/65° dip. Since the line of intersection $I_{\alpha,\beta}$ of planes p_α, p_β dips at an angle ξ_I of 37° according to a measurement on the projection, the 'daylight' condition for potential shear instability is satisfied. Any wedge displacement that might occur against the compounded friction of planes p_α, p_β, would tend to take place under gravity along edge $I_{\alpha,\beta}$ and within the plane $I_{\alpha,\beta}$:W. The directional sense of this movement, as indicated in Figure 10.26, is N 189° azimuth.

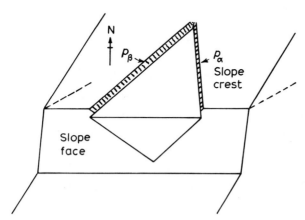

Figure 10.25 Geometry for possible rock wedge failure at the face of a cutting.

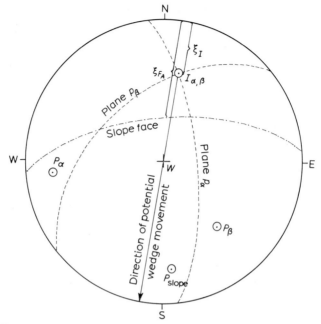

Figure 10.26 Equal area upper hemisphere projection of two intersecting joint planes 'daylighting' at a rock slope. The dip direction of *neither* of the planes fails between the direction of the slope face and the trend of the intersection line, otherwise sliding would have been only along one of the two planes. Note that the apparent dip angle ξ_{FA} for the slope face is measured along the line of wedge movement.

Delimitation of planes P_α, P_β provides information that is really redundant with respect to the location of critical point $I_{\alpha,\beta}$. *The line of intersection of any two planes is defined by the pole to the great circle which passes through the poles to those two planes.* Thus, $I_{\alpha,\beta}$ is the pole to the great circle P_α, P_β as shown on Figure 10.27.

For an appreciation of the stability mechanics, it is useful to recognise that the weight of the rock wedge will be apportioned vectorially as normal forces directed along P_α and P_β and applied to p_α and p_β respectively. But since the constituent shear forces in the Coulomb equation are required to operate in the planes containing $P_\alpha : I_{\alpha,\beta}$ and $P_\beta : I_{\alpha,\beta}$ these planes must be drawn on the projection by arranging first P_α and $I_{\alpha,\beta}$ on the same meridian (great circle) and then repeating the operation for P_β and $I_{\alpha,\beta}$.

Calculation of the factor of safety against shearing down the

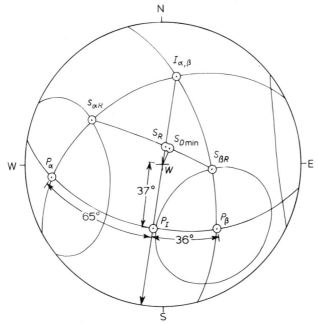

Figure 10.27 Equal area upper hemisphere projection of rock wedge stability analysis.

intersection line $I_{\alpha,\beta}$ requires the resolution, from first principles, of all the normal forces on the two planes which, when modified by the appropriate apparent friction angles and cohesions, are then balanced against the disturbing shear force down the intersection line $I_{\alpha,\beta}$. Hoek and Bray (1974) have noted that each of the normal forces P_α and P_β contributes a component force to the other unless they are orthogonal. We designate the resolved component from P_β along P_α as $P_{\alpha(\beta)}$ and the component from P_α along P_β as $P_{\beta(\alpha)}$. The respective normal forces, in the absence of any pore pressure uplift components, are therefore $P_\alpha + P_{\alpha(\beta)}$ and $P_\beta + P_{\beta(\alpha)}$. One of the first problems is to obtain a general expression for $P_{\alpha(\beta)}$ and $P_{\beta(\alpha)}$ in terms of the dip and dip directions of the two planes.

Let ξ_α, ξ_β be the dip angles of the planes p_α and p_β respectively. Also write ψ_α, ψ_β for the corresponding dip directions of the planes. Resolving P_α into horizontal and vertical components, we have (Figure 10.28):

$$P_{\alpha h} = P_\alpha \cos(90 - \xi_\alpha) = P_\alpha \sin \xi_\alpha$$
$$P_{\alpha v} = P_\alpha \sin(90 - \xi_\alpha) = P_\alpha \cos \xi_\alpha \quad (10.9)$$

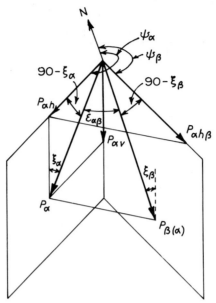

Figure 10.28 Resolution of forces from one direction into another (in the manner of Hoek and Bray, 1974).

Projecting component $P_{\alpha h}$ across to plane p_β,

$$P_{\alpha h \beta} = P_\alpha \sin \xi_\alpha \cos (\psi_\alpha - \psi_\beta) \qquad (10.10)$$

Finally, the projection $P_{\beta(\alpha)}$ is the sum of horizontal component $P_{\alpha h \beta}$ and vertical component $P_{\alpha v}$ in the p_β plane. Hence,

$$P_{\beta(\alpha)} = P_\alpha \sin \xi_\alpha \sin \xi_\beta \cos (\psi_\alpha - \psi_\beta) + P_\alpha \cos \xi_\alpha \cos \xi_\beta$$

or

$$P_{\beta(\alpha)} = k_{\alpha \beta} P_\alpha \qquad (10.11)$$

where

$$k_{\alpha \beta} = \sin \xi_\alpha \sin \xi_\beta \cos (\psi_\alpha - \psi_\beta) + \cos \xi_\alpha \cos \xi_\beta \qquad (10.12)$$

Reversing the process and resolving from p_β to p_α we have,

$$P_{\alpha(\beta)} = P_\beta \sin \xi_\beta \sin \xi_\alpha \cos (\psi_\alpha - \psi_\beta) + P_\beta \cos \xi_\beta \cos \xi_\alpha$$

or

$$P_{\alpha(\beta)} = k_{\beta \alpha} P_\beta \qquad (10.13)$$

It follows that $k_{\alpha \beta}$ must equal $k_{\beta \alpha}$, and substitution of dip and dip direction angles into the appropriate equations will check this point out. Note also that

$$P_{\alpha(\beta)} = P_\beta \cos \epsilon_{\alpha\beta}$$

and

$$P_{\beta(\alpha)} = P_\alpha \cos \epsilon_{\alpha\beta}$$

where $\epsilon_{\alpha\beta}$ is the angle between the two normals to p_α and p_β. Thus,

$$\epsilon_{\alpha\beta} = \cos^{-1} \{\sin\xi_\alpha \sin\xi_\beta \cos(\psi_\alpha - \psi_\beta) + \cos\xi_\alpha \cos\xi_\beta\} \tag{10.14}$$

To these component forces will be added the direct forces due to the weight of the wedge. These are, respectively, $W \cos \xi_\alpha$ and $W \cos \xi_\beta$ on planes p_α and p_β. Writing R_α and R_β for the normal reactions to these forces on the two planes, the constitutive equations are:

$$R_\alpha = W \cos \xi_\alpha - k_{\beta\alpha} R_\beta \tag{10.15}$$

and,

$$R_\beta = W \cos \xi_\beta - k_{\alpha\beta} R_\alpha \tag{10.16}$$

Solving for R_α and R_β, we have that

$$R_\alpha = mW \tag{10.17}$$

and

$$R_\beta = nW \tag{10.18}$$

where

$$m = \frac{\cos \xi_\alpha - k_{\alpha\beta} \cos \xi_\beta}{1 - k_{\alpha\beta}^2} \tag{10.19}$$

and

$$n = \frac{\cos \xi_\beta - k_{\alpha\beta} \cos \xi_\alpha}{1 - k_{\alpha\beta}^2} \tag{10.20}$$

The resisting shear force may therefore be written:

$$S_R = mW \tan \phi_\alpha + nW \tan \phi_\beta + c_\alpha A_\alpha + c_\beta A_\beta \tag{10.21}$$

where c is the cohesion and A is the area of the plane along which the cohesive strength is mobilized.

It is now necessary to calculate the disturbing force, S_D, down the line $I_{\alpha,\beta}$. If ξ_I is the dip of this line to the horizontal, then the disturbing force is simply:

$$S_D = W \sin \xi_I \text{ acting in a direction } \psi_I \tag{10.22}$$

The problem is one of expressing ξ_I in terms of $\xi_\alpha, \xi_\beta, \psi_\alpha, \psi_\beta, \psi_I$. We first resolve from the directions of the normal reactions to each plane into the $I_{\alpha,\beta}$ line. As in equation 10.12, but this time using coefficients $k_{n\alpha I}$ and $k_{n\beta I}$ to resolve from R_α and R_β respectively to $I_{\alpha,\beta}$ we have:

$$k_{n\alpha I} = \sin \xi_\alpha \cos \xi_I \cos(\psi_\alpha - \psi_I) + \cos \xi_\alpha \sin \xi_I \tag{10.23}$$

Similarly,

$$k_{n\beta I} = \sin \xi_\beta \cos \xi_I \cos(\psi_\beta - \psi_I) + \cos \xi_\beta \sin \xi_I \tag{10.24}$$

But since the angle between the normal to the plane and any line (including $I_{\alpha,\beta}$) in the plane must, by definition, be 90° then, by reference to equation 10.14 and noting the cosine function, both equation 10.23 and equation 10.24 must be equal to zero. By manipulation of these two equations it is easily shown that

$$\tan \xi_I = - \tan \xi_\alpha \cos(\psi_\alpha - \psi_I) = - \tan \xi_\beta \cos(\psi_\beta - \psi_I) \tag{10.25}$$

For complete solution of equation 10.25 it is necessary to solve for the *directional* parameter ψ_I of the intersection line $I_{\alpha,\beta}$. From equation 10.25 we find that

$$\tan \psi_I = \frac{\tan \xi_\beta \cos \psi_\beta - \tan \xi_\alpha \cos \psi_\alpha}{\tan \xi_\alpha \sin \psi_\alpha - \tan \xi_\beta \sin \psi_\beta} \tag{10.26}$$

(note that both ψ_α and ψ_β must lie within the range of ±90°)

It is now possible to write down the complete expression for the factor of safety against wedge sliding knowing that most of the elements of the equation are solved:

$$F = \frac{mW \tan \phi_\alpha + nW \tan \phi_\beta + c_\alpha A_\alpha + c_\beta A_\beta}{W \sin \xi_I} \tag{10.27}$$

But it will also be obvious that the weight of the wedge must be evaluated unless the cohesion terms can be suppressed or absorbed into an apparent friction angle. The expressions for *apparent friction angles* ϕ_α^* and ϕ_β^* are:

$$\tan \phi_\alpha^* = \frac{c_\alpha A_\alpha}{P_\alpha} + \tan \phi_\alpha$$

and (10.28)

$$\tan \phi_\beta^* = \frac{c_\beta A_\beta}{P_\beta} + \tan \phi_\beta$$

In order to quantify P_α and P_β in terms of the weight vector W it is first necessary, on Figure 10.27, to construct a great circle passing through P_α and P_β and to note the point P_I at which this plane cuts the line of intersection of planes p_α and p_β. Measurement of the angles between P_α and P_I and between P_β and P_I on the stereographic projection and construction of an appropriate force diagram (Figure 10.29) allows equations 10.28 to be re-written in terms of W, vis:

$$\tan \phi_\alpha^* = \frac{c_\alpha A_\alpha}{W \cos 37° \cos 65°} + \tan \phi_\alpha = \frac{c_\alpha A_\alpha}{0.337W} + \tan \phi_\alpha$$

$$\tan \phi_\beta^* = \frac{c_\beta A_\beta}{W \cos 37° \cos 36°} + \tan \phi_\beta = \frac{c_\beta A_\beta}{0.646W} + \tan \phi_\beta$$

(10.29)

The influence of W, c and A on the problem in hand will be considered later, but at this stage let us assume that the first term in

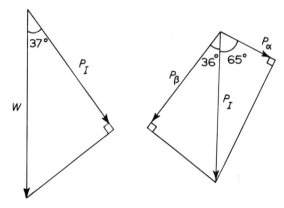

Figure 10.29 Supplementary force triangles for resolution of the weight of the wedge.

equations 10.29 can be 'nested' in apparent friction angles ϕ_α^*, ϕ_β^* by estimation. Suppose that ϕ_α and ϕ_β, as measured in the field or derived from realistic laboratory tests, are 35° and 30° respectively. Assume also that the cohesion contributions, again as measured or estimated, serve to increase ϕ_α and ϕ_β to ϕ_α^* and ϕ_β^* values of 38° and 34° respectively. Given these latter two values, we can now *calculate* a factor of safety.

First, from equation 10.26, and with values $\xi_\alpha = 70°$, $\xi_\beta = 50°$, $\psi_\alpha = 265°$, $\psi_\beta = 140°$, ψ_I is calculated as 10.88°. This azimuthal direction for plane intersection line $I_{\alpha,\beta}$ checks out on the stereographic projection (Figure 10.27). Next, using this ψ_I value, equation 10.25 produces an angle of 36.93° for the dip of $I_{\alpha,\beta}$. Using this ψ_I angle together with the *m* and *n* parameters specified in equations 10.19, 10.20, the factor of safety for the wedge according to equation 10.27 is

$$F = \frac{m \tan \phi_\alpha^* + n \tan \phi_\beta^*}{\sin \xi_I}$$

$$= \frac{0.48415 \tan 38° + 0.73625 \tan 34°}{\sin 36.93°} = 1.456$$

Consider, now, how the factor of safety can be assessed directly from further construction on the stereographic projection. It is first necessary to set off friction cones, having semi-apical angles of 38° and 34°, about P_α and P_β. The lines of intersection of the upper inclined surfaces of each cone with the planes (great circles) containing $P_\alpha : I_{\alpha,\beta}$ and $P_\beta : I_{\alpha,\beta}$ will define respectively $S_{\alpha R}$ and $S_{\beta R}$ on the projection. Furthermore, the plane $S_{\alpha R} : S_{\beta R}$ which is generated by rotating $S_{\alpha R}$ and $S_{\beta R}$ to a common meridian defines the resultant $S_{\alpha R} + S_{\beta R}$ and represents a condition of limiting equilibrium for a factor of safety of unity.

Just as a great circle passing through poles P_α and P_β cuts through the line of intersection of planes p_α and p_β at the point marked P_I, in a similar way, the great circle containing $S_{\alpha R}$ and $S_{\beta R}$ cuts the same line of intersection at S_R. It then follows that the 'effective' friction angle ϕ_i^*, compounded by the two planes as the wedge is constrained to move downslope, will be the angular dispacement between P_I and S_R, which is equal to 48°. In a similar manner, the dip angle ξ_I of the intersection line $I_{\alpha,\beta}$, previously calculated at 36.93° and measured at 37°, is clearly the same as the angular displacement between P_I and W. Thus as with equation 10.7

$$F = \frac{\tan \phi_I^*}{\tan \xi_I} \qquad (10.30)$$

which, from measurement on the projection, produces a factor of safety of 1.47.

The calculated and the 'measured' factors of safety are so close to one another that, in view of the speed of construction on the stereonet, the geologist is recommended to use that technique.

There are several observations that can usefully be made. With an upper hemisphere projection, the directional sense of gravitational down-dip failure is *from* $I_{\alpha,\beta}$ *to* W (with a lower hemisphere projection, the reverse is true). If W is located within one or both of the friction cones or, as in the present example, within an *effective* friction zone of the projection, then there can be no shear failure and the factor of safety must exceed unity. Strictly, the friction cones need not be drawn for the example as discussed, since all that is required is for ϕ_α^* and ϕ_β^* to be struck off along each great circle $P_\alpha : I_{\alpha,\beta}$ and $P_\beta : I_{\alpha,\beta}$. Points ϕ_α^*, ϕ_β^* need only be defined on the $I_{\alpha,\beta}$ side of P_α, P_β, although for complete delimitation of the safe zone these points may be fixed for the dip-side of the wedge as shown in Figure 10.27. This latter construction could conceivably be of some value under circumstances which might promote sliding up the line of intersection. Where one or both of the friction cones overlap the periphery of the projection, further construction is required to delimit the safe zone on the down-dip side. The following procedure, which relates to the current example (Figure 10.30) may be applied when both cones intersect the periphery.

It is first necessary to locate the mirror image of the plane $P_\alpha : I_{\alpha,\beta}$ and to mark the point a at which this plane intersects the friction cone about P_α. Next, place points a and b (the latter denoting the intersection of plane $P_\beta : I_{\alpha,\beta}$ with the down-dip surface of the P_β friction cone) on the same great circle and intersect the primitive circle at point c. Transfer c across the diameter and complete the plane c:a on the same great circle. The lower boundary of the safe zone is defined by plane acb.

A further matter of interest concerns the necessary conditions that must be imposed to promote failure against what is otherwise

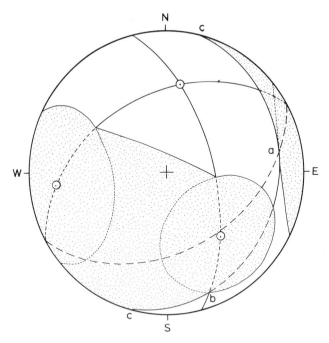

Figure 10.30 Construction for delimitation of stable areas on the upper hemisphere projection.

gravitational stability. The minimum force for failure will, in fact, be the one which acts normal to plane $S_{\alpha R}:S_{\beta R}$ through the point $S_{D\,min}$. Letting the angular distance between $S_{D\,min}$ and W be χ (equal to 10° in Figure 10.27), it follows that the angular sum of ξ_I and χ must be the same as angle ϕ_I^*. It will be noted that the azimuthal orientation (26°) of the minimum force vector is not concordant with the strike of the intersection line of the two planes. As a general comment, the two lines will be compatible only when the two coefficients of friction are the same for both planes and the wedge is acted on by its own weight alone. The angular difference is in most cases small, particularly under self-weight conditions and a factor of safety only a little greater than unity.

Possible external forces acting on the block may take several forms. Seismic forces due to blasting or an earthquake may affect stability, and if the acceleration can be predicted, the influence of these forces can be expressed on the stereographic projection. Similarly, stabilizing forces exerted on the block by cable anchors or rockbolts, or disturbing forces applied by foundation loads or, for example, dam abutment thrusts, may be analytically accommodated on the projection. But, as indicated earlier, probably the most common external force is that due to porewater influences creating an uplift effect at the block. The analytical approach in the wedge situation is little different from that considered for the planar slide problem, but the procedures are rather tedious.

It is really only possible to begin analysing for porewater pressure effects if the areas of the joint: wedge contact interfaces are known. But if these analyses are to be performed using linear measurements in the field together with inter-planar angular measurements from the stereographic projection, then this same information may well be utilized to *calculate* a more accurate apparent friction angle ϕ^*. Given the lengths and areas of the sides of the sliding tetrahedron together with the interfacial angles, the weight of the wedge can be evaluated from the calculated volume and the measured density of the rock. In this way, equation 10.7 becomes tractable. The detailed procedures have been very clearly set out by Hoek *et al* (1973) and would be followed by the engineer responsible for rock slope design.

Since angular measurements on the stereographic projections are involved, and since the principles associated with those measurements should be known by geologists, it is useful to consider the cohesion and pore pressure question further. The symbolism and procedure of Hoek *et al* (1973) will be applied as far as possible to the present problem in order to facilitate any check-out against their original work.

We first re-sketch the wedge from Figure 10.25 and insert a possible tension crack which, if filled with water, would enhance the shear forces driving the rock down-slope. On the wedge drawn in Figure 10.31 have been numbered the intersection lines, or edges, formed by adjacent planes. These lines are also marked on the Figure 10.32 projection containing the complete family of planes involved in the problem. Dips and dip directions of these lines, as measured from Figure 10.32, are listed in Table 10.10. In order to be able fully to describe the wedge in terms of both volume and shape, the angles between edges 1 to 9 must be known and from the initial limited information on slope height and slope crest distance to the tension crack (Figure 10.31) both of the above conditions can be satisfied. Clearly, if all inter-edge angles are calculated or measured, some of the information will be redundant. Nevertheless, by completing the measurements on the stereographic projection, an

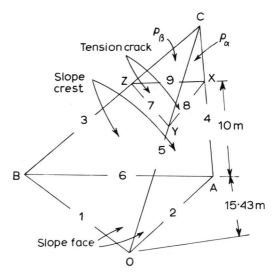

Figure 10.31 Numbering of intersection lines on the wedge.

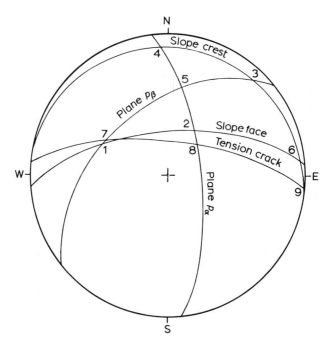

Figure 10.32 Upper hemisphere projection of the planes constituting the sliding wedge problem.

Table 10.10 Dip directions, dips and lengths of the plane intersection lines in the wedge problem.

Line	Planes	Direction	Dip	Length
1	Slope face : p_β	295°	47°	20.5 m
2	Slope face : p_α	31°	59°	18 m
3	Slope crest : p_β	43°	9°	25 m
4	Slope crest : p_α	356°	10°	17 m
5	p_α : p_β	9°	37°	31 m
6	Slope face : Slope crest	84°	3°	17.5 m
7	Tension crack : p_β	299°	47°	7 m
8	Tension crack : p_α	44°	65°	6 m
9	Tension crack : Slope crest	95°	1°	6 m

accuracy check can be made to ensure that the internal angles of each constitutive triangle sum to 180°.

Angular measurements between poles (lines) have been made earlier on the stereographic projection. In the present instance, as before, lines 1 to 9 are placed in pairs and in turn on the same great circle and the included angles measured.

The projections in Figure 10.33 are developed progressively so that wedge face areas, porewater forces, and the wedge weight can be calculated. From a direct measurement on the slope exposure along the OA line (18 m) and from angular measurements of $\widehat{24}$ (56°) and $\widehat{25}$ (27°) on the projection (Figure 10.32), point C and the lengths of lines 4 (17 m) and 5 (31 m) are defined. Angular measurements of $\widehat{15}$ (53°) and $\widehat{35}$ (41°) on the stereographic projection set off respectively from O and C allow point B to be defined in the triangle OBC and hence the lengths of lines 1 (20.5 m) and 3 (25 m) to be similarly represented. Measurements of angles $\widehat{46}$ (90°) and $\widehat{43}$ (46°), if then developed from line 4 (Figure 10.30c), provide a check on the line 3 length.

Figure 10.33d shows how the position X of the tension crack from the slope crest A (10 m as measured in the field) can be marked off on line 4 (AC). Since line 3 (BC) length is now known, a set-off angle $\widehat{49}$ (98°) measured from the stereographic projection fixes both the length of line 9 (ZX = 6 m) and that of CZ (8.5 m). In order to define the shape of the tension crack, the p_α plane projection (Figure 10.33a) is reproduced in Figure 10.33e. From point X, line 8 (XY) is set off at an angle $\widehat{48}$ of 64° and measures of 6 m in length. The length of CY is 11 m. From point Y just determined, line 7 is

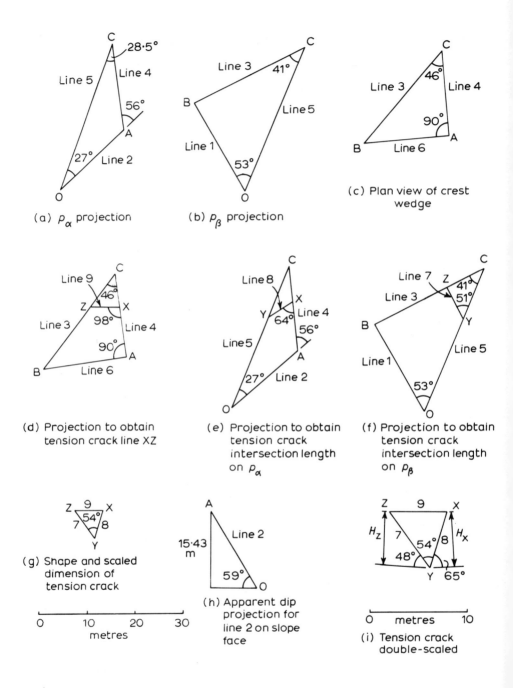

Figure 10.33 Generation of wedge-face and tension crack projections.

Rock Slope Stability   779

projected at an angle $\widehat{57}$ to 51° to intersect line 3 at point Z on the p_β plane projection in Figure 10.33f. Line 7 is found to be 7 m long and CZ is 8.5 m long. Having derived the side lengths of the tension crack and after reading angle $\widehat{78}$ (54°) from the stereographic projection, Figure 10.33 g can be drawn. Measurements of angles $\widehat{89}$ (73°) and $\widehat{79}$ (53°) check out on the stereographic projection.

For the calculation of porewater force, the average height of line 9 above the base (Y) of the tension crack must be known. To facilitate the calculation, Figure 10.33i is drawn to double-scale, the dips of lines 7 (48°) and 8 (65°) being taken from Figure 10.32. The height of point X is 6 sin 65° (5.4 m) and that of point Z is 7 sin 48° (5.2 m). It follows that the average depth of the tension crack is 5.3 m.

Areas of the constituent triangles are calculated from the formula: area of triangle = one half the product of any adjacent sides multiplied by the sine of the included angle. Hoek *et al* (1973) have also noted that the volumes of tetrahedra ACBO and CXYZ are calculated from the formula: volume = one sixth of the product of any three adjacent sides multiplied by a factor K, where in the present case,

$$K = (1 - \cos^2 \widehat{34} - \cos^2 \widehat{35} - \cos^2 \widehat{45} + 2 \cos \widehat{34} \cdot \cos \widehat{35} \cdot \cos \widehat{45})^{1/2}$$
$$= 0.311$$

Based on these relationships, the relevant areas and volumes are calculated in Table 10.11.

Using the information from Table 10.11, the forces acting on the wedge are now calculated. Assuming a unit weight of 24.3 kN m^{-3}

Table 10.11 Calculation of areas and volumes of block

Plane p_α :	area ACO	= ½ x 17 x 31 x sin 28.5°	= 125.7 m²
	area CXY	= ½ x 7 x 11 x sin 28.5	= 18.4 m²
	area AXYO	= 125.7 − 18.4	= 107.3 m²
Plane p_β :	area BCO	= ½ x 25 x 31 x sin 41°	= 254.2 m²
	area CZY	= ½ x 11 x 8.5 x sin 41°	= 30.7 m²
	area BZYO	= 254.2 − 30.7	= 223.5 m²
Tension crack :	area XYZ	= ½ x 6 x 7 x sin 54°	= 17 m²
Whole block :	volume ACBO	= ⅙ x 25 x 17 x 31 x 0.3115	= 684 m³
Above crack :	volume CXYZ	= ⅙ x 7 x 11 x 8.5 x 0.3115	= 34 m³
Below crack :	volume AXZBOY	= 684 − 34	= 650 m³

(155 lb ft^{-3}), the weight of the total wedge ACBO = 684 × 2480 = 1.70 × 10^6 kgf (17 MN). Similarly, the *weight of rock wedge on the dip-side of the tension crack* (AXZBOY) = 650 × 2480 = 1.61 × 10^6 kgf (15.8 MN). The down-dip driving force due to a *water pressure in the tension crack* XYZ (and assuming a linear increase in pressure over the average vertical depth of 5.3 m) = ⅓ × 1000 × 5.3 × 17 = 3.00 × 10^4 kgf or 0.3 MN (note that the force of water is the pyramidal volume formed by pressure at Y on area XYZ). The *uplift force due to water* pressure on area AXYO (p_α) = ⅓ × 1000 × 5.3 × 107.3 = 0.189 × 10^6 kgf (1.9 MN). Similarly, the uplift force due to water pressure on area BZYO (p_β) = ⅓ × 1000 × 5.3 × 223.5 = 0.395 × 10^6 kgf (3.9 MN). It should be noted that the pressure in the tension crack is assumed not to dissipate down the contact interfaces AXYO and BZYO.

We are now in a position to consider the cases of shear stability with and without a water-filled tension crack. The stability assessment of wedge ACBO permits a re-appraisal of the earlier analysis and estimate of apparent friction angles ϕ_α^* and ϕ_β^*. The second assessment, that of wedge AXZBOY, will be a new one which will certainly reduce the factor of safety.

Apparent friction angles are calculated in Table 10.12 according to equation 10.28. Cohesions are entered as $c_\alpha = c_\beta = 0.1$ kgf cm^{-2} (1.42 lbf in^{-2} or 9.81 kN m^{-2}). Appropriate wedge weights and contact areas are taken from Table 10.11.

It will be noted from Table 10.12 that the ϕ_α^* and ϕ_β^* angles, estimated quite arbitrarily for the earlier stability projections, were marginally pessimistic. If the friction cones are re-plotted on the projection in Figure 10.34, the following factor of safety results for the whole wedge:

$$F = \frac{\tan 52.5°}{\tan 37°} = 1.73$$

A slightly lower factor of safety would apply with respect to the lower wedge, but since there is a difference of only about 0.5° in the apparent friction angles, it would be difficult to resolve a ϕ_i^* angle that differed from 52.5°. What should be noted, however, is that the somewhat tedious calculations have had the effect of raising the factor of safety from its earlier 'estimated' level to a value that now exceeds 1.5. A factor of safety of 1.5 would normally be regarded as

Table 10.12 Calculation of apparent friction angles

Wedge condition	Equation	Apparent friction angle
No tension crack	$\tan \phi_\alpha^* = \tan 35° + \dfrac{1000 \times 125.7}{1.7 \times 10^6 \times \cos 60°}$	$\phi_\alpha^* = 40.38°$
	$\tan \phi_\beta^* = \tan 30° + \dfrac{1000 \times 254.2}{1.7 \times 10^6 \times \cos 40°}$	$\phi_\beta^* = 37.69°$
With tension crack	$\tan \phi_\alpha^* = \tan 35° + \dfrac{1000 \times 107.3}{1.61 \times 10^6 \times \cos 60°}$	$\phi_\alpha^* = 39.81°$
	$\tan \phi_\beta^* = \tan 30° + \dfrac{1000 \times 223.5}{1.61 \times 10^6 \times \cos 40°}$	$\phi_\beta^* = 37.18°$

the minimum level at which major concern for the stability of the slope would be abated.

It is now necessary to consider the influence of any water pressure in the tension crack and possible uplift effects at the base of the wedge. Figure 10.35 is used for this analysis and the point T, representing the pole to tension crack, is derived from the appropriate great circle in Figure 10.32. This crack dips at 70° in a direction of 185° and it follows that the water force vector will be directed through T *upwards* at a dip angle of −20°. Uplift forces $u_{f\alpha}$ (= 0.189 x 10^6 kgf or 1.8 MN on plane p_α) and $u_{f\beta}$ (= 0.395 x 10^6 kgf or 3.9 MN on plane p_β) act against the normal wedge forces P_α and P_β respectively and are located diametrically across the projection from poles P_α and P_β. From measurement on the great circle containing $u_{f\alpha}$ and $u_{f\beta}$, it is found that the included angle formed by these two uplift force vectors is 102° and from construction of the ancillary triangle of forces in Figure 10.36a, the resultant force u_{fR} is found to be 0.4 x 10^6 kgf (3.9 MN) acting at an angle of 75° to $u_{f\alpha}$ in the $u_{f\alpha}$: $u_{f\beta}$ plane. By placing u_{fR} and T on the same great circle (Figure 10.35) the included angle is found to be 35° and from the construction in Figure 10.36b, the resultant uplift force u_{fT} is found to be 0.43 x 10^6 kgf (4.2 MN) directed 2.5° from the u_{fR} line in the plane u_{fR} : T. The final resolution of forces is between u_{fT} and W. Measuring along the appropriate great circle in the usual way produces an included angle of 39° and, from Figure 10.36c, the effective weight W_e of the lower wedge is 1.39 x 10^6 kgf (13.6 MN) and acts at an angle of 11° to W in a direction 355°. If W_e and P_I are

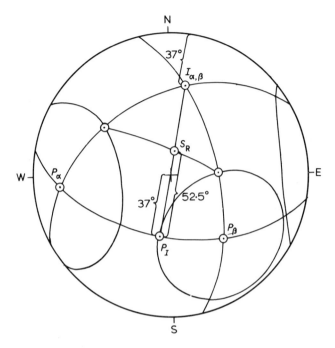

Figure 10.34 Equal area upper hemisphere projection for determination of factor of safety against wedge sliding on the basis of calculated apparent friction angles.

placed on the same great circle, the factor of safety is defined as

$$F = \frac{\tan 51.5°}{\tan 48°} = 1.13$$

This value is unacceptably low and immediate measures would have to be taken to raise it, either by drainage, grouting or, more likely, by cable or bolt anchors.

The factor of safety of the slope can be raised by increasing the normal compressive force on the shear planes. By post-tensioning cable anchors, the W_e vector can be drawn closer to W within the friction 'cone' envelopment. The method, described by Hoek *et al* (1973) involves first of all the measurement of the effective friction angle between P_I and the plane containing $S_{\alpha R}$ and $S_{\beta R}$. If the overlay to the stereonet is rotated about W, a family of 'effective' friction angles ϕ_e^* is generated as the point P_I traverses different great

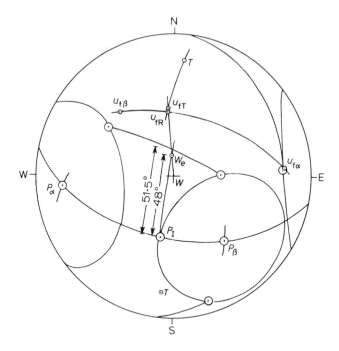

Figure 10.35 Equal area upper hemisphere projection for the analysis of the water pressure influence upon wedge stability.

circles. Now, for a particular great circle and measured value of ϕ_e^*, and for a factor of safety of 1.5, the appropriate angular relationship from P_I along that same great circle will be $\tan^{-1}(\tan \phi_e^*/1.5)$. Thus, in this way a 1.5 factor of safety locus can be expressed on the projection in the manner of the broken line in Figure 10.37. Irrespective of the cable positioning, the resultant of the cable tension CT and the effective weight W_e must lie on the $F = 1.5$ locus.

There is a range of options for the eventual location of CT, but in order to demonstrate the procedure, a maximum cable tension of 0.25×10^6 kgf is specified. It is first arranged for a great circle to pass through W_e and some point W_{CT} on the $F = 1.5$ locus. W_{CT} is the resultant of W_e and the cable tension CT and the included angle between W_e and W_{CT} is measured along the great circle. For example, the triangle of forces in Figure 10.38a is the special case of the plane $W_e:W_{CT}$ lying normal to the $F = 1.5$ locus, the included angle being 7.5°. The included angle $\widehat{W_{CT}\ W_e}$ (52°) is then obtained by setting off vector CT from W_e so that its scaled

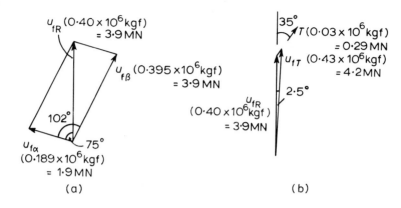

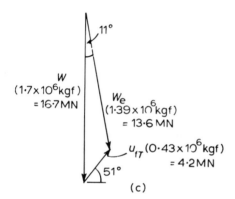

Figure 10.36 Resolving the different force components in the wedge problem.

length closes exactly with vector W_{CT} at 0.25×10^6 kgf (2.5 MN). Amplitude W_{CT} of the resultant may also be measured from the vector diagram. The process is then repeated for several other points W_{CT} in order to provide a range of possible drill hole orientations for the post-tensioning. It is useful to have available such a range of options since some of the orientations may have to be ruled out on the grounds of practical inconvenience at the site. In a similar manner, ranges of orientations may be derived by construction on the stereonet for different cable tensions. Choice of allowable tension would be determined by the thickness and strength of the cable and by the quality of the rock anchorage, these controls being the subject of specialist decisions.

Cable anchor types of reinforcement are usually only a practical proposition for slopes of restricted height since the necessary forces that have to be applied by these restraining devices may be as high as

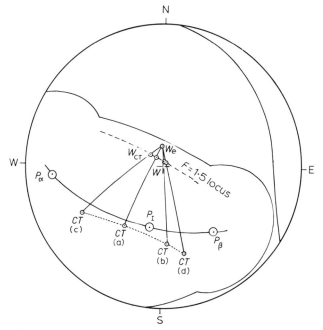

Figure 10.37 Upper hemisphere equal area projection for determining necessary cable anchor orientations for enhancing rock wedge shear stability (note that this particular construction is for a specified cable tension of 0.25×10^6 kgf (2.5 MN) and must be interpreted in conjunction with Figure 10.38).

20 per cent of the rock weight (Hoek and Londe, 1974). If dilation of the mass can be prevented at an early stage following excavation, then any reinforcement will be much more effective. On the other hand, a certain amount of shear displacement is always necessary in order to mobilize maximum shear strength. Dilation will be an almost inevitable consequence of that displacement.

Tension crack porewater pressures exert a considerable influence on possible shear instability. Hoek and Londe (1974) recommend that any cracks be filled with a suitable porous material such as gravel and the tops be sealed-off with a re-moulded clay. Fissures should be drained by drilling holes into the slope for a horizontal distance approximately equal to the height of the slope. Drainage by the insertion of vertical boreholes from the crest of the slope and the use of down-the-hole pumps is a more expensive option because a natural, gravitational facility is not utilized but, on the other hand,

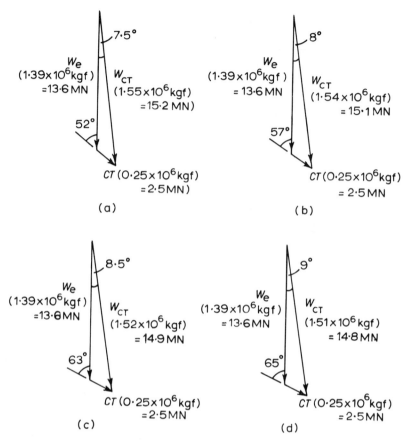

Figure 10.38 Resolving forces in order to determine a locus of cable anchor orientations for a particular cable tension of 0.25×10^6 kgf or 2.5 MN.

the holes can be inserted and drainage started before a slope is actually created. In some circumstances, this latter facility could be important. Drainage galleries could be used in large excavations should the expense be justified. A final point to remember is that clay overburden material should not be allowed to accumulate at the toe of a slope. With its low permeability, it acts as a seal against natural drainage and thereby promotes instability.

If both planes which form the potential failure surface for the wedge can be assumed to be *completely drained* and if the stability is *dependent on frictional forces only*, then analysis on the stereographic projection for a factor of safety has been shown not to be very time-consuming. But for rapid calculation of a factor of safety without the need for stereonet construction, Hoek (1973) has produced stability charts. Based on the dips of two planes and the difference in dip direction, two dimensionless ratios, A and B, can be read off directly, or by interpolation, on the charts. The factor of safety is determined from the equation:

$$F = A \tan \phi_\alpha + B \tan \phi_\beta \qquad (10.31)$$

which is a truncated form of the general equation of Hoek *et al* (1973) and of equation 10.27 earlier. From the charts, it is found that the factor of safety of the earlier wedge situation is

$$F = 0.53 \tan 38° + 0.82 \tan 34° = 0.97.$$

This factor of safety, at critical equilibrium, is significantly lower than that read off from the stereonet earlier. Hoek suggests that the charts be used for the rapid assessment of those slopes which can be considered safe ($F > 2$) and which do not require further consideration. Those which are potentially hazardous ($F < 2$) require more detailed investigations.

10.9 Influence of discontinuity orientation distributions

In the analyses pursued thus far in the present chapter it has been assumed that shear takes place along true joint faces with a displacement continuity facility. But even with a well-defined joint system, orientation distributions of joint faces are never mapped as singular concentrations. Rather it is found that there appears a spread of orientations, each having a concentration kernel which may be used to specify a unique directionality for planar-or wedge-slide analysis.

If terminated discontinuities of the type discussed in Chapter 6 are mapped in overconsolidated clays, clay shales or in weaker rocks such as chalk, then the orientation distributions will be much more attenuated and there may be less evidence of central concentrations. Under those circumstances, and for a rapid non-analytical appraisal, it will be necessary to prepare a transparent overlay for the stereonet, the overlay incorporating the great circle for the slope face, the pole to that face, and a friction circle or circles representative of the discontinuities. The contoured plots and the overlay would then be rotated on top of a base net so that suites of great circles could be drawn through pairs of concentrations. A similar suite of poles to these great circles would then represent a range of possible

Table 10.13 Location and description of coastal sites studied in the south of England by Reeves (1973) and Gavshon (1974)

Site	Location	Lithology	Stratigraphical position	Exposure Inclination	Exposure Height	Nature of Exposure
A	Furzy Cliff, near Weymouth SY 698817	Fissured clay shales	Oxford Clay, Upper Jurassic	78°	–	Cliff section
B	Furzy Cliff, near Weymouth SY 698817	Fissured clay shales	Oxford Clay, Upper Jurassic	73°	16 m	Cliff section
C	Kimmeridge Bay, Dorset SY 906793	Fissile bituminous shale	Kimmeridge Clay, Upper Jurassic	86°	16 m	Cliff section
D	Kimmeridge Bay, Dorset SY 907786	Fissile bituminous shale	Kimmeridge Clay, Upper Jurassic	90°	48 m	Cliff section
E	Kimmeridge Bay, Dorset SY 902790	Fissile bituminous shale	Kimmeridge Clay, Upper Jurassic	82°	–	Cliff section
F	Near Corfe Castle SY 970824	Heavily fissured chalk	Middle Chalk, Upper Cretaceous	66°	–	Hill side cutting
G	Alum Bay, I. of Wight SZ 305853	Sandy fissured clay shale	Base of London Clay, Middle Eocene	74°	8 m	Cliff section
H	Shippards Chine, Compton Bay, I. of Wight SZ 377842	Fissured marl clay	Weald Clay, Lower Cretaceous	75°	8 m	Cliff section
I	Atherfield Point, South I. of Wight SZ 451791	Fissured clay shale	Weald Clay, Lower Cretaceous	68°	–	Cliff section
J	Whitecliff Bay, I. of Wight SZ 639858	Sandy fissured clay shale	Base of London Clay, Middle Eocene	82°	6 m	Cliff section
a	Charmouth, West Dorset	Fissured marls	Lower Lias, Lower Jurassic	90°	15 m	Cliff section
b	Eypesmouth, West Dorset	Fissured marls & silts	Lower Lias, Lower Jurassic	74°	6 m	Cliff section
c	Black Ven, West of Charmouth	Fissured marl/shale	Lower Lias, Lower Jurassic	68°	10 m	Cliff section
d & e	As A					
f	White Nothe, East of Weymouth	Heavily fractured chalk	Lower Chalk, Cretaceous	up to 90°	13 m	Cliff section

intersection lines for potential wedge sliding. The location of these poles with respect to a stippled, unstable zone as in Figure 10.23 would indicate the degree of hazard that could arise from this mode of failure. Alternatively, if the object of the exercise is, for example, to orientate a road cutting and suggest its maximum safe dip, then it would simply be necessary to ensure that the area on the projection bounded by the trace of the slope face and a friction circle: a) was not cut by any great circle representing a concentration of discontinuities that could condition a planar shear failure, and b) did not contain intersection edges of the $I_{\alpha,\beta}$ form.

Following the work of Reeves (1973) on cliff sections along the south coast of England and the Isle of Wight (outlined in Chapter 3), further stability analyses were conducted in the same general area by Gavshon (1974). One of the locations, in a chalk cliff section at Ringstead Bay near Weymouth, is marked 'f' in Figure 3.26 and is used to illustrate the general approach. Other locations specifically studied for slope stability, together with those earlier designated in Figures 3.26 and 3.27, are marked a,b,c,d,e. All locations are specified in Table 10.13.

The discontinuity fabric in Figure 10.39 results from field measurements taken in the Lower Chalk (Cenomanian) horizon of the Cretaceous at a section 200 metres west of White Nothe and 5 miles east of Weymouth. For the most part, the chalk is very unfossiliferous and is rather siltier than that of the Turonian or Senonian. There was evidence of cliff failures in the locality, shear plane development and numerous fallen blocks lying at the foot of the slope.

Since the strata in question occur in the northern limb of the Weymouth Anticline, the bedding dips at approximately 20° in a northerly direction. Four discontinuity concentrations are picked out in Figure 10.40. These comprise a set of bedding discontinuities and three other sets which dip at high angles in a southerly and easterly direction. Also marked on Figure 10.40 are two slope faces I and II, their poles, two friction circles which result from *in situ* shear box tests on chalk discontinuities and which also bracket Priest's Lower Chalk friction angles in Section 10.6, and the discontinuity intersection lines $I_{\alpha,\beta}$ (α,β = 1 ... 4).

Planar sliding along discontinuities 2 seems to be a possibility since the dip of kernel 2 is less than the dip of P_I and P_{II} and a 25° or 35° friction cone about 2 would not take in W. Even considering the

Figure 10.39 Equal area upper hemisphere projection of discontinuity poles in Lower Chalk near Weymouth, England. Maximum concentration 'A' = 11% per 1% area. Discontinuity data *after* Gavshon (1974).

nature of the distribution about 2 at the 5 per cent per 1 per cent area level, this latter condition would not be satisfied. However, the steeper-dipping discontinuities in this set cease to daylight and are therefore safe from shear. Concentration 3 does not daylight and is therefore safe from shear. Concentrations 1 and 4 dip into the slope and can therefore be regarded as safe from planar shear.

Field observations tended to suggest that sliding occurs on two intersecting discontinuity sets and it will be noted that intersection $I_{2,4}$, just fulfills the requirement for wedge failure with respect to slope face I. A situation is envisaged where sliding would take place along $I_{2,4}$ dipping at 60° in a direction of N 160°.

Where wedge shear is deemed to be a live issue, and particularly when the material of immediate concern is a stiff clay or clay shale,

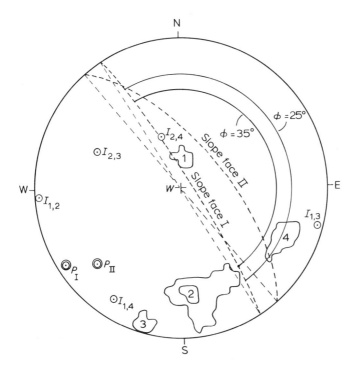

Figure 10.40 Equal area upper hemisphere projection of chalk slope stability problem in Ringstead Bay, near Weymouth, England. 117 poles to joint planes contoured at 5% and 10% per 1% area.

it is recommended that the intersection line $I_{\alpha,\beta}$ be subjected to a two-dimensional back-figuring analysis of the type applied to the clay slopes in Section 9.3. It is usual to construct a range of shear profiles having different dips but to include the dip (in the present example: 60°) of specific interest. The profiles are then analysed by computer program on the basis of Fellenius, Bishop and Janbu (see Section 9.2) and the safety factor determined for different input values of c' and ϕ' and for different assumed piezometric surfaces.

Computations on the chalk slope generate a friction angle of 37° to 38° for a shear plane angle of 60° (Figure 10.41). This friction angle slightly exceeds the upper limit of the bracketed values used in Figure 10.40. If a modest cohesion (10 kN m^{-2}) is entered, then the friction angle ($F = 1$) falls to 32°. The presence of water will act to

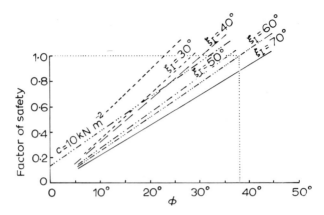

Figure 10.41 Shear stability of chalk slope for different shear plane and friction angles (*after* Gavshon, 1974).

reduce the safety factor or to demand a reduction in shear plane angle for stability. Free flowing water was not observed at the face and the overlying sandy top-soil appeared to be very well drained.

10.10 Seismic influences on stability with respect to sliding

Supplementary forces can be applied to a rock slope as a result of natural seismicity or blasting. As considered subsequently in Section 12.15, an acceleration ng is usually regarded as generating an equivalent static force nW or, more generally, nW_e through the centre of gravity where W_e, as earlier, is the resultant of several internally-imposed loadings.

In order to show how dynamic forces may be accommodated, in an arguably over-simplified manner, on the stereographic projection and to re-capitulate on some of the earlier analysis, a different example will be used (Attewell and Farmer, 1974c). This relates to a limestone quarry in northern England (Figure 10.42) in which the controls on stability are predominantly through three near-vertical sets of joints and a bedding plane. The distribution of poles to joint planes is as shown in Figure 10.43 and the bedding plane, since it is so well-defined, is marked with a single concentrated pole. Although the main tendency is for block rotation out of the face and generally down the bedding plane, a problem to be considered subsequently, there is also, seemingly, the possibility of shear movement down the bedding planes which dip at an angle of 39° to the south. Resistance to shear is either through direct limestone-to-limestone contact or through a thin layer of interbedded shale, the respective shear

Rock Slope Stability 793

Figure 10.42 Showing unstable blocks of Carboniferous Limestone at the north east corner of Helbeck Quarry in Northern England. (Line of major fault runs horizontally across background of long distance telephotograph). Photograph: P. B. Attewell.

strength parameters being $c' = 0$, $\phi' = 44°$ and $c' = 0.0913$ kgf cm^{-2} (8.9 kN m^{-2}), $\phi' = 22°$ from laboratory shear box tests. In the latter case, the cohesive contribution is ignored. Field observations confirmed that the rock was well drained and that possible influences of porewater pressure in the analysis could also be ignored. Problems are also exacerbated by the fact that, in the area of greatest instability, the slope face is vertical and that the 'daylighting' condition is potentially satisfied even for the near-vertical joint planes.

With a distribution of poles to joint planes, the first step is to prepare a transparent overlay projection of the dip of the slope and the two friction cones of 22° and 44°. This overlay is rotated with the contoured stereoplot over the stereonet in order to locate great circles passing through pole concentrations. Figure 10.43 is a 'locked' composite of overlay and stereonet on which have been marked the several planar intersection points (lines). It can be seen that there is a remote possibility of shear failure with respect to the $p_2 : p_3$ wedge and the $p_1 : p_4$ wedge, the latter only if the frictional angle for the shale should drop to 18°. The obvious hazard could take the form of

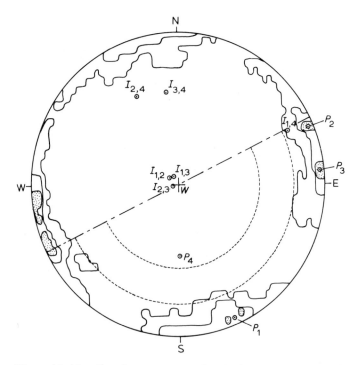

Figure 10.43 Equal area upper hemisphere projection of limestone quarry stability problem. 107 poles to joint planes contoured $\geqslant 10\%$, $\geqslant 5\%$, $\geqslant 1\%$ per 1% area.

a bedding plane slide against which the factors of safety are tan $44°/\tan 39°$ (limestone) = 1.19 and $\tan 22°/\tan 39°$ (shale) = 0.5.

An interpretation of the stability picture in Figure 10.44 confirms the vulnerability of the bedding planes under a self-weight loading condition. If, however, a horizontal force nW is applied to the slope, then for any arbitrary direction this can be represented on the stereonet as a resultant vector describing a locus about W and generating a semi-apical cone angle equal to $\tan^{-1} n$. Temporarily, therefore, the resultant of W and nW will be directed sub-vertically and, for example, should a seismic inertia force be directed from north to south (direction of wave travel from south to north) there would be the possibility of shear instability, with respect to limestone friction, along the bedding planes. The factor of safety for planar or wedge sliding, on the basis of this concept, would be measured off on the projection in exactly the same manner as described earlier, with the angular measurements being made along

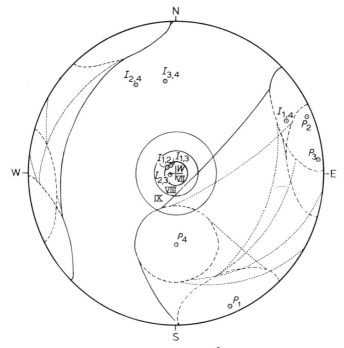

Figure 10.44 Limestone quarry: 22° friction cones and delimitation of safe zones on equal area upper hemisphere projection. Roman numerals define seismic intensities (Figure 12.38).

the great circle containing P or P_I and the resultant position of the $W + nW$ vector.

It must be stated that the meaning of a factor of safety calculated for dynamic loading is rather arbitrary because dynamic forces are never 'static' forces acting in one direction only. The actual factor of safety varies with time and movements may only occur when F falls very temporarily below unity. There is no means in the present type of analysis whereby the magnitude of the necessary displacements may be incorporated.

Hendron *et al* (1971) have used a Newmark (1956) type analysis. This involves calculation of the minimum force applied through the centre of gravity of the sliding mass for which the mass will just begin to move, in the present case along a joint or bedding plane or down two intersecting planes. Such minimum force, or 'dynamic resistance', is compared with the maximum acceleration force, each cycle of which will often occur for only a very short period of time (see Figure 12.32) in order to obtain an equation incorporating the component of ground displacement. It is shown that the slope displacement is proportional to the square of the maximum particle velocity for a given ratio of dynamic resistance (Ng) to acceleration force (Ag), the relevant equations being:

$$\text{Displacement} = \frac{v^2}{2gN}\frac{A}{N} \text{ for } 0.2 < \frac{N}{A} < 0.4 \qquad (10.32)$$

(conservative upper limit)

$$= \frac{v^2}{2gN}\left(1 - \frac{N}{A}\right)\frac{A}{N} \text{ for } \frac{N}{A} > 0.4 \qquad (10.33)$$

(reasonable upper bound)

$$= \frac{6v^2}{2gN} \text{ for } \frac{N}{A} < 0.2 \qquad (10.34)$$

(reasonable upper bound)

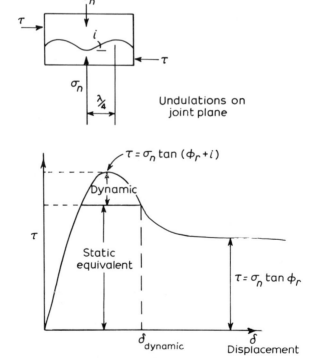

Figure 10.45 Influence of displacement and undulation angle upon the dynamic shear strength of a joint plane (*after* Hendron et al., 1971).

where g is the acceleration due to gravity. Using the earlier factor of safety concept, N would be determined on the stereonet from the necessary angular rotation to reduce F to unity. Parameter A would be determined in the same general way.

Having estimated the displacement, Hendron *et al* (1971) then compare it with the distribution of wavelengths, or, more specifically, quarter wavelengths of the undulations on the joint contact surfaces. If the displacement exceeds that necessary to ride over surface asperities then the effective shear strength of the slope is likely to be reduced (Figure 10.45) and the stability impaired. Of course, the *form* of the frictional interface is not simple-harmonic and there will be a range of 'wavelengths'; nor will it really be a viable proposition to extrapolate the measured morphology of an exposed surface to that concealed between joint faces. But the maximum displacement concept is a useful addition to slope stability thinking.

10.11 Instability caused by block overturning

In certain instances, high friction along a potential shear plane may inhibit block sliding but a possibility may exist for rotational instability. The analytical technique involves taking moments about hinge lines and being able to specify centres of gravity of rotational elements. These latter elements, the blocks bounded by joint planes, can be regarded as being compounded of several standard geometrical solids, and with progressive construction it would be possible to specify centres of gravity for some irregular blocks. A *necessary* condition for rotation is a sufficient resistance to shear movement.

In most systematically-jointed sedimentary rocks, the joint planes are near-vertical and sensibly orthogonal. If the rock has been subjected to later earth movements, the strata may have been upturned; the possibility of shear or rotation on the bedding planes must then be considered. The limestone quarry in the north of England that was considered a little earlier also offers an example of rotational instability with respect to the bedding planes, and for other descriptive examples of this type of failure, reference may be made to de Freitas and Watters (1973).

Consider, first, the simple rectangular block situation sketched two-dimensionally in Figure 10.46. A third plane normal to the plane of section will not affect the analysis. It is also assumed that no cohesion acts along the bedding plane to enhance stability. Taking

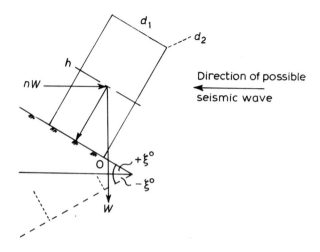

Figure 10.46 Analysis for stability against rotation of a rectangular block on a slope.

moments about 0, the intersection point of the rotation axis with the plane of section, we have

$$W \frac{h}{2} \sin \xi \leqslant W \frac{d_1}{2} \cos \xi \qquad (10.35)$$

or

$$\frac{h}{d_1} \tan \xi \leqslant 1 \qquad (10.36)$$

for stability with respect to overturning.

Returning to the limestone quarry problem (Figure 10.42), it can be seen, from Figure 10.47 that the apparent dip $\xi_{1,4}$ of the bedding plane p_4 in the direction $P_4:P_1$ is 34°. Similar constructions generate apparent dips $\xi_{2,4}$ and $\xi_{3,4}$ of 16° and 4° in directions $P_2:P_4$ and $P_3:P_4$ respectively. It follows, therefore, that with respect to the $P_1:P_4$ plane, a block 1 m wide measured along that plane (d_1) would be unstable if its height exceeded 1.48 m. For a 3 m wide block, the critical height would be 4.45 m, and in a similar manner the critical heights can be quickly calculated for the other two apparent dips. For practical quarry work, or for controls on highway cutting construction, bench design curves can be drawn up and, if feasible, optimum extraction face alignments specified. The curves would tend to be inherently conservative, not only because of a possible cohesion contribution to rotational stability, but also because a

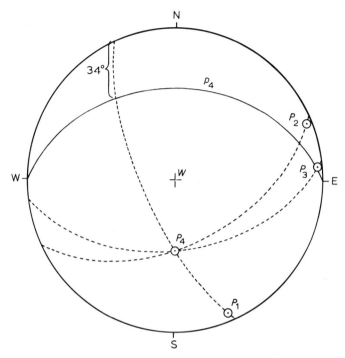

Figure 10.47 Construction on an upper hemisphere projection for apparent dips on the p_4 plane.

certain degree of interlocking might be expected in certain directions when the inter-planar angles are less than 90°. Opposing these stabilizing factors would again be any porewater pressure within the joint and bedding planes arising from inadequate drainage. A cohesion contribution to stability on one or both faces of the block sketched in Figure 10.46 is quite easily accommodated by subtracting a term equal to $k_1 c_{(h)} h^2 d_2$ from the left hand side of equation 10.35 and adding a term equal to $k_2 c_{(d1)} d_1^2 d_2$ to the right hand side ($0 \leqslant k_1, k_2 \leqslant \frac{1}{2}$). Pore pressure influences will be considered shortly.

The stereographic projection can be used to delimit orientation regimes within which rotation can and cannot take place for different aspect (h/d_1) ratios of block or face height. Examination of Figure 10.48 shows how this may be done for just three different block aspect ratios. In this practical joint situation, the joint faces are not exactly orthogonal and so, for example the plane of section sketched in Figure 10.46, when reproduced in Figure 10.48 as I:W, is not quite compatible with the plane of rotation about the $I_{1,4}$:W

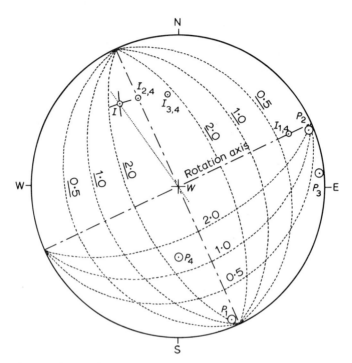

Figure 10.48 Some possible rotation stability regimes for blocks having different aspect ratios (block lengths are taken in the p_1 and p_2 planes; the numbers on the de-limiting planes refer to different h/d_1 and h/d_2 ratios, with those underlined referring to the latter; point I defines the intersection line formed by the plane p_4 and the plane containing p_1 and p_4). The projection is equal area, upper hemisphere.

axis. With an apparent dip ($\xi = I_{2,4}$) of 36°, the limiting aspect ratio of the jointed limestone is 1.38 with respect to rotation about the $I_{1,4}$ axis.

Suppose, now, that there is a head, h_w, of ponded water behind the block sketched in Figure 10.46. As shown in Figure 10.49 this water will exert an additional overturning moment about 0, the point of application of this driving force being $h_w/3 \cos \xi$ from the base of the block.

If there is no seepage through line A to the base of the block, then for stability,

$$W\left(\frac{h}{2}\right) \sin \xi + \tfrac{1}{2}\gamma_w h_w^2 d_2 \cdot \tfrac{1}{3} \frac{h_w}{\cos^2 \xi} < W\left(\frac{d_1}{2}\right) \cos \xi \quad (10.37)$$

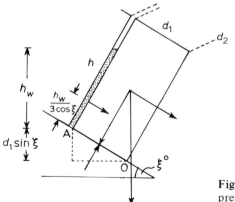

Figure 10.49 Influence of water pressure on rotational stability of block.

or

$$h_w < \left\{ \frac{3W}{\gamma_w d_2} \cos^2 \xi (d_1 \cos \xi - h \sin \xi) \right\}^{1/3} \qquad (10.38)$$

where γ_w is the unit weight of the water, and W can be written as $\gamma h d_1 d_2$, with γ being the rock unit weight. If, on the other hand, while the head h_w is maintained, seepage occurs through line A along the base of the block and drains at 0 (that is, a full head h_w at A and zero head at 0), then for stability,

$$W\left(\frac{h}{2}\right) \sin \xi + \frac{1}{2} \gamma_w h_w^2 d_2 \cdot \frac{1}{3} \frac{h_w}{\cos^2 \xi}$$

$$< W(d_1/2) \cos \xi - \frac{\gamma_w h_w d_1 d_2}{2} \cdot \frac{2 d_1}{3} \qquad (10.39)$$

or

$$\frac{\gamma_w}{3 \cos^2 \xi} h_w^3 + \frac{2\gamma_w d_1^2}{3} h_w + \gamma h d_1 (h \sin \xi - d_1 \cos \xi) < 0 \qquad (10.40)$$

which is a cubic equation in water head h_w. This second condition imposes an additional instability on the block, but if it would have suffered rotational failure anyway under the first condition, there would be little point in engaging in the added complexity associated with basal seepage. In any case, with uplift pressures a real issue, sliding might take place preferentially before rotation.

Equations 10.38 and 10.40 could be represented on the stereonet for various h_w, ξ and h, d_1 ratios, but there would be little technical merit in so doing. Alternatively, the expressions would not be too unwieldy to express graphically.

It remains only to evaluate the influence of an earthquake force on a potential block rotation situation. For convenience we again assume a horizontal inertial component of force nW while still remembering that such forces are usually multi-directional, albeit temporarily applied, within a seismic wave train.

Reference to Figure 10.46 indicates that

$$(W \sin \xi + nW \cos \xi)h/2 \leqslant (W \cos \xi - nW \sin \xi)d_1/2 \tag{10.41}$$

or

$$n \leqslant \frac{(d_1 - h \tan \xi)}{(h + d_1 \tan \xi)} \tag{10.42}$$

for stability against overturning in the presence of a horizontal component of seismic force (nW). From the equilibrium equation 10.42 the chart in Figure 10.50 can easily be constructed. It will be noted, for example, that on a dip slope of 39° under a horizontal, outward-directed inertia force from an intensity $\overline{V}$ earthquake, the blocks would be prone to overturning if their aspect ratio exceeded 1.2. Negative values of n represent an inertia force directed into the slope to enhance stability (perhaps a reversal of the seismic acceleration) but they also tell us something about the necessary bolt anchoring tensions that would have to be applied in order to ensure stability of a cut face using such mechanical restraints. Suppose, for example, that the characteristic block length (d_1) in the plane of the potential rotation is 3 m on the same slope angle of 39° and we want to estimate the tension requirement in a single horizontal bolt for a factor of safety of 1.5 against overturning if the slope height is 9 m. The aspect ratio is 3 and, from the chart, $-n$ for a factor of safety of one is about 0.38. Thus, for $F = 1.5$, the bolt tension would have to be $0.57W$. The calculation, from equation 10.42, is

$$n = \frac{1.5(3 - 9 \tan 39°)}{(9 + 3 \tan 39°)}$$

$$= -0.563$$

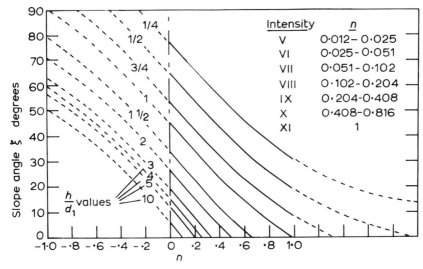

Figure 10.50 Limiting horizontal force nW for block rotation on a plane dipping at ξ degrees expressed in terms of the block aspect ratio h/d_1.

Assuming a third block dimension d_2 also to be 3 m, and taking the unit weight of the rock as 23.5 kN m^{-3}, the bolt tension would then be 0.563 x 23962 x 3 x 3 x 9 = 1.09 x 10^6 kg force (10.7 MN). The calculation could quite easily be reversed to produce a maximum slope height for a permissible bolt tension.

10.12 General rock slope design curves

A preliminary aim in the assessment of rock slope stability is to be able to relate, in a quite general manner, a safe slope height to a particular slope angle. Since, as will be shown later, a regular joint structure imposes a very close control on the stability mechanics, it follows that a generalized design curve approach, in which the premises upon which the eventual curves are based take little or no account of particular rock mass structural detail, is more appropriate to a weaker rock. Such a weaker rock will be more fragmented and could probably be analysed using accepted soil mechanics methods, or it will be replete with a more randomly-distributed field of terminated discontinuities. However, since the stability of any individual slope is sensitive to the presence of porewater pressures, any set of general design curves must embody information on the water table geometry.

Shuk (1965) performed a field study of the stability of natural

804 *Principles of Engineering Geology*

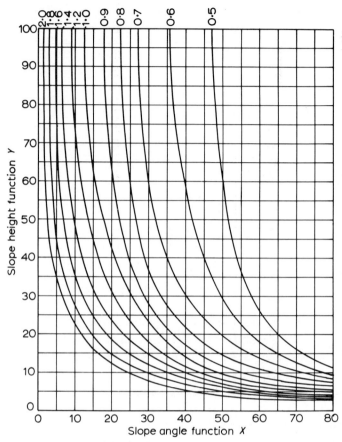

Figure 10.51a Slope design chart for plane failure (*from* Hoek, 1970). Factors of safety are marked for each curve.

slopes in ferruginous shales in Columbia and proposed a relation between slope height and slope angles. Kley and Lutton (1967) and Lutton (1970), using the same field approach for slopes in opencast mines, also suggested relations between these two variables. Lane (1961, 1966) presented slope design charts based on his work on shales at the Fort Peck dam (Montana, U.S.A.), the soil mechanics limit equilibrium, circular failure approach being used. Slope failure in opencast mines have also been studied by Coates *et al* (1963, 1965) and by Rana and Bullock (1969). But a later and most valuable contribution is due to Hoek (1970) whose slope stability design charts are drawn in Figures 10.51a and 10.52a, with the control equations listed in Figures 10.51b and 10.52b. The bases

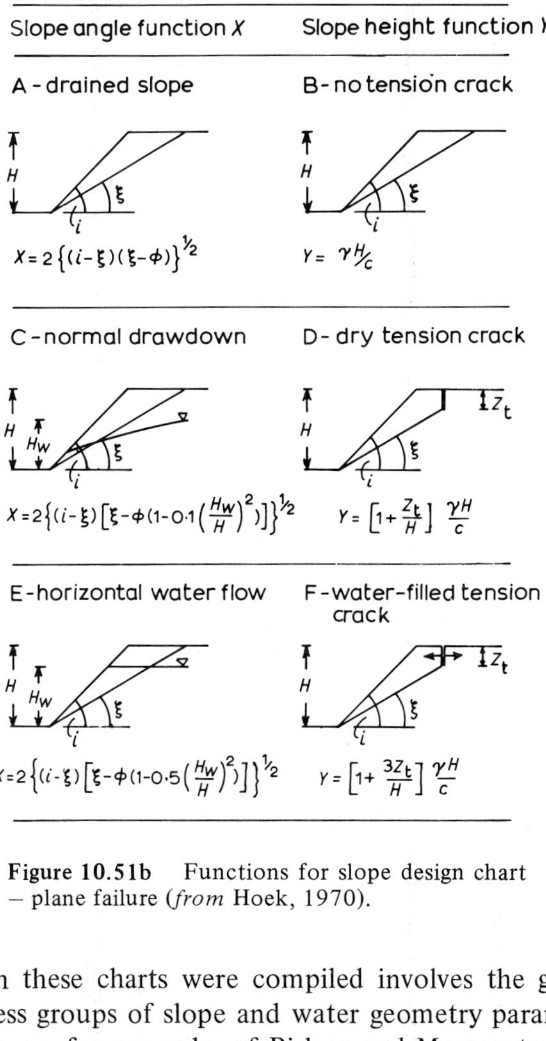

Slope angle function X	Slope height function Y
A - drained slope	B - no tension crack
$X = 2\{(i-\xi)(\xi-\phi)\}^{1/2}$	$Y = \gamma H/c$
C - normal drawdown	D - dry tension crack
$X = 2\{(i-\xi)[\xi-\phi(1-0.1(\frac{H_w}{H})^2)]\}^{1/2}$	$Y = [1 + \frac{Z_t}{H}] \frac{\gamma H}{c}$
E - horizontal water flow	F - water-filled tension crack
$X = 2\{(i-\xi)[\xi-\phi(1-0.5(\frac{H_w}{H})^2)]\}^{1/2}$	$Y = [1 + \frac{3Z_t}{H}] \frac{\gamma H}{c}$

Figure 10.51b Functions for slope design chart – plane failure (*from* Hoek, 1970).

upon which these charts were compiled involves the generation of dimensionless groups of slope and water geometry parameters in the general manner, for example, of Bishop and Morgenstern (1960) for soil slopes.

The manner in which the charts are used will readily be apparent. Having measured up his slope parameters, the observer must then attempt to determine the nature of a possible shear surface, whether a circular type of failure is more likely than a planar failure (the latter occurring in rock containing more well-defined structural features), and the characteristic form of the water table if evidence of such is available from borehole records. On this latter point, a value for H_w (the water table height relatively uninfluenced by the

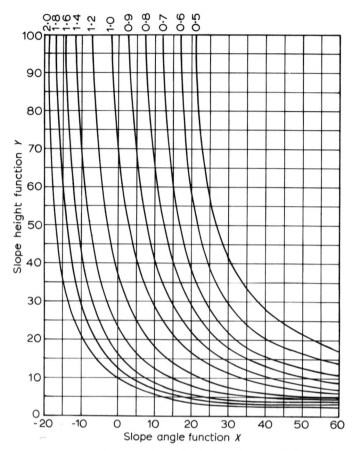

Figure 10.52a Slope design chart for circular failure (*from* Hoek, 1970). Factors of safety are marked for each curve.

free slope surface) should be obtained from measurements made at a distance of approximately four-times the slope height behind the toe of the slope (Sharp, 1970). Without such evidence, Hoek (1970) gives curves which allow H_w to be assessed from the vertical height h_w of seepage from the face of the slope above the toe (Figure 10.53). Actual slope angles are difficult to specify when the slope face is irregular, and in such cases an average slope angle would have to be taken. In a tension crack situation, it is not usually possible to obtain a measured crack depth. If its position can be located at the ground surface then, in the case of a potential planar shear failure along a well-defined structural feature, its depth can be calculated from an assumed intersection with the planar feature. In

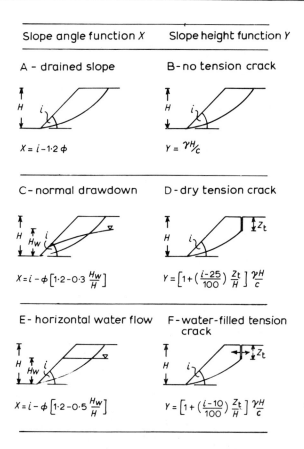

Figure 10.52b Functions for slope design chart — circular failure (*from* Hoek, 1970).

the case of a potential circular failure, and on the assumption that the shear surface passes through the toe of the slope, Hoek (1970) has suggested that the depth of the tension crack may again be estimated on the basis of an intersection, the critical circle being located by construction from the graph in Figure 10.54. If there is no obvious evidence of a tension crack, but it is felt that the influence of a possible concealed crack on the stability of the slope should be checked, then Hoek (1970) suggests using the equation (Terzaghi, 1943; Spencer, 1968):

$$z_t = \frac{2c}{\gamma}\left\{\frac{1-\sin\phi}{1+\sin\phi}\right\} \qquad (10.42)$$

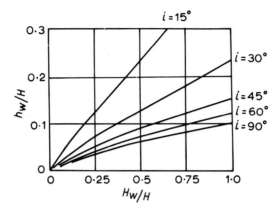

Figure 10.53 Relationship between the height of water seepage h_w from a slope face and the height H_w of the water table, both parameters being expressed as a ratio of the slope crest height H (*from* Hoek, 1970).

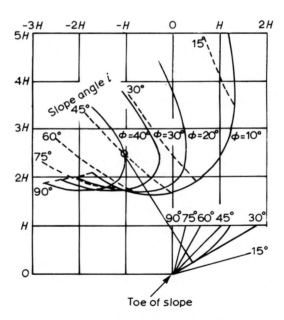

Figure 10.54 Curves for the approximate location of the centre of the critical toe circle in soil and soft rock slopes; the centre is determined by the intersection of the appropriate i and ϕ curves (*from* Hoek, 1970). The critical circle that is illustrated is for $i = 45°$, $\phi = 40°$

Tension crack depths in cohesive soils are usually limited to a maximum depth of half the slope height (Terzaghi, 1943).

10.13 Slopes in highway cuttings and embankments

It is only quite recently that the slopes of highway embankments and cuttings in Britain have received *ad hoc* design attention. Slope angles have usually been based on prior engineering experience in the same or similar materials. Reference to the relevant British Code of Practice on Earthworks (CP 2003) proposes, for example, blanket slopes of 38° to 42° for a hard sandstone and 34° to 38° for a shale. Over-conservative flattening of a slope may not incur the magnitude of financial penalty that would arise if the slopes of a major open-pit mine are under-steepened, but cost-savings can be made with relatively little engineering geological investigation and analysis of the type discussed elsewhere. Some examples of case histories will be outlined in this section.

The example taken from Arrowsmith (1971a) — see Figure 10.55 — of rock slope instability illustrates how a bedding plane slide can develop as a result of sub-vertical joint relaxation. Although the west face was cut cleanly using pre-splitting blast techniques (Langefors and Kihlström, 1963) and stood safely at 70° with little subsequent engineering work required, the east face had to be cut back to about a 34° slope. Evidence on jointing patterns in the rock was, in fact, available before the excavation in an abandoned quarry behind the east slope.

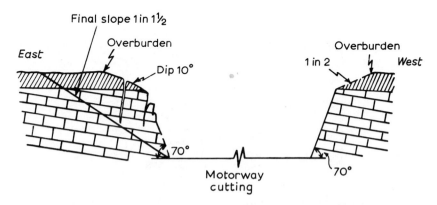

Figure 10.55 50 ft deep cutting in Lower Coal Measures for M61 Manchester-to-Preston motorway at Chorley (*from* Arrowsmith, 1971a).

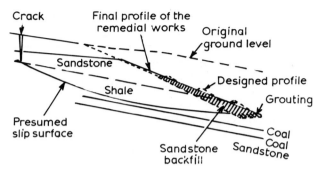

Figure 10.56 Cutting and remedial works at Longden End Moor on the M62 Motorway (*from* Arrowsmith, 1971b).

Arrowsmith (1971b) also describes three other interesting rock-cut and rock-fill situations on the M62 Lancashire-Yorkshire motorway. All were problems involving a sub-vertical tension-cracking facility along pre-existing joint planes and shear development in a shale bed lying between sandstone strata. In each case, unrestricted water intrusion through the joints in the sandstone had softened-up the shale at the base. Designed slopes of from 1 in 1½ to 1 in 3 in the rock and 1 in 1½ in the shale had to be reduced in steepness, the shale bands towards the toes of the slopes had to be cut back and backfilled with rock aggregate from the tops of the cuttings, water ingress at the crests of the slopes reduced, and both piezometers in the rock and target blocks on the slopes were installed for monitoring purposes. Two of the situations are shown in cross-section in Figures 10.56 and 10.57.

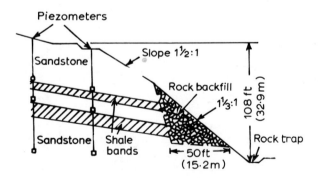

Figure 10.57 Cross-section of north slope of cutting on the M62 motorway at Windy Hill (*from* Arrowsmith, 1971b).

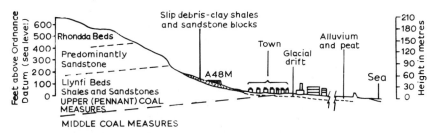

Figure 10.58 Section through the Port Talbot area, South Wales showing location of by-pass road (*from* Rogers, 1970).

Both Woodland (1968) and Rogers (1970) have discussed the slope stability problems associated with the construction of the Port Talbot (South Wales) Bypass (A48M). There, the engineer was faced with alternative choices of route, a coastal route with abundant bridge foundation problems in recent sands, alluvial deposits and peat or a route on the landside of the town but which would encounter rock slope stability problems. These latter instabilities were caused, as in the earlier examples, by weathering and softening of weak shales within a Coal Measures cyclothemic sequence leading eventually to undercutting and failure of the strong, massive Pennant sandstones (see also Knox, 1927 for an early appreciation of these problems). The hill-side route (Figure 10.58) was eventually chosen and Woodland (1968) criticizes the engineers for being aware of the landslipping but for not considering it to be a special hazard. Removal of the toes of the landslips initiated more slipping and some property was affected. Remedial works involved regrading and draining the hillside.

It is appropriate to mention other rock stabilization measures that have considerable practical merit. Grout treatment can render planes of weakness much less susceptible to separation and shear, but if the void volume is large the treatment may then become prohibitively expensive. Masonry or concrete retaining walls are often either too expensive or too obtrusive. In some instances, rock bolting is a highly practical solution, sometimes in conjunction with a limited grouting programme. Price and Knill (1967) have described some research directed towards this solution for the basalt block instability problems at Edinburgh Castle rock and Edwards (1970) has described in some detail the rock bolting at Jeffrey's Mount, a cutting in the M6 motorway near Tebay in Cumbria, England. This latter bolting (the heads of the bolts are strikingly visible from

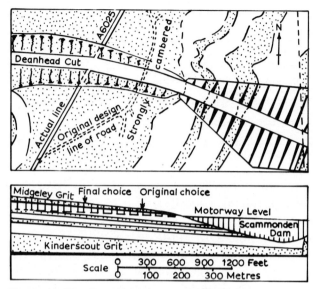

Figure 10.59 Deanhead cut in the Midgeley Grit on the Lancs–Yorks motorway (*from* Woodland, 1968).

the motorway) was into the very strongly structurally disturbed and partially metamorphosed Silurian Coniston Grits which comprised massive-to-open jointed strongly cemented fine-grained greywackes and thin laminations of closely-jointed-to-shattered well-cemented shaly mudstones (Edwards, 1970). The rock surface was glacially shattered to a depth of up to 1½ metres and it was often difficult to de-limit the rock from the over-lying glacial debris. Askey (1970) makes the point that when bolting is to take place in highly-fractured rock, it may be necessary to drill short holes for pressure cement grouting, the holes later to be re-drilled for permanent anchor installation.

When rock blasting for highway cuttings has to take place, a pre-splitting technique (Langefors and Kihlström, 1963) is best adopted if at all possible for two reasons. First, the vibration levels for the subsequent major blasting are reduced, and this can be an important factor in sensitive residential areas through which the highway route might pass. It also reduces the possibility of other block instabilities caused by vibration inertia forces. Second, there is a reduction in fragmentation at the walls of the cutting and therefore a reduced facility for progressive degradation from water ponding and frost heave. This point was made by Attewell and Taylor (1970) with respect to the Al(M) motorway cuttings in Lower Magnesian Limestone of County Durham. In this example, the walls were initially trimmed-back to a slope angle of $45°-50°$, an angle which accorded reasonably well with safe slope angles predicted from laboratory experiments. But this re-grading was later abandoned over lengths of the cutting presumably on the basis of the argument that the material from any minor slumping would not project beyond a generous 'hard-shoulder' bordering the motorway pavement proper.

Under certain circumstances of incipient slope instabilities, it may be desirable to re-route and/or re-grade a highway to take advantage of higher ground. Woodland (1968) quotes the example of the A6025 road crossing the Lancashire-Yorkshire motorway at Deanhead adjoining the Scammonden reservoir (Figure 10.59). Cambering and block slipping in the Midgeley Grit could have affected the A6025 but the problem was alleviated by locating the road on higher ground.

11 ~ Ground Improvement

Ground engineering (or geotechnical) processes cover a very wide area of civil and mining engineering technology. Table 11.1 gives some indication of the number of processes involved. In the narrowest sense, engineering geology might be defined as covering merely the specifically geological aspects of exploration. In the widest sense it might be considered to include, as well as geological aspects (such as soil and rock fabric, structure and moisture), an analysis of soil and rock properties through in-situ and laboratory testing, and this, as we have seen in the consideration of drainage in Section 8.4, might also include recommendations for improvement of these properties. Ideally, of course, an engineering geologist should have a reasonably intimate knowledge of foundation engineering and construction procedures in order to be able to estimate the effect of construction upon the pre-existing distribution of ground pressure and ultimately upon the geotechnical properties of the ground. He would, however, be expected to acquire this knowledge and experience outside any text on engineering geology. *Ground improvement* techniques do, nevertheless, illustrate some fundamental properties of rocks and soils. For their successful application they also require a sound knowledge of site geology, and it is therefore reasonable to include them as a specialist aspect of any formal treatment of engineering geology.

The purpose of ground improvement is essentially to alter the natural properties of a soil or rock and specifically to reduce soil *compressibility* (in order to reduce foundation settlement), to increase *strength* (in order to improve structural stability or bearing capacity) or to reduce *permeability* (to restrict ground water flow). The type of treatment may be considered under two main headings, *densification* and *stabilisation*, with the main methods under these headings being listed in Table 11.2.

There are, in addition, several fairly exotic methods which can be

Table 11.1 Ground engineering processes

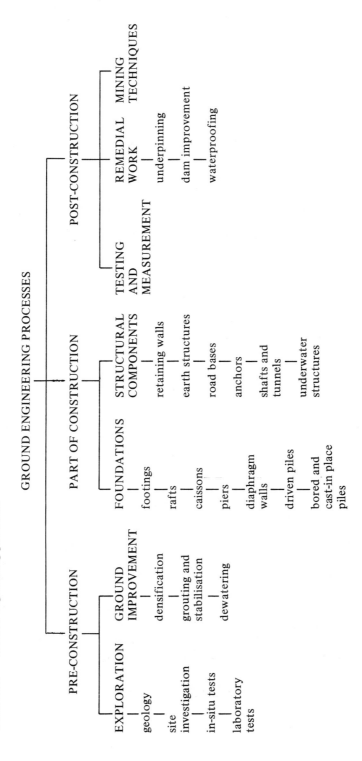

used to change the state of the soil, rock or groundwater, such as the formation of an ice wall by freezing (Sanger, 1968) to give temporary strength and water resistance, or driving off adsorbed and chemical water in clay soils by the application of temperatures up to 600 °C (Beles and Stanculescu, 1958) in order to form stable brick columns. The former is important and is considered in some detail later. The latter is of minimal use and is ignored.

Table 11.2 Methods of ground improvement

Process	Method	Ground type	Property improved
DENSIFICATION			
SHALLOW COMPACTION	vibratory roller	cohesionless and mixed soils	density
	pneumatic (tyre) roller;	mixed and cohesive soils	density
	sheep's foot roller	cohesive soil	density
DEEP COMPACTION	vibroflotation	all types of soil	strength/density
	explosive or dynamic compaction	cohesionless soil	density
CONSOLIDATION	pre-loading; sand drains	saturated and organic clays	strength/density
STABILISATION			
DE-WATERING	deep well and well point pumping	clean sands, gravels	strength/density
	surface drainage	all types soil and rock	strength
PERMEATION GROUTING	cement, clay and chemical grouts	gravels, sand, porous rock	permeability/strength
FISSURE GROUTING	cement grout	fissured rocks	permeability/strength
FRACTURE GROUTING	cement, clay and chemical grouts	fine sands, silts, fissured rocks	permeability/strength
ELECTRO-CHEMICAL STABILISATION	electro-osmosis	fine silty clays with low (10^{-7} m sec^{-1}) permeability	temporary strength
REINFORCEMENT	ground anchors	all types rock and and soil	strength
	nylon mesh	all types soil	strength

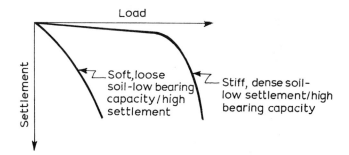

Figure 11.1 Typical load-settlement characteristics of dense and loose soils.

11.1 Shallow compaction

The simplest and most obvious method of reducing the settlement of a structure is to increase the *density* (Figure 11.1) of the soil or fill in which it is founded, or in the case of an earth structure (such as a dam or embankment) the soil or fill from which it is constructed. In the case of placed or shallow fills, densification can be achieved most simply by rolling thin layers with a heavy, possibly vibrating, roller. This method is used to compact fills during highway subgrade and earth dam or core construction. A typical example of an earth or rockfill dam construction is illustrated in Figure 11.2, and other examples may be found in Section 12.13.

The effectiveness of *shallow* compaction depends on four main factors: (a) the *layer thickness*, (b) the *type and grading* of the soil,

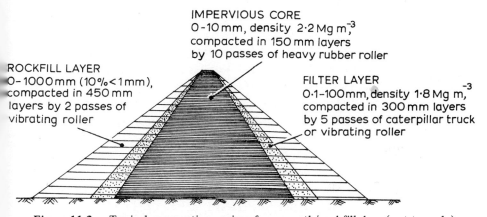

Figure 11.2 Typical compaction regime for an earth/rockfill dam (not to scale).

818 *Principles of Engineering Geology*

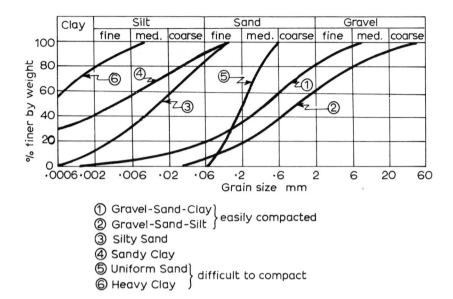

① Gravel-Sand-Clay ⎫
② Gravel-Sand-Silt ⎬ easily compacted
③ Silty Sand
④ Sandy Clay
⑤ Uniform Sand ⎫ difficult to compact
⑥ Heavy Clay ⎭

Figure 11.3 Compactability of typical soil grades.

(c) the *moisture content* and (d) the *energy applied*. Most readily-compactable soils are a mixture of cohesive and cohesionless types, and are well graded. Figure 11.3 illustrates the compactability of the typical size ranges likely to be encountered in practice.

Soft, saturated organic clays cannot be compacted by rolling because they drain poorly. Some cohesive soils and single-size cohesionless soils can only be compacted with difficulty, the former because of drainage and sometimes suction pressure difficulties, the latter because the low shear resistance is often insufficient to support the weight of a roller which rapidly becomes bogged down.

All soils have an *optimum (dry) density*, strongly dependent on the *moisture content*. The effect of increasing moisture content is to enhance the interparticle repulsion forces, thus separating the particles and allowing a more orderly re-arrangement for a given amount of effort. However, as saturation is approached, porewater pressure effects will counteract the effectiveness of the compactive effort. Each soil therefore has an *optimum* moisture content (Figure 11.4) at which the soil will have maximum γ_d.

Optimum density can be determined in the laboratory using a *Proctor dynamic* compaction test (see Lambe, 1951) which involves subjecting a soil to a standard compactive effort by dropping a

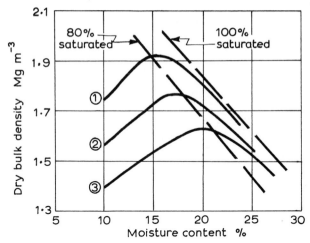

Figure 11.4 Compacted density/moisture content curves for dynamic compaction of silty clay subjected to (1) high energy, (2) medium energy and (3) low energy compaction (*after* Turnbull, quoted in Lambe and Whitman, 1969). See also Figure 12.27.

hammer onto a sample through a fixed height for a given number of times. The sample is usually compacted in three or five layers and the moisture content is varied for each sample. Following the test, the sample is dried and the dry bulk density plotted against the moisture content. A *static* compaction test in which the soil is stressed at a given magnitude for a given number of times may also be carried out.

The empiricism of the tests is demonstrated by the fact that the optimum density/moisture content data vary with the compactive energy, and it is indeed possible to derive for a given soil an approximate log-linear relationship between energy input and density (Figure 11.5 based on Wong, 1971; Tschebotarioff, 1951; Road Research Laboratory, 1952; Remillon, 1955) which makes most of the quoted index texts appear rather arbitrary. Considerable experience combined with rigid field control and testing is therefore required to equate laboratory compactability tests to field compaction. In fact, laboratory tests merely indicate how easily a soil can be compacted. Field samples are required in order to assess the actual resultant density from compaction.

Three types of roller are used on a large scale to compact soils — *vibratory, pneumatic tyre,* and *sheepsfoot* — subjecting the soil respectively to a cyclic, even and point load. The characteristics

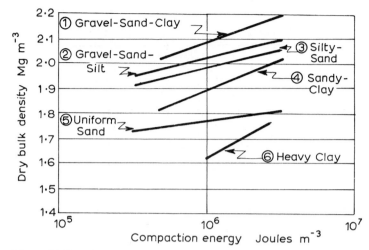

Figure 11.5 Compacted density – compaction energy relationship.

of various rollers are described in detail by Fischer (1970) and Road Research Laboratory (1952) among others.

Vibratory rollers, either towed or self-propelled, can be used to compact cohesionless, mixed or cohesive soils. In cohesionless soils, a *light* roller vibrating at high frequency and low amplitude may be used to liquefy the soil, allowing compaction at relatively low input energy levels. In cohesive soils, a *heavy* (towed) roller vibrating at low frequency and high amplitude can be used to give an impulsive effect. Such a roller will tend to leave the ground between cycles, whilst a light roller will maintain contact with and transmit vibrations into the ground. An extreme example of impulsive rolling is a vibratory *sheepsfoot* roller used in cohesive and mixed soils. This type of roller imparts high loads to relatively small areas of ground and is useful in compacting cohesive soil.

Static steel rollers are effective in mixed soils, but are of limited value in uneven ground due to their relatively restricted contact area. *Pneumatic tyre* rollers, comprising a row of smooth rubber tyres, allow a heavy load to be spread evenly over a large and variable contact area, with the area and pressure exerted depending on the tyre pressures. The stress situation is similar to that imposed by a transient rectangular shallow foundation, and the compactive effect is more uniform than with a steel roller.

The effect of *layer thickness* on compaction is obviously significant; for instance, note the rapid fall-off in vertical stress increments beneath a shallow foundation (Figure 7.10). The effect of increasing layer thickness is less if higher surface pressures are exerted. A similar pattern exists with vibratory rollers and the implication is that soil layers should be as thin as economically practicable during compaction if uniform densification is required.

Compaction is a remoulding process and apart from increasing the soil density and rearranging soil particles in a closer and more orderly arrangement it also causes significant changes in the stress situation in the soil and in its engineering properties. Thus, the lateral stress, and hence K_0, are increased by compaction, and permeability is reduced. Strength and compressibility, however, are only improved significantly at water contents less than *optimum* (see Lambe and Whitman, 1969). At higher moisture contents, rapid deterioration of strength and compressibility can occur. The importance of not exceeding optimum moisture content during compaction should be stressed, but there is a possible exception in dam core material where the increased workability associated with higher moisture contents may justify compaction on the wet side of optimum (Section 12.13).

11.2 Deep compaction

Shallow compaction methods are only effective when used to compact thin layers and are therefore virtually limited to placed fills. In order to compact deeper layers of fill or loose in-place soils, alternative methods involving the application of subsurface compaction forces, usually along a vertical axis, may be considered. Available methods include:

(a) *vibroflotation* in which a Vibroflot machine, essentially a large laterally-vibrating poker, penetrates the ground, compacting by liquefaction of cohesionless ground or by the formation of compacted stone columns in cohesive soils;
(b) *compaction piling* in which a driven steel mandrel compacts by displacement;
(c) *explosive compaction* in which explosive charges are detonated at depth;
(d) *dynamic compaction* in which large weights are dropped onto the ground surface; and
(e) *compaction grouting* (to be considered later) in which dense grouts are injected at high pressures and at depth.

All these methods, and others, are considered by Mitchell (1970)

822 *Principles of Engineering Geology*

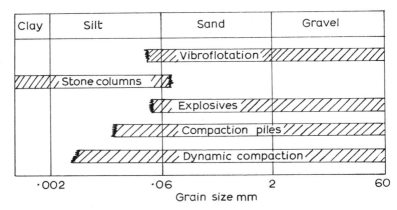

Figure 11.6 Applicability of deep compaction methods (*after* Mitchell, 1970).

in a comprehensive state-of-the-art review. Figure 11.6, which summarizes the principal soil grades in which each method can be used, is based on Mitchell's summary.

Deep *blasting* is probably the most effective method of applying a large amount of compacting energy at depth. The major technical stipulations are that the soil should be loose, cohesionless and free running and preferably saturated. Application and subsequent design data are limited, but Lyman (1942) shows that a simple spherical scaling law, $r = W^{1/3}/C$, where r is the radial distance, W is the charge weight of explosive and C is a charge factor depending on the type of cover and of explosive, will specify the influence zone and hence the depth of cover required to prevent cratering (see also Section 7.13). Kummoneje and Eide (1961) quote a case where, after firing 6 x 1.2 kg charges at a depth of 7 m in loose sand, a total settlement of 400 mm was obtained above the charge, with the total affected area having a radius of 10 m. A series of holes at, say, 3 to 4 m centres would therefore have the effect of lowering the ground level by 0.4 m, which is equivalent to about 5 percent increase in relative density of the compacted soil.

Whilst blasting may have some uses in isolated areas, the more conventional methods of compaction in *cohesionless* soils are vibroflotation and compaction piling. In the former instance, the soil is compacted by dispersion (vibration) and displacement (Mitchell, 1970; Greenwood, 1970) around a heavy vibrating poker which penetrates the soil vertically under its own weight (Figure 11.7), usually assisted by water or air flushing from a leading nose cone.

Ground Improvement 823

Figure 11.7 Vibroflotation in cohensionless soil (*photograph by courtesy of Cementation Ground Engineering Ltd.*).

The diameter of the poker is 30–40 cm, and the hole created by penetration of the poker is backfilled with sand, which is in turn compacted to form a replacement column up to 1 m. diameter.

Compaction piles usually comprise a hollow steel mandrel driven into the ground, so compacting the surrounding soil by displacement. The resultant holes are then backfilled with sand. In both cases, the backfill should comprise surface soil, and the result should be a lowering of surface levels as soil is compacted to the treated depth.

The effectiveness of deep compaction is usually expressed in terms of increased relative density (see Section 2.3) of cohesionless soils, assessed on the basis of dynamic or static penetrometer tests (usually cone penetrometer tests described in Section 7.8). Table 11.3

Table 11.3 Effect of vibroflotation in cohesionless soils (*after* Mitchell, 1970)

Case no.	Soil type	Depth m	Column spacing m	Relative density % Before	Relative density % After
1	Sand	7	2.0	43	80
2	Sand and gravel	4.5	–	63	85–95
3	Well-graded sand	6	2.4	47	79
4	Sand and gravel	4–5	2.1–2.4	7–58	70–100
5	Clean loose sand	4	2.3	33	78
6	Fine sand	5–6	2.0–2.4	0–40	75–93
7	Gravelly sand	6–9	–	0	80
8	Sandy gravel	–	–	33–80	85–95
9	Glacial sand and gravel	5	1.8–2.3	40–60	85–90
10	Well-graded till	3	3.0	50	75
11	Loose sand	7	1.9	loose	80
12	Loose fine sand	6	1.8–2.3	loose	80

(Mitchell, 1970) summarizes some case histories of vibroflotation and indicates significant increases in relative density in a variety of loose soils. There are no hard and fast rules on column (or compaction) spacing, but an average of 2 m between centres is usually chosen. In Figure 11.8, Basore and Boitano (1969) compare the effect of vibroflotation and compaction piling as a means of compacting medium-dense sand fill with an initial average relative density of about 70 percent. They conclude that while both methods are effective in compacting sand fill, vibroflotation produces a more even compaction and more uniform density, and is also effective at greater spacing. Compaction piling was, however, cheaper.

It must be stressed that densification of *all* sand fills or in-place cohesionless soils by deep densification is not essential. The bearing capacity of even loose cohesionless soils is not normally critical and the basic aim of densification in cohesionless soils is to reduce settlement by increasing the relative density of the soil. This procedure has a particular importance in earthquake zones, where post-construction seismic shocks may lead to liquefaction and settlement of cohesionless soils with consequent structural damage (Seed and Lee, 1966; see also Section 12.15). This mechanism and the vibroflotation compaction mechanism have considerable dynamic affinities. However, in a reasonably dense or overconsolidated soil (say, relative density ~80 per cent), attempts at deep densification,

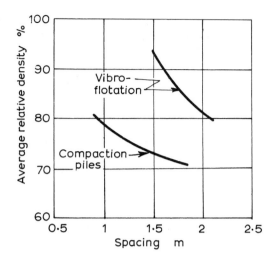

Figure 11.8 Effect of spacing on compaction for vibroflotation and 0.35 m dia. compaction piles (*after* Basore and Boitano, 1969).

particularly by vibroflotation, may in fact reduce the relative density. This possibility has been pointed out by Rowe (1970).

In cohesive soils, deep compaction as such is difficult to justify since the optimum moisture content cannot be controlled accurately using available methods and the amount and rate of energy application is limited by the equipment. If some form of compaction or strengthening of the soil is considered desirable, deep compaction equipment may be used to install stone columns in the soil (Watt *et al*, 1967; Webb, 1969). These columns should effectively improve the load-settlement characteristics of the soil, partly by densification and partly by replacement, acting as compressible load-bearing columns which will tend to yield laterally under load and thereby induce a passive pressure in the soil at the soil-column interface. The logical next step is the installation of concrete piles.

A recent alternative method of deep compaction developed by Ménard (1972), although described earlier by Hobbs (1970) among others, involves dropping a heavy weight from a considerable height onto the ground surface. Weights that are used are between 10 to 20 tonnes, drop heights are 5 to 15 m and resultant compaction depths 5 to 15 m. The method apparently works in everything except thick clays, and in low permeability materials it evidently owes its effectiveness to induced hydrofractures resulting from the induction of high impulsive porewater pressures. These pressures create drainage paths which allow accelerated porewater pressure dissipation with increases in density during subsequent attack.

11.3 Pre-loading and consolidation

A more conventional approach to densification of clay soils is the use of pre-loading or controlled construction techniques to allow porewater dissipation at a rate in line with *consolidation* theory (see Section 2.12). This point will be made again in Section 12.13 in the context of earth dam, and particularly clay core, construction. We have already seen that the rate at which a layer of low-permeability compressible soil subjected to a surface load increases in density is governed by:

(a) The rate at which porewater pressures induced by a surface load are dissipated. If this dissipation occurs principally through vertical flow to a free draining layer, then the rate of porewater pressure dissipation can be estimated from Terzaghi's classical theory of one–dimensional consolidation.
(b) The magnitude of surface load which can be applied to the layer without causing shear failure and subsequent plastic flow.

Any pre-loading or the application of consolidation pressure must therefore start with the condition that the initial stresses applied to the soil are less than the existing shear strength of the soil. Then, since consolidation rates under this stress decrease with time, a second condition must be that the stresses applied to the soil must increase with increasing densification of the soil. The basic design theory has been discussed previously (Section 2.12).

Suppose that the existing vertical and horizontal (principal) effective stresses σ'_1, σ'_3 where $\sigma'_3 = K_0 \sigma'_1$ are to be increased by increments $\delta\sigma'_1$ and $\delta\sigma'_3$ through the application of a surface load. Then provided that the assumption is correct in consolidation theory that all additional total stress increments are transmitted initially from the soil skeleton as porewater pressure, the new effective stresses will be equal to $\sigma_1 - \delta u$ and $\sigma_3 - \delta u$, where δu is obtained from Skempton's pore pressure equation (Section 2.11) using total stress increments estimated from Boussinesq curves (see Figure 7.10).

Then,

$$\delta u = \delta\sigma_3 + A(\delta\sigma_1 - \delta\sigma_3)$$

where A is the pore pressure parameter defining the relative proportions of the applied stress increment carried by the *soil skeleton* and by the porewater in a *saturated* soil. This particularly depends on the relative compressibilities of water and the soil skeleton. In the case of a soft, sensitive clay, A may be as high as 2.5 and δu will be high, but in a heavily overconsolidated clay, A may be negative and δu will be low.

It is evident, therefore, that consolidation theory is only justified in soils where A is large and $\delta u \gg \delta\sigma_3$. In this case, immediately following surface loading, $\delta\sigma'_1 = \delta\sigma_1 - \delta u$, and $\delta\sigma'_3 = \delta\sigma_3 - \delta u$. $\delta\sigma'_3$ will have large negative values and the loading stress path drawn in Figure 11.9 will approach the K_f line or failure envelope. When the soil is fully consolidated, then $\delta u = 0$, $\delta\sigma'_1 = \delta\sigma_1$ and $\delta\sigma'_3 = \delta\sigma_3$, and the point $\sigma'_3 + \delta\sigma_3 = K_0(\sigma'_1 + \delta\sigma_1)$ representing the consolidated effective stress condition will be at a more stable point on the K_0 line.

Since 30 percent of porewater dissipation takes place in 10 percent of the time required for 90 percent consolidation (Figure

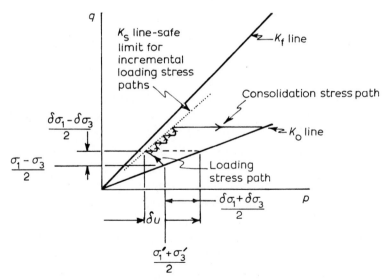

Figure 11.9 $p-q$ representation of pre-loading consolidation.

2.36), it is obviously preferable to load the surface in a continuing series of small increments prior to complete consolidation, rather than to allow full consolidation under a larger increment. This can be arranged by maintaining the stress situation in the ground so that it may be represented graphically as a series of small steps close to the K_f line. In this way, each small increase in shear strength due to consolidation is utilised immediately by the addition of an increased consolidating load.

Clearly, this type of design approach is not wholly satisfactory since in practice most of the initial assumptions regarding consolidation theory must be modified. In particular, the properties (undrained) of the soil will change during consolidation, and this may allow higher stress increments during the later consolidation stages. In addition, the ground is not laterally confined and plastic flow, at the sub-shear failure levels may occur. This means, in effect, that a design for pre-loading and consolidation must be regarded at best as a rough guide, and observations of ground reactions during loading are essential. These observations would normally be obtained from piezometers to measure porewater pressures at points beneath the loaded surface, and inclinometers around the perimeter of the structure to detect possible plastic flow and/or failure states (see Section 7.11).

A further incomplete variable may be the assumed bed thickness. Rowe (1968b) has pointed out that the real drainage or consolidation characteristics of a clay deposit depend on its geological formation and that actual porewater pressure dissipation can often bear no resemblance to dissipation rates based on laboratory or *in situ* tests. For instance, thin layers, lenses or fissures containing sands, silts or organic inclusions can increase the mass permeability of the clay, particularly in a horizontal direction, and can transform the consolidation characteristics, effectively reducing the bed thickness to the thickness of a series of thin clay laminae between drainage layers. Figure 11.10 illustrates this type of structure in split cores and Table 11.4 (Rowe, 1968b) lists the effect of this type of structure upon experimentally-derived values of c_v (using 0.1 m and 0.3 m diameter oedometers) and c_v values estimated from *in-situ* permeability tests. In extreme cases laboratory and field values differ by several orders of magnitude.

Rowe recommends the adoption of consolidation tests on 0.3 m diameter samples to estimate consolidation characteristics,

Figure 11.10 Buried Valley lower laminated clay from Tyneside.

combined with *in-situ* permeability tests (Section 8.6) and some laboratory permeability tests on organic clays. He also recommends careful examination of split cores and suggests the following guide lines for treatment (see Table 11.4):

(a) *Well-point dewatering* (see Section 8.4) — coarsely layered clays containing beds and fissures of sand and silt; boulder clays with gravel pockets, having $c_v > 50$ m² year⁻¹.
(b) *Pre-loading, consolidation* — finely-layered clays with thin sand and silt seams having 10 m² year⁻¹ $< c_v < 50$ m² year⁻¹.
(c) *Sand drains* — homogenous organic clays and uniform clays with $c_v < 10$ m² year⁻¹.

Table 11.4 Laboratory and *in-situ* c_v values for clay soils (*after* Rowe, 1968b)

Site	Stratum	Effective stress range $kN\ m^{-2}$	$c_v\ m^2\ year^{-1}$ Lab. 0.1 m	Lab. 0.3 m	In-situ	Remarks
Staunton Harold	Fissured weathered shale	96–575	5.6h	–	6130	well point drainage recommended
Covenham	Stony clay with thin silt and sand layers	48–192	1.8	–	1860	,,
Derwent	Lower Lake Clay, coarsely layered with sand and silt	240–480	1.1v 2.6h	–	836	,,
Frodsham	Estuarine clay b_2 with vertical rootlets	24	9.3v	180–1800v	929	,,
Derwent	Upper Lake Clay, coarsely layered	240–480	2.6h	–	121	,,
Frodsham	Estuarine clay, b_2 very silty	48	18.6v	464	483	,,
Frodsham	Estuarine clay, b_1 layered with silt and sand	62	0.46v 1.1h	46h	41	pre-loading consolidation recommended
Staunton Harold	Fissured weathered mudstone	96–575	5.6h	–	62	,,
Derwent	Middle Lake Clay, laminated with occasional silt layers	240–480	1.4h	–	11	Sand drains recommended
Frodsham	Estuarine clay, b_2 with vertical rootlets	287	0.65v	2.8	2.5	,,
Selset	Uniform boulder clay	–	1.7h	–	5.1	,,
Skå-Edeby	Swedish post-glacial clay	24–96	0.74h	–	about 0.8	,,

Letters h and v refer to the direction (horizontal/vertical) of flow

11.4 Sand drains

Most sediments have a higher horizontal than vertical permeability, due partly to their near-horizontal stratification and partly to the presence of permeable layers or inclusions of sand. It follows, therefore, that if the main direction of water flow can be changed from the vertical to the horizontal during consolidation of a compressible soil, then the rate of consolidation or porewater pressure dissipation will be substantially increased. If, in addition, the drainage path (the bed thickness in the case of vertical flow) can be reduced, then consolidation of very low-permeability sediments becomes feasible. These conditions can be obtained by inserting equidistant sand drains, comprising circular, vertical, sand-filled holes formed in the compressible layer, in order to introduce radial horizontal flow.

Consolidation by radial flow has been analysed by Baron (1948) through solution of the governing equation for one-dimensional radial consolidation in terms of cylindrical co-ordinates:

$$\frac{\partial u}{\partial t} = c_h \left[\frac{\partial^2 u}{\partial r^2} + \frac{1}{r}\frac{\partial u}{\partial r} \right] \tag{11.1}$$

where

$$c_h = \frac{k_h}{\gamma_w m_v} = c_v \frac{k_h}{k_v}$$

is the coefficient of consolidation due to radial flow,

k_h/k_v is the ratio between horizontal and vertical permeability,

u is the porewater pressure,

and

r is the radial cylindrical co-ordinate.

For the two-dimensional case:

$$\frac{\partial u}{\partial t} = c_h \left[\frac{\partial^2 u}{\partial r^2} + \frac{1}{r}\frac{\partial u}{\partial r} \right] + c_v \frac{\partial^2 u}{\partial z^2} \tag{11.2}$$

First consider radial flow only. Then assume that the cylindrical boundaries of the drain-well influence cylinder shown in Figure 11.11 are impermeable ($r = r_e$; $\partial u/\partial r = 0$), or that settlement during consolidation is uniform (that is, $n^* > 5$) and that the pore pressure at $t = 0$ is uniformly equal to u_0 and at $t > 0$ is equal to zero at the sand drain surface.

Baron's solution can then be obtained in the form:

$$u = u_0 \exp\left(\frac{-8T_h}{d_e^2 F(n^*)}\right) \tag{11.3}$$

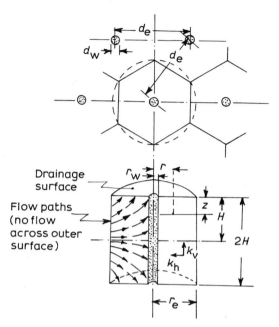

Figure 11.11 Plan and section of drain well (*after* Richart, 1957).

where $n^* = \left(\dfrac{d_e}{d_w}\right)$, effectively the ratio between drain spacing (d_e) and drain diameter (d_w), and

$$F(n^*) = \frac{n^{*2}}{n^{*2} - 1} \ln(n^*) - \frac{(3n^{*2} - 1)}{4n^{*2}}$$

which, since n^* is by definition large, becomes

$$F(n^*) = \ln(n^*) - \frac{3}{4} \tag{11.4}$$

T_h is a dimensionless factor related to consolidation time t, in much the same way that T_v is related to t for vertical consolidation. Thus:

$$t = T_h \frac{d_e^2}{c_h} \text{ compared with } t = T_v \frac{H^2}{c_v} \text{ (as in equation 2.62)} \tag{11.5}$$

where H is the bed half-thickness.

The magnitude of T_h for $u/u_0 = 0.1$ (that is, 90 percent consolidation where $T_v = 1$ and $t = H^2/c_v$) can be estimated from Baron's solution:

$$T_h = -\tfrac{1}{8} [\ln(u/u_0)] [\ln(n^*) - \tfrac{3}{4}] \simeq \tfrac{1}{4}[\ln(n^*) - \tfrac{3}{4}] \qquad (11.6)$$

Thus, for most realistic values of n^*, T_h will have a value between $\tfrac{1}{2}$ and $\tfrac{1}{4}$ for 90 per cent consolidation, so emphasising the conceptual similarity between vertical and radial consolidation (Figure 11.12).

It is immediately evident that the *drain diameter* is not as important as *drain spacing* in determining the rate of consolidation. Thus $t \propto d_e^2$ whereas $t \propto \ln d_e/d_w$. It follows that d_e will have a far greater effect on drain well efficiency than will d_w.

This fact is illustrated in Figure 11.12a which shows that reducing d_e by a factor of 2 causes a reduction in t by a factor of 6, whereas increasing d_w by a factor of 20 (Figure 11.12b) only reduces t by a factor of 4. Bearing in mind the obvious expense and potential disturbance inherent in producing larger holes, it is obviously more efficient to drill smaller holes provided that a continuous drainage column can be maintained. This is the main criterion in installing drains, and to be assured of vertical continuity, a minimum diameter of 150 to 220 mm is usually chosen. However, if the sand can be contained previously in a jute bag (for instance, sandwicks – see Dastidar *et al*, 1969), the continuity is assured and much smaller diameter drains may be considered.

Another major factor controlling sand-drain effectiveness is the *horizontal permeability*. Since the real drainage behaviour of the soil depends on its geological formation, then in the same way that the effective permeability of a rock is determined by its fissure structure, small layers of silt and organic inclusions can dominate the permeability of a soil body, so making it quite different from that of small samples. Unfortunately, *in-situ* permeability measurement (by falling head or constant head methods) usually involves disturbance of the (sensitive) soil in low c_v soils. Whilst this may not affect the relevance of data insofar as it affects a geotechnical process involving similar disturbance, it has affected the collection of comparative data. In fact, k_h/k_v ratios appear to fall in the range 5 to 10 for most low-permeability layered soils. The relative effect is illustrated in Figure 11.13.

The *thickness (H)* of the half layer also influences the sand drain efficiency, as shown in Figure 11.14. Clearly, the smaller the

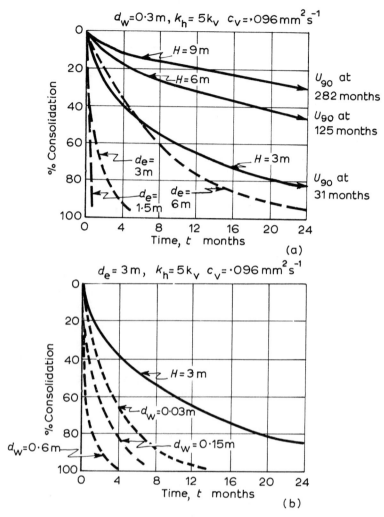

Figure 11.12 Effect of drain installation and drain spacing and diameter on consolidation time for a clay soil (*after* Richart, 1957).

value of H, the greater will be the effectiveness of vertical drainage. The efficiency of drains increases with the increasing vertical thickness of the drained layer and Christie (1959) shows that their presence is only really justified where H is greater than 5 m.

Installation is also an important factor in sand drain performance. Discontinuities due to arching, subsequently filled with impermeable soil, will prevent drainage of the porewater. Similar discontinuities caused by plastic flow or failure of the ground due to excessive

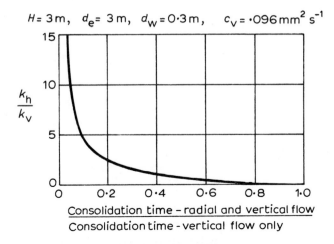

Figure 11.13 Effect of permeability ratio on sand drain efficiency (*after* Richart, 1957).

loading will have an identical effect. More important, however, is the occurrence of *smear* due to remoulding or blinding of permeable layers during drilling or by driving of the drain hole. Where augered holes are used for drains, the properties of the soil surrounding the hole will be largely unaltered. However, the use of a driven mandrel (a common and cheap method of drain installation) will tend to displace and remould adjacent soil, so creating a zone of reduced permeability. In layered soils, the resistance of pervious layers may be increased by smearing from impervious layers.

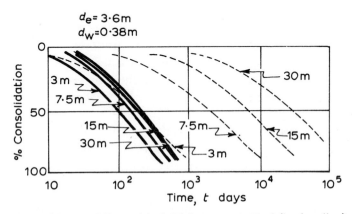

Figure 11.14 Effect of bed thickness on vertical (broken line) and radial (solid line) consolidation (*after* Christie, 1959). $c_v = 0.067$ mm^2 s^{-1}.

The actual effect of smear probably depends on the initial density and porewater pressure in the soil. In loose saturated soils with a high porewater content, it is even possible to visualize an increasing permeability created by the impact of the driver. In denser, layered soils, the effect may be serious.

The extent of the smear zone may be quantified in terms of the effectively altered zone. Thus, the smear ratio SR may be defined:

$$SR = \frac{\text{radius of smear zone}}{\text{radius of drain}}$$

Usually:
if SR = 1 there will be no effect due to smear,
but if SR = 1.2, there will be an effective increase in drainage time of up to 200 per cent, equivalent to a significant reduction in d_w or increase in d_e,
and if SR = 1.5, then vertical drainage will probably be as effective as sand drains.

Quantitative data on smear is, however, rare (Berry and Wilkinson, 1969) and, in practical terms, steps to reduce it are more important than actual quantification. Hence, auger drilling or water-flush methods are often used for the installation of sand drains.

11.5 Grout treatment

Grout treatment is usually designed either to reduce the *permeability* or *compressibility* of rocks or soils, or to increase their *strength* or *stability*. This is usually achieved by filling cavities, fissures and porespaces in the rock or soil with a set grout of variable strength. The grout is injected at pressure through a standpipe which may be driven into the ground or fixed in a borehole by grouting or packers, depending both upon the type of ground treated and the grouting pressure.

Choice of grout and the mechanism of grout penetration are determined by the properties of the ground material and the purpose for which the treatment is designed. Some of the basic mechanisms are:

(a) *Permeation grouting* in which the grout is injected evenly and comprehensively into soil or rock porespace, forming a series of near-symmetrical interconnecting spheres or cylinders around

grout sources. This method is suitable in pervious sands or gravels and porous rocks.
(b) *Fissure grouting* in which a well-dispersed water-cement grout is injected into fissures in rocks or layered soils having a low intrinsic permeability. The purpose of grouting is to seal the fissures, often using high pressures to widen them slightly in order to ease the passage of the grout, and relying on subsequent deposition of cement particles to 'silt up' the fissures.
(c) *Fracture grouting* may be used in low- or variable-permeability rocks and soils, usually to increase penetration rates by inducing limited and controlled zones of hydrofracture in the vicinity of the injection source.
(d) *Compaction (or consolidation) grouting* is used to compact loose soils or sands, usually by high pressure injection over a limited zone of a high solids-content grout with low penetration characteristics.
(e) *Bulk grouting* in which the purpose of the grouting is to stabilise natural cavities or old mine workings.

A detailed description of the types of grout used in ground engineering is outside the scope of the present work but they are briefly summarized in Table 11.5. Grouts are most simply divided into two main groups (and four sub-groups), as *particulate* grouts (sub-divided into *cement* grouts and *clay* grouts) and *non-particulate* grouts (subdivided into *silicate* grouts and *organic polymers*). This division is important in grouting, since the *size of particle* limits the types of ground which the grout can effectively permeate. Other factors limiting grout flow are its *viscosity* and its *shear strength* in the case of non-Newtonian fluids or suspensions, since the pressures which can be used to inject grouts are normally limited by the necessity to avoid fracturing the ground. Once the grout is in place, its *set properties*, particularly its strength and the strength of the grouted soil, are important. Some indication of grouted soil strengths is given in Table 11.5. In fact, there is often a relationship between strength and viscosity in grouts of a particular subgroup since a grout with a high strength is invariably denser and more viscous in a fluid state than is a low strength grout. Some of these interrelationships have been considered by Hilton (1967) in developing a classification systems for chemical grouts (Figure 11.15).

Typical examples of permeation grouting are illustrated in Figures

Table 11.5 Types of grout

Group	Type	Composition	Approximate unconfined compressive strength of grouted sand (kN m^{-2})
Cement Grouts	Cement Suspension	water, cement (ratio > 1)	3 000
	Cement slurry	water, cement (ratio < 1)	15 000
	Sand–cement	water, cement, sand	12 000
	Flyash–cement	water, cement, flyash	4 800
	Clay–cement	water, cement, bentonite	3 000
	Alum–cement	water, cement, aluminium sulphate	600
Clay Grouts	Bentonite Suspension	water, bentonite (10 per cent)	very low
	Bentonite–silicate	water, bentonite, sodium silicate, sodium phosphate	120
	Bentonite–diesel	bentonite, diesel oil, water	very low
Silicate Grouts	Joosten	sodium silicate solution, calcium chloride solution	3 600
	Guttman	sodium silicate, calcium chloride, sodium carbonate solutions	2 400
	Silicate–Bicarbonate	sodium silicate, sodium bicarbonate solutions	400
	Silicate–Ethylacetate	sodium silicate solution, ethylacetate	600
	Silicate–Aluminate	sodium silicate solution + sodium aluminate	1 500
Organic Polymers (Two-shot)	Epoxy Resin	non-aqueous resin	35 000 depending on concentration
	Polyester Resin	non-aqueous resin	35 000
	AM-9	acrylamide	600
	Chrome-lignin	calcium lignosulphonate and sodium dichromate	1 200
	Urea-Formaldehyde	urea and formaldehyde in acid solution	3 600
	Polythixon	polyurethane	3 500
	Resorcinol–Formaldehyde	resorcinol and formaldehyde in aqueous solution	3 500

11.16 and 11.17. Figure 11.16 illustrates a classic case of tunnel grouting quoted by Glossop (1968). The ground for the large excavation of a near-surface underground station tunnel was grouted in three stages from three headings.

The first heading A was used initially to give access for stabilizing ground around two other headings, B and C. These were then used with heading A to establish a complete grouted zone around the proposed excavation. First-stage grouting used a bentonite-cement

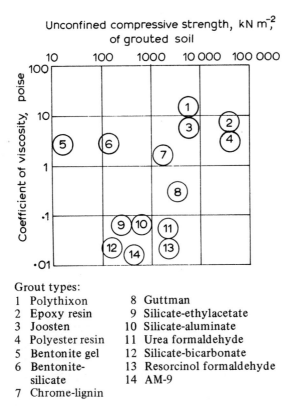

Grout types:
1 Polythixon
2 Epoxy resin
3 Joosten
4 Polyester resin
5 Bentonite gel
6 Bentonite-silicate
7 Chrome-lignin
8 Guttman
9 Silicate-ethylacetate
10 Silicate-aluminate
11 Urea formaldehyde
12 Silicate-bicarbonate
13 Resorcinol formaldehyde
14 AM-9

Figure 11.15 Chemical grout classification on a strength-viscosity basis (*after* Hilton, 1967).

treatment for filling the larger fissures. This was followed by treatment with silicate based grouts with final sealing, particularly in the fine sand, using resin (probably resorcinol-formaldehyde) grouts.

Figure 11.17 illustrates the drilling and injection pattern for waterproofing 27 m of porous sandstone during the sinking of a shaft for a coal mine at Monktonhall in Scotland. The waterbearing sandstone, yielding 700 l min^{-1}, was treated by primary injection with sodium silicate-bicarbonate grout to fill large openings and then with low viscosity AM–9 to permeate pores (see Neelands and James, 1963).

Permeation grouting of soils or porous rocks is limited in the case of particulate grouts partly by the size of the grout particles in relation to the pore size of the soil and partly by the interfacial shear resistance between the pore sidewalls and the fluid grout.

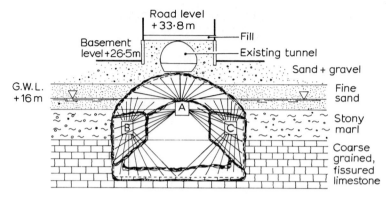

Figure 11.16 Ground treatment for underground station — Paris (*after* Glossop, 1968).

It can be shown (Taylor, 1948) that in a bed of spherical grains packed in the tightest possible way (that is, face-centred cubic packing, having a porosity of 0.26), the maximum size of particle capable of passing in suspension through the minimum pore cross-section will have a diameter in the region of $0.15\,d_0$, where d_0 is the diameter of the spherical particles. In the case of an irregular granular bed, d_0 will be the diameter of equivalent spherical grains having the same surface area as the bed. In the case of a dense suspension, where trapping of particles may lead to a build-up of

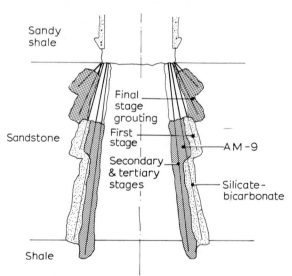

Figure 11.17 Section through a grouted shaft (*after* Neelands and James, 1963).

filter cake and subsequent limitation of flow, the maximum particle diameter d_p may be taken as $0.1\,d_0$, and d_0 may be equated to the D_{10} soil grain size (Section 2.2) in the case of an irregular bed.

d_0 may be related to other properties of granular materials through the empirically modified Kozeny-Carman equation (see Raffle and Greenwood, 1961 and Section 2.4) for laminar flow through a porous material (a simpler form is the totally empirical Hazen formula quoted by Terzaghi and Peck, 1967):

$$k = \frac{\gamma_g}{5\eta} \frac{n^3}{(1-n)^2} \frac{1}{S_o^2} \qquad (11.7)$$

where $S_o = \dfrac{6}{d_o}$ is the specific surface area per unit volume of particles,

k is the coefficient of permeability

n is the porosity,

and η, γ_g are the viscosity and unit weight of the suspension.

Thus, the maximum size of particle (d_p) capable of flowing through a granular bed as part of a suspension in water ($\gamma_g = 9.81$ kN m^{-3}, $\eta = 0.01$ poise) is given by:

$$d_p = 0.1 d_o = \frac{1-n}{n} \left(\frac{1.8 \eta k}{\gamma_g n} \right)^{\frac{1}{2}} \qquad (11.8)$$

The limiting grout particle size is therefore governed by the *porosity* and *permeability* of the ground which, since porosity varies little in sands and gravels, effectively means the permeability (Table 11.6). The particle size will, in effect, be the maximum particle size, since a relatively small proportion of coarse particles (1 to 2 per cent) can rapidly form a filter cake near to the injection source, severely limiting flow. Ordinary portland cements normally contain particles up to 100 μm. This size range therefore limits the permeation of o.p.c. grouts to relatively coarse sands and fine gravels, as shown in Table 11.6. High early strength cements having a maximum grain size of 20 μm might be used to grout sands with a permeability as low as 10^{-4} m s^{-1}, although in practice inhomogenous pore distribution leading to variable permeabilities, particularly in vertical directions, would render this impracticable. In the case of bentonite grouts, a maximum particle size of less than 10 μm should allow

Table 11.6 Permeation size limits in typical sands and gravels

Material	k $m\ s^{-1}$	n	d_p μm	α μm
Coarse gravel	1	.40	1000	2260
Medium gravel	10^{-1}	.40	320	713
Fine gravel	10^{-2}	.40	100	226
Coarse sand	10^{-3}	.35	45	76
Medium sand	10^{-4}	.30	18	26
Fine sand	10^{-5}	.30	6	8

permeation of medium sands. This information is represented graphically in Figure 11.18.

Particulate grouts normally act as Bingham fluids, having a measurable yield (or shear) strength which must be overcome before flow commences. Chemical grouts are solutions and normally act as Newtonian fluids. Typical flow characteristics are illustrated in Figure 11.19.

The permeation characteristics of both types are limited by the increasing shear resistance at the extending interfacial contact area between the grout and the soil being penetrated. An estimate of the total shear resistance, and hence the limiting extent of grout permeation, can be obtained by considering the ground as a series of capillaries each having a radius (α) equal to the hydraulic or effective pore radius of the ground material, where:

$$\alpha = \frac{2 \times \text{Volume per unit length of pore}}{\text{Surface area per unit length}} = \frac{2n}{S_o(1-n)} = \frac{nd_o}{3(1-n)}$$

(11.9)

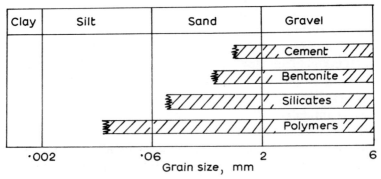

Figure 11.18 Soil size limitations on grout permeation (*after* Mitchell, 1970).

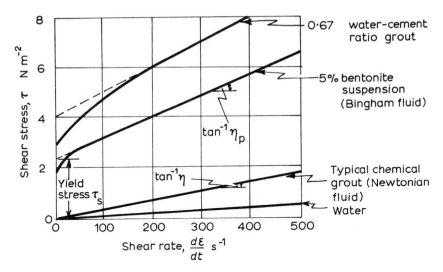

Figure 11.19 Flow properties of typical grouts (*after* Scott, 1963a).

Substituting in equation 11.7, this gives:

$$\alpha = \left(\frac{20\eta k}{\gamma_g n}\right)^{\frac{1}{2}} \quad (11.10)$$

Magnitudes of α for typical soils show an effective pore radius of about twice the minimum particle size capable of free passage through the soil (Table 11.6).

Resistance to flow through a single capillary may be obtained from the Buckingham-Reiner equation for flow of (Bingham) fluids through a pipe:

$$Q = \frac{\pi \alpha^4}{8\eta_p}\left[\frac{p}{r} - \frac{2\tau_s}{\alpha}\right] \quad (11.11)$$

where η_p, τ_s are the (Bingham) plastic viscosity and the shear strength respectively of the grout (Figure 11.19),

and p is the pressure drop over a pipe length r (or, in terms of injection from a grout source, injection pressure p and radial penetration r).

In the case of a Newtonian fluid ($\tau_s = 0$), flow will gradually diminish linearly as $r \to \infty$. In the case of a Bingham fluid, flow will cease when $p/r = 2\tau_s/\alpha$. Thus, since in practice p will be limited by the desirability of reducing hydrofracture, the extent of permeation will be limited in the case of high shear strength grouts. This

(a) Uniform penetration from a cylindrical borehole source.

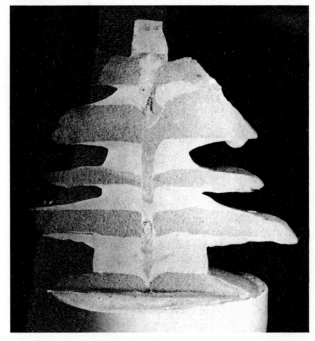

(b) Non-uniform penetration due to stratification.

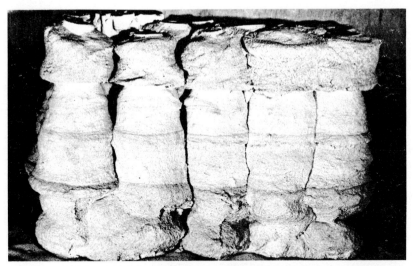

(c) Grout curtain formed by contiguous cylinders.

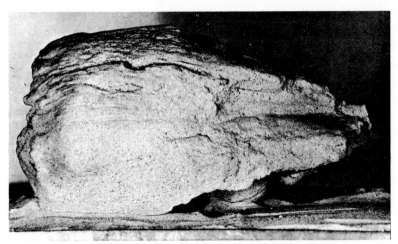

(d) Distortion of grouted mass due to groundwater flow.

Figure 11.20 AM9 Chemical Grout penetration into saturated dense medium sand (*reproduced from* Karol, 1968, *by permission of* the American Society of Civil Engineers).

emphasizes the importance of dispersed particulate grouts (where the flow properties will approximate to those of water) when attempting to permeate even the coarsest granular material if significant radial penetration from a single grout source is required.

Raffle and Greenwood (1961) provide an expression for the rate of radial permeation in the case of spherically-divergent two-phase (grout displacing water) flow of Newtonian fluids in the form:

$$t = \frac{ea_o^2}{kh_g}\left[\frac{\eta_g}{3\eta_w}\left(\frac{r^3}{a_o^3}-1\right) + \frac{\eta_s/\eta_w + 1}{2}\left(\frac{r^2}{a^2}-1\right)\right] \quad (11.12)$$

where t is the time taken for the flow to reach a radial distance r,
a_o is the source radius, or equivalent source radius in the case of a cylindrical source of length L and radius a, when (see Hvorslev, 1951)

$$a_o = \frac{m^{1/2}}{\ln[(L+m^{1/2})(L-m^{1/2})]}$$

where $m = L^2 - 4a^2$,
e, k are the void ratio and coefficient of permeability of the soil,
h_g is the hydraulic injection head,
and η_g/η_w is the ratio between grout and water viscosity.

In the case of a Bingham fluid, the effect of shear strength in introducing an additional pressure gradient which has to be overcome will be to reduce still further the rate of expansion of the grouted zone, eventually reducing it to zero. The effect of shear strength may be included in equation 11.12 as a correction to the hydraulic head h_g, based on equation 11.11. Thus, the shear resistance may be expressed in terms of hydraulic head h_s as:

$$h_s = \frac{2\tau_s}{\alpha} \quad (11.13)$$

The new hydraulic injection head will be equal to $h_g - h_s$. Flow will cease at $r = r_{max}$ when $h_g = h_s$.

This approach to spherically divergent (or even cylindrical) flow is rather idealistic since soil formations, in which permeation grouting is most commonly used, are rarely homogenous deposits with constant permeability. Permeability is likely to change significantly in a horizontal direction in layered soils, and horizontal permeability is likely to be significantly greater than vertical permeability. Any grout treatment is therefore likely to be uneven, and considerable care will be required to ensure that finer soils are adequately permeated (Figure 11.20).

Another cause of uneven grouting may be instability and mixing at the grout/groundwater interface. It is unlikely to occur if the grout viscosity is greater than that of the water being displaced, and the

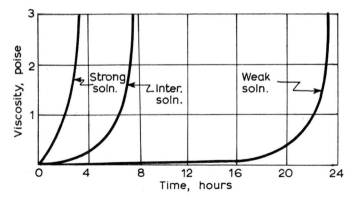

Figure 11.21 Setting characteristics of a typical organic polymer, chemical grout.

velocity (due to injection) of the lower boundary of the injection zone is greater than the rate of flow or sinking of the grout under its own weight. Scott (1963) quotes a criterion for stability under the combined influence of injection and sinking as:

$$\frac{\eta_g - \eta_w}{\eta_w} \cdot \frac{v}{k_w} + \frac{\gamma_g - \gamma_w}{\gamma_w} \geq 0 \qquad (11.14)$$

where v is the downward velocity of the grout boundary and the subscripts g and w refer to grout and water respectively.

The problem of grout stability when subjected to rather higher hydraulic gradients, as is often the case when grouting for water-stopping, is more severe. It is evident that flow of a Newtonian grout (equation 11.11, $\tau_s = 0$) displacing water at a specific hydraulic gradient will be reversed as soon as pumping pressures are removed. To prevent reverse flow it is desirable either that the grout has some shear resistance (implying the use of a particulate grout) or that pumping is maintained until the grout develops some set strength. Strength should ideally develop rapidly in a chemical grout following a period during which the grout retains a uniform and low viscosity (see Figure 11.21).

The hydraulic gradient required to overcome the shear resistance of particulate grout may be equated to $2\tau_s/\alpha$ (equation 11.11), and through this to the permeability and porosity of the ground. In the case of a coarse gravel subjected to a 10 per cent hydraulic gradient, flow will be resisted by a grout having a shear strength of

1 N m^{-2}. Normally the gradient would be lower. However, in grouting wide fissures and cavities and particularly those containing running water, very high shear strength grouts may be required.

Grouts for permeation grouting are selected primarily on their ability to permeate a particular soil or rock. Thus, *particulate* grouts will only be used in gravels, and the more *viscous* grouts in the coarser sands. The least permeable soils having a permeability of about 10^{-6} ms^{-1} will only be capable of treatment with the most *fluid* grouts having the lowest set strengths. Nevertheless, injection of very fluid chemical grouts into cohesionless soils can give them considerable strength. Some results of *undrained* tests on sands of various sizes injected with five chemical grouts of low, medium and high strength are included in Tables 11.7 and 11.8 (Farmer, 1974). The results in Table 11.8, each the average of 10 tests at different confining pressures, show that the grouts, in addition to imparting cohesion, have influenced the angle of internal friction of the sample, so tending to reduce this parameter in the case of the weaker grouts and to increase it in the case of the stronger grouts (Figure 11.22). The c values are roughly equivalent to those of clay shales and tend to increase with decreasing sand size.

Table 11.7 Set grout properties

Grout	(a) AM9	(b) Geoseal MQ4	(c) Chrome lignin	(d) Geoseal MQ9	(e) Urea–formaldehyde
percentage dissolved solids	11	13	30	25	45
unconfined compressive strength kN m^{-2}	gel	gel	680	1740	2800

(a) AM9 is a proprietary grout supplied by Cyanamid International and based on the solution polymerization of acrylamide.
(b) Geoseal MQ4 is a proprietary resin-forming grout supplied by Borden Chemical Company.
(c) A chrome-lignin grout (CL) using sodium dichromate to gel calcium lignosulphonate in sulphite lye.
(d) Geoseal MQ9 is a proprietary resorcinol–formaldehyde grout supplied by Borden Chemical Company.
(e) A urea–formaldehyde (UF) grout formed by reaction of urea with formaldehyde in acidic solution.

Table 11.8 Chemically grouted soil properties

Grout	Particle size mm	c kN m^{-2} (cohesion)	$\phi°$ (friction angle)
AM9	0.06–0.2	181	33.8
	0.2 –0.6	170	29.5
	0.6 –2	126	23.9
	2 –6	89	19.8
GEOSEAL MQ4	0.06–0.2	160	32.8
	0.2 –0.6	159	29.6
	0.6 –2	103	28.5
	2 –6	95	21.6
CHROME LIGNIN	0.06–0.2	149	38.7
	0.2 –0.6	161	34.9
	2 –6	103	32.6
GEOSEAL MQ9	0.06–0.2	278	45.6
	0.2 –0.6	224	41.0
	0.6 –2	185	37.7
	2 –6	158	34.3
UREA–FORMALDEHYDE	0.06–0.2	376	51.6
	0.2 –0.6	341	38.2
	2 –6	212	30.3
DRY SAND	0.06–0.2	—	36.5
	0.2 –0.6	—	35.4
	0.6 –2	—	33.2
	2 –6	—	31.0

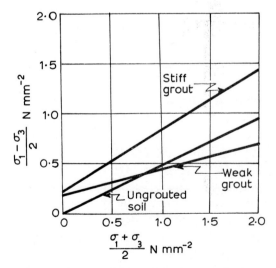

Figure 11.22 Failure envelopes for grouted and ungrouted sand samples 0.6–2 mm size range.

The apparent reduction in ϕ for the weaker grouts can be explained relatively simply. The presence of a low strength, low solids polymer gel or resin in the pore structure, replacing the air in the original porespace with a high water content Bingham type fluid having a high viscosity and shear resistance, will create a modified effective stress condition in the grouted soil. Thus, the total stress applied to the soil will be distributed between the original soil skeleton and the plastic grout infill. This will reduce the shear resistance of the original cohesionless soil skeleton, represented by the angle of internal friction, whilst the limitation on particle movement imposed by the grout, particularly at lower stress levels, will create cohesion. Introduction of stiffer grouts having a higher solids content and increased shear strength will tend to create a void-bound composite or cemented structure in which the grout, apart from limiting interparticle movement, will contribute to the overall strength of the composite. This is evidently the case with the urea-formaldehyde and Geoseal MQ9 grouts, whilst the chrome lignin grout has an intermediate effect. Ultimately, as in the case of an epoxy or polyester resin-grouted soil, the grouted soil will develop similar properties to those of hard continuous rock.

The influence of particle size on the shear resistance of samples of similar relative density reflects the reducing porosity and increasing number of interparticle contacts with finer (or more evenly graded) sand fractions. This will reduce the tendency for particle translation and rotation and interparticle crushing, with consequent increase in the sand stability. The obvious inference is that the shear resistance of denser soils and aggregates will be less affected by grout penetration.

The results may also be affected by two other factors. First, the index tests were carried out on dry sand, whereas the grouted samples had high aqueous contents. Bishop and Eldin (1958) report consistently higher ϕ values for dry sand than for saturated sand, the difference ranging from 2° (loose sand) to 5° (dense sand) at low confining pressures. Second, the method of sample preparation ensured reasonable sample homogeneity, particularly with regard to grout penetration. In practice, the degree of grout penetration will be variable, and relatively minor reductions in penetration may have considerable effect on c values.

The practical significance of the results should not be overstated. An extreme effect may be illustrated by examining the worst case of a uniform coarse sand (2 to 6 mm) injected with AM9 grout. On the basis of the results, the effect of complete injection with grout would be to reduce the shear resistance of the grouted soil at confining pressures greater than 350 kN m^{-2}. This reduction will be of critical importance in a minority of civil engineering stabilisation applications, and may be of minor significance when compared with other factors such as the efficiency of penetration.

11.6 Fissure grouting

Except in the case of porous rocks, different criteria apply to the grouting of rocks and soils. For instance, we have seen that many rock *materials* have low permeabilities and it would obviously be unrealistic to expect significant grout permeation through the rock material. Any significant water percolation and grout penetration must therefore occur along joints, bedding planes and fissures of variable width and extent, although analytical treatment of a homogeneously fissured rock may use the analogy of discharge flow through a soil (Chapter 8).

It is generally accepted (Vaughan, 1963) that rocks with an effective permeability coefficient as low as 10^{-7}ms^{-1} can be successfully treated with cement grouts. This indicates that cement particles will penetrate fissures rather more efficiently than the analogy with permeation might suggest. The reason for this can be demonstrated by comparing equivalent pore sizes and crack thicknesses for soils and homogeneously fissured rocks of similar permeability.

Table 11.9 summarizes hydraulic pore diameter (2α) data extracted from Table 11.6, with thickness (δ) data calculated from equation 8.35 on the basis of Snow's (1968) prediction of a minimum fissure separation of 1 m and on more conservative estimations of 0.1 m and 0.01 m.

It is evident that, except in the case of the closest fissure separations, a limiting coefficient of permeability of about 10^{-3}ms^{-1} for permeation cement grouting of soils is entirely compatible with a limiting effective permeability of 10^{-6} ms^{-1} for fissure cement grouting of rocks.

Initial rates of flow of a Newtonian liquid into a series of fissures in a borehole are given by equations 8.32 or 8.33. Cement grouts for fissure injection normally have a high water-cement ratio and may therefore be considered to have negligible shear strength. In the case

Table 11.9 Comparison of pore and fissure sizes (N is the number of fissures per metre and δ is the fissure thickness) in porous and fissured materials having the same nominal permeability

Permeability (effective) coefficient m s^{-1}	2α (μm)	$1/N = 1$ m	δ microns $1/N = 0.10$ m	$1/N = 0.01$ m
10^{-1}	1426	5340	2480	1150
10^{-2}	452	2480	1150	530
10^{-3}	152	1150	530	250
10^{-4}	52	530	250	115
10^{-5}	16	250	115	53
10^{-6}		115	53	25
10^{-7}		53	25	12

of a grout having significant shear strength, adjustment of injection pressures p_a in the form suggested by equation 11.13 would be required.

The initial rate of flow of the grout is governed principally by the ability of the fissures to accept the grout particles. This normally presupposes an effective permeability coefficient greater than 10^{-6} ms^{-1}, equivalent to a fissure width greater than 100 μm in jointed competent rock (Table 11.9). Rocks with an effective permeability less than 10^{-6} ms^{-1} (equivalent to one Lugeon unit) are considered by many engineers to be practically impermeable (see Benko, 1966) for dam foundation purposes.

Rate of flow of grout into cracks after overcoming initial shear resistance is dependent to a large extent on the grouting pressure. There is increasing evidence that, following initial penetration, fissures are expanded by grout fluid pressures in excess of the constraining pressure on the rock. At moderate excess pressures in competent rock, this expansion will result from quasi-elastic compression of the rock, which may be analysed in terms of the Boussinesq equations for deformation of a circular area of elastic material subjected to uniform loading (Figure 11.23a), although p_r reduces with r (Figure 11.23b). Adapted for radial injection (see Sabarly, 1968) of a crack, these equations give a deformation at the centre of the loaded area of

$$\delta_o = \frac{4(1 - \nu^2)}{E} r p_a \qquad (11.15)$$

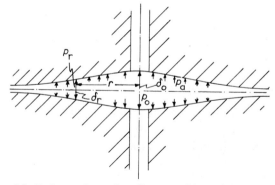

(a) Deformation of a fissure subjected to a uniform pressure p_a.

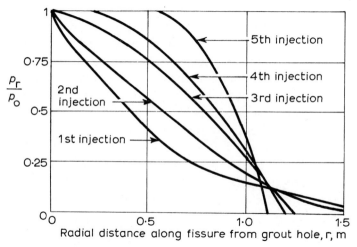

(b) Pressure distribution along a fissure (*after* Houlsby, 1969).

Figure 11.23 Deformation and pressure distribution along a fissure.

and at the edge of the injection front:

$$\delta_r = \frac{8(1 - v^2)}{\pi E} r p_a \qquad (11.16)$$

where E, v are respectively the modulus of elasticity and Poisson's ratio of the rock

and p_a is the average pressure acting in the fissure (Figure 11.23a).

It can be shown that for average values of $E = 3 \times 10^4$ N mm^{-2} $v = 0.25$, $p_a = 1$ N mm^{-2}, the deformation due to radial injection to a radial distance of 0.5 m will be equal to:

$$\delta_o = 0.062 \text{ mm}; \quad \delta_r = 0.04 \text{ mm}$$

and for a radial distance of 1 m:

$$\delta_o = 0.124 \text{ mm}; \quad \delta_r = 0.08 \text{ mm}$$

This effect can have two important implications for the fissure injection process. First, the effective permeability of, and hence rate of flow in, rock around a borehole is increased (equation 8.35) according to the relationship:

$$k_e = \frac{N\gamma(\delta + \delta_a)^3}{15\eta} \tag{11.17}$$

where δ_a is the average fissure width extension
and N is the number of fissures per unit length of borehole.

This equation is particularly important in the interpretation of Lugeon test results and may lead to indications of greater effective permeability and crack thickness than is in actual fact the case.

Second, increasing crack thickness with increasing radial penetration will allow continued absorption of large particles of cement which in turn will effectively seal the fissure when injection ceases.

A typical example of fissure grouting is the cementation process sketched in Figure 11.24 and developed in the nineteenth century in continental Europe, partly for stabilisation but mainly to reduce water flow into shafts and adits during construction. In this process, grout is pumped (using a compressed air piston pump) at relatively high pressures into the rock through the borehole extension of a grouted standpipe. The grout initially may comprise a 2:1* water/cement ratio suspension and the basic grouting mechanism involves forcing this fluid suspension through widening fissures until refusal at

---

*This ratio will depend upon the permeability measured during a pre-grout water test. Rule of thumb design mixes (Houlsby, 1973) are:

Permeability (Lugeons)	w/c ratio
10	3:1
10–30	2:1
30–60	1:1
60	0.8:1

This mix will then be thickened or diluted depending on subsequent grout penetration experience.

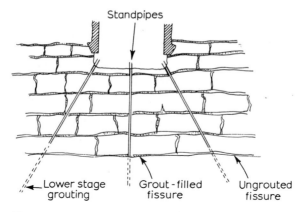

Figure 11.24 Cementation grouting from a shaft in fissured rock.

a specified upper limit pressure or pressure-volume flow combination.

Some of the grouting methods used most commonly in practice are summarised by Houlsby (1973) in an excellent practical grouting manual. There are two basic methods – *full depth* grouting for near-surface blanket treatment, in which holes are grouted to their full depth in one operation, and *stage* grouting for deep curtain or waterproofing treatment in which holes are grouted in stages.

Although circuit grouting of a number of holes through a ring main is feasible, holes are usually grouted individually in a closure sequence (Figure 11.25). This means that following the grouting of an initial grid, further intermediate holes are drilled, tested and grouted until the whole grouting zone reaches the desired level of impermeability.

11.7 Hydrofracture

Permeation and fissure grouting aim to inject grouts into pore-space or fissures in order to reduce permeability or increase strength. This is best achieved if the original physical structure of the rock is not disturbed by inducing inter-granular movement or creating new fissures. It is, however, often necessary to exceed optimum pressures in order to obtain economical injection rates and penetrations. If pressures significantly exceed overburden pressure some movement of the rock fabric may occur, the resultant disturbance or fissuring being called *hydrofracture* (see Cambefort, 1964). The term was

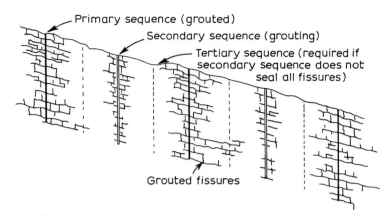

(a) Full-length blanket grouting of fissured slope.

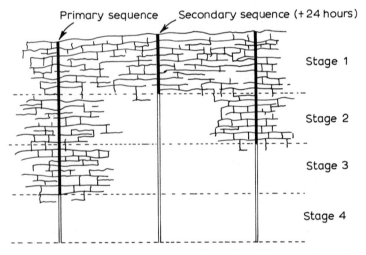

(b) Down-stage curtain grouting in fissured rock [up-stage grouting would start from the bottom of the holes].

Figure 11.25 Grouting methods (*after* Houlsby, 1973).

originally used to describe a similar process utilized widely to rejuvenate non-productive oil wells by injection of water at high pressures. Hydrofracture is caused when fluid (grout) pressures induce tensile stresses greater than the strength of the ground. It can be illustrated most simply by comparing the stress situation around a borehole in an elastic continuum subjected to hydrostatic overburden pressure (γz), and initially to zero grout pressure, with the

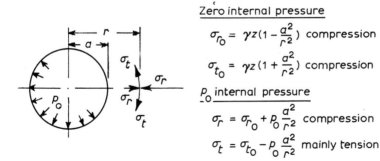

Figure 11.26 Stress distribution around a borehole.

stress situation around the borehole when it is subjected to an internal grout pressure p_0 (shown in Figure 11.26).

If failure occurs, fractures will develop in a radial direction when the tensile stress σ_t caused by the grout pressure exceeds (or if a tensile stress negative convention is used — is less than) the tensile strength S_t of the ground. Thus the conditions for initiation of hydrofracture at the surface of the hole ($r = a$) are:

$$\sigma_t = 2\gamma z - p_0 \leqslant -S_t \tag{11.18}$$

or

$$p_0 \geqslant 2\gamma z + S_t \tag{11.19}$$

where S_t is a positive quantity.

If $S_t = 0$ as in most soils or near-surface, fissured, massive rocks, p_0 must be greater than twice the overburden pressure to initiate hydrofracture. This approach is extremely simple and, in practice, rules of thumb limiting grouting pressure are based on similar analyses (see, for instance, Hubbert and Willis, 1957; Morgenstern and Vaughan, 1963; and Fairhurst, 1964). These rules suggest safe pressures of 1 to 5 times overburden pressures. It is useful to refer to Table 11.10 which summarizes some case histories quoted by Wong and Farmer (1973).

The simple approach, however, assumes the presence of totally impermeable ground in which all the fluid force is expended in expanding the grout source. In fact, in a permeable rock or soil, much of the fluid pressure is expended in forcing grout through the ground, inducing at the same time a *seepage* force (see Section 8.3) acting radially outwards. In other words, only a fraction of the available internal pressure (say) Ξp_0 will be available for source

858   *Principles of Engineering Geology*

Table 11.10   *Grouting case histories*

Structure	Type of ground	Grout used	Depth of injection m	Injection pressure kN m^{-2}	Reference
Mangla Dam, West Pakistan; grout cut-off	22 m alluvial gravel	clay–cement (cement 135 kg, clay 415 kg, water 800 l m^{-3} of grout)	3–22	3000–8000	Skempton and Cattin (1963)
Kotah Dam, India; grout cut-off	about 9 m of black clay and fine sand underlain by about 6 m of sandy cobbles	clay–cement (clay–cement ratio = 14)	9–15	1000	Greenwood and Raffle (1963)
Great Cumberland Place; underpinning	fine sand and gravel to 5 m	clay–cement and chrome-lignin	1–5	350 (up to 3 m)	Neelands and James (1963)
Blackwall Tunnel	alluvial sand and gravel	clay-chemical chrome-lignin	5–6 (minimum depth)	350	Perrot (1965)
Shimouke Arch Dam, Japan; consolidation grouting	Andesite bedrock	cement		90–130 per linear metre of cover	Soejima and Shidomoto (1970)
Balderhead Dam; water injection test	4 m alluvial gravel underlain by shale and sandstone	water	13–35	maximum 50 per metre cover	Morgenstern and Vaughan (1963)
30 dams of world wide location	various	mainly cement		40–80 per metre cover	Grundy (1955)
Oldbury Power Station; grout curtains for excavation work	4 m alluvial clay underlain by horizontal beds of calcareous sandstone and marl	cement	6–15	150	Perrot and Lancaster-Jones (1963)

expansion while the remainder $(1 - \Xi)\, p_0$ of the original pressure will be expended as a seepage force. Ξ may be termed a dimensionless *area factor* and will be equal to 1 in impervious rocks and equal to zero in infinitely porous soils. For intermediate cases (Pulpan and Scheidegger, 1965), such as increasingly fractured rocks, or soils and fractured rocks permeated by increasingly viscous or granular grouts, Ξ will assume values between 1 and 0. Numerical values of Ξ are, however, extremely difficult to state because of the varying structure of the massive rock or soil. Thus, in sound rocks with small porosity n, the value of Ξ is close to $(1-n)$. In this case, Ξ is equivalent to the porosity factor of Lubinski (1954) and Biot and Willis (1957). However, with increasing porosoty, Ξ will decrease much more rapidly than $(1-n)$ to a value approaching zero in soils permeated by water, where n is normally about 0.4. On the other hand, permeation of a particulate grout (especially one containing large particles) through a soil or fractured rock will tend to increase the effective contact area between particles, and hence the value of Ξ. In practice, Ξ will probably lie in the 0 to 0.5 range for soils, 0.5 to 1.0 range for rocks. As such, it will have importance in the calculation of uplift pressures.

In terms of Ξ, the seepage force per unit volume at radius r around a borehole will be given by:

$$-(1-\Xi)\frac{dp_r}{dr} = \frac{(1-\Xi)p_0}{r\ln(R/a)} \tag{11.20}$$

where p_r is the fluid pressure at radius r,
      a is the borehole radius and
      R is the radius of grout penetration.

This equation has been solved by Sauvage de St. Marc et al (1960) to give stresses around the hole due to grouting in an elastic continuum as:

$$\sigma_R = \Xi p_0 \left(\frac{a}{r}\right)^2 - (1-\Xi)p_0 \left[\frac{p_r}{2(1-v)p_0} - F(r)\right] \tag{11.21}$$

$$\sigma_T = -\Xi p_0 \left(\frac{a}{r}\right)^2 - (1-\Xi)p_0 \left[\frac{p_r}{2(1-v)p_0} + F(r)\right] \tag{11.22}$$

where

$$F(r) = \frac{1}{2(1-v)}\left(\frac{a}{r}\right) + \frac{(1-2v)}{4(1-v)\ln(R/r)}\left(1-\frac{a^2}{r^2}\right) \tag{11.23}$$

and the subscripts R and T denote the *seepage* case.
Assuming zero vertical strain, the vertical stress σ_z is given by:

$$\sigma_z = \frac{v(1-\Xi)}{(1-v)}p_r \tag{11.24}$$

The original stresses (assuming a vertical borehole) are given by the zero internal pressure equations in Figure 11.26 modified to allow for non-hydrostatic conditions, and equations 11.21 – 11.24 replace the pressurized hole equations. Then, assuming cracking to take place principally in the radial stress direction, the conditions for vertical hydrofracture initiation at the borehole sidewall ($r = a$) are written according to Wong and Farmer (1973) as:

$$\sigma_{t_0} + \sigma_T \leqslant - S_t \qquad (11.25)$$

or

$$2K\gamma z - \frac{(1 - \Xi v)}{(1 - v)} p_0 \geqslant - S_t \qquad (11.26)$$

whence

$$\frac{p_0}{\gamma z} = \frac{(1 - v)}{(1 - \Xi v)} \left(2K + \frac{S_t}{\gamma z} \right) \qquad (11.27)$$

with fractures initiated mainly in a vertical mode and radial direction. K is the original horizontal to vertical stress ratio.

Criteria for fracture initiation where $S_t = 0$ are illustrated in Figure 11.27. The $K = 0$, $\Xi = 1$, $p_0/\gamma z = 2$ case (equation 11.19) is obviously an optimum condition for hydrofracture initiation. Otherwise, hydrofracture in soils will be initiated at pressures close to or below overburden pressure and in rocks at higher pressures depending upon their tensile strengths. In addition, porous rocks and soils subjected to seepage forces will be more susceptible to hydrofracture. This possibility is of some significance in the context of permeability measurement in soils, as noted by Bjerrum *et al* (1972). In grouting, hydrofracture initiation may be tolerable in most rocks and soils provided that it does not *propagate* uncontrollably from the injection source (see Houlsby, 1969). In fact, some hydrofracture may actually be desirable as a means of speeding injection in tight or variable ground. If the previous analysis is extended for values of $r > a$, it is immediately obvious that *very high* pressures are required to cause fractures in zones remote from the injection source. Even if the analysis is adapted to allow for a fractured Mohr-Coulomb zone surrounding the injection source instead of a completely elastic rock, very high pressures are still required (see Wong and Farmer, 1973). This indicates that the approach may be unrealistic, for *surface uplift* has been noted in grout injection when high pressures

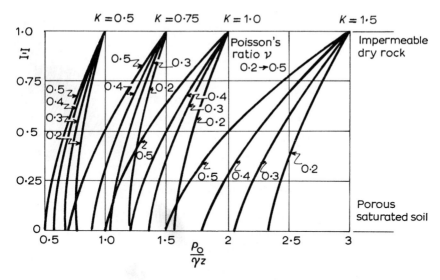

Figure 11.27 Pressure required to initiate cracks at a pressurized borehole surface (*after* Wong and Farmer, 1973).

are used. Two simple approaches may be suggested to assess the degree of fracture and the prospects of large scale ground failure associated with grouting. These assume respectively the existence of a truncated cone failure zone in the case of soils, and a series of expanding stress-induced fissures in the case of rocks.

The former case is analysed in Figure 11.28. The latter case may be analysed through an energy balance approach (see Perkins and Krech, 1966, 1968) equating grout energy input $\delta E = p_0 \delta V$ (where p_0 is the average injection pressure and δV the volume of fluid injected during a specified time) to the stored energy δE_s (elastic strain energy in the rock and fluid) and irrecoverable energy δE_r (mainly work done in hydrofracturing the soil and in overcoming fluid friction drag and shear resistance). The implications of this analysis, discussed in Wong and Farmer (1973), are that an increasingly small amount of energy is available for hydrofracture propagation with increasing source radius, and that ultimately a rapid extension of the input energy will be required in order to maintain the hydrofracture process. This conclusion is demonstrated by Morgenstern and Vaughan, (1963). The result is that hydrofracture can be controlled more readily by input energy than by arbitrary hydrofracture *initiation* pressure limitations.

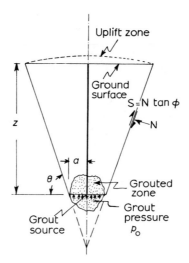

Weight of truncated cone

$$W_c = \frac{\pi \gamma z}{3 \tan^2 \theta} [z^2 + 3az \tan \theta + 3a^2 \tan^2 \theta]$$

Shear resistance at cone surface

$$S = 2W_c \frac{(1 - \sin \phi)\cos(\overline{180 - \phi + \theta})}{\cos \phi \sin \theta}$$

Upward force of grout

$$F_G = \pi a^2 p_0$$

Then for uplift to occur

$$F_G \geqslant W_c + S$$

or

$$\frac{p_0}{\gamma z} \geqslant \left[\frac{(z/a)^2 + 3(z/a)\tan \theta + 3 \tan^2 \theta}{3 \tan^2 \theta} \right] C'$$

where

$$C' = 1 + \frac{2(1 - \sin \phi)\cos(\overline{180 - \phi + \theta})}{\cos \phi \cos \theta}$$

Then if $\phi = 30°$ and $\theta = 45 + (\phi/2) = 60°$ then $C' = 2.3$ and

$$\frac{p_0}{\gamma z} \geqslant 0.11 \left[\frac{z^2}{a^2} + 5.2 \frac{z}{a} + 9 \right]$$

Figure 11.28 Conditions for ground surface uplift based on truncated cone analogy. Any cohesion in the ground is neglected. (*After* Wong, 1971).

11.8 Cavity grouting

The existence of cavities such as abandoned mineral workings or natural openings beneath new development land can create considerable difficulties during site investigation and design. The origin and detection of these cavities has been considered in Chapter 7. From a technical point of view, the actual grouting is relatively simple and, provided that the cavity can be accurately defined, it can always be filled. However, with extensive mineworkings, this may involve considerable cost which can only be reduced by a careful assessment of the short-and long-term interactions between the cavity and the structure.

It is important, therefore, during the site investigation process to identify and define any potential cavities (see Section 7.15). This may involve rather more than a cursory preliminary investigation, particularly in areas where minerals of economic interest or soluble rocks are found. Since mining operations can include such unlikely materials as chalk, limestone and basalt (for roadstone), and extensive cave systems can exist in limestones, some examination of local historical, archaeological and caving sources is often required.

The actual investigation will depend on the geology and available knowledge of the underground workings or caves. If these are well defined, or if underground access is feasible, relatively minor drilling works will be required. If the geology is complex and the workings partially collapsed, detailed exploration and accurate and careful core recovery and logging will be needed.

The most common type of underground cavity in Britain results from old coal mine workings at shallow depths in the North of England. These are often extensive and well-preserved comprising say 5 m x 3 m pillars supporting an area from which approximately 70 per cent of the coal has been extracted. The potential stability of these workings can range from good to totally unstable when subjected to the type of foundation loading associated with medium to large buildings — a loading of the order of 100 kN m^{-2} (1 ton ft^{-2}).

Typical of the type of damage associated with these workings is that of partial roof collapse between the pillars (Figure 11.29) leading to controlled collapse of ground and eventual migration of the cavity to the surface. Wardell and Wood (1966), who have studied this type of collapse in detail, suggest that a depth of cover equal to at least five times seam height is required for stability in this

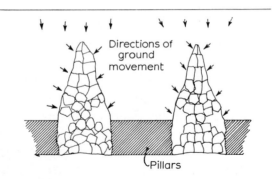

Figure 11.29 Collapse zones migrating towards ground surface over bord and pillar workings.

type of situation, and it is likely that in weaker rocks a greater thickness than this would be required. The problem is considered in detail in Section 7.15, and an estimate of safe cover depths can be obtained from equation 7.28.

In fact, it is doubtful whether any undergound cavity within 30 to 40 m of a structure can be guaranteed stable once its stress regime has been altered by the imposition of large structural loads. Butler (1974), referring particularly to mine workings, suggests a series of questions which would be useful in determining treatment needs for any cavity situation. These include:

(a) *Structure:* Is the structure permanent or temporary? Is it flexible, rigid, simply-supported or continuous? What are the costs of failure and restoration?
(b) *Cover:* What is the thickness of cover? Does it comprise competent or incompetent rocks? Are the rocks jointed or fissured?
(c) *Pillar or sidewall support:* What is the strength of pillar or sidewall rocks? What is the stress acting on them? What is the additional stress from the structure? Are the supporting rocks wet or subject to periodic flooding? If flooded, what is the piezometric level and what changes occur in this level?
(d) *Drainage:* Will development affect surface drainage? Where will storm water be diverted? Will landscaping affect surface drainage?

If the answers to these questions raise sufficient doubt as to stability or economy then it is likely that some form of treatment will be

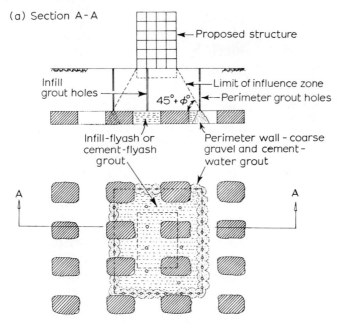

Figure 11.30 Treatment of extensive uncollapsed bord and pillar mineworkings below a foundation.

required. An obvious economy is that treatment be confined to the area of influence of the structure and utilize relatively cheap materials. In isolated cavities this will mean totally infilling the cavities beneath the structure. In extensive mine workings, an essential prerequisite is the construction of a high repose angle perimeter wall from gravel and cement or stiff grout in order to confine the infilling material. This material will usually comprise, because of its relative cheapness, availability and good pumpability, a flyash (pfa) and water slurry. A typical example of a perimeter wall infill treatment is sketched in Figure 11.30. Figure 11.31 illustrates a chalk mine infill situation at Woolwich as described by Leonard and Moller (1963), where 450 000 m^3 of flyash were used to fill abandoned chalk mines when collapse of numerous cavities 10 to 20 m below ground surface was causing extensive surface subsidence. In this case, ready access allowed infilling to proceed from behind bulkheaded underground galleries.

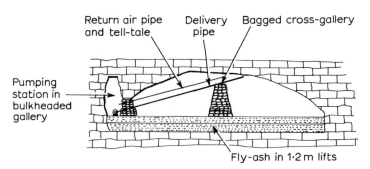

Figure 11.31 Fly-ash infilling of chalk mine (*after* Leonard and Moller, 1963).

11.9 Electro-chemical stabilisation

Although grout penetration is effectively limited by permeation rates to soils with permeabilities greater than 10^{-7} ms^{-1}, many important ground stabilisation problems concern low permeability saturated clays and particularly weakly bonded active clays with a large, low-valency cation content. As is seen in Chapter 1, clays containing significant proportions of sodium (Na$^+$) and lithium (Li$^+$) cations have a tendency to adsorb large quantities of water onto their particle surfaces. On the other hand, the presence of particles containing higher valency cations, such as calcium (Ca^{2+}), magnesium (Mg^{2+}) and particularly ferric iron (Fe^{3+}) and aluminium (Al^{3+}), with higher bond strength, will create a clay with a low adsorbed water content.

Since cations are interchangeable, the introduction into a saturated active clay porespace of a solution or solid containing an excess of high valency cations leads to a reduction of adsorbed water and a consequent improvement in soil strength — a feature that is particularly marked in the case of montmorillonite clays. This property has in fact been widely utilised in the stabilisation of montmorillonite clays, particularly the Black Cotton soils widely distributed in Africa and Asia, by ploughing cement or lime into the exposed soil surface (Bulman, 1972). It also forms the basis for electrochemical stabilisation of soils where dissolved additives can be dispersed with reasonable efficiency into low permeability clays by a process of *electro-osmotic* flow (Farmer, 1975).

If an *electric potential* is applied to a saturated porous material, water will flow to the *cathode*. This phenomenon, known as electro-osmosis, can be explained in terms of the Helmholtz double

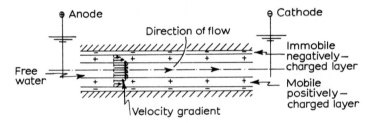

Figure 11.32 Representation of electro-osmotic flow mechanism.

layer theory which postulates that, in a capillary full of electrolyte, two layers are adsorbed onto the capillary surface (Figure 11.32). A thin, negatively-charged layer rigidly attached to the capillary surface attracts a positively-charged layer of soluble cations and associated molecules which is thicker and less rigidly attached.

When a potential difference is applied to the ground, the positively-charged layer moves toward the cathode, carrying with it any free electrolyte in the capillary. A mathematical treatment, based on the analogy between the double layer structure and a condenser, gives a flow rate Q through a capillary of:

$$Q = \frac{Z\,Da^2}{4\,\eta}\frac{E}{L} \qquad (11.28)$$

where Z is the potential across the double layer, termed the *zeta* or electrokinetic potential,
D is the dielectric constant and η the viscosity of the electrolyte,
a is the capillary radius,
and $E/L = E_i$ is the potential gradient.

If equation 11.28 is multiplied by

$$\frac{1}{An} = \frac{1}{\pi a^2}$$

(where A is the discharge area and n the soil porosity) then the discharge flow through a bundle of capillaries, equivalent to a soil mass, is given by:

$$\frac{Q}{An} = \frac{ZD}{4\eta}\frac{a^2}{\pi a^2} E_i \qquad (11.29)$$

or,

$$\frac{Q}{A} = k_e E_i \qquad (11.30)$$

which is directly analogous with Darcy's law (Section 2.4) for hydraulic gradient-dominated discharge flow through a porous medium where $k_e = ZDn/4\pi\eta$ may be termed the *electro-osmotic coefficient of permeability* having units of ms^{-1} per volt cm^{-1} of potential gradient equal to 10^4 mm^2 volt^{-1} s^{-1}.

The importance of k_e is that it is independent of a. Of its related variables, D and η are constant, and although Z is constant in saturated soils, it does reduce significantly in unsaturated soils. Parameter n depends mainly on the adsorbed water content of the soil and may be very high in some montmorillonite clays. In most saturated soils, however, n and k_e will be reasonably constant. Casagrande (1952) quotes an average value of $k_e = 5 \times 10^{-7}$ ms^{-1} per volt cm^{-1} (5×10^{-3} mm^2 volt^{-1} s^{-1}) for a series of fine-grained soils of widely varying permeability and moisture content (see Table 11.11). The implication is that in soils having a hydraulic coefficient of permeability less than 5×10^{-7} m s^{-1}, improved drainage and/or grout permeation can be obtained using electro-osmosis.

Although claims have been made to the contrary, various factors militate against the use of electro-osmosis simply for drainage. The most important of these is the fact that electro-osmosis is only really effective in saturated soils with a high water content, the efficiency of the process decreasing radically with reduction in water content.

Table 11.11 Values of k_e (*after* Casagrande, 1952)

Material	Moisture content percentage	k_e mm^2 volt^{-1} s^{-1}
London Clay	52.3	5.8 x 10^{-3}
Boston Blue Clay	50.8	5.1 x 10^{-3}
Kaolin	67.7	5.7 x 10^{-3}
Clayey Silt (U.K.)	31.7	5.0 x 10^{-3}
Rock flour (Hartwick, N.Y.)	27.2	4.5 x 10^{-3}
Red Marl (Scotland)	23.7	1.7 x 10^{-3}
Bentonite	170	2.0 x 10^{-3}
Bentonite	2000	12.0 x 10^{-3}
Mica powder	49.7	6.9 x 10^{-3}
Fine sand	26.0	4.1 x 10^{-3}
Quartz powder	23.5	4.3 x 10^{-3}

There is therefore a minimal water content reduction with every probability of rehydration unless some other agency is involved. In fact, where electro-osmosis has been found to be effective in the past, its efficacy has probably been due to stabilisation of the soil by solution and dispersion of the anode material rather than by reduction of the water content.

Electrochemical stabilisation resulting from *anode solution* relies on two main processes. These comprise initial base exchange reactions on the surface-active area of the clay mineral particles, replacing weakly with strongly bonding cations, and second, the formation of cementing compounds by reaction between the electrolyte and the dissolved anode. In the simplest cases, these reactions are obtained by hydrolysis and oxidation of the anode material where this is readily oxidizable, as with the iron or aluminium electrodes invariably used in electro-osmosis. In the case of iron, ions released from the anode react with water to form ferrous hydroxide, at the same time increasing the hydrogen ions and hence acidity in the water:

$$Fe^{2+} + 2H_2O = Fe(OH)_2 + 2H^+ \qquad (11.31)$$

The ferrous hydroxide is in turn readily oxidized to form ferric hydroxide:

$$4\,Fe(OH)_2 + O_2 + 2H_2O = 4\,Fe(OH)_3. \qquad (11.32)$$

In addition, both ferrous, ferric and aluminium hydroxides are amorphous colloids, acting as cementing compounds and binding the soil particles together. Release of hydrogen ions in the anode zone will tend to retard the reactions, encouraging the movement of the metal cations towards the cathode and creating a uniform spread of cementing material over the stabilisation zone.

This type of process can be utilized to give permanent stability in wet clays with a relatively low reduction in water content. Casagrande (1952) shows that there can be a 60 per cent increase in the undrained strength of clay strata for a decrease in water content of only 2.5 per cent. Although this increase was attributed to dewatering, a more logical explanation would include base exchange and cementation as contributing factors.

Addition of dissolved additives at the anode may be designed to increase the base exchange reaction, and to improve or catalyse cementation. Most of the additives that are used are organic or inorganic compounds of aluminium or calcium, with iron usually being present as the electrode material. The most commonly used additive (see Adamson *et al*, 1966) is calcium chloride which intensifies dissolution of the iron anode by formation of ferrous chlorides while at the same time producing calcium hydroxide which acts as a *cementing agent*. Another useful group of additives are polyelectrolytes combining electrolytic properties with an ability to gel. Phosphoric acid has also been used (Gillott, 1968) with aluminium electrodes; the mechanism of stabilisation involving a release of amorphous aluminium hydroxide and aluminium phosphate.

Quantitative data on additives is rare, but laboratory experiments described by Adamson *et al* (1966) indicate the creation of — and increase in — strengths greater than the strengths associated with the use of electrodes alone, and with specific indications of changes in the clay mineral structure and the formation of cementing compounds.

Owing to the high cost and relative sophistication of the process, practical applications have been limited. A summary of case histories given by Farmer (1975) outlines its uses in silts, loess, loam and clays together with various mixed soils. Cathode layouts are similar to well pointing (Section 8.4) arrays in coarser soils and in many cases they utilize similar equipment. Some successful applications have been seen in the vicinity of river or harbour works where electro-osmotic flow reinforces and utilizes pre-existing hydraulic gradients to ensure maximum solute penetration in a soil of uniformly high water content (Figure 11.33).

Power requirements for the system may be estimated by substituting for $A = \rho L/R$ in equation 11.30, whence:

$$Q = k_e \frac{E}{L} \rho \frac{L}{R} = k_e \, \rho \frac{E}{R} = k_e \rho I \qquad (11.32)$$

where ρ is the resistivity of the soil, R is the resistance over length L, and I is the current.

In fact, in a typical saturated clay where $\rho = 1$ ohm–m, a potential of about 100 V and separation of about 3 m will give a potential gradient of about 30 Vm^{-1}, and will require a direct current output of about 150 amps in order to maintain a reasonable flow rate.

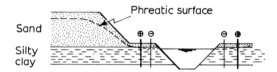

Figure 11.33 Typical electrode placement for electro-chemical stabilisation, utilizing direction of groundwater flow (*after* Casagrande, 1952).

11.10 Groundwater freezing

The properties of a saturated soil or rock will be altered if the state of the soil water phase is changed from liquid to solid (ice). This change of state imposes on the soil two main effects which have some importance in engineering geology:

(a) the strength of the soil will be improved, and
(b) depending on the particle size of the soil, groundwater flow will be eliminated.

The mechanical properties of frozen soil, particularly strength, depend upon the *bond* between soil particles and the ice, and also upon the amount of unfrozen water in the soil. In clay soils, this can be high since a substantial amount of porewater adsorbed onto the surface of the clay minerals will have a lower freezing point than the freezing point of free water. In quartzitic soils containing a minimum of adsorbed water and a relatively lower specific surface area, more ice and stronger bonds are developed. The strength of the ice and the soil-ice interfacial bonds will also increase with decreasing temperature, so giving a frozen soil-temperature relationship of the type illustrated in Figure 11.34 (see Jumikis, 1966). The data are based on the short-term strength of saturated frozen soils. However, since ice has well-documented creep properties (see Robertson, 1963) it is evident that long-term strengths will be rather less than those quoted and Sanger (1968) suggests that it is reasonable to take the ultimate long-term strength as being about one-quarter the short-term strength (Figure 11.35). Larger reductions in strength tend to occur in unsaturated soils.

Nevertheless, the strength of frozen soils is relatively higher than the strength which could be obtained by using some geotechnical processes, and for this reason groundwater freezing can be utilized to prevent groundwater flow. The most common application is in shafts or tunnels (see Sanger, 1968; Collins and Deacon, 1972; Braun, 1972)

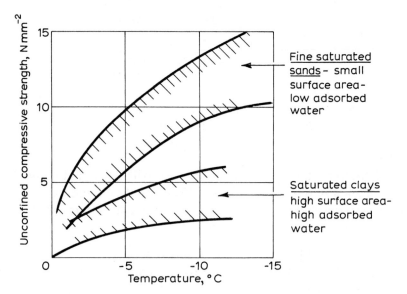

Figure 11.34 Relationship between compressive strength and temperature of frozen soils (*after* Tsytovich *quoted by* Jumikis, 1966).

where the limited geometry, sometimes combined with extreme instability and groundwater problems, lends itself to what is essentially an expensive and difficult geotechnical exercise.

Design and application of freezing processes are considered in detail by Sanger (1968) and by Jumikis (1966). Stability computations are usually based on plastic failure of a $c - \phi$ material, sometimes on the basis of elastic theory with allowances for a safety factor and estimates of long-term deformation. The computed ice wall thickness is then obtained by forming cylinders of frozen soil around cased boreholes through which a coolant of suitable type is circulated. Liquid nitrogen may also be used. These cylinders eventually join up with one another and subsequent cooling then extends the required wall to the design thickness.

The freezing process can be analysed in two ways. In the transient state while ice is forming, the rate of advance of the ice will depend on the *thermal diffusivity* (ψ)* and on the moisture content. It is evident that the rate of extension of the frozen zone will reduce with increasing diameter, but this will be offset to a certain extent by the

---

$$*\psi = \frac{\text{Thermal Conductivity}}{\text{Density} \times \text{Specific heat}}$$

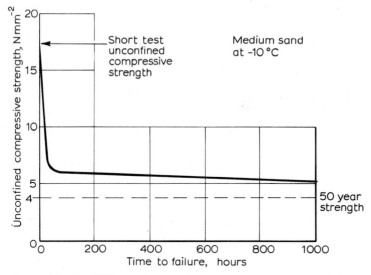

Figure 11.35 Effect of creep on compressive strength of frozen sand (*after* Sayles, *quoted by* Sanger, 1968).

reduced ψ of the frozen soil. In the *steady state* condition after the ice wall is formed, the thermal conductivity of the soil will determine heat outflow. Obviously, both transient and steady state freezing will be affected by groundwater flow and the method is not feasible except in reasonably quiescent conditions.

The expansion of water during freezing is not a serious problem in free-draining soils, or even in poorly-draining soils where the effect of freezing will probably be improved by induced consolidation. However, migration of water from the unfrozen soil into the frozen zone can occur in fine soils. This action is caused through the attraction of adsorbed water by the growing ice crystals, and subsequent replenishment of the adsorbed water by capillary action. Formation of lenses of ice normal to the direction of heat flow and in the plane of freezing by this process can lead to expansion of the soil, a phenomenon known as *frost heave*. Frost heave is a feature of soils with, according to Casagrande (1932c), a clay size fraction (<0.002 mm) greater than 3 per cent in non-uniform soils and greater than 10 per cent in uniform soils. During the formation of ice walls, heave and associated high heave pressures can be reduced by rapid freezing rates.

Naturally-frozen soils or *permafrost* cover about 25 per cent of the surface land of the earth, principally the polar regions, but also in isolated mountainous regions at quite southerly latitudes. Permafrost occurs where the nett groundheat flow over a period of years is outwards, and it is usually and simply defined as perennially frozen ground. It is thickest in non-glaciated zones, reaching maximum thicknesses up to 1600 m in the Yakutskcaya region of the USSR, 400–600 m in Northern Siberia and 300–500 m in Alaska and Canada (Black, 1954). In these zones, the temperature at depths from 10–30 m will be approximately equal to the mean surface temperature, reducing thereafter with depth. In glaciated zones, ice thicknesses range from 50–300 m, and in isolated permafrost zones they may be 10–30 m thick.

Calculations of heat flow capacities suggest that much of the permafrost has been deposited in a frozen state. Engineering problems occur where the ice component of the permafrost exceeds the porespace or where the presence of ice maintains an artificially high porespace which predisposes the ground to settlements on thawing. There is indeed a serious settlement problem associated with thawing of the ground which may, for example, occur during hot oil pumping or transportation by pipeline. Suggested remedies range from refrigeration to maintain the permafrost, small earth-contact structures to reduce thaw, and piles coated with bitumen to combat negative draw-down friction.

11.11 Bentonite suspension

Bentonite, although in the narrowest sense the name of a pure montmorillonite clay found near Fort Benton, Wyoming (see Gillott, 1968) is widely used as a general term to describe strongly *colloidal* clays which disperse in suspension with water to form a colloidal gel. Usually, but not exclusively, these are montmorillonite clays with a high sodium cation content, and when fully hydrated form a thixotropic, colloidal gel at bentonite concentrations of between 3 percent and 12 percent by weight of water.

The structural mechanisms creating these properties have been discussed previously (see Section 1.4). Most simply, it is found that on dispersion, by *efficient* mixing in a colloidal or high shear mixer, the bentonite clay breaks down into small plate-like particles in the sub-micron size range. These plates are negatively charged on the surfaces and positively charged at the broken edges. Mutual attraction within the dispersed suspension forms a typical, dispersed edge-to-face card-house gel structure (Section 3.1) which increases in strength with time. Subsequent re-agitation will break down this structure into a randomly orientated fluid suspension. Dehydration will lead to closer packing and ultimately to restoration of an orientated structure.

The fluid properties of bentonite suspensions are essentially those of a Bingham fluid (Figure 11.19), the shear stress τ to initiate flow being defined in terms of yield strength τ_s or gel strength and plastic viscosity η_p ($\tau = \tau_s + \eta_p \, d\epsilon/dt$), and sometimes apparent viscosity η_a ($\tau = \eta_a \, d\epsilon/dt$). These parameters are related to the solids content of the suspension, to the surface area of the solids and to the base exchange capacity of the bentonite. In effect, the solids content is determined by measurement, the base exchange capacity can be optimised by pre-treatment to raise the level of sodium cations in suspension, and the surface area can be maximized by efficient mixing and *hydration*. Simple relationships based on Fann viscometer measurements are illustrated in Figure 11.36 for commercial Berkbent CE (Berk Ltd., 1971), a type of bentonite commonly used

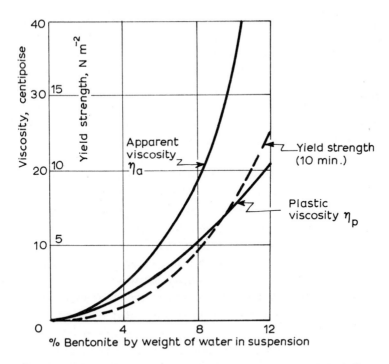

Figure 11.36 Effect of bentonite concentration on initial flow properties — Berkbent CE (*after* Berk Ltd., 1971).

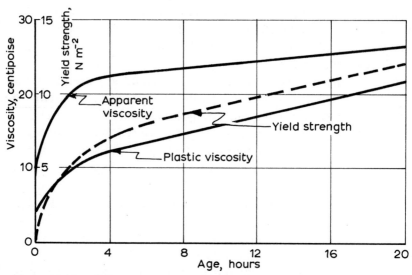

Figure 11.37 Effect of ageing on suspension flow properties. (6 per cent suspension, Berkbent CE – *after* Berk Ltd., 1971).

in civil engineering. These relationships show the effect of solids concentration on initial apparent viscosity at 600 rpm (Fann viscometer) and on initial plastic viscosity (at 300 rpm and 600 rpm) and ten minute gel strength (at 3 rpm). Figure 11.37 shows the effect of ageing on the same parameters for a 6 per cent solids suspension by weight of water. In this case, ageing is essentially an indication of continuing hydration and break-up of particles, with a consequent increase in solid-phase surface area following initial mixing. If re-agitated, therefore, the suspension flow properties would rapidly return to levels near those represented by the hydration state prior to agitation.

The uses of bentonite suspensions in ground engineering are numerous. Strictly speaking, the only application aimed at ground improvement is when bentonite suspensions are used alone or in combination with cement or chemicals as a grout to improve stability or reduce water flow in permanently saturated ground. Combined with cement, sodium silicate or phosphates, the suspension forms, partly through cement hydration or chemical reaction, and partly through cation exchange, a stiff non-thixotropic gel which can be used in saturated soils to strengthen the soil. Used alone, —although bentonite suspensions can improve soil shear resistance (see Farmer *et al*, 1971) —they are not considered to have sufficient strength or permanence for widespread use. They have, however, been used widely to waterproof reservoirs, ponds or canals or to form waterproof cut-offs or membranes. The effectiveness of the injected suspension as a waterproofing medium depends on the yield strength of the bentonite to prevent, in effect, its expulsion from porespaces by moderate hydraulic gradients. The design parameters exist in the zero flow condition for the Buckingham-Reiner equation (equation 11.11) where:

$$\frac{P}{r} = 2\frac{T_s}{\alpha} \qquad (11.34)$$

Here, α is the effective hydraulic radius of the soil and $P/r = i$ is the hydraulic gradient.

A 6 per cent bentonite suspension at 4 h (Figure 11.37) having a yield strength in the region of 7 N m^{-2} and injected into (say) a fine sand having an effective hydraulic radius of 0.01 mm (Table 11.6) would therefore be able to resist a hydraulic gradient of approximately 10^3-10^4 kN m^{-2} m^{-1}, a gradient substantially higher than that existing in any feasible groundwater flow situation. The normal method of waterproofing varies with the site and the manner in which the material is applied. In a reservoir, pond or canal, a layer of dry bentonite (10–50 kg m^{-2}) spread on the pervious floor material would be ploughed into the surface, in a similar way to lime or cement stabilisation of clay soils, and covered with a further layer of the original surface material prior to flooding. In the case of a water cut off, material excavated from a trench (probably using a suspension support system) would be backfilled with some compaction into the bentonite suspension-filled trench.

Probably the most common uses of bentonite suspensions are in the *support* of deep trench or pile excavations — formed for the construction of cast-in-place concrete diaphragm walls or piles — and as a *drilling mud*. Although the actual support mechanisms are different, the principle is the same in each case. The bentonite suspension and a certain amount of the excavation products or drill cuttings are used to form a slurry which possesses sufficient hydrostatic head to prevent collapse of the hole sidewalls, the stability possibly being assisted by a certain amount of sidewall penetration or filtercake formation. In addition, part or all of the cuttings and excavation products may be removed from the hole in the circulating suspension from which they can be rejected by

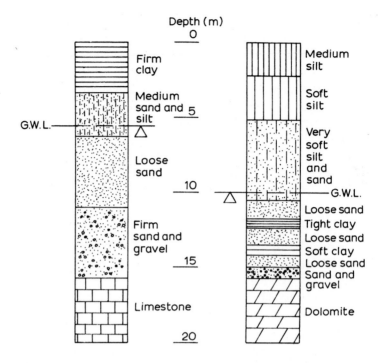

Figure 11.38 Typical ground sequences where bentonite support has proved effective (*after* McKinney and Gray, 1963).

sieving or settlement. Typical 'collapsing' strata in which slurry trenching or piling might be utilized are dry or water bearing sands, silts, gravels, boulder or soft clays. Typical sequences as suggested by McKinney and Gray (1963) are reproduced in Figure 11.38. Clearly, in the case of deep boreholes, a rather more sophisticated mud or slurry design is required, and detailed recipes are given in specialist publications (see Cumming, 1951). The basic requirements for a drilling mud, as outlined by Hetherington (1963) are:

(a) sufficient density (possibly involving artificial weighting with barytes or galena) for support of the borehole sidewall;
(b) sufficiently high viscosity to carry cuttings from depth;
(c) sufficiently low viscosity for effective screening and circulation;
(d) capacity to produce a thin filter cake on the sidewall;
(e) low water or filtrate loss;
(f) thixotropy to keep cuttings in suspension and the hole open when rods are withdrawn.

From a ground engineering point of view, the most interesting, and least stable, type of slurry-supported excavation is a trench, or diaphragm wall excavation. Various workers have studied the stability of this type of excavation, usually by equating the supporting forces acting on the excavation sidewall (principally hydrostatic) and the disturbing forces acting on an assumed wedge failure surface (see, for instance Elson, 1968). On this basis, it can be shown that bentonite trench support is viable in all except loose cohesionless and weakly-cohesive soils provided that the excavation has a reasonably low width-depth ratio. The analysis in Figure 11.39 presents a simplified view of horizontal arching around the trench, and various workers (see Prater, 1973 and Costet and Sanglerat, 1969) have shown that in trenches with a high width-depth ratio, the resultant curved failure surface in a horizontal plane can give significant additional stability.

In fact, because of the varied strata through which bentonite slurry–supported excavations pass, and because of the varied piezometric levels which may be present, overall stability is often considered to be less of a problem than is localized overbreak. This problem has been studied by the authors (Farmer and Attewell, 1973) who have suggested that by equating the active earth pressure in the ground to the hydrostatic pressure of the slurry at the bentonite sidewall interface, it is possible to predict the position of pressure peaks in the sidewall.

11.12 Ground anchors

Design and installation of ground anchors is, strictly speaking, the province of the foundation engineer rather than the engineering geologist. However, the dependence of successful anchor construction on the properties of the ground and on the utilization of grouting techniques does merit a brief mention along with other geotechnical processes.

Ground anchors are essentially a form of tension pile comprising a rod or cable grouted into a borehole and stressed after the grout has set. They can be used to support diaphragm or sheet pile walls in soft ground excavations and (usually in the form of short bolts) to stabilise exposed hard rock surfaces. In the case of soft ground excavations, they are used mainly as structural 'ties', but in the case of hard rock they can be used to improve rock stability by creating a compressed surface zone (see Obert and Duvall, 1967).

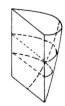

Probable failure wedge allowing for horizontal arching

Ground water level at surface

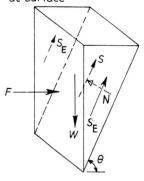

(a) Cohesive wedge

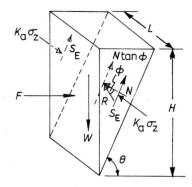

(b) Cohesionless wedge

Shear resistance = $S + 2S_E$

$$= c_u H \left[\frac{L}{\sin \theta} + H \cot \theta \right]$$

Disturbing force = $W \sin \theta$

$$= \tfrac{1}{2} \gamma_b H^2 L \cos \theta$$

Hydrostatic resisting force

$$= F \cos \theta$$
$$= \tfrac{1}{2} (\gamma_g - \gamma_w) H^2 L \cos \theta$$

Thus conditions for stability are

(a) $H \leqslant \dfrac{4 c_u (1 + (H/L) \cos \theta)}{(\gamma_b + \gamma_w - \gamma_g) \sin 2\theta}$

Shear resistance at ends = $2 S_E$

$$= K_a \gamma_b \frac{H^3}{3} \tan \phi \cot \theta$$

Conditions for equilibrium of the wedge are

$$F = W \tan(\theta - \phi) - 2 S_E \sin \theta$$

where

$$F = \tfrac{1}{2} (\gamma_g - \gamma_w) H^2 L$$

and

$$W = \tfrac{1}{2} \gamma_b H^2 L \cot \theta$$

Thus conditions for stability are

(b) $H \geqslant \dfrac{3}{2} \dfrac{L}{K_a} \left[\dfrac{\tan(\theta - \phi)}{\tan \phi \sin \theta} - \dfrac{\gamma_g - \gamma_w}{\gamma_b \tan \phi \cos \theta} \right]$

Figure 11.39 Conditions for stability of a bentonite supported sidewall. γ_b is the submerged density of the ground and γ_g is the bentonite suspension density.

Calculation of permissible cable anchor tensions and locations in rock wedge stability problems has been considered in Chapter 10 and an example of the use of steel tendons for gravity dam stabilisation is given in Figure 12.18.

Design of soil and *soft* ground anchors is covered comprehensively by Littlejohn (1970). The capacity or pull-out resistance F_R of a grouted anchor depends on the amount of skin friction resistance R_s and end bearing resistance R_p generated by the anchor as it is loaded (Figure 11.40a). In the case of granular soils these resistances in turn will depend upon the depth, density and ϕ-value of the soil and upon the anchor dimensions:

$$F_R = R_s + R_p \tag{11.35}$$

$$R_s = \tfrac{1}{2} K \gamma H \tan \phi_s (\pi \, d \, L) \tag{11.36}$$

$$R_p = \gamma H \frac{\pi d^2}{4} (N_q - 1) \tag{11.37}$$

where N_q is the bearing capacity factor (see p. 480), ϕ_s is the angle of skin or wall friction $\simeq 0.7\,\phi$, and K is an earth pressure coefficient approximating to K_0.

These are essentially bored pile equations and their derivation is covered by Tomlinson (1963), among others. The principal weakness in adapting these equations for anchors is the difficulty, particularly in low density soils, of obtaining the full end-bearing capacity. This difficulty is exacerbated in the case of anchors since the presence and subsequent collapse of the borehole in the end-bearing zone will further weaken the soil. The situation is ameliorated in two ways. First, the anchor capacity is mainly frictional, and second, (in contrast to the behaviour of piles) if the anchor is pre-stressed, considerable extension may be tolerated during the stressing provided that the final load on the compacted end is maintained.

In the above equations, the magnitude of d (the diameter of the anchorage) will depend on the type of soil and the method of construction. Clearly, in an anchor of the length and diameter required to generate a reasonable resistance of about 60 tons (600 kN) in soft ground, a cement grout is the only suitable material for anchor construction on the grounds of economy – and probably efficiency also. This requirement can effectively limit d in many

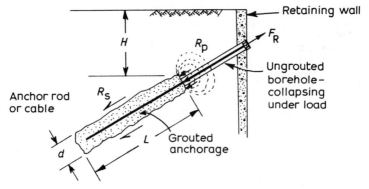

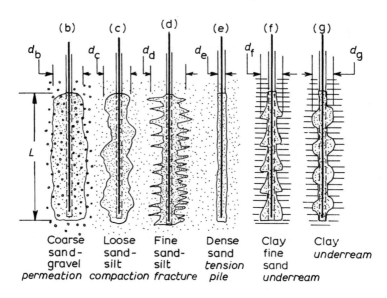

Figure 11.40 Ground anchors.

soils, for only in the case of coarse sands and gravels will cement particles be able to permeate the soil. In this case, radial penetration of the grout will facilitate the formation of a large anchor from quite a small borehole source (Figure 11.40b).

In low density materials, a larger anchor may be obtained by compaction of the soil under pressure. This process will create a large and uneven anchor, even if penetration is limited. In this case, shown in Figure 11.40c, the grouting will have the double effect of increasing ϕ and γ as well as forming an anchorage.

In addition to permeation or compaction, pressure may also be used to cause hydrofracture in stronger soils. In order to obtain controlled hydrofracture in a lateral direction, some form of packer is normally used and the effect is to increase d by radial fissuring (Figure 11.40d). The well known tube-à-manchette system is an example of this type of anchor formation.

Where neither permeation, compaction nor hydrofracture is considered feasible, the remaining alternatives are tension piles (Figure 11.40e) or underreaming (Figure 11.40f,g). With tension piles, a large diameter borehole or an extremely long anchor may be required to obtain a realistic anchor resistance, but in dense, deep sands, sufficient resistance may often be generated from a 4 in (100 mm) or 6 in (150 mm) diameter anchor. Where neither alternative is feasible, underreaming in sands and underreaming or 'belling' in clays (see Littlejohn, 1970) will spread the anchor load and effectively increase d.

In the case of clay anchors, underreaming of some form is normally considered essential, bearing in mind the reduced adhesion at the clay–anchor interface resulting from softening of the clay in contact with drilling water or bleed water from grouting. Since the soil parameters are different, the design approach must be modified and Littlejohn suggests that:

$$R_p = 9c_u \frac{\pi d^2}{4} \qquad (11.38)$$

and

$$R_s = \pi d\, L\, c_u \qquad (11.39)$$

where d is the diameter of the belled length (Figure 11.40g) and the undrained clay strength c_u is not adjusted for softening since the boundary is considered to be largely in unaffected clay.

In dealing with *rock* anchors, although some underreaming may be feasible in weaker rocks, the 'tension pile' approach is normally used. The capacity for a given length and diameter of anchor will, however, be higher because of the potentially greater surface resistance at the much stronger grout-rock interface. A simple approach to design recommended by most authorities such as Coates (1965) is to assume that load transfer from the grouted anchorage to the rock is evenly distributed as shear stress along the cylindrical grout/rock interface. Failure will then occur if this shear stress ($F_R/\pi dL$) exceeds the shear strength c of the weaker substrate. In weak rocks where $c < 3$ N mm^{-2}, the weaker substrate will be the rock, whilst in stronger rock it will be the grout ($c = 3 - 4$N mm^{-2} for 0.4 water/cement ratio grout at 7 days).

A safe assumption that c is approximately one tenth of the magnitude of the unconfined compressive strength of the rock produces a simple equation for anchor resistance in which any end-bearing is ignored:

$$F_R = \pi d L \frac{S_c}{10} \quad (11.40)$$

where S_c is the unconfined compressive strength of the weaker material.

The inherent assumption in the foregoing analyses that load transfer from the anchorage to the surrounding ground is evenly distributed as shear stress along the anchorage length can be challenged in a situation where full skin-friction mobilization is not obtained. For instance, work on reinforced concrete bars (Hawkes and Evans, 1951) has shown that shear stress may decay exponentially along a reinforcement bar according to the relationship:

$$\frac{\tau_x}{\tau_0} = \exp\left(\frac{-Ax}{d}\right) \quad (11.41)$$

where τ_0 is the shear stress at the top of the anchor (Figure 11.41),
τ_x is the shear stress at length x from that end,
and A is the ratio between the shear stress in the receiving substrate and the axial stress in the anchor rod.

If the strains in the anchor are small and the stress-strain relationships linear, A is by definition directly proportional to and arguably similar to the modulus ratio E_a/E_R, where the subscripts a and R refer to the anchoring material and the rod. Further analysis by Farmer (1975c) has shown that for a rod of radius r anchored in a hole of radius a in a rigid material by a shearable grout, at small strains:

$$\frac{\tau_x}{\sigma_0} = \tfrac{1}{2} r \alpha \exp - [\alpha x] \quad (11.42)$$

where

$$\alpha^2 = \frac{E_a/E_R}{r(a-r)} \quad \text{with } (a-r) < r,$$

and σ_0 is the normal stress at the free end of the anchor.

Table 11.12 Range of modulus ratios for typical anchor materials

Anchorage material	Rod material	E_a N mm^{-2}	E_R N mm^{-2}	$A = E_a/E_R$
Cement grout	steel	20 000	200 000	0.1
Polyester resin	steel	3 000	180 000	0.02
Hard rock	cement grout	90 000	20 000	4.5
Soft rock	cement grout	5 000	20 000	0.25
Hard rock	polyester resin	90 000	3 000	30
Soft rock	polyester resin	5 000	3 000	1.7
Dense soil	cement grout	800	20 000	0.04

Figure 11.41 shows that the implications of this analysis are quite important. If A is high, the major part of the anchor load will be transferred as a peak shear stress at the top of the anchor. If A is low, the shear stress at the top of the anchor will be comparatively low and will be uniformly distributed over the anchor length.

A rock or soil anchor is, of course, a three phase system and the anchored rod analysis that has been used refers to a two phase system. If, however, the stress distribution along the *rod* in a three phase grouted rod system is reasonably uniform then there is some justification for considering the grouted anchor as a two phase system. Table 11.12 gives typical A values for conventional grouting

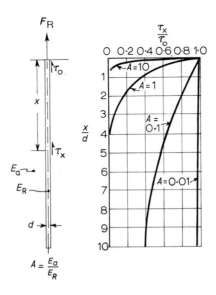

Figure 11.41 Stress distribution along an anchored rod.

materials— cement grout for soil and rock anchors, and polyester resin, which is the grout base in a commonly-used commercial rock anchor. In either material, the A value at the rod-anchor interface is sufficiently low to ensure uniform distribution of stress. An interesting anomaly is the high E_a/E_R value for a resin-hard rock interface, which would indicate a high stress concentration at the top of the fixed anchorage in this particular type of rock anchor.

We may conclude by noting a possible developing trend in soft-ground tunnels towards the use of forward roof bolting (or 'spiling' — see Brekke and Korbin, 1974) to replace fore-poling or breast-boarding at the face. In rock tunnels there is some quite firm evidence of a support trend towards the use of bolts, both with and without tensioning (Deere *et al*, 1974) and it is also likely that there will be an increased use of shotcrete in tunnels — alone or with bolts or steel sets — as steel costs increase.

12 ~ Water Resources, Reservoirs and Dams

Although the world's store of water is finite it is in a state of constant re-distribution through mechanisms of evapotranspiration, condensation and precipitation, and transportation both surface and underground. The major oceans of the world cover approximately 70 per cent of the surface of the earth and they comprise the primary element in the dynamic hydrological cycle. Vast quantities of water are involved, as is shown in Figure 12.1 which is adapted from the information provided by Clarke (1974). But in spite of these figures, the world-wide distribution of water on land is not uniform in space and time. Even in temperate climates, where in the past water has hardly been regarded as a valuable commodity, it is now being realised that resources must be treated with care and be conserved. Advances in civilization and industrial expansion make increasingly heavy demands on available water resources and it is found, for example, that the *total* consumption in the north of England expressed *per capita* was about 275 litres d^{-1} in 1967, the equivalent figure for the south of the Country being 255 litres d^{-1}*. Household consumption in Britain as a whole was running at about 135 litres d^{-1} day in 1974 with the equivalent figure projected to the end of the century more likely to be in the region of 250 litres – an 85 per cent increase. Indeed, a *per capita* estimated demand forecast for the north of England in the year 2001 is 455 litres d^{-1}. In both north and south of the Country, the growth rate of both industrial and domestic demands is estimated at 2 per cent compounded. Electricity generation and manufacturing industry together account for more than three-quarters of the total available surface and underground water in Britain and, as the figures imply, this fraction will tend to increase in the future.

The problem is therefore formidable, but it is one in which the engineering geologist can participate in its solution.

---

*Shiell (1975) quotes *per capita* figures for Scotland as 263 litres day^{-1} (domestic) and 140 litres day^{-1} (industrial).

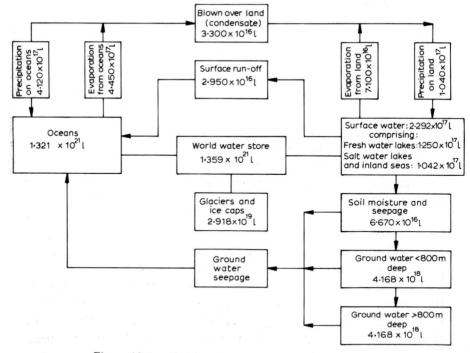

Figure 12.1 Estimated worldwide hydrological cycle.

12.1 Water requirements in England and Wales

It is useful to have some idea of the global quantities that are involved in a national water management problem, and for this purpose some very basic and general figures relating to England and Wales are quoted.

In 1971, the total consumption of water in the two countries averaged out at about 3000 million gallons per day (13.6 x 10^9 l d^{-1}). With an average annual rainfall of 41 in (1041 mm), the total run-off comprising immediate or 'flood flow' (~13 in, or 330 mm) and shallow percolation or groundwater flow (~ 5 in, or 127 mm) amounts to about 18 in (~457 mm). This represents a long-term average run-off of about 200 million m³ per day (2 x 10^{11} l d^{-1})* but this figure is subject to a multiplying factor, the magnitude of which depends upon the volumes of natural and

---

*With smaller mainland area of about 6.8 million hectares and an average annual (1916–1950) precipitation (unevenly distributed spatially) of 56 inches (1422 mm), the run-off for Scotland is just about the same at 1.9 x 10^{11} litres day^{-1} or about 40 inches (1016 mm per annum), and of this run-off only a little over 1% is actually used (Schiell, 1975).

artificial storage that are available, the ways in which those storage volumes are utilized, the pattern of water use and re-use, and the reliability of supply that the consumers deem it worth while to pay for (Rydz, 1971). For the two countries, there is a predicted water deficiency at the year 2000 of about 11 million gallons per day (50×10^6 l d^{-1}). Only 10 per cent of this deficiency could be met by a second use of river flows for the public water supply.

The British Water Resources Board has planned an inventory of between 50 and 100 major storage sites aggregating over 5000 million m³ of water (5×10^{12} litres) from which selection can be made to meet the needs of conurbations and demand districts generally. It has been estimated that the requirements to around the end of the century will be met in England and Wales by about 20 major reservoirs, each of about 10^8 m³ (10^{11} l) average capacity, some possibly being estuary barrage schemes. Any upgrading of potential resources can also be based on improved and new river regulation schemes including second-use of resources as with the River Thames. Rydz (1971) has estimated that there is an aggregate of some 30 million cubic metres per day (3×10^{10} l d^{-1}) based on a limitation of primary yield, from each of the major rivers in the countries, to about one third of mean flow.

Some specific aspects of water resource management will be considered further in the subsequent text.

12.2 Planning of water resources

Water supplies can be categorized in terms of:

(1) Public water supply,
(2) Direct industrial supply, and
(3) Agricultural supply.

Two methods can most generally be adopted to forecast demand:

(a) Project past trends of consumption into the future, or
(b) Use past trends for per capita consumption and project forward on future population estimates.

A typical water resource:water demand forward projection curve is shown in Figure 12.2.

Since the growth rate of domestic and industrial demands on the public water supply in England has been a compound interest rate of about 2 per cent (Ineson, 1970), this is reflected in a necessary

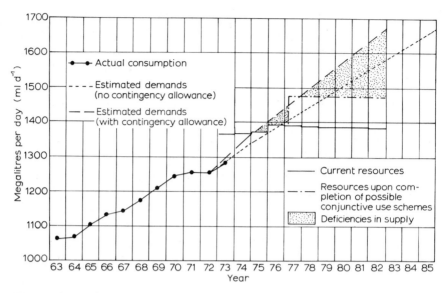

Figure 12.2 Typical water resource: water demand relationship for an area in England.

increase in the construction of impounding reservoirs. There is usually adequate water theoretically available to match inevitable deficiencies on forward projections, but this water is often distributed unevenly with respect to the populated and industrialized areas that it must serve and it is also potentially available to different degrees depending upon the time of the year. Development of increasing amounts of reservoir storage will be more restricted in the future as catchment areas are more and more taken over for agricultural purposes, and it may also be found that resistance to valley flooding becomes stronger as environmental lobbies become more vocal. On the question of direct river abstraction by industry (and in Great Britain, the Central Electricity Generating Board was responsible for 80 per cent of the industrial abstraction in 1967; Ineson, 1970) there must obviously be an increasing reliance upon water re-use through re-circulation. The C.E.G.B., for example, must re-use their cooling water to an increasing extent and this will mean either the construction of more, or bigger, cooling towers having a bigger environmental intrusive impact or, alternatively, the development of smaller heat exchangers having equivalent throughputs. Increased evaporation losses through cooling towers must be tolerated because unless power stations are located adjacent to rivers having high flow volumes, large temperature rises in through-flow water would be unacceptable. An increasing general industrial demand for river-abstracted water usually requires the construction of more regulating reservoirs to provide a reliably-even supply throughout the year.

Developments in water resource management in Great Britain, in addition to established programmes of impounding and regulating reservoir construction, can be expected to take the form of:

(a) Balanced ground water abstraction from aquifer-penetrating wells. Particular attention will have to be paid to such factors as: avoidance of over-pumping both with respect to an aquifer as a whole and also with respect to individual wells; spacing of wells to optimize exploitation of the aquifer but to avoid excessive overlap of cones of depression; avoidance of water pollution in shallow wells near to the gathering grounds and before natural filtration mechanisms have become fully effective; avoidance of overpumping from boreholes near to rivers at times of low flow, otherwise extraction will be at the expense of potentially impounded water and the time constant between the cause and effect will be low; avoidance of saline intrusion.

(b) Use of re-charge techniques. This involves the replenishment of a depleted aquifer in times of an abundance of surface water flow, the re-charge being achieved, using river water pumped through boreholes into a suitably pervious rock horizon which, in the extreme, might be wholly confined. Re-charge methods are used in the Federal Republic of Germany and the Netherlands mainly because of the contaminated water in the River Rhine. The method is also effective in preventing or reducing the effects of saline intrusion into aquifers. Other obvious advantages mentioned by Ineson (1970) are:

(i) Counteracting the over-development of aquifers by arresting the fall in ground water levels,
(ii) Temperature and quality of groundwater is usually much better than in the case of surface water supplies,
(iii) Less treatment is required because of natural filtration effects,
(iv) An aquifer can be used more fully for storage than surface impounded water,
(v) Surface works for a re-charge scheme should be less extensive and obtrusive than for equivalent surface reservoir storage,

(vi) The aquifer also has ducting properties and as a conduit may move the recharged water from an under-developed area having surplus water to a new urban development having a potential nett deficit.

The technical attractions of such a scheme must not be allowed to remove the requirement that it should be proved economically viable. In England, there have been schemes for artificially recharging the Bunter sandstones and the Trent river gravels of Nottinghamshire, the Chalk in the Great Ouse river basin, and the Chalk and overlying sands of the Thanet Beds in the London Basin.

There are problems arising from chemical and bacterial effects resulting from interaction between the existing ground water and re-charge water. Precipitation of calcium carbonate, and iron and manganese salts will tend to reduce the permeability of the aquifer as will the production of sludges from bacterial action. Increased seepage of sewage effluents into an aquifer as urbanization develops may be detrimental to ground water quality through the build up of nitrate ions (see Chapter 8) and it would seem that the shorter the retention time of re-charged water, the less significant will the problems be.

(c) Fresh water flooding of low lying, low-cost land near to river estuaries through the construction of barrages across the mouths of rivers. Typical proposals in Britain have been related, for example, to the Rivers Dee and Solway, Kent and Leven, the latter two rivers draining into Morecambe Bay. It is doubtful if the Morecambe Bay scheme will materialize in the foreseeable future and it would now appear that barrage schemes previously proposed for Britain will enjoy a lower priority rating in the future in view of the environmental and possibly economic attractions of conjunctive use schemes.

(d) In the late summer, pumping to rivers from fairly remote wells in order to augment direct run-off (Ineson, 1970; Richards, 1970). The depleted aquifer would be re-charged naturally during the coming winter, and if the area happened to be one that was prone to flooding during the wet months, then such procedures would go some way towards offsetting more major flood relief schemes.

Any river re-charge scheme is limited to some extent by the degree of permeability of the aquifer rocks. As the hydraulic connection between the ground water in the aquifer and the river water

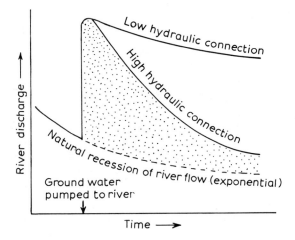

Figure 12.3 Augmentation of river flow by pumped abstraction from an aquifer (*from* Ineson, 1970).

increases, the improvement to the river flow is less significant. This point is illustrated schematically in Figure 12.3.

London and the south-east of England will require an additional 460 million gallons of water per day by the year 1981 and one suggestion to offset this nett deficit would involve such an augmentation of the River Thames flow by groundwater pumping from permeable strata at the catchment in times of need. A pilot study into this proposal conducted in the Chalk of the Lambourn valley has been described by Jones (1971). Stage I of the scheme is equivalent to supplying the Thames Basin with a 34,100 million l reservoir capable of being drained, under conditions of extreme drought, at a rate of 183 million l d^{-1}. This supply would be achieved without exploiting the Chalk aquifer to the extent that it would not recover during the following winter. Figure 12.4 shows how natural recharge to the water table in the Chalk eventually discharges to rivers which flow along an impermeable Tertiary cover and how the aquifer exploitation can be accelerated by pumped extraction and piping to the rivers. An idea of the influence that such a scheme would have imposed on the (statistical 50 year) drought hydrograph for the years 1943 to '45 can be derived from Figure 12.5. The drilling of production wells in the Chalk brings into operation some interesting geological and geotechnical techniques. For example, in addition to the basic geological assessment of

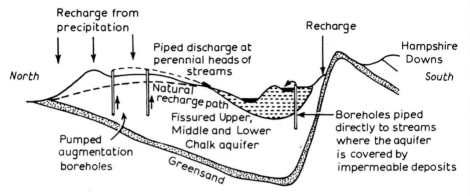

Figure 12.4 North–south idealized section of Stage 1 in the Berkshire groundwater exploitation and river supplementation scheme for the Thames Basin (*from* Taylor, 1973).

structural and lithological units, it is also possible, by taking samples of the Chalk, to differentiate between fossil water in the interstices of the Chalk and which is not available to supply and the more mobile water in the larger fissures, this water being available for abstraction. Geophysical resistivity logs are used down the hole to confirm strata changes. Measurements of temperature variations and in-flow changes with depth serve also to locate those levels where the major fissuring facilitates water movement into the hole. It is also the usual practice to clean the borehole and widen those fissures in the

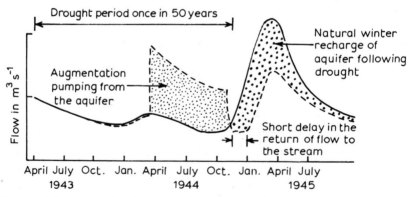

Figure 12.5 A synthesized groundwater hydrograph for a typical drought period. Hydrograph is for the River Lambourn, Berkshire, England. Continuous line represents the natural river flow actually recorded in the 1943–46 drought. Broken line shows the flow that would have taken place then if stage 1 of the river supplementation scheme had been in operation (*from* Taylor, 1973).

Chalk that intersect the borehole (and thereby to improve water yields) by a process of hydrochloric acid injection (sometimes, many tonnes per borehole).

The geological character of many countries, including Britain, is not generally such as to provide an abundance of land areas where the rocks are sufficiently porous to permit rapid abstraction and yet, at the same time, sufficiently widespread to store large quantities of water. The Chalk in the south of England is one such situation and it is really only logical that these desirable qualities of the aquifer should be exploited in a managed way.

12.3 Conjunctive use schemes

In these schemes, sometimes known as water transfer schemes, the idea is that a number of sources of water are used conjunctively (meaning one with the other) so as to utilize them in the most efficient manner. A typical conjunctive use scheme might involve the abstraction of a percentage of the water from one river having a relatively high dry weather flow and transferring it by aqueduct (open channel, pipeline or tunnel) to another river in order to augment the flow of the latter. Such a transfer scheme might be used in conjunction with pumped abstraction from the same command area.* At the same time, it is an essential feature of a conjunctive use scheme that certain constituent parts of it should each be capable of bearing on their own for short periods of time the whole demand load or at least a large proportion of it. This characteristic ensures that the scheme as a whole retains operational flexibility but it does also result in a duplication or overlapping of the capital expenditure necessary to bring it to fruition. Greater capital expenditure must be viewed as one of the prices to be paid for avoiding new reservoir development on the scale that would be required in a more traditional scheme offering the same nett increase in yield to supply. If conjunctive use schemes are deemed to be costly, then this is the most obvious reason and the environmental element bears a large proportion of the responsibility. However, the additional first-stage costs may well be more than recovered subsequently if a regional water transfer scheme eventually forms part of a national water grid.

With a high degree of water resource management in a country,

---

*A 'command area' is, by definition, the area within which a source can conveniently and economically supply its water.

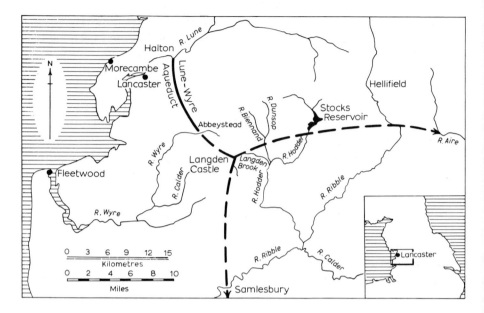

Figure 12.6 Proposed and projected water transfer links in the Lancashire River Authority area.

numerous transfer schemes of this type would be established within a broader framework of reservoir and barrage construction and water re-use strategems. The attraction of this type of water-transfer scheme is that inter-catchment water movement is theoretically possible over very large distances mainly through existing river channels and therefore at low cost. One disadvantage arises from possible eco-system changes which can accompany variations in volumes, velocities, temperatures and chemistry of imported water.

An example of a water transfer, conjunctive use scheme is the proposed River Lune – River Wyre link in Lancashire, England (Anon. 1974). Figure 12.6 shows the generalized line of transfer from Halton to Abbeystead with possible further extensions to the south and east for future consideration. Works at Stocks reservoir and at the tidal weir on the Lune are included in the short-term programme (see also the papers by Law, 1965; Walsh, 1971 and Burrow, 1971, which consider the conjunctive use of Stocks reservoir and abstraction from the Bunter sandstone aquifer of the Fylde).

The River Lune has a catchment of about 1000 km^2 to the point of transfer, an average annual rainfall of 1.520 m, a long-term average annual flow of 3040 megalitres per day (Ml d^{-1} = 666 mgd) and on several occasions each year discharges exceed 15 000 Ml d^{-1} (3300 mgd). The lowest recorded river flow has been 180 Ml d^{-1} (40 mgd). Under the conjunctive scheme, 280 Ml d^{-1} (62 mgd) would be abstracted at Halton under the first stage of the scheme provided that the flow at Skerton tidal weir at Lancaster near to the river mouth did not fall below 365 Ml d^{-1} (100 mgd). Discharge to the Wyre at Abbeystead from the pumping of the Lune will cease when the natural flow of the Wyre exceeds 2300 Ml d^{-1} (500 mgd). Supplemented Wyre flow will subsequently be re-abstracted, treated, and pumped away to the public supply.

12.4 Flood and dam design parameters

Some dams, particularly earth embankments, are constructed primarily for purposes of flood control. Others have a primary impounding or power generation function. But provision for expected flood must be incorporated into the design of any dam.

Design is usually based on the *maximum probable flood* concept. Such a determination is based on a rational consideration of the chances of a simultaneous occurrence of the maximum of several factors which contribute to the flood. The conjunction of these factors would lead to a maximum rate of run off and/or total run off from the catchment or catchments, but putting a figure to a possible maximum is really a statistical exercise with the *quality* of the estimate being dependent upon the *quantity* of real prior information.

The factors which influence flooding can broadly be grouped under the headings of precipitation, topography and geology. Any given storm will be distributed in a given manner over an area, the total precipitation that is received being related to position under the storm. The spatial form of the precipitation distribution is therefore important. But even with this concept of a stationary storm, there will also be a time-dependent distribution of precipitation over any area as the storm builds-up and later abates. These distributions are likely to be complex and variable but superimposed on them will be a dynamic element as the storm changes its position with time. Furthermore, the overlapping of storms adds further complications to any attempt at analysis.

The topography will control the extent to which, and the rate at which the precipitation runs off the catchment and contributes to the immediate flood flow. This run-off is also conditioned by the form of the catchment and its area, the latter therefore exerting a control both with respect to precipitation and with respect to run-off. Short steep, narrow catchments reduce the period of concentration up to maximum flood flow. Wide, shallow catchments or long narrow catchments increase the period.

The extent of immediate run-off is also controlled by the nature of the rock or soil at the ground surface and the pre-existing moisture deficit If a rock exposure is strongly jointed or if the *infiltration capacity* and the *moisture deficit* of a soil are high then a high proportion of the precipitation will be required to prime the watershed. Both the period of concentration and the period of recession of the flood hydrograph will be attenuated, and there will be a significant groundwater flow contribution to the terminal stages of the recession curve as it decays towards base flow of the river.

If precipitation conditions can be established over a long period of time and if the run-off and evapo-transpiration characteristics of the catchment can be approximated then, in theory, flood flows can be estimated, albeit rather imprecisely. However, the hydrological data most directly useful in predicting flood flows are actual stream flow records taken over a considerable period of time (see, for example, mean daily discharge records published by Crawford, 1967). These records of actual flows and flow rates incorporate the flood hydrographs the elements of which have been referred to above. Hydrographs of exceptional flood events can be analysed and used with precipitation data to provide run-off factors for the purpose of determining the maximum probable flood.

In some areas, large floods can originate from rapid snowmelt on a frozen watershed. Those instances where snow run-off provides a major proportion of a maximum probable flood usually involve major streams and large drainage basins (Miller and Clark, 1965).

The dam designer will also require information on the magnitude of floods having different frequencies of occurence. 50 or 100 year floods, although rare events, must be provided for in the design since the period is simply an expression of frequency and such a long period flood could statistically occur at any time during completion of construction and the next 100 years. But again, quality of prediction is very much related to quantity of reliable prior experience data and, of course, the longer the period, the more sparce these data are likely to be.

Rivers develop a naturally economic channel size which is small enough to be self-maintaining and yet large enough for the water to remain within the banks most of the time; overtopping of the banks probably occurs every two or three years on average (Butler, 1972) and in some areas it has been the practice in the past to build *levees* or flood banks to prevent such overtopping in the flood seasons. With increased urbanization and greater paved areas, higher rates of run-off can increase the tendency towards flooding in a natural channel. *Flood routing* is the description of the process adopted to alleviate such problems. It is a mechanism for attenuating a flood (strictly a flood wave) between points along a river by the use of various relief measures such as dam construction, overflow channels, storage basins, channel deepening and widening and so on.

There is usually a relationship between flood magnitude and the probability of its recurrence at a particular level. For example, by suitably scaling the flood *return period* (defined as the average interval in years between events of equal or greater magnitude occurring in any one year), a linear relationship can be obtained with the flow rate of the River Trent at Nottingham, England (Figure 12.7). Thus, rather than simply designing against the most severe flood known, the engineer is able to assess the degree of risk and quantify the possible flood damage. Wilson (1974) demonstrates the use of probability analysis in this context.

Problems of prediction can to some extent be alleviated through the use of empirical formulae which relate flood discharge to catchment area. There is, for example, the British Institution of Civil Engineers formula, the Myers formula from the United States (which allocates a percentage rating to a catchment, the rating being expressive of its flooding capability), and the Creager formula, again of the United States origin, which is a little more complicated than the other two (see Brown, 1964). All these formulae (Table 12.1) and others (see Wilson, 1974) are expressive of a consensus of actual measured and observed data and would only have application at the early design stage. Probability formulae for intensity evaluation could also be used. But in all cases, stream-flow measurements and unit hydrograph development for the actual catchment would always be required.

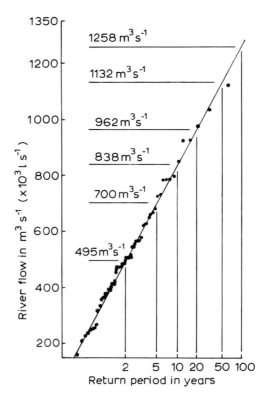

Figure 12.7 Flood frequency relationships for the River Trent at Nottingham: annual flood records for the period 1904–1966 (*from* Butler, 1972).

For more complete reading, reference may be made to De Wiest (1965), Miller and Clark (1960) and Twort *et al* (1974).

12.5 Channel protection

Overall slope adjustment of a river channel is a very slow process, involving as it does the re-distribution of huge quantities of solids in levelling higher ground and the infilling of complete alluvial plains. On an engineering time scale, change can generally be ignored. Most catchment areas have certain run-off and rainfall characteristics which, again over the long term do not change appreciably. As stated by Ackers (1972), river flows are statistically definable even if they vary stochastically, and the long term channel geometry responds to the long term pattern of flow.

Table 12.1   Some empirical intensity formulae for estimating discharge.

Formula	Source	Comments
$Q = 750\,M^{0.67}$ Q (discharge) in cusecs M (catchment area) in miles^2	British Institution of Civil Engineers	For normal maximum floods on upland catchments in Britain. In certain areas, the design flood should be not less than double the normal maximum.
$Q = qa = 10\,000\,(a)^{1/2}$ Q (peak discharge) in cusecs. q (peak discharge mile^{-2}) in cusecs mile^{-2} a (catchment area) in miles^2	Myers formula (United States)	For an area < 9 miles^2. Note that for a 1 percent flood $Q' = q'a = 100\,(a)^{1/2}$ or, maximum flood $Q_{max} = q'Pa = 100\,P\,(a)^{1/2}$ Myers rating (P) for 100 U.S. rivers varies from 0.14 to 289 Catchments < 1000 miles^2 have $P < {\sim}50$ For British catchments, $P \simeq 16$
$Q = 46\,Ca^{0.894a^{-0.048}}$ Q and a as above, C is a coefficient	Creager (1929)	A value for C of 100 envelopes all U.S. major flood records (Natural Resources Committee, 1938)

In Figure 12.8, three channel parameters — width (W), depth (D) and cross-sectional area (A) — are plotted against the discharge rate for a number of rivers in England and Wales. The graphs are taken from Ackers (1972) who used the original data of Nixon (1959a,b). Discharges are taken under full bank conditions (maximum possible water level with respect to the banks).

It is useful to compare the equations linking the channel parameters to discharge for the English and Welsh rivers to the equations compounding data from British and American rivers (Nash, 1959). These equations are listed in Table 12.2.

Quite apart from providing an albeit crude first approximation of discharge rate from known channel geometries, the graphs in Figure 12.8 are also useful to the engineer charged with designing flood protection schemes. A question which exercises the minds of water engineers is concerned with the concept of a dominant discharge. Within a range of river flows, is the geometry of the river channel controlled, or dominated, by the long-term average flow, the

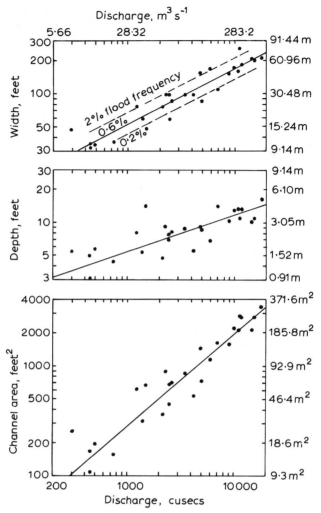

Figure 12.8 River channel parameters for England and Wales related to the full bank discharge (*from* Ackers, 1972 *after* Nixon, 1959a,b).

mean annual flood flow, the flow in the wettest month, week or day, by a rare flood, by the flow exceeded n per cent of the time, or by what, if any, other parameter (Ackers, 1972)? Which of these possibilities generates the best measure of the formative or dominant discharge?

It would appear that the stream geometry is a function of peak flow rather than long term average flow and Nixon (1959a)

Table 12.2 River channel geometry parameters related to discharge

	English and Welsh rivers	British and American rivers
Width W (ft) (at water surface)	$W = 1.65\, Q_{bf}^{0.5}$	$W = 1.32\, Q_{bf}^{0.54}$
Depth D (ft) (mean, $= A/W$)	$D = 0.545\, Q_{bf}^{0.33}$	$D = 0.93\, Q_{bf}^{0.27}$
Cross-sectional Area A of channel (ft²)	$A = 0.9\, Q_{bf}^{0.83}$	$A = 1.7\, Q_{bf}^{0.76}$

Note that Q_{bf} is the full bank discharge in cusecs (cubic feet per second).

concluded that the dominant condition coincided with an average frequency of 0.6 per cent (slightly greater than 2 days per year) for English and Welsh rivers*. Therefore, in order to adjust an existing river channel, perhaps by raising the banks, or even to create a new one, the procedure would seem to be: first, draw up a flow/persistance curve and note the discharge having a 0.6 per cent frequency; next, for that discharge, read off the width and depth parameters from Figure 12.8, or if one of them is constrained then evaluate the other from the area parameter; third, calculate the channel slope required to give the velocity necessary to pass that discharge through the channel area parameters selected; fourth, choose the most economical route on the ground for the channel at the calculated slope.

A problem arises during the application of the third stage of the procedure. There appears to be no generally acceptable channel slope/flow velocity formula.

Nixon (1959a) has noted that, from the relative scatter of the data points used to generate the relationships in Table 12.2, rivers would seem to adjust themselves more readily in width than in depth and that the adjustment in velocity and therefore slope is a much slower process. Many British rivers are 'obstructed' by weirs and sluices to further restrict this adjustment.

Slope is a function not only of basic topography and topographic changes but it is also related to the bed load characteristics. The nature of the sediment – the type of source rock and the 'silt

---

*The prevalent concern over the River Thames towards the mouth stems from the fact that the frequency figure for the river is higher at 2.91% and this poses a flood hazard for London.

factor — is an important factor and the sediment transport mechanics is a problem in fluvial hydraulics.

Nixon offers the following equation to link stream slope(S) and water discharge rate:

$$S = 0.00296 \, Q_{bf}^{-0.17} \qquad (12.1)$$

with Q_{bf} again in cusecs.

This equation gives a somewhat steeper slope than that used in the popular Manning formula and also slopes that are steeper than the slopes of the actual rivers used in generating the empirical equation. He justifies the constants by suggesting that some of the rivers may not yet have reached a stable slope and that others may not even be able to achieve such a slope. Ackers has suggested a limiting equation applicable to channel routings which vary between a straight (s) and meandering (m) condition. The channel slope, S, is expressed in terms of the bank-full discharge through the expression:

$$_s S_m = 0.0018 \, Q_{bf}^{-0.21} \qquad (12.2)$$

These expressions imply that a decreased discharge will be associated with an increased slope, which is contrary to what might be expected and is also contrary to the general trend of uncorrelated data given by Ackers and in which the slope is expressed in feet per 1000 ft.

The term 'bank storage' describes that fraction of the total run-off in a rising flood which is absorbed by the permeable banks of a river or stream above the normal phreatic surface and it is a phenomenon which serves to smooth somewhat the total flood flow. It will readily be noted that the recession portion of the flood hydrograph will have a base flow contribution which reflects movement of groundwater out of relatively temporary storage in the flanks and into the stream or river.

12.6 Design capacity of a storage reservoir

Two things must be known. First, it is necessary to find the sequence of years with the lowest annual river flows over a chosen period of years, usually 50 to 100 years if records are available over such a period of time. If records have been taken for a short period of time only, then either extrapolations must be made or comparative matches attempted with similar rivers that have been measured over longer periods of time. Second, upon these flow curves must be superimposed other curves depicting the supply of and demand for water over the sequence of dry years.

These demand/supply curves are best presented on a cumulative basis. In Figures 12.9a,b, the mass demand line is drawn tangentially to the mass reservoir supply curve at its lowest point. If a parallel line is drawn through the supply curve at a point representing a full lake, the vertical displacement between the two limiting lines represents the reservoir storage capacity needed to satisfy an assumed uniform demand. The safe yield of a reservoir is expressed as the maximum possible uniform flow. In order to determine the safe yield for a given storage, a series of tangents to the cumulative supply curve may be drawn so that the maximum ordinate displacement between a tangent and the curve is equal to the necessary storage (De Wiest, 1965). The tangent having the shallowest slope fixes the safe yield.

It is usual to express the capacity of a reservoir in terms of a uniform depth of water coverage over the reservoir area that is actually covered. The basic unit most commonly used hitherto is the acre-foot but the equivalent S.I. unit is the hectare-metre. When capacity is related to water level at the dam, the actual form of the relationship tells us something about the shape of the basin. A typical water level/capacity curve is drawn in Figure 12.10.

After calculating flood water levels at a dam, the design of the structure based on a maximum probable flood will usually incorporate a specified maximum head of water at an overflow weir with the dam having a further specified margin of height to allow for any catastrophic flood. The spillway is a vital appurtenance of a dam because on streams with a large flood potential the spillway can be the dominant structure, with the actual choice of type of dam playing a secondary role. In some cases, therefore, the cost of the spillway can represent a high proportion of the total cost of the dam.

There are two general forms of spillway. The most obvious, that of a concrete overflow spillway, can pose problems when used with earth or rockfill embankments prone to differential settlement. The second type of arrangement utilizes a conduit under or through the dam or a channel routed around one of the dam abutments. In the latter case the tunnel or channel used for the overflow water would have been used earlier for river diversion.

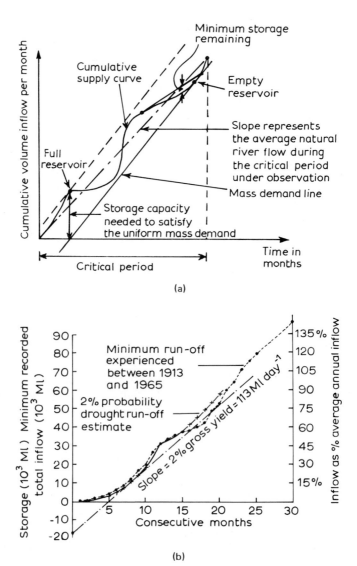

Figure 12.9 (a) Cumulative inflow/demand curves useful in the preliminary design of a reservoir (*after* de Wiest, 1965). (b) Cumulative minimum run-off diagram showing that the critical reservoir drawdown period lengthens as the storage is increased (*from* Twort *et al*, 1974). Note that the yield of a direct supply reservoir may be expressed as:

$$\text{Yield} = \frac{\text{Inflow volume over the critical period} + \text{Effective storage}}{\text{Length of critical period from full to empty}}$$

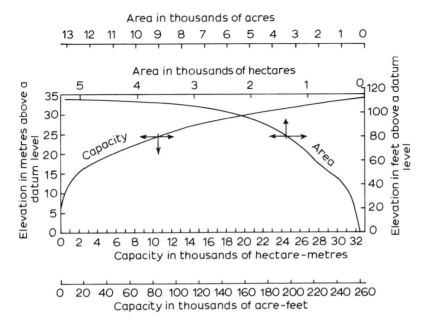

Figure 12.10 Area capacity curve for a reservoir in N. America.

12.7 Air photo-interpretation for catchment development

Aerial photography and terrain evaluation have been considered in Chapter 7 in the context of road communication, but the same techniques may be adopted to identify land factors that have a particular bearing upon the hydrological characteristics of a catchment and which also may assist in the choice of dam type and location. These factors include: drainage patterns; rocky and steep land allowing immediate run-off and flood-flow; areas of natural weakness (swamps, peat and reed beds) which pre-condition the ground for flooding; slope breaks, the flatter slopes of which will be mantled with top soil and vegetation to retard the immediate run-off; crests and ridges along which access roads could be constructed. Evaluation of the factors might suggest, among other things, possible changes in the vegetation cover along the catchment to control erosion and encourage infiltration, the location of storm drain and waterway construction works (including levees) for protecting land against storm-water run-off erosion of river banks, and aquifer/aquiclude contact zones with the possible presence of groundwater supplies.

Jones (1964) has made specific reference to the siting of small dams on the basis of air photo-interpretation. The function of these small dams is for irrigation, silt detention and water conservation, or for the detention and storage of floodwater peak discharges for their later release at a rate less than the capacity of the downstream channel

Gradients are important elements determining run-off and stream flow. There are many pointers to gradient severities on an air photograph. Some of these are: areas of natural wetness (generally flat ground); meandering streams and oxbows, silt bars, near right-angle junctions of tributaries with main rivers, wide terrace intervals in cultivated land (shallow gradients). Jones has suggested that the photo-interpreter will also be on the lookout for stream junctions (maximising storage), potential borrow areas (for earth dam construction, for concrete aggregate, and for rip-rap), rocky areas (foundations for concrete gravity or buttress dams), and positions on stream flows where the gradient changes. This last point is important because dams can most usefully be located at places where the gradient changes such that, for maximum pool area, the upstream bed gradient will not exceed about 4 per cent. Water storage capacities of areas suitable for impounding can be calculated from stereoscopic examinations for possible dam heights and from appropriate formulae.

12.8 Geological influences upon the selection of reservoir sites

Geology will never be the prime factor governing the choice of site for a reservoir. The major controls will be as follows:—

(a) *Storage potential*, which will be related most directly to the topography of the valley;
(b) *Hydrological regime* of the catchment, of which the degree and persistance of precipitation are the parameters of major concern;
(c) Proximity to major conurbations;
(d) Environmental constraints.

The importance of the latter two factors should not be over-emphasized. Water is already conveyed by pipeline over considerable distances (leakages and frictional losses are quantifiable) and although the environmental lobby is vocal, particularly when areas of natural parkland are scheduled for flooding, there is also a body of opinion which holds the view that the introduction of a lake to an otherwise largely wild and inaccessible area actually improves the environment and certainly the potential amenity for many people (see, for example, Knill, 1970).

Some major geological considerations are as follows:

(a) The basin should generally be watertight.

(b) Hard, impermeable (low joint/fissure density) rocks satisfy condition *a* and one would therefore look particularly favourably on igneous rocks outcropping in a suitable valley topography.

(c) Since, however, all reservoirs leak to some degree (there is a steep hydraulic gradient at the dam as the impounded water passes from the reservoir, through and around the cut-off, where it meets and constrains the existing groundwater to flow downstream of the dam) and provided that the leakage can be calculated for the water balance equations and will not increase with time, then the water tightness requirement is less important particularly in the case of a regulating reservoir. Bedrock requirements are consequently much less stringent. In the case of a pumped storage reservoir, however, leakage is a direct cost against pumping — these reservoirs must therefore be chosen with water tightness as a priority and this means that they will usually be constructed on igneous rock. Leakage paths can be checked using radioactive tracers or fluorescein dye.

(d) Since sedimentary rocks are jointed, they are potentially leaky. But since the joints tighten up at depth, leakage through the base of the reservoir may not be a live issue except at the dam itself where suitable cut off grouting will reduce the losses if this is considered desirable. The sides of a valley may, however, pose a problem. Rocks may be prone to slabbing due to stress relief. There may be fractures resulting from sandstone cambering (as in the Pennines in Northern England) and any shaly horizons may have slabbed and fractured due to seasonal variations in temperature and precipitation. Avoid anticlinal sites with associated outward leakage.

(e) Reservoirs constructed in Coal Measures rocks could leak through channelling into old mine workings in the vicinity. The pattern of subsidence above and adjacent to the old workings will have created zones of lateral compressive and tensile strain in the ground together with bed separation. These latter two effects are particularly conducive to leakage from the reservoir but they can be evaluated using standard curves published by the British National Coal Board (Subsidence Engineers' Handbook). The subsidence-strain problem is also considered in Chapter 7.

(f) Limestone can be unsuitable because of the possible presence of solution cavities (Kennard and Knill, 1969) and their exploitation by low pH groundwater. Knill (1970) has pointed out, however, that there are many examples of successful reservoirs on limestone outside Great Britain and has implied that dam construction is often feasible provided that suitable exploration and ground treatment techniques are used. Remove peat from reservoir area to reduce acidity.

(g) Leakage could take place through residual soils which can be of great thickness and which would therefore require deep cut-off trenches at damsites.

(h) If a dam is to be built on a granite, there must be an investigation into the degree of feldspar decomposition. The rock will probably remain acceptably impermeable if the decomposition is at an advanced stage. On the other hand, if there is quartz veining in the granite, fracturing and jointing of the quartz may render a foundation permeable.

(i) There could be leakage through transported superficial deposits. Leakage from Sautet Reservoir on the River Drac near Grenoble, France, is an example of this and is considered later in this chapter.

(j) Leakage through faults is generally unlikely but should not be dismissed without investigation. An obvious source of leakage is through an aquifer dipping away from the reservoir and into an adjacent valley. Major faulting parallel to the valley side could reduce the loss significantly but otherwise the site would have to be abandoned as a reservoir or extensive ground treatment resorted to.

(k) Should extensive landslipping occur after impounding, the water table could be raised and a leakage path provided to more permeable material downstream of the dam. This is one reason why slope stability analysis can be very important in the context of dams and reservoirs. On the other hand, tensile relief of joints and cracking of rock at valley sides without landslipping constitutes potential leakage channels from a reservoir beyond the abutments of a dam and may require grout treatment (a cut-off trench was required at the Derwent Dam in Derbyshire, England). Infilling of joints with clay and silt could reduce leakage flow with time. However, probably the most important example of the disastrous effect of reservoir landslipping can be found in the case history of the Vaiont reservoir where on October 9, 1963 a short distance upstream from the dam, a mass of steeply inclined limestone strata, 200 m high and 1800 m long with a volume estimated at 250 to 300 million m^3, slid into the water along clay seams. About 25–30 million m³ of water overtopped the dam and with a wave front at least 70 m high engulfed the town of Longarone and other populated areas both upstream and downstream of the dam. Very careful checks had been made for possible landslipping prior to the event but there was no indication or warning of the actual catastrophic movement that occurred so suddenly. On the other hand, it appears that a major landslip had previously taken place while the dam was being built, and there were also other historical precedents in Italy such as the landslip of Mount Spitz in the Alps in 1772 which blocked a river valley and formed a lake which itself, a few months later, received a further landslip and from which the wave destroyed the town of Alleghe.

(1) Perched water tables may occur where beds of well-jointed rocks are separated by relatively impermeable shale or mudstone bands. If the perched water table is interpreted as the natural ground water table and the top water level of a reservoir established on this basis, considerable leakage could ensue below the aquiclude. Borehole packer tests must be instituted to detect such a condition, but even during the actual drilling the presence of perched water will be manifest by a drop in the water level. After the borehole has been drilled beyond the perched water table, the new water level would remain static until more perched water is encountered. Accurate historical records of water levels during exploratory drilling are therefore essential.

(m) Seepage losses through joints and fissures may decrease with time as the inflow of silt and clay reduce the permeability. Conversely, erosion of clay-filled joints and cavities may take place under the greater water pressures (Binnie, 1969 quotes two examples of this).

(n) An adequate amount of suitable material must be available near to the dam site for construction purposes. In the case of an earth embankment, material will be required for the impermeable clay core, rolled fill, drainage blanket, rip rap, and as aggregate for concrete works.

12.9 Foundation investigations

Having established a social and economic need for a reservoir, either impounding, regulating or balancing, and having confirmed a location from hydrological, topographic, and preliminary geological information, it is then necessary to examine in detail the proposed site for the dam proper In England it is usually found that about 1 or 1½ per cent of the capital cost of the works is allocated for the site investigation.

Sometimes, foundation conditions at the damsite are revealed, or can be inferred from visual inspection of outcrops, previous excavations for roadways and so on. Groundwater information can often be derived from local wells. Specific geological maps and sections should then be drawn to show the boundaries of all deposits, fault zones, dips and strikes of joints and bedding. Using either separate maps or appended sheets there should be reports classifying the soils and rocks on the basis of both observation and tests from boreholes and trial pits. Specific reference should be made to the type of cementing material in the rocks and the origin and mode of deposition of the soils. A reconnaissance report should consider the relationship between the mass permeability of the foundation material and the stability of the dam when fully impounded. Recommendations should be made for the acquisition of further geological information.

A row of boreholes will usually be required along the line of a proposed dam in order to determine the bedrock profile, if feasible, and the character of the rock and soil. A few additional holes upstream and downstream of this axis will allow for any adjustments in alignment. Although the actual number of holes required will vary with geological complexity, Hilf (1960, p. 72) has suggested that the spacing should not exceed 500 ft (152 m) and that the depth of the holes should be at least equal to crest height of the dam. Boreholes will also be required in the areas of appurtenant structures, spillways, outlets, cut-off trenches, tunnel portals, and also to resolve any outstanding problems of geological interpretation. For these, Hilf (1962, p. 72) has suggested a maximum spacing of 100 ft (30 m) and a depth below the foundation of at least 1½ times the base width of the structure. For an investigation of river-diversion dams, he suggests a 100 ft maximum spacing with at least one hole located at each pier site.

As is particularly important with tunnel site investigation boreholes, it is vitally necessary to interpret the groundwater behaviour in boreholes during and after drilling at dam sites in order to build up a picture of the pre-existing groundwater regime and which will prove to be significant in the assessment of reservoir watertightness. Groundwater levels should be recorded in each borehole at the beginning of each shift on the assumption that equilibrium has been achieved during the previous stand period. According to Knill (1971), when the ground permeability exceeds 10^{-6} ms^{-1}, a time period of about 12 hours is needed for stability but in rocks of lower permeability and in deeper boreholes, greater time is required. It may be necessary to hold off the drilling to see if water levels in boreholes continue to fall. He also stresses that a standing water level in the ground is a visible manifestation of a possible *distribution* of ground water pressures along the length of the boreholes. Localized ground water pressures can be determined by using packer tests in which limited sections of the borehole are isolated and sealed off in turn. Alternatively, piezometers can be installed in short sections. But the requirement of overriding importance is for 3-dimensional information on the permeability distribution in the ground, particularly at the site of the dam proper and along the flanks of the reservoir. It should also be appreciated that if the borehole investigations indicate a natural groundwater table higher than the proposed top water level of the reservoir and that the pressures at depth below the water table are also above the proposed reservoir level, then, in general, after allowing for some re-elevation after impounding, there should be no leakage from the flanks. In theory, and irrespective of expense, a depressed flank water table could be raised by pumping water in.

It should not be assumed that permeability will necessarily decrease with depth because ground pressure increases will tend to close rock discontinuities. Nor should it be assumed that a relatively impermeable clay blanket lining the base of a reservoir will necessarily retain the impounded water with minimal seepage loss. Continuity of the reservoir base cut-off might be suspect from the outset, and even if continuous, it may be of variable thickness. Super-adjacent impounded water pressure could cause disruption of the blanket on fractured or cavernous rock or the cover could be lifted and fractured by artesian pressure from below (Knill, 1972). On the other hand, if a geological investigation should reveal the widespread presence of glacial lake deposits over the whole impounding area then one might feel, with some justification, that what has remained watertight in the Glacial past might perform equally well in the future.

If glacial-fluvial lacustrine alluvium infills a buried channel at a dam site then since such buried channels represent deep erosion of rock head and consequently contain coarsely-granular bed-load deposits from the original stream, there is a potential leakage path through the less permeable clays and into the granular material unless ground treatment processes are resorted to. Conversely, water at artesian pressure may be present beneath a clay layer in a buried valley (for example, Derwent reservoir in Northumberland, England, see Early, 1971) and create problems. The profile of a buried channel must be determined using a combination of soft ground boring and seismic methods.

12.10 Water movement into and out of a reservoir

An impounding reservoir acts as a smoothing device in the natural yield-demand equation of a catchment. Analysis of any cumulative yield curve will highlight a potential nett deficiency of yield over demand at any particular time and thereby will prescribe the required capacity of the reservoir.

The two major factors which control the retention of water in reservoirs are the piezometric conditions in, and the natural permeability of the base and flanks of the basin (Kennard and Knill, 1969).

The presence of a reservoir will obviously change the groundwater system over an area that is much larger than that occupied by the reservoir itself. If the maximum height of the phreatic surface in the flanks of the reservoir after impounding exceeds the top water level, then there should be a nett flow of groundwater towards the reservoir. The flanks, therefore, offer a concealed increase in potential storage which in total volume could approach that of the reservoir itself. On the other hand, the phreatic surface adjacent to the reservoir will almost inevitably have risen somewhat in response to the impounding, the extent of the rise, the form of the phreatic surface and the time constant with respect to the latter being dependent upon the characteristics of the flank rocks. One result of this rise will be an increase in the groundwater seepage losses to the opposite side of the flank from the reservoir.

If the groundwater table in the flanks is depressed with respect to reservoir top water level then deep seepage flow will take place from the reservoir to an adjacent catchment.

The rise of a phreatic surface from an existing river can usually be regarded as being quasi-parabolic in form (see Section 8.2 for the contruction of this surface), but it is the steepness of this rise that can not only reflect the permeability of the rock and the extent of the groundwater re-charge facility but also the potential influence of a reservoir upon the ultimate hydrogeology of the region. If the rise is gentle, then the long term presence of even a shallow reservoir may be to create only a slight increase in the groundwater head but which on a relatively widespread areal basis could create quite important hydrological changes. Knill has referred to Ahmed's (1960) arguments concerning the possibility of increased discharges from oases, fed from the Nubian sandstone, as a result of impounding behind the Aswan High Dam in Egypt.

With the current accelerated trend towards the construction and use of pumped storage schemes as a means of meeting peak electricity demands, the choice of upper reservoir site, its watertightness at a maximum efficiency head of about 300 metres (which implies that it will be positioned at a level considerably above the natural water table in the rock), and its bonus possibilities with respect to a natural catchment capability to slightly improve the efficiency of the scheme, all become important factors to be investigated by the engineering geologist. In a pumped storage scheme, an existing lake will usually be adopted for the lower reservoir and this will be subjected respectively to daily rises and falls in water level as the upper reservoir feeds the turbines for power generation and as it is replenished at night by reverse pumping. The smaller the lake, the greater the drawdown and the more prone are the banks to erosion. Larger lakes are to be preferred and ultimately, in suitable circumstances, the sea would be used as the lower 'reservoir' water source. Clearly, it would then be even more necessary to prevent a contaminating seepage from the upper reservoir. This restriction on seepage would still apply if the sea were to be used as the upper reservoir and a suitable underground cavern as the lower reservoir.

There are numerous pumped storage schemes throughout the world, operative, under construction, or in the design stage and it can be noted that the necessary topographical and geological conditions for a scheme are most likely to be found where metamorphic or plutonic rocks are relatively fault–free and have been subjected to strong glaciation (Anderson, 1971). A number of schemes satisfying these conditions in Great Britain and Ireland have been mentioned by Anderson, and Learmonth (1972) has performed a detailed joint survey and pressure tunnel finite element analysis on the Ferris

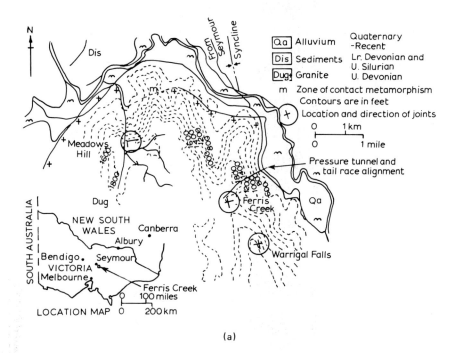

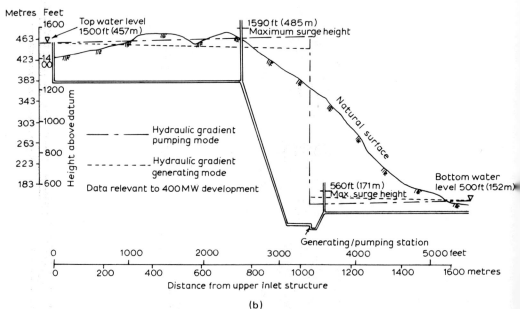

Figure 12.11 (a) Ferris Creek Pumped Storage Hydroelectric scheme location maps.
(b) Tunnel profile for Ferris Creek Pumped Storage Hydroelectric scheme. (*after* Learmonth, 1972).

Creek pumped storage hydro-electric scheme 60 miles (97 Km) north of Melbourne in Australia. Location geology and a cross-sectional sketch of this latter scheme taken from Learmonth (1972) in order to illustrate the general principal of pumped storage are given in Figures 12.11a and 12.11b.

12.11 Synthetic flow generation techniques

In order to determine the water potential of a catchment area, long term run-off data usually in the form of streamflow records, are required. In the absence of such information, short term data can be used with synthesised data in order to generate predictive flows.

Of the statistical models most commonly used for this purpose, the rainfall/run-off model is probably best-known. If Q_n is the estimate of run-off in month n, R_n is the rainfall in month n, a_n is a constant of proportionality, and b is a constant defining input:output losses, then the input:output relationship can be expressed most simply in a linear manner as:

$$Q_n = a_n R_n - b \qquad (12.3)$$

Many variations on this simple relationship have been proposed in order to improve the statistical curve fitting to known precipitation and run off data. For example, Midgley and Pitman (1969) adopted an equation of the form

$$\log Q_n = a_0 + a_1 P_n + a_2 P_{n-1} \qquad (12.4)$$

where P_n and P_{n-1} are the catchment precipitation indices for months n and $n - 1$ respectively. However, the intrinsic complexity of catchment response can only be quantifiably modelled with equal complexity (Crawford and Linsley, 1966).

Irrespective of the logarithmic term in equation 12.4, there is a fundamental improvement created by the incorporation of data from a single preceding month. The introduction of further successive antecedent terms into equation 12.3 improves the accuracy of the model until the number of terms exceeds the total period of run-off, surface and groundwater, for a single input. Hamlin (1970) has used a generalized equation 12.5 in the form

$$Q_n = a_n R_n + a_{n-1} R_{n-1} + a_{n-2} \ldots + a_{n-k} R_{n-k} + b \qquad (12.5)$$

where $R_{n-1} \ldots R_{n-k}$ are the rainfalls in months $n - 1 \ldots n - k$ respectively.

918 *Principles of Engineering Geology*

He found, however, that this form of equation approximated only crudely to real flow mainly because run-off is less sensitive to rainfall than it is to the nett intake, that is, to rainfall less evaporation/transpiration losses.

If the 'effective rainfall' R'_n is expressed as

$$R'_n = R_n - cE_n \qquad (12.6)$$

where E_n is the evaporation parameter for month n and c is the ratio of actual to potential evaporation, and if antecedent conditions are also applied to a_n to take account of the degree of saturation (soil moisture deficit) of the catchment:

$$a_n = d_{n-1} R'_{n-1} + d_{n-2} R'_{n-2} + \ldots d_{n-k} R'_{n-k} + F \qquad (12.7)$$

then the run-off equation becomes

$$Q_n = d_{n-1} R'_n R'_{n-1} + d_{n-2} R'_n R'_{n-2} \ldots d_{n-k} R'_n R'_{n-k} + fR'_n + a_{n-1} R'_{n-1} + a_{n-2} R'_{n-2} \ldots a_{n-k} R'_{n-k} + b \qquad (12.8)$$

Synthetic flows may also be obtained by the technique of serial correlation of monthly flow data followed by a so-called 'random circular walk' technique. Alternatively, Markov models may be adopted. Reference should be made to appropriate textbooks for further reading on this subject. See also Bloomer and Sexton (1972).

12.12 Dam foundations

The really large site investigation programmes will usually be associated with the field of dam technology. This is because some of the greatest peace-time disasters have occurred as a direct result of dam failures, failures less directly in the dam structure itself as in the adjacent ground as a result of the structure-ground interaction. The interaction effect is most acute in the case of arch dams which have experienced the most spectacular failures.

A cursory perusal of the Monar Dam profile and analysis in Chapter 6 will show that the abutment rocks are required to withstand the main hydrostatic thrust of the water. The nature of these loadings is much more clearly expressed in Figure 12.12 taken from Judd (1964) and they emphasize very forcibly the requirement for preliminary and on-going (during excavation) structural and mapping surveys of the abutment and foundation rocks at, and in the vicinity of the line of the dam, together with large scale field jacking, permeability, and geophysical tests. It will be noted, for example,

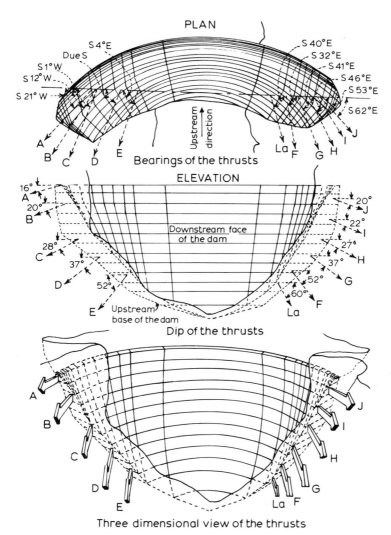

Figure 12.12 Abutment thrusts resulting from an arch dam loading (Judd and Wantland, 1956; *re-drawn from* Judd, 1964).

that in concrete dams there should be as near a compatibility as possible between the Young's moduli of concrete and rock in order to reduce stresses in the concrete caused by foundation deflection. Whereas a quite large systematic difference between $E_{concrete}$ and E_{rock} can be tolerated provided that it is uniformly spread, abrupt local changes in strata, or faulting, can create problems due to differential changes in the modulus ratio: $E_{concrete}/E_{rock}$. The results of such observations and tests will determine the volume of rock that has to be removed to achieve the required degree of structural competence and also the extent and complexity of any grout treatment programme. $E_c/E_{r(mass)}$ should preferably exceed 4 (Rocha, 1974).

Methods of presentation of discontinuity data have been considered in Chapter 6 and these techniques can be used at exposures and excavations. Exploratory shafts, tunnels and adits, probably not exceeding 3 m in width or height will also be driven and along which linear mapping can be conducted. An example of this technique, taken from Knill and Jones (1965), is shown in Figure 12.13. Changes in rock types and lithologies will be spatially correlated with the same variations observed in surface exposures, particular attention being paid to the thickness and orientation of any weaker beds or faults. Water seepages will also be measured. Knill and Jones (1965) have suggested that tunnels can be so logged at a rate of about 20 m h^{-1}.

Another major factor of importance in dam site investigations concerns the specification, logging and interpretation of borehole information. Frequency and spacing of boreholes will be determined by the finance that is available and by the complexity of the site geology. An original specification will allow sufficient reserve funds for additional boreholes to be put down in order to resolve indeterminate evidence from earlier boreholes.

The actual presentation of borehole information in the form of a borehole log has been considered in Chapter 7 but it is worth listing, after Knill and Jones (1965), the basic data that should appear on a comprehensive log:

Entire hole information
 (i) Name of site and borehole number as per master plan;
 (ii) Coordinates of borehole and topographic elevation at the top;
 (iii) Angular inclination with respect to the vertical;
 (iv) Date of start and end of drilling.

Down-the-hole log
 (i) Hole diameter, type of drilling, bit and bit size, length of casing;
 (ii) Both geological and engineering description of the rock cores making particular note of the degree of weathering;
 (iii) Percentage core/sample recovery;
 (iv) Points of inflow or loss of water, level of any standing water,

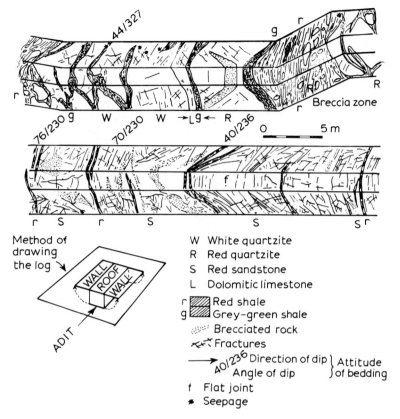

Figure 12.13 Record of geological conditions in exploratory adits and tunnels at the Latiyan dam (*from* Knill and Jones, 1965).

the colour of the water and the presence of any sediment in it; before and after any stand-time (e.g. beginning and end of a shift) water levels must be carefully monitored,
(v) Results of any tests made down the borehole (e.g. packer permeability, dilatometer).

It will then be the duty of the engineering geologist to gather together all his ground surface, tunnel and borehole information and to draw up a three dimensional picture of the geology and hydrology of the dam site and adjacent areas. Superimposed on this will be the basic geotechnical information which will complete the presentation upon which any ground treatment and design decisions will be based.

A useful technique is to transfer all the information on to a number of parallel vertical cross-sections through the valley for a distance upstream and downstream of the dam. Ideally, the spacing of the cross-sections will be closer in the immediate vicinity of the dam abutments. The next step is to trace the geological boundaries, joint orientations and fault lines on to thin perspex (plexiglas/lucite) sheets and colour in the different geological horizons. These sheets can then be slotted into a frame, the distance between the sheets preferably having the same scales as the horizontal dimension of the cross-section. If necessary, mouldable bridging sheets can be used to complete the visual impact of the physical model by creating structural and geological continuity between cross-sections.

12.13 Classification of dam types according to their purpose, construction and foundation geology

Both *impounding or storage* dams and *regulating or detention* dams act as smoothing elements in the hydrological run-off cycle. Impounding dams hold back water in times of plenty for use at times when there would otherwise be a nett deficiency to supply. This supply is required to meet both industrial and household demand and for consistency and continuity it usually involves impounding the winter and spring run-off for use in the drier summer months. Regulating dams are designed to retard flood run-off and minimize the effects of sudden flooding. They also serve to equalize wet-weather and dry-weather flow so that, for example, industry towards the mouth of a river will not be deprived of water during the summer season (this was the criterion behind the construction of the Cow Green reservoir in Upper Teesdale, England). Another type of detention dam, while holding the water as long as possible, allows it to seep away into pervious banks or foundation strata and so artificially re-charges an underground water supply. The third primary function of a dam is for the generation of electricity. *Hydroelectric power* is obtained by permitting restricted access of water to a turbine system usually housed in the body of the dam.

Dams are classified by the type of construction and the materials used in the construction. There are many variations on each theme and some dams are composite structural types. But broadly, the types are:

Gravity dam

A gravity or mass concrete dam is suitable where there is a rock foundation within 20 to 30 ft of the surface. In this way, the amount of necessary excavation work is limited. According to Walters

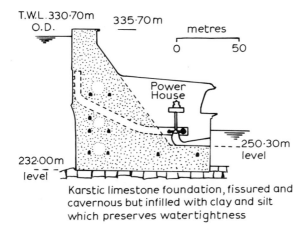

Figure 12.14 Génissiat straight massive concrete gravity dam on the River Rhône, France, 31 miles S.W. of Geneva. Construction completed 1948. Maximum height: 104 m. Reservoir extends 14 miles up-river (*after* Walters, 1971).

(1971), the bearing capacity of the rock should be at least 124 to 155 lbf in^{-2} (8.68–10.85 kgf cm^{-2} = 0.855–1.069 MN m^{-2}). A gravity dam will be selected when the length of the crest of the dam would be five-times or more the height from the foundation; this type of dam is therefore suited to a wide, shallow valley. Two examples of gravity dam are shown in cross-section in Figures 12.14 and 12.15. As an additional requirement in view of the vast quantities of concrete involved in construction, there must be plentiful supplies of suitable sand and gravel or crushed aggregate within a few miles of the dam site.

A gravity dam takes its name from the fact that it relies on its own weight alone to resist the sliding and overturning forces imposed by the water. It is triangular in cross-section, the width of the base being at least two-thirds of the crest height. Since foundation pore pressures cannot be totally dissipated by the drainage that is always provided, it is usual to double the head of water for calculating the design of the structure. This, of course, has the effect of increasing the width of the base with respect to crest height (Figure 12.16). Walters (1971) has noted that, although not a design requirement, some gravity dams have been built slightly curved along their line, probably for aesthetic reasons. But this curvature does increase the stability of the structure by the arching effect. It would also appear

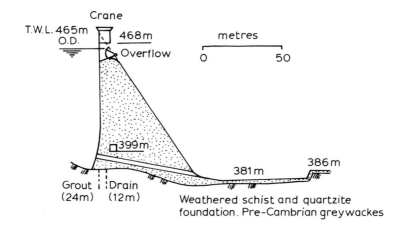
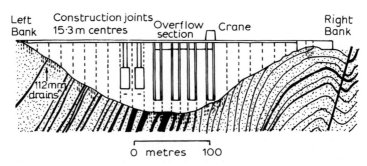

Figure 12.15 Hiwassee straight concrete gravity dam on River Hiwassee, South Carolina, U.S.A., 60 miles south of Knoxville. Hydro-electric (reversible turbine) dam completed 1940. 94 m high. Development of uplift pressures in late '40's relieved by drilling drainage holes (*after* Walters, 1971).

to accommodate expansions and contractions due to temperature effects more easily, and the overall watertightness would seem to be greater since the water pressure on the arch tends to close up any cracks.

Sautet Dam on the River Drac, near Grenoble, France, is of the gravity arch type, 126 m high and a crest width of 65 m. It rests on Lower Lias limestone of variable composition, from the hard crystalline variety to alternations of soft argillaceous clays. A great deal of grouting was required in the dam foundations but a much more severe problem was encountered through leakage along the old buried river course of the Drac (see Figure 12.17). The dam is used for power generation, but since the leakage water supplements the

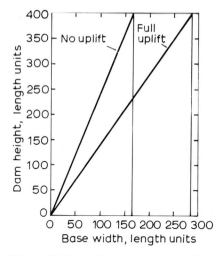

Figure 12.16 Concrete gravity dam width-to-height ratios assuming a condition of no hydrostatic pressure for uplift at the base and also a condition of full uplift according to the head of impounded water (*from data given by* Walters, 1971).

flow along the River Sézia it is further available for doing work downstream from the Sautet Dam. Various schemes for reducing the permeability of the reservoir, schemes involving grouting and clay lining, have been considered and rejected but Walters notes that natural silting processes are achieving this over the period of time since the dam was constructed.

It should also be noted that concrete gravity dams are not of exact triangular configuration. The crest will be flattened to take a road or pathway for access between banks and the upstream face will often be slightly raked to add to the stability through a downward component of the water pressure.

If there is an extremely strong rock foundation in a wide valley then there might be a cost saving in using a pre-stressed gravity dam. This would involve the anchoring of steel cables deep in the bedrock, taking them through the body of the dam and anchoring them either at the crest of the dam (for vertical tendons) or on the downstream face (inclined tendons). Costs would be saved through a reduction in concrete quantities. A cross-section of such a system is shown in Figure 12.18 and Walters (1971) quotes the Allt-na-Lairige dam on

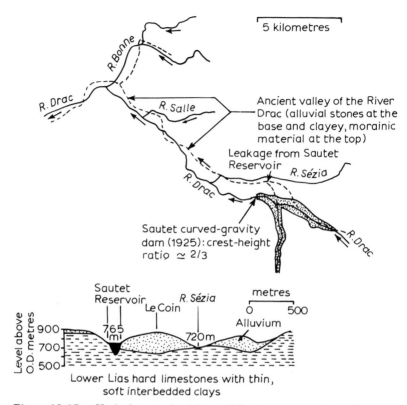

Figure 12.17 Hydrology and geology of Sautet reservoir, near Grenoble, France (*after* Walters, 1971). (Note that at the actual damsite, the gorge conditions are really more suited to a cupola-type dam.)

igneous rocks capable of sustaining a pressure of 155 lbf in^{-2} (10.85 kgf cm^{-2}; 1.069 MN m^{-2}) and which were subjected to a pull of 18,290 lbf in^{-2} (1280 kgf cm^{-2}; 126 MN m^{-2}) from steel cables bedded in concrete within the rocks.

Post-stressed tendons have also been used to reinforce older and suspect gravity dams. It was also one of the options open to the engineers faced with the necessity to strengthen the foundation of the Mequinenza Dam on the Ebro River in north-east Spain. The concrete gravity dam for hydroelectric generation, which rises 77.4 m above the foundations and has a crest length of 451 m, was almost completed (except for a small flood-flow length) in the early 1960's when serious doubts developed concerning the stability of the foundations. These comprised an alternating succession of horizontally-bedded limestones with intermediate layers of lignite and marl laid

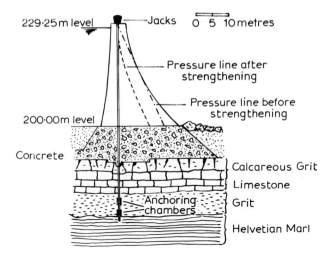

Figure 12.18 Use of steel tendons to enhance stability of Les Cheurfas gravity dam in Algeria. Dam built 1882–1885 with the left-bank on sound foundations but with the right-bank on poor Quaternary conglomerates. In 1885, right-bank was swept away and there was a subsequent history of re-building. In 1935, the foundations were grouted and the tendons added. (*After* Walters, 1971).

down in Oligocene times. The possibility of sliding along these layers now seems obvious, and both *in situ* and laboratory test programmes were set in motion (Salas and Uriel, 1964; Gete-Alonso, 1968; Vallarino and Alvarez, 1971).

The tendon solution was quickly abandoned on the basis of cost and anchorage difficulties. Another possibility was to construct a *rastrillo* (an extension of the toe under the dam) into the bedrock layers under the dam but after model experiments in Spain and England (Sheffield University and Imperial College, London) this was overruled because of excavation problems under the dam, the predicted introduction of strong tensions to the dam proper, and the possibility of heavy stress concentrations in the rock.

The solution which was finally adopted involved the thickening and strengthening of the apron of the dam as a toe extension and the construction of a vertical *rastrillo* at the toe. The effect, as shown in Figure 12.19, is to lend added stability to the dam through the dead loading of the concrete, the reaction at the *rastrillo* to horizontal thrust, and the added length of concrete along which resistance to lateral motion can be mobilized.

928 *Principles of Engineering Geology*

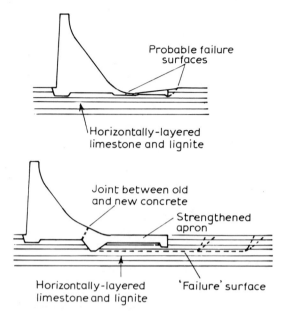

Figure 12.19 Cross-sections of unstrengthened (upper) and reinforced (lower) Mequinenza Dam (*after* Vallarino and Alvarez, 1971).

Although gravity structures can sit on sedimentary rocks, it will be noted that some rocks such as, for example, compacted shales and some tuffs tend to slake. This behaviour could pose problems during construction. Cemented shales are less prone to this type of disintegration.

Buttress dam
Whereas a gravity dam relies on its total weight to resist the sliding and overturning forces imposed by the water, a buttress dam cuts down on the total weight of concrete by mobilizing a lateral component of force in the foundation to resist the hydrostatic thrust of the water. The foundation rock must therefore be very strong and capable of sustaining pressures of 311 to 467 lbf in^{-2} (21.77–32.69 kgf cm^{-2} = 2.14–3.22 MN m^{-2}) but where the valley is wide and the dam is high (over 40 ft) the savings in materials and the reduction in total uplift force due to the lower foundation-contact area offset the higher costs involved in the detailed formwork and reinforcement of the structure and the greater attention to concrete quality as a safeguard against deterioration of the thinner structure. If the rock is

Figure 12.20 Sketch of Lubreoch massive gravity buttress dam in Scotland. View to north-west. Dam is 530 m long, just over 30 m high and the foundation is in quartz mica schist (drawn *from* photograph in Walters, 1971, courtesy N. of Scotland Hydro-Electric Board).

too weak, then the buttresses could punch into the ground and cause heave in the intervening ground. Dam sites underlain by fresh igneous intrusions of granite, monzonite, syenite, gabbro and other such strong rocks will support the buttress loadings quite readily, but granites showing evidence of feldspar decomposition must be investigated cautiously and more thoroughly. With layered sedimentary structures — and as noted earlier — checks and calculations must be made against sliding along bedding planes.

Walters (1971) quotes examples of buttress dams in Scotland and a 2000 ft (610 m) long combined buttress-gravity dam in Italy. In Figure 12.20 is a sketch of a Scottish buttress dam.

Multiple arch dam

This type of concrete dam is a variation on the buttress theme and is again suitable for wide valleys where foundation rock strengths are again of the order $311-467$ lbf in^{-2} ($21.77-32.69$ kgf cm^{-2} = $2.14-3.22$ MN m^{-2}). There is a requirement for probably an even higher standard of formwork and construction than in a buttress dam. Walters (1971) points out that whereas in a pure buttress type of dam each half of the arch between buttresses acts as a cantilever with an expansion joint on the crown of the arch, in the case of a multiple arch dam each arch is a monolithic structure transferring its thrust to the 'buttress' intercept lines between adjacent arches. Thus, although some settlement between the buttresses of a buttress dam might be tolerated by the design of the mid-buttress elements, this would certainly not be the case with the monolithic multiple arches.

Thick arch dam

This type of concrete dam is a possible option in wide valleys (width-at-crest to height ratio of between 3 and 5) where the foundation and abutment rock is very strong (at least 467 lbf in^{-2}; 32.82 kgf cm^{-2} or 3.22 MN m^{-2}). Although the greater thickness is demanded by the width of the structure and there are more formidable design problems than in the case of a concrete gravity dam, an important saving in materials costs would swing the decision in its favour. With confidence in treatment techniques for weak foundations it is likely that engineers will, in the future, express rather greater preferences for the thick arch dam.

Thin arch dam

These concrete dams require narrow valleys to have a crest chord-height ratio of less than 3, with a radius of under 500 ft or 152 m (Walters 1971). They are curved about a vertical axis only, the axis passing through the centre of the dam. As a result of this curvature, the hydrostatic thrust of the impounded water is transferred to the abutments on the valley sides. These abutments must, in consequence, have a resistance capacity of 775– 1085 lbf in^{-2} (54.25–75.95 kgf cm^{-2}; 5.35–7.49 MN m^{-2}) and the thrust of the dam must be evaluated in the context of any major joint set directions. A cross section example of this type of dam is shown in Figure 12.21.

Cupola or dome dam

These dams have both a horizontal and a vertical axis of curvature and are ideal for a gorge having a crest chord-height ratio of less than 3. Pressures at least as high as those from an arch dam are transmitted to the abutments which must obviously be formed of very sound rock. Sometimes the actual bearing pressures are reduced by thickening the structure at the abutments, as, for example, with the Monar Dam considered elsewhere. But whereas the chord-height

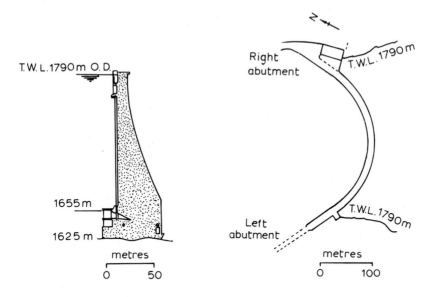

Figure 12.21 Tignes thin arch concrete dam on the River Isère 100 miles east of Lyons, France. Dam is nearly 170 m high, the highest in France and sits on hard, compact Lower Triassic quartzite in a gorge having a chord-height ratio of 2. The radius of the dam is about 45.7 m and it was completed in 1952 (*after* Walters, 1971).

ratio of the Monar Dam is 3, this ratio for the Vaiont Dam in Northern Italy is only 0.7. The Vaiont cupola dam (Figure 12.22), built 1957–59 for hydroelectric purposes in a deep gorge of Middle Jurassic dolomitic limestones, is one of - if not *the* highest dam in the world at 265.5 m, and in 1963 withstood heavy overstressing when about 25–30 million m^3 (25–30 x 10^9 litres) of water were forced over the crest by major slope failures of the limestone into the reservoir (Mencl, 1966, and see also p. 910).

Earthfill dams
Over the years, this has been the most common type of gravity dam for wide valleys having soft floors. Cross-sections of some modern earthfill dams under construction at the time of writing are given in Figures 12.23–12.26. The earthfill dam utilizes materials in the natural state needing a minimum of processing and it is particularly suitable for relatively small gathering grounds for which the impounding depths do not exceed 30 m and are usually less than 15 m. However, such dams can be more difficult to construct on soft

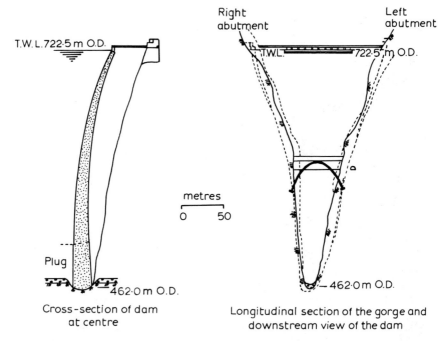

Figure 12.22 Sections of the Vaiont concrete cupola arch dam, Northern Italy. Dam built 1956–1960. Maximum height 265.5 m. Foundation is in dolomitized Middle Jurassic limestones (*after* Walters, 1971).

foundations than are concrete gravity dams on their suitable foundations. There will be an allowance for continuing dam settlements and this will take the form of an overbuilding to the extent of the computed terminal settlement (note that negative settlement, or foundation rise, occurring during excavation must also be taken into account. The original design will also take account of the maximum allowable bearing capacity of the foundation material – for example, at Pipestem Dam (Figure 12.25) the appropriate figure was 20 000 lbf ft^{-2} (958 kN m^{-2}).

Depending upon the foundation material excavation, drainage and/or ground treatment will be necessary to achieve the required cut-off or controlled leakage. In close proximity to the damsite, there must be an adequate supply of clay for filling up the excavation trench (or concrete can be used as a substitute), puddled clay or a satisfactory substitute for the core (or, again, the concrete would be carried through from the trench), graded filter material adjacent to the core and capable of supporting a clay core without

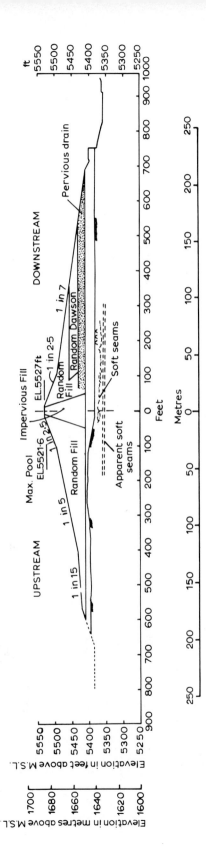

Figure 12.23 Cross-section of Chatfield Dam on the South Platte River, nr. Denver, Colorado, U.S.A. (*re-drawn after* U.S. Army Corps of Engineers, Omaha District, Design Memorandum PC-24, 1968).

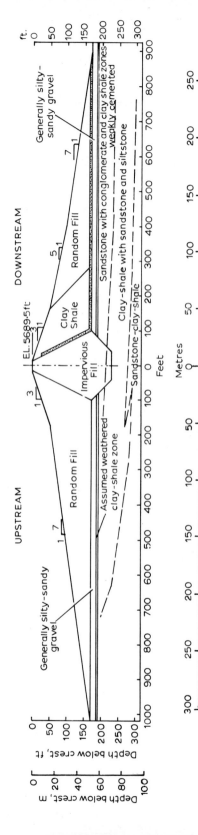

Figure 12.24 Cross-section of Bear Creek earth dam, South Platte River Basin, nr. Denver, Colorado, U.S.A. (*after* U.S. Army Corps of Engineers, Ohama District, Design Memorandum PB-6, 1972).

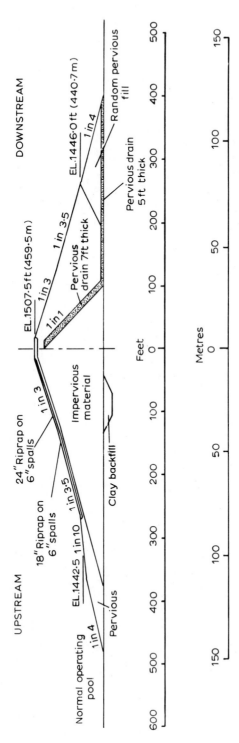

Figure 12.25 Cross-section of Pipestem Dam on the James River, 3 miles west of Jamestown, N. Dakota, U.S.A. (*after* U.S. Army Corps of Engineers, Omaha District, Design Memorandum JP-3, 1969).

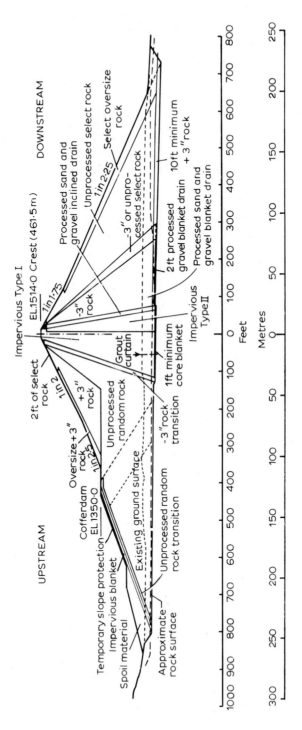

Figure 12.26 Cross-section of Bloomington Dam on the North Branch of the Potomac River, Maryland/West Virginia border, U.S.A. (*after* U.S. Army Corps of Engineers, Baltimore District, Design Memorandum No. 3, 1969).

allowing it shear and crack in tension, and an outer protective blanket of quarry-run rock faced with a larger rock fraction forming a rolled, protective skin.

Although this type of zoned construction is the most usual, homogenous dams comprising just one type of rolled/vibrated material plus a drain are to be found where relatively high leakages are not intolerable. It is often difficult to acquire large quantities of very low permeability clay for puddling over the whole embankment so many successful embankments have been built from relatively permeable sand and sand/gravel mixtures.

Earthfill embankments can also be constructed with manufactured impermeable membranes to replace a clay core. They can be constructed from concrete, steel or asphaltic concrete and are usually placed either as a core inside the embankment or on the upstream face of the dam. In the latter case it remains more accessible and also enhances the overall stability of the embankment by removing seepage pore pressures from the fill and by ensuring that the total water force on the dam is directed downwards into the foundation at as steep an angle as possible. An impervious membrane is also free from progressive leakage phenomena ('piping') that can be a feature of earth embankments (see Chapter 8 plus Terzaghi and Peck, 1967; Zaslavsky and Kassiff, 1965). There are a number of other advantages associated with the use of upstream face impermeable membranes. They permit some economy in design in that the embankment side slopes can be much steeper than is the case with rolled core embankments and, of course, in some upland areas where suitable earth core material may not exist they would be the only answer. They also offer good protection against severe wave action in the reservoir and are not prone to optimum density/moisture content considerations and controls that are a feature of rolled cores. On the disadvantage side, there would be deterioration of the membrane and a shortening of its life, although with good quality materials and careful on–site control a life of at least 30 years should be expected. There is a higher initial cost but this disadvantage is reduced as the crest height of the dam increases. Initial leakages may develop due to poor welds in steel plate, cracks in concrete, or tears in asphalt blankets; repairs may require the reservoir to be partially drawn down, or divers may have to be used, but such repairs are much more easily accomplished than repairs of equivalent leaks in a concealed core. Rigid membranes are also a little more prone to cracking if the dam base spreads during construction or if there is any post-construction differential settlement.

Many earth dams constructed pre-1925 in the U.S.A. were of the concrete impermeable membrane type but they went out of favour when it was established that rolled earth cores could do the job just as well. A history of membrane cracking, together with inadequate drainage facilities, militates against their continued acceptance. But with improved materials and construction methods they may enjoy a new lease of life in the future.

There are a number of problems that can arise with earthfill dams that are essentially structural and material and which are generally independent of foundation conditions. A fuller discussion of these problems can be found in Sherard *et al* (1963).

Although the detailed design of earthfill dams is a specialist subject outside the scope of this present work, it is nevertheless useful for the engineering geologist to be aware of some of the problems which will influence any pre-conceived standard designs. Most of these problems have obvious material affinities and thereby create additional engineering geology interest.

First of all through its own weight, the dam must be capable of resisting the force of the water; this means that the resultant thrust of the dam weight and water force must be directed well inside the base of the dam cross-section. Relatively shallow side slopes add mass to the dam and increase the cross-sectional base length. A raked upstream face also enhances stability at the heel by accepting a downward component of water force. Weights may often be calculated from a buoyant density up to top water level on the upstream side of the core and from either a saturated or dry density from top water to the crest. Seepage forces must also be considered.

A major portion of the design effort will be directed towards analyses of embankment stability during construction and at various stages as the reservoir is filled. The embankment slopes must also be checked out against shear rotation should the reservoir be drawn-down at a faster rate than the rate at which the internal pore pressures can be dissipated. Excess pore pressures on a critical shear surface reduce the effective normal pressure and thereby lower the factor of safety. During the stability design procedures several methods of equilibrium analysis would be applied to numerous trial circles and wedges within the structure, but the factors of safety derived from these would be critically dependent upon the material shear strength parameters entered into each analysis. Detailed shear strength parameters using triaxial and direct shear techniques would have to be carefully considered before the most likely values were selected from a quite scattered range, but since the material being placed in the embankment will be quite closely defined and controlled, the choice of parameters will be less exacting than in the case of a foundation stability analysis where a consensus strength must be selected from what must usually be a very wide spectrum of results. Embankment rock fill will also usually be of quite a large size range beyond the means of laboratory testing and so it will be necessary to allocate shear strengths from test results on a much smaller size range.

The relatively impervious cut-off core poses problems both during construction and when the reservoir is full.

During construction, and before there is any impounded water to generate seepage pressures at the core, it is well known that very high pore water pressures can develop. Sherard *et al* (1963) have noted that pressures equal to 50 per cent of the total pressure of the overlying embankment are common in the central part of the core and pressures in excess of 80 per cent have actually been measured. These pressures, which reduce the effective shear strength of the core, can be predicted from laboratory triaxial and consolidation tests on the material, but the only really satisfactory control is for the embankment to be fully instrumented with piezometers and settlement devices and for the rate of emplacement not to exceed the rate at which excess pore pressures are dissipated (see Section 11.3). There will, in fact, be two controls. The first will take the form of an embankment height/time curve based on the results of laboratory tests on the material which relate the degree of consolidation to the overburden pressure and the rate of pore pressure dissipation. Superimposed on this primary, guide control are the *in situ* instrumentation records which show immediately the extent to which the pore pressure has risen above the primary control line and the need or otherwise to cease construction until the pore pressures have reduced to a satisfactory lower level.

Pore pressures are not uniform throughout an impervious core either during or after construction. They will usually be highest towards the base of the central portion* and decrease towards the

---

*If the foundation is relatively permeable, the zone of maximum pore pressure will be located higher up the core.

crest as the vertical pressure reduces. Pressures will also reduce steadily towards the edges of the core as the drainage facility increases. But a major cause of pressures becoming higher than predicted can be attributed to soil emplacement with moisture contents on the wet-side of optimum. Sherard *et al.* (1963) have noted that in almost every dam in which the average moisture content exceeded 0.6 per cent below Standard Proctor Optimum, substantial pore pressures developed but that no significant pressures developed in embankment materials compacted a few per cent drier. On the other hand Bjerrum (1968) has noted that cracking may develop with moisture contents both below and above Proctor optimum. But a narrow clay core with a high moisture content is much more compressible than the supporting filter members and it may hang on them and never fully consolidate. A close control on the moisture content of the core material as it is being rolled-in is therefore necessary but difficult in view of climatic variations and the greater ease of working with slightly damper material. Nevertheless, the material must be placed dry of optimum, so permitting, incidentally, with fine impermeable soils, the mobilization of quite significant negative construction pore pressures which create surface tensions to add to the mass stability.

Optimum moisture contents will be related to maximum densities via the standard compaction test (see Section 11.1). Some typical curves for the assessment of embankment material are shown in Figure 12.27 and some relationships between material and compaction parameters for four N. American earthfill dams are given in Figure 12.28. See Table 2.4 for the U.C.S. symbols in Figure 12.27.

A further problem can arise from differential post-construction settlements—core with respect to graded filter and rockfill. The core must not 'hang' with respect to the filter (that is, it must not settle more rapidly) or it will fracture easily in tension. On the other hand, if the surrounding fill settles more rapidly, drag on the outer side of the core can cause sub-horizontal fractures. The use of a transition zone graded filter, in addition to its permeability protective function, also helps to offset differential settlement. Nevertheless, Sherard (1967) states that for earthquake design a single well-graded gravel mixture should be used in preference to a graded transition since it would be able to withstand more severe conditions if the core were to be ruptured or offset.

Sherard (1967) provides a general discussion on core material

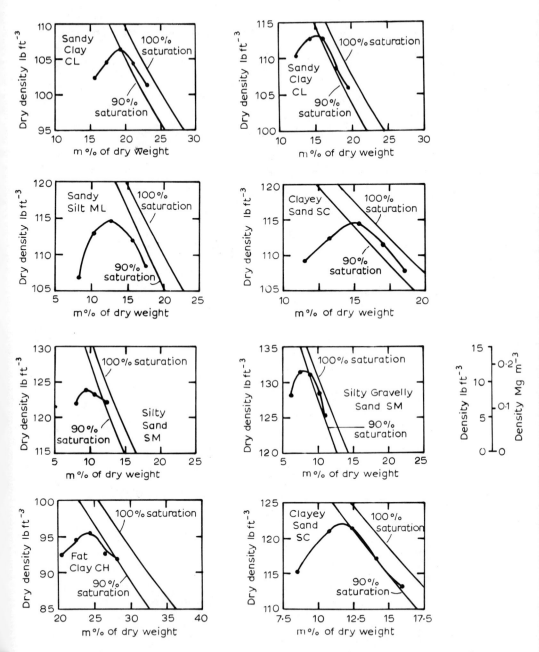

Figure 12.27 Some A.A.S.H.O. compaction test results on remoulded soils for Chatfield Dam, Littleton, Colorado. Soil types are graded according to the Unified Soil Classification System (Casagrande, 1948). Curves *after* Chatfield Dam and Reservoir Design Memorandum No. PC–24, Embankment and Excavation, Vol. II, U.S. Army Corps of Engineers, Omaha District, 1968).

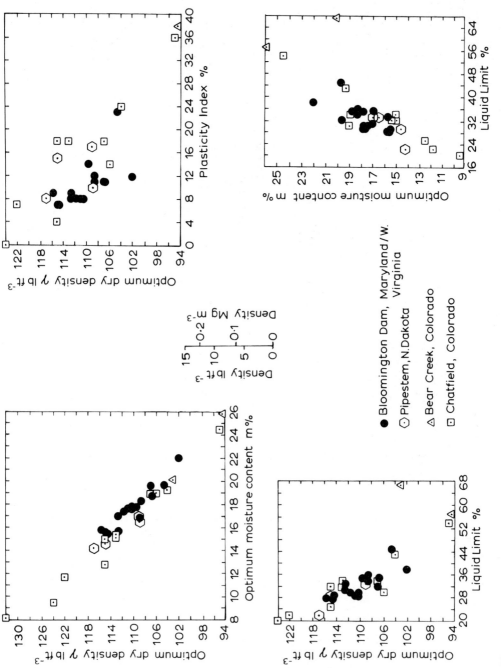

Figure 12.28 Material properties influencing compaction; data from four N. American earth fill dams under construction in 1974 (after U.S. Army Corps of Engineers Design Memoranda).

Table 12.3 Sherard's (1967) classification of dam core materials* on the basis of resistance capability to concentrated leaks

(a) *Very good material*
Very well-graded coarse mixtures of sand, gravel and fines. D_{85} coarser than 2 in; D_{50} coarser than 0.25 in. If fines are cohesionless, not more than 20 per cent finer than the No. 200 sieve.

(b) *Good materials*
- (i) Well-graded mixture of sand, gravel and clayey fines. D_{85} coarser than 1 in. Fines consisting of inorganic clay (CL) with plasticity index > 12.
- (ii) Highly plastic tough clay (CH) with plasticity index > 20.

(c) *Fair materials*
Fairly well-graded, gravelly, medium-to-coarse sand with cohesionless fines. D_{85} coarser than 0.75 in; 0.5 mm $< D_{50} <$ 3.0 mm. Not more than 25 per cent finer than the No. 200 sieve.

(d) *Poor materials*
- (i) Clay of low plasticity (CL and CL–ML) with little coarse fraction. Plasticity index between 5 and 8. Liquid limit > 25.
- (ii) Silts of medium-to-high plasticity (ML or MH) with little coarse fraction. Plasticity index > 10.
- (iii) Medium sand with cohesionless fines.

(e) *Very poor materials*
- (i) Fine, uniform, cohesionless silty sand. $D_{55} <$ 0.3 mm.
- (ii) Silt from medium plasticity to cohesionless (ML). Plasticity index < 10.

*A general summary would be that all silts (ML and MH in the Unified Soil Classification System) are relatively poor materials. For inorganic clays, the higher the liquid limit and the higher the position above the A-line on the plasticity chart, the higher the leakage resistance (Sherard, 1959). See also Table 2.4.

gradings and these are reproduced on Table 12.3. It will be realised that the strength of a compacted clay is roughly related to the liquid limit which should be restricted to some maximum value. Sherard *et al* (1963, p. 607) quote an approximate decrease in shear strength (in effective stress terms) of 10 to 15 per cent through an increase in the liquid limit from 50 to 60 per cent. There is therefore a basic conflicting requirement for core material in that a high liquid limit will usually be associated with relative impermeability but lower liquid limits will usually be expressive of the higher shear strengths necessary to resist distortion. The materials in Table 12.3 are listed and rated on a compromise basis.

In general, the foundation requirements for an earth embankment are less demanding than for other types of concrete structure although this should not be taken to imply that problems do not exist nor that the site investigation should be anything less than thorough. But the loading is usually more distributed and differential settlements caused by changes in lithology within the foundation area can be more rapidly accepted by the structure without incurring high shear stresses and possible tensions in the fabric.

Although dams are usually constructed across existing river courses, settlements can occur when dry foundations become saturated as impounding proceeds. A logical procedure for dry foundations would be to saturate them before construction begins and to keep them saturated during construction. Arthur (1965), for example, quotes the case of Medicine Creek Dam in Nebraska, U.S.A. founded on loess.

In view of its dryness, the loess was pre-irrigated to raise the water content on average to 28 per cent, thereby reducing the settlement that would otherwise have been experienced.

12.14 Long term stability of earth dams

It seems to be generally held that if dams remain stable and there is no increase in leakage during the short-term following construction, then having equilibriated they are likely to remain stable throughout their life-time unless the operating conditions are changed. Kennard (1972) has described the condition of a number of old British dams and has concluded that weakening can be attributed to such factors as (a) Overtopping (design height inadequate with respect to flood flows, probably exacerbated by urbanization, and insufficient spillway capacity); (b) Erosion of upstream face; (c) Settlement of the dam as a whole; (d) Slipping of one or both of the faces; (e) Cavities at the base of the clay core; (f) Construction rates too high and compaction insufficiently rigorous; (g) Differential settlements of the clay core with respect to the filter leading to arching across the core and cracking. In those areas, the total earth pressure is reduced to below the full reservoir water pressure, so making the core susceptible to hydraulic fracturing by the full reservoir pressure.

Another possible cause for concern could arise from the bonding characteristics between the clay minerals comprising the core of a dam. Vaughan (1972) has pointed out that if the clay, when placed, contained a high proportion of salts in its pore water, and if it was subsequently leached by water having a lower saline concentration, then an intercrystalline structure that was initially flocculated and open would change to a dispersed one of less open character. The edge-to-face bonding between clay mineral particles would be considerably reduced and the clay *en masse* would consequently be more prone to internal erosion.

Although it has been customary to leave pillars of coal under reservoirs as 'key points' in mining areas, there is always the possibility that substantial tensions (see Section 7.15) could be projected forward beyond the last line of mine workings, so affecting the water tightness of the dam and basin. Cochrane (1972) has suggested that in some cases an old dam has possibly been saved only by the existence of a substantial thickness of mantle or drift material over the area of the reservoir.

The most obvious way of monitoring for leakage through the dam is by relating the downstream flow to variations in top water level adjusted for yield, run-off and evaporation effects. But if the existence of a developing leakage is confirmed it is then necessary to locate the sources of that leakage on the basis that if it is associated with only a few discrete locations in the core, erosion will tend to increase rapidly and the leakage would become unacceptable. The only available action would then be to draw down the reservoir and commence remedial works. If the percolation is of a more general type, then remedial works may not be necessary in the short term.

Remedial works might involve one or more of the following procedures: (a) Replacement of puddle clay in the core. This may be relatively straightforward and inexpensive near the top of the core but it will be much more expensive to deal with leakage at the lower levels in this way; (b) Grouting, although puddle clay is difficult to grout and only low injection pressures can be used with the holes closely spaced; (c) Sheet piling to replace, in effect, the original clay as the cut-off medium.

12.15 Dam seismicity

(a) *Factors in dam design*

The forces which a dam is designed to resist are:– (i) Self weight of the structure, (ii) Water pressure, (iii) Silt pressure, (iv) Ice pressure, (v) Earthquake acceleration, (vi) Resonance.

In many areas, (iv) can be ignored, and resonance in low dams up to 100 ft is not usually taken into account since the vibration period of the dam is normally much less than the range 0.2 to 1.0 second, the period associated with severe earthquake shocks. Dams must resist with a reasonable factor of safety the following main causes of failure:— (i) Rotational failure of the side slopes of earth and rock-fill dams; (ii) Rotation of gravity dams; (iii) Shearing and sliding on a horizontal plane (taken as a construction joint); (iv) Settlements in earth embankments; (v) Tensile cracking of concrete; (vi) Stresses in the foundation strata.

The first four causes will be considered in a rather superficial manner here, but specifically with respect to earthquake attack. Examples of (vi) are given elsewhere.

(b) *Effect of seismic activity on dams*

Dams in earthquake-prone areas of the world will embody within the design concept a resistance capacity to dynamic forces which can be applied during earthquake attack. It is necessary, therefore, to have some understanding of the amplitudes and nature of these forces. This can best be achieved before construction by checking past seismological records for the area or areas nearest to the proposed damsite, by locating major fault lines, and more particularly by undertaking an instrumentation programme on site as soon as preliminary decisions on location have been made. The programme would take the form of long-term seismic monitoring using geophones in orthogonal modes. Recording would be on magnetic tape for subsequent processing with the aim of resolving amplitude, frequency content and directional properties of any incoming waves.

As a general point, it should be noted that probably the earliest method of determining the dynamic lateral pressure on a retaining structure is that due to Mononobe (1929) and Okabe (1926). The Mononobe-Okabe analysis, upon which are based many national building code requirements for lateral pressure tolerance during earthquakes, is discussed in Seed and Whitman (1970). Westergaard's early curves (Figure 12.29) are also still referred to but modern methods of design and analysis using digital computation are more relevant.

(c) *Nature of ground movements*

Shallow depth earthquakes can be assumed to originate from the sudden release of strain energy caused by fracture or slip along a pre-existing fault plane (Ambraseys and Sarma, 1967). The slip-stick brittle character of the shear movement releases distortional strain

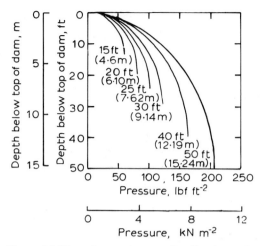

Figure 12.29 Approximate distribution of pressure at a dam due to an earthquake of amplitude $0.1g$ and a period of one second. Curves are for dams of different heights and were calculated from Westergaard's (1933) approximate formulae (*after* National Resources Committee, 1938).

energy rapidly in the form of stress waves, but the actual energy released is only a fraction of the total potential energy that is stored in the volume of rock within which the shearing takes place Taking some very approximate comparative amplitudes, an earthquake would probably involve a stress drop at source of less than 100 bars (1450 lbf in^{-2} or 1 MN m^{-2}) but since it is probable that an area of about 10^5 km^2 could be involved in the dissipation of energy (releasing about 10^{25} ergs) there is the suggestion of a semi-brittle, deep-seated origin.* Most large earthquakes occur at depths of 20 to 30 km (pressures of 6000–9000 kgf cm^{-2} or $589 - 883$ MN m^{-2}) with accompanying ambient temperatures of 500–600°C. It is claimed that the phenomenon of after-shocks trailing large earthquakes is the result of removing residual stresses within the focus of the main shock The presence of residual deformation is characteristic of plastic deformation, although it is possible to visualize

---

*A sandstone having a 20 per cent porosity and confined at 1 kbar at room temperature deforms in a ductile manner. It ceases to shear in a discrete manner and there is no sudden stress drop, failure being defined only in terms of a strain criterion. Under geo-thermal temperature gradients, ductility would develop at lower confining pressures (shallower depths).

progressive movement along an elastic shear plane as protrusions are planed down.

Dilatancy theory may be invoked to explain seismic precursor deformation. Although accumulated strain energy causes a decrease in rock volume and an increase in its strength, there is a later increase in volume and decrease in porewater pressure as more strain energy is accumulated. Water enters the stressed rock, again builds up the porewater pressure and so reduces the effective stress.

With an increase in rock volume there is an associated reduction in seismic velocity which can be detected in an environment favourable for a future earthquake. Changes in the ratio of P-wave to S-wave velocity have also been used to predict approximate magnitudes and timings of earthquakes.

A further earthquake origin could possibly be attributed to a sudden density change (density metastability) associated with a polymorphic transition in the rock material. For upper mantle materials, a 3 per cent density change is possible, while in the middle mantle, 20 per cent density changes could occur*. The possibility is then raised of fault development being the result of the earthquake rather than actually creating it.

There is some evidence that the onset of an earthquake is preceded by changes in the rate of ground surface deformation which may be measured using techniques of precise levelling, precision distance-measuring equipment, and tilt meters. Pre-earthquake elastic deformation is approximately sinusoidal with its wavelength decreasing and its amplitude increasing as the onset time for the earthquake approaches (Oborn, 1974). The greatest movements will be found at the earthquake epicentre.

(d) Source movement

Frictional displacement along a pre-existing fault plane creates imaginary nodal planes across which the algebraic stress is zero (Figure 12.30). These planes separate dilations from compressions, the direction of wave motion changing as it passes through them, and they can sometimes be defined by plotting the distribution directions of compressions and tensions over a wide area at ground surface. In a

---

*Note that eclogite is 15 per cent denser than gabbro and that it is claimed that polymorphic transformations have been achieved under laboratory shock conditions when the time for completing a reaction is of the order of a few microseconds (Hughes and McQueen, 1958).

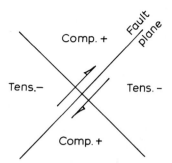

Figure 12.30 Movement along a pre-existing fault plane.

similar manner, a force in excess of hydrostatic in the three-dimensional case can create a fold system and give rise to a quadrant distribution of compressions and dilations if the release of excess strain is sudden (Figure 12.31). On the other hand, a sudden cavitation due to a density change would produce a tensile first arrival independent of direction with respect to source.

(e) *Measured parameters of movement*

The motions produced by deep focus earthquakes result predominantly from surface reflections and refractions of P and S body waves and their derivatives. Shallow earthquakes generate surface waves and it is the directional properties of the surface motion, particularly the horizontal components of the motion that should be estimated for design.

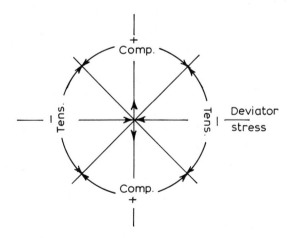

Figure 12.31 Generation of a fault plane due to the presence of a deviator stress.

Seismic disturbance can be recorded and/or expressed in terms of surface displacement, velocity or acceleration. Single event ground *displacements* of up to 12 inches (0.3 m) have been recorded, ground *velocities* of about 12 in s^{-1} (0.3 ms^{-1}) have been measured indirectly near the epicentre, and while accelerations of about 0.3 g have actually been measured a few miles from epicentre, accelerations exceeding 1 g have been inferred (see reference quoted by Ambraseys and Sarma, 1967). From record analysis, Ambraseys and Sarma have drawn several important conclusions concerning the generality of strong ground motion from seismic sources.

Ground dispacement takes place usually over a period of 2 to 10 seconds. Larger displacements develop slowly and do not induce significant accelerations. Ground velocities have much shorter periods of between 0.5 and 2 seconds and the amplitudes of both velocity and accelerations are magnified in the presence of superficial deposits overlying bedrock. Ground accelerations fluctuate much faster, with half-periods of 0.05 to 0.4 second and with pulse shapes that are nearly triangular in configuration (¼ period rise and fall-off time). Peak accelerations persist for only a few milliseconds, but it is during this time that a non-compliant structure experiences its maximum inertia force.

The nature of an arch dam and the critical design concept behind it implies that its seismic response requires careful evaluation. For chord-height ratios up to 3, and with respect to horizontal excitation, the first 5 or 6 natural frequencies fall generally in the range 1 to 10 Hz. However, as the chord-height ratio becomes greater than 3, there is a significant increase in the number of natural frequencies which fall within this 1 to 10 Hz frequency range.

In terms of a pure sinusoid there should, of course, be a quarter wavelength phase difference between displacement, velocity, and acceleration respectively. Vibration frequencies of the three modes should be the same. Departure from these conditions can be attributed to the wide spectral band of the pulse frequency composition. A typical record and its derivatives are shown in Figure 12.32.

The question arises as to which mode of vibration — displacement, velocity or acceleration — to accept as a specification for safe structural design. Although there is a complex spectrum of input frequencies to a dam, and the internal resonances of the dam itself will be complex, it is probably most conveniently simple to consider

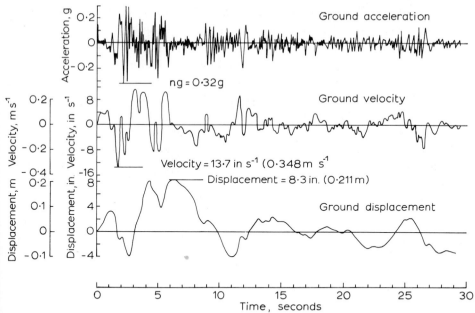

Figure 12.32 Seismic record of El Centro, California, earthquake of May 18, 1940. N–S polarized component traces (*after* Newmark, 1965).

a basic forcing frequency (say p) and a fundamental structural frequency (say f). If p is sufficiently high compared with f, then horizontally-vectored movement could be described in very approximate terms by saying that the dam foundation accompanies the ground while the crest remains relatively still. Distortion of the dam structure would then be represented by the amplitude of *ground movement* itself, the correlation between the two becoming more refined the larger the ratio p/f. It is not realistic to suggest damage threshold levels with respect to vibration displacement on the lines, for example, of the 0.008 in* (0.02 cm) displacement suggested by Morris (1950) as being just not capable of causing distress to a building. This same criterion could be expressed in terms of acceleration quite simply as $d \simeq u \simeq a/(2\pi p)^2$, where d is the induced distortion, u is the ground displacement and $a (= \partial^2 u/\partial t^2)$ is the peak acceleration for a monochromatic, or single frequency,

*This is a very conservative figure that has at times been doubled or quadrupled; for example, the State of Pennsylvania adopted a displacement of 0.030 inch (0.012 cm) as fixing a safe stand-off distance for property from an explosive blast source (Duvall and Fogelson, 1961).

excitation wave. We find that a displacement of 0.008 in (0.2 mm) is associated with an acceleration of 1 g at a frequency of 35 Hz (or 0.030 inch (0.76 mm) at 18 Hz); these frequencies are more related to the excitation imposed on building structures quite close to a man-made blast vibration source and, in fact, an acceleration of 1 g and upwards has been used as a damage criterion in America (Morris, 1950; Leet, 1960).

At natural earthquake frequencies (for example, 4 Hz quoted by Ambraseys and Sarma for the El Centro records of the Imperial Valley earthquake of 18 May 1940) p will often be less than f, leading to larger permissible displacements and correspondingly smaller accelerations. Under these conditions,

$$d \simeq \frac{up^2}{f^2} \simeq \frac{a}{(2\pi f)^2}$$

with distortion being less than ground displacement and the structure attempting to follow the ground in a more or less rigid manner. Acceleration is then a more obvious quantity to correlate with damage by distortion and it will be seen that smaller accelerations, although larger displacements, will be required for damage than in the case of high frequency waves. As an example, if a structure of natural frequency $f = 5$ Hz is subjected to a seismic frequency of 1 Hz, then $d \simeq u/25 \simeq a/900$. Structural distortions would then require ground displacements of 0.75 in (19 mm) but low level accelerations of 0.07 g.

It is possible to conclude from the above somewhat over-simplified arguments that structural damage is generally going to be controlled by ground displacement at high input frequencies and by acceleration at low seismic frequencies. Threshold values for typical structures could be taken at about 0.03 in (1.2 mm) and 0.1 g respectively at the two extremes. It should be pointed out, however, that displacement values are not easily obtained directly and that double integration errors from acceleration records arise from an uncertainty with respect to the true base line of the source trace.

A third criterion, that of velocity, has been used for structural damage from blast vibrations. In fact the choice between displacement, velocity and acceleration for blast damage was tested at the U.S. Bureau of Mines (Duvall and Fogelson, 1961) through the use of double-log graphs of critical displacement u against forcing frequency p. A slope of zero corresponds to the simple displacement criterion, a slope of -1 to velocity, and a slope of -2 to acceleration. Their results favoured a criterion based on critical ground velocity, and suggested that 2 in s^{-1} (50 mm s^{-1}) would be a conservative value. Other workers, for example Langefors and Kihlström (1963), have confirmed this with values varying between 2 to 4 in s^{-1} (50–100 mm s^{-1}). There is logic in the adoption of a velocity criterion since it is related to both strain and energy and it can be shown that the most important measure of earthquake intensity is the maximum ground velocity reached at any time during the earthquake (Newmark, 1965). But in any simple analysis of dam stability involving force balances, then obviously the acceleration approach is most germane.

(f) *Factors influencing the response of a dam to seismic motion*

The response of a dam to earthquake vibration is a function of the amplitude and duration of the acceleration that is applied to it since these control the time-distribution of the inertia forces. If these forces are low, the dam will tend to respond in a reasonably elastic manner and that response can be computed albeit through the adoption of simplifying assumptions. In any case, even if the level of acceleration takes the structure beyond the elastic regime, elastic calculations will produce conservative results that err on the side of safety (Ambraseys and Sarma, 1967). A coda of smaller amplitude accelerations may also impose greater structural damage than will a single, but higher, maximum acceleration. If analyses are to be performed on a two-dimensional section of, say, an earth embankment, the section being taken normal to the long, horizontal axis, then a one-dimensional wave propagation situation would be assumed, with the component of acceleration being resolved from the resultant of three orthogonal accelerations, the latter being either measured on site or predicted.

(g) *General foundation awareness*

Much more earthquake damage tends to be created in unconsolidated deposits than in solid rock exposures at ground surface.

The susceptibility of superficial deposits to seismicity is a function of the degree of strain energy acquired by the material. Writing u for the seismic displacement amplitude and ω_h for the circular frequency in radians, then the horizontal velocity and acceleration of the ground can be written as $u\omega_h \cos \omega_h t$ and $-u\omega_h^2 \sin \omega_h t$

respectively. The maximum amplitudes of horizontal surface velocity and acceleration can then be written:

$$v_{max} = u\omega_h \qquad (12.9)$$

and

$$ng = u\omega_h^2 \qquad (12.10)$$

Thus, if f is the ground frequency or spectrum of frequencies,

$$v_{max} = \frac{ng}{\omega} = \frac{ng}{2\pi f} \qquad (12.11)$$

and the kinetic energy applied in the horizontal direction to the foundation per unit time and per unit volume at maximum velocity is

$$K.E \text{(units of force per unit area)} = \tfrac{1}{2} m v_{max}^2 = \frac{n^2 \rho g^2}{8\pi^2 f^2} \qquad (12.12)$$

where ρ is the density of the foundation material.

As an example of vibration magnification in unconsolidated superficial deposits, in the San Fransisco 1906 'quake the maximum acceleration recorded in marshy ground was about 10 ft s^{-2} (3 m s^{-2}) but in corresponding rock outcrops was as low as 0.89 ft s^{-2} (0.27 m s^{-2}). Therefore, for civil engineering work in earthquake-prone terrain, foundations should be in solid bedrock whenever possible. It is useful also to consider a raft-type foundation so designed that it will withstand upward forces in addition to the usual downward loads. Filled ground will often consolidate drastically under earthquake forces. In the New Zealand earthquake of 1929, for example, one fill that had been in place for 6 years sank 3 ft (1 m). In addition to filled ground, undulating and hilly ground should be avoided whenever possible and it is also advisable to keep away from river beds, canals, and coastlines. A further point which should receive attention at the design stage, and which has been the subject of a number of scientific papers over the last decade, concerns the possibility of liquefaction phenomena developing in the saturated fill material of an earth/rockfill dam and even in the foundation material. This is a specialist subject and the interested reader is referred to any appropriate publications, for example: Seed, 1968 (Sheffield Dam, Santa Barbara, 1925); Kishida, 1969

(Tohnankai earthquake, 1944; Fukui earthquake, 1948, Japan); Marsal, 1961 (Jaltipan earthquake, Veracruz, Mexico, 1959); Steinbrugge and Clough (1960), Duke and Leeds (1963), Dobrey and Alvarez (1967), Seed (1968) – (Chilean earthquake of 1960); McCulloch and Bonilla (1967), Seed and Wilson (1967), Seed (1968), Ross *et al.* (1969) – (Alaska earthquake of 1964); Seed and Idriss, 1967 (Niigata earthquake, west coast of Japan, 1964); Kishida, 1970 (Tokachioki earthquake, Japan, 1968). Geotechnical maps of earthquake prone areas should indicate those places which might be prone to failure through liquefaction.

(h) *Possible dam failure modes*

It is possible to visualize at least three types of earth or rock fill dam failure: circular shear failure of the embankment; planar base failure in shear along the embankment-foundation interface; rotational failure of the embankment as a whole. Of these, the first is most likely to occur. Concrete gravity dams may be subjected to these latter two modes, and in earthquake-prone areas they would incorporate appropriate provisions in their structure, for example, panel construction with asphalt-filled groove-joint connections.

(i) *Circular shear failure of embankment*

The first type of failure, taking the form of a series of slides on the upstream and downstream faces of a dam due to the motions from several shocks originating from different directions, was illustrated by Ambraseys (1958), see Figure 12.33. Ultimate crest settlements of the form shown have been observed in older dams having no inbuilt resistance to seismic forces.

Acceleration forces are usually expressed as a fraction of g, and if W is the weight of a shearing element or slice of circularly-sheared mass then nW is the magnitude of the earthquake force. The simplest specification of the problem is as shown in Figure 12.34 with a seismic wave directed towards the embankment to decrease the stability. Taking W and nW to act through the centre of gravity of the rotating mass, the factor of safety against rotational shear can be written

$$F = \frac{S_s r^2 \theta}{Wa + nWb} \qquad (12.13)$$

The shear strength S_s can include either or both of the shear

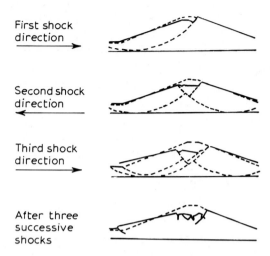

Figure 12.33 Seismic deformation effects on an earth embankment (*after* Ambraseys, 1958; Newmark, 1965).

strength parameters. Very simple, first approximation solutions can be derived in this way if the centres of gravity can be located for a series of trial circles.

Newmark's (1965) method involves determining the maximum possible acceleration ng that can be transmitted to a potential shear failure surface having a known shear strength, and if an actual earthquake acceleration $n'g$ exceeds this, then the material deforms 'plastically' (there is displacement along the shear surface) and the factor of safety with respect to shear falls below unity.

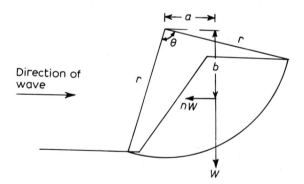

Figure 12.34 Parameters for an analysis of circular slope failure under the influence of seismic forces.

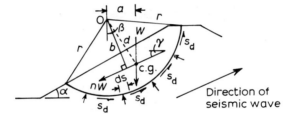

(a) Arbitrary direction of seismic force

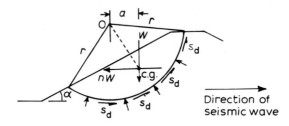

(b) Horizontal seismic force

Figure 12.35 Rotational slide on face of dam (*after* Newmark, 1965).

Consider a static condition ($n' = 0$). Equating the disturbing moment and the restoring moment on the failure surface shown in Figure 12.35 and writing τ for the shear stresses,

$$Wa = r\Sigma \tau ds \qquad (12.14)$$

Now the moment of the maximum possible resisting forces to dynamic shearing under seismic acceleration is $r\Sigma s_d ds$. Thus the dynamic factor of safety is:

$$F_d = \frac{r\Sigma s_d ds}{r\Sigma \tau ds} = \frac{\Sigma s_d\, ds}{\Sigma \tau\, ds} \qquad (12.15)$$

In order to find the necessary n for sliding just to occur,

$$Wa + nWb = r\Sigma s_d\, ds \qquad (12.16)$$

or, by taking equation 12.14 from equation 12.16,

$$nWb = r\Sigma s_d\, ds - r\Sigma \tau\, ds \qquad (12.17)$$

Dividing equation 12.17 by equation 12.14 and multiplying by a/b,

$$n = \frac{a}{b}\left(\frac{\Sigma s_d\, ds}{\Sigma \tau\, ds} - 1\right)$$

or if average values of s_d and τ are taken

$$n = \frac{a}{b}\left(\frac{s_d}{\tau} - 1\right) \qquad (12.18)$$

This equation can be written in terms of factor of safety:

$$n = \frac{a}{b}(F_d - 1) \qquad (12.19)$$

Since b at its maximum becomes equal to d for a particular shear surface, the minimum value of n arises for a slope perpendicular to d or

$$n = \frac{a}{d}(F_d - 1) = (F_d - 1)\sin\beta \qquad (12.20)$$

If the seismic force is applied horizontally, then from equation 12.19

$$n = (F_d - 1)\tan\beta \qquad (12.21).$$

If, as Newmark suggests, soils have almost the same static and dynamic shear resistance, then the factor of safety symbol F can be written unsubscripted.

Now if n' is not zero, the same type of analysis leads to:

$$r\,\Sigma\,s_d\,ds = F'_d(Wa + n'Wb) \qquad (12.22)$$

and equating equation 12.22 with equation 12.16 gives

$$n = n'F'_d = (F'_d - 1)\frac{a}{b} \qquad (12.23)$$

Newmark has pointed out that the use of equation 12.23 incorporating a trial value of n' leads to more accurate results than the use of the earlier equations 12.17 and 12.18 in which no assumed value of accelerating force is included in the calculation. The most accurate results arise when the dynamic factor of safety under an arbitrary seismic acceleration (F'_d) is nearly equal to unity.

For $n' \neq 0$ and n perpendicular to d,

$$n = n'F'_d + (F'_d - 1)\sin\beta \qquad (12.24)$$

and for $n' \neq 0$ and n horizontal,

$$n = n'F'_d + (F'_d - 1)\tan\beta \qquad (12.25)$$

(j) Stability with respect to horizontal base shear

During an earthquake, and considering the horizontal component of surface wave motion only, it is assumed that there will be a sudden lateral translation of the base of the structure as a function of the shear stresses acting along that horizontal cross-section. This has the effect of creating a shear wave in the dam structure. The velocity of the shear wave (determined by the material composition and its elasticity) together with the crest height will set the dam in oscillation, and it is one of the design requirements to attempt to estimate the fundamental and higher modes of vibration characteristic of the dam. Concrete buttress, arch and gravity dams, being relatively stiff, will oscillate at higher fundamental frequencies than earth and rockfill dams, and in areas of continuous seismicity the question of harmful resonances (Dungar and Severn, 1971), structural fatigue and possibly tensile failure at critical interfaces in concrete dams could perhaps arise under circumstances where single oscillations might be well below damaging levels. There will also be a tendency for stress magnification towards the crest of a dam; in the case of an arch dam, a computed magnification of 7 or 8 in the maximum crest acceleration relative to the maximum input at the base of the dam. As a general point, larger overall forces seem to be developed in smaller dams (although they persist for rather shorter periods of time) than in higher dams. Engineers might also consider the point that a stiff structure, when set into resonance is only one lumped component in a potential oscillating system and that there is always the possibility of vibratory energy feedback from the structure to the foundation after the disturbing force has passed. Reference might be made to Chakrabarti and Chopra (1973) and to Petrovski *et al* (1974) for further reading on this subject.

Consider a triangular cross-section concrete gravity dam as sketched for example in Figures 12.36, 12.37 and take the impounded condition. H_1 is the crest height of the dam and H_2 is the full head of water. W is the weight of the dam structure per unit run of its longitudinal axis. Using a simple analysis, the static water force is $\gamma_w H_2^2/2$ and the dynamic water force also acting against the upstream face of the dam for a siesmic wave directed upstream is $n\gamma_w H_2^2/2$, assuming that a body of water confined between a certain parabola and the upstream face of the dam were forced to vibrate with the dam. To these two forces must be added the seismic component of the weight of the dam, nW. These disturbing forces are

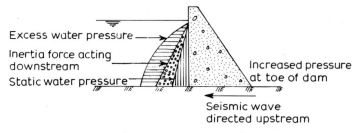

(a) Reservoir full; seismic wave travelling upstream.

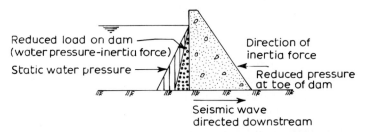

(b) Reservoir full; seismic wave travelling downstream.

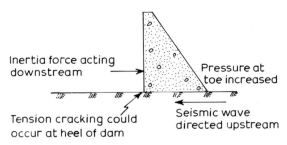

(c) Reservoir empty; seismic wave travelling upstream.

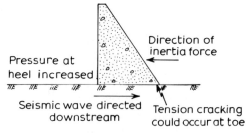

(d) Reservoir empty; seismic wave travelling downstream.

Figure 12.36 Various stress conditions created at a gravity dam through the horizontal motion of a seismic wave.

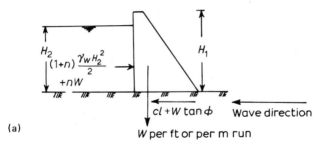

(a)

Possible uplift seepage forces acting against W are not taken into account

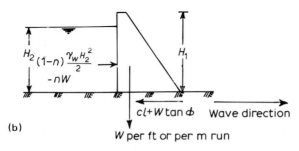

(b)

Figure 12.37 Simple analysis for factors of safety against planar sliding at the base of the dam.

resisted along the foundation area by the shear strength of interface, $cl + W \tan \phi$, where l is the base width of the dam. It follows that the factor of safety F_U (upstream wave) with respect to basal shear is:

$$F_{US} = \frac{cl + W \tan \phi}{\frac{\gamma_w H_2^2}{2}(1+n) + nW} \qquad (12.26)$$

If the seismic wave is directed horizontally downstream, then the disturbing forces on the upstream face of the dam will be reduced and the factor of safety F_D becomes:

$$F_{DS} = \frac{cl + W \tan \phi}{\frac{\gamma_w H_2^2}{2}(1-n) - nW} \qquad (12.27)$$

In theory, as n increases, the second term in the denominator of equation 12.27 rapidly becomes dominant and the resisting shear force in the foundation will be required to change direction. But generally the factor of safety for a downstream wave will be higher. Safety margins will also be enhanced with the reservoir empty.

There are one or two other points to bear in mind. The equivalent lateral force method assumes the dam to be infinitely rigid, which of course it is not. The simple analysis also ignores any restraining influence offered by the dam abutments; the shorter the dam, the more important these influences become. The base of the dam and its foundation contact is, of course, never planar; the shear strength parameters will be supplemented by a component of reaction. Wave action would have an influence on the factor of safety, but this has been ignored It should also be mentioned that a vertically-directed component of seismic wave motion would enhance the shear stability of the structure by increasing the gravitational load through the imposition of inertia forces. On the other hand, water pressures on the upstream face of a gravity dam during vertical seismic motions have been found by analysis (Chopra, 1967; Nath, 1969, 1971) to exceed those on the same face due to horizontal seismic motions.

With the pseudo-static method of analysis, a coefficient of $0.1\,g$ is often taken for regions of expected moderate-to-severe shocks. Alternatively, a coefficient of $0.05\,g$ can be taken for regions of minor seismic risks. An earthquake acceleration of up to approximately $0.3\,g$ is only about half as effective in the silt that will have accumulated at the base of the upstream wall as in water. Since the density of water is about half that of silt, the increase in pressure on the dam due to an upstream earthquake is approximately the same for either silt or water (Lewis *et al*, 1960). The presence of silt can therefore be ignored in calculations.

With a full reservoir, the resultant force at any horizontal plane must fall within the limiting confines of the structure so long as allowable horizontal stresses are not exceeded. Although a factor of safety of 4 with respect to sliding will usually be built in to the design of the dam, it will be permissible to see this factor drop to about 2 for the short periods during which it will be subjected to seismicity.

In certain areas, ice pressure produced by thermal expansion in the ice sheet and by wind drag can create a problem. Rose's (1947) curves relating thrust against the dam to the thickness of the ice sheet and air temperature have been used in design, and Lewis *et al* (1960) have quoted evidence which suggests that a snow covering provides a good insulation effect against progressive sheet thickening. Such thrust would act more effectively in the overturning mode.

(k) Stability with respect to overturning

If the gravity dam considered earlier is assumed to be of triangular configuration in cross-section then a vertical line through the centre of gravity of the dam will pass through the base at a distance equal to $2l/3$ from the toe. Otherwise, this distance must be determined for the actual shape.

Reference to Figure 12.37 indicates that the factor of safety against rotation about the toe of the dam is:

$$F_{UO} = \frac{l(4W + 3cl)}{H_2[(1 + n)\gamma_w H_2^2 + 2nW]} \tag{12.28}$$

if cohesion is assumed to act as a stabilizing force against overturning.

In the case of a downstream wave there are, in theory, two possible overturning modes – about the toe and the heel of the dam respectively. The appropriate equations are:

$$F_{DOT} = \frac{l(4W + 3cl)}{H_2[(1 - n)\gamma_w H_2^2 - 2nW]} \tag{12.29}$$

(overturning about the toe) $\quad \left\{(1 - n)\dfrac{\gamma_w H_2^2}{2} > nW\right\}$

$$F_{DOH} = \frac{l(2W + 3cl)}{H_2[2nW - (1 - n)\gamma_w H_2^2]} \tag{12.30}$$

(overturning about the heel) $\quad \left\{(1 - n)\dfrac{\gamma_w H_2^2}{2} < nW\right\}$

As before, application of a seismic-equivalent static horizontal force to an earth embankment provides a rather less valid solution since it takes no account of the quasi-elastic response of the compliant fill material to the seismic accelerations.

(l) Criteria of failure for arch dams

There seem to be no definite criteria of failure for concrete arch dams but there are a number of factors which bear on the problem. It is useful to attempt an estimation of the extent of tensile cracking on both upstream and downstream faces of the dam, based on the amplitude of the calculated stresses. From these calculations, it may also be possible to specify those parts of the dam where cracking can propagate from one face to the other. If the dam is cracked, it is still possible for it to retain some strength, but it would be difficult to assess for how long a cracked dam would retain any structural competence under subsequent stress reversals. Following a large earthquake, design engineers should immediately engage in a detailed inspection of all dam structures in the vicinity in order to log any damage and make decisions as to whether a particular dam is unsafe.

New and more realistic evaluations of the effects of seismic loading can be made using dynamic response methods which combine the interactive effects of structural and ground vibration. The pseudo-static design procedure earlier described is empirical and is not based on rational analysis. But the factors of safety so derived tend to be conservative to compensate for any inaccuracies; the structural dynamic method of analysis generates appreciably larger concrete stresses than the equivalent stresses computed by the 'g' method.

(m) *Seismic stability of earth dams*

Under moderate earthquake attack, some earth dams have settled between 0.5 per cent and 1.0 per cent of their height (Lane *et al*, 1974) It is therefore necessary in seismically-prone areas to provide ample freeboard above top water level and to design for shallower slope angles at the crest. The material used for the construction of earth dams in seismically-active areas should have 'self-healing' properties and be able to adjust readily to differential settlements. One design criterion, for example, limits the maximum strain at any point in the embankment to 5 per cent. A well-graded sandy gravel possesses this characteristic, it has good damping properties, and it would be suitable not only for transition zones but could also be sufficiently impermeable for the cores of earth dams. From a seismic stability viewpoint, earth dam cores should be vertical and central, but a core that is sloped upstream is likely to have a higher self-sealing capability.

(n) *Relationship between seismic intensity and wave parameters*

Figure 12.38 expresses the relationship between a known seismic intensity level on either the MM (Modified Mercalli) or MSK (Medvedev, Sponheuer, Karnik) scales and the acceleration, n, velocity and displacement parameters.

Intensity must be estimated from observations of the seismic influence on natural and man-made structures and objects. From such observations and reports of witnesses, tentative contours of

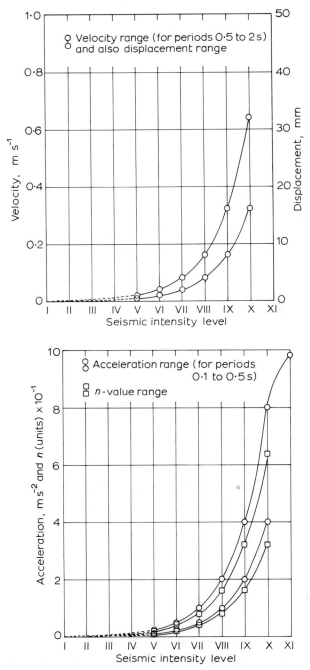

Figure 12.38 Seismic force parameters related to seismic intensity.

seismic intensity may sometimes be drawn on geographical maps of the affected area and the epicentre thereby located. In contrast to seismic *intensity*, earthquake *magnitude* for P-waves and surface waves is computed directly from the wave amplitude on seismograms. Logarithmic relationships between magnitude and intensity are proposed from time to time for different areas, but they can be misleading and may even be an order of magnitude in error when used in a predictive role.

By analysis of seismic frequency on a geographical basis, maps of maximum intensity and maps of a given return period may be constructed in order to express the degree of seismic risk. Although Britain is deemed to be relatively aseismic (in spite of the fact that the return period for low intensity seismicity is surprisingly short) certain areas such as the Great Glen of Scotland and a diffuse area stretching from South Wales, Hertfordshire and into the North Midlands may expect to experience an intensity 7 'quake once every 200 years (Dollar et al, 1975). For a specified lower intensity level, the return period may be expected to decrease. A fundamental problem arises from the paucity of data upon which to base viable predictive analyses, but such predictions must be made in order to assist the structural designer. The terms 'Design Earthquake', 'Operating Basis Earthquake' and 'Safe Shut-down Earthquake' have been used in this context (Oborn, 1974).

(o) *Investigation*

At the reservoir design stage, it is necessary to attempt a prediction of the maximum probable earthquake which can occur in the general area of the dam site. Unfortunately, there is usually a lack of strong-motion recordings for areas which will generally be rather remote from population centres and upon which predictions must be based. Other investigations are often of equal value. Geologically, all the major fault zones within about 100 miles of the damsite must be located and the direction, length, possible age and seismic history (from ancient records) of the main and branch faults determined. Aerial reconnaissance will be used at the primary stage, with early morning or late afternoon colour photography or sidelooking airborne radar in order to highlight fault scarplets and ground undulations generally. It should be assumed that if a fault has slipped in the past, it is also likely to slip in the future especially if it can be shown that it has not slipped for a long time. Fault displacements can be measured by geomorphological and stratigraphical observation. Reference may be made to Sherard *et al.* (1974) for an exposition of the problems related to faults actually at damsites — faults which may undergo active movement through seismic triggering. In a similar, parallel investigation of old data, the intensity, depth and epicentre of all historical earthquakes must be evaluated. From this geological and seismic information it might then be possible to estimate the magnitude of the maximum probable earthquake at each of the faults and fault systems located. In addition to this magnitude, it might also be possible, having located the epicentre, to estimate the duration, number and magnitude of any aftershocks which can add to the main shock effect and even prove critical. Data will be available on the variation of ground acceleration with distance from source, and from this information it will be possible to estimate the peak ground acceleration at the damsite for each maximum probable earthquake at the fault location. This acceleration information provides a framework from around which the maximum probable earthquake for the damsite proper can be estimated.

(p) *Reservoir-induced seismicity*

There has been a number of well-documented cases of seismic activity development or intensification correlating with reservoir impounding. These include lakes Koyna (India), Kariba (Zambia), Kremasta (Greece), Bileća (Yugoslavia), Talbingo (Australia), Mead and Cabin Creek, Colorado (U.S.A.) and a large number of high dams in Japan. There are two main explanations:

(a) increased vertical loading and localized crustal depression caused by the impounded water, and
(b) seepage to and/or pore pressure elevation along a potential shear plane with a consequent reduction in the effective normal stress on the plane and a diminishing shear resistance along it.

Condition a) by itself would be unlikely to initiate shear motion of the stick-slip kind unless the stability of the ground was critically balanced prior to impounding. With respect to condition b), it is known that prior to shear failure under all-round compression, a rock at first decreases in volume and then dilates just prior to failure (see Section 4.3 and also Oborn, 1974). This would lead to temporary pore pressure elevations (conducive the shear) followed by pore pressure reductions (shear resistance). But the greatest justification for a) arises from evidence such as was provided by the Rangely oil field (Colorado, U.S.A) and the subsequent water injection experiments* (see Snow 1968; Hollister and Weimer, 1968; Scopel, 1970; Evans, 1968, 1970; Bardwell, 1970; Munson, 1970; Carder, 1970).

There is often a time-delay between impounding and the onset of statistically significant seismic activity (for example, Green, 1973, has mentioned 6 months delay at the Hendrik Verwoerd dam in South Africa). Howells (1973) has used the solution of the one-dimensional diffusion equation to suggest that surface pore water pressure may be realized at 5 to 10 km depth after some hundreds of days and at a depth greater than 15 km after some thousands of days. The consensus of evidence also points to the fact that such seismic activity is quite shallow in origin (1 to 7 km) and is confined to the immediate vicinity of the reservoir. 'Plastic' rocks tend to display more temporary seismicity than do brittle rocks (Snow, 1972). Further reference may be made to Guha *et al* (1973), Muirhead (1973) and Okamoto (1973).

A great deal of effort is being directed towards acquiring a better understanding of the detailed causes of this type of seismic activity (Oborn, 1974). For example, it may well be confirmed that the impounding rate with respect to the consequential density of activity is just as important in its own way as is the rate of construction of an earthen dam with respect to shear failure through pore pressure elevation. Dams on the margins of shields seem to experience more post-impounding seismicity than do dams in the centres of shield areas. Joint/bedding/fault plane densities and orientations at the reservoir site may directly influence the development or otherwise of seismic activity. If vertical loading is a critical factor, then a steep catchment, high dam and concentrated lake might be considered least desirable. Furthermore, given a seepage facility, higher heads of water would reduce the time constant between cause and event.

Extensive dam instrumentation programmes underway throughout the world (seismometers, tiltmeters and so on) will throw much more light on this problem in the not-too-distant future.

*These field trials were conducted when correlations were established between 150 million (U.S.) gallons of effluent injection from the U.S. Rocky Mountain arsenal outside Denver into a 12,045 ft. deep Pre-Cambrian reservoir of highly-fractured, low permeability granite gneiss and an increased tremor intensity. Between 1882 and 1962, there were only 3 earth tremors recorded at Denver. Since injection began in 1962 and up to 1968, there were 610 tremors of which 18 registered 3 or above on the Richter scale.

References

Aas, G. (1965) A study of the vane shape and rate of strain on the measured values of *in situ* shear strength of soils, *Proc. 6th Int. Conf. Soil Mech.*, Montreal, 1, 141–145.
Ackers, P. (1972) River regime: research and application, *J. Inst. Water Engrs.*, 26, 257–274.
Ackroyd, T. N. W. (1954) 'Laboratory testing in soil engineering,' Soil Mechanics Ltd., London.
Adamson, L. G., Chilingar, C. V., Beeson, C. M. and Armstrong, R. A. (1966) Electrokinetic dewatering consolidation and stabilisation of soils, *Eng. Geol.*, 1, 291–304.
Adamson, L. G., Quigley, D. W., Ainsworth, H. R. and Chilingar, C. V. (1966) Electrochemical strengthening of clayey sandy soils, *Eng. Geol.*, 1, 451–499.
Agarwal, K. B. (1967) 'The influence of size and orientation of sample on the undrained strength of London Clay,' Ph.D. Thesis, University of London.
Ahmed, A. A. (1960) An analytical study of the storage losses in the Nile Basin with special reference to Aswan Dam Reservoir and the High Dam Reservoir (Sadd-el-Ali), *Proc. Inst. Civ. Engrs.*, 17, 181–200.
Akai, K., Yamamoto, K. and Arioka, M. (1970) Experimental research on the structural anisotropy of crystalline schists. *Proc. 2nd Int. Congress, International Soc. Rock Mechanics*, Belgrade, 2, 181–186.
Allum, J. A. E. (1966) 'Photogeology and regional mapping.' Pergamon, Oxford.
Alpan, I. (1970) The geotechnical properties of soils, *Earth Science Rev.*, 6, 5–49.
Alpan, I. and Meidav, Ts. (1963) The effect of pile driving on adjacent buildings, a case history, *Proc. RILEM Symp. on Measurement and Evaluation of Dynamic Effects and Vibrations of Constructions*, Budapest, 2, 171–181.
Alty, J. L. (1968) Recording and analysis of pole figures by computer, *J. Appl. Physics*, 39, 4189–4192.
Ambraseys, N. N. and Hendron, A. J. (1968) 'Dynamic behaviour of rock masses', *Rock Mechanics in Engineering Practice*, Eds. K. G. Stagg and O. C. Zienkiewicz, Wiley & Sons, London, 203–236.
Ambraseys, N. N. and Sarma, S. K. (1967) The response of earth dams to strong earthquakes. *Géotechnique*, 17, 181–213.
American Geological Institute (1962) 'Dictionary of Geological Terms,' Dolphin Reference Books Edition, New York.
American Society for Testing and Materials (1970) 'Standard method for classification of soils for engineering purposes.' ASTM Designation D 2487–1969.
Anderson, J. G. C. (1971) Reservoirs for pumped storage, *Quart. J. Engrg. Geol.*, 4, 370–371.

Anderson, W. (1941) *Water supply from underground sources of north-east England.* Wartime pamphlet No. 19, Part III, Geological Survey of Great Britain.

D'Andrea, D. A., Fischer, R. L. and Fogelson, D. E. (1965) 'Prediction of compressive strength of rock from other rock properties,' U.S. Bureau of Mines, Report of Investigation No. 6702.

Anon (1974) £23 million scheme to link the Wyre and the Lune, *Water Services,* **78**, No. 938, 108.

Antevs, E. (1925) Conditions of formation of varved glacial clays, *Bull. Geol. Soc. Amer.,* **36**, 171–172.

Antsyferov, M. S. (Ed) (1966) 'Seismo-acoustic methods in mining,' Consultants Bureau, New York.

Arrowsmith, E. J. (1971a) 'Slope stability,' Lecture delivered at Symp. on Rock Mechanics in Highway Construction, Univ. of Newcastle-upon-Tyne, England, 6 and 7 April, 1971.

Arrowsmith, E. J. (1971b) Slope stability, *Trans. Inst. Highway Engrs.,* **18**, No. 12, 13–15.

Arthur, H. G. (1965) Earthfill Dams, *Design of small Dams,* (3rd Printing), U.S. Bureau of Reclamation, 157–230.

Askey, A. (1970) Rock bolting with polyester resins, *J. Inst. Highways Engrs.,* **18**, No. 12, 28–32.

Attewell, P. B. (1970) Triaxial anisotropy of wave velocity and elastic moduli in slate and their axial concordance with fabric and tectonic symmetry, *Int. J. Rock Mech. Min. Sci.,* **7**, 193–207.

Attewell, P. B. (1971) Geotechnical properties of the Great Limestone in Northern England, *Engineering Geology,* **5**, 89–116.

Attewell, P. B. (1973) 'Soft ground tunnelling in urban areas.' Paper to: 9th Regional Meeting of Eng. Group of Geol. Soc. of London, University of Durham, 35–38.

Attewell, P. B. and Boden, J. B. (1969) 'Site investigation practice for shallow tunnels,' Report for External Distribution, Department of Geology, University of Durham.

Attewell, P. B. and Brentnall, D. (1964) Attenuation measurements on rocks in the frequency range 12 kc/s to 51 kc/s and in the temperature range $100°K$ to $1150°K$, *6th Symp. on Rock Mechanics,* Eds. E. M. Spokes and C. R. Christiansen, Rolla, Missouri, 330–337.

Attewell, P. B. and Cripps, J. C. (1975) *Economic design of site investigation in engineering geology and geotechnics,* Internal Report: Enginering Geology Laboratories, University of Durham, England.

Attewell, P. B. and Farmer, I. W. (1964a) Ground vibrations from blasting: their generation, form and detection, *Quarry Man. J.,* **48**, 191–198.

Attewell, P. B. and Farmer, I. W. (1964b) Attenuation of ground vibrations from blasting, *Quarry Man. J.,* **48**, 211–215.

Attewell, P. B. and Farmer, I. W. (1973a) Fatigue behaviour of rock, *Int. J. Rock Mech. Min. Sci.,* **10**, 1–9.

Attewell, P. B. and Farmer, I. W. (1973b) Attenuation of ground vibrations from pile driving, *Ground Engineering,* **6**, No. 4, 26–29.

Attewell, P. B. and Farmer, I. W. (1974a) Ground deformation resulting from shield tunnelling in London Clay, *Can. Geotech. J.,* **11**, 380–395.

Attewell, P. B. and Farmer, I. W. (1974b) Ground disturbance caused by shield tunnelling in a stiff, overconsolidated clay, *Engineering Geology,* **8**, 361–381.

Attewell, P. B. and Farmer, I. W. (1974c) Analysis of structural controls on mass stability of a jointed limestone, *Proc. 2nd Int. Congress of Engineering Geology*, Sao Paulo, Brazil, Theme V − 2.1 to 2.13.

Attewell, P. B., Farmer, I. W. and Haslam, D. (1965) Prediction of ground vibration parameters from major quarry blasts, *Mining and Minerals Engrg.*, 1, 621−626.

Attewell, P. B., Farmer, I. W., Glossop, N. H. and Kusznir, N. J. (1975) 'A case history of ground deformation caused by tunnelling in laminated clay,' *Proc Conference on Subway Construction*, Balatonfüred-Budapest, Hungary, 165−178.

Attewell, P. B., Hirst, D. M. and Taylor, R. K. (1969) Diagenetic re-crystallization and orientation of two carbonate species, *Sedimentology*, 11, 237−247.

Attewell, P. B. and Ramana, Y. V. (1966) Wave attenuation and internal friction as functions of frequency in rocks, *Geophysics*, 31, 1049−1056.

Attewell, P. B. and Sandford, M. R. (1971) On an energy criterion of failure for intrinsically anisotropic rock, *Symposium, Soc. Internat. Mécanique des Roches*, Nancy, Paper 11-4.

Attewell, P. B. and Sandford, M. R. (1974a) Intrinsic shear strength of a brittle anisotropic rock − I Experimental and mechanical interpretation, *Int. J. Rock Mech. Min. Sci., & Geomech. Abstr.*, 11, 423−430.

Attewell, P. B. and Sandford, M. R. (1974b) Intrinsic shear strength of a brittle anisotropic rock − II Textural data acquisition and processing, *Int. J. Rock Mech. Min. Sci., & Geomech. Abstr.*, 11, 431−438

Attewell, P. B. and Sandford, M. R. (1974c) Intrinsic shear strength of a brittle, anisotropic rock − III Textural interpretation of failure, *Int. J. Rock Mech. Min. Sci., & Geomech. Abstr.*, 11, 439−451.

Attewell, P. B. and Taylor, R. K. (1969) A microtextural interpretation of a Welsh slate, *Int. J. Rock Mech. Min. Sci.*, 6, 423−428.

Attewell, P. B. and Taylor, R. K. (1970) 'Foundation engineering − some geotechnical considerations,' (Dewdney, J. C. Ed.) *Durham County and City with Teesside*, British Association for the Advancement of Science, Durham Meeting, 89−107.

Attewell, P. B. and Taylor, R. K. (1971a) Jointing in Robin Hood's Bay, North Yorkshire Coast, England, *Int. J. Rock Mech. Min. Sci.*, 8, 477−481.

Attewell, P. B. and Taylor, R. K. (1971b) 'Investigations, tests and experiments on the mechanical strength and breakdown characteristics of certain overconsolidated clay shales.' Interim and Final Reports on DA−ERO−591−70−G0002, European Research Office, U.S. Army.

Attewell, P. B. and Taylor, R. K. (1973) 'Clay shale and discontinuous rock mass studies.' Final Report to European Research Office, U.S. Army on Contract No. DA−ERO−591−72−G0005.

Attewell, P. B. and Woodman, J. P. (1971) Stability of discontinuous rock masses under polyaxial stress systems; *Stability of Rock Slopes*, 13th Symposium on Rock Mechanics, Ed. E. J. Cording, Publ. ASCE, University of Illinois, Urbana, Ill., U.S.A. Aug/Sept 1971, pp. 665−683.

Awojobi, A. O. and Sobayo, O. A. (1974) Ground vibrations due to seismic detonation in oil exploration. *Earthquake Engineering and Structural Dynamics*, 3, 171−181.

Ayyar, T. S. R. (1968) Discussion: Paper on 'Microscopic structures in kaolin subjected to direct shear,' *Géotechnique*, 18, No. 3, 379−382.

Badger, C. W., Cummings, A. D. and Whitmore, R. L. (1956) The disintegration of shales in water, *J. Inst. Fuel*, 29, 417−423.

Badgley, P. C. (1965) 'Structural and tectonic principles,' Harper and Row, New York.
Bagnold, R. A. (1940) Beach formation by waves: some model experiments in a wave tank, *J. Inst. Civ. Engrs.*, **15**, 27–52.
Bailey, S. W., Bell, R. A. and Peng, C. J. (1958) Plastic deformation of quartz in nature. *Bull. Geol. Soc. Am.*, **69**, 1443–1466.
Baker, D. W. and Wenk, H. R. (1969) Spherical harmonic analysis of X-ray pole figure data for specimens with low symmetry. *Trans. Am. Geophys. Un.*, **50**, 323 (abstr.).
Baker, D. W., Wenk, H. R. and Christie, J. M. (1969) X-ray analysis of preferred orientation in fine-grained quartz aggregates, *J. Geol.*, **77**, 144–172.
Baker, J. F. and Heyman, J. (1969) 'Plastic design of frames,' Cambridge University Press.
Balla, A. (1960) Stress conditions in triaxial compression. *J. Soil Mech. Found. Div.*, Am. Soc. Civ. Engrs., **86**, SM6, 57–84.
Banks, D. C. (1972) Study of clay shale slopes. *Stability of Rock Slopes*, 13th Symp. on Rock Mechanics, Univ. of Illinois, Ed. E. J. Cording, Publ. A.S.C.E., 303–328.
Barden, L. (1968) Primary and secondary consolidation of clay and peat, *Géotechnique*, **18**, 1–24.
Barden, L. (1972a) The relation of soil structure to the engineering geology of clay soil. *Quart. J. Eng. Geol.*, **5**, 85–102.
Barden, L. (1972b) The influence of structure on deformation and failure in clay soil. *Géotechnique*, **22**, 159–163.
Barden, L. and Sides, G. R. (1970) The influence of weathering on the microstructure of Keuper Marl. *Quart. J. Eng. Geol.*, **3**, 259–261.
Barden, L. and Sides. G. R. (1971) Sample disturbance in the investigation of a clay structure. *Géotechnique*, **21**, 211–222.
Bardwell, G. E. (1970) Some statistical features of the relationship between Rocky Mountain Arsenal Waste disposal and frequency of earthquakes, *Engineering Geology Case Histories*, No. 8, Engineering Seismology: The Works of Man, Ed. W. M. Adams, Publ. Geol. Soc. Am., 33–37.
Baron, R. A. (1948) Consolidation of fine-grained soils by drain wells, *Trans. Amer. Soc. Civ. Eng.*, **113**, 718–742.
Barrett, C. S. (1943) 'Structure of metals; crystallographic methods, principles, and data,' McGraw-Hill, New York.
Barrett, C. S. and Massalski, T. B. (1966) 'Structure of metals,' 3rd Edition, New York, McGraw-Hill.
Barron, K. (1965) Glass insert stressmeters, *Trans. Amer. Soc. Min. Engrs.*, **232**, 287–297.
Barron, K. (1971a) Brittle fracture initiation in and ultimate failure of rocks, Part 1 – Isotropic rock, *Int. J. Rock Mech. Min. Sci.*, **8**, 541–551.
Barron, K. (1971b) Brittle fracture initiation in and ultimate failure of rocks, Part II – Anisotropic rocks: theory, *Int. J. Rock Mech. Min. Sci.*, **8**, 553–563.
Barron, K. (1971c) Brittle fracture initiation in and ultimate failure of rocks, Part III – Anisotropic rocks: experimental results, *Int. J. Rock Mech. Min. Sci.*, **8**, 565–575.
Barshad, I. (1955) Adsorptive and swelling properties of clay-water systems, *Clays and Clay Technology*, Dept. of Natural Resources, State of California, Bulletin **169**, 70–77.
Barton, N. (1973) Review of a new shear-strength criterion for rock joints, *Engineering Geology*, **7**, 287–332.

Barton, N., Lien, R. and Lunde, J. (1974a) Analysis of rock mass quality and support practice in tunnelling, and a guide for estimating support requirements, Internal Report to Norwegian Geotechnical Institute (54206, dated 19 June).
Basore, C. E. and Boitano, J. D. (1969) Sand densification by piles and vibroflotation. *J. Soil Mech. Found. Div., Amer. Soc. Civ. Engrs.,* **95**, 1303–1323.
Bazett, D. J., Adams, J. L. and Matyas, E. L. (1961) An investigation of a slide in a test trench excavated in fissured sensitive clay, *Proc. 5th Int. Conf. Soil Mech.*, Paris, **1**, 431–435.
Beard, C. N. (1959) Quantitative study of columnar jointing, *Bull Geol. Soc. Am.*, **70**, 379–382.
Beer, De. E. (1953) Donées concernant la résistance au cisaillement déduites des essais de penetration en profondeur. *Géotechnique*, **1**, 22–39.
Beer, De. E. (1965) Influence of the mean normal stress on the shear strength of sand, *Proc. 6th Int. Conf. on Soil Mech. Found. Engrg.*, Montreal, **1**, 165–169.
Beles, A. A. and Stanculescu, I. I. (1958) Thermal treatment as a means of improving the stability of earth masses. *Géotechnique*, **18**, 158–165.
Benjamin, J. R. and Cornell, C. A. (1970) 'Probability, statistics and decision for civil engineers,' Publ. McGraw-Hill, New York.
Benko, K. F. (1966) Instrumentation in rock grouting for Portage mountain dam, *Water Power*, **18**, 407–415.
Berger, L. and Gnaedinger, J. (1949) Thixotropic strength of clays, *Bull. Amer. Soc. for Testing Materials,* **106**, 64–68.
Berger, P. R. (1971) Blasting seismology, recent developments in criteria, regulations, and instrumentation, *Quarry Man. J.*, **55**, 187–190.
Berk Ltd. (1971) 'Bentonite for civil engineering,' Berk Ltd., Basingstoke.
Berkovitch, I., Manackerman, M. and Potter, N. M. (1959) The shale breakdown problem in coal washing, Part I – Assessing the breakdown of shales in water, *J. Inst. Fuel,* **32**, 579–589.
Berman, H. (1937) Constitution and classification of natural silicates. *Amer. Mineralogist*, **22**, 342–408.
Berry, P. L. and Poskitt, T. J. (1972) The consolidation of peat, *Géotechnique*, **22**, 27–52.
Berry, P. L. and Wilkinson, W. B. (1969) The radial consolidation of clay soils, *Géotechnique*, **19**, 253–284.
Bevan, O. M. (1972) In 'Municipal tunnelling: a contractor's viewpoint,' *Tunnels and Tunnelling*, **4**, 237–243.
Bhatia, H. S. (1969) In: Engineering properties and behaviour of clay shales (Reporter S. J. Johnson), *7th Int. Conf. on Soil Mechanics and Foundation Engineering*, Mexico City, **3**, 483–488.
Bhattacharyya, K. K. and Boshkov, S. H. (1970) Determination of the stresses and the displacements in slopes by the finite element method. *Proc. 2nd Congress, Int. Soc. for Rock Mech.*, Belgrade, **3**, 339–344.
Bieniawski, Z. T. (1967) Stability concept of brittle fracture propagation in rock, *Engineering Geology*, **2**, 149–162.
Bieniawski, Z. T. (1968) 'Mechanisms of brittle rock fracture.' D.Sc.(Eng) Thesis, University of Pretoria.
Bieniawski, Z. T. (1970) Time-dependent behaviour of fractured rock. *Rock Mechanics*, **7**, 123–137.
Bierbaumer, A. (1913) 'Die Dimensionierung des Tunnelmauerwerks.' Englemann, Leipzig.
Billings, M. P. (1962) 'Structural geology,' Prentice-Hall, New York.

Binnie, G. M. (1969) Discussion on paper 'Reservoirs on limestone, with particular reference to the Cow Green Scheme,' *J. Inst. Water Engrs.*, 23, 128–129.

Biot, M. A. and Willis, D. G. (1957) The elastic coefficients of the theory of consolidation, *Journal of Applied Mechanics*, 24, 594–601.

Bishop, A. W. (1952) 'The stability of earth dams,' Ph.D. thesis, University of London.

Bishop, A. W. (1955) The use of the slip circle in stability analysis of earth slopes, *Géotechnique*, 5, 7–17.

Bishop, A. W. (1966a) The strength of soils as engineering materials, *Géotechnique*, 16, 91–128.

Bishop, A. W. (1967) Progressive failure – with special reference to the mechanism causing it, Panel discussion. *Proc. Geotech. Conf.*, Oslo, 2, 142.

Bishop, A. W. (1972) Shear strength parameters for undisturbed and remoulded soil specimens, *Proc. Symp. on Stress-strain Behaviour of Soils*, Ed. R. H. G. Parry, G. T. Foulis, Oxford, 3–58.

Bishop, A. W., Alpan, I., Blight, E. E., and Donald, I. B. (1960) Factors controlling the strength of partly saturated cohesive soils, *Proc. Res. Conf. on Shear strength of cohesive soil, Amer. Soc. Civ. Engrs.*, 503–532.

Bishop, A. W. and Eldin, A. K. G. (1958) The effect of stress history on the relation between ϕ and porosity in sand, *Proc. 3rd Int. Conf. Soil Mechs., Found. Engrg.*, Zurich, 1, 100–105.

Bishop, A. W., Green, G. E., Garga, V. K., Andresen, A. and Brown, J. D. (1971) A new ring shear apparatus and its application to the measurement of residual strength, *Géotechnique*, 21, 273–328.

Bishop, A. W. and Henkel, D. J. (1962) 'The triaxial test,' 2nd Edition, Edward Arnold, London.

Bishop, A. W. and Little, A. L. (1967) The influence of size and orientation of the sample on the apparent strength of the London Clay at Maldon, Essex, *Proc. Geotech. Conf.*, Oslo, 1, 89–96.

Bishop, A. W. and Morgenstern, N. (1960) Stability coefficients for earth slopes, *Géotechnique*, 10, 129–150.

Bishop, A. W., Webb, D. L. and Lewin, P. I. (1965) Undisturbed samples of London Clay from the Ashford Common Shaft: Strength-effective stress relationships, *Géotechnique*, 15, 1–30.

Bjerrum, L. (1954a) Stability of natural slopes in quick clay, *Proc. European Conf. on Stability of Earth Slopes*, Stockholm, 2, 16–40, and *Géotechnique*, 5, 101–119.

Bjerrum, L. (1954b) Geotechnical properties of Norwegian Marine Clays, *Géotechnique*, 4, 49–69.

Bjerrum, L. (1967) Mechanism of progressive failure in slopes of over-consolidated plastic clays and clay-shales. *J. Soil Mechanics and Foundations Div.*, A.S.C.E., 93, SM5, 1–49.

Bjerrum, L. (1967) Engineering geology of Norwegian normally-consolidated marine clays as related to settlements of buildings, *Géotechnique*, 17, 81–118.

Bjerrum, L. (1968) Discussion. *Proc. 9th Int. Congress on Large Dams*, Istanbul, VI, 456, 457.

Bjerrum. L. (1973) Geotechnical problems involved in foundations of structures in the North Sea. *Géotechnique*, 23, 319–358. .

Bjerrum, L., Nash, J. K. T. L., Kennard, R. M. and Gibson, R. E. (1972) Hydraulic fracturing in field permeability testing. *Géotechnique*, 22, 319–332.

Bjerrum, L. and Rosenqvist, I. Th. (1956) Some experiments with artificially sedimented clays, *Géotechnique*, **6**, 124–136.
Bjerrum, L. and Simons, N. E. (1960) Comparison of shear strength characteristics of normally consolidated clays, *Proc. Res. Conf. on Shear Strength of Cohesive Soils, Amer. Soc. Civ. Engrs.*, 711–726.
Black, R. F. (1954) Permafrost: A review, *Bull. Geol. Soc. Amer.*, **65**, 839–56.
Blair, B. E. and Duvall, W. I. (1954) Evaluation of gauges for measuring displacement, velocity and acceleration of seismic pulses, *U.S. Bureau of Mines, Report of Investigations*, No. **5073**.
Boast, C. W. and Kirkham, D. (1971) Auger hole seepage theory, *Proc. Soil Sci. Soc. America*, **35**, 365–373.
Boden, J. B. (1969) 'Site investigation and subsequent analysis for shallow tunnels,' Dissertation, M.Sc., Advanced Course in Engineering Geology, University of Durham.
Bohor, B. F. and Hughes, R. E. (1971) Scanning electron microscopy of clays and clay minerals, *Clays and clay minerals*, **19**, 49–54.
Bolt, G. H. and Miller, R. D. (1951) Compression studies of illite suspensions, *Proc. Soil Sci. Soc. Amer.*, **19**, 285–288.
Bostrom, R. C. and Sherif, M. A. (1974) Negative strength regions in offshore construction, *Proc. 1st Conf. Int. Assoc. Eng. Geol.*, **1**, Th III, 7.1–7.10.
Boulton, N. S. (1963) Analysis of data from non-equilibrium pumping tests allowing for delayed yield from storage, *Proc. Inst. Civ. Engrs.*, **26**, 469–482.
Bourgin, A. (1953) 'The design of dams,' (translation by F. F. Fergusson), Pitman, London, 344 pp.
Boussinesq, J. (1885) 'Applications des potentiels à l'étude de l'equilibre et du mouvement des solides élastiques,' Ganthier-Villard, Paris.
Bowden, F. P. and Tabor, D. (1965) 'Friction and Lubrication,' Methuen, London.
Brace, W. F. (1960) An extension of the Griffith theory of fracture to rocks, *J. Geophys. Res.*, **65**, 3477–3480.
Brace, W. F. (1965) Relation of elastic properties of rocks to fabric, *J. Geophys. Res.*, **70**, 5657–5667.
Brace, W. F. and Orange, A. S. (1968) Further studies of the effects of pressure on electrical resistivity of rocks, *J. Geophys. Res.*, **73**, 5407–5420.
Bradshaw, R. and Phillips, F. C. (1970) The use of X-rays in petrofabric studies, 'Experimental and natural rock deformation,' (P. Paulitsch, Ed.,), Springer-Verlag, Berlin, 75–97.
Brady, B. T. (1969a) A statistical theory of brittle fracture for rock materials – I: Brittle failure under homogeneous axisymmetric states of stress, *Int. J. Rock Mech. Min. Sci.*, **6**, 21–42.
Brady, B. T. (1969b) A statistical theory of brittle fracture for rock materials – II: Brittle failure under homogenous triaxial states of stress, *Int. J. Rock Mech. Min. Sci.*, **6**, 285–300.
Braun, W. (1973) Aquifer re-charge to lift Venice, *Underground Services*, **1**, 16–19.
Bray, J. W. (1967) A study of jointed and fractured rock, *Rock Mechanics and Engineering Geology*, **5**, 119–136, 197–216.
Braybrooke, J. C. (1966) 'The strength effects and measurement of discontinuities in rock masses,' M.Sc. thesis, University of London.
Brekke, T. L. and Howard, T. R. (1972) Stability problems caused by seams and faults, *Proc. North American Rapid Excav. and Tunnelling Conf.*, Illinois, U.S.A., (Eds. K. S. Lane and L. A. Garfield,) ASME, **1**, 25–64.
Brekke, T. L. and Korbin, G. (1974) Some comments on the use of spiling in

underground openings, *Proc. 2nd Int. Cong., Int. Assoc. of Eng. Geology, São Paulo, Brazil,* 2, Th. VII, PC 4.1–4.6.

Bretschneider, C. L. (1952) The generation and decay of wind in deep water, *Trans. Amer. Geophys. Union,* **33**, 381–389.

Brewer, R. C. (1964) 'Fabric and mineral analysis of soils,' Wiley, New York.

Brink, A. B., Mabbut, J. A., Webster, R. and Beckett, P. H. T. (1966) 'Report of the working group on land classification and data storage,' Military Engineering Experimental Establishment, Christchurch, England, Report **940**.

British Standards Institution (1957) 'Site Investigation,' British Standard Code of Practice, CP 2001.

British Standards Institution (1967) 'Methods for sampling and testing of mineral aggregates, sands and fillers,' British Standard B.S. 812.

British Standards Institution (1967) 'Methods for testing soils for civil engineering purposes,' British Standard B.S. 1377.

Brooker, E. W. and Ireland, H. O. (1965) Earth pressures at rest related to stress history, *Canadian Geotech. J.,* **2**, 1–15.

Brooker, E. W. (1967) Strain energy and behaviour of overconsolidated soils, *Canadian Geotech. J.,* **2**, 326–333.

Broscoe, A. J. and Thomson, S. (1969) Observations on an alpine mudflow, Steele Creek, Yukon, *Can. J. Earth Sci.,* **6**, 219–229.

Brown, E. T. (1970a) Modes of failure in jointed rock masses, *Proc. 2nd Congress of Int. Soc. for Rock Mechanics,* Belgrade, **2**, 293–298.

Brown, E. T. (1970b) Strength of models of rock with intermittent joints, *J. Soil Mech. Found. Div.,* ASCE, **96**, SM2, 685–704.

Brown, G. (1961) The X-ray identification and crystal structures of clay minerals, *Mineralogical Society,* London.

Brown, J. G. (Ed.) (1964) *Hydro-electric Engineering Practice,* 2nd Edition, **1**. *Civil Engineering,* Blackie & Sons, Ltd., London.

Bruckl, E. and Scheidegger, A. E. (1973) Application of the theory of plasticity to slow mud flows, *Géotechnique,* **23**, 101–107.

Brunsden, D. and Jones, D. K. C. (1972) The morphology of degraded landslide slopes in South West Dorset, *Quart. J. Eng. Geol.,* **5**, 205–222.

Buchan, S., McCann, D. M. and Smith, D. T. (1972) Relations between the acoustic and geotechnical properties of marine sediments, *Quart. J. Eng. Geol.,* **5**, 265–284.

Buessem, W. R. and Nagy, B. (1953) Mechanisms of the deformation of clay, *Proc. 2nd Nat. Conf. Clays and Clay Minerals,* Un. of Missouri, 480–491.

Building Research Establishment (1973) 'Concrete: materials,' *Digest No. 150,* Dept. of the Environment, London.

Bulman, J. N. (1972) 'Soil stabilization in Africa,' *Transport and Road Research Laboratory Report LR 476,* Dept. of the Environment, London.

Bundred, J. (1969) 'Basic geology for engineers,' Butterworths, London.

Burdine, N. T. (1963) Rock failure under dynamic loading conditions, *J. Soc. Petrol. Engrs.,* **3**, 1–8.

Burgess, A. S. (1970) 'Engineering geology and geohydrology of the Magnesian Limestone of Northern England,' Ph.D. thesis, University of Durham, England.

Burnett, A. D. and Fookes, P. G. (1974) A regional engineering geological study of the London Clay in the London and Hampshire Basins, *Quart. J. Eng. Geol.,* **7**, 257–296.

Burrow, D. C. (1971) Conjunctive use of water resources (computation), *J. Inst. Water Engrs.,* **25**, 381–396.

Burton, A. N. (1965) Classification of rocks for rock mechanics, *Int. J. Rock Mech. Min. Sci.*, **2**, 105.
Burwash, E. M. J. (1938) The deposition and alteration of varved clays, *Trans. Roy. Canadian Inst.*, **22**, pt. 1, no. 47, 3–6.
Burwell, E. B. and Nesbitt, R. H. (1954) The NX borehole camera, *Trans. Amer. Inst. Min. Engrs.*, **199**, 805–808.
Butler, F. G. (1972) Treatment of unstable ground arising from old mine workings, *Symposium on Methods of Treatment of Unstable Ground*, Sheffield Polytechnic, U.K. (unpublished preprint).
Butler, R. M. J. (1972) Water as an unwanted commodity: some aspects of flood alleviation, *J. Inst. Water Engrs.*, **26**, 311–322.
Cambefort, H. (1964) 'Injection des sols,' Publ. Eyrolles, Paris.
Capper, P. L. and Cassie, W. F. (1969) 'The mechanics of engineering soils,' 5th Edition, E. & F. N. Spon, London.
Capper, P. L., Cassie, W. F. and Geddes, J. D. (1966) 'Problems in engineering soils,' E. & F. N. Spon, London.
Carder, D. S. (1970) Reservoir loading and local earthquakes, *Engineering Geology Case Histories, No. 8, Engineering Seismology: The Works of Man*, Ed. W. M. Adams, *Geol. Soc. Amer.*, 51–61.
Carson, M. A. and Kirkby, M. J. (1972) *Hillside form and process*, Cambridge Univ. Press, Cambridge, U.K.
Carson, M. A. and Petley, D. J. (1970) The existence of threshold hillslopes in the denudation of the landscape, *Trans. Inst. of British Geographers*, **45**, 71–95.
Casagrande, A. (1932a) Research on the Atterberg Limits of soils, *Public Roads*, **13**, 121–136.
Casagrande, A. (1932b) The structure of clay and its importance in foundation engineering, *Contributions to Soil Mechanics, 1925–1940, Boston Soc. Civ. Engrs.*, 72–112.
Casagrande, A. (1932c) Discussion on frost heaving, *Proc. Highway Research Board*, **12**, 169.
Casagrande, A. (1936) The determination of the preconsolidation load and its practical significance, *Proc. 1st Int. Conf. Soil Mechs. Found. Eng.*, Cambridge, Mass., **3**, 60–64.
Casagrande, A. (1937) Seepage through Earth Dams, In *Contributions to Soil Mechanics 1925–1940, Boston Soc. Civ. Engrs.*, 295–296 and *J. New England Water Works Assoc.*, **51**, 136–137.
Casagrande, A. (1948) Classification and identification of soils, *Trans. Am. Soc. Civ. Engrs.*, **113**, 901–992.
Casagrande, A. and Carrillo, N. (1944) Shear failure of anisotropic materials, In: *Contributions to Soil Mechanics, 1941–53, Boston Soc. Civ. Engrs.*, **31**, 122–135.
Casagrande, L. (1947) Structures produced in clays by electric potentials and their relation to natural structure, *Nature*, **160**, 470–471.
Casagrande, L. (1952) Electro-osmotic stabilisation of soils, *J. Boston Soc. Civil Engrs.*, **39**, 51–83.
Cashman, P. M. and Haws, E. T. (1970) 'Control of groundwater by water lowering.' *Ground Engineering*, Ed. J. S. Davis, *Inst. of Civ. Engrs.*, London.
Chakrabarti, P. and Chopra, K. (1973) Earthquake analysis of gravity dams including hydro-dynamic interaction, *Earthquake Engineering and Structural Dynamics*, **2**, 143–160.
Chandler, R. J. (1969) The effect of weathering on the shear strength properties of Keuper Marl, *Géotechnique*, **19**, 321–334.

Chandler, R. J. (1970) Shallow slab slide in the Lias Clay near Uppingham, Rutland, *Géotechnique,* **20**, 253–260.

Chandler, R. J. (1972a) Lias Clay: weathering processes and their effect on shear strength. *Géotechnique,* **22**, 403–431.

Chandler, R. J. (1972b) Periglacial mudslides in Vestspitsbergen and their bearing on the origin of fossil 'solifluction' shears in low angled clay slopes, *Quart. J. Eng. Geol.,* **5**, 223–241.

Chandler, R. J. (1974) Lias clay: the long-term stability of cutting slopes, *Géotechnique,* **24**, 21–38.

Chandler, R. J., Pachakis, M., Mercer, J. and Wrightman, J. (1973) Four long-term failures of embankments founded on areas of landslip, *Quart. J. Engrg. Geol.,* **6**, 405–422.

Chandler, R. J. and Skempton, A. W. (1974) The design of permanent cutting slopes in stiff fissured clays, *Géotechnique,* **24**, No. 4, 457–466.

Chappell, B. A. (1974) Deformational response of blocking models, *Int. J. Rock Mech. Min. Sci., and Geomech. Abstr.,* **11**, 13–19.

Chayes, F. (1949) Statistical analysis of three-dimensional fabric diagrams, 'Structural Petrology of Deformed Rocks,' Ed., H. W. Fairbairn, Addison-Wesley Press, Cambridge, Mass., U.S.A.

Chen, J. C. (1966) Petrofabric studies of some fine-grained rocks by means of X-ray diffraction, *Bull. Houston Geol. Soc.,* **85**, 16 (Abstr.).

Chenevert, M. E. and Gatlin, C. (1965) Mechanical anisotropies of laminated sedimentary rocks, *J. Soc. Petroleum Engrs.,* **5**, 67–77.

Childs, E. C., Cole, A. H. and Edwards, D. H. (1953) The measurement of the hydraulic permeability of saturated soil in-situ. *Proc. Royal Soc. Lond.,* Series A, **216**, 72–89.

Chopra, A. K. (1967) Hydraulic pressures in dams during earthquakes, *J. Engrg. Mechs. Div., Proc. Amer. Soc. Civ. Engrs.,* **93**, No. EM6, 205–223.

Chowdhury, R. N. (1970) Discussion on paper 'Slopes in stiff-fissured clays and shales' by J. M. Duncan and P. Dunlop, *J. Soil Mech. Foundations Div.,* ASCE, **96**, 336–338.

Christensen, D. M. (1964) A theoretical analysis of wave propagation in fluid-filled drill holes for the interpretation of the 3-dimensional velocity log, *Soc. Professional Well Log Analysts,* 5th Annual Symposium, Tulsa, Oklahoma, K1–K29.

Christian, C. S. (1958) The concept of land units and land systems, *Proc. 9th Pacific Sci. Cong.,* **20**, 74–81.

Christian, C. S. and Stewart, G. A. (1953) General report on survey of Katherine-Darwin region, 1946. *C.S.I.R.O. Aust. Land Res. Series,* No. 1.

Christiansen, F. W. (1970) Isotropic tensional and compressional strain patterns in tabular shaped bodies, *Proc. 2nd Congress, Int. Soc. Rock Mechanics,* Belgrade, **1**, 265–269.

Christie, I. F. (1959) Design and construction of vertical drains to accelerate the consolidation of soils. *Civ. Eng. Pub. Works Rev.,* **54**, 197–200, 339–342, 473–476.

Clark, S. P. Jr. (1966) Handbook of physical constants, *Geol. Soc. Amer. Memoir* 97.

Clarke, D. D. (1904) A phenomenal landslide, *Trans. A.S.C.E.,* **53**, 322–397, and Discussion, 398–412.

Clarke, F. (1924) The data of geochemistry, *U.S. Geol. Survey Bull.,* 770.

Clarke, P. (1974) Turning on the £1500 m water tap, *The Sunday Times,* August 11, 44–45.

Clarkson, C. G. (1974) 'An investigation of the nature of the laminated clays and

their involvement in landslips around Durham City,' Dissertation: M.Sc. Advanced Course in Engineering Geology, University of Durham.
Clevenger, W. A. (1956) Experiences with loess as a foundation material. *Proc. Soil Mech. Found. Div.*, A.S.C.E., **82**, SM3, 1025, 1–26.
Clough, R. W. and Woodward, R. J. (1967) Analysis of embankment stresses and deformations, *J. Soil Mechanics and Found. Div.*, A.S.C.E., **93**, 529–549.
Coates, D. F. (1964) Classification of rocks for rock mechanics, *Int. J. of Rock Mech. Min. Sci.*, **1**, 421–431.
Coates, D. F. (1965) Rock mechanics principles, *Canadian Department of Energy, Mines and Resources, Mines Branch Monograph*, **874**..
Coates, D. F., Gyenge, M. and Stubbins, J. B. (1965) Slope stability studies at Knob Lake, *Proc. Rock Mech. Symp.*, Toronto, Queen's Printer, Ottawa, 35–46.
Coates, D. F., McRorie, K. L. and Stubbins, J. B. (1963) Analyses of pit slides in some incompetent rocks, *Trans. Am. Soc. Min. Engrs.*, **226**, 94–101.
Coates, D. F. and Parsons, R. C. (1966) Experimental criteria for classification of rock substances, *Int. J. Rock Mech. Min. Sci.*, **3**, 181–189.
Cochrane, N. J. (1972) Discussion: Examples of the internal conditions of some old earth dams, by M. F. Kennard, *J. Inst. Water Engrs.*, **26**, 153–154.
Cole, K. W. and Burland, J. B. (1972) Observations of retaining wall movements associated with a large excavation, *Proc. 5th European Conf. on Soil Mech. and Found. Engineering*, Madrid, **1**, 445–453.
Collin, A. (1846) 'Glissements spontanes des terrains argileux,' Paris, Carilian-Goeury and Dulmont (English trans. publ. 1956 by University of Toronto Press).
Collins, K. and McGown, A. (1974) The form and function of microfabric features in a variety of natural soils, *Géotechniquee*, **24**, no. 2, 223–254.
Collins, S. P. and Deacon, W. G. (1972) Shaft sinking by ground freezing: Ely Ouse – Essex scheme, *Proc. Inst. Civil Engrs., Supplementary Volume*, paper 75065, 129–156.
Conlon, R. J. (1966) Landslide on the Toulnustouc River, Quebec, *Can. Geotech. J.*, **3**, 113–144.
Connelly, R. J. (1970) 'X-ray fabric analysis of laboratory sedimented muscovite clay,' M.Sc. Advanced Course Dissertation, University of Durham, England.
Cooper, H. H. and Jacob, C. E. (1946) A generalized graphical method for evaluating formation constants and summarizing well-field history, *Trans. Amer. Geophys. Un.*, **27**, 526–534.
Cooling, L. F. (1940) Discussion on 'Cliff stabilization works in London Clays,' *Journ. Inst. Civ. Engrs.*, **17**, 251–276.
Costet, J. and Sanglerat, G. (1969) 'Mécanique des sols,' Dunod, Paris.
Coulson, J. H. (1970) The effects of surface roughness on the shear strength of joints in rock, *Technical Report MRD–2–70 to the Missouri River Division*, U.S. Corps. of Engineers, Contract DACA–45–67–C–0009.
Courchée, R. (1970) 'Properties of boulder clay slopes and their influence on coastal erosion at Robin Hood's Bay, Yorkshire,' M.Sc. Advanced Course Dissertation, Durham University.
Cramér, H. (1954) 'The elements of probability theory and some of its applications,' Almquist and Wisksell, Stockholm.
Crampton, C. B. (1958) Muscovite, biotite, and quartz fabric reorientation, *J. Geol.*, **66**, 28–34.
Crandell, D. R. and Varnes, D. J. (1961) Movement of the Slumgullion earthflow near Lake City, Colorado, *U.S. Geol. Survey, Prof. Paper*, 424 B, 136–139.

Crandell, F. J. (1949) Ground vibrations due to blasting and its effect upon structures, *Journ. Boston Soc. Civ. Engrs.*, **36**, 222–245.

Crawford, C. B. and Eden, W. J. (1957) Report on the Nicolet landslide of November, 1955, *National Research Council Canada, Div. of Building Research*, Internal Report No. 128, Ottawa.

Crawford, N. H. (1967) Some observations on rainfall and run-off, *Water Research*, Kneese, A. V. and Smith, S. C. (Eds.), Publ. for Resources for the Future Inc., by the Johns Hopkins Press, Baltimore, 343–353.

Crawford, N. H. and Linsley, R. K. (1966) 'Digital Simulation in hydrology – Stanford Watershed Model IV,' *Tech. Rept. no. 39*, Dept. Civil Engineering, Stanford University, U.S.A.

Creager, W. P. (1929) 'Masonry dams,' 2nd Edition, John Wiley & Sons, New York.

Cripps, J. C. (1971) 'Anisotropy in Lumley Mudstone,' M.Sc. Advanced Course, Dissertation, University of Durham.

Cripps, J. C. (1977) 'Urban planning: storage and processing of geotechnical data,' Ph.D. Thesis, University of Durham.

Croft, J B. (1967) The influence of soil mineralogical composition on cement stabilization, *Géotechnique*, **17**, 119–135.

Croney, D. and Coleman, J. D. (1961) Pore pressure and suction in soil, *Pore Pressure and Suction in Soils*, Butterworths, London, 31–38.

Croney, D. and Lewis, W. A. (1949) The effect of vegetation on the settlement of roads, *Proc. Biology and Civ. Eng. Conf.* 1948, Inst. of Civ. Engrs., London, 195–202.

Crosby, W. O. (1882) On the classification and origin of joint structures, *Proc. Boston Soc. of Nat. History*, **22**, 72–85.

Crosby, W. O. (1893) The origin of parallel and intersecting joints, *Technology Quarterly*, **VI**, 230–236.

Cumming, J. D. (1951) 'Diamond drill handbook,' Smit, Toronto.

Cundall, P. A. (1974) 'A computer model for rock mass behaviour using interactive graphics for the input and output of geometrical data,' Report prepared under Contract No. DACW–45–74–C–006 for the Missouri River Division, U.S. Army Corps of Engineers by University of Minnesota, Dept. of Civil and Mineral Engineering, July 31st.

Curry, R. R. (1966) Observation of Alpine mudflows in the Ten mile Range, Central Colorado, *Bull. Geol. Soc., America*, **77**, 771–776.

D'Appolonia, E. Ed.) (1972) Underwater soil sampling, testing and construction control, *Special Tech. Pubn. No. 501, Amer. Soc. Testing and Matl.*, Philadelphia.

D'Appolonia, E., Alperstein, R. and D'Appolonia, D. J. (1967) Behavior of a colluvial slope, *J. Soil Mech. Found. Div., A.S.C.E.*, **93**, SM4, 447–473.

Darbyshire, M. and Draper, L. (1963) Forecasting wind-generated sea waves, *Engineering*, **195**, 482–484.

Das Ries, (1969) 'Geological Bavarica', No. 61, Bayer, Geologischen Landesant, Munchen.

Dastidar, A. G., Gupta, S. and Ghosh, T. K. (1969) Application of sandwicks in a housing project. *Proc. 7th Int. Conf. Soil Mech. Found. Eng.*, Mexico City, **2**, 59–64.

Davies, B., Farmer, I. W. and Attewell, P. B. (1964) Ground vibration from shallow sub-surface blasts, *The Engineer*, **217**, 553–559.

Davis, A. G. (1968) The structure of Keuper Marl. *Quart. Jl. Eng. Geol.*, **1**, 145–153.

Davis, G. H, Small, J. B. and Counts, H. B. (1964) 'Land subsidence related to decline of artesian pressure,' *Engineering Geology Case Histories, Nos. 1–5*, Eds. Trask, P.D. and Kiersch, G. A., *Geol. Soc. Amer.*, 185–192.

Davis, S. N. and De Wiest, R. J. M. (1966) 'Hydrogeology,' Wiley & Sons, New York.

Daw, G. P. (1971) A modified Hoek Franklin triaxial cell for rock permeability measurements, *Géotechnique*, **21**, 89–91.

Dearman, W. R. (1974) Presentation of information on engineering geological maps and plans, *Quart. J. Eng. Geol.*, **7**, 317–320.

Dearman, W. R. and Fookes, P. G. (1974) Engineering geological mapping for civil engineering practice in the United Kingdom, *Quart. J. Eng. Geol.*, **7**, 223–256.

Decker, B. F., Asp, E. T. and Harker, D. (1948) Preferred orientation determination using a Geiger counter X-ray diffraction goniometer, *J. Appl. Phys.*, **19**, 388–392.

Deer, W. A., Howie, R. A. and Zussman, J. (1966) 'An introduction to the rock forming minerals,' Longman, London.

Deere, D. U. (1966) Contribution to discussion, *Proc. 1st Cong., Int. Soc. Rock Mech.*, Lisbon, **3**, 156–158.

Deere, D. U. (1968) Geological considerations, In: 'Rock mechanics in engineering practice,' Eds., Stagg, K. G. and Zienkiewicz, O. C., Wiley, London, 1–20.

Deere, D. U., Coon, R. F. and Merritt, A. H. (1969) Engineering classification of in situ rock, *Technical Report No. AFWL–TR–67–144 to Air Force Weapons Laboratory*, U.S.A.

Deere, D. U., Hendron, A. J., Patton, F. D. and Cording, E. J. (1967) 'Design of surface and near-surface construction in rock,' *Failure and Breakage of Rock, Proc. 8th Symp. on Rock Mech., Minneapolis, AIME*, 237–303.

Deere, D. U., Merritt, A. M. and Cording, E. J. (1974) Engineering Geology and Underground Construction, *Proc. 2nd Int. Congr., Int. Assoc. of Eng. Geol.*, São Paulo, Brazil, **2**, Th. VII GR 1–27.

Deere, D. U. and Miller, R. P. (1966) Engineering classification and index properties for intact rock, *Air Force Weapons Lab. Tech. Report, AFWL–TR–65–116*, Kirtland Base, New Mexico.

Deere, D. U., Peck, R. B., Monsees, J. E. and Schmidt, B. (1969) Design of tunnel liners and support systems, *Final Report on Contract No. 3–0152 to Office of High Speed Ground Transportation*, U.S. Dept. of Transportation, Washington, D.C.

Deklotz, E. J. and Stemler, O. A. (1966) Anisotropy of a schistose gneiss, *Proc. 1st Congress, Int. Soc. of Rock Mech.* Lisbon, **1**, 465–470.

Denisov, N. Ya. (1953) Stroitel 'nie svoistva lessa i lessovidnykh suglinkov (The engineering properties of loess and loess-like clay soils), Moscow (Gosstroiizdat) 133.

Denness, B. (1969) 'Fissure and related studies in selected Cretaceous rocks of S.E. England.' Ph.D. thesis, Imperial College, London University.

Desai, C. S. and Abel, J. F. (1972) 'Introduction to the finite element method,' Van Nostrand, New York.

Devine, J. F. (1962) Vibration levels from multiple holes per delay quarry blasts, *Earthquake Notes*, **33**, No. 3, 32–39.

Dobrey, R. and Alvarez, L. (1967) Seismic failures of Chilean tailings dam. *J. Soil Mechs. Found. Div., Amer. Soc. Civ. Eng.*, **93**, 237–260.

Dobrin, M. B. (1960) 'Introduction to geophysical prospecting,' McGraw-Hill, New York.

Dodd, J. S. (1967) Morrow Point underground power plant: *Rock Mechanics*

Investigations, *U.S. Bureau of Reclamation Technical Publication*, Denver, Colorado, U.S.A.

Dollar, A. T. J., Abedi, S. M. H., Lilwall, R. C. and Willmore, P. L. (1975) Earthquake risk in the U.K., *Proc. Inst. Civ. Engrs.*, **58**, pt. I, 123–124.

Donath, F. A. (1961) Experimental study of shear failure in anisotropic rocks, *Geol. Soc. Amer. Bull.*, **72**, 985–990.

Donath, F. A. (1964) Strength variation and deformation behaviour in anisotropic rock, *State of stress in the earth's crust*, Ed. W. R. Judd, Elsevier, New York, 281–298.

Doornkamp, J. C. and King, C. A. M. (1971) 'Numerical analysis in geomorphology; an introduction,' Arnold, London.

Douglass, P. M. and Voight B. (1969) Anisotropy of granites: a reflection of microscopic fabric, *Géotechnique*, **19**, 376–398.

Dowling, J. W. F. (1968) The classification of terrain for road engineering purposes, *C.E.P.O.Conference*, Session 5, Paper **16**, 33–57.

Dowling, J. W. F. and Beaven, P. J. (1969) Terrain evaluation for road engineers in developing countries, *J. Inst. Highway Engineers*, **16**, 5–15.

Dreyer, W. (1972) 'The science of rock mechanics – Part 1,' *Trans. Tech. Publications*, Cleveland, Ohio, U.S.A.

Dudley, J. H. (1970) Review of collapsing soils, *J. Soil Mechs. Found. Div., Amer. Soc. Civ. Engrs.*, **96**, 925–947.

Duke, C. M. and Leeds, D. J. (1963) Response of soils, foundations and earth structures to the Chilean earthquakes of 1960. *Bull. Seism. Soc. Amer.*, **53**, 309–357.

Dumbleton, M. J. and West, G. (1968) Soil suction by the rapid method: an apparatus with extended range, *J. Soil Science*, **19**, no. 1, 40–46.

Dumbleton, M. J. and West, G. (1970a) The suction and strength of remoulded soils as affected by composition, *Road Research Laboratory Report*, LR 306. Dept. of the Environment, London.

Dumbleton, M. J. and West, G. (1970b) Air-photograph interpretation for road engineers in Britain, *Road Research Laboratory Report*, LR 369. Dept. of the Environment, London.

Dumbleton, M. J. and West, G. (1971) Preliminary sources of information for site investigations in Britain, *Road Research Laboratory Report* LR 403, Dept. of the Environment, London.

Dumbleton, M. J. and West, G. (1972) Preliminary sources of site information for roads in Britain, *Q. J. Eng. Geol.*, **5**, 7–14.

Dumbleton, M. J. and West, G. (1974) Guidance on planning, directing and reporting site investigations, *Report* LR 625, *Transport & Road Research Laboratory*, Dept. of the Environment, London.

Duncan, J. M. and Goodman, R. E. (1968) Finite element analyses of slopes in jointed rock, *U.S. Corps of Engineers*, Contract Report S–63–3.

Duncan, J. M. and Seed, H. B. (1966) Anisotropy and stress re-orientation in clay, *J. Soil Mech. Found. Div.*, A.S.C.E., **92**, SM5, 21–49.

Dungar, R. and Severn, R. T. (1971) Experimental vibration and response analysis of water-retaining structures, 'Dynamic Waves in Civil Engineering,' Eds. D. A. Howells, I. P. Haigh and C. Taylor, Wiley-Interscience, London, 407–422.

Dunn, C. G. (1959) On the determination of preferred orientations, *J. Appl. Phys.*, **30**, 850–857.

Dunn, C. G. and Walter, J. L. (1959) Synthesis of a (110) 001 type torque curve in silicon iron, *J. Appl. Phys.*, **30**, 1067–1072.

Dupuit, J. (1863) 'Etudes théoriques et pratiques sur le mouvement des eaux dans les canaux découverts et à travers les terrains perméables,' Dunod, Paris.
Duvall, W. I., Devine, J. F., Johnson, C. F. and Meyer, A. V. C. (1963) Vibrations from blasting at Iowa limestone quarries, *U.S. Bureau of Mines, Report of Investigations*, No. 6270.
Duvall, W. I. and Fogelson, D. E. (1962) Review of criteria for estimating damage to residences from blasting vibrations, *U.S. Bureau of Mines, Report of Investigations*, No. 5968.
Duvall, W. I., Johnson, C. F., Meyer, A. V. C. and Devine, J. F. (1963) Vibrations from instantaneous and millisecond-delayed quarry blasts, *U.S. Bureau of Mines, Report of Investigations*, No. 6151.
Early, K. R. (1971) Buried channels, *Quart. J. Engrg. Geol.*, **4**, 368–370.
Early, K. R. and Skempton, A. W. (1972) Investigations of the landslide at Walton's Wood, Staffordshire, *Quart. J. Engrg. Geol.*, **5**, 19–41.
Eden, W. J. (1955) A laboratory study of varved clay from Steep Rock Lake, Ontario, *Am. J. Sci.*, **253**, 659–674.
Edwards, A. T. and Northwood, T. D. (1960) Experimental studies of the effects of blasting on structures, *The Engineer*, **210**, 538–546.
Edwards, R. J. G. (1970) The practical application of rock bolting – Jeffrey's Mount rock cut on the M.6, *J. Inst. Highway Engrs.*, **18**, No. 12, 21–27.
Edwards, R. J. G. (1971) The engineering geologist in project reconnaissance and feasibility studies, *Quart. J. Eng. Geol.*, **4**, 283–289.
Eide, O. and Bjerrum, L. (1954) The slide at Bekkelaget, *Proc. European Conf. on Stability of Earth Slopes*, Stockholm, **2**, 1–15, and *Géotechnique*, **5**, 88–100.
Elson, W. K. (1968) An experimental investigation of the stability of slurry trenches, *Géotechnique*, **18**, 37–49.
Elvery, R. H. (1973) Estimating strength of concrete in structures, *Concrete*, **7**, 49–51.
Endlich, E. M. (1876) Annual Report (for 1874) of the United States geological and geographical survey of the territories embracing Colorado and parts of adjacent territories, *U.S. Geological Survey*, Washington, 181–239.
Ernst, L. F. (1950) 'A new formula for the calculation of the permeability factor with the auger hole method,' *Agricultural Experiment Station I.N.O. Groningen*. (Translation from Dutch by H. Bouwer, Cornell University, New York State.)
Esu, F. (1966) Short-term stability of slopes in unweathered jointed clays, *Géotechnique*, **16**, 321–328.
Eswaran, H. (1972) Morphology of Allophane, Imogolite and Halloysite, *Clay Minerals*, **9**, 281–285.
Evans, D. M. (1968) The Denver area earthquakes and the Rocky Mountain Arsenal disposal well, *The Mountain Geologist*, **3**, no. 1, 23–36.
Evans, D. M. (1970) 'The Denver area earthquakes and the Rocky Mountain Arsenal disposal well,' *Engineering Geology Case Histories, no. 8, Engineering, Seismology: The works of Man*, Ed. W. M. Adams, Geol. Soc. Amer. 25–32.
Evans, I. and Pomeroy, C. D. (1966) 'Strength fracture and workability of coal,' Pergamon, Oxford.
Ewing, W. M., Jardetzky, W. S. and Press, F. (1957) 'Elastic waves in layered media,' McGraw-Hill, New York.
Faber, J. and Mead, F. (1965) 'Reinforced concrete,' Spon, London.
Fairburn, H. W. (1949) 'Structural petrology of deformed rocks,' Addison-Wesley Press, Cambridge, Mass., U.S.A.

Fairhurst, C. (1964) 'Measurement of in situ rock stresses with particular reference to hydraulic fracturing, *Rock Mech. and Eng. Geol.*, **2**, 130–155.

Falkiner, R. H. and Tough, S. G. (1968) The Tyne Tunnel: Construction of the main tunnel, *Proc. Inst. Civ. Engrs.*, **39**, 213–234.

Farmer, I. W. (1968) 'Engineering properties of rocks,' Chapman and Hall, London.

Farmer, I. W. (1970) Leaching of cement from a concrete pile by groundwater flow, *Civ. Eng. Pub. Wks. Rev.*, **64**, 457–458.

Farmer, I. W. (1975a) Undrained strengths of chemically grouted cohesionless soils, *Proc. 2nd Int. Cong., Int. Assoc. of Eng. Geol.*, São Paulo, Brazil, **2**, Th VI, 1.1–1.6.

Farmer, I. W. (1975b) 'Electro-osmosis and electro-chemical stabilisation,' In: *Methods of treatment of unstable ground*, Butterworths, London, 26–36.

Farmer, I. W. and Attewell, P. B. (1973) The effect of particle strength on the compression of crushed aggregate, *Rock Mechs.*, **5**, 237–248.

Farmer, I. W. and Attewell, P. B. (1973) Ground movements caused by a bentonite supported excavation in London Clay, *Géotechnique*, **23**, 576–581.

Farmer, I. W. and Attewell, P. B. (1975) A note on the similarities between ground movement around soft ground tunnels and longwall mining excavations, *Mining Engineer*, **134**, 392–402.

Farmer, I. W., Buckley, P. J. C. and Sliwinski, A. (1971) 'The effect of bentonite on the skin friction of cast-in-place piles,' In: *Behaviour of Piles*, Inst. Civ. Engrs., London, 115–120.

Fellenius, W. (1936) Calculation of the stability of earth dams, *Trans. 2nd Congress on Large Dams*, Washington, **4**, 445–459.

Feng, C. (1965) Determination of relative intensity in X-ray reflection study, *J. Appl. Phys.*, **36**, 3432–3435.

Field, M. and Merchant, M. E. (1949) Reflection method of determining preferred orientation on the Geiger counter spectrometer, *J. Appl. Phys.*, **20**, 741–745.

Fischer, F. (1970) 'The technical aspects of compaction in earthmoving and road construction,' *Clark Equipment Co.*, Michigan, U.S.A.

Fish, B. G. (1951a) Fundamental considerations of seismic vibrations from blasting, *Mine and Quarry Engrg.*, **17**, 111–114.

Fish, B. G. (1951b) Seismic vibrations from blasting – reduction by means of short delay initiation, *Mine and Quarry Engrg.*, **17**, 189–192.

Fish, B. G. (1951c) Seismic vibrations from blasting – solution of the problem in specific cases, *Mine and Quarry Engrg.*, **17**, 217–222.

Fish, B. G. and Hancock, J. (1949) Short delay blasting, *Mine and Quarry Engrg.*, **15**, 339–344.

Fleming, R. W., Spencer, G. S. and Banks, D. C. (1970) 'Empirical study of behaviour of clay shale slopes,' *N.C.G., Technical Report No. 15 (2 vols), U.S. Army Engineer Nuclear Cratering Group*, Livermore, California.

Flinn, D. (1958) On tests of significance of preferred orientation in three-dimensional fabric diagrams, *J. Geol.*, **66**, 526–539.

Flinn, D. (1962) On folding during three-dimensional progressive deformation, *Q. Jl. Geol. Soc., Lond.*, **118**, 385–433.

Flinn, D. (1965) 'Deformation in metamorphism,' In: *Controls of Metamorphism*, W. S. Pitcher and G. W. Flinn (Eds.), Oliver and Boyd, London, 48–72.

Focardi, P., Gandolfi, S. and Mirto, M. (1970) Frequency of joints in turbidite sandstone, *Proc. 2nd Congress, Int. Soc. Rock Mechanics,* Belgrade, **1**, 97–101.

Folayan, J. I., Hoeg, K. and Benjamin, J. R. (1970) Decision theory applied to settlement predictions, *J. Soil Mech. Found. Div.*, A.S.C.E., **96**, 1127–1141.

Fookes, P. G. (1965) Orientation of fissures in stiff, overconsidated clay of the Siwalik system, *Géotechnique*, **15**, no. 2, 195–206.

Fookes, P. G. (1969) Geotechnical mapping of soils and sedimentary rock for engineering purposes with examples of practice from the Mangla Dam project, *Géotechnique*, **19**, 52–74.

Fookes, P. G. and Best, R. (1969) Consolidation characteristics of some late Pleistocene metastable soils of east Kent, *Q. Jl. Eng. Geol.*, **2**, 103–128.

Fookes, P. G., Dearman, W. R. and Franklin, J. A. (1971) Some engineering aspects of rock weathering with field examples from Dartmoor and elsewhere, *Q. Jl. Eng. Geol.*, **4**, 139–185.

Fookes, P. G. and Denness, B. (1969) Observational studies on fissure patterns in Cretaceous sediments of south-east England, *Géotechnique*, **19**, 453–477.

Fookes, P. G., Hinch, L. W. and Dixon, J. C. (1972) Geotechnical considerations of the site investigation for stage IV of the Taff Vale trunk road in South Wales, *Proc. 2nd British Regional Congress, Permanent Int. Assoc. of Road Congresses*, Cardiff, 20, 21 March, 1972.

Fookes, P. G. and Parrish, D. G. (1969) Observations on small-scale structural discontinuities in the London Clay and their relationship to regional geology, *Q. J. Eng. Geol.*, **1**, 217–240.

Fookes, P. G. and Wilson, D. D. (1966) The geometry of discontinuities and slope failures in Siwalik clay, *Géotechnique*, **16**, no. 4, 305–320.

Forbes, A. C. (1971) 'The problem of failure in rock mass,' Dissertation: M.Sc. Advanced Course in Engineering Geology, University of Durham, England.

Foster, R. H. and De, P. K. (1971) Optical and electron microscopic investigation of shear induced structures in lightly consolidated (soft) and heavily consolidated (hard) kaolinite, *Clays and Clay Minerals*, **19**, 31–47.

Francis, E. A. (1970) Geology of Durham County, Ch. 12, *Trans. Nat. Hist. Soc. Northumberland, Durham and Newcastle*, **41**, 134–153.

Franklin, J. A. (1974) Rock quality in relation to the quarries and performance of rock construction materials, *Proc. 2nd Int. Cong., Int. Assoc. of Eng. Geol.*, São Paulo, Brazil, **1**, Th IV-PC-2.1–2.11.

Franklin, J. A., Broch, E., and Walton, G. (1971) Logging the mechanical character of rock, *Trans. Inst. Min. Metall*, **80**, A1–9.

Freitas, M. H. De and Watters, R. J. (1973) Some field examples of toppling failure, *Géotechnique*, **23**, 495–514.

French, B. M. and Short, N. M. (eds.) (1968) 'Shock metamorphism of natural materials,' Mono Book Corp. Baltimore.

Freyberg, B. von (1973) Geologie des Isthmus von Korinth, *Erlanger Geologische Abhandlungen*, Heft 95.

Friedman, M. (1963) Petrofabric techniques for the determination of principal stress directions in rocks, in *State of Stress in the Earth's Crust*, (Ed. W. R. Judd), Elsevier, 451–552.

Fukuoka, M. (1953) Landslides in Japan, *Proc. 3rd Int. Conf. Soil Mech. and Found. Engrg.*, Zurich, **2**, 234–238.

Gavshon, R. A. C. (1974) 'Discontinuities and their effect on the stability of soft rock slopes,' Dissertation: M.Sc. Advanced Course in Engineering Geology, University of Durham.

Gehlen, K. von (1960) Die Röntgenographische und Optische Gefugeanalyse von Erzen, *Beitr, Mineralogy, Petrog.*, **1**, 340–388.

Geological Society of London (1970) The logging of cores for engineering purposes – Engineering Group Working Party Report, *Q. J. Eng. Geol.*, **3**, 1–24.

Geological Society of London (1972) The preparation of maps and plans in terms of engineering geology – Engineering Group Working Party Report, *Q.J. Eng. Geol.*, **5**, 293–382.

Gete-Alonso, A. (1968) Applications des concepts de sûreté dans les fondations et culées des Rives au barrage de Mequinenza, *Proc. 9th Int. Cong. on Large Dams*, Istanbul, Report C9, **5**.

Geuze, E. C. A. W. and Rebull, P. M. (1966) Mechanical force fields in a clay mineral particle system, *Proc. 14th Nat. Conf. on Clays and Clay Minerals*, 103–116.

Gibbs, H. J. and Bara, J. P. (1962) 'Predicting surface subsidence from basic soil tests,' ASTM Special Technical Publication, **322**, 231–246.

Gibbs, H. J. and Holland, W. Y. (1960) Petrographic and engineering properties of loess, *Eng. Monograph*, **28**, (U.S. Bureau of Reclamation, Denver), 1–37.

Gibson, R. E. (1966) A note on the constant head test to measure permeability in-situ, *Géotechnique*, **16**, 256–259.

Gibson, R. E. (1974) The analytical method in soil mechanics, *Géotechnique*, **24**, no. 2, 115–140.

Gilbert, G. K. (1882) Post-glacial joints, *Amer. J. Sci.*, 3rd series, **23**, 25–27.

Gillott, J. E. (1968) 'Clay in engineering geology,' Elsevier, Amsterdam.

Glossop, R. (1968) The rise of geotechnology and its influence on engineering practice, *Géotechnique*, **18**, 107–150.

Gold, L. (1960) The cracking activity in ice during creep, *Canadian J. Physics*, **38**, 1137–1148.

Goodman, R. E. (1963) Subaudible noise during compression of rocks, *Bull. Geol. Soc. Amer.*, **74**, 487–490.

Goodman, R. E. and Blake, W. (1964) *Microseismic detection of potential earth slumps and rock slides*, Report No. MT–64-6, Inst. of Engineering Research, University of California, Berkeley, U.S.A.

Goodman, R. E. and Blake, W. (1965) 'An investigation of rock noise in landslides and cut slopes,' *Rock Mechanics and Engineering Geology*, Supplement No. 2, 89–93.

Grant, K. (1974) The composition of some Australian laterites and 'lateritic' gravels, *Proc. 2nd. Int. Congr. Int. Assoc. of Eng. Geology*, São Paulo, Brazil, **1**, Th IV–32.1–32.16.

Gray, D. A., Mather, J. D. and Harrison, I. B. (1974) Review of groundwater pollution from waste sites in England and Wales, with provisional guidelines for future site selection, *Quart. J. Engrg. Geol.*, **7**, 181–196.

Green, H. W. (1966) Preferred orientation of quartz due to re-crystallization during deformation, *Trans. Am. Geophys. Un.*, **47**, 491.

Green, H. W. (1967) Quartz; Extreme preferred orientation produced by annealing, *Science*, N.Y., **157**, 1444–1447.

Green, R. W. E. (1973) Seismic activity associated with the Hendrik Verwoerd Dam, Summary, Int. Colloquium on Seismic Effects of Reservoir Impounding, (COSERI) Royal Soc., 27–29 March.

Greenwood, D. A. (1970) Mechanical improvement of soils below surface. In: *Ground Engineering*, Inst. Civ. Engrs, London, 11–22.
Greenwood, D. A. and Raffle, J. F. (1963) Formulation and application of grouts containing clay, In *Grouts and Drilling Muds in Engineering Practice*, Butterworths, London, 127–130.
Greer, G. de (1910) A geochronology of the last 12,000 years, *Compte Rendus, Cong. Geol. Intern.*, session 11, 241–253.
Grice, R. H. (1969) 'Test procedures for the susceptibility of shale to weathering,' In: *Engineering properties and behaviour of clay shales*, (Johnson, S. J. Reporter), Speciality Session No. 10, Proc. 7th Int. Conf. on Soil Mech. and Found. Engrg., Mexico City, 3, 884.
Griffin, O. G. (1954) A new internal standard for the quantitative X-ray analysis of shales and mine dusts, *Research Report No. 101*, Safety in Mines Research Establishment, Ministry of Fuel and Power, Sheffield, England.
Griffith, A. A. (1921) The phenomenon of rupture and flow in solids, *Phil. Trans. Roy. Soc., London*, A221, 163–198.
Griffith, A. A. (1925) Theory of rupture, *Proc. First Int. Cong. for Applied Mechanics*, Waltham Int. Press, Delft, 53–64.
Griffiths, D. H. and King, R. F. (1965) 'Applied Geophysics for Engineers and Geologists,' Pergamon Press, London.
Griggs, D. and Bell, J. F. (1938) Experiments bearing on the orientation of quartz in deformed rocks, *Geol. Soc. Amer. Bull.*, 49, 1723–1746.
Griggs, D. T., Turner, F. J. and Heard, H. C. (1960) Deformation of rocks at 500°C–800°C. *Rock Deformation*, Geological Society of America, Memoir 79.
Grim, R. E. (1953) 'Clay mineralogy,' (2nd Edn. 1968) McGraw-Hill Book Co., Inc., New York.
Grim, R. E. (1962) 'Applied clay mineralogy,' McGraw-Hill Book Co., Inc. New York.
Grobbelaar, C. (1970) 'A theory for the strength of pillars.' Voortrekkers B Pk., Johannesburg.
Grundy, C. F. (1955) The Treatment by grouting of permeable foundations of dams, *Proc. Fifth International Congress on Large Dams*, Paris, 1, 647–674.
Guha, S. K., Gosavi, P. D., Agrawal, B. N. P., Padale, J. G. and Marwadi, S. C. (1973) Case histories of some artificial crustal disturbances, *Summary, Int. Colloquium on Seismic Effects of Reservoir Impounding* (COSERI), Royal Soc., 27–29 March, and *Eng. Geol.*, 8, 59–78.
Haefeli, R. (1948) The stability of slopes acted upon by parallel seepage, *Proc. 2nd Int. Conf. on Soil Mechanics*, Rotterdam, 1, 57–62.
Haimson, B. C. and Kim, C. M. (1971) Mechanical behaviour of rock under cyclic failure, *13th Symp. Rock Mechs.*, Univ. of Urbana, Illinois, 845–862.
Haimson, B. and Fairhurst, C. (1967) Initiation and extension of hydraulic fractures in rocks, *J. Soc. Petroleum Engrs.*, 7, 310–318.
Haimson, B. and Fairhurst, C. (1969) In-situ stress determination at great depth by means of hydraulic fracturing, *Eleventh Symposium on Rock Mechanics*, Berkeley, 559–584.
Hall, E. B. and Gordon, B. B. (1963) Triaxial testing with large-scale pressure equipment, *Symp. on laboratory Shear Testing of Soils*, Ottawa, ASTM STP 361, 315–328.
Hambly, E. C. (1972) Plane strain behaviour of remoulded normally consolidated kaolin, *Géotechnique*, 22, no. 2, 301–317.
Hamlin, M. J. (1970) 'Extension of run-off data,' Paper to Convention on Water for the Future, Republic of South Africa.

Hamrol, A. (1961) A quantitative classification of the weathering and weatherability of rocks, *Proc. 5th Int. Conf. Soil Mech. Found. Eng.* Paris, 2, 771–774.

Hanna, T. H. (1973) 'Foundation Instrumentation,' Trans. Tech. Pubs. Cleveland.

Hanrahan, E. T. (1954) An investigation of some physical properties of peat, *Géotechnique,* 4, 108–123.

Hanrahan, E. T. (1964) A road failure on peat, *Géotechnique,* 14, 185–202.

Hansen, B. (1961) Shear box tests on sand, *Proc. 5th Int. Conf. on Soil Mech. Found. Engrg.,* Paris, 1, 127–131.

Hantush, M. S. (1956) Analysis of data from pumping tests in leaky aquifers, *Trans. Am. Geophys. Union,* 37, 702–714.

Hantush, M. S. (1964) Hydraulics of wells, *Advances in Hydroscience,* Ed. V. T. Chow, 1, 281–442.

Hantush, M. S. and Jacob, C. E. (1954) Plane potential flow of groundwaters with linear leakage, *Trans. Am. Geophys. Union,* 35, 917–936.

Hardy, H. R. (1969) Microseismic activity and its application in the natural gas industry, *Proc. Amer. Gas Assoc., Operating Section,* A.G.A. Cat. No. X59969, New York, pp. T–147–T–156.

Hardy, H. R. (1971) Application of acoustic emission techniques to rock mechanics research, *Internal Report RML–IR/71–13.,* Dept. of Mineral Engineering, Pennsylvania State University, U.S.A.

Hardy, H. R. (1973) Microseismic techniques, basic and applied research, *Rock Mechanics,* Suppl. 2, 93–114.

Hardy, H. R., Kim, R. Y., Stefanko, R. and Wang, Y. J. (1970) Creep and microseismic activity in geologic materials, *Rock Mechanics – Theory and Practice,* AIME, New York, 377–413.

Hardy, H. R. and Chugh, Y. P, (1970) Failure of geologic materials under low-cycle fatigue, Presented at *6th Canadian Symp. Rock Mech.,* Montreal, 1970.

Hartshorne, N. H. and Stuart, A. (1964) 'Practical Optical Crystallography,' Edward Arnold, London.

Hast, N. (1958) The measurement of rock pressure in mines, *Sveriges Geologiska Undersokming,* Arsbok 52, Ser.C., no. 560, Stockholm.

Hawkes, I. and Mellor, M. (1970) Uniaxial testing in rock mechanics laboratories, *Engineering Geology,* 4, 177–286.

Hawkes, I. and Moxon, S. (1966) The measurement of *in situ* rock stress using the photoelastic biaxial gauge with the core relief technique, *Int. J. Rock Mech. Min. Sci.,* 2, 405–419.

Hawkes, J. M. and Evans, R. H. (1951) Bond stresses in reinforced columns and beams, *Structural Engineer,* 29, 323–327.

Heard, H. C. (1963) The effect of large changes in strain rate in the experimental deformation of rocks. *J. Geol.,* 71, 162–195.

Hearmon, R. F. S. (1961) 'An introduction to applied anisotropic elasticity,' Oxford Univ. Press.

Heckler, A. J., Elias, J. A. and Woods, A. P. (1967) Automatic computer plotting of pole figures and axis density figures, *Trans. Metal Soc., AIME,* 239, 1241–1244.

Heerden, W. L. van (1969) Potential failure zones around boreholes with flat and spherical ends, *Int. J. Rock Mech. Min. Sci.,* 6, 453–463.

Helfrich, H. K., Hasselstrom, B. and Sjogren, B. (1970) Geoscience site investigations for tunnels, *The Technology of Tunnelling,* 1, 65–69.

Hem, J. D. (1959) Study and interpretation of the chemical characteristics of natural water, *U.S. Geol. Survey Water Supply Paper 1473.*

Hendron, A. J., Cording, E. J. and Aiyer, A. K. (1971) Analytical and graphical methods for the analysis of slopes in rock masses, *U.S. Army Engineer Waterways Experiment Station*, NCG Technical Report No. 36.

Henkel, D. J. (1970) The role of waves in causing submarine landslides, *Géotechnique,* **20**, 75–80.

Henkel, D. J., Knill, J. L., Lloyd, D. G. and Skempton, A. W. (1964) Stability of the foundations of Monar Dam, *Trans. 8th Int. Congress on Large Dams,* Edinburgh, Scotland, **1**, 425–441.

Henkel, D. J. and Skempton, A. W. (1954) A landslide at Jackfield, Shropshire, in an over-consolidated clay, *Proc. European Conf. Stability of Earth slopes,* Stockholm, **1**, 90–101, and *Géotechnique,* **5**, 131–137.

Herbert, R. (1965) Well hydraulics – 1., *Water Research Association Technical Paper,* TP 40, Water Research Association, Medmenham, Bucks, England.

Herbert, R. and Rushton, K. R. (1966) Groundwater flow studies by resistance networks, *Géotechnique,* **16**, 53–75.

Hetherington, H. A. (1963) 'Drilling muds for mineral drilling and water well construction', In: *Grouts and drilling muds in engineering practice,* Butterworths, London, 206–210.

Heuze, E. E., Goodman, R. E. and Bornstein, A. (1971) Numerical analyses of deformability tests in jointed rock – 'Joint Perturbation' and 'No Tension' finite element solutions, *Rock Mechanics,* **3**, 13–24.

Hibbert, E. S. (1956) The hydrogeology of the Wirral Peninsula, *J. Inst. Water Engrs.,* **10**, 441–469.

Higginbottom, I. C. (1971) Superficial structures in reconnaissance and feasibility studies, *Q. Jl. Eng. Geol.,* **4**, 307–310.

Higgs, D. V., Friedman, M. and Gebhart, J. E. (1960) Petrofabric analysis by means of the X-ray diffractometer, *Mem. Geol. Soc. Am.,* **79**, 275–292.

Hilf, J. W. (1960) Foundations and Construction Materials, In: *Design of Small Dams,* Publ. U.S. Dept., of the Interior, Bureau of Reclamation, 71–155.

Hill, R. (1950) 'The mathematical theory of plasticity,' Clarendon Press, Oxford.

Hill, R. A. (1941) Salts in irrigation water, *Proc. Amer. Soc. Civ. Engrs.,* **67**, 975–990.

Hilliard, J. E. (1962) Specification and measurement of microstructural anisotropy, *Trans. Metall. Soc., AIME.,* **224**, 1201–1211.

Hilton, I. C. (1967) Grout selection: a new classification system, *Civ. Eng. Pub. Works Rev.,* **62**, 993–995.

Hinckley, D. N. (1963) Variability in 'crystallinity' values among the kaolin deposits of the coastal plain of Georgia and South Carolina, *Proc. 11th Nat. Conf. on Clays and Clay Minerals,* Natl. Acad. Sci., Natl. Research Council Publ., Pergamon Press, Oxford, 229–235.

Hiramatsu, Y. and Oka, Y. (1966) Determination of the tensile strength of a rock by compression of an irregular test piece. *Int. J. Rock. Mech. Min. Sci.,* **3**, 89–99.

Hirschfield, R. C., Whitman, R. V. and Wolfskill, L. A. (1965) Engineering properties of nuclear craters, Review and analysis of available information on slopes excavated in weak shale, *Tech. Report, No. 3–699,* Report 3; Prepared by Massachusetts Inst. of Tech., for U.S. Army Engineer Waterways Experiment Station, C. E., Vicksburg, Miss., U.S.A.

Hobbs, D. W. (1962) Strength of irregularly shaped specimens of rock crushed between parallel platens. *MRE Report No. 2216,* National Coal Board, U.K.

Hobbs, N. B. (1970) Contribution to discussion, *Symp. Ground Engineering*, Institution of Civil Engineers, London, 81–82.

Hodgson, E. A. (1943) Recent developments in rock burst research at lakeshore mines, *Trans. Can. Inst. Mining Metal*, 46, 313–324.

Hodgson, E. A. and Gibbs, Z. E. (1945) Seismic research program, rock burst problem – lakeshore mines, *Department Mines, Resources and Surveys Engrg. Branch Report*, No. 14, Ottawa, Canada.

Hodgson, R. (1961) Regional study of jointing in Comb Ridge – Navajo, Mountain Area, Arizona and Utah, *Bull. Amer. Assoc. Petrol. Geol.*, 45, No. 1, 1–38.

Höeg, K., Andersland, O. A. and Rolfsen, E. N. (1969) Undrained behaviour of quick clay under load tests at Äsrum, *Géotechnique*, 19, 101–115.

Hoek, E. (1964) Fracture of anisotropic rock, *J. South Africa Inst. of Mining and Metallurgy*, 64, 501–518.

Hoek, E. (1966) Rock Mechanics – an introduction for the practical engineer, *Min. Mag.*, 114, 236–243.

Hoek, E. (1970) Estimating the stability of excavated slopes in opencast mines, *Trans. Inst. Min. Metall.*, Sect. A, 79, 109–132.

Hoek, E. (1971) Rock slope stability – how far away are reliable design methods?, *Proc. First Australia-New Zealand Conf. on Geomechanics*, Melbourne, 307–313.

Hoek, E. (1973) Methods for the rapid assessment of the stability of three-dimensional rock slopes, *Q. Jl. Eng. Geol.*, 6, 243–255.

Hoek, E. and Bray, J. W. (1974) 'Rock Slope Engineering,' Inst. of Mining and Metall., London.

Hoek, E., Bray, J. W. and Boyd, J. M. (1973) The stability of a rock slope containing a wedge resting on two intersecting discontinuities, *Q. Jl. Eng. Geol.*, 6, 1–55.

Hoek, E. and Franklin, J. A. (1968) Simple triaxial cell for field or laboratory testing of rock. *Trans. Inst. Min. Metall.*, 77, A22–26.

Hoek, E. and Londe, P. (1974) Surface workings in rock, *Proc. 3rd Congress, Int. Soc. for Rock Mech.*, Denver, Colorado, U.S.A., 1, pt A, 613–654.

Hoek, E. and Pentz, D. L. (1968) 'The stability of open pit mines: review of the problems and of methods of solution,' Imperial College (London) Inter-departmental Rock Mechanics Project Report No. 5.

Holden, W. S. (1970) *Water treatment and examination*, Ch. 2 and 3, J. & A. Churchill, London.

Hollister, J. C. and Weimer, R. J. (1968) Geophysical and geological studies of the relationship between the Denver earthquakes and the Rocky Mountain Arsenal well, Part A, General Summary and Conclusions, *Q. Jl. Colorado Sch. of Mines*, 63, no. 1, 1–8.

Holtz, W. G. (1960) Effect of gravel particles on friction angle, *Proc. Amer. Soc. Civ. Engrs., Research conf. on Shear Strength*, 1000–1001.

Holtz, W. G. and Gibbs, H. J. (1951) Consolidation and related properties of loessial soils, *Special Technical Publ. 126*, American Society for Testing Materials, 9–33.

Holtz, W. G. and Hilf, J. W. (1961) Settlement of soil foundations due to saturation, *Proc. 5th Int. Conf. Soil Mech. Found. Eng.*, Paris, 1, 673–679.

Hooper, J. A. and Butler, F. G. (1966) Some numerical results concerning the shear strength of London Clay, *Géotechnique*, 16, 282–304.

Horder, R. I. (1960) Road designing from the air, *The New Scientist*, 7, 589–591.

Horn, H. M. and Deere, D. U. (1962) Frictional characteristics of minerals, *Géotechnique,* **12**, 319–335.
Horsley, R. V. (1952) Oily collectors in coal flotation, *Trans. Inst. Min. Engrs.,* **111**, 886–894.
Hoskins, E. R. (1967) An investigation of strain-relief methods of measuring rock stress, *Int. J. Rock Mech. Min. Sci.,* **4**, 155–164.
Hough, B. K. (1957) 'Basic soils engineering,' Ronald Press Co., New York.
Houlsby, A. C. (1969) Rock movements during grouting. *Rock Mechanics Symposium,* Sydney, 116–120.
Houlsby, A. C. (1973) Private communication.
Howells, D. A. (1973) The time for a significant change in pore pressure, *Summary. Int. Colloquium on Seismic Effects of Reservoir Impounding,* (COSERI), Royal Soc., 27–29 March, and *Eng. Geol.,* **8**, 135–138.
Hubbert, M. K. and Willis, D. G. (1957) Mechanics of hydraulic fracturing, *Trans. Amer. Inst. Min. Eng.,* **210**, 153–168.
Hudson, J. A., Brown, E. T. and Fairhurst, C. (1971) Optimising the control of rock failure in servo-controlled laboratory tests, *Rock Mechs.,* **3**, 217–224.
Hughes, D. S. and McQueen, R. G. (1958) Density of basic rocks at very high pressures, *Trans. Amer. Geophys. Union,* **39**, 959–965.
Hurlbut, C. S. (1971) 'Dana's manual of mineralogy,' 18th Ed., Wiley, New York.
Hutchinson, J. N. (1961) A landslide on a thin layer of quick clay at Furre, Central Norway, *Géotechnique,* **11**, 69–94.
Hutchinson, J. N. (1965) The landslide of February, 1959 at Vibstad in Hamdalen, *Publ. Norwegian Geotech. Inst.,* No. 61, 1–16.
Hutchinson, J. N. (1967a) The free degradation of London Clay cliffs, *Proc. Geotech. Conf.,* Oslo, **1**, 113–118.
Hutchinson, J. N. (1967b) Discussion on Session 2, *Proc. Geotech. Conf.,* Oslo, **2**, 183–184.
Hutchinson, J. N. (1968) Field meeting on the coastal landslides of Kent, 1–3 July, 1966, *Proc. Geol. Assoc.,* **79**, 227–237.
Hutchinson, J. N. (1969) A reconsideration of the coastal landslides at Folkestone Warren, Kent, *Géotechnique,* **19**, 6–38.
Hutchinson, J. N. (1970) A coastal mudflow on the London Clay cliffs at Beltinge, north Kent, *Géotechnique,* **20**, 412–438.
Hutchinson, J. N. and Bhandari, R. K. (1971) Undrained loading, a fundamental mechanism of mudflows and other mass movements, *Géotechnique,* **21**, 353–358.
Hutchinson, J. N. and Hughes, M. J. (1968) The application of micro-palaeontology to the location of a deep-seated slip surface in the London Clay, *Géotechnique,* **18**, 508–510.
Hvorslev, M. J. (1951) Time lag and soil permeability in ground water observations, *Bull. No. 36, Waterways Exper. Stn.,* U.S. Army Corps of Engrs., Vicksburg.
Hvorslev, M. J. (1960) Physical components of the shear strength of saturated clays. *A.S.C.E. Research Conf. on Shear Strength of Cohesive Soils,* Boulder, Colorado, 169–273.
Iida, K. (1938) The mudflow that occurred near the explosion crater of Mt Bandai on May 9 and 15, 1938, and some physical properties of volcanic mud, *Bull. Earthquake Res. Inst. Tokyo Univ.,* **16**, 658–681.
Iliev, I. G. (1970) An attempt to estimate the degree of weathering of intrusive rocks from their physico-mechanical properties, *Proc. 2nd Cong. Int. Soc. Rock Mechanics, Belgrade,* **1**, 109–114.

Inderbitzen, A. L. (1971) Empirical relationships between physical properties for recent marine sediments of Southern California. *Marine Geol.*, 9, 311–329.

Ineson, J. (1970a) Water resource planning and development in Northern England, *Symposium Report on Geological Aspects of Development and Planning in Northern England*, Ed. P. T. Warren, Organized and Published by the Yorkshire Geological Society, 4–11.

Ineson, J. (1970b) Development of ground water resources in England and Wales, *J. Inst. Water Engrs.*, 24, 155–177.

Inman, D. L. (1952) Measures for describing the size distribution of sediments, *J. Sed. Petrol.*, 22, 125–145.

Insley, A. E. and Hillis, S. F. (1965) Triaxial shear characteristics of a compacted glacial fill under unusually high confining pressures, *Proc. 6th Int. Conf. on Soil Mech. Found. Eng.*, Montreal, 1, 244–248.

Institute of Geological Sciences (1967) 'Characteristic symbols for use on six-inch and one-inch maps of the Geological Survey of Great Britain.' Inst. Geological Sciences, London.

International Geographical Union (1968) 'The unified key to the detail geomorphological map of the world. 1:25,000 to 1:50,000 scale,' Int. Geographical Union, University of Chicago, U.S.A.

Ireland, H. O. (1954) Stability analysis of the Congress Street open cut in Chicago, *Géotechnique*, 4, 163–168.

Irwin, G. R. (1957) Analyses of stresses and strains near the end of a crack traversing a plate, *J. Appl. Mech.*, 24, 361–364.

Iverson, N. L. (1969) In: Engineering properties and behaviour of clay shale, (Reporter S. J. Johnson), *7th Int. Conf. on Soil Mechanics and Foundation Engrg.*, Mexico City, 3, 483–488.

Jackson, M. L. (1964) 'Chemical composition of soils,' In: *Chemistry of Soil*, Ed. F. E. Bear, Reinhold, New York, 71–135.

Jacob, C. E. (1950) 'Flow of ground water,' *Engineering Hydraulics*, Ed. H. Rouse, Wiley, New York, 321–386.

Jaeger, J. C. (1960) Shear failure of anisotropic rocks, *Geol. Mag.*, 97, 65–72.

Jaeger, J. C. (1969) 'Elasticity, Fracture and Flow: with engineering and geological applications,' Chapman and Hall, London.

Jaeger, J. C. (1970) The behaviour of closely jointed rock, *Proc. 11th Symp. Rock Mech.*, Berkeley, California, U.S.A., 57–68.

Jaeger, J. C. (1971) Friction of rocks and stability of rock slopes, *Géotechnique*, 21, 97–134.

Jaeger, J. C. and Cook, N. G. W. (1969) 'Fundamentals of Rock Mechanics,' Chapman and Hall, London.

Janbu, N. (1954) Application of composite slip circles for stability analysis, *Proc. European Conf. on Stability of Earth slopes*, Stockholm, 4, 43–49.

Janbu, N. (1957) Earth pressure and bearing capacity by generalized procedure of slices, *Proc. 4th Int. Conf. Soil Mech. Found. Engrg.*, London, 2, 207–212.

Jenkins, J. D. and Szecki, A. (1964) The properties of some rock materials and their behaviour in pillars, *Unpublished report*, Newcastle University, England.

Jennings, J. E. (1970) A mathematical theory for the calculation of the stability of slopes in open cast mines, *Symposium on Planning Open Pit Mines*, Johannesburg, Ed. D. W. J. Van Rensburg, Publ. A. A. Balkema, Cape Town, 87–102.

Jenny, H. (1941) 'Factors of soil formation,' McGraw-Hill, New York.

Jetter, L. K. and Borie, B. S. (1953) Method for the quantitative determination of preferred orientation, *J. Appl. Phys.*, 24, 532–535.

Jetter, L. K., McHargue, C. J. and Williams, R. O. (1956) Method of representing preferred orientation data, *J. Appl. Phys.*, **27**, 368–374.

John, K. W. (1962) An approach to rock mechanics, *J. Soil Mech. Found. Div., ASCE*, **88**, pt. 1, 1–30.

John, K. W. (1968) Graphical stability analysis of slopes in jointed rock, *J. Soil Mech. Found. Div., ASCE*, **94**, 497–526.

John, K. W. (1974) Geologists and civil engineers in the design of rock foundations of dams, *Proc. 2nd Int. Congr., Int. Assoc. of Eng. Geol.*, São Paulo, Brazil, **2**, Th. VI–PC–3.1–3.9.

Johnson, A. I. (1963) Selected references on analog models for hydrologic studies, *Proc. Symp. Transient Ground Water Hydraulics*, Appendix F, Colorado State University, U.S.A.

Johnson, A. I. (1968) 'An outline of geophysical logging methods and their uses in hydrologic studies,' U.S. Geol. Survey, Water Supply Papers, No. 1892, 158–164.

Johnson, S. J. (1969) 'Engineering properties and behaviour of clay shales,' Speciality Session 10, *Proc. 7th Int. Conf. on Soil Mechanics & Foundation Engrg.*, Mexico City, **3**, 483–488.

Johnson, S. J., Krinitzsky, E. L. and Dixon, N. A. (1973) Preliminary review of reservoir filling and associated earthquakes, *Summary, Int. Colloquium on Seismic Effects of Reservoir Impounding* (COSERI), Royal Soc., 27–29 March.

Jones, G. P. (1971) Management of underground water resources, *Quart. J. Eng. Geol.*, **4**, 317–328.

Jones, F. O., Embody, D. R. and Peterson, W. L. (1961) 'Landslides along the Columbia River valley, north eastern Washington,' U.S. Geol. Survey, Prof. Paper 367.

Jones, P. H. (1952) 'Electrical logging methods; principles of interpretation on applications in groundwater studies.' U.S. Geol. Survey.

Jones, R. C. B. (1964) Some engineering aspects of air photo-interpretation in catchment development programmes, *The Photogrammetric Record*, **4**, 466–475.

Jorgensen, P. (1965) Mineralogical composition and weathering of some late Pleistocene marine clays from the Kongsvinger area, Southern Norway, *Geologiska Foreningen i Stockholm Forhandlinger*, **87**, no., 520, 62–83.

Judd, W. R. (1964a) 'Effect of the elastic properties of rocks on civil engineering design,' *Engineering Geology Case Histories, No's. 1–5*, Eds. P. D. Trask and G. A. Kiersch, Geol. Soc. Amer., 163–184.

Judd, W. R. (1964b) Rock stress, rock mechanics and research, 'State of stress in the Earth's crust', (Ed. W. R. Judd), Elsevier, New York, 5–53.

Judd, W. R. and Huber, C. (1962) Correlation of rock properties by statistical methods, *Int. Symp. on Mining Research* (Ed. G. Clarke), Pergamon, Oxford, **2**, 621–648.

Jumikis, A. R. (1962) 'Soil Mechanics,' D. van Nostrand, & Co. Inc., Princetown, New Jersey.

Jumikis, A. R. (1969) 'Thermal Soil Mechanics,' Rutgers Univ. Press, New Jersey.

Jungst, H. (1934) Zur Geologischen Bedentung der Synarese, *Geologische*, **25**, 321–325.

Jurgenson, L. (1934) 'The application of theories of elasticity and plasticity to foundation problems,' *Contributions to Soil Mechanics (1925–1940)*, Boston Soc. Civ. Engrs., 148–183.

Kamb, W. B. (1959) Theory of preferred crystal orientation developed by crystallization under stress, *J. Geol.*, **67**, 153–170.

Kamb, W. B. (1959) Ice petrofabric observations from the Blue Glacier, Washington, in relation to theory and experiment, *J. Geophys. Res.*, **64**, 1891–1909.

Kankare, E. (1969) Failures at Kimola floating canal in southern Finland, *Proc. 7th Int. Conf. Soil Mech. and Found. Engrg.*, Mexico City, **2**, 609–616.

Karplus, W. J. (1958) 'Analog simulations,' McGraw-Hill, New York.

Karplus, W. J. (1968) 'Analog simulation – solution of field problems,' McGraw-Hill, New York.

Kassiff, G. and Shalom, A. B. (1971) Experimental relationship between swell pressure and suction. *Géotechnique*, **21**, no. 3, 245–255.

Kazi, A. and Knill, J. L. (1969) The sedimentation and geotechnical properties of the Cromer Till between Happisburgh and Cromer, Norfolk, *Q. Jl. Eng. Geol.*, **2**, 63–86.

Keller, G. H. (1969) Engineering properties of some sea floor deposits, *J. Soil Mech. Found. Div., Am. Soc. Civ. Engrs.*, **95**, 1379–1392.

Kellogg, C. E. (1941) 'The soils that support us,' Macmillan Co., New York.

Kennard, M. F. (1972) Examples of the internal conditions of some old earth dams, *J. Inst. Water Engrs.*, **26**, 135–147.

Kennard, M. F. and Knill, J. L. (1969) Reservoirs on limestone, with particular reference to the Cow Green scheme, *J. Instn. Water Engrs.*, **23**, 87–113.

Kennard, M. F., Knill, J. L. and Vaughan, P. R. (1967) The geotechnical properties and behaviour of Carboniferous Shale at Balderhead Dam, *Q. Jl. Eng. Geol.*, **1**, 3–24.

Kennard, M. F., Penman, A. D. M. and Vaughan, P. R. (1967) Stress and strain measurements in the clay core at Balderhead Dam, *Proc. 9th Int. Conf. Large Dams*, Istanbul, **3**, 129–151.

Kenney, T. C. (1967a) 'Shearing resistance of natural quick clays,' Ph.D. thesis, Univ. of London.

Kenney, T. C. (1967b) The influence of mineral composition on the residual strength of natural soils, *Proc. Geotech. Conf.*, Oslo, **1**, 123–129.

Kenney, T. C., Moum, J. and Berre, T. (1968) An experimental study of bonds in a natural clay, *Norwegian Geotechnical Institute Publ.*, no. 76.

Kenyon, W. J. and Hibbitt, J. A. (1974) The Upper Wylye investigation, *Water Services*, **78**, no. 939, 154–159.

Kinahan, G. H. (1875) 'Valleys and their relation to fissures, fractures and faults,' Trabner & Co., London.

Kindle, E. M. (1923) Note on mud-crack and associated joint structure, *Amer. Journ. Sci.*, 5th Series, **28**, 329–330.

Kindle, E. M. (1926) Contrasted types of mud cracks, *Trans. Roy. Soc., Canada*, 3rd series, **20** (4), 71–76.

Kirkby, M. J. and Chorley, R. J. (1967) Throughflow, overland flow and erosion, *Bull. Int. Assoc. for Scientific Hydrology*, **12**, 5–21.

Kirkham, D. (1955) Measurement of the hydraulic conductivity of soil in place, *Symp. on Permeability of Soils, Amer. Soc. Testg. Mat.*, S.T.P. No. 163, 80-97.

Kirkham, D. and Bavel, C. H. M. von (1948) The theory of seepage into auger holes, *Proc. Soil Sci. Soc. Amer.*, **13**, 78–82.

Kirkpatrick, W. M. and Rennie, I. A. (1972) Directional properties of consolidated kaolin, *Géotechnique*, **22**, 166–169.

Kishida, H. (1969) Characteristics of liquefied sands during Mino-Owari, Tohnankai and Fukui earthquakes, *Soils and Foundations*, **9**, 75–92.

Kishida, H. (1970) Characteristics of liquefaction of level sandy ground during the Tokachioki earthquake, *Soils and Foundations*, **10**, 103–118.

Kley, R. J. and Lutton, R. J. (1967) *Engineering properties of nuclear craters: a study of selected rock excavations as related to large nuclear craters*, U.S. Army Corps of Engineers, Report PNE 5010.

Klug, H. P. and Alexander, L. E. (1954) 'X-ray diffraction procedures for poly-crystalline and amorphous materials,' Wiley and Sons, New York.

Knights, C. (1971) 'Small amplitude ground vibrations,' Dissertation M.Sc. Advanced Course in Engineering Geology, University of Durham.

Knights, M. C. (1974) Exploratory shafts for in-town surveys, *Ground Engineering*, **17**, no. 1, 43–45.

Knill, J. L. (1969) The application of seismic methods in the prediction of grout take in rock, *Proc. Conf. on In Situ Investigation in Soils and Rocks*, British Geotechnical Soc., 93–100.

Knill, J. L. (1970) Environmental, economic and engineering factors in the selection of reservoir sites, with particular reference to Northern England, *Symp. Report on Geological Aspects of Development and Planning in Northern England*, Ed: P. T. Warren, Yorkshire Geological Society, 124–141.

Knill, J. L. (1972) Assessment of reservoir feasibility, *Q. Jl. Eng. Geol.*, **4**, 355–372.

Knill, J. L. (1973) Rock conditions in the Tyne Tunnels, North eastern England, *Bull. Assoc. Engrg. Geologists*, **10**, no. 1, 1–19.

Knill, J. L. (1974) Engineering Geology related to dam foundations, *Proc. 2nd Int. Congr. Int. Assoc. of Eng. Geology*, São Paulo, Brazil, **2**, Th VI, PC–1.1–1.7.

Knill, J. L., Franklin, J. A. and Malone, A. W. (1967) A study of acoustic emission from stressed rock, *Int. J. Rock Mech. Min. Sci.*, **5**, 87–121.

Knill, J. L. and Jones, K. S. (1965) The recording and interpretation of geological conditions in the foundations of the Roseires, Kariba, and Latiyan Dams, *Géotechnique*, **15**, 94–124.

Knox, G. (1927) Landslides in South Wales valleys, *Proc. South Wales Inst. of Engineers*, **43**, 161–247; 259–290.

Krigbaum, W. R. and Roe, R. J. (1964) Crystallite orientation in materials having fibre texture, II; A study of strained samples of cross-linked polyethelene, *J. Chem. Physics*, **41**, 737–748.

Krigbaum, W. R. and Balta, Y. I. (1967) Pole figure inversion for triclinic crystal class, polyethelene teraphthalate, *J. Phys. Chemistry*, **71**, 1770–1779.

Krinitzsky, E. L. (1970) The effect of geological features on soil strength, *U.S. Army Engineer Waterways Experimental Station*, Vicksburg, Miss., U.S.A., Misc. Paper S–70–25.

Krinitzsky, E. L. and Turnbull, W. J. (1967) Loess deposits of Mississippi, *Geol. Soc. Amer.*, Boulder, Colorado, Special Paper no. 94.

Krumbein, W. C. and Graybill, F. A. (1965) 'An introduction to statistical models in geology,' McGraw-Hill, New York.

Krynine, D. P. (1948) The megascopic study and field classification of sedimentary rocks. *J. Geol.*, **56**, 130–165.

Krynine, D. P. and Judd, W. R. (1957) 'Principles of engineering geology and geotechnics,' McGraw-Hill, New York.

Kummeneje, O. and Eide, O. (1961) Investigations of loose sand deposits by blasting, *Proc. 5th Int. Conf. Soil Mech. and Found. Eng.*, Paris, **1**, 491–497.

Lafeber, D. (1968) Discussion: Paper on 'Microscopic structures in kaolin subjected to direct shear,' *Géotechnique*, **18**, 379–382.

Lafeber, D. and Kurbanovic, M. (1965) Photographic reproduction on soil fabric patterns, *Nature,* **208**, 609–610.
Lajtai, E. Z. (1969) Strength of discontinuous rocks in direct shear, *Géotechnique,* **19**, 218–233.
Lam, P. W. H. (1969) Computer method for plotting beta diagrams, *Am. J. Sci.,* **267**, 1114–1117.
Lambe, T. W. (1951) 'Soil testing for engineers,' Wiley, New York.
Lambe, T. W. (1953) The structure of inorganic soil, *Proc. Amer. Soc. Civ. Engrs.,* **79**, Paper no. 315.
Lambe, T W. (1960) A mechanistic picture of the shear strength of clays, *Proc. Conf. on Shear Strength of cohesive soils, Amer. Soc. Civ. Engrs.,* 555–580.
Lambe, T. W. (1967) Stress path method, *J. Soil Mech. and Found. Div. Amer. Soc. Civ. Engrs.,* **93**, 309–331.
Lambe, T. W. and Whitman, R. V. (1969) 'Soil Mechanics,' Wiley & Sons, New York.
Lamplugh, G. W. (1890) On the larger boulders of Flamborough Head and other parts of the Yorkshire coast — Part 4, *Proc. Yorks. Geol. Soc.,* **XI**, 397–408.
Lane, K. S. (1960) Garrison Dam: Evaluation of test results, *Trans. Amer. Soc. Civ. Engrs.,* **125**, 268–306.
Lane, K. S. (1961) Field slope charts for stability studies, *Proc. 5th Int. Conf. Soil Mech. Found. Engrg.,* Paris, **2**, 651–655.
Lane, K. S. (1966) Stability of reservoir slopes. In *Failure and breakage of rock (Proc. 8th Symp. Rock Mechanics,* 1966), AIME, New York, 1967, 321–336.
Lane, K. S. (1969) In: 'Engineering properties and behaviour of clay shales.' (Reporter, S. J. Johnson), *7th Int. Conf. on Soil Mech. and Found. Engineering, Mexico City,* **3**, 483–488.
Lane, K. S. and Washburn, D. E. (1946) Capillary tests by capillarimeter and by soil-filled tubes, *Proc. Highway Res. Board 26th Ann. Meeting,* **26**, 460–473.
Lane, R (1966) The interpretation of rock test results for the design of structures, *Proc. 1st Congress Int. Soc. of Rock Mech.,* Lisbon, **2**, 563–566.
Langefors, U. and Kihlström, B. (1963) 'The Modern Technique of Rock Blasting,' John Wiley & Sons., Inc., New York.
Langefors, U., Westerberg, H. and Kihlström, B. (1958) Ground vibrations in blasting, *Water Power,* **10**, no. 2, 335–338, 390–395, 421–424.
Larsson, L. (1952) Graphic testing procedure for point diagrams, *Amer. J. Sci.,* **250**, 586.
Lauffer, H. (1958) Gebirgsklassifizierung, fur der Stollenbau, *Geologie und Bauwesen,* **24**, 46–51.
Law, F. (1965) Integrated use of diverse resources, *J. Inst. Water Engineers,* **19**, 413–461.
Leach, B. (1973) 'The effects of planar anisotropy on the consolidation and drained shear strength characteristics of a laminated clay.' M.Sc. Advanced Course Dissertation, University of Durham.
Learmonth, A. P. (1972) 'An engineering geology investigation of a pressure tunnel,' M.Sc. Advanced Course Dissertation, University of Durham.
LeConte, J. (1882) Origin of jointed structures in undisturbed clay and marl deposits, *Amer. Jl. Sci.,* 3rd series, **23**, 233–234.
Leeman, E. R. (1964a) The measurement of stress in rock, Part 1: The principles of rock stress measurements, *Symp. on Rock Mechanics and Strata Control in Mines, South African Inst. Min. Metall.,* 248–284.
Leeman, E. R. (1964b) The measurement of stress in rock, Part II: Borehole

rock stress measuring instruments, *Symp. on Rock Mechanics and Strata Control in Mines, J. South African Inst. Min. Metall.*, 285–317.

Leeman, E. R. (1964c) The measurement of stress in rock, Part III: The results of some rock stress investigations, *Symp. on Rock Mechanics and Strata Control in Mines, J. South African Inst. Min. Metall.*, 318–349.

Leet, L. D. (1946) Vibrations from blasting, *Explosives Eng.*, **24**, 41–44, 55–59, 85–89.

Leet, L. D. (1949) Vibrations from delay blasting, *Seismol. Soc. Amer. Bull.*, **39**, 9–20.

Leet, L. D. (1960) 'Vibrations from blasting rock,' Harvard Univ. Press., Mass.

Legget, R. F. and Bartley, M. W. (1953) An engineering study of glacial deposits at Steep Rock Lake, Ontario, Canada, *Economic Geology*, **48**, 513–540.

Leonard, D. R., Grainger, J. W. and Eyre, R. (1974) 'Loads and vibrations caused by eight commercial vehicles with gross weights exceeding 32 tons (32.5 Mg),' Dept. of the Environment, TRRL Report LR 582, Crowthorne.

Leonard, M. W. and Moller, K. (1963) 'Grouting for support, with particular reference to the use of some chemical grouts,' *Grouts and drilling muds in engineering practice*, Butterworths, 156–164.

Leslie, D. D. (1963) Large scale triaxial tests on gravelly soils, *Proc. 2nd Amer. Conf. Soil Mech. Found. Eng.*, Brazil, **1**, 181–202.

Lewis, A. T., Conrad, J. S. and Watson, E. L. (1960) Concrete gravity dams. In: *Design of Small Dams*, Publ. U.S. Dept. of the Interior, Bureau of Reclamation, 231–245.

Linell, K. A. and Shea, H. F. (1960) Strength and deformation characteristics of various glacial tills of New England, *Proc. Res. Conf. on Shear Strength of Cohesive Soils*, A.S.C.E., 275–314.

Little, J. A. (1972) 'Slope stability in the Kimmeridge Clay,' Dissertation: M.Sc. Advanced Course in Engineering Geology, University of Durham, England.

Little, A. L. and Price, V. E. (1958) The use of an electronic computer for slope stability analysis, *Géotechnique*, **8**, 113–120.

Littlejohn, G. S. (1970) Soil anchors, 'Ground Engineering,' Inst. Civil Engrs., London, 33–45.

Litwiniszyn, J. (1956) Application of the equation of stochastic processes to mechanics of loose bodies, *Archiwum Mechaniki Stosowanej*, **8**, 393–411.

Litvinov, I. M., Rzhanitzin, B. A. and Bezruk, V. M. (1961) Stabilisation of soils for constructional purposes, *Proc. 5th Int. Conf. Soil Mech. Found. Engrg.*, Paris, **2**, 775–780.

Lo K. Y. (1965) Stability of slopes in anisotropic soils, *Proc. Amer. Soc. Civ. Engrs.*, **91**, No. SM4, 85–106.

Lo, K. Y. (1970) The operational strength of fissured clays, *Géotechnique*, **20**, 57–74.

Lo K. Y. and Morin, J. P. (1972) Strength anisotropy and time effects of two sensitive clays, *Canadian Geotech. J.*, **9**, 261–277.

Londe P. (1973a) The role of rock mechanics in the reconnaissance of rock foundations, *Quart. J Eng. Geol.*, **6**, no. 1, 57–74.

Londe, P. (1973b) Water seepage in rock slopes, *Quart. J. Eng. Geol.*, **6**, no. 1, 75- 92.

Loughnan, F. C. (1969) 'Chemical weathering of the silicate minerals,' Elsevier Publ. Co., Amsterdam.

Love, A. E. H. (1892) 'A treatise on the mathematical theory of elasticity,' 2 vols., Cambridge Univ. Press, Cambridge, U.K.

Lowe, J. (1964) Shear strength of coarse embankment dam materials, *Proc. 8th Int. Congress on Large Dams*, Edinburgh, **3**, 745–761.

Lubinski, A. (1954) The theory of elasticity for porous bodies displaying a strong pore structure, *Proc. of the 2nd U.S. Nat. Cong. of Applied Mechanics*, 247–256.

Lueder, D. R. (1959) 'Aerial photographic interpretation,' McGraw-Hill, New York.

Lumb, P. (1962) The properties of decomposed granite, *Géotechnique*, **12**, 226–243.

Lumb, P (1965) The residual soils of Hong Kong, *Géotechnique*, **15**, no. 2, 180-194.

Luthin, J. N. (1961) 'Drainage of agricultural lands,' Amer. Soc. Agronomy, Madison, Wisconsin.

Lutton, R. J. (1969) Fractures and failure mechanics in loess and applications to rock mechanics, *U.S. Army Engineer Waterways Experiment Station Research Report S–69–1*, Vicksburg, Miss., U.S.A.

Lutton R. J. (1970) Rock slope charts from empirical slope data, *Trans. Am. Soc. Min. Engrs.*, **247**, 160–162.

Lutton, R. J and Banks, D. C. (1970) Study of clay shale slopes along the Panama Canal, Report 1: East Culebra and West Culebra Slides and the Model Slope, *U.S. Army Engineer Waterways Experiment Station Technical Report*, S–70–9.

Lykoshin, A G., Yaschenko, A. D., Mikhailov, A. D., Savitch, A. I. and Kop, V. J. (1971) Investigation of rock jointing by seismo-acoustic methods, *Symp. Soc. Int. Mécanique des Roches*, Nancy, Paper 1.19.

Lyman, A. K. B. (1942) Compaction of cohesionless foundation soils by explosives, *Trans. Amer. Soc. Civ. Engs.*, **107**, 1330–1342.

Lynn, B. C. (1973) 'An investigation of the slope failure affecting the Carrville link road.' Dissertation, M.Sc. Advanced Course in Engineering Geology, University of Durham, England.

Maasland, M. and Haskew, H. C. (1975) The auger hole method of measuring the hydraulic permeability of soil and its application to drainage problems, *3rd Cong. Int. Comm. Irrig. Drainage*, Question 8.69–8.114.

MacDonald, D. F. (1915) Some engineering problems of the Panama Canal in their relation to geology and topography, *U.S. Bureau of Mines Bulletin*, No. 86.

MacDonald, G. J. E. (1960) Orientation of anisotropic minerals in a stress field, *Geol. Soc. Am. Mem* **79**, 1–8.

Mahtab, M. A. and Goodman, R. E. (1970) Three dimensional finite element analysis of jointed rock slopes, *Proc. 2nd Congress, Int. Soc., Rock Mech.*, Belgrade, **3**, 353–360.

Marachi, H. D., Chan, G. K. and Seed, H. B. (1972) Evaluation of properties of rockfill materials, *J. Soil Mech. Found. Div. ASCE.*, **98**, SM1, 95–112.

Marsal, R. J. (1961) Behaviour of a sandy uniform soil during the Jaltipan earthquake, Mexico, *Proc. 5th Int. Conf. Soil Mech. Found. Eng.*, Paris, **1**, 229–234.

Marshall, A. (1969) 'Glacial deposits of the Team Wash,' Dissertation, M.Sc. Advanced Course in Engineering Geology, University of Durham.

Marsland, A. (1957) The design and construction of earthen flood banks, *J. Inst. Water Engs.*, **11**, 236–258.

Marsland, A. (1971a) Large in-situ tests to measure the properties of stiff fissured clays, *Proc. 1st Australia-New Zealand Conf. Geomechanics,* **1**, 180–189.(Building Res. Estab., U.K. Current paper No. 1/73).

Marsland A. (1971b) 'Laboratory and in-situ measurements of deformation moduli of London Clay.' In: *Interaction of Structure and Foundation, Midland Soil Mech. Found. Eng. Soc.,* 7–17 (Building Res. Estab., U.K. Current Paper, no. 24/73.)

Marsland, A. (1972a) Clays subjected to in-situ plate tests. *Ground Engineering,* **5**, no. 6, 24–31.

Marsland, A. and Butler, M. E. (1967) Strength measurements on stiff, fissured Barton Clay from Fawley, Hampshire, *Proc. Geotech. Conf.,* Oslo, **1**, 131–138.

Martin, G. (1849) *The 'Undercliff' of the Isle of Wight; its climate, history and natural productions,* John Churchill, London.

Martin, G. R. and Kayes, T. J. (1971) Effects of anisotropy and sample disturbance on the "$\phi_u = 0$" stability analysis, *Proc. 1st Australia – New Zealand Conference on Geomechanics,* Melbourne, **1**, 349–354.

Martna, J. (1970) Engineering problems in rocks containing pyrrhotite, 'Large Permanent Underground Openings,' *Proc. Int. Symp.,* Oslo, 1969, Eds., T. L. Brekke and F. A. Jorstad, 87–92.

Masure, P. (1970) Comportement mécanique des roches à anisotropie planaire discontinué, *Proc. 2nd Cong. Int. Soc. Rock Mechanics,* Belgrade, **1**, 197–208.

Mayer, A. (1954) Joseph Boussinesq 1842–1929, *Géotechnique,* **4**, 3–5.

McCann, D. N. and Smith, D. T. (1973) Geotechnical mapping of sea floor sediments by the use of geophysical techniques. *Trans. Instn. Min. Metall.,* **81**, B155–B167.

McClintock, F. A. and Walsh, J. B. (1962) Friction on Griffiths cracks in rocks under pressure, *Proc. 4th Nat. Congr. Applied Mech.,* Berkeley, Calif., U.S.A., 1015–1021.

McCracken, D. D. (1967) 'Fortran with engineering applications,' Wiley, New York.

McCulloch, D. S. and Bonilla, M. G. (1967) Railroad damage in the Alaska earthquake, *Jl. Soil Mech. Found. Div., Amer. Soc. Civ. Engrs.,* **93**, 89–100.

McDowell, P. W. (1970) The advantages and limitations of geophysical methods in the foundation investigation of the tracked hovercraft experimental site in Cambridgeshire, *Jl. Eng. Geol.,* **3**, 119–126.

McGown, A. (1971) The classification for engineering purposes of tills from moraines and associated land forms, *Q. Jl. Eng. Geol.,* **4**, 115–130.

McGown A., Saldivar-Sali, A. and Radwan, A. M. (1974) Fissure patterns and slope failures in till at Hurlford, Ayrshire, *Q. Jl. Eng. Geol.,* **7**, 1–26.

McGregor, K. (1968) 'Drilling of rocks.' McLaren, London.

McHargue, C. J. and Jetter, L. K. (1960) Use of axis distribution charts to represent cold rolled thorium sheet textures, *Trans. Metal. Soc., Amer. Inst. Min. Eng.,* **218**, 550–553.

McKinlay D. G. (1969) Engineering properties of boulder clays, *Geol. Soc. of London Engineering Group Regional Meeting,* Glasgow, 25th September, 1969.

McKinney, J. R. and Gray, G. R. (1963) 'The use of drilling mud in large diameter construction borings.' In: *Grouts and drilling muds in engineering practice,* Butterworths, London, 218–221.

McWilliams, J. R. (1966) The role of microstructure in the physical properties of rock, Testing Techniques for Rock Mechanics, ASTM Special Tech. Publ., **402**, *Amer. Soc. Testing Materials,* 175–189.

Mellis, O. (1942) Gefugediagramme in Stereographischer Projecktion, *Zeitschr. Min. Pet. Mitt.,* **53**, 330–353.
Ménard, L. (1966) Use of pressuremeter to study rock masses, *Rock Mechanics and Engineering Geology,* **4**, 160–171.
Ménard, L. (1972) La consolidation dynamique des remblais récents et sols compressibles, *Travaux,* **54**, No. 452, 56–60.
Mencl, V. (1966) Mechanics of landslides with non-circular slip surfaces with special reference to the Vaiont slide, *Geotechnique,* **16**, 329–337.
Mendes, F. de M. (1966) About the anisotropy of schistose rocks, *Proc. 1st Congress Int. Soc. of Rock Mech.,* Lisbon, **1**, 607–611.
Merriam, R. (1960) Portuguese Bend Landslide, Palso Verdes Hills, California, *Journal Geol.,* **68**, 140–153.
Merriam, R., Rieke, H. H., and Young, C. K. (1970) Tensile strength related to mineralogy and texture of some granitic rocks, *Engineering Geol.,* **4**, 155–160.
Merrill, R. H., Williamson, J. V., Ropchan, D. M. and Kruse, G. H. (1964) Stress determinations by flatjack and borehole deformation methods, *U.S. Bureau of Mines, Report of Investigations,* RI 6400.
Mesri, G. and Olson, R. E. (1970) Shear strength of montmorillonite, *Géotechnique* **20**, 261–270.
Meyerhof, G. G. (1956) Penetration tests and bearing capacity of cohesionless soils, *J. Soil Mech. and Found. Div.,* Amer. Soc. Civil Engrs., **82**, SM1 paper 866.
Middlebrooks, T. A. (1936) Foundation investigation of the Fort Peck Dam closure section, *Proc. 1st Int. Conf. Soil Mech. Found. Engrg.,* Cambridge, Mass. **1**, 135–145.
Midgley, D. C. and Pitman, W. V. (1969) 'Surface water resources of South Africa,' Hydrological Res. Unit. University of Witwatersrand, Rept. 2/69.
Mielenz, R. C. (1948) 'Petrography and engineering properties of igneous rocks,' U.S Bureau of Reclamation, Eng. Monograph No. 1, Denver, Colorado.
Millard, R. S. (1962) Road building in the tropics. *J. Applied Chemistry,* **12**, 342–357.
Millard, R. S. (1967) Road Planning, *Road International,* **66**, 12–18.
Miller, D. L. and Clark, R. A. (1965) Flood studies, *Design of Small Dams,* Publ. U.S. Bureau of Reclamation, 19–61.
Milligan, V., Soderman, L. G. and Rutka, A. (1962) Experience with Canadian varved clays, *J. Soil Mech. Found. Div., ASCE,* **88**, SM4, 31–67.
Mirata, T (1969) A semi-empirical method for determining stresses beneath embankments, *Géotechnique,* **19**, 188–204.
Mitchell, J. K. (1956) The fabric of natural clays and its relation to engineering properties, *Proc. Highway Research Board,* Washington, D.C. (NAS-NRC Publ. 426), **35**, 693–713.
Mitchell, J. K. (1969) Structural and physico-chemical basis for the mechanical properties of clays, *Proc. 7th Int. Conf. Soil Mech. Found. Engrg.,* Mexico City, **3**, 458–461.
Mitchell, J. K. (1970) In place treatment of foundation soils, *J. Soil Mech. Found. Div., Amer. Soc. Civ. Engrs.,* **96**, 73–110.
Mitchell, R. T. (1956) The fabric of natural clays and its relation to engineering properties, *Proc. Highway Res. Board,* **35**, 693–713.
Mogilevskaya, S. E. (1974) Morphology of joint surfaces in rock and its importance for engineering geological examination of dam foundations, *Proc. 2nd Int. Congr. Int. Assoc. of Eng. Geology,* São Paulo, Brazil, **2**, Th VI, 17.1–17.8.

Mohr, E. C. J. and Van Baren, F. A. (1954) *Tropical soils,* Interscience, London.
Monahan, C. J. (1963) Discussion 'An approach to rock mechanics,' *Proc. Amer. Soc. Civ. Eng.,* **89**, SM1, 306–307.
Mononobe, N. (1929) Earthquake-proof construction of masonry dams, *Proc. World Engineering Conf.,* **9**, 275.
Mooney, H. M. and Wetzel, W. W. (1956) *The potentials about a point electrode and apparent resistivity curves for a two-, three- and four-layer earth,* Univ. of Minnesota Press, Minneapolis, U.S.A.
Moore, C. A. (1968) 'Mineralogical and pore fluid influences on deformation mechanisms in clay soils,' Ph.D. thesis, Univ. of California, Berkeley.
Moore, J. F. A. (1974) Mapping major joints in the Lower Oxford clay using terrestrial photogrammetry, *Quart. Jl. Eng. Geol.,* **7**, 57–67.
Moos, A. von, (1953) The subsoil of Switzerland, *Proc. 3rd Int. Conf. Soil Mech. Found. Engrg.,* Zurich, **3**, 252–264.
Morgenstern, N. R. (1969) Structural and physico-chemical effects on the properties of clays, *Proc. 7th Int. Conf. Soil Mech. Found. Eng.,* Mexico, **3**, 455–471.
Morgenstern, N. R. (1969) Shear strength of stiff clay, *Proc. Geotechnical Conf.* Oslo, **2**, 59–69.
Morgenstern, N. R. and Phukan, A. L. T. (1966) Non-linear deformation of a sandstone, *Proc. 1st Cong. Int. Soc. Rock Mechs.,* Lisbon, **1**, 543–548.
Morgenstern N. R. and Price, V. E. (1965) The analysis of the stability of general slip surfaces, *Géotechnique,* **15**, 79–93.
Morgenstern, N. R. and Tchalenko, J. S. (1967a) The optical determination of preferred orientation in clays and its application to the study of microstructure in consolidated kaolin, I. *Proc. Roy. Soc.,* **A300**, 218–234.
Morgenstern, N. R. and Tchalenko, J. S. (1967b) The optical determination of preferred orientation in clays and its application to the study of microstructure in consolidated kaolin, II, *Proc. Roy. Soc.,* **A300**, 235–250.
Morgenstern, N. R. and Tchalenko, J. S. (1967c) Microscopic structures in kaolin subjected to direct shear, *Géotechnique,* **17**, no. 4, 309–328.
Morgenstern, N. R. and Tchalenko, J. S. (1967d) Microstructural observations on shear zones from slips in natural clays, *Proc. Geotechn. Conf. on shear strength properties of natural soils and rocks,* Oslo, **1**, 147–152.
Morgenstern, N. R. and Vaughan, P. R. (1963) 'Some observations on allowable grouting pressures,' In: *Grouts and Drilling Muds in Engineering Practice,* Butterworths, London, 36–43.
Morris, G. (1950a) Vibrations due to blasting and their effects on building structures, *The Engineer,* **190**, 394–395, 414–418.
Morris, G. (1950b) The reduction of ground vibrations from blasting operations, *Engineering,* **169**, 430–433.
Mossman, R. W., Heim, G. E., and Dalton, F. E. (1973) Vibroseis applications to engineering work in an urban area, *Geophysics,* **38**, 489–499.
Muchowski, J. and Sztyk, Z. (1974) The contribution of internal erosion processes in the development of morphology of loess slopes, *Proc. 2nd Int. Congr. Int. Assoc. of Eng. Geology,* São Paulo, Brazil, **2**, Th V, 25.1–25.10.
Muirhead, K. J., Cleary, J. R. and Simpson, D. W. (1973) Seismic activity associated with the filling of Talbingo reservoir, *Summary, Int. Colloquium on Seismic Effects of Reservoir Impounding* (COSERI), Royal Soc., 27–29 March.
Muller, L. (1963) Discussion on 'An approach to rock mechanics,' *Proc. Amer. Soc. Civ. Engrs.,* **89**, SM2, 137–139.

Mullineaux, D. R. and Crandell, D. R. (1962) Recent lahars from Mount St. Helens, Washington, *Bull. Geol. Soc. Amer.*, **73**, 855–869.

Munson, R. C. (1970) Relationship of effect of water flooding of the Rangeley oil field on seismicity, *Engineering Geology Case Histories, No. 8, Engineering Seismology: The works of Man*, Ed. W. M. Adams, *Geol. Soc. Amer.*, 39–49.

Murrell, S. A. F. (1966) The effect of triaxial stress systems on the strength of rocks at atmospheric temperatures, *Geophys. J. Roy. Astron. Soc.*, **10**, 231–281.

Muskat, M. (1947) 'The flow of homogeneous fluids through porous media,' Edwards, Ann Arbour, Michigan, U.S.A.

Nadai, A. (1952) 'Theory of fracture and flow of solids,' 2nd edn. Vol. 1, McGraw-Hill, New York.

Nakano, R. (1967) On weathering and change of properties of Tertiary mudstone related to landslide, *Soils and Foundations*, **7**, 1–14.

Nash E. A (1959) Discussion on: 'A study of the bank-full discharges of rivers in England and Wales,' *Proc. Inst. Civ. Engrs.*, **14**, 403–406.

Nasmith, H. (1964) Landslides and Pleistocene deposits in the Meikle River Valley of Northern Alberta, *Canadian Geotech. Journ.*, **1**, 155–166.

National Coal Board, (1963) 'Subsidence Engineers' Handbook,' National Coal Board, London.

National Resources Committee, (1939) *Low Dams: A manual of design for small water storage projects* U.S. Govt., Printing Office, Washington

Nath, B. (1969) Hydrodynamic pressures on high gravity dams during vertical earthquake motions, *Proc. Inst. Civ. Engrs.*, **42**, 413–422.

Nath, B. (1971) Structural and hydrodynamic coupling for a gravity dam during vertical earthquake motions, *Dynamic Waves in Civil Engineering*, Eds. D. A. Howells, I. P. Haigh and C. Taylor, Wiley Interscience, London, 421–440.

Neelands, R. J and James, A. N. (1963) Formulation and selection of chemical grouts with typical examples of their field use, *Grouts and drilling muds in engineering practice*, 150–155, Butterworths, London.

New Civil Engineer (1974) Esso's giant oil tanks – a question of more haste less speed, *New Civil Engineer*, no. 81, 28 February, 1974.

Newland, D. H. (1916) Landslides on unconsolidated sediments, *Bull. N.Y. State Museum*, **187**, 79–105.

Newmark, N. M. (1942) Influence charts for computation of stresses in elastic foundations, *Univ. of Illinois, Eng. Experimental Stn.*, Bull. No. **338**.

Newmark, N. M. (1947) Influence charts for computation of vertical displacements in elastic foundations, *Univ. of Illinois, Eng. Experimental Stn., Bull. No. 367*.

Newmark, N. M. (1965) Effects of earthquakes on dams and embankments, *Géotechnique*, **15**, 139–159.

Nixon, M. (1959a) A study of the bank-full discharges of rivers in England and Wales, *Proc. Inst. Civ. Engrs.*, **12**, 157–173.

Nixon, M. (1959b) Discussion on: 'A study of the bank-full discharges of rivers in England and Wales,' *Proc. Inst. Civ. Engrs.*, **14**, 416–425.

Nonveiller, E. (1965) The stability analysis of slopes with a slip surface of general shape, *Proc. 6th Conf. Soil Mech. Found. Engng.*, Montreal, **2**, 522–525.

Nonveiller, E. (1968) Shear strength of bedded and jointed rock as determined from Zalesina and Vajont slides, 'Shear strength properties of natural soils and rocks,' *Proc. Geotech. Conf.*, Oslo, **1**, 289–294.

Noorany, I. (1972) Underwater sampling and testing – a state-of-the-art review. 'Underwater sampling, testing and construction control,' *Amer. Soc. Testing and Matl.,* STP No. **501**.

Noorany, I. and Gizienski, S. F. (1970) Engineering properties of submarine soils: state of the art review, *J. Soil Mech. Found. Div., Amer. Soc. Civ. Engrs.*, **96**, 1735–1762.

Norman, J. W. (1969) Photo-interpretation of boulder clay areas as an aid to engineering geological studies, *Q. Jl. Eng. Geol.*, **2**, 149–157.

Norton, J. T. (1948) A technique for quantitative determination of texture of sheet metals, *J. Appl. Phys.*, **19**, 1176–1178.

Nur, A. (1971) Effects of stress on velocity anisotropy in rocks with cracks, *J. Geophys. Res.*, **76**, 2022–2034.

Nye, J. (1951) The flow of glaciers and ice sheets as a problem in plasticity, *Proc. Roy. Soc., Series A*, **207** 554–572.

Obert, L. (1939) Measurement of pressures on rock pillars in underground mines, Part I, *U.S. Bureau of Mines, R.I.* **3444**.

Obert, L. (1940) Measurement of pressures on rock pillars in underground mines, Part II, *U.S. Bureau of Mines, R.I.* **3521**.

Obert, L. (1941) Use of subaudible noise for prediction of rock bursts, *U.S. Bureau of Mines, R.I.* **3555**.

Obert, L. and Duvall, W. I. (1942) Use of subaudible noise for prediction of rock bursts, Part II *U.S. Bureau of Mines, R.I.* **3654**.

Obert, L. and Duvall, W. I. (1945a) Microseismic methods of predicting rock failure in underground mining, Part I, General Method, *U.S. Bureau of Mines, R.I.* **3799**.

Obert, L. and Duvall, W. I. (1945b) Microseismic method of predicting rock failure in underground mining, Part II, Laboratory experiments, *U.S. Bureau of Mines, R.I.* **3803**.

Obert, L. and Duvall, W. I. (1967) *Rock Mechanics and the Design of Structures in Rock*, Publ. Wiley & Co., New York.

Obert, L. and Rich, C. (1971) Classification of rock for engineering purposes, *Proc. 1st Australia-New Zealand Conf. on Geomechanics*, Melbourne, **1**, 435–441.

Obert, L., Windes, S. L. and Duvall, W. I. (1946) Standardised tests for determining the physical properties of mine rocks, *U.S. Bureau of Mines, Report of Investigation No.* **3891**.

Oborn, L. E. (1974) Seismic phenomena and engineering geology, *Proc. 2nd Int. Congr. Int. Assoc. of Eng. Geology*, São Paulo, Brazil, **1**, Th II, GR–1–41.

Odenstad, S. (1951) The landslide at Skottorp on the Lidan River, February 2, 1946, *Proc. Roy. Swedish Geotech. Inst.*, No. **4**.

Okabe, S. (1926) General theory of earth pressure, *Journ. Japanese Soc. of Civil Engineers*, **12**, 1.

Okamoto, S. (1973) Relationship between the earthquake occurrence and the reservoir impounding in Japan, *Summary, Int. Colloquium on Seismic Effects of Reservoir Impounding (COSERI), Royal Soc.*, 27–29 March.

Okeson, C. J. (1964) Geologic requirements of the foundations of large dams, *Proc. 8th Int. Conf. on Large Dams*, Edinburgh, **1**, 73–85.

Olphen, H. van (1963a) Compaction of clay sediment in the range of molecular particle distances, *Proc. 11th Natl. Conf. Clays and Clay Minerals*, Pergamon Press, New York, **13**, 178–189.

Olsen, H. W. (1962) Hydraulic flow through saturated clay, *Proc. 9th Nat. Conf. Clays and Clay Minerals*, 131–161.

Olson, R. E. (1962) Shear strength properties of calcium illite, *Géotechnique*, **12**, 23–43.
Olson, R. E. (1963) Shear strength properties of a sodium illite, *J. Soil Mech. Found. Div., ASCE.*, **89**, SM1, 183–208.
Olson, R. E. and Hardin, J. (1963) Shearing properties of remoulded sodium illite, *Proc. 2nd Pan. Am. Conf. Soil Mech.*, **1**, 203–218.
Olson, R. E and Mitronovas, F. (1962) Shear strength and consolidation characteristics of calcium and magnesium illite, *Proc. 9th Nat. Conf. on Clays and Clay Minerals*, 185–209, Pergamon Press, Oxford.
O'Neill, A. L. (1963) Slope failures in foliated rocks, Bukte County, *California Natl. Research Council – Highways Research Board, Research Record*, no. 17 40–42.
Onodera, T. (1963) Dynamic investigation of foundation rocks 'in situ', *Proc. 5th Symp. Rock Mech.*, Minnesota, Pergamon, New York, 517–533.
Onodera, T., Yoshinaka, R. and Kazama, H, (1974) Slope failures caused by heavy rainfall in Japan, *Proc. 2nd Int. Congr. Int. Assoc. of Eng. Geology*, São Paulo, Brazil, 2, Th V–11.1–11.10.
Otto, G. H. (1938) The sedimentation unit and its use in field sampling, *J. Geol.*, **46**, 509–582.
Over, V. G. (1969) 'X ray fabric analysis of laboratory sedimented and consolidated monomineralic clay,' M.Sc. Advanced Course Dissertation, University of Durham, England.
Palache, C., Berman, H. and Frondel, C. (1944) *Dana's system of mineralogy*, 7th Ed., **1**, Wiley, New York.
Palit, R. (1953) Determination of swelling pressure of Black Cotton soil – a method, *Proc. 3rd Int. Conf. Soil Mech. Found. Eng.*, Zurich, **1**, 170–172.
Palladino, D. J. (1972) Slope failures in an overconsolidated clay, Seattle, Washington, *Géotechnique*, **22**, no. 4, 563–595.
Palmer, C. (1911) The geochemical interpretation of water analyses, *U.S. Geol. Surv. Bulletin* **479**.
Panek, L. A. (1961) Methods of determining rock pressure, *Proc. 4th Symp. on Rock Mechanics*, Pennsylvania State Univ., Ed. H. L. Hartman, 181–184.
Paolo, B. and Armando, S. (1970) Investigations on the characteristics of rock masses by geophysical methods, *Proc. 2nd Cong., Int. Soc. of Rock Mechanics*, Belgrade, 489–500.
Papageorgiou, S. A. (1974) 'A laboratory investigation of the shear strength and compressibility of rockfill,' M.Sc. Advanced Course Dissertation, University of Durham.
Parnasis, D. S. (1972) 'Principles of Applied Geophysics,' 2nd edition, Chapman & Hall, London.
Patterson, M. S. and Weiss, L. E. (1961) Symmetry concepts in the structural analysis of deformed rocks, *Geol. Soc. Amer. Bull.*, **72**, 841–882.
Patton, F. D. (1966a) 'Multiple modes of shear failure in rock and related materials,' Ph.D. thesis, University of Illinois, Urbana, U.S.A.
Patton, F. D. (1966b) Multiple modes of shear failure in rock, *Proc. 1st Congress, Int. Soc. Rock Mechanics*, Lisbon, **1**, 509–513.
Patton, F. D. and Deere, D. U. (1970) Significant geologic factors in rock slope stability, *Proc. Symp. on Planning Open Pit Mines*, Ed. D. W. J. Van Rensburg, Publ. A. A. Balkema, Cape Town, 143–151.
Peck, R. B. (1969a) Deep excavations and tunnelling in soft ground, State of the Art Report, *7th Int. Conf. on Soil Mech. and Found Engrg.*, Mexico City, 225–290.

Peck, R. B. (1969b) Advantages and limitations of the observational method in applied soil mechanics, *Géotechnique,* **19**, no. 2, 171–187.
Peck, R. B., Deere, D. U., Monsees, J. E., Parker, H. W. and Schmidt, B. (1969) Some design considerations in the solution of underground support systems, *Final Report for U.S. Dept. of Transportation,* Washington, D.C., Contract No. 3-0152.
Peck, R. B., Hanson, W. E. and Thorburn, T. H. (1953) 'Foundation Engineering,' Wiley, New York.
Pellegrino, A. (1965) Geotechnical properties of coarse-grained soils, *Proc. 6th Int. Conf. on Soil Mech. and Found. Engrg.,* **1**, 87–92.
Peng, S. and Johnson, A. M. (1972) Crack growth and faulting in cylindrical specimens of Chelmsford granite, *Int. J. Rock Mech. Min. Sci.,* **9**, 37–86.
Penman, A. D. M. (1972) Effect of the position of the instrumentation for embankment dams subjected to rapid drawdown, *Current Paper CP 1/72.* Building Research Establishment, U.K.
Penman, A. D. M. and Charles, J. A. (1972a) Effect of the position of the core on the behaviour of two rock fill dams. *Current Paper CP 18/72,* Building Research Establishment, U.K.
Penman, A. D. M. and Charles, J. A. (1972b) Constructional deformations in a rockfill dam. *Current Paper CP 19/72,* Building Research Establishment, U.K.
Penny, L. F., Coope, G. R. and Catt, J. A. (1969) Age and insect fauna of the Dimlington Silts, East Yorkshire, *Nature,* **224**, 65–66.
Pentecost, J. L and Wright, C. H. (1963) 'Preferred orientation in ceramic materials due to forming techniques,' *Advances in X-ray Analysis, Proc. 12th Amer. Conf. on Applications of X-ray Analysis,* Eds. W. M. Mueller, G. Mallet and M. Fay, 174–181.
Perkins, T. K. and Krech, W. W. (1966) Effect of cleavage levels and stress level on apparent surface energies of rocks, *J. Soc. Pet. Engrs.,* AIME, **6**, 308–314.
Perkins, T. K. and Krech, W. W. (1968) The energy balance concept of hydraulic fracturing, *J. Soc. Pet. Engrs.,* AIME, **8**, 1–12.
Perrin, R. M. S. (1971) 'The clay mineralogy of British sediments,' Mineralogical Society (Clay Mins. Group), London.
Perrott, W. E. and Lancaster-Jones, P. F. F. (1963) Case Records of Cement Grouting, *Grouts and Drilling Muds in Engineering Practice,* Butterworths, London, 80–84.
Peterson, R. (1958) Rebound in the Bearpaw Shale, Western Canada, *Bull. Geol. Soc. America,* **69**, 1113–1124.
Peterson, R., Jaspar, J. L., Rivard, P. J. and Iverson, N. L. (1960) Limitations of laboratory shear strength in evaluating stability of highly plastic clays, *A.S.C.E., Research Conference on Shear Strength of Cohesive Soils,* Boulder, Colorado, 765–791.
Petrovski, J., Paskalov, T. and Jurukovski, D. (1974) Dynamic full-scale test of an earthfill dam, *Géotechnique,* **24**, 193–206.
Pettijohn, F. J. (1957) 'Sedimentary rocks,' 2nd Edn., Harper Bros. New York.
Pettijohn, F. J. and Potter, P. E. (1964) 'Atlas and glossary of primary sedimentary structures,' Springer-Verlag, Berlin.
Phillips, F. C. (1971) 'The use of stereographic projection in structural geology,' 3rd Edition, Edward Arnold, London.
Philofsky E. M. and Hilliard, J. E. (1969) The measurement of the orientation distribution of lineal and areal arrays, *Quart. J. Appl. Math.,* **27**, 79–86.
Philpott, K. D. (1970) 'Suction and swelling pressures of certain argillaceous materials,' Dissertation, M.Sc. Advanced Course in Engineering Geology, University of Durham

Pickering, D. J. (1970) Anisotropic elastic parameters for soil, *Géotechnique*, **20**, 271–276.
Pincus, H. J. (1966) Optic processing of vectorial rock fabric data, *Proc. 1st Congress, Int. Soc., of Rock Mech.*, Lisbon, **1**, 173–177.
Pincus, H. J. (1969a) Sensitivity of optical data processing to changes in rock fabric, Part I, Geometric patterns, *Int. J. Rock Mech. Min. Sci.*, **6**, 259–268.
Pincus, H J. (1969b) Sensitivity of optical data processing to changes in rock fabric, Part II, Standardized grain patterns, *Int. J. Rock Mech. Min. Sci.*, **6**, 269–275.
Pinder, G. F. and Bredehoeft, J. D. (1968) Application of the digital computer for aquifer evaluation, *Water Resources Res.*, **4**, 1069–1093.
Pinto, J. L. (1970) Deformability of schistous rocks, *Proc. 2nd Congress, Int. Soc. Rock Mechanics*, Belgrade, **1**, 491–496.
Piper, A. M. (1953) A graphic procedure in the geochemical interpretation of water analyses, *U.S. Geol. Survey, Ground Water*, Note 12.
Pirsson, L. and Knopf, A. (1948) 'Rocks and rock minerals,' Wiley & Sons, New York.
Pitcher, W. S , Shearman, D. J. and Pugh, D. C. (1954) The loess of Pegwell Bay, Kent and its associated frost soils, *Geol Mag.*, **91**, 308–314.
Piteau, D. R. (1970) Geological factors significant to the stability of slopes cut in rock, *Proc. Symp. on Planning Open Pit Mines*, Ed. P. W. J. Van Rensburg, Publ. A. A. Balkema, Cape Town, 33–53.
Pitty, A. F. (1970) 'Introduction to geomorphology,' Methuen, London.
Poland, J. F. and Green, J. H. (1962) Subsidence in the Santa Clara Valley, California, A progress report, *U.S. Geological Survey, Water Supply Paper* 1619–C.
Pomeroy, C. D., Hobbs, D. W. and Mahmoud, A. (1970) The effect of weakness-plane orientation on the fracture of Barnsley Hards by triaxial compression, *Int. J. Rock Mech. Min. Sci.*, **8**, 227–238.
Poskitt, T J. (1970) Settlement charts for anisotropic soils, *Géotechnique*, **20**, 325–330.
Potts, E. L. J. (1956) A scientific approach to strata control, *Trans. Inst. Min. Engrs.*, **116**, 114–129.
Potts, E. L. J. (1964a) The in situ measurement of rock stress based on deformation measurements, 'State of stress in the earth's crust,' 397–409, Elsevier, Amsterdam.
Potts, E. L. J. (1964b) Current investigations in rock mechanics, *4th Int. Conf. Strata Control and Rock Mechs,* Columbia University, New York, 29–45.
Poulos, H. G. (1967) The use of the sector method for calculating stresses and displacements in an elastic mass, *Proc. 5th Australian-New Zealand Conf. on Soil Mech.*, Auckland, 198–204.
Prater, E. G. (1973) Die Gewölbewirkung der Schlitzwände, *Der Bauingenieur*, **48**, 125–131.
Price, D. G. and Knill, J. L. (1967) The engineering geology of Edinburgh Castle rock, *Géotechnique*, **17**, 411–432.
Price, N. J. (1959) Mechanics of jointing in rocks, *Geol. Mag.*, **96**, 149–167.
Price, N. J. (1960) Compressive strength of Coal Measure rocks, *Colliery Engineering*, **37**, 283–289.
Price, N. J. (1966) 'Fault and joint development in brittle and semi-brittle rock,' Pergamon Press, Oxford.
Priest, S. D. (1974) Private communication.

Priest, S. D. and Hudson, J. A. (1976) Discontinuity spacings in rock (In Press).
Prokopovich, N. P. (1974) Land subsidence and pollution, *Proc. 2nd Int. Congr. Int. Assoc. of Eng. Geology*, São Paulo, Brazil, 1, Th III–2.1–2.10.
Prosser, J. R. and Grant, P. A. St.C. (1968) The Tyne Tunnel I – Planning of the scheme, *Proc. Inst. Civ. Engrs.*, **39**, 193–212.
Pruška, L. and Thú, L. (1974) The genesis of slip lines in slopes. *Proc. 2nd Int. Congr. Int. Assoc. of Eng Geology*, São Paulo, Brazil, 2, Th V–12.1–12.4.
Pryor, E. J. (1965) 'Mineral Processing' (3rd edn.), Elsevier, Amsterdam.
Pugh, S. F. (1967) The fracture of brittle materials, *British J. Appl. Phys.*, **18**, 129–161.
Pusch, R. (1966) Quick clay microstructure. *Engineering Geology*, **3**, 433–443.
Pusch, R. (1964) On the structure of clay sediments, *Nat. Swed. Council Build. Res.*, Handlingar, No. 48.
Rabcewicz, L. von (1959) Die Ankerung im Tunnelbau ersetzt bisher gebräuchliche Einbaumethoden, *Schweizerische Bauzeitung*, no. 9, 123.
Raffle, J. F. and Greenwood, D. A. (1961) Relations between the rheological characteristics of grouts and their capacity to permeate soil, *Proc. 5th Int. Conf. Soil Mech. and Found. Engrg.*, **2**, 789–794.
Rana, M. H. and Bullock, W. D. (1969) The design of open pit slopes, *Can. Min. J.*, **90**, 58–62.
Ramez, M. R. H. and Attewell, P. B. (1963) Shock deformation of rocks, *Geophysics*, **28**, 1020–1036.
Randell, P. A. (1972) 'The modelling of discontinuous rock masses,' M.Sc. Advanced Course Dissertation, University of Durham, England.
Rankilor, P. R. (1974) A suggested field system of logging rock cores for engineering purposes, *Bull. Assoc. Eng. Geol.*, **11**, 247–258.
Ranganatham, B. V. (1961) Soil structure and consolidation characteristics of Black Cotton soil, *Géotechnique*, **11**, 333–338.
Rankine, W. J. M. (1857) On the stability of loose earth, *Phil. Trans. Roy. Soc. Lond.*, **147**, 9–28.
Rapp, A. (1960) Talus slopes and mountain walls at Templefjorden, Spitzbergen, *Norks Polarinstitutt Skrifter*, No. **119**.
Raw, G. (1940) 'Proposed crossing under the River Tyne in the neighbourhood of Jarrow and Howdon,' Mining report presented to the Durham County Council and to the Northumberland County Council.
Reeve, R. C., and Kirkham, D. (1951) Soil anisotropy and some field methods for measuring permeability, *Trans. Amer. Geophys. Union*, **32**, 582–590.
Reeves, G. M. (1973) 'Discontinuities in rock and clay shale masses,' Dissertation, M.Sc. Advanced Course in Engineering Geology, University of Durham.
Rehbinder, P. A., Schreiner, L. A. and Zhigach, K. (1948) 'Hardness reducers in drilling.' Translation – C.S.I.R., Melbourne.
Remillon, A. (1955) Stabilisation of granular materials, *Highway Research Board, Bull.*, **108**, 96–101.
Reséndiz, D. and Zonana, J. (1969) The short term stability of open excavations in Mexico City clay, *Nabor Carrillo Volume, Secretaria de Hacienda y Credito Publico*, Mexico, 203–227.
Ricceri, G. and Butterfield, R. (1974) An analysis of compressibility data from a deep borehole in Venice, *Géotechnique*, **24**, 175–192.
Richards, A. F. (Ed.) (1967) 'Marine Geotechnique,' Univ. of Illinois Press, Urbana.

Richards, A. F., McDonald, V. J., Olson, R. E. and Keller, G. H. (1972) In-place measurement of deep sea soil shear strength. *Underwater soil sampling, testing and construction control,* Amer. Soc. Testing and Matls., S.T.P. No. 501, 55–68.

Richards, H. J. (1970) Ground-water resources in Northern England, *Symposium Report on Geotechnical Aspects of Development and Planning in Northern England,* Ed: P. T. Warren, Yorkshire Geological Society, 14–21.

Richart, F. E. (1957) Review of the theories for sand drains. *J. Soil Mechs. and Found. Div., Amer. Soc. Civ. Engrs.,* 83, SM3, Paper 1301.

Rinehart, J. S. (1962) Effect of transient stress waves in rock, *Int. Symp. on Mining Research* (ed. G. Clarke), Pergamon, Oxford, 713–726.

Rinehart, J. S. and Pearson, J. (1954) 'Behaviour of metals under impulsive loads,' Amer. Soc. Metals, Cleveland.

Ringheim, A. S. (1964) Experiences with Bearpaw shales at the South Saskatchewan, River Dam, *Proc. 8th Int. Cong. on Large Dams,* Edinburgh, 1, 529–550.

Ritchie, A. M. (1963) The evaluation of rockfall and its control, *Highway Record,* 17, 13–28.

Road Research Laboratory (1952) 'Soil mechanics for road engineers,' H.M.S.O., London.

Roberts, A. (1965) The photoelastic glass insertion stress meter, *Engineer,* 220, 164–171.

Roberts, A. (1968) 'Measurement of strain and stress in rock masses,' In: *Rock Mechanics in Engineering Practice,* Wiley, London, 157–202.

Roberts, A., Hawkes, I. and Williams, F. T. (1965) Some field applications of the photoelastic stress meter, *Int. J. Rock Mech. Min. Sci.,* 2, 93–103.

Roberts, A., Hawkes, I., Williams, F. T. and Murrell, S. A. F. (1964) The determination of the strength of rock in situ, *Proc. 8th Int. Cong. on Large Dams,* Edinburgh, 1, 167–186.

Roberts, A., Hawkes, I., Williams, F. T. and Dhir, R. K. (1964) A laboratory study of the photoelastic stressmeter, *Int. J. Rock Mech. Min. Sci.,* 1, 441–457.

Robertson, A. MacG. (1970) The interpretation of geological factors for use in slope theory, *Symp. on Planning Open Pit Mines,* Ed. P. W. J. van Rensburg, Publ. A. A. Balkema, Cape Town, 55–71.

Robertson, E. C. (1964) Viscoelasticity of rock, *State of Stress in the Earth's Crust,* ed. W. R. Judd, Elsevier, New York, 181–233.

Robinson, W. (1969) 'The engineering geology of Durham City,' Dissertation M.Sc. Advanced Course in Engineering Geology, University of Durham, England.

Rocha, M., Seraphim, J. L., Da Silveira, A. F. and Neto, J. M. R. (1955) Deformability of foundation rocks, *Proc. 5th Int. Cong. on Large Dams,* Paris, 3, 531–559.

Rocha, M., Da Silveira, A., Grossmann, N. and De Oliveira, E. (1966) Determination of the deformability of rock masses along boreholes, *Proc. 1st Cong. Int. Soc. Rock Mech.,* Lisbon, 1, 697–704.

Rocha, M. and Silvério, A. (1969) A new method for the complete determination of the state of stress in rock masses, *Géotechnique,* 19, 116–132.

Rodrigues, F. P. (1970) Anisotropy of rocks. Most probable surfaces of the ultimate stresses and the moduli of elasticity, *Proc. 2nd Int. Cong. Int. Soc. Rock Mech.,* Belgrade, 1, 133–142.

Rogers, S. H. (1970) The engineering geology of motorway location and design,

Symp. Report on Geological Aspects of Development and Planning in Northern England, Ed. P. T. Warren, Yorkshire Geol. Soc., 144–157.
Rosa, S. A., Fedorenko, A. N., Eristov, V. S. and Tokachirov, V. A. (1964) Studies of deformation properties of rock foundations of high arch and gravity dams in the USSR, *Proc. 8th Int. Cong. on Large Dams,* Edinburgh, 1, 1023–1040.
Roscoe, K. H. (1970) The influence of strains in soil mechanics, 10th Rankine Lecture, *Géotechnique,* 20, 129–170.
Roscoe, K. H., Schofield, A. N. and Wroth, C. P. (1958) On the yielding of soils, *Géotechnique,* 8, 22–53.
Roscoe, K. H., Schofield, A. N. and Wroth, C. P. (1959) On the yielding of soils, Discussion, *Géotechnique,* 9, 72–83.
Rose, E. (1947) Thrust exerted by expanding ice sheet, *Trans. Amer. Soc. Civ. Engrs.,* 112, 871–900.
Rosenblad, J. L. (1971) Geomechanical model study of the failure modes of jointed rock masses, *Technical Report, MRD 1–71. Dept of the Army, Corps of Engineers, Missouri River Division,* Omaha, Nebraska.
Rosenblueth, A. (1969) Contribution to discussion, *Main Session 4, Proc. 7th Int. Congress on Soil Mechanics and Found. Engrg.,* Mexico City, 3, 230–241.
Rosengren, K. J. (1968) Rock mechanics of the Black Star open cut, Mount Isa, Ph.D. thesis, The Australian National University.
Rosengren, K. J. and Jaeger, J. C. (1968) The mechanical properties of an interlocked low-porosity aggregate, *Géotechnique,* 18, 317–328.
Rosenqvist, I. Th. (1955) Investigations in the clay-electrolyte-water system, *N.G.I. Publ.* No. 9, 83–87.
Rosenqvist, I. Th. (1958) Physical-chemical properties of soil water system (discussion), *J. Soil Mech. Found. Div., Amer. Soc. Civ. Engrs.,* 85, 31–53.
Rosenqvist, I. Th. (1960) The influence of physico-chemical factors upon the mechanical properties of clays, *Proc. 9th Nat. Conf. on Clays and Clay Minerals,* Indiana, U.S.A. (Ed. A. Swineford), 12–27.
Ross, G. A., Seed, H. B. and Migliaccio, R. R. (1969) Bridge foundation damage in Alaska earthquake. *J. Soil Mech. Found. Div., Amer. Soc. Civ. Engrs.,* 95, 1007–1036.
Ross-Brown, D. M. and Atkinson, K. B. (1972) Terrestrial photogrammetry in open-pits: 1 – description and use of the photo-theodolite in mine surveying, *Trans. Inst. Min. Metal.,* 81, A205–A213.
Ross-Brown, D. M., Wickens, E. H. and Markland, J. T. (1973) Terrestrial photogrammetry in open-pits: 2 – an aid to geological mapping, *Trans. Inst Min. Metal.,* 82 A115–A130.
Roussel, J. M. (1968) Theoretical and experimental considerations of the dynamic modulus of rock masses, *Revue de l'Industrie Minerale,* 80, 373–600.
Rowe, P. W. (1959) Measurement of the coefficient of consolidation of a lacustrine clay, *Géotechnique,* 9, 107–118.
Rowe, P. W. (1962) The stress-dilatancy relation for static equilibrium of an assembly of particles in contact, *Proc. Roy. Soc., London,* Series A., 269, 500–527.
Rowe, P. W. (1963) Stress dilatancy, earth pressures and slopes, *J. Soil Mech. Found. Div., Amer. Soc. Civ. Engs.,* 89, No. SM3, 37–61.
Rowe, P. W. (1964) Calculation of consolidation rates of laminated, varved or

layered clays with particular reference to sand drains, *Géotechnique*, **14**, 321–340.

Rowe, P. W. (1968a) Failure of foundations and slopes on layered deposits in relation to site investigation, *Proc. Instn. Civ. Engrs.* Supplementary Paper 7057S. (Summary in **39**, 465–466).

Rowe, P. W. (1968b) The influence of geological features of clay deposits on the design and performance of sand drains, *Proc. Inst. Civ. Engrs.*, Supplementary Paper, 7058S (Summary in **39**, 581).

Rowe, P. W. (1970) Contribution to discussion, *Ground Engineering, Institution of Civil Engineers*, London, 79.

Rowe, P. W. (1971) Representative sampling in location quality and size, In *Sampling of soil and rock, ASTM Spec. Tech. Pub. 483*, 77–106.

Rowe, P. W. (1972) The relevance of soil fabric to site investigation practice, *Géotechnique*, **22**, 195–300.

Ruddock, E. C. (1967) Residual soils of the Kumasi district in Ghana, *Géotechnique*, **17**, 359–377.

Ruiz, M. D. and de Camargo, F. P. (1966) Large scale field test on rock. *Proc. 1st Cong. Int. Soc. Rock Mech.*, Lisbon, **1**, 257–262.

Rummell, F. and Fairhurst, C. (1970) Determination of post failure behaviour of brittle rock using a servo-controlled machine, *Rock Mechs.*, **2**, 189–204.

Russam, K. (1959) Climate and moisture conditions under road pavements, *Regional Conf. for Africa on Soil Mech. and Found. Eng.*, Laurenco Marques, 263–271.

Rydz, B. (1971) Regional water resources analysis, *Proc. Inst. Civ. Engrs.*, **49**, 129–143.

Rzhevsky, V. and Novik, G. (1971) 'The physics of rocks,' Mir, Moscow.

Salamon, M. G. D. and Munro, H. (1967) The study of the strength of coal pillars, *Jl. S. African Inst. Min. Metall.*, **68**, 57–67.

Salas, A. J. and Uriel, S. (1964) Some recent rock mechanics testing in Spain, *Proc. 8th Int. Congress on Large Dams*, Edinburgh, **1**, 995–1021.

Sanger, F. J. (1968) Ground freezing in construction, *J. Soil Mech. Found. Div. Amer. Soc. Civ. Engrs.*, **94**, 131–158.

Sangrey, D. A. (1972) Naturally cemented sensitive soils, *Géotechnique*, **22** 139–152.

Sankaran, K. S. and Bhaskaran, R. (1973) Deformation and failure pattern in an anisotropic kaolinite clay, *Géotechnique*, **23** 113–117.

Sapegin, D. D. and Shiryaev, R. A. (1966) Deformability characteristics of rock foundations before and after grouting, *Proc. 1st Cong. Int. Soc. of Rock Mech.*, Lisbon, **1**, 755–760.

Sarma, S. K. (1973) Stability analysis of embankments and slopes, *Géotechnique*, **23**, 423–433.

Saul, T. and Higson, G. R. (1971) The detection of faults in coal panels by a seismic transmission method, *Int. J. Rock Mech. Min. Sci.*, **8**, 483–499.

Sauvage de Sainte Marc, G., Bouvard, M. and Ma Min-Yuan (1960) Préssions interstitielles dans les galeries en charge. *La Houille Blanche*, **15**, 173–193.

Savanick, G. A. and Johnson, D. L. (1974) Measurement of the strength of grain boundaries in rock, *Int. J. Rock Mech. Min. Sci.*, **11**, 173–180.

Scheidig, A. (1934) *Der Löss und seine geotechnischen Eigenschaftan, Dresden und Leipzig*, Verlag von Theodor Steinkopf, Berlin.

Schlaifer, R. (1959) 'Probability and Statistics for Business Decisions' McGraw-Hill, New York.

Schmidt. B. (1969) 'Settlements and ground movements associated with tunnelling in soil,' Ph.D. thesis, University of Illinois, Urbana.

Schmidt, W. (1925) Gefugestatistik, *Min. Petrog. Mitt.*, **38**, Ch. 23, 1928, 392–423.
Schmidt. W. E. (1967) Field determination of permeability by the infiltration test, *Permeability and Capillarity of Soils, Amer. Soc. Test Mat. S.T.P.* No. **417**, 141–157.
Schofield, A. N. and Wroth, C. P. (1968) 'Critical state soil mechanics,' McGraw-Hill, London.
Schofield, R. K. (1935) The pF of the water in soil, *Trans. 3rd Int. Cong. Soil Sci.*, **2**, 37–48 (and Discussion, **3**, 182–186).
Schultz, E. and Horn, A. (1965) The shear strength of silt, *Proc. 6th Int. Conf. on Soil Mech. and Found. Eng.*, Montreal, **1**, 350–353.
Schulz, L. G. (1949a) A direct method of determining preferred orientation of a flat reflection sample using a geiger counter X-ray spectrometer, *J. Appl. Phys.*, **20**, 1030–1033.
Schulz, L. G. (1949b) Determination of preferred orientation in flat transmission samples using a Geiger counter X-ray spectrometer, *J. Appl. Phys.*, **20**, 1033–1035.
Schulz, L. G. (1964) Quantitative interpretation of mineralogical composition from X-ray and chemical data for the Pierre Shale, *Geol. Survey Professional Paper* 391–C, U.S. Govt. Printing Office, Washington.
Schuster, R. L. (1965) 'Minor structures in the London Clay,' D.I.C. thesis, University of London.
Scopel, L. J. (1970) Pressure injection disposal well, Rocky Mountain arsenal, Denver, Colorado, *Engineering Geology Case Histories No. 8, Engineering Seismology: The Works of Man*, Ed. W. M. Adams, *Geol. Soc. Amer.*, 19–24.
Scott, J. S. and Brooker, E. W. (1968) Geotechnical and engineering aspects of Upper Cretaceous shales in Western Canada, *Paper 66-37, Geological Survey of Canada*, Dept. of Energy, Mines and Resources, Ottawa.
Scott, R. A. (1963a) 'Fundamental considerations governing the penetrability of grouts and their ultimate resistance to displacement,' In: *Grouts and Drilling Muds in Engineering Practice*, Butterworths, London, 10–14.
Scott, R. F. (1970) In place ocean soil strength by acceleration, *J. Soil Mech. Found. Div., Amer. Soc. Civ. Engs.*, **96**, 199–211.
Scrivenor, J. B. (1929) The mudstream ('Lahars') of Gunong Keloet in Java, *Geol. Mag.*, **66**, 433–434.
Seed. H. B. (1968) Landslides during earthquakes due to soil liquefaction, *Journ. Soil Mech. Found. Div., ASCE*, **94**, SM5, 1053–1122.
Seed, H. B. and Idriss, I. N. (1967) Analysis of soil liquefaction – Niigata earthquake, *J. Soil Mech. Found, Div., Amer. Soc. Civ. Engs.*, **93**, no. SM3, 83–108.
Seed, H. B. and Lee, K. L. (1966) Liquefaction of saturated sands during cyclic loading, *J. Soil Mech. Found. Div., Amer. Soc. Civ. Engrs.*, **92**, SM6 105–134.
Seed, H. B. and Whitman, R. V. (1970) Design of earth retaining structures for dynamic loads, *Proc. Speciality Conf. on Lateral Stresses in the Ground and Design of Earth-Retaining Structures*, A.S.C.E., 103–147.
Seed, H. B. and Wilson, S. D. (1967) The Turnagain Heights landslide, Anchorage, Alaska, *Jl. Soil Mech. and Found. Div., ASCE*, **93**, 325–353.
Sevaldson, R. A. (1956) The slide at Lodalen, October 6th, 1954, *Géotechnique*, **6**, 167–182.
Shannon and Wilson Inc. (1964) Report on in-situ rock tests, Divoshak Dam Site, *For U.S. Army Engineer District, Walla Walla*, Corps of Engineers, Seattle, Washington.

Sharp, J. C. (1970) 'Fluid flow through fissured media,' Ph.D. thesis, University of London.
Sharp, R. P. and Nobles, L. H. (1953) Mudflow of 1941 at Wrightwood, Southern California, *Bull. Geol. Soc. America*, **64**, 547–560.
Sherard, J. L. (1959) Discussion of: 'Tractive resistance of cohesive soils' by I. S. Dunn, *J. Soil Mech. Found. Div.*, ASCE, **185**, Paper No. 2325, 157–162.
Sherard, J. L. (1967) Earthquake considerations in earth dam design, *J. Soil Mech. Found. Div.*, ASCE, **93**, SM4, 377–410.
Sherard, J. L., Cluff, L. S. and Allen, C. R. (1974) Potentially active faults in dam foundations, *Géotechnique*, **24**, no. 3, 367–428.
Sherard, J. L., Woodward, R. J., Gizienski, S. F. and Clevenger, W. A. (1963) 'Earth and Earth-Rock Dams,' John Wiley and Sons. Inc., New York.
Shergold, F. A. (1955) Results of tests on single sized roadmaking aggregates, *Quarry Managers Journal*, **38**, 636–642.
Shergold, F. A. (1960) The classification, production and testing of roadmaking aggregates, *Quarry Managers Journal*, **44**, 47–54.
Shiell, J. W. (1975) The water resources of Scotland: their use and abuse, *Proc. Inst. Civ. Engrs.*, **85**, pt. I, 107–109.
Short, N. M. (1966) Effect of shock pressures from a nuclear explosion on mechanical and optical properties of granodiorite, *J. Geophys. Res.*, **71**, 1195–1215.
Shuk, T. (1965) Discussion of paper by Langejan, A. entitled 'Some aspects of the safety factor in soil mechanics considered as a problem of probability,' *Proc. 6th Int. Conf. Soil Mech. Found. Engng.*, Montreal, **2**, 500–502; **3**, 576–577.
Shuk, T. (1970) Optimization of slopes designed in rock, *Proc. 2nd Cong. Int. Soc. Rock Mechanics*, Belgrade, **3**, 275–280.
Silveira, A. F. Da, Rodrigues, F. P., Grossmann, N. F. and Mello Mendes, F. (1966) Quantitative characterization of the geometric parameters of jointing in rock masses, *Proc. 1st Cong. Int. Soc. of Rock Mech.*, Lisbon, Portugal, **1**, 225–233.
Simmons, G. and Richter, D. (1974) 'Microcracks in rocks,' Distributed preprint.
Simmons, G., Richter, D., Todd, T. and Wang, H. (1973) Microcracks: their potential for obtaining P-T history of rocks, *Geol. Soc. Amer.*, **5**, abstract only, 810.
Simmons, G., Todd, T. and Baldridge, W. S. (1974) Toward a quantitative relationship between elastic properties and cracks in low porosity rocks, Publication preprint.
Simmons, G., Wang, H., Richter, D. and Todd, T. (1973) Microcracks created by P-T changes, *EOS*, **54**, abstracts 451.
Sitter, L. U. de (1956) 'Structural Geology,' Publ. McGraw-Hill, London.
Skempton, A. W. (1945) A slip in the West bank of the Eau Brink Cut, *Journ. Inst. Civ. Eng.*, **24**, 267–287.
Skempton, A. W. (1946) Earth pressure and the stability of slopes. In: *The Principles and Application of Soil Mechanics*, Inst. Civ. Engrs., London, 31–61.
Skempton, A. W. (1953a) Soil mechanics in relation to geology, *Proc. Yorks. Geol. Soc.*, **29**, 33–62.
Skempton, A. W. (1953b) Discussion on: Soil stability problems in road engineering, *Proc. Inst. Civ. Eng.*, **2**, 265–268.
Skempton, A. W. (1953c) The colloidal activity of clay, *Proc. 3rd Int. Conf. Soil Mech. Found. Eng.*, Zurich, **1**, 57–61.

Skempton, A. W. (1954) The pore-pressure coefficients A and B, *Géotechnique*, **4**, 143–147.
Skempton, A. W. (1961a) Horizontal stresses in an over-consolidated Eocene clay, *Proc. 5th Int. Conf. on Soil Mechanics and Found. Eng.*, Paris, **1**, 351– 387.
Skempton, A. W. (1961b) 'Effective stress in soils, concrete and rocks,' *Pore Pressure and Suction in Soils*, Butterworths, London, 4–16.
Skempton, A. W. (1964) Long-term stability of clay slopes, *Géotechnique*, **14**, 77–101.
Skempton, A. W. (1966) Some observations on tectonic shear zones, *Proc. 1st Congress, Int. Soc. Rock Mech.*, Lisbon, **1**, 329–335.
Skempton, A. W. (1970) First-time slides in over-consolidated clays, *Géotechnique*, **20**, 320–324.
Skempton, A. W. and Bjerrum, L. (1957) A contribution to the settlement analysis of foundations on clay, *Géotechnique*, **7**, 168–178.
Skempton, A. W. and Brown, J. D. (1961) A landslide in boulder clay at Selset, Yorkshire, *Géotechnique*, **11**, 280–293.
Skempton, A. W. and Cattin, P. (1963) A full scale alluvial grouting test at the site of the Mangla Dam, *Grouts and Drilling Muds in Engineering Practice*, Butterworths, London, 131–135.
Skempton, A. W. and DeLory, F. A. (1957) Stability of natural slopes in London Clay, *Proc. 4th Int. Conf. on Soil Mech. Found. Engrg.*, London, **2**, 378–381.
Skempton, A. W. and Henkel, D. J. (1953) The post glacial clays of the Thames Estuary at Tilbury and Shellhaven, *Proc. 3rd Int. Conf. Soil Mech. and Found. Eng.*, Zurich, **1**, 302–308.
Skempton, A. W. and Henkel, D. J. (1957) Tests on London Clay from deep borings at Paddington, Victoria and the South Bank, *Proc. 4th Int. Conf. on Soil Mech. and Found. Engrg.*, London, **2**, 100–106.
Skempton, A. W. and Hutchinson, J. N. (1969) Stability of natural slopes and embankment foundations, State of the Art Report, *7th Int. Conf. on Soil Mech. and Found. Engrg.*, Mexico City, 291–340.
Skempton, A. W. and LaRochelle, P. (1965) The Bradwell Slip, a short term failure in London Clay, *Géotechnique*, **15**, 221–242.
Skempton, A. W. and Petley, D. J. (1967) The strength along structural discontinuities in stiff clays, *Proc. Geotech. Conf.*, Oslo, **2**, 29–46.
Skempton, A. W. and Northey, R. D. (1952) The sensitivity of clays, *Géotechnique*, **2**, 30–54.
Skempton, A. W. and Petley, D. J. (1970) Ignition loss and other properties of peats and clays from Avonmouth, Kings Lynn and Cranberry Moss, *Géotechnique*, **20**, 343–356.
Skempton, A. W., Schuster, R. L. and Petley, D. J. (1969) Joints and fissures in the London Clay at Wraysbury and Edgeware, *Géotechnique*, **19**, 205–217.
Skibitzke, H. E. (1963) The use of analogue computers for studies in groundwater hydrology, *J. Inst. Water Engrs.*, **17**, 216–230.
Skipp, B. O. and Tayton, J. W. (1970) Blasting vibrations – ground and structure response, *Dynamic Waves in Civil Engineering*, Eds. D. A. Howells, I. P. Haigh, C. Taylor, Wiley – Interscience, 181–210.
Sloane, R. L. and Kell, T. F. (1966) The fabric of mechanically compacted kaolin. *Proc. 14th Natl. Conf. Clays and Clay Minerals*, 289–296.
Smalley, I. J. and Cabrera, J. A. (1969) Particle association in compacted kaolin, *Nature*, **222**, 80–81.

Smart, P. (1966) Particle arrangements in kaolin, *Proc. 15th Natl. Conf. Clays and Clay Minerals*, 241–254.
Smedley, M. I. (1974) 'A geotechnical investigation of laminated clays from Northern England,' Dissertation: M.Sc. Advanced Course in Engineering Geology, University of Durham.
Smiles, D. E. and Young, E. G. (1965) Hydraulic conductivity determinations by several methods in a sand tank, *Soil Sci.*, **99**, 83–87.
Smith, D. B. and Francis, E. A. (1967) 'The geology of the country between Durham and West Hartlepool,' Mem. Geol. Survey, Great Britain, H.M.S.O. London.
Smith, R. E. (1970) Guide for depth of foundation exploration, *J. Soil Mechs. Found. Div. Amer. Soc Civ. Engrs.*, **96**, 377–384.
Snow, D. T. (1968a) Rock fracture spacings, openings and porosities, *J. Soil Mechs. Found. Div., Amer. Soc. Civ. Engrs.*, **94**, 73–91.
Snow, D. T. (1968b) Hydraulic character of fractured metamorphic rocks of the Front Range and implications to the Rocky Mountains Arsenal well, *Quart. Jl. Colorado Sch. of Mines*, **63**, 167–199.
Snow, D. T. (1972) Geodynamics of seismic reservoirs, *Proc. Int. Symp. on Percolation through fissured rocks, Int. Soc. Rock Mech.*, Germany, Paper T2–J.
Søderblom, R. (1966) Chemical aspects of quick clay formation, *Eng. Geol.*, **1**, 415–431.
Soejima, T. and Shidomoto, Y. (1970) Foundation improvement of an arch dam by special consolidation grouting, *Proc. 2nd Cong. Int. Soc. for Rock Mech.*, Belgrade, **3**, 167–174.
Southwell, R. V. (1946) 'Relaxation methods in theoretical physics,' Oxford Univ. Press, England.
Sowers, G. B and Sowers, G. F. (1971) 'Introductory soil mechanics and foundations,' (3rd Edition), Collier – McMillan, London.
Sparks, B. W. (1960) 'Geomorphology,' Longmans, London.
Spears, D. A. (1969) A laminated marine shale of Carboniferous age from Yorkshire, England, *J. Sed. Petrol.*, **39**, 106–112.
Spears, D. A. (1971) The mineralogy of the Stafford Tonstein, *Proc. Yorks. Geol. Soc.*, **38**, 497–516.
Spears, D. A., Taylor, R. K. and Till, R. (1971) A mineralogical investigation of a spoil heap at Yorkshire Main Colliery, *Q. Jl. Eng. Geol.*, **3**, 239–252.
Spencer, A. B. and Clabaugh, P. S. (1967) Computer program for fabric diagrams, *Amer. Journ. Sci.*, **265**, 166–172.
Spencer, E. (1967) A method of analysis of the stability of embankments assuming parallel inter-slice forces, *Géotechnique*, **17**, 11–26.
Spencer, E. (1968) Effect of tension on stability of embankments, *J. Soil Mech. Found. Div.*, A.S.C.E., **94**, 1159–1173.
Spencer, E. (1973) Thrust line criterion in embankment stability analysis, *Géotechnique*, **23**, 85–100.
Spencer, E. W. (1959) Geologic evolution of the Beartooth Mountains, Montana, and Wyoming; Part 2, Fracture Patterns, *Bull. Geol. Soc. America*, **70**, 467–508.
Sridharan, A. and Rao, G. V. (1973) Mechanisms controlling volume change of saturated clays and the role of the effective stress concept, *Géotechnique*, **23**, 359–382.

Stacey, T. R. (1969) Application of the finite element method in the field of rock mechanics with particular reference to slope stability, *The South African Mechanical Engineer,* **19**, 131–134.

Stacey, T. R. (1970) The stresses surrounding open-pit mine slopes, *Proc. Symp. on Planning Open Pit Mines,* Ed. P. W. J. van Rensburg, Johannesburg, 199–207.

Stallman, R. W. (1963) Electric analog of 3-dimensional flow to wells and its application to unconfined aquifers, *U.S. Geol. Water Supply,* Paper No. 1536–H.

Stapledon, D. H. (1968) Discussion – Classification of rock substances, *Int. Jl. Rock Mech. Min. Sci.,* **5**, 371–373.

Starkey, J. (1964) An X-ray method for determining orientations of selected planes in a polycrystalline aggregate, *Am. J. Sci.,* **262**, 735–752.

Stauffer, M. R. (1966) An empirical-statistical study of three-dimensional fabric diagrams as used in structural analysis, *Canadian Jl. Earth Sci.,* **3**, 473–498.

Steinbrugge, K. V. and Clough, R. W. (1960) Chilean earthquake of May 1960. A brief trip report, *Proc. 2nd World Conf. on Earthquake Eng.,* **1**, 629–637.

Stimpson, B., Metcalfe, R. G. and Walton, G. (1970) A new field technique for sealing and packing rock and soil samples, *Q. Jl. Eng. Geol.,* **3**, 127–133.

Stini, J. (1950) 'Tunnel baugeologic,' Springer-Verlag, Vienna.

Stoker, J. J. (1957) 'Water waves,' Interscience, New York.

Straud, T. (1945) A method of counting out petrofabric diagrams, *Norsk, Geol Tidsskr.,* **24**, 112–113.

Suits, C. G. (1951) New instruments from the research laboratory, *Gen. Elect. Review,* **54**, No. 11, 32–33.

Suklje, L. and Vidmar, S. (1961) A landslide due to long-term creep, *Proc. 5th Int. Conf. on Soil Mech.,* Paris, **2**, 727–735.

Sultan, H. A. and Seed, H. B. (1967) Stability of sloping core earth dams, *Jl. Soil Mech. and Found. Eng., A.S.C.E.,* **93**, SM4, 45–67.

Svatoš, A. (1974) Identification of gravitational slope deformations on aerial photographs, *Proc. 2nd Int. Congr. Int. Assoc. of Eng. Geology,* São Paulo, Brazil, 2 Th V 13.1–13.9.

Symons, I. F. and Booth, A. I. (1971) Investigation of the stability of earthwork construction on the original line of Sevenoaks By-Pass, Kent, *Road Research Laboratory Report LR 393.* Department of the Environment, London.

Tagg, G. F. (1934) Interpretation of resistivity measurements, 'Geophysical Prospecting, 1934,' *Trans. Am. Inst. Mining Met. Engrs.,* **110**, 183–200.

Takano, M. and Shidomoto, Y. (1966) Deformation test on mudstone enclosed in a foundation by means of tubedeformeter, *Proc. 1st Cong. Int. Soc. Rock Mech.,* Lisbon, **1**, 761–764.

Talbot, C. J. (1970) The minimum strain ellipsoid using deformed quartz veins, *Tectonophysics,* **9**, 47–76.

Talobre, J. A. (1957) 'La méchanique des roches,' Dunod, Paris.

Taniguchi, T. and Watari, M. (1965) Landslide at Yui and its countermeasure, *Soils and Foundations,* **5**, 7–25.

Taylor, D. W. (1948) 'Fundamentals of soil mechanics,' John Wiley and Sons, New York.

Taylor, K. (1973) Borrowing water for thirsty London, *New Civil Engineer,* **68**, 22 November, 32–33.

Taylor, R. K. (1971) The functions of the engineering geologist in urban development, *Q. Jl. Eng. Geol.*, **4**, 221–240.
Taylor, R. K. and Spears, D. A. (1970) The breakdown of British Coal Measure rocks, *Int. J. Rock Mech. Min. Sci.*, **7**, 481–501.
Taylor, R. K. and Spears, D. A. (1972) The influence of weathering on the composition and engineering properties of *in situ* Coal Measures rocks, *Int. J. Rock Mech. Min. Sci.*, **9**, 729–756.
Tchalenko, J. S. (1968) The microstructure of London Clay, *Q. Jl. Eng. Geol.*, **1**, 145–155.
Teichmann, G. A. and Hancock, J. (1951) Blasting vibrations and the householder, *Quarry Managers Journal*, **34**, 397–405.
Teichmann, G. A. and Westwater, R. (1957) Blasting and associated vibrations, *Engineering*, **183**, 460–465.
Terzaghi, K. (1922) Der Grundbruch an Stanwerken und seine Verhutung, *Die Wasserkraft*, **17**, 445–449.
Terzaghi, K. (1936) Stability of slopes in natural clay, *Proc. 1st Int. Conf. on Soil Mechanics and Foundations Eng.*, **1**, 161–165.
Terzaghi, K. (1943) 'Theoretical soil mechanics,' Wiley, New York.
Terzaghi, K. (1946) In: *Rock tunnelling with Steel Supports*, by R. V. Proctor and T. L. White, The Commercial Shearing and Stamping Co., Youngstown, Ohio.
Terzaghi, K. (1950) Mechanism of landslides. In: *Application of Geology to Engineering Practice (Berkey volume)*, *Geol. Soc. Amer.*, 83–123.
Terzaghi, K. (1962) Stability of steep slopes on hard, unweathered rock, *Géotechnique*, **12**, 251–270.
Terzaghi, K. and Peck, R. B. (1967) 'Soil mechanics in engineering practice,' (2nd Edn.) John Wiley and Sons, New York.
Terzaghi, R. D. (1965) Sources of error in joint surveys, *Géotechnique*, **15**, 287–304.
Theis, C. V. (1935) The relation between the lowering of the piezometric surface and the rate and duration of discharge of a well using groundwater storage, *Trans. Am. Geophys. Union*, **16**, 519–524.
Thill, R. E., Willard, R. J. and Bur, T. R. (1969) Correlation of longitudinal velocity variation with rock fabric, *J. Geophys. Res.*, **74**, 4897–4909.
Thoenen, J. R. and Windes, S. L. (1938) House movement caused by ground vibrations, *U.S. Bureau of Mines*, Report of Investigations No. 3431.
Thoenen, J. R. and Windes, S. L. (1942) Seismic effects of quarry blasting, *U.S. Bureau of Mines*, Bull. 442.
Thoenen, J. R., Windes, S. L. and Ireland, A. T. (1940) House movement induced by mechanical agitation and quarry blasting, *U.S. Bureau of Mines*, Report of Investigations No. 3542.
Thompson, T. F. (1947) Origin, nature and significance of the slickensides in the Cucaracha clay shales, *Isthmian Canal Studies Memorandum*, **245**, Panama Canal Co., Diablo Heights, Canal Zone.
Thorn, R. B. (1960) 'The design of sea defence works,' Butterworths, London.
Thurber, W. C. and McHargue, C. J. (1960) Deformation textures in aluminium-uranium alloys, *Trans. Metal Soc., AIME*, **218**, 141–144.
Tirey, G. B. (1972) Recent trends in underwater soil sampling methods, *Underwater soil sampling testing and control, Amer. Soc. Testing and Matls.*, S.T.P. No. **501**, 42–54.
Todd, D. K. (1959) 'Ground water hydrology,' Wiley, New York.

Todd, T., Wang, H., Baldridge, W. S. and Simmons, G. (1972) Elastic properties of Apollo 14 and 15 rocks, *Proc. 3rd Lunar Sci. Conf.*, **3**, 2577–2586.
Todd, T., Richter, D., Simmons, G. and Wang, H. (1973) Unique characterization of lunar samples by physical properties, *Proc. 4th Lunar Sci. Conf.*, **3**, 2639–2662.
Tomlinson, M. J. (1963) 'Foundation design and construction,' (2nd Edn. 1969), Pitman & Sons, London.
Toms, A. H. (1948) The present scope and possible future development of soil mechanics in British railway civil engineering construction and maintenance, *Proc. 2nd Int. Conf. Soil Mech.*, Rotterdam, **4**, 226–237.
Toms, A. H. (1953) Recent research into coastal landslides at Folkestone Warren, Kent, England, *Proc. 3rd Int. Conf. Soil Mech.*, Zurich, **2**, 288–293.
Torrance, J. K. (1974) A laboratory investigation of the effect of leaching on the compressibility and shear strength of Norwegian marine clays, *Géotechnique*, **24**, 155–173.
Tourtelot, H. A. (1962) Preliminary investigation of the geologic setting and chemical composition of the Pierre shale, Great Planes Region, *U.S. Geol. Survey Professional Paper 390*, Washington, D.C.
Troeh, E. R. (1965) Landform equations fitted to contour maps, *Am. J. Sci.*, **263**, 616–627.
Trostel, L. J. and Wynne, D. J. (1940) Determination of quartz (free silica) in refractory clays, *Am. Ceram. Soc. Journ.*, **23**, 18–22.
Tschebotarioff, G. P. (1951) 'Soil mechanics, foundation and earth structures,' McGraw-Hill, New York.
Turnbull, W. J. (1948) Utility of loess as a construction material, *Proc. 2nd Int. Conf. on Soil Mech. Found. Eng.*, Rotterdam, **5**, IVd, 97–103.
Turner, F. J. (1957) Lineation, symmetry, and internal movement in monoclinic tectonite fabrics, *Bull. Geol. Soc. Amer.*, **68**, 1–17.
Turner, F. J. and Weiss, L. E. (1963) 'Structural analysis of metamorphic tectonites,' McGraw-Hill, New York.
Turner, M. J. (1967) 'The influence of geological structures on rock slope stability,' M.Sc. Advanced Course Dissertation, University of Durham.
Twenhofel, W. H. (1939) 'Principles of sedimentation,' McGraw-Hill, New York.
Twort, A. C., Hoather, R. C. and Law, F. M. (1974) 'Water Supply,' Arnold, London.
U.S. Army Corps of Engineers (1968) South Platte River, Colorado, Chatfield Dam and Reservoir, *Design Memorandum No. DC–24*, Embankment and Excavation, U.S. Army Engineer District, Omaha, Corps of Engineers Omaha, Nebraska, **1**, 2–5.
U.S. Bureau of Reclamation (1947) Laboratory tests on protective filters for hydraulic and static structures, *Earth Materials Laboratory Report, EM–132*, Denver, Colorado, U.S.A.
U.S. Bureau of Reclamation (1953) Physical properties of some typical foundation rocks, *Engineering Laboratories Branch, Concrete Laboratory Report*, No. SP–39.
U.S. Coastal Engineering Research Center (1966) *Shore protection planning and design*, Technical Report No. 4, 3rd Edition.
Ulmer, G. C. and Smothers, W. J. (1967) Application of mercury porosimetry to refractory materials, *Ceramic Bull.*, **46**, 649–656.
Underwood, L. B. (1967) Classification and identification of shales, *J. Soil Mech. and Found. Div., A.S.C.E.*, **93**, no. SM6, 97–116.

Vallarino, E. and Alvarez, A. (1971) Strengthening the Mequinenza dam to prevent sliding, *Water Power*, **23**, 104–108, 121–126.

Vargas, M. and Pichler, E. (1957) Residual soil and rock slides in Santos (Brazil), *Proc. 4th Int. Conf. Soil Mech. Found. Engrg.*, London, **2**, 394–398.

Vargas, M., Silva, F. P. and Tubio, M. (1965) Residual clay dams in the state of São Paulo, Brazil, *Proc. 6th Int. Conf. on Soil Mech. and Found. Eng.*, Montreal, **2**, 579–582.

Varnes, D. J. (1958) Landslide types and processes, In: *Landslides and Engineering Practice* (Ed. E. B. Eckel), Highway Research Board Special Report 29, Washington, D.C., U.S.A. 20–47.

Vaughan, P. R. (1963) Contribution to discussion: *Grouts and Drilling Muds in Engineering Practice*, Butterworths, London, 54.

Vaughan, P. R. (1972) Discussion: Examples of the internal conditions of some old earth dams, by M. F. Kennard, *J. Inst. Water Engrs.*, **26**, 152–153.

Vaughan, P. R. and Walbancke, H. J. (1973) Pore pressure changes and the delayed failure of cutting slopes in over-consolidated clay, *Géotechnique*, **23**, 531–539.

Verwey, E. J. W. and Overbeek, J. Th. G. (1948) 'Theory of the stability of lyophobic colloids,' Elsevier, New York.

Voight, B. (1973) Correlation between Atterburg plasticity limits and residual shear strength of natural soils, *Géotechnique*, **23**, 265–267.

Voight, B. and Pariseau, W. (1970) State of the predictive art in subsidence engineering, *J. Soil Mechs. Found. Eng. Div., Amer. Soc. Civ. Engrs.*, **96**, 721–750.

Vucetik, R. (1958) Determination of shear strength and other characteristics of coarse, clayey schist material compacted by pneumatic wheel roller, *Proc. 6th Int. Cong. on Large Dams*, New York, **4**, 465–473.

Wagner, A. A. (1957) The use of the Unified Soil Classification system by the Bureau of Reclamation, *Proc. 4th Int. Conf. Soil Mech. and Found. Eng.*, London, **1**, 125–134.

Waldorf, W. A., Veltrop, J. A. and Curtis, J. J. (1963) Foundation modulus tests for Karadj arch dam, *J. Soil Mech. Found. Div., ASCE*, **89**, No. SM4, 91–126.

Walker, P. E. (1971) 'The shearing behaviour of a block jointed rock model,' Ph.D. thesis, Queens University, Belfast, N. Ireland.

Walker, W. H. (1974) Monitoring toxic chemical pollution from land disposal sites in humid regions, *Ground Water*, **12**, no. 4, 213–218.

Wallace, K. B. (1973) Structural behaviour of residual soils of the continually wet Highlands of Papua, New Guinea, *Géotechnique*, **23**, 203–218.

Waller, R. A. (1969) 'Building on Springs,' Pergamon Press, London.

Walsh, J. B. and Brace, W. F. (1964) A fracture criterion for brittle anisotropic rock, *J. Geophys. Res.*, **69**, 3449–3456.

Walsh, J. B. and Decker, E. R. (1966) Effect of pressure and saturating fluid on the thermal conductivity of compact rock, *J. Geophys. Res.*, **71**, 3053–3061.

Walsh, P. D. (1971) Designing control rules for the conjunctive use of impounding reservoirs, *J. Inst. Water Engrs.*, **25**, 371–380.

Walters, R. C. S. (1971) 'Dam Geology,' Butterworths, London.

Walthall, S. (1970) 'Ground vibrations from vehicular traffic,' Dissertation, M.Sc. Advanced Course in Engineering Geology, University of Durham.

Walton, W. C. (1960) *Leaky artesian aquifer conditions in Illinois*, Report of Investigation No. 39, Illinois State Water Survey.

Walton, W. C. (1962) *Selected analytical methods for well and aquifer evaluation*, Illinois State Water Survey, Bulletin 49.
Walton, W. C. (1970) 'Groundwater resource evaluation,' McGraw-Hill, New York.
Walton, W. C. and Prickett, T. A. (1963) Hydrogeologic electric analog computers, *Proc. Am. Soc. Civ. Engrs., Hydraul. Div.*, 89, (HY6), 67–91.
Wang, H., Todd, T., Weidner, D. and Simmons, G. (1971) Elastic properties of Apollo 12 rocks, *Proc. 2nd Lunar Sci. Conf.*, 3, 23–32.
Ward, W. H., Burland, J. B. and Gallois, R. W. (1968) Geotechnical assessment of a site at Mundford, Norfolk, for a large proton accelerator, *Géotechnique*, 18, 399–431.
Ward, W. H., Marsland, A. and Samuels, S. G. (1965) Properties of the London Clay at the Ashford Common Shaft: *In situ* and drained strength tests, *Géotechnique*, 15, 321–344.
Ward, W. H., Samuels, S. G. and Butler, E. (1959) Studies of the properties of London Clay, *Géotechnique*, 9, 33–58.
Wardell, K. and Wood, J. C. (1966) Ground instability problems arising from the presence of old shallow mine working, *Proc. Mid. Soc. Soil Mech. Found. Eng.*, 7, 5–30.
Watt, A. J., de Boer, B. B. and Greenwood, D. A. (1967) Loading tests on structures founded on soft cohesive soils strengthened by compacted granular columns, *Proc. 3rd Asian Reg. Conf. on Soil Mech. Found. Eng.*, Haifa, 1, 248–251.
Wawersik, W. R. and Brace, W. F. (1971) Post failure behaviour of a granite and diabase, *Rock Mechs.*, 3, 61–85.
Wawersik, W. R. and Fairhurst, C. (1970) A Study of brittle rock fracture in laboratory compression experiments. *Int. J. Rock Mech. Min. Sci.*, 7, 561–575.
Weaver, R. and Call, R. D. (1965) Computer estimation of oriented fracture set intensity, *Symposium on Computers in Mining and Exploration*, Tucson, Arizona, U.S.A., March 1965.
Webb, D. L. and Hall, R. I. (1969) Effects of vibroflotation on clayey sands, *J. Soil Mech. Found. Div., Amer. Soc. Civ. Eng.*, 95, 1365–1378.
Weeks, A. G. (1969) The stability of natural slopes in south-east England as affected by periglacial activity, *Q. J. Eng. Geol.*, 2, 49–62.
Weibull, W. (1939) 'A statistical theory of the strength of materials,' Ingvetensk, Acad. Handl., no. 151.
Welham, K. P. (1969) A study of the shear strength parameters of dolomite and microconcrete, B.Sc. Honours in Engineering Science, Project Dissertation, University of Durham.
Wenk, H. R., Baker, D. W. and Griggs D. T. (1967) X-ray fabric analysis of hot-worked and annealed flint, *Science*, 157, 1447–1449.
Wenk, H. R. and Kolodny, Y. (1968) Preferred orientation of quartz in a chert breccia, *Natl. Acad. Sci. Proc.*, 59, 1061–1066.
Wenzel, L. K. (1942) Methods for determining permeability of water-bearing materials with special reference to discharging well methods, *U.S. Geol. Survey, Water – Supply Paper 887*, Washington, D.C., U.S.A.
Wesley, L. D. (1973) Some basic engineering properties of halloysite and allophane clays in Java, Indonesia, *Géotechnique*, 23, 471–494.
Westergaard, H. M. (1933) Water pressure on dams during earthquakes, *Trans. Amer Soc. Civ. Engrs.*, 98, 418–472.

Whiffin, A. C. and Leonard, D. R. (1971) 'A survey of traffic-induced vibrations,' Department of the Environment (U.K.) TRRL Report LR418.

Whipkey, R. Z. (1965) Subsurface storm flow from forested slopes, *Bull. Int. Assoc., for Scientific Hydrology*, **10**, 74–85.

White, W. A. (1955) 'Water sorption properties of homoionic clay minerals,' Ph.D. thesis, University of Illinois.

White, W. A. (1961) Colloid phenomena in sedimentation of argillaceous rocks, *J. Sed. Petr.*, **31**, 560–570.

Whitman, R. V. and Bailey, W. A. (1967) Use of computers for slope stability analysis, *J. Soil Mech. Found. Div., ASCE*, **93**, SM4, 475–498.

Wiebols, G. A. and Cook, N. G. W. (1968) An energy criterion for the strength of rock in polyaxial compression, *Int. J. Rock Mech. Min. Sci.*, **5**, 529–549.

Wiegel, R. L. (1964) 'Oceanographical Engineering,' Prentice-Hall, New York.

Wiest, De R. J. M. (1965) 'Geohydrology,' Wiley & Co., New York.

Wilkinson, W. B. (1967) Discussion of Gibson, *Géotechnique*, **16**, 68–71.

Wilkinson, W. B. (1968) Constant head *in-situ* permeability tests in clay strata, *Géotechnique*, **18**, 172–194.

Willard, R. J. and McWilliams, J. R. (1969) Microstructural techniques in the study of physical properties of rock, *Int. J. Rock Mech. Min. Sci.*, **6**, 1–12.

Willis, D. E. and Wilson, J. T. (1960) Maximum vertical ground displacement of seismic waves generated by explosive blasts, *Seismological Soc. Amer. Bull.*, **50**, 455–459.

Wilson, A. H. (1961) A laboratory investigation of a high modulus borehole plug gauge for the measurement of rock stress, *Proc. 4th Symp. on Rock Mech.*, Pennsylvania State Univ., Ed. H. L. Hartman, 185–195.

Wilson, E. M. (1974) 'Engineering hydrology,' 2nd edn. Macmillan, London.

Wilson, G. and Grace, H. (1942) The settlement of London due to under-drainage of the London Clay, *J. Inst. Civ. Engrs.*, **19**, 100–127.

Wilson, S. D. (1970) Observational data on ground movements related to slope instability, *J. Soil Mechanics and Found. Div., A.S.C.E.*, **96**, SM5, 1521–1543.

Winchell, H. (1937) A new method of interpretation of petrofabric diagrams, *Amer. Mineralogist*, **22**, 15–36.

Winslow, N. M. and Shapiro, J. (1959) An instrument for the measurement of pore size distribution by mercury penetration. *Bull., American Soc. Test. Mat.*, **236**, 39–44.

Wisecarver, D. W., Merrill, R. H., Rausch, D. O. and Hubbard, S. J. (1964) Investigation of in situ rock stresses, Ruth Mining District, Nevada, with emphasis on slope design problems in open-pit mines, *U.S. Bureau of Mines, Report of Investigations*, R.I. **6541**.

Wittke, W. (1970) Three-dimensional percolation of fissured rock, *Symp. on Planning Open Mines*, Johannesburg, S. Africa, P. W. J. van Rensburg (ed.) A. A. Balkema, Cape Town, 181–191.

Wittke, W. and Louis, C. (1966) Determination of the influence of ground water flow on the stability of slopes and structures in jointed rock, *Proc. 1st Int. Cong. Rock Mechs.*, Lisbon, **2**, 201–206.

Wong, H. Y. (1971) Unpublished communication.

Wong, H. Y. and Farmer, I. W. (1973) Hydrofracture mechanisms in rock during pressure grouting, *Rock Mech.*, **5**, 21–41.

Wood, A. N. Muir (1969) 'Coastal Hydraulics,' Macmillan and Co., Ltd., London.

Wood, C. C. (1958) 'Shear strength and volume change characteristics of compacted soil under conditions of plane strain,' Ph.D. thesis, University of London.
Wood, W. O. (1923) The Permian formation in East Durham, *Trans. Inst. Min. Engrs.*, **65**, 178–186.
Woodland, A. W. (1968) Field geology and the civil engineer, *Proc. Yorks. Geol. Soc.*, **36**, 531–578.
Woodruff, S. (1966) 'Methods of working coal and metal mines,' **1**, Pergamon, Oxford.
Wright, G., Kulhawy, F. H. and Duncan, J. M. (1973) Accuracy of equilibrium in slope stability analysis, *J. Soil Mech. Found. Div., A.S.C.E.*, **99**, SM10, 783–791.
Wright, S. G. (1969) 'A study of slope stability and the undrained shear strength of clay shales,' Ph.D. thesis, University of California, Berkeley, California, U.S.A.
Wright, S. G. and Duncan, J. M. (1969) Anisotropy of clay shales, *Paper prepared for Speciality Session No. 10*, 'Engineering Behaviour of Clay Shales,' 7th Int. Conf. on Soil Mech. and Found. Eng., Mexio City.
Wroth, C. P. (1972) Some aspects of the elastic behaviour of overconsolidated clay, *Stress-strain behaviour of soils, Proc. Roscoe Memorial Symp.*, Cambridge Univ. England, March, 1971, Ed. R. H. G. Parry, Foulis, Oxford, 347–361.
Wroth, C. P. and Hughes, J. M. O. (1973) An instrument for the in-situ measurement of the properties of soft clay, *Proc. 8th Int. Conf. Soil Mech. Found. Engrg.*, Moscow, **2**(1), 487–494.
Wu, T. H. (1958) Geotechnical properties of glacial lake clays, *J. Soil Mech. Found. Div.*, ASCE., **84**, SM3, 1–34.
Wu, T. S. and Kraft, L. M. (1967) The probability of foundation safety, *J. Soil Mech. Found. Div.*, ASCE., **93**, SM5, 213–231.
Wyckoff, R. D. and Reed, D. W. (1935) Electrical conduction models for the solution of water seepage problems, *Physics*, **6**, 395–401.
Young, E. G. and Smiles, D. E. (1963) The pumping of water from wells in unconfined aquifers: a note on the applicability of Theis' formula, *J. Geophys. Res.*, **68**, 5905–5907.
Zaruba, Q. and Mencl, V. (1969) 'Landslides and their control.' Academia and Elsevier, Prague.
Zaslavsky, D. and Kassiff, G. (1965) Theoretical formulation of piping mechanism in cohesive soils, *Géotechnique*, **15**, 305–316.
Zee, C. H., Peterson, D. F. and Beck, R. O. (1957) Flow into a well by electric and membrane analogy, *Trans. Am. Soc. Civ. Eng.*, **122**, 1086–1112.
Zienkiewicz, O. C. (1971) 'The finite element method in engineering science,' McGraw-Hill, London.
Zienkiewicz, O. C. and Cheung, Y. K. (1967) 'The finite element method in structural and continuum mechanics,' McGraw-Hill, London.
Zienkiewicz, O. C. and Stagg, K. G. (1966) The cable method of in situ testing, *Proc. 1st Congress, Int. Soc., of Rock Mech.*, Lisbon, **1**, 667–672.
Zienkiewicz, O. C., Valliappan, S. and King, I. P. (1968) Stress analysis of rock as a 'no tension' material, *Géotechnique*, **18**, 1, 56–66.
Zwart, H. J. (1951) Breuken en Diaklazen in Robin Hood's Bay (England), *Geologie Mijnb.*, **13**, 1–4.

Supplementary references

Alden, W. C. (1928) Landslide and flood at Gros Ventre, Wyoming, *Trans. Amer. Inst. Min. Met. Eng.,* **76**, 347–360.

Bannister, A. and Raymond, S. (1972) 'Surveying,' 3rd Edn. Pitman, London.

Barton, N., Lien, R. and Lunde, J. (1974b) Engineering classification of rock masses for the design of tunnel support, *Rock Mechanics,* **6**, 189–236.

Bendel, L. (1948) 'Ingenieurgeologie,' Springer Verlag, Vienna.

Bieniawski, Z. T. (1973) Engineering classification of jointed rock masses, *Trans S. Afr. Inst. Civ. Engrs.,* **15**, 335–344.

Bishop, A. W. (1966) Soils and soft rocks as engineering materials, *Inaugural Lectures,* 1964–65 and 1965–66. Imperial College of Science and Technology, London, 289–313.

Bloomer, R. J. G. and Sexton, J. R. (1972) 'The generation of synthetic river flow data.' Publication No. 15, Water Reserves Board, Reading, England.

Bornitz, G. (1931) 'Uber die Ausbrietung der von Groszkolbenmaschinen erzeugten, Bodenschwingungen in die Tiefe, Springer, Berlin.

Boulton, N. S. (1951) Flow pattern near gravity well in uniform water-bearing medium, *J. Inst. Civil Engineers,* **36**, 534–550.

Bridgeman, P. W. (1952) 'Studies in large plastic flow and fracture with special emphasis on the effects of hydrostatic pressure.' Harvard Univ. Press. Cambridge, Mass.

Cedergren, H. R. (1967) 'Seepage, drainage and flownets,' Wiley and Sons, N.Y.

Chatwin, C. P. (1960) 'The Hampshire Basin and adjoining areas,' British Regional Geology, Inst. Geological Sciences, London.

Dalrymple, J. B., Blong, R. J. and Conacher, A. J. (1968) A hypothetical nine unit land surface model, *Zeit. fur Geomorph.,* **12**, 60–76.

Davison, C. (1889) On the creeping of the soil cap through the action of frost, *Geol. Mag.* **6**, 255–261.

Deere, D. U. and Patton, F. D. (1971) Slope stability in residual soils, *Proc. 4th Pan American Conf. Soil Mech. Found. Eng.,* Puerto Rico, **1**, 87–170.

De Wiest, R. J. M. (1965) 'Geohydrology,' Wiley and Sons, New York.

Edvokinov, P. D. and Chiriaev, R. A. (1966) Quelques lois de la résistance au cisaillement des ouvrages de retenue en beton sur fondations rocheuses, *Proc. 1st Cong. Int. Soc. Rock Mech.,* Lisbon, **2**, 661–666.

Edwards, A. G. (1970) 'Scottish aggregates: rock constituents and suitability for concrete.' Current Paper CP 28/70, Building Research Establishment, London.

Edwards, A. W. F. (1972) 'Likelihood', Cambridge Univ. Press., Cambridge.

Farmer, I. W. (1975c) Stress distribution along a resin grouted rock anchor. *Int. Jl. Rock Mech. Min. Sci. Geomech. Abstr.,* **12**, 347–351.

Fayed, L. M. and Attewell, P. B. (1965) A simplified, non-rigorous, tabular classification of clay minerals, with some explanatory notes, *Int. J. Rock Mech. Min. Sci.,* **2**, 271–276.

Gerber, E. and Scheidegger, A. E. (1974) On the dynamics of scree slopes, *Rock Mechanics,* **6**, 25–38.

Gilboy, G. (1940) Mechanics of hydraulic-fill dams, *Boston Soc. Civ. Engrs,* Contributions to Soil Mechanics 1925–1940, **21**, no. 3 (July 1934) 127–145.

Golder, H. Q., Gould J. P., Lambe T. W., Tschebotarioff G. P., Wilson S. D. (1970) Predicted performance of braced excavations, *Jl. Soil Mech. Found. Div.,* Am. Soc. Civ. Engrs., **96**, 810–815.

Griggs, D. T. (1941) An experimental approach to dynamic metamorphism, *Trans. Am. Geophys. Un.*, **22**, 526–528.

Gruner, E. and Gruner, G. (1953) Moraine and decomposed rock as construction materials for earth dams, *Proc. 3rd Int. Conf. Soil Mech. Found. Eng.*, Zurich, **2**, 245–249.

Halstead, P. N., Call, R. D. and Rippere, K. H. (1968) 'Geological structural analysis for open pit slope design, Kimberly Pit, Ely, Nevada.' Paper to *Am. Inst. Min. Eng.*, Annual Meeting, New York.

Handin, J. and Hager, R. V. (1958) Experimental deformation of sedimentary rocks under confining pressure – tests at high temperature, *Bull. Amer. Assoc. Petr. Geol.*, **42**, 2892–2934.

Hardy, R. M. (1969) In: 'Engineering properties and behaviour of clay shales', (Reporter S. J. Johnson), *Proc. 7th Int. Conf. Soil Mech. Found. Engrg*, Mexico City, **3**, 483–488.

Hobbs, D. W. (1960) Strength and stress-strain characteristics of Oakdale coal under compression, *Geol. Mag.*, **97**, 422–435.

Janbu, N., Bjerrum, L. and Kjaernsli, B. (1956) Veiledning ved løsning qv fundamenteringsoppgaver (Soil mechanics applied to some engineering problems), *Norwegian Geotechnical Institute*, Publ. No. 16.

Judd, W. R. and Wantland, D. (1965) Influence of geotechnical factors on arch dam design, *Proc. 20th Int. Geol. Congress*, Mexico, Sect. 13, 191–214.

Karol, R. H. (1968) Chemical grouting technology, *J. Soil Mech. Found. Div.*, Amer. Soc. Civ. Engs., **94**, 175–204.

Leeman, E. R. and Hayes, D. J. (1966) A technique for determining the complete state of stress in rock using a single borehole, *Proc. 1st Cong. Int. Soc. Rock Mechs.*, Lisbon, **2**, 17–24.

Lenczner, D. (1973) 'Movement in buildings,' Pergamon, Oxford.

Lewis, W. A. and Croney, D. (1965) The properties of chalk in relation to road foundations and pavements, *Proc. Symp. on Chalk in Earthworks and Foundations,* Paper 3, Inst. Civ. Engrs, London, 27–41.

Marsland, A. (1972b) The shear strength of stiff fissured clays, *Proc. Symp. on Stress Strain Behaviour of Soils,* Ed. R. H. G. Parry, Foulis, Oxford, 59–68.

Murrell, S. A. F. (1958) The strength of coal under triaxial compression, In *Mechanical Properties of Non-metallic Brittle Materials,* Butterworths, London, 123–146.

Murrell, S. A. F. (1963) A criterion for brittle fracture of rocks and concrete under triaxial stresses, and the effect of pore pressure on the criterion, *Rock Mechanics, Proc. 5th Symp. Rock Mech.*, University of Minnesota, Ed. C. Fairhurst, Pergamon, Oxford, 563–577.

Nicholls, G. D. (1963) Environmental studies in sedimentary geochemistry, *Science Progress*, **51**, 12–31.

Noble, B. (1964) 'Numerical methods: 2 differences, integration and differential equations,' Oliver and Boyd, Edinburgh.

Obermeier, S. F. (1974) Evaluation of laboratory techniques for measurement of swell potential of clays, *Bull. Assoc. Engrg. Geol.*, **11**, No. 4, 293–314.

Priest, S. D. (1975) 'Geotechnical aspects of tunnelling in discontinuous rock, with particular reference to the Lower Chalk,' Ph.D. Thesis, Univ. of Durham, England.

Pulpan, H. and Scheidegger, A. E. (1965) Calculation of tectonic stresses from hydraulic well fracturing data, *Jl. Inst. Petroleum*, **51**, 169–176.

Raiffa, H. and Schlaifer, R. (1961) 'Applied statistical decision theory', Boston

Division of Research, Graduate School of Business Administration, Harvard University, **28**.

Rocha, M. (1974) Recent possibilities of studying foundations of concrete dams, *Proc. 3rd Cong. Int. Soc. for Rock Mech.*, Denver, **1**, pt A 879–897.

St John, C. M. (1972) 'Numerical and observational methods of determining the behaviour of rock slopes in opencast mines,' Ph.D. Thesis, Imperial College, London.

Sabarly, F. (1968) Les injections et les drainages de fondation de barrages. *Géotechnique*, **18**, 229–249.

Scott, R. F. (1963b) 'Principles of soil mechanics,' Addison-Wesley, Reading, Mass.

Siegfried, R. W., Simmons, G., Schatz, J. and Richter, D. (1974) Effect of shock waves on elastic properties of rocks, *EOS*. 5S, ABS, 418.

Shiell, J. W. (1975) The water resources of Scotland, their use and abuse, *Proc. Inst. Civ. Engrs.*, **85**, pt. 1, 107–109.

Silvestrii, T. (1961) Determinazioni sperimentale de resistenza meccanica del materiale constitutente il corpo di una diga del tipo 'Rockfill', *Geotechnica*, **8**, 186–191.

Skempton, A. W. (1944) Notes on the compressibility of clays. *Qt. Jl. Geol. Soc.*, London, **100**, 119–135.

Smith, R. C. (1973) 'A comparison of some *in situ* methods for hydraulic conductivity measurement.' Dissertation: M.Sc. Advanced Course in Engineering Geology, University of Durham.

Sowers, G. F. (1963) Engineering properties of residual soils derived from igneous and metamorphic rocks, *Proc. 2nd Pan.Amer. Cong. Soil Mech. Found. Eng.*, Brazil, **1**, 39–61.

Steers, J. A. (1964) 'The coastline of England and Wales.' 2nd Edn. Cambridge Univ. Press, Cambridge.

Steffens, R. J. (1974) 'Structural vibration and damage,' Building Research Establishment Report H.M.S.O., London.

Talwani, M. and Ewing, M. (1960) Rapid computation of gravitational attraction of 3-dimensional bodies of arbitrary shape, *Geophysics*, **25**, 203–235.

Taylor, R. K. (1974) Private communication.

Tourtelot, H. A. (1974) Geologic origin and distribution of swelling clays, *Bull. Assoc. Engrg. Geologists*, **11**, no. 4, 259–275.

Valentin, H. (1954) Den Landverlust in Holderness, Ostengland von 1852 bis 1952, *Die Erde*, **3**, 295–315.

Van Olphen, H. (1963b) 'An introduction to clay colloid chemistry,' Interscience, New York.

Willis, B. and Willis, R. (1934) 'Geological Structures,' 3rd Edn. McGraw-Hill, New York.

Author Index

Aas, G., 482
Abel, J. F., 746
Ackers, P., 900–902
Ackroyd, T. N. W., 96
Adamson, L. G., 870
Agarwal, K. B., 306
Ahmed, A. A., 915
Akai, K., 286, 289, 291, 300
Alden, W. C., 724
Alexander, L. E., 26
Allum, J. A. E., 439
Alpan, I., 101, 516
Alty, J. L., 257
Alvarez, A., 927, 928
Alvarez, L., 955
Ambraseys, N. N., 519, 522, 946, 952, 953, 955, 956
Anderson, J. G. C., 915,
Anderson, W., 615
Andrea, D. A. D., 193, 194
Antevs, E., 307
Antsyferov, M. S., 205
Armando, S., 513
Arrowsmith, E. J., 706, 809, 810
Arthur, H. G., 944
Askey, A., 812
Atkinson, K. B., 337
Attewell, P. B., 19, 68, 69, 153, 155, 157, 166, 168, 170, 173–181, 200, 216, 218, 233, 236, 238, 241, 244, 256, 259, 266, 269, 271, 273–277, 279–281, 283, 289, 290, 292, 294–297, 299, 300, 308, 318, 320, 356, 365, 366, 381, 382, 385, 407, 426, 460, 503, 506–508, 514, 517, 523, 528, 535, 541, 552, 654, 700, 792, 812, 879
Awojobi, A. O., 518
Ayyar, T. S. R., 251

Badger, C. W., 165, 166
Badgley, D. C., 739
Bagnold, R., 715
Bailey, S. W., 264
Bailey, W. A., 639
Baker, D. W., 253, 254, 256, 257, 260
Baker, J. F., 198
Balla, A., 190

Balta, Y. I., 253
Banks, D. C., 307, 661, 663–665, 667–669, 686
Bannister, A., 503
Barden, L., 111, 113–122, 305
Bardwell, G. E., 968
Baren, F. A. Van, 696
Baron, R. A., 831, 833
Barrett, C. S., 253, 256, 272, 436
Barron, K., 289, 290, 295, 501
Barshad, I., 163, 164,
Bartley, M. W., 307
Barton, N., 326, 338, 734, 735, 753, 754
Basore, C. E., 824, 825
Bavel, C. H. M. von, 592
Bazett, D. J., 693
Beard, C. N., 740
Beavan, P. J., 443, 444
Beer, E. de, 480
Beles, A. A., 816
Bell, J. F., 282
Bendel, L., 733
Benjamin, J. R., 364
Benko, K. F., 852
Berger, L., 318
Berger, P. R., 518
Berkovitch, I., 166
Berman, H., 12
Berry, P. L., 121, 836
Best, R., 118
Bevan, O. M., 461
Bhandari, H. K., 692
Bhaskaran, R., 305
Bhatia, H. S., 676
Bhattacharyya, K. K., 748
Bieniawski, Z. T., 201, 203, 209, 210–212, 215, 545, 728, 729, 735
Bierbaumer, A., 733
Billings, M. P., 364, 739
Binnie, G. M., 911
Biot, M. A., 859
Bishop, A. W., 57, 153, 229–233, 307, 308, 399, 426, 558, 639, 644, 649, 658, 661, 666, 673–675, 682, 701, 702, 791, 805, 850
Bjerrum, L., 90, 92, 113, 130, 133, 134, 146, 150, 151, 153, 327, 532, 533, 693–695, 860, 940

Author Index

Black, R. F., 874
Blair, B. E., 523
Blake, W., 206
Bloomer, R. J. G., 918
Boast, C. W., 594
Boden, J. B., 128, 460, 537–539
Bohor, B. F., 18, 20
Boitano, J. D., 824, 825
Bonilla, M. G., 955
Booth, I. N., 709
Borie, B. S., 254
Bornitz, G., 241
Boshkov, S. H., 748
Bostrom, R. C., 533
Boulton, M. S., 604, 610–614
Bourgin, A., 413
Boussinesq, J., 92, 458
Bowden, F. P., 61
Brace, W. F., 199, 208, 250, 282, 290, 292, 295
Bradshaw, R., 254
Brady, B. T., 290, 294
Braun, W., 600, 871
Bray, J. W., 316, 723, 743, 752, 753, 762, 763, 767, 768
Braybrooke, J. C., 385
Bredehoeft, J. D., 619
Brekke, T. L., 326, 886
Brentnall, D., 503
Bretschneider, C. L., 714
Brewer, R. C., 112
Bridgeman, P. W., 222
Brink, A. B., 443
Brooker, E. W., 146, 152, 153, 177
Broscoe, A. J., 692
Brown, E. T., 316, 749
Brown, G., 172
Brown, J. D., 656, 681, 693
Brown, J. G., 899
Brückl, E., 692
Brunsden, D., 705
Buchan, S., 531
Buessem, W. R., 305
Bullock, W. D., 804
Bulman, J. N., 866
Bundred, J., 561
Burdine, N. T., 216
Burgess, A. S., 614, 616–619, 623, 624
Burrow, D. C., 896
Burton, A. N., 730, 731
Burwash, E. M. J., 307
Burwell, E. B., 475
Butler, F. G., 385, 864
Butler, M. E., 317, 337
Butler, R. M. J., 899, 900
Butterfield, R., 600

Cabrera, J. A., 305

Call, R. D., 337
Cambefort, H., 855
Capper, P. L., 157, 458, 543, 544, 574, 576
Carrillo, N., 308
Carder, D. S., 968
Carson, M. A., 692, 698–700, 702, 704, 720–722
Casagrande, A., 23, 35–41, 84, 110, 127, 129, 308, 317, 572, 574, 656, 873, 941
Casagrande, L., 868, 869, 871
Cashman, P. M., 583
Cassie, W. F., 157, 458
Cattin, P., 858
Cedergren, H. R., 568, 597
Chakrabarti, P., 959
Chandler, R. J., 121, 124, 130, 131, 146, 170, 676, 679, 680, 682–687, 688, 692, 701
Chappell, B. A., 749
Charles, J. A., 504–506
Chatwin, C. P., 136, 138, 140
Chayes, F., 365
Chen, J. C., 254
Chenevert, M. E., 294, 304
Cheung, Y. K., 746
Childs, E. C., 591
Chiriaev, R. A., 700
Chopra, A. K., 962
Chopra, K., 959
Chorley, R. J., 711
Chowdhury, R. N., 746
Christensen, D. M., 475
Christian, C. S., 443
Christiansen, D. W., 156
Christie, I. F., 834, 835
Chugh, Y. P., 216
Cladbaugh, P. S., 366, 368
Clark, R. A., 898, 900
Clark, S. P., Jr., 182
Clarke, D. D., 695
Clarke, P., 887
Clarkson, C. G., 654
Clevenger, W. A., 118
Clough, R. W., 399, 401, 955
Coates, D. F., 182, 183, 185, 723, 730, 731, 804, 881
Cochrane, N. J., 945
Coleman, J. D., 158
Collin, A., 693
Collins, K., 108
Collins, S. P., 871
Conlon, R. J., 308
Connelly, R. J., 305
Cook, N. G. W., 50, 191, 207, 231, 232, 290–292, 294, 371, 374, 377, 502
Cooper, H. S., 609
Cornell, C. A., 364

Costet, J., 879
Coulson, J. H., 749
Courchée, R., 314, 654, 655, 657–661, 718
Cramér, H., 547
Crampton, C. B., 282
Crandell, D. R., 692, 694
Crandell, F. J., 516, 517
Crawford, C. B., 695, 917
Crawford, N. H., 898
Creager, W. P., 901
Cripps, J. C., 294, 301, 302, 552, 557
Croney, D., 158–160
Crosby, W. O., 317, 318
Cumming, J. D., 878
Curry, R. R., 694
Cundall, P. A., 749

Dalrymple, J. B., 704
D'Appolonia, E., 531, 695
Darbyshire, M., 715,
Das Ries, 200
Dastidar, A. G., 833
Davies, B., 240, 517
Davis, A. G., 120
Davis, G. H., 460
Davis, S. N., 561, 562, 591, 605, 607–610
Davison, G., 702
Daw, G. P., 585
De, P. K., 305
Deacon, W. J., 871
Dearman, W. R., 431, 436
Decker, B. F., 254
Decker, E. R., 250
Deer, W. A., 8
Deere, D. U., 63, 65, 80, 185, 197, 198, 361, 449, 466, 471, 475, 486, 487, 489, 502, 503, 512–514, 700, 730, 732, 737, 751, 753, 884
Deklotz, E. J., 300
DeLory, F. A., 634, 659, 676, 687, 710
Denisov, N. Ya., 118
Denness, B., 138, 139, 141, 142, 315, 317–319, 322, 323, 338, 755, 756
Desai, C. S., 746
Devine, J. F., 523
Dobrey, R., 955
Dobrin, M. B., 449, 453, 454
Dodd, J. S., 475
Dollar, A. T. J., 966
Donath, F. A., 289, 291, 294, 308
Doornkamp, J. C., 704, 716
Douglass, P. M., 250
Dowling, J. W. F., 442–444
Draper, L., 715
Dreyer, W., 213
Dudley, J. H., 119
Duke, C. M., 955
Dumbleton, M. J., 158, 159, 161, 429, 437,

Dumbleton, M. J. (*continued*)
444, 476, 477
Duncan, J. M., 302–306, 308, 735, 736
Dungar, R., 959
Dunn, C. G., 253
Dupuit, J., 581
Duvall, W. I., 30, 183, 205, 241, 502, 516, 517, 523, 879, 951, 952

Early, K. R., 181, 708, 709, 914
Eden, W. J., 307, 695
Edvokimov, P. D., 700
Edwards, A. G., 245
Edwards, A. T., 516, 517
Edwards, A. W. F., 556
Edwards, R. J. G., 447, 811, 812
Eide, O., 693, 822
Eldin, A. K. G., 850
Elson, W. K., 879
Elvery, R. H., 236
Ernst, L. F., 593
Esu, F., 317, 337, 676, 693
Eswaran, H., 19
Evans, D. M., 968
Evans, R. H., 884
Ewing, M., 236
Ewing, W. M., 238

Fairburn, H. W., 364
Fairhurst, C., 198, 207, 208, 210, 857
Falkiner, R. H., 536
Farmer, I. W., 48, 68, 69, 183, 216, 218, 223, 233, 236–238, 241, 356, 426, 504–506, 517, 523, 535, 792, 848, 857, 866, 870, 876, 879, 879, 884
Fayed, L. M., 19
Fellenius, W., 639, 649, 791
Feng, C., 256
Field, M., 254
Fischer, F., 820
Fish, B. G., 517
Fleming, R. W., 146, 147, 150, 151
Flinn, D., 263–266, 365, 366, 368
Focardi, P., 318
Fogelson, D. E., 516, 517, 951, 952
Folayan, J. I., 550
Fookes, P. G., 109, 118, 138, 139, 141, 142, 316, 321–323, 337, 338, 431, 436, 447, 448, 755, 756,
Forbes, A. C., 492–494, 741
Foster, R. H., 305
Francis, E. A., 615, 654
Franklin, J. A., 192–194, 219, 220, 476, 477
Freitas, M. H. De, 797
French, B. M., 200
Freyburg, B. von, 671
Friedman, M., 365
Fukuoka, M., 694

Gatlin, C., 304
Gavshon, R. A. C., 397, 788–790, 792
Gehlen, K. von, 254
Gerber, E., 724
Gete-Alonso, A., 927
Geuze, E. C. A. W., 305
Gibbs, H. J., 118
Gibbs, Z. E., 205
Gibson, R. E., 307
Gilbert, G. K., 317
Gillott, J. E., 27, 111, 179, 251, 273, 870, 874
Gizienski, S. F., 532
Glossop, R., 708, 709, 838
Gnaedinger, J., 318
Gold, L., 205
Golder, H. Q., 103
Goodier, J., 236
Goodman, R. E., 206, 735, 736, 748
Gordon, B. B., 700
Grace, H., 599
Grant, K., 109
Grant, P. A. St. C., 536
Gray, D. A., 629
Gray, G. R., 878
Graybill, F. A., 355, 364
Green, H. W., 282
Green, J. H., 599
Green, R. W. E., 968
Greenwood, D. A., 822, 841, 844, 858
Greer, G. de, 307
Grice, R. H., 177
Griffin, O. G., 168
Griffith, A. A., 199, 201–203, 209, 213, 226, 228, 290, 295, 297
Griffiths, D. H., 449
Griggs, D. T., 221–223, 282
Grim, R. E., 16, 22–24, 175, 176
Grobbelaar, C., 546
Grundy, C. F., 858
Gruner, E., 701
Gruner, G., 701
Guha, S. K., 968

Haefeli, R., 634
Hager, R. V., 222
Haimson, B. C., 217
Hall, E. B., 700
Halstead, P. N., 337
Hambly, E. C., 305
Hamlin, M. J., 917
Hamrol, A., 754
Hancock, J., 517
Handin, J., 222
Hanna, T. H., 503, 511, 512, 595
Hanrahan, E. T., 121
Hantush, M. S., 565, 604, 609
Hardin, J., 305

Hardy, H. R., 206, 216
Hardy, R. M., 680
Hartshorne, N. H., 260
Haskew, H. C., 592
Hast, N., 499
Hawkes, I., 184, 191, 499
Hawkes, J. M., 884
Haws, E. T., 583
Hayes, D. J., 501
Heard, H. C., 220
Hearmon, R. F. S., 250
Heckler, A. J., 253
Heerden, W. L. van, 501
Helfrich, H. K., 513
Hem, J. D., 628
Hendron, A. J., 101, 519, 522, 795–797
Henkel, D. J., 130, 132, 133, 327, 409–411, 413, 414, 417, 694
Herbert, R., 604, 619
Hetherington, H. A., 878
Heuzé, E. E., 746
Heyman, J., 198
Hibbert, E. S., 630
Higginbottom, I. C., 438, 439
Higgs, D. V., 254
Higson, G. R., 514
Hilf, J. W., 912
Hill, R., 231
Hill, R. A., 629
Hilliard, J. E., 390
Hillis, S. F., 700
Hilton, I. C., 837, 839
Hinckley, D. N., 168
Hiramatsu, Y., 194
Hirschfield, R. C., 686
Hobbs, D. W., 194, 222, 476
Hobbs, N. B., 825
Hodgson, E. A., 205
Hodgson, R., 317, 739
Hoeg, K., 118
Hoek, E., 219, 220, 226, 227, 290, 295, 316, 337, 495, 496, 585, 686, 723, 743, 744, 748, 750, 752, 753, 762, 763, 767, 768, 775, 779, 782, 785, 787, 804, 806–808
Holden, W. S., 628
Holland, W. Y., 118
Hollister, J. C., 968
Holtz, W. G., 118
Hooper, J. A., 385
Horder, R. I., 441
Horn, H. M., 63, 64, 80
Horsley, R. V., 166
Hoskins, E. R., 501
Hough, B. K., 353
Houlsby, A. C., 853–856, 860
Howard, T. R., 326
Howells, D. A., 968

Hubbert, M. K., 857
Huber, C., 187, 196, 236
Hudson, J. A., 198, 207, 363
Hughes, D. S., 948
Hughes, J. M. O., 497, 510
Hughes, M. J., 468, 693
Hughes, R. E., 18, 20
Hurlbut, C. S., 3, 10
Hutchinson, J. N., 111, 468, 639, 646, 668, 680, 687–689, 691–696, 702
Hvorslev, M. J., 595, 596, 844

Idriss, I. N., 533, 955
Iida, K., 694
Iliev, I. G., 513
Inderbitzen, A. L., 532
Ineson, J., 889–893
Inman, D. L., 716
Insley, A. E., 700
Ireland, H. O., 177
Irwin, G. S., 203, 209
Iverson, N. L., 680

Jacob, C. E., 581, 605, 609, 618
Jaeger, J. C., 50, 191, 195, 207, 224, 227, 231, 232, 289, 294, 308, 316, 371, 374, 502, 749
James, A. N., 839, 840, 858
Janbu, N., 641, 649, 791
Jardetzky, W. S., 238
Jenkins, J. D., 545
Jennings, J. E., 122, 316, 339, 385, 394, 738
Jenny, H., 697
Jetter, L. K., 253, 254
John, K. W., xi, 316, 326, 730
Johnson, A. I., 471, 568
Johnson, A. M., 251
Johnson, D. L., 187
Johnson, S. J., 146, 150, 679
Jones, D. K. C., 705
Jones, F. O., 693
Jones, G. P., 893
Jones, K. S., 735, 737, 920, 921
Jones, P. H., 449
Jones, R. C. B., 908
Jørgenson, P., 27
Judd, W. R., xi, 6, 182, 187, 196, 236, 918, 919
Jumikis, A. R., 574, 632, 871, 872
Jungst, H., 318
Jurgenson, L., 400

Kamb, W. B., 282, 365
Kankare, E., 681
Karol, R. H., 846, 847
Karplus, W. J., 568
Kassiff, G., 162, 937

Kayes, T. J., 307
Kell, T. F., 305
Keller, G. H., 125, 532
Kellogg, C. E., 698
Kennard, M. F., 177, 910, 914, 944
Kenney, T. C., 113, 173, 175, 176, 178, 179, 308, 699
Kihlström, B., 517, 809, 812, 953
Kim, C., 217
Kinahan, G. H., 317
Kindle, E. M., 318
King, C. A. M., 704, 716
King, R. F., 449
Kirkby, M. J., 692, 698–700, 702, 704, 711, 720–722
Kirkham, D., 592, 594, 595
Kirkpatrick, W. M., 305
Kishida, H., 954, 955
Kley, R. J., 686, 804
Klug, H. P., 26
Knights, C., 515
Knights, M. C., 463
Knill, J. L., 205, 411, 412, 459, 513, 537, 735, 737, 741, 743, 811, 909, 910, 913–915, 920, 921
Knopf, A., 8
Knox, G., 811
Kolodyny, Y., 282
Korbin, G., 884
Kraft, L. M., 550
Krech, W. W., 861
Krigbaum, W. R., 253
Krinitzky, E. L., 118, 125, 126
Krumbein, W. C., 355, 364
Krynine, D. P., 6
Krynine, P. D., 5
Kummeneje, O., 822
Kurbanovic, M., 251

Lafeber, D., 251
Lajtai, E. Z., 724
Lam, P. W. H., 366
Lambe, T. W., 19, 21, 45, 49, 55, 63, 72, 80, 81, 86, 87, 90, 96, 100, 101, 105, 106, 111, 112, 127, 130, 132, 263, 303, 578, 584, 632, 639, 720, 818, 819, 821
Lamplugh, G. W., 654
Lancaster-Jones, P. F. F., 858
Lane, K. S., 58, 59, 153, 679, 804, 964
Lane, R., 487
Langefors, U., 516, 517, 809, 812, 953
La Rochelle, P., 315, 317, 385, 693
Larsson, L., 365
Lauffer, H., 733
Law, F., 896
Leach, B., 310–313
Learmonth, A. P., 915–917

Le Conte, J., 318
Lee, K. L., 824
Leeds, D. J., 955
Leeman, E. R., 501
Leet, L. D., 517, 952
Legget, R. F., 307
Leonard, D. R., 515
Leonard, M. W., 865, 866
Lewis, A. T., 962
Lewis, W. A., 158-160
Linell, K. A., 129
Linsley, R. K., 917
Little, J. A., 341, 347
Little, A. L., 327, 644, 687
Littlejohn, G. S., 881
Litwiniszyn, J., 356, 536
Lo, K. Y., 307-309, 327, 328
Londe, P., 352, 509, 510, 513, 514, 743, 744, 748, 750, 752, 753, 785
Loughnan, F. C., 155
Louis, C., 570, 572
Love, A. E. H., 250
Lowe, J., 700
Lubinski, A., 859
Lueder, D. R., 439
Lumb, P., 109, 693, 699, 701
Luthin, J. N., 593
Lutton, R. J., 118, 307, 661, 663-665, 667-669, 686, 712, 713, 804
Lykoshin, A. G., 513
Lyman, A. K. B., 822
Lynn, B. C., 645, 651

Maasland, M., 592
MacDonald, D. F., 173, 686, 725
MacDonald, G. J. E., 282
Mahtab, M. A., 748
Marachi, H. D., 72, 73
Marsal, R. J., 955
Marshall, A., 129
Marsland, A., 130, 131, 144-146, 317, 337, 701, 718
Martin, G., 317
Martin, G. R., 307
Martna, J., 155
Massalski, T. B., 253, 256, 272
Masure, P., 286
Mayer, A., 458
McCann, D. N., 531
McClintock, F. A., 190, 199, 228, 290
McCracken, D. D., 621
McCulloch, D. S., 955
McDowell, P. W., 451, 452
McGown, A., 108, 122-124, 328, 337
McGregor, K., 461
McHargue, C. J., 253
McKinlay, D. G., 656
McKinney, J. R., 878

McQueen, R. G., 948
McWilliams, J. R., 250
Meidav, Ts., 516
Mellis, O., 364, 366, 368
Mellor, M., 184, 191
Ménard, L., 497-499, 825
Mencl, V., 696, 931
Mendes, F. de M., 300
Merchant, M. E., 254
Merriam, R., 5, 189, 694
Merrill, R. H., 499
Mesri, G., 305
Meyerhof, G. G., 480
Middlebrooks, T. A., 399
Midgley, D. C., 917
Mielenz, R. C., 6
Millard, R. S., 444
Miller, D. L., 898
Miller, R. P., 197, 198
Milligan, V., 307, 308
Mirata, T., 399, 400
Mitchell, J. K., 105, 821, 822, 824, 842
Mitchell, R. T., 251, 303
Mitronovas, F., 305
Mogilevskaya, S. A., 354
Mohr, E. J. C., 697
Moller, K., 865, 866
Monahan, C. J., 365
Mononobe, N., 946
Mooney, H. M., 451
Moore, C. A., 105
Moore, J. F. A., 337
Moos, A. von, 694
Morgenstern, N. R., 77, 105, 186, 251, 305, 558, 673, 675, 679, 805, 857, 858, 861
Morin, J. P., 308
Morris, G., 517, 951, 952
Mossman, R. W., 457
Moxon, S., 499
Muchowski, J., 118
Muirhead, K. J., 968
Muller, L., 316
Mullineaux, D. R., 692
Munro, H., 545
Munson, R. C., 968
Murrell, S. A. F., 199, 228, 230
Muskat, M., 567

Nadai, A., 224
Nagy, G., 305
Nakano, R., 150, 166, 167
Nash, E. A., 901
Nasmith, H., 694
Nath, B., 962
Neelands, R. J., 839, 840, 858
Nesbitt, R. H., 475
Newland, D. H., 695

Newmark, N. M., 458, 795, 951, 953, 956–958
Nicholls, G. D., 9
Nixon, M., 901–904
Noble, B., 619
Nobles, L. H., 694
Nonveiller, E., 586
Noorany, I., 531, 532
Norman, J. W., 437
Northey, R. D., 318
Northwood, T. D., 516, 517
Norton, J. T., 254
Novik, G., 237, 451, 452
Nur, A., 251
Nye, J., 692

Obert, L., 30, 183, 190, 205, 502, 735, 879
Oborn, L. E., 948, 966, 968
Odenstad, S., 695
Oka, Y., 194
Okabe, S., 946
Okamoto, S., 968
Okeson, C. J., 318
Olphen, H. van, 108
Olsen, H. W., 112
Olson, R. E., 305
O'Neill, A. L., 352
Onodera, T., 512, , 513, 686
Orange, A. S., 250
Otto, G. H., 318
Over, V. G., 305
Overbeek, J. Th. G., 105

Palache, C., 7
Palit, R., 122
Palladino, D. J., 677
Palmer, C., 629
Panek, L. A., 499
Paolo, B., 513
Papageorgiou, S. A., 72
Pariseau, W., 536
Parish, D. G., 139
Parnassis, D. S., 449
Parsons, R. C., 185, 730, 731
Patterson, M. S., 260, 261, 271
Patton, F. D., 700, 750, 751, 753
Pearson, J., 243
Peck, R. B., 33, 36, 48, 49, 58, 68, 92, 166, 318, 345, 356; 479, 481, 547, 578, 627, 632, 640, 642, 645, 676, 684, 841, 937
Pellegrino, A., 700
Peng, S., 251
Penman, A. D. M., 504–506, 597
Penny, L. F., 654
Pentecost, J. L., 253
Pentz, D. Y., 337
Perkins, T. K., 861

Perrin, R. M. S., 181
Perrott, W. E., 858
Peterson, R., 153, 327
Petley, D. J., 121, 317, 682, 694, 695, 700
Petrovski, J., 959
Pettijohn, F. J., 31, 32, 318, 321
Phillips, F. C., 254, 328, 332, 364, 758
Philofsky, E. M., 390
Philpott, K. D., 159–161, 163–165
Phukan, A. L. T., 186
Pichler, E., 693
Pickering, D. J., 307
Pincus, H. J., 252
Pinder, G. F., 619
Pinto, J. L., 286, 294, 300
Piper, A. M., 630
Pirsson, L., 8
Pitcher, W. S., 118
Piteau, D. R., 339, 394
Pitman, W., 917
Polond, J. F., 599
Pomeroy, C. D., 300
Poskitt, T. J., 121, 307
Potts, E. L. J., 499, 509
Potter, P. E., 318
Poulos, H. G., 458
Prater, E. G., 879
Press, F., 238
Price, D. G., 741, 743, 811
Price, N. J., 189, 221, 224, 318, 319, 338, 738
Price, V. E., 644
Prickett, T. A., 619, 623
Priest, S. D., 342–344, 363, 426, 754–757
Prokopovich, N. P., 599
Prosser, J. R., 536
Pruška, L., 677
Pryor, E. J., 578
Pugh, S. F., 189
Pusch, R., 251, 305
Pulpan, H., 858

Rabcewicz, L. von, 733
Raffle, J. F., 48, 841, 844, 858
Raiffa, H., 556
Ramana, Y. V., 514
Ramez, M. R. H., 200
Rana, M. H., 804
Randell, P. A., 749
Rankilor, P. R., 194, 465
Ranganatham, B. V., 121
Rankine, W. J. M., 102
Rao, G. V., 305
Rapp, A., 697
Raw, G., 537
Raymond, S., 503
Rebull, P. M., 305
Reed, D. W., 619

Reeve, R. C., 595
Reeves, G. M., 129, 135, 138, 397, 788, 789
Rehbinder, P. A., 81, 189
Remillon, A., 819
Rennie, I. A., 305
Resendiz, D., 746, 748
Ricceri, G., 600
Rich, C., 735
Richards, A. F., 531, 532
Richards, H. J., 892
Richart, F. E., 832, 834, 835
Richter, D., 200, 251
Rinehart, J. S., 211, 243
Ritchie, A. M., 727, 729
Roberts, A., 499
Robertson, A. MacG., 349, 351, 354, 363, 394
Robertson, E. C., 212, 871
Robinson, W., 654
Rocha, M., 486, 497, 920
Rodrigues, F. P., 294, 300
Roe, R. J., 253
Rogers, S. H., 460, 811
Rosa, S. A., 483
Roscoe, K. H., 681
Rose, E., 962
Rosenblad, J. L., 749
Rosenblueth, A., 547
Rosengren, K. J., 337
Rosenqvist, I. Th., 105, 113, 303, 318
Ross, G. A., 955
Ross-Brown, D. M., 337
Roussell, J. M., 513
Rowe, P. W., 64, 67, 80, 302, 308, 828–830
Ruddock, E. C., 109
Rummel, F., 207
Rushton, K. R., 619
Russam, K., 158
Rydz, B., 888
Rzhevsky, V., 237, 451, 452

Sabarly, F., 588, 852
St. John, C. M., 736
Salamon, M. G. D., 545
Salas, A. J., 927
Sandford, M. R., 256, 259, 289, 290, 292 294–297, 299, 300, 308, 700
Sanger, F. J., 816, 871–873
Sanglerat, G., 879
Sangrey, D. A., 308
Sankaran, K. S., 305
Sapegin, D. D., 485
Sarma, S. K., 946, 952, 953
Saul, T., 514
Sauvage de St. Marc, G., 859
Savanick, G. A., 187
Scheidegger, A. E., 692, 724, 858

Scheidig, A., 712
Schlaifer, R., 551, 556
Schmidt, B., 356, 535
Schmidt, W., 364
Schmidt, W. E., 595
Schofield, A. N., 98–100, 681
Schofield, R. K., 157, 158
Schulz, L. G., 27, 254, 256, 257
Schuster, R. L., 318
Scopel, L. J., 968
Scott, J. S., 146
Scott, R. A., 48, 843
Scott, R. F., 49, 530
Scrivenor, J. B., 694
Seed, H. B., 304, 305, 306, 308, 533, 663, 695, 824, 946, 954, 955
Sevaldson, R. A., 681, 693
Severn, R. J., 959
Sexton, J. R., 918
Shalom, A. B., 162
Shapiro, J., 586
Sharp, J. C., 806
Sharp, R. P., 694
Shea, H. F., 129
Sherard, J. L., 399, 939, 940, 943, 967
Shergold, F. A., 245, 248, 249
Sherif, M. A., 533
Shiell, J. W., 887, 888
Shidomoto, Y., 497, 858
Shirayaev, R. A., 485
Short, N. M., 200
Shuk, T., 686, 803
Sides, G. R., 111, 120
Siegfried, R. W., 200
Silveira, A. F. Da, 316, 337
Silvério, A., 497
Silvestrii, T., 725
Simmons, G., 200, 251, 390
Sitter, L. U. de, 318
Skempton, A. W., 35, 56, 58, 82, 84, 88, 90, 111, 121, 130, 132, 133, 151, 153, 154, 181, 183, 315–318, 321, 327, 337, 385, 426, 634, 639, 646, 649, 656, 659, 668, 676, 677, 680, 681–685, 687–689, 691–696, 701, 702, 708, 710
Skibitzke, H. E., 619
Skipp, B. O., 519, 527
Sloane, R. L., 305
Smalley, I. J., 305
Smart, P., 305
Smedley, M. I., 307, 308
Smiles, D. E., 591, 596
Smith, D. B., 615
Smith, D. T., 531
Smith, R. C., 595
Smith, R. E., 459
Smothers, W. J., 586

Snow, D. T., 589, 590, 851, 968
Søderblom, R., 113
Soejima, T., 858
Southwell, R. B., 618
Sowers, G. B., 585
Sowers, G. F., 585, 701
Spears, D. A., 155, 156, 166, 167, 171, 177, 698, 701
Spencer, A. B., 366, 368
Spencer, E., 639, 807
Spencer, E. W., 365
Sridharan, A., 305
Stacey, T. R., 746
Stagg, K. G., 490
Stallman, R. W., 619
Stanculescu, I. I., 816
Stapledon, D. H., 185
Starkey, J., 260
Stauffer, M. R., 364, 365
Steers, J. A., 655
Steffens, R. J., 515
Steinbrugge, K. V., 955
Stemler, O. A., 300
Stewart, G. A., 433
Stimpson, B., 478
Stini, J., 732
Stoker, J. J., 715
Straud, T., 364
Stuart, A., 260
Suits, C. G., 254
Suklje, L., 693
Sultan, H. A., 663
Symons, I. F., 709
Szecki, A., 545
Sztyk, Z., 118

Tabor, D., 61
Tagg, G. F., 450
Takano, M., 497
Talbot, C. J., 263–266
Talobre, J. A., 483
Talwani, M., 236
Taniguchi, T., 695
Taylor, D. W., 46, 49, 72, 95, 511, 672, 673
Taylor, K., 894
Taylor, R. K., 153, 155–157, 166–168, 170, 173–181, 271, 279–281, 318, 381, 382, 401, 541, 646, 651, 654, 698, 701, 812
Tayton, J. W., 519, 527
Tchalenko, J. S., 77, 251, 305
Teichmann, G. A., 517
Terzaghi, K., 33, 36, 48, 49, 58, 68, 92, 93, 146, 166, 316, 318, 345, 480, 481, 544, 545, 547, 578, 627, 632, 640, 642, 645, 676, 684, 702, 723, 724, 730, 732, 735, 807, 809, 826, 841
Terzaghi, R. D., 349, 351, 937

Theis, C. V., 605
Thill, R. E., 250
Thoenen, J. R., 516, 517
Thompson, T. F., 173
Thomson, S., 692
Thorn, R. B., 714, 718
Thú, L., 677
Thurber, W. C., 253
Timoshenko, S., 236
Tirey, G. B., 531
Todd, D. K., 591
Todd, T., 200
Tomlinson, M. J., 480, 481, 483, 707, 881
Toms, A. H., 693, 694
Torrance, J. K., 113
Tough, S. G., 536
Tourtelot, H. A., 176
Trostel, L. J., 168
Tschebotarioff, G. P., 819
Tsytovich, N. A., 872
Turnbull, W. J., 118, 712, 819
Turner, F. J., 364, 365
Turner, M. J., 540–542
Twenhofel, W. H., 318
Twort, A. C., 900, 906

Ulmer, G. C., 586
Underwood, L. B., 146–150
Uriel, S., 927

Valentin, H., 655, 657
Vallarino, E., 927, 928
Vargas, M., 693
Varnes, D. J., 693–695
Vaughan, P. R., 588, 676, 851, 857, 858, 861, 944
Verwey, E. J. W., 105
Vidmar, S., 693
Voight, B., 84, 251, 536
Vucetik, R., 698

Wagner, A. A., 31, 35–41
Walbancke, H. J., 676
Waldorf, W. A., 485
Walker, P. E., 749
Walker, W. H., 629
Wallace, K. B., 109, 700
Waller, R. A., 518, 519
Walsh, J. B., 190, 199, 228, 250, 290, 292, 295
Walsh, P. D., 896
Walter, J. L., 253
Walters, R. C. S., 424, 579, 923–927, 930–932
Walthall, S., 515
Walton, W. C., 605, 610, 611, 619, 623, 630
Wang, H., 200
Wantland, D., 919

Ward, W. H., 318, 321, 488
Wardell, K., 863
Washburn, D. E., 58, 59
Watari, M., 695
Watt, A. J., 825
Watters, R. J., 797
Wawersik, W. R., 208, 210
Weaver, R., 337
Webb, D. L., 825
Weeks, A. G., 694
Weibull, W., 191
Weimer, R. J., 968
Weiss, L. E., 260, 261, 271, 364, 365
Wenk, H. R., 253, 282
Wenzel, L. K., 605, 607
Wesley, L. D., 109
West, G., 158, 159, 161, 429, 437, 444, 476, 477
Westergaard, H. M., 946, 947
Westwater, R., 517
Wetzel, W. W., 451
Whiffin, A. C., 515
Whipkey, R. Z., 711
White, W. A., 24, 318
Whitman, R. V., 19, 21, 45, 49, 55, 63, 72, 80, 81, 86, 87, 90, 96, 100, 101, 111, 127, 130, 132, 263, 578, 632, 639, 720, 819, 821, 946
Wiebols, G. A., 290–292, 294, 377
Wiegel, R. L., 714
Wiest, R. J. M. De, 561, 562, 591, 605, 607–610, 900, 905, 906
Wilkinson, W. B., 836
Willard, R. T., 250
Willis, B., 740
Willis, D. E., 517
Willis, D. G., 857, 859
Willis, R., 740

Wilson, A. H., 501
Wilson, D. D., 316
Wilson, E. M., 899
Wilson, G., 599
Wilson, J. T., 517
Wilson, S. D., 695, 696, 748, 955
Winchell, H., 365
Windes, S. L., 516, 517
Winslow, N. M., 586
Wisecarver, D. W., 499
Wittke, W., 352, 570, 572
Wong, H. Y., 819, 857, 860–862
Wood, A. N. Muir, 714–718
Wood, C. C., 232
Wood, J. C., 863
Wood, W. O., 600
Woodman, J. P., 365, 366, 385, 407
Woodland, A. W., 537, 540, 709, 711, 811–813
Woodruff, S., 30
Woodward, R. J., 399, 401
Wright, C. H., 253
Wright, S. G., 302, 303, 663
Wroth, C. P., 69, 98–100, 497, 510, 681
Wu, T. H., 129, 251
Wu, T. S., 550
Wyckoff, R. D., 619
Wynne, D. J., 168

Young, E. G., 591, 596

Zaruba, Q., 696
Zaslavsky, D., 937
Zee, C. H., 604, 619
Zienkiewicz, O. C., 490, 746
Zonana, J., 746, 748
Zwart, H. J., 654

Subject Index

Main entries are indicated in bold type

A-line, **36**, 656
Abstraction rights (water), 601
Acidising, 627
Acidity, 910
Acoustic emission, see Microseismic Activity
Acoustic holography, 439
Acoustic impedance, 242
Activity, 35, 176
Adits, 920
Aerial photographs, 437, 546, 602, 907, 908
Aggregates, **244**, 248, 753, 810, 908, 911, 923
    abrasion test, 247
    bitumen-coated, 244
    crushing strength, 245
    crushing test, 247
    durability, 245
    impact test, 247
    polished stone coefficient, 249
    tests generally, 476
    thermal insulation, 245
Albite, 4, 279
Allophane, 19
Alluvial plains, 900
Alluvium, 701, 913
Alteration index, 754
Anchors,
    cable, 782
    ground, 816, 879
Andesite, 4, 752, 858
Anions, **10**
Anisotropic flow, **570**
Anisotropic index, **266**
Anisotropy,
    intrinsic, **285**
    intrinsic strength, 289
Ankerite, 155
Anticline, 909
Apatite, 171
Aqueduct, 895
Aquiclude, 565, 907, 911
Aquifer, 468, 564, 892, 907, 910
    confined, 582
    unconfined, 581
    water table, 610
Aquitard, 565

Arching, 879, 944
Area factor, 858
Arid regions, 445
Arrhenius equation, 220
Artesian aquifer, 610
Artesian head, 600
Artesian pressure, 565, 913, 914
Aspect ratio, 190
Asperities, 61
Asphalt blanket, 937
Asphalt jointing, 718, 955
Attapulgite, 19
Attenuation, wave, 513, 524
Atterberg limits, **23**, 476
Atterberg liquid limit, **23**, 35, 42, 82, 161, 168, 175, 179, 180, 353, 943
Atterberg plastic limit, **23**, 35, 168
Auger hole method (permeability testing), **592**
Avalanches, rock, 723

Backwash (tidal), 716
Bacterial action, 892
Bank storage, 904
Barometric pressure, 603
Barrages, 892
Barytes, 878
Basalt, 4, 245, 248, 741, 752, 811, 863
Base exchange, clay minerals, **20**, 178
Base flow (river), 898, 904
Bayesian statistics, 547
Beach, 715
Beam, 542
Bearing capacity, 480, 923
Bed load, 903, 913
Bed separation, 909
Bedding planes (control on slope stability), 661
Bell pits, 439, 546
Bentonite, 20, 22, 753, 868, 874, 879
Berm, 706
Bitumen, 874
Blasting, 809, 812
Block glides, 724
Boehmite, 168
Bolting, rock, 725, 782, 802, 811

1036 Subject Index

Bonding,
  ionic, 10
  in clay shales, 151
Bond strength, 9, 871, 944
Bonds,
  covalent, 10
  diagenetic, 113, 151
Bord, see Pillar and stall
Borehole, 233, 458, 878, 891, 911, 912, 913, 920
Borehole extensometers, 508, 509
Borehole logging, **465**, 920
Borehole measurement devices, **496**, 501
Borehole T.V. cameras, 475
Borrow areas, 908
Bouger anomaly, 449
Boussinesq, 458, 459, 484, 543, 826, 852
Boulders, 31, 654
Breccia, 6, 753
Brittle–ductile transition, 219, 222
Brittleness index, 229
Brucite, 16
Bulk density, **43**, 477
Bulk modulus, 236, 455
Bulkhead, 865

Cable anchors, 782, 925
Cable fencing, 727
Calcite, 63, 154, 326, 712, 753
Calcium, 171, 178
Cambering, 710, 909
Canal, 661, 671, 877, 954
Canyons, 722
Capillaries, 697, 867
Capillarity, 76, 725
Capillary forces, **58**, 561
Capillary zone, 58, 562, 563
Cartesian coordinates, 333
Catchment, 897, 907, 908
Catchment area, 899, 900
Catchment development, 907
Cation exchange (see also base exchange), **20**, 866, 869
Cations, **10**
Cavern, 915
Cavitation, 949
Chalk, 6, 141, 488, 626, 752, 754, 788, 863, 892, 893, 894, 895
Channel,
  buried, 913
  geometry, 902
  protection, **900**
  size, 899
Chemical analysis, 478
Chert, 5
Chlorite, 17, 154, 173, 326
Classification,
  clay slides, **688**

Classification (*continued*)
  clays, **112**
  dams, **922**
  discontinuities, **320**
  discontinuity roughness, 753
  land, see Terrain evaluation
  rock masses, **730**
  rock particles, **30**, 943
  rock slope profiles, **704**
  rock slope morphology, **720**
  rock slopes, **735**
Clay, 6, 31, **104**, 307, 449, 452, 532, 645, 718, 724, 734, 753, 816, 818, 820, 842, 858, 866, 870, 878, 911, 924, 932
  boulder, 6, 36, 122, 314, 538, 633, 654, 657, 701, 718, 824
  brackish water, 115
  fissured, 692
  freshwater, **115**
  laminated, **301**, 597, 633, 653, 692, 829
  marine, 114
  normally consolidated, 84
  organic, 818, 829
  overconsolidated, 84, 113, 117, **127**, 676, 748
  'quick', 113, 309
  sensitive, 113
  varved, 307, 692
  volcanic, 701
Clay,
  Ampthill, 153, 177
  Atherfield, 709
  Boston Blue, 129, 868
  Gault, 139
  Kimmeridge, 127, 344, 357, 360, 379, 682, 685, 688, 788
  Lias, 679
  London, 55, 83, 129, 360, 426, 682, 684, 688, 702, 788, 868
  Lower Lias, 360, 788
  Norwegian marine, 129
  Oxford, 131, 360, 788
  Thames Estuary, 129
  Upper Lias, 129, 682, 684, 688, 702
  Weald, 137, 360, 709, 788
Clay,
  depth/strength relationship, 125
  lining, 925
  puddled, 937, 945
  strength anisotropy, 302
Clay shale, **104**, **127**, **146**, 302, 319
  Bearpaw, 152
  Claggett, 152
  Colorado, 152
  Cucaracha, 173, 178, 661
  Dawson, 402
  Fort Union, 152

Subject Index   1037

Clay shale *(continued)*
  Pierre, 152
Clay shale,
  breakdown of, 155, 166
  geotechnical properties of, 175
Clay minerals, 14, **16**, 167, 676, 677, 871, 944
  orientation density distribution of, **250**
Claystone, 6
Cliffs,
  erosion of, 655
  recession rate of, 655
Coal, 6, 708, 724, 945
Coastal erosion, **714**
Coastal protection, 716
Cobble, 31
Cohesion, 65, 76, 77, 354
Collapsing soils, 119
Colloids, 105
Colluvial mantle, 697
Colluvial mass, 661
Columns,
  brick, 816
  stone, 821, 822, 825
Command area, 895
Compaction, 674, 714, 816, 822
  deep, **816**, 825
  explosive or dynamic, 816, 821, **822**, 825
  shallow, 817
Compaction grouting, 821
Compaction piling, 821, 822, **823**
Compaction test, 940
Compliance, axis of, 281
Composition of earth, **2**
Composition of rocks, **1**
Compressibility,
  axis of linear, 281
  coefficient of, 83
  coefficient of volume, 83
Compression index, 82
Cone of depression, 617, 891
Confining pressure, 73
Conglomerate, 6, 670
Conjunctive use, 895
Consolidation, 150, 478, 534, 597, 816, 826, 873, 939, 954
  coefficient of, 94
  one-dimensional, 92
Conurbations, 908
Cores, 465
Cracks, 199, 292
  classification in rocks, 200
  tension, 541
Creep, 212, 655, 657, 702, 751
Creep rate, 658
Critical state (soil mechanics), **97**
Crust (earth), 1

Crystallinity factor, 168
Cumulative yield curve, 905, 914
Curve fitting, statistical (synthetic flow generation), 917
Cut-off, 909, 910, 912, 913, 932
Cuttability, 513
Cyclic loading, 215

Dacite, 4
Dam,
  arch, 858, **929**, **930**, 950, 963
    double (cupola), **930**
    multiple, **929**
    thick, **930**
    thin, **930**
  buttress, 908, **928**
  concrete gravity, 567, 908, **922**, 946, 955, 959
  earth/rockfill, **931**, 937, 944, 946, 954
  sheet pile coffer, 567
Dam
  hydroelectric, 922
  impounding, 922
  regulating, 922
Dam
  arch, foundation, 409
  core, 579
  earth or rockfill, foundation, 398, 817
  earth, water flow through, 573
  fill, 674
  foundation investigations, **911**
  leakage, 910
  loadings, 918
  measurement of movements in, **504**
  seismic stability, **955**
    base shear, **959**
    overturning, **963**
  sites, 735, 908, 911
Darcy's law, 46, 94, 567, 585, 587, 591, 868
Deformation, granular soils, 66
Deformation domain, 263
Deformation ellipsoid, **263**
Deformation modulus, 487, 490
Deformation paths, **263**
Delta function, 366
Desiccation, 156
Diaphragm wall, 877, 879
Diatomaceous earth, 36
Diatomite, 6
Differential thermal analysis, 27
Dilatancy, 37, 73, 948
Dilation, 64, 72, 202, 749, 785, 967
Dilatometer, 921
Dimensional analysis, 520
Diorite, 686, 752
Direction cosines, 335
Discharge, 898

1038   Subject Index

Discharge (*continued*)
  dominant, 901
Discontinuites, 127, 143, 321
  character of, **326**
  classification of, **320**
  continuity of, **394**
  critical equilibrium in shear, **369**
  fabric, 136, 138
  genesis, **317**
  orientation, 346, 349
  orientation density distribution of, **364**, 365
  probability of propagation, 387
  spatial density distribution of, **355**
  survey of, **336**
  water flow through, 587
Distribution,
  a priori, 550
  negative exponential, **357**
  normal probability, **356**
  orientation density, **364**, 787
  Poisson, **357**
  posterior, **550**
  sampling, 550
  spatial density, **355**
Dolerite, 4, 245, 493, 740
Dolomite, 5, 6, 171, 275
Drain,
  counterfort, 652, 707, 709
  horizontal, 652, 670
  interceptor, 652
  sand, 816, 829, **830**
  vertical, 652
Drain wells, **580**
Drainage, 59, 444, **580**, 715, 785, 811, 864, 907, 923, 932, 940
  surface, 816
  tributary, 712
Drainage blanket, 911
Drainage galleries, 786
Draw-down,
  reservoir, 575, 915
  water table, 602
Drilling mud, 877, **878**
Dry bulk density, 43
Durability, 477
Dutch cone test, 480
Dykes, 445
Dynamic moduli, 513

Earth pressure
  cell, 510, 511
  coefficient of active, 102, 721
  coefficient of, at rest, 51, 55, 57, 66, 100, 153, 721
  coefficient of passive, 102, 676
Earthflow, **689**
Earthquake, see Seismic intensity, Seismicity

Eclogite, 948
Effluent injection, 968
Elastic modulus, see Modulus of Elasticity
Elastic storage (aquifer), 610
Electro-osmosis, 23, 706, 816, 866
Electrolyte, 449
Electron microscope, 251
Elements, composition of rocks, 3
Environmental constraints, 908
Epicentre, 948, 966, 967
Equipotential lines, 566
Erosion, 907, 911, 915, 944, 945
Errors, sampling, 338
Escarpments, 444
Eulerian angles, 336
Evapo-transpiration, 898, 918
Evaporites, 5
Exploration, groundwater, 601
Explosive, 233, 523

Fabric, clay mineral, 107
Failure criteria, **224**
  Extended Tresca, 231
  Extended Von Mises, 231
  Maximum shear stress (Coulomb), 224
  Tresca, 230
  Von Mises–Huber–Henky, 231
Failure envelope, 77, 78
  peak and residual, 79
Fatigue of rock, 216
Faulting, 445
Faults, 321, 601, 661, 724, 910, 912, 920, 946, 948, 966
Feldspar, 4, 5, 15, 32, 154, 910, 929
Feldspathoids, 4
Fetch (waves), 714
Field monitoring, **503**
Fill, 821, 911
Filled ground, 954
Filter, 584, 652, 932, 940, 944
  inverted, 706
Filter cake, 841, 877
Filter design, 626
Filter screen, 627
Filtration, 891
Finite element analyses, 352, 746, 915
Fissility, 127
Fissures, see Discontinuities
Flat jacks, 485
Flint, 248
Flood, maximum probable, 897
Flood banks, 899
Flood flow, 888, 898, 907, 908
Flood relief, 892
Flood plains, 444, 460
Flow,
  grout, 843
  synthetic, generation, **917**

Subject Index 1039

Flow lines, 567
Flow net, **566**
Flow slide in rock, 724
Fluorescein dye, 909
Flyash, 865
Foundation investigation, 458
Fracture intensity, 513
Freebord, 964
Freeze-thaw cycles, 697, 703, 722
Freezing, 816, 871
Frequency,
    fundamental, of dam, 959
    wave transmission, 513, 950, 966
Frequency spectrum, 950
Friction,
    coefficient of, 61, 64, 225
    negative skin, 874
    peak angle, 80
    residual angle, 80, 84
    skin, 884
Friction angle for discontinuity, 354, 414
Friction angle of breakdown products, 699
Friction circle method, **635**
Friction cone, **758**
Frost action, 657, 697, 725
Frost heave, 812, 873
Frost shattering, 439
Frozen soil, 871

Gabbro, 4, 248, 929, 948
Galena, 878
Gel strength, 874
Geochemical analyses, **167**
Geophysical exploration, **448**, 602
Geophysical gravity method, 449, 602
Geophysical magnetic method, 449
Geophysical marine exploration, **531**
Geophysical resistivity method, 449, 602, 894
Geophysical seismic refraction method, **453**, 602, 914
Gibbsite, 16
Glaciated landforms, 447
Glaciated zones, 874
Glaciation, 915
Gliding, twin and translation, 199, 210, 223
Glycolation, 26, 155
Gneiss, 7, 445, 701, 735
Gouge, 352, 495, 725, 749, 753
Graben, 696
Grading curves, 34
Granite, 4, 221, 248, 686, 702, 752, 910, 929, 968
Granodiorite, 4
Granular soils, 66, 82
Granularity value, 397
Granulite, 411
Graphite, 326
Gravel, 6, 31, 47, 59, 73, 74, 452, 583, 597, 626, 645, 701, 718, 816, 818, 820, 824, 841, 842, 847, 858, 878, 882, 892, 937, 964
Greywacke, 5, 245, 752
Griffith cracks, 290
Griffith theories, 201, 226, 228
Gritstone, 6, 248
Groundwater, 465, 478, **560**, 816, 898, 904, 907, 914
    economics, **598**
    freezing, **871**
    level, 657
    lowering, 583
    quality, **627**
Grout,
    cut-off, 858
    setting time, 845
    treatment, **836**, 910, 932
Grout,
    AM-9, 838, 848
    cement, 837, **842**, 847, 848, 859
    chrome lignin, 838, 848
    clay, 837, 838
    epoxy resin, 838
    Guttman, 838
    Joosten, 838
    non-particulate, 837
    organic polymer, 837, 842, 847, 484, 859
    particulate, 837 **842**, 847, 848, 859
    polyester resin, 838
    polythixon, 838
    silicate, 837, 838
Grouting, 579, 811, 812, 836, 884, 909, 925, 945
    bulk, 837
    cavity, **863**
    compaction, 821, 837, 858
    fissure, 816, 837, **851**
    fracture, 816, 837
    full depth, 855
    permeation, 816, 836, 837, 839
    seepage force, 857
    stage, 855
Groynes, 717
Gumbo, 36
Gypsum, 170, 171, 326

Haematite, 281
Halloysite, 17
Hardness (Schmidt hammer), 9
Headings, 461
Heat flow, 874
Heat treatment of clay minerals, 26
Hooke's law, 195
Horst, 671, 696
Histograms, 358
Hornfels, 7, 248
Hydration, 875

1040 Subject Index

Hydraulic conductivity, see Permeability
Hydraulic connection, 892
Hydraulic fracturing, 944
Hydraulic gradient, 93, 568, 578, 847, 870, 877, 909
Hydraulic radius, 47
Hydrofractures, 826, 837, **855**, 883
Hydrograph, 893, 898, 904
Hydrological regime, 908

Ice, 871
    crystals, 873
    cylinders, 872
    wall, 816, 872, 873
Illite, 17, 169, 171, 172, 173, 178, 274, 295, 301, 753
Inclinometer, 504, 645, 703, 828
Inertia forces, 953
Infiltration, 561
Infiltration capacity, 898
Interlocking, 61
Internal friction, 514
Interparticle,
    dispersion, 107
    double (Stern) layer, 106
    electrolyte, 105
    flocculation, 107
    forces, **105**
    Van der Waal's–London forces, 107
Intrusion,
    igneous, 601
    salt water, 599, 617
Inverse pole figure, 253
Inverted pendulum, 510
Iron, 154
Irrigation, 908

Jarosite, 169
Joints, 318, 321, 411, 536, 571, 585, 697, 722, 724, 734, **738**, 758, 810, 894, 909, 910, 911, 912, 913, 915, 920, 922, 930

Kalsite, 4
Kandites, 17
Kaolin, 868
Kaolinite, 17, 35, 154, 169, 171, 172, 173, 274, 296, 301
Kentledge, 485
Keuper Marl, 120
Kronecker delta, 335

Laminar fluid flow, 45
Laminations, 155, 321
Land
    elements, 443
    facets, 443
    slipping, 910, 911
    systems, 443

Landform, 697
Laplace's equation of flow, 570, 619
Laterites, 36, 109
Lattice compliances, 280
Leached soils, 445
Leakance, 610
Legendre polynomial, 294, 295
Lignite, 6, 926
Limestone, 3, 5, 6, 204, 248, 493, 564, 699, 709, 724, 740, 750, 752, 753, 792, 797, 863, 910, 924, 926, 931
    Magnesian, 541, 614, 813
    Solenhofen, 221
Limits, see Atterberg limits,
Liquefaction, 824, 954, 955
Liquidity index, **35**
Liquid limit, see Atterberg limits
Liquid nitrogen, 872
Lithosphere, 1
Littoral drift, 716
Loam, 36, 870
Loess, 36, 118, 712, 870, 944
Logarithmic spiral, 581
Logger, borehole, 471
Logging,
    borehole, 471
    borehole core, 735
Longwall workings, 535
Loss function, 550
Lugeon, 588, 852

Magma, 3
Magnetostrictive transducer, 471
Mapping, 920
Maps,
    geological, 430, 735, 912
    geotechnical, 431
    hydrological, 564
    topographic, 429, 601
Marble, 7, 210
Marine geotechnical exploration, 529
Marl, 6, 36, 858, 868, 926
    Keuper, 702
Marlstone, 6
Mass demand line, 905
Mass reservoir supply line, 905
Membrane, dam, 938
Mesh, nylon, 816
Mica, 154, 868
Microseismic,
    activity, 205
    monitors, 662, 703
Mine workings, 837, 863, 909, 945
Mineralogical
    analyses, 167
    compostion, 26
    identification, **25**
Minerals,
    detrital, 154

Subject Index

Minerals (*continued*)
   non-detrital, 154
   rock-forming, 7
   wave velocity parameters of, 238
Modulus of deformation, 487, 490
Modulus of elasticity, 145, 195, 235, 455, 502, 513, 918
   dynamic, 737
Modulus of rigidity, 236, 455
Moh, scale of hardness, 8
Moisture,
   content, **43**, 477
   deficit, 898
Montmorillonite, 17, 35, 45, 156, 162, 163, 164, 165, 169, 171, 173, 175, 178, 179, 180, 326, 753, 866, 868, 874
Monzonite, 752, 929
Mudflow, 537, 655, **692**, 703
Mudstone, 5, 6, 155, 298, 301, 449, 708, 812, 911
Muscovite, 165, 279
   hydro-, 178
Muskeg, see Peat

Nepheline syenite, 4
Nodal planes, 948
Non-detrital minerals, 154
Non-steady state flow, 605

Oases, 915
Oedometer, 66, 89, 164
Orthoclase, 4
Overconsolidation ratio, 80, 84, 101, 681
Overcoring, 499
Ownership of water, 601
Oxbows, 908

Palaeontology, 468
Parabola, construction for phreatic surface, **574**, 675
Particle,
   average diameter, 68
   friction, **60**, 699
   size, 47, 80, 476, 656, 716
   strength tests, 476
Peat, 6, 12, 36, 449, 452, 907, 910
Penetration test, 478
Penetrometer, 823
Percentage rating, 899
Percolation, 888
Peridotite, 4
Permafrost, 873
Permeability, 45, 59, 468, 477, 585, 618, 715, 735, 830, 841, 851, 852, 892, 911, 912, 913, 914
   coefficient of, 46, 47, 93, 610
   electro-osmotic coefficient of, 868
   of particulate systems, **45**
   rock, 585
   soils, **591**

Permeability test, packer, 587, 911, 913, 921
Perspex (plexiglas/lucite) model, 922
Petrographic microscope, 251
pH, 23, 628, 910
Phonolite, 4, 741
Phosphoric acid, 870
Photogeology, 439
Photogrammetry, 439
Phreatic,
   maps, 563
   surface, 562, 914
   zone, 562
Phyllite, 7
Picrite, 4
Piezoelectric transducers, 503
Piezometer, 511, 594, 645, 705, 810, 828, 939
Piezometric conditions, 914
Piles, 652, 874, 877
   concrete, 825
   tension, 879, 883
Piling, sheet, 707, 879, 945
Pillar and stall, 542, 863
Piping, 578, 706, 937
Plagioclase, 4
Plains, 444
Planimeter, 168
Plastic limit, see Atterberg limits
Plasticity index, **35**, 85, 100, 101
Plate bearing tests, 144, 478, 484, 488, 514
Plateaux, 444
Point load test, 192
Poisson's ratio, 196, 202, 234, 455
Polymorphic transition, 948
Pore pressure,
   parameters, **87**
   rock, 223, 750
   ratio, 666, **674**
Porewater pressure, 51, **56**, 538, 646, 670, 923, 967
   measurement, 511
   negative, 65, 87, 90
   suction, 697
Porosimeter, mercury, 586
Porosity, **42**, 477, 859
Porphyrite, 4
Porphyry, 248, 752
Precipitation, 897, 898, 900, 908
Precipitation indices, 917
Pre-loading, **826**, 829
Pressure arch, 544
Pressure chamber test, 489
Pressuremeter, Ménard, **497**
Probability,
   discontinuity orientation, 365
   flooding recurrence, 897
   theory, **547**
Proctor dynamic test, 818

Profilograph, 751
Projection,
 equal angle, 333
 equal area, 329, 330, 331
 stereographic, **328**, **368**
Pyrite, 154, 170, 171
Pyrrhotite, 155

Quartz, 4, 5, 15, 32, 63, 154, 167, 171, 172, 282, 283, 868, 910
 conglomerate, 5
 monzonite, 4
 porphyry, 4
 undulatory extinction in, 224
Quartzite, 5, 7, 248, 737, 752
Quartzitic soils, 871
Quicksand, 578

RQD, 361, 465, **466**, 514, 734
Radioactive tracers, 909
Radio-carbon dating, 654
Raft, apron, 707
Randomness of discontinuity spacings, 363
Rastrillo, 927
Recession curve, 898
Reconnaissance, site, 438
Regression analysis, 79
Reinforcement, 816
Relative density, **44**, 64, 480, 822, 823, 850
Reliability (statistical), 556
Reports, site investigation, 483
Reservoir,
 design capacity, **904**
 leakage, 909, 910, 911, 913
 pumped storage, 909, 915, 916, 917
 regulating, 891, 909
 sites, 889, 908
 storage, 914
 waterproofing, 877, 925
 yield, 905
Residual factor, 649, 677
Residual shear strength, 72, 73, 178, 687, 752
Residual soils, 109, 699, 910
Resonance, 945
Retaining wall, 103, 811
Return period, 899
Revetment, 718
Rheosphere, 1
Rhyolite, 4
Rigidity modulus, see Modulus of Rigidity
Rippability, 513
Rip-rap, 706, 908, 911
River abstraction, 890
Rivers, 892, 899
Rock,
 anisotropy, 188
 dynamics, **232**

Rock (*continued*)
 classification, 196
 constructional material, 244
 cores, 189, 920
 density, 186
 grain shape, 187
 grain size, 187
 mass strength, 315
 mineralogy, 189
 particles,
  classification, 30
  shape, 32
  size, 31
  size distribution, 33
 porosity, 186
 strength, **182**, 190, 194
 testing (laboratory), 191, 192
 wave velocity, parameters of, 237
Rocks,
 acid, 4
 basic, 4
 biofragmental, 6
 clastic, 6
 crystalline, 6
 extrusive, 4
 igneous, 2, 5, 286, 449, 452, 564, 738, 909
 intermediate, 4
 metamorphic, 2, 6, 445, 738, 915
 plutonic, 4, 915
 schistose, 286
 sedimentary, 2, 5, 286, 626, 738, 909, 928
Rockfill, 73, 753
Roller, 816
 pneumatic (tyre), 816, 819, **820**
 sheep's foot, 816, 819, **820**
 vibratory, 816, 819, **820**
Room, see Pillar and stall
Roughness, surface of discontinuities, **352**, 749
Roughness index, 756
Run-off, 888, 897, 900, 904, 907, 917, 922
Rutile, 155, 171, 172

Safety, factor of, **557**, 634, 638, 641, 649, 651, 661, 759, 770, 772, 773, 780, 782, 787
Saline intrusion, 891
Sand, 6, 31, 47, 59, 72, 449, 532, 583, 597, 645, 715, 718, 816, 818, 820, 822, 824, 841, 842, 850, 858, 868, 873, 878, 882, 892, 923, 937
Sand layer, 660, 692
Sandstone, 3, 5, 6, 68, 69, 449, 564, 585, 645, 670, 699, 708, 709, 719, 722, 724, 737, 748, 751, 752, 810, 811, 839, 858, 892, 896, 909, 947

Subject Index 1043

Sandwick, 833
Saturation, **43**, 59
   partial, 57
Scaled distance law, 519
Scan line, **339**, 361
   discontinuity statistics of, 388
Scanning electron microscope, 111
Schist, 7, 248, 445, 702, 752, 929
Schmidt hammer, 194
Scour, 718
Sea defence, 715
Sea wall, 717
Seatearth (underclay), 702
Secant modulus, **75**, 487
Sediment
   fabric symmetry, 110
   formation, 109
   marine, 533
   transportation mechanisms, 110
Sedimentation, 300
Seepage,
   forces, **577**, 857, 937, 939
   losses, 911, 915, 920
Seismic,
   aftershocks, 967
   body waves (P) and (S), 234
   boundary reflection, **242**
   forces, 751, 792, 802, 803, 940
   intensity, 964
   one-dimensional transmission, 234
   site investigation, methods for, **512**
   surface waves (R) and (Q), 239
   surveys, 735
   wave attenuation, **239**
   waves, **234**
   zones, 824
Seismicity, 966
   dam, **945**
   reservoir induced, 967
Seismographs, 457
Seismological records, 946
Seismometers, 968
Sensitivity, clay, 113, **117**
Serpentinite, 7
Settlement (see also Subsidence), 439, 459, 503, 580, 599, 600, 874, 905, 938, 940, 944, 946, 964
   gauging, 645, 939
   negative, 932
Shafts, 463, 547, 854, 871, 920
Shale, 3, 5, 6, 73, 74, 268, 304, 449, 654, 702, 708, 748, 752, 810, 858, 911, 928
Shale grit, 699, 701, 724, 793, 811
Shear,
   box, 75, 76, 77, 490, 496, 751, 752, 793
   modulus, see Rigidity modulus
   strength,

Shear (*continued*)
   discontinuities, 369
   parameters (clay slopes), 675, 939
   residual, 325, 647, 653, 676, 703, 754
   tests, 478, 490
Shield margins, 968
Shingle, 715, 718
Shock wave, 233
Shrinkage, 161
   cracks, 660
Siderite, 154, 171, 173, 277
Sieve analyses, 34
Silicates,
   framework, 63
   sheet, 63
Silt, 6, 31, 47, 59, 645, 752, 816, 818, 842, 868, 870, 878, 882
   bars, 908
   detention, 908
Siltstone, 6, 702
Simulation,
   analogue,
      non-steady state, 624
      steady state, 622
   digital methods, 619
   groundwater regimes, **618**
Site exploration, **457**
Site investigation, **427**
Slab failure (rock slope), 720
Slabbing, 909
Slake durability test, 477
Slaking, 166, 176, 676, 684, 697, 725, 928
Slate, 7, 259, 266, 267, 282, 289, 293, 297, 299, 300, 701
   Marl, 273, 276
Slices,
   Conventional Method of, **639**
   Simplified Method of, **639**
Slickensides, 318, 321
Slides,
   block, 689
   bottle neck, 696
   colluvium, 696
   compound, 688
   first time, 677, 696, 710
   reactivation, 677, 709
   rotational, 688
   slab, 689
   successive, 692
   translational, 688
Slip detectors, 645
Slopes, 444, 571, **632**, 907
   design curves, 672, 686, **803**
   height-angle relationships, **683**
   highway, 708, 809
   planar failure of soil, **633**, 655, 677
   profile development, **704**

1044   *Subject Index*

Slopes (*continued*)
   regrading, 653
   rock, **720**
   rotational failure of soil, 576, **635**, 655
   soil, **632**
   submerged, 641
   wedge method of analysis, **661**
Sluices, 903
Smear zone, **836**
Smearing, 835
Snowmelt, 898
Softening, 676, 677, 722
   critical, 711
Soil probe, 510
Soil moisture deficit, 918
Solifluction, 124, 438, 703, 709, 710
   lobes and sheets, 692
Solution cavities, 439, 910
Specific capacity, 604, 618
Specific gravity, **43**, 478
Spillway, 905, 912
Spreading failures, 696
Springs, 561
Stabilisation, 178, 816
   de-watering, 816
   electro-chemical, 816, **866**, 869
Stability fabrics, 380, 404, 418
Standpipe, 512
Standards, clay mineral, **27**, 168
State paths, 99
Static sounding, 480
Stiff testing, **206**
   complete stress-strain curves, 208
Storage
   coefficient, 605
   potential, 908
Storativity, see Storage coefficient
Storms, 711, 897
Strain,
   energy, 177, 285, 291, 861, 946, 953
      diagenetic bonds, 152, 153
   gauge rosette, 501
   ground, 537, 909
   measurement, 509
   volumetric, 69, 93, 202
   strain energy, shear, 376
Strength,
   critical state, 681
   fully softened, 681
   material, 477
   peak, 327
   tensile, 541
Stress field, polyaxial, 369
Stress path, 55, 91, 92
Stress relief, 438, 499, 909, 910
Stress,
   deviator, 69, 73
   effective, 51, **56**, 69, 578

Stress (*continued*)
   interparticle, 49
   lateral, 51
   normal, **53**
   principal, **50**, 69, 334
   shear, **53**, 72
   total, 59
   vertical, 51
Striations, 338
Structural damage, 516, 951
Sub-grade, 708
Subsidence, mining (see also Settlement), **534**, 865, 909
Subsidence profile, 535, 536
Suction,
   plate, 159
   pressure, **157**, 597, 818
      membrane, 159
Sulphur, 155
Surface area, minerals, 161, **164**
Swamps, 907
Swash (tidal), 715
Swell, tidal, 715
Swell index, 82
Swelling, 155, 157, 676
   interlayer, 163
   intramicellar, 163
   pressure, **162**, 326
Syenite, 4, 929
Symbols,
   for rocks and soils, 472, 473
   for geotechnical maps and plans, 434, 435
Symmetry, 368
Symmetry concepts, **260**
Syneresis, 150

Talc, 326
Talus, 697
Talysurf, **62**, **63**, 354
Tangent modulus, **75**
Temperature, effect of, 222
Tendons, see Cable anchors
Tension crack, 643, 658, 676, 688, 722, 744, 746, 762, 775, 777, 785, 806
Terracettes, 699
Terrain evaluation, **442**, 692, 907
Thermal conductivity, 872
Thermal diffusivity, 872
Tides, 715
Till, see Clay, boulder
Tilt meters, 948, 968
Time lag (*in situ* permeability tests), 512, 595
Tip, colliery, 538
Topography, 897, 908, 909
Toughness, 37
Transformations,

Transformations (*continued*)
  angular, 329
  linear orthogonal, 333
Transition zone (filter), 940
Transmissibility coefficient, 605
Transmissivity, see Transmissibility coefficient
Trench, 877, 879
Trial pit, 458, 912
Triaxial testing,
  cell, 71, 219, 220
  rock, 145, **218**
  soil, 60, 66, 89, 939
Trilinear diagrams, 630
Tube-à-machette, 883
Tuff, 6, 928
Tunnel, 426, 536, 730, 838, 871, 920
  ground measurements around, 506
Tunnel face, discontinuity survey of, 342
Tunnel pressure, 915
Tunnelling, 460, 600
Type curve (groundwater pumping test), 607

Ultimate shear strength, see Residual shear strength
Underclay, see Seatearth
Underdrain, 573
Underground openings, 730, 863
Underpinning, 858
Underreaming (piles), 883
Uniaxial compressive strength (rock), 194
Uniformity coefficient, 33, 627
Uplift, 584, 662, 762, 801, 859, 861
Utility, see Loss function

Vadose zone, 562
Valence, **9**
  electrons, 9
Valley sides, 909, 910
Vane tests, **481**
Variance, 359, 551
Vegetation cover, 907
Velocity index, 513, 737
Vermiculite, 19, 163, 172, 181
Vibration,
  ground, **514**, 516, 812
  human perception, 518
  quarry blast, 517
  traffic, **515**
Vibroflotation, 816, 821, 822
Viscometer, 875
Void ratio, **43**, 84, 477

Voids, 5
Volcanic ash, 36
Volcanic rocks, 626

Water,
  adsorption, clay minerals, 20
  balance of a catchment, 909
  conservation, 908
  consumption, 887
  distribution, 887, 888
  fossil, 894
  level, 921
  perched, 561, 711, 911
  proofing, 877, 913
    ponds, 877
  resource planning, **889**
  table, 562, 647, 659, 915
  tightness of a reservoir, 909
  transfer scheme, see Conjunctive use
Watershed, 898
Wave (water) height, 714
Weathering, 2, 348, 655, 657, 684, 697, 751, 754, 920
  chemical, 697
  classification, 431, 698
  rates, 698
  tropical, 109, 697
Wedge, 661
  active and passive, 661
  analysis, rock slope, 765
  neutral, 663
Wedging, ice, 751
Weirs, 897, 903, 905
Well, 891, 912
  function, 605
  image method, 606
  losses, **626**
  point, 583, 707, 816, 829
  tests, 602
Wenner electrode configuration, 450

X-ray,
  diffraction, 25, 252
    'spiking-up', 27, 168
    standards, 27, 167, 168
  fluorescence, 167
  texture goniometry, **252**

Yield criteria, see Failure criteria
Yield surface, 99

Zeta potential, 867